Springer Series in Geomechanics and Geoengineering

Series editor

Wei Wu, Universität für Bodenkultur, Vienna, Austria
e-mail: wei.wu@boku.ac.at

Geomechanics deals with the application of the principle of mechanics to geomaterials including experimental, analytical and numerical investigations into the mechanical, physical, hydraulic and thermal properties of geomaterials as multiphase media. Geoengineering covers a wide range of engineering disciplines related to geomaterials from traditional to emerging areas.

The objective of the book series is to publish monographs, handbooks, workshop proceedings and textbooks. The book series is intended to cover both the state-of-the-art and the recent developments in geomechanics and geoengineering. Besides researchers, the series provides valuable references for engineering practitioners and graduate students.

More information about this series at http://www.springer.com/series/8069

Wei Wu · Hai-Sui Yu
Editors

Proceedings of China-Europe Conference on Geotechnical Engineering

Volume 1

Editors
Wei Wu
Institut für Geotechnik
Universität für Bodenkultur
Vienna, Austria

Hai-Sui Yu
Faculty of Engineering
University of Leeds
Leeds, UK

ISSN 1866-8755   ISSN 1866-8763 (electronic)
Springer Series in Geomechanics and Geoengineering
ISBN 978-3-319-97111-7   ISBN 978-3-319-97112-4 (eBook)
https://doi.org/10.1007/978-3-319-97112-4

Library of Congress Control Number: 2018949380

© Springer Nature Switzerland AG 2018
This work is subject to copyright. All rights are reserved by the Publisher, whether the whole or part of the material is concerned, specifically the rights of translation, reprinting, reuse of illustrations, recitation, broadcasting, reproduction on microfilms or in any other physical way, and transmission or information storage and retrieval, electronic adaptation, computer software, or by similar or dissimilar methodology now known or hereafter developed.
The use of general descriptive names, registered names, trademarks, service marks, etc. in this publication does not imply, even in the absence of a specific statement, that such names are exempt from the relevant protective laws and regulations and therefore free for general use.
The publisher, the authors, and the editors are safe to assume that the advice and information in this book are believed to be true and accurate at the date of publication. Neither the publisher nor the authors or the editors give a warranty, express or implied, with respect to the material contained herein or for any errors or omissions that may have been made. The publisher remains neutral with regard to jurisdictional claims in published maps and institutional affiliations.

This Springer imprint is published by the registered company Springer Nature Switzerland AG
The registered company address is: Gewerbestrasse 11, 6330 Cham, Switzerland

Foreword

Nigh on a century after Karl Terzaghi published his epoch-making book "Erdbaumechanik auf bodenphysikalischer Grundlage" in 1925, geotechnical engineering has developed from its infancy to a full-fledged engineering discipline. Europe is the cradle of modern soil mechanics and geotechnical engineering. The old continent still hosts the finest researchers, engineers, contractors, and manufacturers in geotechnical engineering. However, the construction activities, as the driving force for research and innovation, have subsided considerably in Europe. A major player in the construction sector is China with its huge domestic market as well as its ambitious Belt and Road Initiative abroad. For many years, China has the most construction activities of the world with such impressive infrastructure projects such as the Three Gorges Dam, South–North Water Transport, High-Speed Railway. This conference will link the birthplace of modern soil mechanics with the country with most construction and research activities in geotechnical engineering. It offers a welcoming opportunity to take stock of the state-of-the-art practice and the current research trends in China, Europe, and beyond.

The responses following our call for papers were overwhelming and echoed well beyond China and Europe. We have received about 400 papers from 35 countries, which make this conference a truly international event. The contributions in this proceedings cover virtually all areas of geotechnical engineering including constitutive model; numerical simulation; micro–macro relationship; laboratory testing; monitoring, instrumentation, and field test; foundation engineering; underground construction; innovative geomaterials; environmental geotechnics; cold regions geotechnical engineering; transportation and hydraulic engineering; unsaturated soils in waste management and CO_2 storage; geohazards–risk assessment, mitigation, and prevention. The proceedings provide an excellent overview of the current geoengineering research and practice in China, Europe, and beyond.

I am indebted to the plenary speakers, session organizers, and authors. The generous support from the City of Vienna and PORR, Austria, represented by Dr. Schön Harald (CEO), is gratefully acknowledged. My co-workers in Vienna, Dr. Wang Shun, Dr. He Xu-Zhen, and Dr. Lin Jia, spent many hours reading and correcting the papers. Lastly, I am grateful to my dear wife Ms. Wang Jing-Xiu, who made every effort to make this conference a success.

April 2018   Wei Wu

Contents

Part I: Constitutive Model

A Simple Anisotropic Mohr-Coulomb Strength Criterion
for Granular Soils 3
Wei Cao, Rui Wang, and Jian-Min Zhang

The Equivalent Stress of Soil Skeleton 8
Tong Dong, Liang Kong, Likun Hua, Xing Wang, and Yingren Zheng

Constitutive Model for Granular-Fluid Flows Based
on Stress Decomposition 13
Xiaogang Guo, Wei Wu, Liangtong Zhan, and Ping Chen

The Hypoplastic Model Expressed by Mean Stress and Deviatoric
Stress Ratio 17
Xuzhen He, Wei Wu, Dichuan Zhang, and Jong Kim

Multi-scale Modelling of Crushable Granular Materials 21
Pierre-Yves Hicher

Study on the Constitutive Model of Sandy Pebble Soil 26
Jizhi Huang and Guoyuan Xu

Isothermal Hyperelastic Model for Saturated Porous Media
Based on Poromechanics 31
Yayuan Hu

Review of N Models in Simulating Mechanical Behavior of Clays 35
Jianhong Jiang, Hoe I. Ling, and Victor N. Kaliakin

Strain-Rate and Temperature Dependency for Preconsolidation
Pressure of Soft Clay 39
Ting Li, Qi-Yin Zhu, Lian-Fei Kuang, and Li Gang

A Nonlinear Elastic Model for Dilatant Coarse-Grained Materials 43
Si-hong Liu, Yi Sun, and Chao-min Shen

Influence of Multidirectional Principal Stress Axes Rotation Sequences on Deformation of Sand 48
Long Xue, Rui Wang, and Jian-Min Zhang

Transversely Isotropic Strength Criterion for Soils Based on the Characteristic Mobilized Plane 52
Dechun Lu, Jingyu Liang, Chao Ma, and Xiuli Du

A Multi-Phase-Field Anisotropic Damage-Plasticity Model for Crystalline Rocks 57
SeonHong Na and WaiChing Sun

Simulation of the Undrained Behavior of Particulate Assemblies Subjected to Continuous Rotation of Lode Angle 61
Mohammadjavad Salimi and Ali Lashkari

On Effective Stress and Effective Stress Equation 65
Longtan Shao, Xiaoxia Guo, and Boya Zhao

Hypoplastic Simulation of Axisymmetric Interface Shear Tests in Granular Media 69
Hans Henning Stutz and Alejandro Martinez

A Cam-Clay-Based Fractional Plasticity Model for Granular Soil 74
Yifei Sun and Yufeng Gao

Modelling Root-reinforced Soils with Nor-Sand 79
Barbara M. Świtała and E. James Fern

Experience of Parameter Optimization of the High-Cycle Accumulation Model for Undrained Triaxial Tests on Sand 84
Yining Teng, Hans Petter Jostad, and Youhu Zhang

Hardening Soil Model - Influence of Plasticity Index on Unloading - Reloading Modulus 89
Wojciech Tymiński, Tomasz Kiełczewski, and Hubert Daniluk

A Visco-Hypoplastic Constitutive Model and Its Implementation 94
Guofang Xu, Wei Wu, Jilin Qi, Xiaoliang Yao, Fan Yu, and Aiguo Guo

Advance in the Constitutive Modelling for Frozen Soils 98
Guofang Xu, Lingwei Kong, Yiming Liu, Cheng Chen, and Zhiliang Sun

Towards a Better Understanding of the Mechanics of Soil Liquefaction 103
Jun Yang, H. Y. Eric Sze, Liming Wei, and Xiao Wei

UH Model for Granular Soils 108
Yang-ping Yao, Lin Liu, and Ting Luo

**An Optimization-Based Parameter Identification Tool
for Geotechnical Engineering** 112
Zhen-Yu Yin and Yin-Fu Jin

**Constitutive Model for Soaking-Induced Volume Change
of Unsaturated Compacted Expansive Soil** 116
Wei-lie Zou, Xie-qun Wang, Jun-feng Zhang, Zhong Han,
and Liang-long Wan

Part II: Micro-Macro Relationship

**Linking Microstructural Behavior with Macrostructural Observations
on Unsaturated Porous Media** 123
Hiram Arroyo and Eduardo Rojas

**The Effect of Particle Shape Polydispersity on Deformability
of Granular Materials** 128
Yuxuan Che, Fang Liu, Gang Deng, and Jizhong He

Grain **Learning: Bayesian Calibration of DEM Models and Validation
Against Elastic Wave Propagation** 132
Hongyang Cheng, Takayuki Shuku, Klaus Thoeni, Pamela Tempone,
Stefan Luding, and Vanessa Magnanimo

**An Analytical Study of the Bi-rock System
Considering Interface Effect** 136
Xing-wei Chen and Zhong-qi Quentin Yue

Natural State Parameter for Sand 140
Katarzyna Dołżyk-Szypcio

**Micro-structure Evolution in Layered Granular Materials
During Biaxial Compression** 144
Longlong Fu, Peijun Guo, Shunhua Zhou, and Yao Shan

**Dilatancy Phenomenon Study in Remolded Clays –
A Micro-Macro Investigation** 148
Qian-Feng Gao, Mohamad Jrad, Lamine Ighil Ameu, Mahdia Hattab,
and Jean-Marie Fleureau

DEM Simulation of Wave Propagation in Anisotropic Granular Soil ... 153
Xiaoqiang Gu and Shuocheng Yang

On Collapse of 2D Granular Columns: A Grain-Scale Investigation ... 157
Xuzhen He, Wei Wu, Dichuan Zhang, and Jong Kim

DEM Simulation of the Behavior of Rockfill Sheared with Different Intermediate Principle Stress Ratios 161
Juntian Hong, Ming Xu, and Erxiang Song

DEM Simulation of Sand Liquefaction Under Partially Drained Conditions 165
Qian-Qian Hu, Rui Wang, and Jian-Min Zhang

A Hard-Sphere Model for Wet Granular Dynamics 169
Kai Huang

Distinct Element Modelling of Constant Stress Ratio Compression Tests on Pore-Filling Type Methane Hydrate Bearing Sediments 174
Mingjing Jiang, Wenkai Zhu, and Jie He

Investigating the Mechanical Behavior of Pore-Filling Type Methane Hydrate Bearing Sediments by the Discrete Element Method 178
Mingjing Jiang, Jun Liu, and Jie He

A New Bottom-Up Strategy for Multiscale Studying of Clay Under High Stress 182
Lianfei Kuang, Guoqing Zhou, and Yazhou Zou

Evaluation of Grain Shape Parameters of Fujian Standard Sand 186
Shuai Li, Wan-Huan Zhou, Xue-Ying Jing, and Hua-Xiang Zhu

DEM Simulation of Oedometer Tests with Grain Crushing Effects 191
Jia Lin, Erich Bauer, and Wei Wu

Mechanical Behaviour of Meso-Scale Structure at Critical State for Granular Soils 195
Yang Liu, Xiaoxiao Wang, and Shunchuan Wu

Calibrating the Strength Ratio and the Failure Envelope for DEM Modeling of Quasi-Brittle Materials 199
Haiying Huang and Yifei Ma

Investigation on Stress-Fabric Relation for Anisotropic Granular Material 203
Weiyi Li, Jiangu Qian, and Xiaoqiang Gu

Micromechanical Analysis of the At-Rest Lateral Pressure Coefficient of Granular Materials 207
Chaomin Shen and Sihong Liu

Shear Modulus Degradation and Its Association with Internal Damage in Cemented Granular Material 211
Zhifu Shen, Mingjing Jiang, Zhihua Wang, and Shengnian Wang

Induced Crack Network Evolution in Geomaterials: μct Examination and Mathematical Morphology Based Analysis 216
Maciej Sobótka, Michał Pachnicz, and Magdalena Rajczakowska

Kinematic Nature of Sand Strength 220
Zenon Szypcio

Shortcomings of Existing Scan Line Void Fabric Tensors 225
A. I. Theocharis, E. Vairaktaris, and Y. F. Dafalias

DEM Simulation of Coarse-Grained Soil in Slip Zone 229
Shun Wang, Wei Wu, Wei Xiang, and Deshan Cui

Discrete Element Experiment and Simulation 233
Zhaofeng Li, Jun Kang Chow, Yu-Hsing Wang, Quan Yuan, Xia Li, and Yan Gao

Analytical Data Based Time-Dependent Mechanical Responses of an Intergranular Bond Considering Various Bond Sizes 238
H. N. Wang, N. Che, H. Gong, and M. J. Jiang

Visualization of Geogrid Reinforcing Effects Under Plane Strain Conditions Using DEM 242
Zhijie Wang, Martin Ziegler, and Guangqing Yang

DEM Investigation of Fracture Characteristic of Calcareous Sand Particles Under Dynamic Compression 247
Lei Wang, Xiang Jiang, Hanlong Liu, Zhichao Zhang, and Yang Xiao

Validation of Synthetic Images for Contact Fabric Generated by DEM ... 252
Max Wiebicke, Václav Šmilauer, Ivo Herle, Edward Andò, and Gioacchino Viggiani

Particle-Scale Observations of the Pressure Dip Under the Sand Pile 256
Qiong Xiao and Xia Li

Experimental Study on Clay Pore Microstructure Based on SEM and IPP 261
Ri-qing Xu, Li-yang Xu, and Xiao-ran Yan

Investigating the Effect of Intermediate Principal Stress on Granular Materials via DEM 265
Dunshun Yang, Xia Li, and Hai-Sui Yu

DEM Analysis of Cone Penetration Test in Sand 269
Fuguang Zhang

DEM Investigation of Particle Crushing Effects on Static and Dynamic Penetration Tests 274
Ningning Zhang, Marcos Arroyo, Matteo Ciantia, and Antonio Gens

Simulation of One-Dimensional Compression of Sand Considering
Irregular Grain Shapes and Grain Breakage 279
Fan Zhu and Jidong Zhao

Size Effects in Cone Penetration Tests in Sand 283
Pei-Zhi Zhuang and Hai-Sui Yu

Part III: Numerical Simulation

SPH-FDM Boundary Method for the Heat Conduction
of Geotechnical Materials Considering Phase Transition 291
Bing Bai, Dengyu Rao, Nan Wu, and Tao Xu

MPM Simulations of the Impact of Fast Landslides
on Retaining Dams 296
Francesca Ceccato, Paolo Simonini, and Veronica Girardi

Extensions to the Limit Equilibrium and Limit Analysis
in Geomechanics 300
Zuyu Chen

Extrusion Flow Modelling of Concentrated Mineral Suspensions 307
Chuan Chen, Damien Rangeard, and Arnaud Perrot

Enriched Galerkin Finite Element Method for Locally Mass
Conservative Simulation of Coupled Hydromechanical Problems 312
Jinhyun Choo and Sanghyun Lee

Influence of Individual Strut Failure on Performance of Deep
Excavation in Soft Soil 316
Kamchai Choosrithong and Helmut F. Schweiger

Implementation of Advanced Constitutive Models for the Prediction
of Surface Subsidence After Underground Mineral Extraction 320
Yury Derbin, James Walker, Dariusz Wanatowski, and Alec M. Marshall

Effect of Travelling Waves on Tunnels in Soft Soil 324
Stefania Fabozzi, Emilio Bilotta, Haitao Yu, and Yong Yuan

Numerical Simulation of a Post-grouted Anchor and Validation
with In-Situ Measurements 328
Carla Fabris, Václav Račanský, Boštjan Pulko, and Helmut F. Schweiger

Numerical Analysis of Monopile Behavior Under
Short-Term Cyclic Loading 332
Anfeng Hu, Peng Fu, and Kanghe Xie

An Improved Beam-Spring Model for Excavation Support 336
Biao Huang, Mingguang Li, Jinjian Chen, Yongmao Hou,
and Jianhua Wang

Numerical Modelling of a Monopile for Estimating the Natural
Frequency of an Offshore Wind Turbine 340
David Igoe, Luke J. Prendergast, Breiffni Fitzgerald,
and Saptarshi Sarkar

Modelling of Deep Excavation Collapse Using Hypoplastic
Model for Soft Clays 344
Jan Jerman and David Mašín

Simulation of Caisson Foundation in Sand Using a Critical State
Based Soil Model 350
Zhuang Jin, Zhen-Yu Yin, Panagiotis Kotronis, and Ze-Xiang Wu

Analytical Solution for the Transient Response of Cylindrical Cavity
in Unsaturated Porous Medium 355
Wei-Hua Li, Feng-Cui Feng, and Zhao Xu

Seepage and Deformation Analyses in Unsaturated Soils
Using ANSYS 360
Yangyang Li, Harianto Rahardjo, and K. N. Irvine

Implicit and Explicit Integration Schemes of a Double Structure
Hydro-Mechanical Coupling Model for Unsaturated Expansive Clays.... 365
Jian Li, Pengyue Wang, Lu Hai, Yanhua Zhu, and Yan Liu

Influence of Different Soil Constitutive Models on SSI Effect
on a Liquefiable Site...................................... 370
Zheng Li

Performance of Composite Buttress and Cross Walls to Control
Deformations Induced by Excavation 373
Aswin Lim and Chang-Yu Ou

Study on the Envelope of Stress Path During Deep Excavation 377
Li Liu, Hong-ru Zhang, and Jian-kun Liu

New Bond Model in Disk-Based DDA for Rock Failure Simulation 381
Feng Liu, Kaiyu Zhang, and Kaiwen Xia

On the Pseudo-Coupled Winkler Spring Approach for Soil-Mat
Foundation Interaction Analysis............................ 386
Dimitrios Loukidis and Georgios-Pantelis Tamiolakis

Numerical Analysis of Soil Ploughing Using the Particle Finite
Element Method 390
Lluís Monforte, Marcos Arroyo, Maxat Mamirov, and Jong R. Kim

Numerical Analyses with an Equivalent Continuum Constitutive
Model for Reinforced Soils with Angled Bar Components 394
Seyed Mehdi Nasrollahi and Ehsan Seyedi Hosseininia

A GPU-Accelerated Three-Dimensional SPH Solver
for Geotechnical Applications 398
Chong Peng, Wei Wu, and Hai-sui Yu

Numerical Analysis of Three Adjacent Horseshoe Galleries
Subjected to Seismic Load 402
Sid Ali Rafa, Allaoua Bouaicha, Idriss Rouaz, and Taous Kamel

Dynamic Analysis by Lattice Element Method Simulation 405
Zarghaam Haider Rizvi, Frank Wuttke, and Amir Shorian Sattari

Numerical Simulation of Plane Wave Propagation in a Semi-infinite
Media with a Linear Hardening Plastic Constitutive Model 410
Erxiang Song, Abbas Haider, and Peng Li

Construction Simulation and Sensitivity Analysis of Underground
Caverns in Fault Region 415
Chao Su and Yijia Dong

A Stress Correction Algorithm for a Simple Hypoplastic Model 419
Shun Wang, Wei Wu, Xuzhen He, Dichuan Zhang, and Jong Ryeol Kim

Experimental and Numerical Study of the Mechanical Properties
of Granite After High Temperature Exposure 423
Su-ran Wang and You-liang Chen

Conformable Derivative Modeling of Pressure Behavior
for Transport in Porous Media 427
R. Wang, H. W. Zhou, S. Yang, and Z. Zhuo

Boundary Element Analysis of Geomechanical Problems 431
Sha Xiao and Zhongqi Yue

Numerical Simulation of the Bearing Capacity of a Quadrate
Footing on Landfill 435
Dawei Xue and Xilin Lü

How Spatial Variability of Initial Porosity and Fines Content
Affects Internal Erosion in Soils 439
Jie Yang, Zhen-Yu Yin, Pierre-Yves Hicher, and Farid Laouafa

CFD Analysis of Free-Fall Ball Penetrometer in Clay 444
Yuqin Zhang and Jun Liu

Multiscale Modeling of Large Deformation in Geomechanics:
A Coupled MPM-DEM Approach 449
Jidong Zhao and Weijian Liang

Effect of Principal Stress Rotation on the Wave-Induced Seabed
Response Around a Submerged Breakwater 453
Hongyi Zhao, Jianfeng Zhu, and Dong-Sheng Zheng

Numerical Analysis of Spudcan-Footprint Interaction Using Coupled Eulerian-Lagrangian Approach 458
Jingbin Zheng and Dong Wang

Upper Bound Solution for Ultimate Bearing Capacity of Suction Caisson Foundation Based on Hill Failure Mode 463
Wen-bo Zhu, Guo-liang Dai, Wei-ming Gong, and Xue-liang Zhao

Part IV: Laboratory Testing

Soil Liquefaction Tests in the ISMGEO Geotechnical Centrifuge 469
Sergio Airoldi, Vincenzo Fioravante, and Daniela Giretti

ISMGEO Large Triaxial Apparatus............................. 473
Sergio Airoldi, Alberto Bretschnaider, Vincenzo Fioravante, and Daniela Giretti

Experimental Study on Static Liquefaction of Carbon Fiber Reinforced Loose Sand... 478
Xiaohua Bao, Zhiyang Jin, Hongzhi Cui, and Haiyan Ming

Experimental Study on Effect of Temperature and Humidity on Uniaxial Mechanical Properties of Silty-Mudstone 483
Jingcheng Chen and Hongyuan Fu

On Shear Strength of Stabilized Dredged Soil from İzmir Bay 488
İnci Develioglu and Hasan Firat Pulat

Instrument for Wetting-Drying Cycles of Expansive Soil Under Loads... 492
Jungui Dong, Guoyuan Xu, Hai-bo Lv, and Junyan Yang

Liquefaction of Firoozkuh Sand Under Volumetric-Shear Coupled Strain Paths.. 496
S. R. Falsafizadeh and A. Lashkari

Shear Strength of Tailings with Different Void Ratios 501
Tingting Fan

Dynamic Behaviour of Wenchuan Sand with Nonplastic Fines 506
Tugen Feng, Qiuying Qian, Fuhai Zhang, Mi Zhou, and Kejia Wang

Effects of Induced Anisotropies on Strength and Deformation Characteristics of Remolded Loess............................ 510
Yongzhen Feng, Wuyu Zhang, Lingxiao Liu, and Yanxia Ma

Experimental Analysis of Over-Consolidation in Subsoil in Bratislava, Slovakia 514
Zuzana Galliková

Study on the Influence of Moisture Content on Mechanical Properties of Intact Loess in Qinghai, China 519
Anbang Guo, Wuyu Zhang, Lingxiao Liu, and Yanxia Ma

Undrained Shear Behaviour of Gassy Clay with Varying Initial Pore Water Pressures 524
Y. Hong, L. Z. Wang, and B. Yang

Creep of Reconstituted Silty Clay with Different Pre-loading 529
Minyun Hu, Bin Xiao, Shuchong Wu, Peijiao Zhou, and Yuke Lu

Experimental Study on the Influence of Gravel Content on the Tensile Strength of Gravelly Soil 534
Enyue Ji, Shengshui Chen, Zhongzhi Fu, and Jungao Zhu

Direct Simple Shear Tests on Swedish Tailings 538
Qi Jia and Jan Laue

Concrete-Sand Interface in Direct Shear Tests 542
Zihao Jin, Qi Yang, Junzhe Liu, and Chen Chen

Improvement of Unconfined Compressive Strength of Soft Clay by Grouting Gel and Silica Fume 546
Mahdi O. Karkush, Haifa A. Ali, and Balqees A. Ahmed

Shear Modulus and Damping Ratio of Coarse Fused Quartz 551
Gangqiang Kong, Hui Li, Qing Yang, Gang Yang, and Liang Chen

Field and Laboratory Investigation on Non-linear Small Strain Shear Stiffness of Shanghai Clay 555
Q. Li, W. D. Wang, Z. H. Xu, and B. Dai

Introducing a New Test System for the Rock-Machine Interaction 560
S. D. Li, Z. M. Zhou, and W. S. Hou

A Parallel Comparison of Small-Strain Shear Modulus in Bender Element and Resonant Column Tests 564
Xin Liu and Jun Yang

Big Data and Large Volume (BDLV)-Based Nanoindentation Characterization of Shales 569
Shengmin Luo, Yucheng Li, Yongkang Wu, Yuzhen Yu, and Guoping Zhang

Development of a Large Temperature-Controlled Triaxial Device for Rockfill Materials 574
Hangyu Mao and Sihong Liu

Experimental Study of Small Strain Stiffness of Unsaturated Silty Clay 578
Tomáš Mohyla and Jan Boháč

**Effects of Soil Fabric on Volume Change Behaviour of Clay
Under Cyclic Heating and Cooling** 582
Qingyi Y. Mu, C. W. W. Ng, C. Zhou, and Hongjian J. Liao

**Comparing Water Contents of Organic Soil Determined on the Basis
of Different Oven-Drying Temperatures** 586
Brendan C. O'Kelly and Weichao Li

Planar Granular Column Collapse: A Novel Releasing Mechanism 591
Gustavo Pinzon and Miguel Angel Cabrera

**Effect of Non-plastic Fines on Cyclic Shear Strength of Sand
Under an Initial Static Shear Stress** 597
Daniela Dominica Porcino, Valentina Diano, and Giuseppe Tomasello

**Shear Band Between Steel and Carbonate Sand Under Monotonic
and Cyclic Loading** 602
Shengjie Rui, Zhen Guo, Lizhong Wang, Wenjie Zhou, and Kanmin Shen

**Comparative Study on the Compressibility and Shear Parameters
of a Clayey Soil** 607
Erik Schwiteilo and Ivo Herle

**Digital Image Measurement System for Soil Specimens
in Triaxial Tests** 611
Longtan Shao, Xiaoxia Guo, and Boya Zhao

Open Science Interface Shear Device 615
Hans Henning Stutz, Ralf Doose, and Frank Wuttke

Effect of Soil Characteristics on Shear Strength of Sands 619
Danai Tyri, Polyxeni Kallioglou, Kyriaki Koulaouzidou,
and Stefania Apostolaki

**Mechanical Behaviors of GRPS Track-Bed with Changing Water
Levels and Loading Cycles** 624
Han-Lin Wang and Ren-Peng Chen

**Analysing the Permanent Deformation of Cohesive Subsoil Subject
to Long Term Cyclic Train Loading** 628
Natalie M. Wride and Xueyu Geng

Dynamic Tensile Failure of Rocks Under Triaxial Stress State 632
Bangbiao Wu and Kaiwen Xia

**Experimental Study of Dynamic Shear Module and Damping Ratio
of Intact Loess from Xining, China** 636
Wenju Wu, Wuyu Zhang, Lingxiao Liu, and Yanxia Ma

True Triaxial Tests on Natural Shanghai Clay 640
Guan-lin Ye, Shuo Zhang, Yun-qi Song, and Jian-hua Wang

**Study on Strength of Expanded Polystyrene Concrete
Based on Orthogonal Test** 644
Quan You, Linchang Miao, Chao Li, Shengkun Hu, and Huanglei Fang

**An Investigation on the Effects of Rainwater Infiltration in Granular
Unsaturated Soils Through Small-Scale Laboratory Experiments** 648
R. Darban, E. Damiano, A. Minardo, L. Olivares, Lei Zhang, L. Zeni,
and L. Picarelli

**Adaptive Strength Criterion of Sandy Gravel Material
Based on Large-Scale True Triaxial Tests** 653
Yuefeng Zhou, Jiajun Pan, Zhanlin Cheng, and Yongzhen Zuo

Part V: Geotechnical Monitoring, Instrumentation and Field Test

Validation of In-Situ Probes by Calibration Chamber Tests 659
Sergio Airoldi, Alberto Bretschneider, Vincenzo Fioravante,
and Daniela Giretti

Cyclic Tests with a New Pressuremeter Apparatus 663
Soufyane Aissaoui, Abdeldjalil Zadjaoui, and Philippe Reiffsteck

**A New in Situ Measurement Technique for Monitoring the Efficiency
of Expansive Polyurethane Resin Injection Under Shallow
Foundations on Clayey Soils**..................................... 668
Hossein Assadollahi, L. K. Sharma, Anh Quan Dinh,
and Blandine Tharaud

**Monitoring and Early Warning System in Earth Management -
Based on Self-organized Fusion Sensor Networks** 672
Rafig Azzam, Hui Hu, and Herbert Klapperich

**Monitoring the Foundation Soil of an Existing Levee
Using Distributed Temperature Fiber Optic Sensors** 677
Giulia Bossi, Luca Schenato, Alessandro Pasuto, Silvia Bersan,
Fabio De Polo, Simonetta Cola, and Paolo Simonini

Development of IoT Sensing Modulus for Surficial Slope Failures 681
Wen-Jong Chang, An-Bin Huang, Shih-Hsun Chou, and Jyh-Fang Chen

**Evaluation of Free Swelling of Expansive Soil Using Four-Electrode
Resistivity Cone** 685
Ya Chu, Songyu Liu, Guojun Cai, and Hanliang Bian

**Evaluation of the Variation of Deformation Parameters Before
and After Pile Driving Using SCPTU Data** 689
Wei Duan, Guojun Cai, Jun Yuan, and Songyu Liu

Development of an FBG-Sensed Miniature Pressure Transducer
and Its Applications to Geotechnical Centrifuge Modelling 694
An-Bin Huang, Kuen-Wei Wu, Mohammed Z. E. B. Elshafie,
Wen-Yi Hung, and Yen-Te Ho

Evaluation of Engineering Characteristics of South China Sea
Sedimentary Soil in Sanya New Airport Based on CPTU Data 699
Kuikui Li, Wencheng Liang, Guojun Cai, Songyu Liu, Yu Du,
and Liuwen Zhu

Simplified Vibration Control Technique of Monitoring Devices
in Operating Tunnels 703
Bangan Liu and Yongbin Wei

Optimization of Location of Robotic Total Station in Tunnel
Deformation Monitoring 707
Bangan Liu and Yongbin Wei

Life-Cycle Analysis of Foundation Structures Using Sensor-Based
Observation Methods 712
Arne Kindler and Karolina Nycz

From Mobile Measurements to Permanent Integrity Control:
New Geotechnical Monitoring Instruments for Leakage
Detection and Localization 716
Mike Priegnitz and Ernst Geutebrück

Visual and Quantitative Investigation of the Settlement Behavior
of an Embankment Using Aerial Images Under Large-Size
Field Loading 720
Ali Alper Saylan, Okan Önal, Ali Hakan Ören, Gürkan Özden,
and Yeliz Yükselen Aksoy

Real-Time Monitoring of Large-Diameter Caissons 725
Brian Sheil, Ronan Royston, and Byron Byrne

Influences on CPT-Results in a Small Volume Calibration Chamber ... 730
F. T. Stähler, S. Kreiter, M. Goodarzi, D. Al-Sammarraie, T. Stanski,
and T. Mörz

Estimation of Relative Density in Calcareous Sands
Using the Karlsruhe Interpretation Method 734
Franz Tschuchnigg, Johannes Reinisch, and Robert Thurner

Investigation on Precursor Information of Granite Failure
Through Acoustic Emission 739
C. S. Wang, H. W. Zhou, W. G. Ren, S. S. He, H. Pei, and J. F. Liu

An Internet Based Intelligent System for Early Concrete
Curing in Underground Structures 743
Yongbin Wei, Jindi Lin, Wei Zhao, Chao Hou, Zhenhuan Yu,
Maowei Qiao, and Bangan Liu

Application of Multivariate Data-Based Model in Early Warning
of Landslides 747
Hongyu Wu, Mei Dong, and Xiaonan Gong

Establishment of a Health Monitoring Assessment System
for Hong-Gu Tunnel in China 751
Xiangchun Xu, Songyu Liu, Liyuan Tong, Jun Guo, and Xing Long

Monitoring a Flexible Barrier Under the Impact of Large Boulder
and Granular Flow Using Conventional and Optical Fibre Sensors 755
Jian-Hua Yin, Jie-Qiong Qin, Dao-Yuan Tan, and Zhuo-Hui Zhu

Quantifying Fiber-Optic Cable–Soil Interfacial Behavior
Toward Distributed Monitoring of Land Subsidence 759
Cheng-Cheng Zhang, Bin Shi, Su-Ping Liu, Hong-Tao Jiang,
and Guang-Qing Wei

Monitoring the Horizontal Displacement by Soil-Cement
Columns Installation 763
Wei Zhao, Longcai Yang, Binglong Wang, Congcong Xiong,
Ye Zhang, and Xilei Zhang

Long-Term Field Monitoring of Additional Strain of Vertical
Shaft Linings in East China 767
Guoqing Zhou, Lianfei Kuang, Guangsi Zhao, Hengchang Liang,
and Xiaodong Zhao

Part VI: New Geomaterials and Ground Improvement

Sacrificial Anode Protection for Electrodes in Electrokinetic
Treatment of Soils 773
Abiola Ayopo Abiodun and Zalihe Nalbantoglu

The Effects of Colemanite and Ulexite Additives on the Geotechnical
Index Properties of Bentonite and Sand-Bentonite Mixtures 778
Şükran Gizem Alpaydın and Yeliz Yukselen-Aksoy

Strength Development of Lateritic Soil Stabilized by Local
Nanostructured Ashes 782
Duc Bui Van, Kennedy Chibuzor Onyelowe, Phi Van Dang,
Dinh Phuc Hoang, Nu Nguyen Thi, and Wei Wu

Laboratory Carbonation Model of Single Soil-MgO Mixing Column in Soft Ground 787
Guanghua Cai, Songyu Liu, and Guangyin Du

Enhancing the Strength Characteristics of Expansive Soil Using Bagasse Fibre 792
Liet Chi Dang and Hadi Khabbaz

Engineering Properties and Microstructure Feature of Dispersive Clay Modified by Fly Ash 797
HengHui Fan, YingJia Yan, XiuJuan Yang, Lu Zhang, and HaiJun Hu

Effect of Strain Rate on Interface Friction Angle Between Sand and Alkali Activated Binder Treated Jute 801
Shashank Gupta, Anasua GuhaRay, Arkamitra Kar, and V. P. Komaravolu

Enhancement of Mechanical Properties of Expansive Clayey Soil Using Steel Slag 805
Anurag S. Hirapure and R. S. Dalvi

Breakage Effect of Soft Rock Blocks in Soil-Rock Mixture with Different Block Proportions 809
Xinli Hu, Han Zhang, Chuncan He, and Wenbo Zheng

Comparative Experimental Study on Solar Electro-Osmosis and Conventional Electro-Osmosis 814
Jianjun Huang and Huiming Tan

Stress-Dilatancy Relationship for Fiber-Reinforced Soil 818
Yuxia Kong

Improvement of Soft Soils Using Bio-Cemented Sand Columns 822
Aamir Mahawish, Abdelmalek Bouazza, and Will P. Gates

Stabilization of Expansive Black Cotton Soils with Alkali Activated Binders 826
S. Mazhar, A. GuhaRay, A. Kar, G. S. S. Avinash, and R. Sirupa

Formation of Biomineralized Calcium Carbonate Precipitation and Its Potential to Strengthen Loose Sandy Soils 830
Sangeeta Shougrakpam and Ashutosh Trivedi

Experimental Study on Calcium Carbonate Precipitates Induced by Bacillus Megaterium 834
Xiaohao Sun, Linchang Miao, and Chengcheng Wang

Effect of Fly Ash on Swell Behaviour of a Very Highly Plastic Clay ... 838
İ. Süt-Ünver, M. A. Lav, and E. Çokça

Analysis of Pore Size Distributions of Nonwoven Geotextiles Subjected to Unequal Biaxial Tensile Strains 842
Lin Tang, Songtai Sun, Xiaowu Tang, and Ruixuan Zhang

Effect of Fine Particles on Cement Treated Sand 847
Ganapathiraman Vinoth, Sung-Woo Moon, Jong Kim, and Taeseo Ku

A Geochemical Model for Analyzing the Mechanism of Stabilized Soil Incorporating Natural Pozzolan, Cement and Lime 852
Ba Thao Vu, Van Quan Tran, Quoc Dung Nguyen, Anh Quan Ngo, Huu Nam Nguyen, Huy Vuong Nguyen, and Hehua Zhu

Compaction Characteristics and Shrinkage Properties of Fibre Reinforced London Clay 858
Jianye Wang, Andrew Sadler, Paul Hughes, and Charles Augarde

Flow Behavior of Clays Amended by Superhydrophobic Additives 862
Yongkang Wu, Dongfang Wang, Yuzhen Yu, and Guoping Zhang

Mechanical and Thermal Behaviour of Cemented Soil with the Addition of Ionic Soil Stabilizer 866
Fei Xu, Hua Wei, Wenxun Qian, and Yuebo Cai

Experimental Study on the Relation Between EDTA Consumption and Cement Content of Cement Mixing Pile 870
Canhong Zhang, Baotian Wang, and Enyue Ji

Author Index 875

Part I: Constitutive Model

A Simple Anisotropic Mohr-Coulomb Strength Criterion for Granular Soils

Wei Cao, Rui Wang, and Jian-Min Zhang

Department of Hydraulic Engineering, National Engineering Laboratory for Green & Safe Construction Technology in Urban Rail Transit, Tsinghua University, Beijing 100084, China
caow09@foxmail.com, zhangjm@mail.tsinghua.edu.cn

Abstract. Granular soils usually possess inherent fabric anisotropy due to natural deposition or compaction, leading to the dependence of peak shear strength on loading direction. In order to describe this strength anisotropy, a novel aniso-tropic strength variable Λ, which is determined by stress tensor and fabric orientation, is introduced to measure the "distance" between the bedding plane and the maximum shear stress ratio plane. Incorporation of this variable in the traditional Mohr-Coulomb criterion allows for the description of strength anisotropy. The new anisotropic formulation needs only one extra parameter. Validation against test results shows that the formulation has good capability in describing the anisotropic strength.

Keywords: Anisotropy · Strength criterion · Granular soil

1 Introduction

Natural deposition or compaction always lead to the anisotropic characteristics of granular soils. Anisotropy in strength has been observed in both physical experiments [1, 2] and numerical simulation [3, 4].

Classical isotropic strength criteria such as Mohr-Coulomb criterion, Matsuoka-Nakai criterion [5] and Lade-Duncan criterion [6] does not reflect the anisotropic feature. To formulate an anisotropic strength criterion, there are generally two methods. One way is to directly introduce the angle between the bedding plane and a specific stress plane, such as potential shear plane [3] or "spatially mobilized plane" [7]. This approach is straight forward, but the calculation is complicated when the stress tensor is axisymmetric and there are infinite such stress planes. The other disadvantage is that the angle is usually not a continuous function of the stress tensor. The second method is to include joint invariants of stress tensor and a fabric tensor [8, 9] or use a modified stress tensor [10] in the strength criterion. Guaranteeing objectivity automatically, the second method is mathematically elegant, but suffers the disadvantage of lacking physical meaning, and may need more parameters to depict the anisotropic strength accurately.

This study aims at developing an anisotropic Mohr-Coulomb criterion (AMC) with simplicity and clear physical meaning. An anisotropic strength variable Λ is introduced and used to extend Mohr-Coulomb criterion to become anisotropic. The proposed AMC criterion is validated against test results.

2 Anisotropic Variable and Anisotropic Mohr-Coulomb Criterion

Existing test results [1–4] show that the peak strength usually decreases first and then increases with the increase of δ, which is the angle between the axis of the maximum principal stress and the normal direction of the bedding plane, and the peak strength reaches its minimum value when δ is around 60°. Hence, it is postulated that anisotropic granular soil obtains its minimum peak shear strength when the bedding plane coincides with the plane of maximum shear stress ratio. Such a postulate is a reasonable assumption for friction dominated granular soil. To describe the relationship between the two planes, a novel anisotropic strength variable (ASV) Λ is proposed [11] (Fig. 1):

$$\Lambda \equiv \frac{\sqrt{(\sigma_{msr} - \sigma_{bed})^2 + (\tau_{msr} - \tau_{bed})^2}}{\sigma_3} \quad (1)$$

where σ_3 is the minimum principal stress, σ_{msr}, τ_{msr} and σ_{bed}, τ_{bed} denote the normal stress and shear stress on the plane of maximum shear stress ratio and the bedding plane, respectively. Λ measures the "distance" between the two planes and is equal to 0 when the two planes coincide.

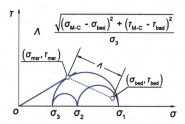

Fig. 1. The anisotropic strength variable Λ's definition

Thus, the anisotropic Mohr-Coulomb criterion (AMC) can be proposed as:

$$\frac{\sigma_1}{\sigma_3} = k_{f0} + a\Lambda \quad (2)$$

where k_{f0} represents the minimum strength and a represents the intensity of anisotropy. Both parameters can be conveniently and accurately calibrated with 3 sets of test data using a least square fit method.

The AMC can be validated against existing test data on granular soils. Figure 2 shows the comparison between the test results and the prediction of AMC criterion. It is shown that the proposed AMC criterion is capable of capturing the strength anisotropy of different anisotropic granular materials.

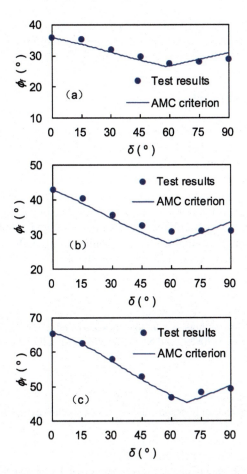

Fig. 2. Test results and the prediction of AMC criterion (a) Elliptical rods (Test data from [12]) (b) 2D DEM simulation (Test data from [3]) (c) Toyoura sand (Test data from [2])

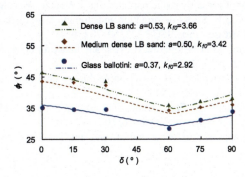

Fig. 3. Different materials' anisotropic parameters (Leighton Buzzard is abbreviated to LB, Test data from [1])

The anisotropic strength criterion is governed by the two parameters. As shown in Fig. 3, the material with a greater k_{f0} also has a greater minimum peak strength, and the material with a greater a has stronger anisotropic intensity.

3 Concluding Remarks

Based on the postulate that anisotropic granular soil obtains its minimum peak shear strength when the bedding plane coincides with the plane of maximum shear stress ratio, a novel anisotropic strength variable (ASV) is introduced to reflect the influence of the loading direction on the peak strength of such material. The ASV is introduced in the Mohr-Coulomb criterion, extending it to become anisotropic, which is referred to as AMC. The proposed formulation only introduces one extra parameter compared to the existing isotropic strength criterion, which can be calibrated conveniently. Validation shows good agreement between the predicted values using AMC and measured results from existing tests. The method of formulating the AMC can also readily be used to extend other classical isotropic criteria to become anisotropic with great simplicity and clarity.

Acknowledgements. The authors would like to acknowledge the National Natural Science Foundation of China (No. 51708332 and No. 51678346) for funding the work in this paper.

References

1. Yang, L.: Experimental study of soil anisotropy using hollow cylinder testing. In: University of Nottingham (2013)
2. Miura, K., Miura, S., Toki, S.: Deformation behavior of anisotropic dense sand under principal stress axes rotation. Soils Found. **26**, 36–52 (1986)
3. Fu, P., Dafalias, Y.F.: Study of anisotropic shear strength of granular materials using DEM simulation. Int. J. Numer. Anal. Met. **35**, 1098–1126 (2011)
4. Wang, R., Fu, P., Tong, Z., Zhang, J.M., Dafalias, Y.F.: Strength anisotropy of granular material consisting of perfectly round particles. Int. J. Numer. Anal. Met. Geomech. (2017)
5. Matsuoka, H., Nakai, T.: Stress-deformation and strength characteristics of soil under three different principal stresses. Proc. JSCE **232**, 59–70 (1974)
6. Lade, P.V., Duncan, J.M.: Elastoplastic stress-strain theory for cohesionless soil. J. Geotech. Eng. Div. **101**, 1037–1053 (1975)
7. Yao, Y., Kong, Y.: Extended UH model: three-dimensional unified hardening model for anisotropic clays. J. Eng. Mech. (2011)
8. Pietruszczak, S., Mroz, Z.: Formulation of anisotropic failure criteria incorporating a microstructure tensor. Comput. Geotech. **26**, 105–112 (2000)
9. Wu, W.: Rational approach to anisotropy of sand. Int. J. Numer. Anal. Met. **22**, 921–940 (1998)

10. Yao, Y., Tian, Y., Gao, Z.: Anisotropic UH model for soils based on a simple transformed stress method. Int. J. Numer. Anal. Met. **41**, 54–78 (2017)
11. Cao, W., Wang, R., Zhang, J.M.: Formulation of anisotropic failure criteria incorporating a microstructure tensor. Comput. Geotech. **26**, 105–112 (2000)
12. Zhang, L.W.: Research on failure mechanism and strength criterion of anisotropic granular materials and its application. Tsinghua University, Beijing (2007). (in Chinese)

The Equivalent Stress of Soil Skeleton

Tong Dong[1,2], Liang Kong[2(✉)], Likun Hua[2], Xing Wang[2], and Yingren Zheng[1]

[1] Army Logistics University of PLA, Chongqing 401311, China
[2] School of Sciences, Qingdao University of Technology, Qingdao 266033, China
qdkongliang@163.com

Abstract. Introducing the fabric tensor to reflect the distribution of the skeleton of the granular material, the real stress and the equivalent stress of the skeleton are obtained. A generalized method of anisotropic transformation of the isotropic constitutive model into an anisotropic one is proposed with the true stress of soil skeleton described by the equivalent stress and the mechanical properties of skeletons described by the isotropic constitutive models. Taking the Cam-Clay model as an example, the anisotropic transformed Cam-Clay model is built and verified by existing test results.

Keywords: True stress · Soil skeleton · Equivalent stress method
Anisotropic transformation

1 Introduction

Geomaterials is a system composed of a large number of discrete particles which usually have a special assembly. Therefore, it often shows the mechanical properties of anisotropy. In general, to describe this property, various kinds of anisotropy parameters are used to modify isotropic constitutive models [1]. However, the theoretical basis of such modification is not rigorous, which made the modified models complicated and poorly applicable. In this paper, based on theoretical analysis of the effective stress, the true stress of soils is discussed. Then, a general form of the true stress is proposed by tensor analysis. Finally, with the idea of stress transformation, an anisotropic constitutive model is established and verified by the existing test data.

2 Effective Stress and Equivalent Stress

2.1 Effective Stress

The saturated soil is a typical two-phase material. The soil-skeleton, which is composed of interconnected soil particles, can bear and pass the effective stress. Soil particles and water are assumed to be incompressible. Therefore, the deformation and strength of soil are the deformation and strength of soil-skeleton. In general, the effective stress is put forward to represent the average stress carried by the soil skeleton [2], which follows

$$\sigma = \sigma' + u \tag{1}$$

where, σ is the total stress, σ' is the effective stress and u is the pore pressure.

On the z-direction in Fig. 1(a), the equation of static equilibrium is

$$\sigma S_z = \sigma' S_z + u S_z \tag{2}$$

where, S_z is the area of the element along z-direction.

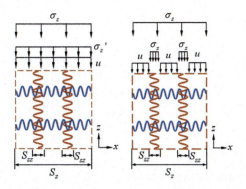

Fig. 1. (a) Averaging model; (b) Area fraction model

It can be found that the total stress, the effective stress and the pore stress are applied on the whole element uniformly. This means that soil is treated as a kind of homogeneous material and the average theory is applied in the definition of the effective stress. Therefore, σ' calculated by Eq. (1) is a virtual stress.

2.2 The True Stress

In fact, as shown in Fig. 1(b), the effective stress only acts on the soil-skeleton while the total stress and the pore pressure act on the whole element. Therefore, the equation of static equilibrium should follow:

$$\sigma_z S_z = \tilde{\sigma}_z S_{sz} + u S_{wz} \tag{3}$$

in which, $\tilde{\sigma}_z$ is the true stress, S_{sz} and S_{wz} are cross areas of soil and water. Therefore, σ is divided into $\tilde{\sigma}$ and u according to the area. The three principal fabric quantities can be defined as

$$F_i = 3 S_{si} / S_i \tag{4}$$

which can be written as the function of the fabric parameter Δ for simplify, as

$$\mathbf{F} = \begin{bmatrix} F_1 & 0 & 0 \\ 0 & F_2 & 0 \\ 0 & 0 & F_3 \end{bmatrix} = \frac{1}{3+\Delta} \begin{bmatrix} 1-\Delta & 0 & 0 \\ 0 & 1+\Delta & 0 \\ 0 & 0 & 1+\Delta \end{bmatrix} \tag{5}$$

Therefore, the true stress can be described by the stress tensor and the fabric tensor by

$$\tilde{\sigma}_{ij} = \sigma'_{il} F_{lj}^{-1} \tag{6}$$

Equation (6) is the logical expression of the true stress since the cross-area of soil-skeleton is involved in the true stress. Thus, for anisotropic granular materials the true stress is more suitable to describe the stress passing through the soil skeleton. If the basis vectors of the effective stress and the fabric tensor are different, transformation of coordinates is necessary. Meanwhile, to make sure the true stress is symmetric, Eq. (6) should be modified as

$$\tilde{\sigma}_{ij} = \left[\sigma'_{il}(T_{lm}T_{jn}F_{mn}^{-1}) + (T_{im}T_{ln}F_{mn}^{-1})\sigma'_{lj} \right] / 2 \tag{7}$$

in which, T is the spinning alternate tensor.

3 Equivalent Modified Cam-Clay Model

Similar to the Modified Cam-Clay model (MCC model) [3, 4], the particle skeleton is taken as the research object when building the Equivalent Modified Cam-Clay model (EMCC model). The yield surface of the EMCC model is given as

$$f = c_p \ln \frac{\tilde{p}}{p_0} + c_p \ln\left(1 + \frac{\tilde{q}^2}{M^2 \tilde{p}^2}\right) - \varepsilon_v^p \tag{8}$$

where, $c_p = (\lambda - \kappa)/(1 + e_0)$, in which λ and κ are parameters of the MCC model. Using the associated flow rule, the plastic potential surface is given as $g = f$. Other parts of the EMCC model are similar with the MCC model.

Only the fabric parameter Δ is added in the EMCC model. However, Δ can hardly be measured directly. In laboratory, triaxial tests on horizontal and vertical specimen can be used to determine Δ. When $\alpha = 0°, T_{ij} = \delta_{ij}$ and the value of internal friction angle is φ_0. Strength of anisotropic soils can be expressed as

$$\left(\frac{\sigma_1}{\sigma_3}\right)_{\alpha=0°} = \frac{1 + \sin \varphi_0}{1 - \sin \varphi_0} = k_0, \quad \frac{\tilde{\sigma}_1 F_1}{\tilde{\sigma}_3 F_3} = k_0 \tag{9}$$

where, k_0 is a strength index. Similarly, when $\alpha = 90°$. The strength of anisotropic soils can be expressed as

$$\left(\frac{\sigma_1}{\sigma_3}\right)_{\alpha=90°} = \frac{1+\sin\varphi_{90}}{1-\sin\varphi_{90}} = k_{90}, \quad \frac{\tilde{\sigma}_1 F_3}{\tilde{\sigma}_3 F_1} = k_{90} \qquad (10)$$

Finally, the fabric parameter can be calculated by

$$\Delta = \frac{1-\sqrt{k_0/k_{90}}}{1+\sqrt{k_0/k_{90}}} \qquad (11)$$

4 Model Verification

With $\lambda = 0.0504$, $\kappa = 0.003$, $e_0 = 0.55$ and $v = 0.3$, the measured and predicted curves of shear tests with a fixed principal stress direction are given in Fig. 2 [5]. The EMCC model shows an acceptable accuracy in predicting the stress-strain curves and the anisotropy of shear strength.

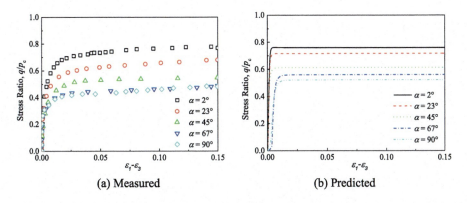

Fig. 2. Curves of normalized deviatoric stress versus shear strain.

5 Conclusions

The true stress of the soil skeleton is studied and expressed using tensor analysis. The stress acting on each component in a soil depends on its area fraction. With tensor analysis, the equivalent stress of the skeleton is obtained using the fabric tensor to reflect the distribution of the skeleton of the granular material and to describe the stress acting on the skeleton. With the idea of stress transformation, the anisotropic constitutive model can be established. As an example, the anisotropic transformed Cam-Clay model is built and verified by existing test results. All these provide the theoretical basis and experimental evidence for the equivalent stress method in building constitutive models which consider the anisotropy of materials and the direction of stress.

Acknowledgement. This paper is financially supported by the National Natural Science Foundation of China (51778311 and 11572165).

References

1. Dafalias, Y.F.: Must critical state theory be revisited to include fabric effects? Acta Geotech. **11**(3), 479–491 (2016)
2. Terzaghi, K.: Erdbaumechanik auf Bodenphysikalischer Grundlage. F. Deuticke (1925)
3. Roscoe, K.H.: On the generalized stress-strain behavior of 'wet' clay. Eng. Plast. 535–609 (1968)
4. Schofield, A.N., Wroth, P.: Critical State Soil Mechanics. McGraw-Hill, New York (1968)
5. Kumruzzaman, M., Yin, J.H.: Influences of principal stress direction and intermediate principal stress on the stress-strain-strength behaviour of completely decomposed granite. Can. Geotech. J. **47**(47), 164–179 (2012)

Constitutive Model for Granular-Fluid Flows Based on Stress Decomposition

Xiaogang Guo[1,2(✉)], Wei Wu[1], Liangtong Zhan[2], and Ping Chen[3]

[1] Institute of Geotechnical Engineering, Universität für Bodenkultur, Feistmantelstrasse 4, 1180 Vienna, Austria
[2] MOE Key Laboratory of Soft Soils and Geoenvironmental Engineering, Zhejiang University, Hangzhou 310058, China
xiaogang_guo@zju.edu.cn
[3] Department of Civil Engineering, Zhejiang Sci-Tech University, Hangzhou 310018, China

Abstract. In Bagnold's description of rheological properties of a solid-fluid suspension, the shear and normal stresses demonstrate linear dependence on the shear rate in the 'macro-viscous' regime and quadratic dependence in the 'grain-inertia' regime. Based on this pioneering work, a framework for the modeling of solid-fluid flows is proposed in this paper. In the framework, the total stress of the solid phase is decomposed into a frictional term contributed from prolonged contact of solid particles, a viscous term due to the viscosity of interstitial fluid and a inertia term results from particle collisions. A constitutive model for simple-shearing granular-fluid flows is developed based on this framework and verified in the element tests of granular-fluid materials under rapid shear.

1 Introduction

Debris flow is one of the most dangerous and destructive natural hazards in the mountainous area. Debris materials show solid-like behaviors before failure and fluid-like behaviors after failure. It is normally simplified to a mixture of monosized particles and Newtonian fluid, and treated as a fluid continuum with microstructural effect in constitutive modeling [1,2]. Based on the study about a gravity-free dispersion of solid spheres sheared in Newtonian liquids, Bagnold [3] proposed a constitutive model in which the shear stress has the relation

$$T_v = 2.25\lambda^{\frac{3}{2}}\mu\frac{dU}{dy} = k_1\frac{dU}{dy} \tag{1}$$

in the so-called 'macro-viscous' regime and

$$T_i = 0.042\rho_s(\lambda d)^2\left(\frac{dU}{dy}\right)^2 \sin\alpha_i = k_2\left(\frac{dU}{dy}\right)^2 \tag{2}$$

in the 'grain-inertia' regime, respectively. In the above expressions, μ is the dynamic viscosity of the interstitial fluid; ρ_s and d are the material density and the mean diameter of the grains, respectively; dU/dy denotes the shear strain rate; the tangent of the angle α_i corresponds to the ratio between the shear and normal stress in the grain-inertia regime; λ is the linear concentration. The normal stress P is observed proportional to the shear stress T with a constant value, 0.75, in the macro-viscous region and 0.32 in the grain-inertia region. In Bagnold's experiments, the density of solid particles is equal to that of the interstitial fluids. Thus, the effect of gravity is eliminated, and no yield stress is observed in the experiment. The constitutive relations (1) and (2) are developed for flow stages rather than the entire flow process from quasi-static state to high-speed shearing state. The main intention of this study is developing framework of constitutive model for granular-fluid mixture based on Bagnold's work.

2 A Constitutive Model for Solid-Fluid Flows

By fitting the experimental results, the total rheological shear stress T_r is assumed to the sum of T_v and T_i, i.e.

$$T_r = T_v + T_i, \tag{3}$$

and the total rheological normal stress P_r is

$$P_r = P_v + P_i. \tag{4}$$

As mentioned before, yield stress was not considered in Bagnold's model. However, the effect of gravity is not negligible for common granular-fluid mixture. The stress state in the quasi-static stage should be described in a complete model. Therefore, the general framework of constitutive model for granular fluid flow can be expressed as

$$P = P_0 + P_r = P_0 + P_v + P_i \tag{5a}$$
$$T = T_0 + T_r = T_0 + T_v + T_i. \tag{5b}$$

P_0 and T_0 are the shear and normal stresses in the quasi-static state and regarded as the static portion of the framework. T_r and P_r are termed the dynamic portion. The idea of stress decomposition is akin to Wu's assumption based on statics [4] and preliminarily validated in the literature [5,6] with the case of dry granular flow. The constitutive model based on this framework is promising to describe the stress state throughout the shearing process from quasi-static to high-speed shearing stage.

The stresses P_0 and T_0 satisfied a generalized Mohr-Coulomb type yield criterion [7,8]. The shear stress in quasi-static stage is expressed as

$$T_0 = -P_0 \, \mathrm{sgn}\left(\frac{dU}{dy}\right) \tan \phi. \tag{6}$$

The specific expression for the flowing stage is determined by modifying Bagnold's models. Based on the framework (5), the dynamic stresses are combined with the static stresses T_0 and P_0 to obtain a simple-shearing model as

$$P = P_0 + \frac{K_1}{\tan \alpha_v} \frac{dU}{dy} + \frac{K_2}{\tan \alpha_i} \left(\frac{dU}{dy}\right)^2 \tag{7a}$$

$$T = -P_0 \, \text{sgn}\left(\frac{dU}{dy}\right) \tan \phi + K_1 \frac{dU}{dy} + K_2 \left(\frac{dU}{dy}\right)^2 \tag{7b}$$

where K_1 is called effective viscosity; K_2 is the modified coefficient. The model (7) is applicable to the whole flow process, from the quasi-static state to rapidly flowing stage.

The new model is used to predict the stress-strain relations in the granular-fluid flow obtained in [9]. The simulation results are shown in Fig. 1. The dynamic portion of the new model (7) shows competent to describe the stress-shear rate relation in the flowing stage. The static portion overestimates the stresses of the loose sample with $C = 0.49$ in the low shear rate stage.

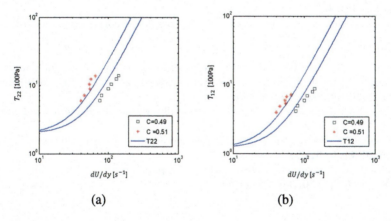

Fig. 1. Element test results for the granular-water flow with different solid volume fraction: (a) normal stress (b) shear stress.

3 Conclusions

Based on the analysis about Bagnold's pioneering work in [3] and the idea of stress decomposition, a framework is proposed for constitutive modeling of granular-fluid mixing materials. The constitutive relations for the quasi-static and the flowing stage were combined to obtain a model which can describe the stress state throughout the shear process from yielding to high-speed shearing. The new model is applied in element tests to simulate stress-strain relationships of granular-fluid flows. The predicted stress-shear rate curves show well agreement with the experimental data in the high shear rate stage. However, the stress in the beginning of the flow is overestimated by the model (7). The employed static portion of the new model is not capable in describing the initiation of granular-fluid flows where shear softening or hardening may happen. It encourages us to find more sophisticated theory to be the static portion of the new model in our future work.

References

1. Fang, C., Wu, W.: On the weak turbulent motions of an isothermal dry granular dense flow with incompressible grains: part I. Equilibrium turbulent closure models. Acta Geotech. **9**, 725–737 (2014)
2. Fang, C., Wu, W.: On the weak turbulent motions of an isothermal dry granular dense flow with incompressible grains: part II. Complete closure models and numerical simulations. Acta Geotech. **9**, 739–752 (2014)
3. Bagnold, R.A.: Experiments on a gravity-free dispersion of large solid spheres in a Newtonian fluid under shear. Proc. R. Soc. Lond. A Math. Phys. Sci. **225**, 49–63 (1954)
4. Wu, W.: On high-order hypoplastic models for granular materials. J. Eng. Math. **56**, 23–34 (2006)
5. Guo, X., Peng, C., Wu, W., Wang, Y.: A hypoplastic constitutive model for debris materials. Acta Geotech. **11**, 1217–1229 (2016). https://doi.org/10.1007/s11440-016-0494-0
6. Peng, C., Guo, X., Wu, W., Wang, Y.: Unified modelling of granular media with Smoothed Particle Hydrodynamics. Acta Geotech. **11**, 1231–1247 (2016). https://doi.org/10.1007/s11440-016-0496-y
7. Cowin, S.C.: Constitutive relations that imply a generalized Mohr-Coulomb criterion. Acta Mech. **20**, 41–46 (1974)
8. Savage, S.B.: Gravity flow of cohesionless granular materials in chutes and channels. J. Fluid Mech. **92**, 53–96 (1979)
9. Hanes, D.M., Inman, D.L.: Observations of rapidly flowing granular-fluid materials. J. Fluid Mech. **150**, 357–380 (1985)

The Hypoplastic Model Expressed by Mean Stress and Deviatoric Stress Ratio

Xuzhen He[1(✉)], Wei Wu[1], Dichuan Zhang[2], and Jong Kim[2]

[1] Institut für Geotechnik, Universität für Bodenkultur, 1180 Wien, Austria
{xuzhen.he,wei.wu}@boku.ac.at
[2] Nazarbayev University, Astana 010000, Kazakhstan

Abstract. Over the past several decades, hypoplasticity has been shwon to be a powerful tool to predict the non-linear behaviour of soils. Early hypoplastic models were developed from trial-and-error procedures and these models are usually expressed in a unique tensorial equation regarding the stress tensor. However, most models for fluid-like soil are expressed in the deviatoric stress ratio and the mean stress and these variables are usually modelled differently. This paper presents the hypoplastic model in a new format and written in these two variables. Additionally, parameters of hypoplastic models usually do not have any clear physical meaning and the authors try to investigate the meaning of parameters in the new equations.

Keywords: Hypoplastic model · Dilatancy

1 Introduction

The soil can behave like a fluid under certain circumstances and it is said to stay at the flowing state then. The critical state [2] after continuous shear deformation is actually a flowing state but at constant volume. For any fluid, its mechanical behaviour is characterised by its viscosity and compressibility. From extensive observations, it is found that the shear stress of fluid-like soil is largely mean stress dependent:

$$\boldsymbol{r} = \frac{\boldsymbol{s}}{p} = r^f \frac{\dot{\boldsymbol{\epsilon}}}{|\dot{\boldsymbol{\epsilon}}|} \qquad (1)$$

where p is the mean stress ($p = \text{tr}\boldsymbol{\sigma}/3$), $\boldsymbol{s}$ is the deviatoric stress tensor $\boldsymbol{s} = \boldsymbol{\sigma} - p\mathbf{I}$, $\boldsymbol{r} = \boldsymbol{s}/p$ is the deviatoric stress ratio tensor, or simply the stress ratio, $\mathbf{I}$ is the isotropic tensor, $\dot{\boldsymbol{\epsilon}}$ is the deviatoric strain rate tensor ($\dot{\boldsymbol{\epsilon}} = \dot{\boldsymbol{\varepsilon}} - \dot{\varepsilon}_v \mathbf{I}/3$), $\dot{\boldsymbol{\varepsilon}}$ is the strain rate tensor, $\dot{\varepsilon}_v$ is the volumetric strain rate and the modulus of a tensor is its Euclidean norm. r^f is the norm of $\boldsymbol{r}$ at the flowing state, or simply called the friction coefficient. In the simplest situation, r^f is constant. In soil mechanics community, r^f is found to depend on the loading direction (i.e. $r^f(\dot{\boldsymbol{\epsilon}})$). Recently, it is also found that r^f should have different values when the soil is flowing at the quasi-static state, the inertia state or the collision state and r^f is a function of the inertia number I.

The compressibity of fluids can be characterised by an equation like $p/\rho^\gamma =$ constant. For a soil, its bulk density is related to its solid fraction $\rho = \rho_{grain}\nu$ and fluid-like soils have a minimum solid fraction ν_m such that p is zero at this solid fraction. Therefore, the compresbility of fluid-like soil is more appropriately expressed as

$$p = C^c(\nu - \nu_m)^\gamma$$
$$\text{or, } \nu = \nu_m + C^{c\prime} p^{1/\gamma} \qquad (2)$$

In soil mechanics community, an equation similar to Eq. 2 is called the critical state $e - p$ relationship where $e = (1 - \nu)/\nu$ is the void ratio. The compresibility also varies when the soil is flowing at different states and C^c and ν_m can be assumed to be functions of I.

The two equations above constitute a complete rheology model for fluid-like granular soils. The left side are the deviatoric stress ratio tensor r and the mean stress p, which indicates the importance of these two variables in the modelling of fluid-like soil. The hypoplastic model is usually expressed in a unique tensorial equation regarding the stress tensor. In this paper, we investigate how it is expressed in r and p and some features of the model.

2 Hypoplastic Model in Mean Stress and Deviatoric Stress Ratio

Wu's Model [1] is

$$\dot{\boldsymbol{\sigma}} = I_e\{-C_1 \text{tr}\boldsymbol{\sigma}\dot{\boldsymbol{\varepsilon}} - C_2 \boldsymbol{\sigma}\text{tr}\dot{\boldsymbol{\varepsilon}} - C_3 \frac{\text{tr}(\boldsymbol{\sigma}:\dot{\boldsymbol{\varepsilon}})}{\text{tr}\boldsymbol{\sigma}}\boldsymbol{\sigma} + C_4 I_d(\boldsymbol{\sigma} + \boldsymbol{s})|\dot{\boldsymbol{\varepsilon}}|\} \qquad (3)$$

where the constants are the same as the original model and the different signs are because the compressive stress and strain are defined positive in this paper. I_d is the critical state function, which is 1 at the critical state, greater than 1 for a loose state, and less than 1 for a dense state. Substitute the identities $\boldsymbol{\sigma} = p\mathbf{I} + \boldsymbol{s}$ and $\dot{\boldsymbol{\varepsilon}} = \dot{\varepsilon}_v \mathbf{I}/3 + \dot{\boldsymbol{e}}$ into Eq. 3, we get

$$\dot{\boldsymbol{\sigma}} = I_e\{-[C_1 + C_2 + C_3/3]p\dot{\varepsilon}_v \mathbf{I} - [C_2 + C_3/3]\dot{\varepsilon}_v \boldsymbol{s} - 3C_1 p\dot{\boldsymbol{e}}$$
$$-[C_3/3]\text{tr}(\boldsymbol{s}:\dot{\boldsymbol{e}})[\mathbf{I} + \boldsymbol{r}] + C_4 I_d(p\mathbf{I} + 2\boldsymbol{s})|\dot{\boldsymbol{\varepsilon}}|\} \qquad (4)$$

Therefore, the rate equations for $\boldsymbol{s}$ and p are:

$$\dot{p} = I_e\{-[C_1 + C_2 + C_3/3]p\dot{\varepsilon}_v - [C_3/3]\text{tr}(\boldsymbol{s}:\dot{\boldsymbol{e}}) + C_4 I_d p|\dot{\boldsymbol{\varepsilon}}|\} \qquad (5)$$

$$\dot{\boldsymbol{s}} = I_e\{-[C_2 + C_3/3]\dot{\varepsilon}_v \boldsymbol{s} - 3C_1 p\dot{\boldsymbol{e}} - [C_3/3]\text{tr}(\boldsymbol{s}:\dot{\boldsymbol{e}})\boldsymbol{r} + 2C_4 I_d \boldsymbol{s}|\dot{\boldsymbol{\varepsilon}}|\} \qquad (6)$$

Using the indentity $\dot{r} = \frac{\dot{s}}{p} - r\frac{\dot{p}}{p}$, the rate equations for r and p are:

$$\frac{\dot{p}}{p} = I_e\{-[C_1 + C_2 + C_3/3]\dot{\varepsilon}_v - [C_3/3]\text{tr}(r:\dot{\epsilon}) + C_4 I_d|\dot{\epsilon}|\} \tag{7}$$

$$\dot{r} = I_e\{C_1\dot{\varepsilon}_v r - 3C_1\dot{\epsilon} + C_4 I_d r|\dot{\epsilon}|\} \tag{8}$$

At the critical state, for any given deviatoric stretch rate $\dot{\epsilon}$, the soil is moving at constant volume ($\dot{\varepsilon}_v = 0$, $\dot{\epsilon} = \dot{\epsilon}$) and the stress ratio is constant $r = r^f = r^f\frac{\dot{\epsilon}}{|\dot{\epsilon}|}$. Also, I_d is one. Therefore, Eqs. 7 and 8 must have

$$0 = -[C_3/3]r^f|\dot{\epsilon}| + C_4|\dot{\epsilon}| \tag{9}$$

$$0 = -3C_1\dot{\epsilon} + C_4 r^f|\dot{\epsilon}| \tag{10}$$

which means that to have the critical state embedded in the model, the constants must have $3C_1 = C_4 r^c$ and $C_3/3 = C_4/r^c$. Substitute these relationship into Eqs. 7 and 8, we have

$$\dot{r} = [-C_4 I_e]\{[r^f\dot{\epsilon} - I_d r|\dot{\epsilon}|] - r^c\dot{\varepsilon}_v r/3\} \tag{11}$$

$$\frac{\dot{p}}{p} = [-C_4 I_e]\{[\text{tr}(\frac{r}{r^f} : \frac{\dot{\epsilon}}{|\dot{\epsilon}|})|\dot{\epsilon}| - I_d|\dot{\epsilon}|] + \frac{C_1 + C_2 + C_3/3}{C_4}\dot{\varepsilon}_v\} \tag{12}$$

After some rearrangements, absorbing $-C_4$ into I_e and defining $C_1 + C_2 + C_3/3/C_4$ as μ, we have the following constitutive equations:

$$\dot{r} = I_e\{[r^f - I_d r]|\dot{\epsilon}| - r^f\dot{\varepsilon}_v r/3 - \chi I_d|\dot{\varepsilon}_v|/\sqrt{3}\} \tag{13}$$

$$\frac{\dot{p}}{p} = I_e\{[\text{tr}(\frac{r}{r^f} : \frac{\dot{\epsilon}}{|\dot{\epsilon}|}) - I_d]|\dot{\epsilon}| + \mu\dot{\varepsilon}_v - \chi I_d|\dot{\varepsilon}_v|/\sqrt{3}\} \tag{14}$$

where $\chi = \frac{|\dot{\epsilon}|-|\dot{\epsilon}|}{|\dot{\varepsilon}_v|/\sqrt{3}} = \frac{\sqrt{|\dot{\epsilon}|^2+\dot{\varepsilon}_v^2/3}-|\dot{\epsilon}|}{|\dot{\varepsilon}_v|/\sqrt{3}} = \sqrt{\lambda^2+1} - \lambda$ and $\lambda = \sqrt{3}|\frac{\dot{\epsilon}}{\dot{\varepsilon}_v}|$. Therefore, $0 \leq \chi \leq 1$ quantifies the relative significance of the volumetric deformation and shear deformation. (a) When the shear deformation dominates ($|\dot{\varepsilon}_v| \ll |\dot{\epsilon}|$) in tests such as conventional shear tests, $\chi \approx 0$ and the third terms in Eqs. 13 and 14 are negligible. (b) When the volumetric deformation dominates ($|\dot{\varepsilon}_v| \gg |\dot{\epsilon}|$) in tests such as oedometer tests or isotropic compression tests, $\chi \approx 1$ and the first terms in Eqs. 13 and 14 are negligible and the last two terms are to model the non-linear behaviour.

3 Discussions

The first term in Eq. 13 means that under shear deformation, the rate of r is proportional to the deformation rate $|\dot{\epsilon}|$, a stiffness coefficient I_e and a distance, which is between the current stress ratio r corrected by the critical state function I_d and the critical state stress ratio r^f. Therefore, this equation models (a) the evolution of stress ratio towards its critical state value; (b) the generally greater rate $\dot{r}$ when the stress is more or less isotropic and the smaller $\dot{r}$ when the stress ratio is close to the critical state value; (c) the peak stress ratio and the strain-softening behaviour of dense soils because for them, $I_d < 1$ and $|r|$ could increase till a peak value greater than r^f.

In Eq. 14, I_e is a stiffness function, therefore, $\frac{\dot{p}}{pI_e}$ is like a definition of 'elastic' volumetric strain $\dot{\varepsilon}_v^e$. In shearing tests, the third term is negligible and therefore, $[\mathrm{tr}(\frac{r}{r^c} : \frac{\dot{\epsilon}}{|\dot{\epsilon}|}) - I_d]|\dot{\epsilon}| = -(\mu \dot{\varepsilon}_v - \dot{\varepsilon}_v^e)$. The right hand side is like a definition of 'plastic' volumetric strain $\dot{\varepsilon}_v^p$. Therefore, the following conceptual equation is obtained.

$$\mathrm{tr}(\frac{r}{r^c} : \frac{\dot{\epsilon}}{|\dot{\epsilon}|}) - I_d = -\frac{\dot{\varepsilon}_v^p}{|\dot{\epsilon}|} = -d \qquad (15)$$

where d is the dilatancy. In conventional laboratory shear tests, r is coaxial with $\dot{\epsilon}$ and Eq. 15 is expressed as

$$d = I_d - \frac{|r|}{r^c} \qquad (16)$$

which is the same as the dilatancy equation proposed by Li [3]. Therefore, the first term of Eq. 14 models the diltancy of soils.

Acknowledgements. The first author acknowledges the financial support from the Otto Pregl Foundation for Geotechnical Fundamental Research, Vienna, Austria and Jiangsu Province natural sciences fund subsidisation project (BK20170677). The research is also partly funded by the project "GEORAMP" within the RISE programme of Horizon 2020 under grant number 645665.

References

1. Wu, W., Lin, J., Wang, X.: A basic hypoplastic constitutive model for sand. Acta Geotech. **12**(6), 1373–1382 (2017)
2. Schofield, A., Wroth, P.: Critical State Soil Mechanics. McGraw Hill, London (1968)
3. Li, X.-S., Dafalias, Y.F.: Dilatancy for cohesionless soils. Geotechnique **50**(4), 449–460 (2000)

Multi-scale Modelling of Crushable Granular Materials

Pierre-Yves Hicher[✉]

UMR CNRS GeM, Ecole Centrale de Nantes, 44321 Nantes, France
`pierre-yves.hicher@ec-nantes.fr`

Abstract. A microstructural model to consider the mechanical behavior of granular materials has been enhanced to incorporate the evolution of the critical state when significantly high grain breakage occurs during loading. The influence of grain size on grain strength is highlighted and its impact on the properties of granular assemblies is shown through the results of oedometer test simulations.

Keywords: Grain breakage · Micromechanical model · Grain size effect

1 Introduction

Grain crushing has been widely recognized as a major factor controlling the performance of geotechnical structures such as loaded piles, rockfill dams, etc.... It is therefore advisable to keep improving our knowledge of grain crushing. This paper presents a new approach to model the mechanical behavior of granular materials subjected to significant grain breakage. The approach will be based on a multi-scale analysis relating the behavior of grains in interaction with neighboring grains to the behavior of a granular assembly.

2 Micromechanical Model for Granular Materials

The numerical method is based on a homogenization technique for deriving the stress-strain relationship of the granular assembly from forces and displacements at the particle level. The microstructural model developed by Chang and Hicher [1] treats a soil as a collection of non-cohesive particles. The deformation of a representative volume of the material has been generated by mobilizing particle contacts along various orientations. The stress-strain relationship can be derived as an average of the deformation behavior of local contact planes in all orientations. The inter-particle behavior is based on an elastoplastic relationship between local forces and displacements. The elastic part is given by the Hertz-Mindlin's formulation, whereas the plastic part is based on a Coulomb criterion with a hardening function of the plastic sliding. Resistance against sliding on a contact plane depends on the degree to which neighboring particles are interlocked. The degree of interlocking is a function of the actual void ratio and of the critical state void ratio at a given stress state.

3 Behavior of Crushable Granular Materials

Grain breakage commonly occurs when a granular material undergoes compression and shearing, especially under high confining stresses. For a given material, the amount of grain breakage increases when stresses and strains increase. In this study, we chose the following quantity corresponding to a modified expression of the plastic work during mechanical loading [2]:

$$w_p = \int p'\langle d\varepsilon_v^p\rangle + q d\varepsilon_d^p \tag{1}$$

where $d\varepsilon_v^p$ and $d\varepsilon_d^p$ are volumetric and deviatoric plastic strain increments, respectively. The evolution of the gradation is expressed by the breakage ratio B_r^* as defined by Einav [3]. Its evolution with grain breakage is given by:

$$B_r^* = \frac{w_p}{a + w_p} \tag{2}$$

where a is a material parameter which controls the evolution of the breakage amount depending on the initial properties of the grains (mineralogy, shape, size, size distribution). The amount of grain breakage depends also on the stress path and the degree of humidity. Various experimental and numerical works have shown that the position of the critical state line in the e-lnp' plane depends on the grain size distribution. If grain breakage occurs during loading, the grain size distribution will evolve; as a consequence, the critical state line is shifted towards lower values of e. in the void ratio–mean effective stress plane with a relatively constant critical friction angle, provided that the particle shape remains relatively unchanged. A simple way to describe the change of the critical state is to change the reference critical point $\left(e_{ref}, p_{ref}\right)$. This reference point is the only variable necessary to control the evolution of the critical state line, if one can assume that the slope remains constant. In order to describe the evolution of the critical state due to grain breakage, we propose the following expression:

$$P_{ref} = p_{ref0} \exp(-b B_r^*) \tag{3}$$

where b is a material parameter which controls the evolution rate of the critical state line with grain breakage. Grain ruptures will reduce the dilatancy of granular assemblies and increase the contractancy. As a consequence, the material will develop more deformation when subjected to oedometric compression and the maximum strength along deviatoric loading paths will be reduced.

4 Considering Grain Size Effect

Large civil engineering works, such as rockfill dams, are increasingly constructed with coarse grained materials. However, the size of the apparatus required, even for small-sized rockfills, has considerably limited the experimental characterization of such materials. We propose here a new modelling approach to evaluate the mechanical properties of assemblies of large particles from the mechanical behavior of assemblies of smaller particles based on size effect in granular materials [4]. The influence of the particle sizes on the mechanical behavior of granular assemblies is related to the amount of particle breakage during loading and the size influence on particle crushing strength. The statistical theory of the strength of brittle materials proposed by Weibull [5] gives the following distribution for the survival probability of a material of size d subjected to a tensile stress σ:

$$P_s(d) = \exp\left(-\left(\frac{d}{d_o}\right)^{n_d}\left(\frac{\sigma}{\sigma_o}\right)^m\right) \qquad (4)$$

where σ_o is the characteristic strength ($P_s = 37\%$ for a sample of size d_o) and m is the control factor for the data scatter on crushing strength. According to Bažant and Planas [6], n_d represents the geometric similarity of the mechanical problem: $n_d = 1, 2$ or 3, for linear, surface or volume similarity, respectively.

The method developed considers two homothetic granular materials (say $G1$ the finer and $G2$ the coarser) under the same loading condition and with the same mineralogy and grain shape. Thus, if an identical survival probability is assumed for two grains under compression of characteristic sizes d_1 and d_2 in $G1$ and $G2$, respectively, the following relation for the particle crushing strength (σ_{Gi}) can be obtained from Eq. 5:

$$\sigma_{G2} = \sigma_{G1}\left(\frac{d_2}{d_1}\right)^{-n_d/m} \qquad (5)$$

It can then be demonstrated that plastic works w_{pG1} and w_{pG2} necessary to create the same amount of grain breakage in Materials $G1$ and $G2$ along the same loading path are linked by the relation [7]:

$$w_{pG2} = w_{pG1}\left(\frac{d_2}{d_1}\right)^{-n_d/m} \qquad (6)$$

and, consequently, the material constants a_{G1} and a_{G2} are linked by a similar relation:

$$a_{G2} = a_{G1}\left(\frac{d_2}{d_1}\right)^{-n_d/m} \qquad (7)$$

Therefore, if we can assume that all the other parameters are identical for materials $G1$ and $G2$, it is possible to calibrate the parameters of coarse material $G2$ from experimental tests performed on fine material $G1$. This assumption is reasonable if materials $G1$ and $G2$ are made of the same grains, having similar shapes for different sizes and homothetic grain size distributions. Subsequently, if we want to model the behavior of a given coarse granular material, we have to select a finer homothetic grain size distribution to prepare a representative finer material whose grain sizes are compatible with the available experimental resources of the laboratory.

As an illustration of our approach, numerical simulations of oedometer tests on two granular assemblies *G1* and *G2* with different grain sizes were performed. *G1* has been assumed to have a mean grain size value $d_{m1} = 1$ mm and *G2* a mean grain size value $d_{m2} = 100$ mm. The same set of parameters was used for the simulations. The size effect was taken into account by selecting the breakage parameter a for the two materials according to Eq. 7. The results of the simulations are presented in Fig. 1, illustrating the influence of the amount of grain breakage on the deformability of granular assemblies.

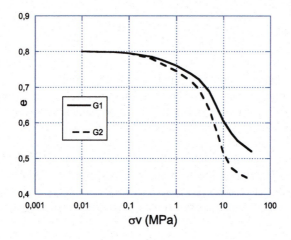

Fig. 1. Influence of grain size on mechanical properties of granular assemblies

5 Conclusions

Modelling the mechanical behavior of crushable granular materials through a novel approach has been suggested in this study. First of all, a constitutive model based on micromechanical considerations able to reproduce the main features of granular materials has been adopted. This model incorporates the critical state which, under our approach, is considered to depend on the amount of grain breakage developed by the material during loading. The grain breakage amount is then related to the plastic work along a given loading path. This relation includes a model parameter which depends on the individual strength of the grains. A size effect is then introduced which takes into account the decrease of the grain strength with the grain size. Numerical simulations of oedometer tests on granular assemblies with different grain sizes were performed. The same set of parameters was used for all simulations. The results illustrate how the model can predict the mechanical behavior of coarse granular materials from test results performed on a finer fraction with a homothetic gradation, provided that the scale effect on the strength of individual grains was taken into account in the modelling of the granular assembly.

References

1. Chang, C., Hicher, P.-Y.: An elastoplastic model for granular materials with microstructural consideration. Int. J. of Solids and Structures **42**(14), 4258–4277 (2005)
2. Yin, Z.Y., Hicher, P.-Y., Dano, C., Jin, Y.F.: Modeling the mechanical behavior of very coarse granular materials. J. Eng. Mech. **143**(1), C401600 (2017)
3. Einav, I.: Breakage mechanics- Part I: Theory. J. Mech. Phys. Solids **55**(6), 1274–1297 (2007)
4. Frossard, E., Hu, W., Dano, C., Hicher, P.Y.: Rockfill shear strength evaluation: a rational method based on size effects. Géotechnique **62**(5), 415–428 (2012)
5. Weibull, W.: A statistical theory of the strength of materials. Proc. Roy. Swedish Inst. Eng. Res. **151**, 1–49 (1939)
6. Bažant, Z.P., Planas, J.: Fracture and Size Effect in Concrete and Other Quasibrittle Materials. CRC Press, Boca Raton and London (1998)
7. Hicher, P.-Y., Jin, Y.F., Dano, C., Yin, Z.Y.: A constitutive model for rockfills considering grain breakage and size effect. Dam Eng. **5**(4), 157–169 (2015)

Study on the Constitutive Model of Sandy Pebble Soil

Jizhi Huang^(✉) and Guoyuan Xu

South China University of Technology, Guangzhou 510641, China
`201610101235@mail.scut.edu.cn`

Abstract. Sandy pebble soil is mainly composed of pebbles, gravel and some clayey soil, the structure of pebble soil is discrete. Different from other soils and rocks, the mechanical properties of pebble soil are complex and easy to be disturbed, shield construction in sandy pebble soil may cause land subsidence easily. Constitutive model plays an important role in better understanding the mechanical properties of sandy pebble soil. This paper reviews two constitutive models (Reuss model and Voigt model) of composite materials. Based on the assumption of stress uniformity and strain uniformity, the analytical formulas for calculating the equivalent elastic modulus of pebble soil are derived. To verify the formulas, a series of numerical experiments are established through ABAQUS.

Keywords: Sandy pebble soil · Equivalent elastic modulus
Constitutive equation · Finite element simulation

1 Introduction

Sandy pebble soil is usually composed of pebbles, gravel, sand and clayey soil. Different from clay and weathered rock, sandy pebble soil is a typical heterogeneous material with special physical and mechanical properties. Sandy pebble soils distribute unevenly in the stratum, the permeability of sandy pebble soil is strong and the self-stability is poor. Since the structure of sandy pebble soil is loose, TBM in this kind of soil layer may cause land subsidence and collapse easily [1]. Investigations on the constitutive model of the pebble soil are significant for underground engineering, especially for shield construction.

Equivalent elastic modulus is a basic property of sandy pebble soil. Currently, much effort has been devoted to modelling the equivalent elastic modulus of multiphase materials, the basic theory includes "Reuss model" and "Voigt model" [2]. But these models are only suitable for composite materials with ideal delamination. Based on the two models, Tian Yunde [3] proposed a mixed-mode method in investigating composite ceramic material. In sandy pebble soil modeling research, Hu Min [4] proposed a constitutive model based on Eshelby tensors and Mori-Tanaka equivalent method. However, this model is only suitable for small deformation condition and the calculation formula is complex. In addition, the accuracy will drop when the pebble content is more than 50%.

2 Basic Constitutive Model for Composite Material

2.1 Reuss Model (Stress Uniformity Model)

Reuss model [5] is also named series model, which is based on the hypothesis of stress uniformity. The basic assumption is the distribution of two materials are horizontal and the stresses in two materials are same under a vertical load. If ω represents the volume fraction of pebble, the equivalent elastic modulus of Reuss model is:

$$\frac{1}{E} = \frac{1}{E_1}(1-\omega) + \frac{1}{E_2}\omega \tag{1}$$

E_1, E_2——Elastic modulus of soil and pebble.

2.2 Voigt Model (Strain Uniformity Model)

Voigt model [5] is also named parallel model, which is based on the hypothesis of strain uniformity. The basic assumption is that the two materials are in a juxtaposition delamination, under the vertical load, the strain in two materials are the same. The equivalent elastic modulus of Voigt model is:

$$E = E_1(1-\omega) + E_2\omega \tag{2}$$

3 Constitutive Model for Sandy Pebble Soil

To investigate the constitutive model of sandy pebble soil, a RVE (Representative Volume Element) was selected from a section of sandy pebble soil [6] (see Fig. 1). In theoretical derivation, the mechanical properties of RVE are considered the same as that of the main body.

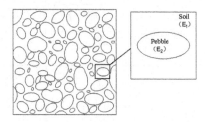

Fig. 1. Sections of sandy pebble soil and RVE

3.1 Stress Uniformity Model

Similar with the Russ model in Sect. 2.1, the basic assumption of stress uniformity model is the distribution of stress in RVE is homogeneous, which means the stress in pebble is the same as the stress in soil (Fig. 2).

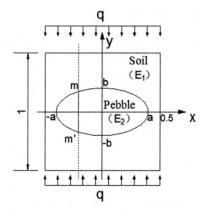

Fig. 2. Schematic diagram of equivalent elastic modulus calculation

By calculation, the equivalent elastic modulus of stress uniformity model is:

$$E = \frac{\sigma}{\varepsilon} = \frac{q}{\delta} = \frac{E_1 E_2}{E_2 + \pi ab(E_1 - E_2)} \tag{3}$$

3.2 Strain Uniformity Model

Similar with the Voigt model in Sect. 2.2, the basic assumption of strain uniformity model is the distribution of strain in RVE is homogeneous, which means the strain in pebble is the same as that in soil.

The equivalent elastic modulus of strain uniformity model is:

$$\frac{1}{E} = \frac{a}{m} \left[\pi - \frac{2E_1 \ln\left(\frac{m + \sqrt{m^2 - E_1^2}}{E_1}\right)}{\sqrt{m^2 - E_1^2}} \right] + \frac{1 - 2a}{E_1} \tag{4}$$

$$m = 2b(E_2 - E_1) \tag{5}$$

3.3 Applicability of Two Models

The applicability of the two models depends on the content of pebbles. If RVE contents fewer pebbles and more soil, the pebble will be surrounded by soil and the stress in pebble will be consistent with that in the soil. In this condition, the "Stress Uniformity Model" is more accurate in calculating equivalent elastic modulus. However, when the pebbles get more, they begin to contact each other and stress will transfer through pebble. In this case, the stress in soil is much smaller, but the strain distribution keeps uniform and "Strain Uniformity Model" will be more effective.

4 Numerical Experiment and Modified Formula

To verify the constitutive models and formulas in Sect. 3, a series of numerical experiments was conducted. Results show that (see Fig. 3 left), when the content of pebble is lower than 30%, formula of "Stress Uniformity Model" matches numerical experiment results well. When the content of pebble is about 60%, the formula of "Strain Uniformity Model" matches numerical experiment results better. The mechanism can be explained by the theory in Sect. 3.3.

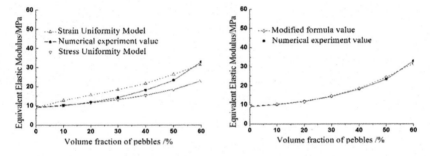

Fig. 3. Comparison between constitutive equation and numerical experiment

For the pebble content lower than 20%, the equivalent elastic modulus could be described by the "Stress Uniformity Model". When the content is 20%–60%, the state of RVE is a transition from "Stress Uniformity" to "Strain Uniformity". Through data analysis, the constitutive model and modified formula could be expressed as:

$$E = E_{stress} + \frac{(E_{strain} - E_{stress})(\omega - 0.2)}{0.4} \quad (20\% < \omega < 60\%) \tag{6}$$

Figure 3(b) illustrates the modified formula value and numerical experiment value.

References

1. Bai, Y.X., Qi, T.Y.: Prediction for surface collapse deformation of shield construction based on LESSVM. Chin. J. Rock Mechan. Eng. **32**(S2), 3666–3674 (2013)
2. Alan, J.M., James, H.S.: A model of structure optimization for a functionally graded material. Mater. Lett. **22**(1), 103–107 (1995)
3. Tian, Y.D., Qin, S.L.: Improved mixed-mode method of valid elastic modulus of composites. J. Southwest Jiaotong Univ. **40**(06), 783–787 (2005)
4. Hu, M., Xu, G.Y.: Study of equivalent elastic modulus of sand gravel soil with Eshelby tensor and Mori-Tanaka equivalent method. Rock Soil Mech. **34**(05), 1437–1442+1448 (2013)
5. Wang, M.R., Pan, N.: Predictions of effective physical properties of complex multiphase materials. Mater. Sci. Eng., R **63**(1), 1–30 (2008)
6. Paul, B.: Prediction of elastic constants of multiphase materials. Trans. Metall. Soc. AIME **218**(1), 36–41 (1960)

Isothermal Hyperelastic Model for Saturated Porous Media Based on Poromechanics

Yayuan Hu

Research Center of Coastal and Urban Geotechnical Engineering, Zhejiang University, Hangzhou 310058, People's Republic of China
huyayuan@zju.edu.cn

Abstract. Assuming that solid and fluid constituents have same temperature and based on the balance equation of energy, the isothermal isotropic hyperelastic theory of saturated porous media at finite strains is formulated from solid and fluid free energy in term of the invariant set of skeleton deformation or porous solid deformation.

Keywords: Saturated porous media · Finite deformation · Free energy Hyperelastic model

1 Introduction

Resulting both from nonlinear and irreversible material responses and from nonlinear geometrical effects, the large-strain theory of saturated porous media is still a challenging problem [1]. Gajo used macroscopic free-energy function to formulate isothermal hyperelastic model of saturated porous medium subjected to finite deformations [1], but his theory is only valid when the hydrostatic pressure of solid phase linear changes with the logarithmic volumetric strain. In order to avoid this shortcoming, the aim of this work is to investigate the isothermal hyperelastic equations in the framework of tensor theory in the continuum mechanics of porous media [2].

2 The Constitutive Framework of Hyper-poroelasticity

2.1 The Potential of Free Energy

A saturated porous medium consists of two constituents: solid (S) and fluid (F). $\alpha = \{S, F\}$ is the constituent index. $n_{\alpha 0}$ and n_α denotes the original and present volume fractions of α constituent, $n_S + n_F = 1$. $\rho_{\alpha 0}$ and ρ_α denotes the original and present partial densities of α constituent and $\rho_{R\alpha}$ satisfies $\rho_\alpha = n_\alpha \rho_{R\alpha}$ is its real material density. $J_{R\alpha} = \rho_{R\alpha 0}/\rho_{R\alpha}$, $\vartheta_\alpha = -\ln(J_{R\alpha})$. $\boldsymbol{\sigma}$ denotes the total stress. $P_T = -\boldsymbol{\sigma} : \mathbf{1}/3$. P_F denotes fluid pore pressure, $P_F^* = P_F$, $P_S^* = (P_T - n_F P_F^*)/n_S$, $\tilde{\boldsymbol{\sigma}} = \boldsymbol{\sigma} + P_F \mathbf{1}$, $\tilde{\sigma}_m = \tilde{\boldsymbol{\sigma}} : \mathbf{1}/3$. Let:
$\boldsymbol{F}_S = \partial \boldsymbol{x}_S / \partial \boldsymbol{X}_S$, $J_S = \det[\partial \boldsymbol{x}_S / \partial \boldsymbol{X}_S]$, $\boldsymbol{C}_S = \boldsymbol{F}_S^T \cdot \boldsymbol{F}_S$, $\boldsymbol{B}_S = \boldsymbol{F}_S \cdot \boldsymbol{F}_S^T$, $\boldsymbol{F}_H = J_{RS}^{-1/3} \boldsymbol{F}_S$,
$J_H = \det \boldsymbol{F}_H = J_S/J_{RS} = n_{S0}/n_S$, $\boldsymbol{C}_H = \boldsymbol{F}_H^T \cdot \boldsymbol{F}_H = J_{RS}^{-2/3} \boldsymbol{C}_S$, $\boldsymbol{B}_H = \boldsymbol{F}_H \cdot \boldsymbol{F}_H^T = J_{RS}^{-2/3} \boldsymbol{B}_S$, $\boldsymbol{E}_H = (\boldsymbol{C}_H - \mathbf{1})/2$, $\tilde{\boldsymbol{T}}_H = J_{RS}^{2/3} \boldsymbol{F}_S^{-1} \cdot \tilde{\boldsymbol{\sigma}} \cdot \boldsymbol{F}_S^{-T}$, $\overset{\backslash}{\boldsymbol{E}}_H = J_{RS}^{-2/3} \boldsymbol{F}_S^T [\text{grad } \overset{\backslash}{\boldsymbol{x}}_S +$

$(1/3)\overset{\backslash}{\vartheta}_S]F_S$, where C_S and B_S is the right and left Cauchy-Green tensor of porous solid, respectively. C_H and B_H is the right and left Cauchy-Green tensor of skeleton solid, respectively. E_H is the Green strain of skeleton. The material time derivative of Γ_α in the direction of α constituent is defined as $\overset{\backslash}{\Gamma}_\alpha = (\partial \Gamma_\alpha / \partial t) + \operatorname{grad} \Gamma_\alpha \cdot \overset{\backslash}{x}_\alpha$, where $\overset{\backslash}{x}_\alpha$ denotes the velocity of α constituent. $W_F = \overset{\backslash}{x}_F - \overset{\backslash}{x}_S$. We can obtain:

$$\sum_\alpha \rho_\alpha \overset{\backslash}{\mathcal{E}}_\alpha = (\tilde{T}_H : \overset{\backslash}{E}_H + n_S P_S^* \overset{\backslash}{\vartheta}_S) + n_F P_F \overset{\backslash}{\vartheta}_F + (P_F^* - P_F)\overset{\backslash}{\lambda}_F + W_F \cdot (P_F \operatorname{grad} n_F - \hat{p}_F) \qquad (1)$$

Let us consider the isothermal process of thermodynamics and the same temperature θ of solid and fluid constituents at the same point, η_α denotes the entropy of unit density for α constituent. The thermodynamic free energy of fluid constituent is assumed to be decoupled from porous solid. Thus, the free energy of fluid and porous solid can be expressed as $\mathcal{E}_F(\eta_F, \vartheta_F)$, and $\mathcal{E}_S(\eta_S, E_H, \vartheta_S)$. We have:

$$\sum_\alpha \rho_\alpha \overset{\backslash}{\mathcal{E}}_\alpha = \rho_S(\frac{\partial \mathcal{E}_S}{\partial \eta_S}\overset{\backslash}{\eta}_S + \frac{\partial \mathcal{E}_S}{\partial E_H}:\overset{\backslash}{E}_H + \frac{\partial \mathcal{E}_S}{\partial \vartheta_S}\overset{\backslash}{\vartheta}_S) + \rho_F(\frac{\partial \mathcal{E}_F}{\partial \eta_F}\overset{\backslash}{\eta}_F + \frac{\partial \mathcal{E}_F}{\partial \vartheta_F}\overset{\backslash}{\vartheta}_F) \qquad (2)$$

According to the thermodynamics and the rule that Eq. (1) is equal to Eq. (2) for any $\eta_S, \eta_F, E_H, \vartheta_S, \vartheta_F$ and λ_F, we can get

$$\theta = \frac{\partial \mathcal{E}_S}{\partial \eta_S} = \frac{\partial \mathcal{E}_F}{\partial \eta_F}, \quad \tilde{T}_H = \rho_S \frac{\partial \mathcal{E}_S}{\partial E_H}, \quad P_S^* = \rho_{RS}\frac{\partial \mathcal{E}_S}{\partial \vartheta_S}, \quad P_F = \rho_{RF}\frac{\partial \mathcal{E}_F}{\partial \vartheta_F}, \quad P_F^* - P_F = 0 \qquad (3)$$

For the isothermal process of thermodynamics ($d\theta = 0$), the Helmholtz free energy of porous solid and fluid material can be defined by $\psi_S = \mathcal{E}_S - \theta \eta_S$ and $\psi_F = \mathcal{E}_F - \theta \eta_F$, respectively, Eq. (3) becomes:

$$\tilde{T}_H = \rho_S \frac{\partial \psi_S}{\partial E_H}, \quad P_S^* = \rho_{RS}\frac{\partial \psi_S}{\partial \vartheta_S}, \quad P_F = P_F^* = \rho_{RF}\frac{\partial \psi_F}{\partial \vartheta_F} \qquad (4)$$

Equation (4) are the general hyperelastic constitutive equations of saturated porous media. After Eq. (2)–(3) are substituted into Eq. (1), entropy production of mechanical diffusion can be obtained:

$$\sum_\alpha \rho_\alpha \overset{\backslash}{\eta}_\alpha = W_F \cdot (P_F \operatorname{grad} n_F - \hat{p}_F)/\theta \qquad (5)$$

Based on Eq. (5), the following dissipative rate potential $\mathcal{D}^* = \mathcal{D}^*(W_F)$ can be assumed

$$P_F \operatorname{grad} n_F - \hat{p}_F = \frac{\partial \mathcal{D}^*}{\partial W_F} \qquad (6)$$

Equation (6) is regarded as the generalized Darcy's law of saturated porous medium.

2.2 Isotropic Constitutive Equations

The Helmholtz free energy of fluid material as a function of ϑ_F is too simple to discuss. Let us investigate the isotropic Helmholtz free energy of porous solid.

Let $I_{H1} = \text{tr}\,\boldsymbol{C}_H$, $I_{H2} = [(\text{tr}\,\boldsymbol{C}_H)^2 - \text{tr}\,\boldsymbol{C}_H^2]/2$ and $I_{H3} = \det \boldsymbol{C}_H$, be the first, second and third invariant of $\boldsymbol{C}_H$. ψ_S can be expressed by $\psi_S(I_{H1}, I_{H2}, I_{H3}, \vartheta_S)$ as

$$\frac{d I_{H1}(\boldsymbol{C}_H)}{d \boldsymbol{C}_H} = \mathbf{1}, \quad \frac{d I_{H2}(\boldsymbol{C}_H)}{d \boldsymbol{C}_H} = I_{H1}\mathbf{1} - \boldsymbol{C}_H, \quad \frac{d I_{H3}(\boldsymbol{C}_H)}{d \boldsymbol{C}_H} = I_{H3}\boldsymbol{C}_H^{-1} \tag{7}$$

As $\tilde{\boldsymbol{T}}_H = J_{RS}^{2/3} \boldsymbol{F}_S^{-1} \cdot \tilde{\boldsymbol{\sigma}} \cdot \boldsymbol{F}_S^{-T}$, the first equation of Eq. (4) becomes:

$$\tilde{\boldsymbol{\sigma}} = 2 J_{RS}^{-2/3} \boldsymbol{F}_S \cdot \tilde{\boldsymbol{T}}_H \cdot \boldsymbol{F}_S^T = 2\rho_S \left[\frac{\partial \psi_S}{\partial I_{H3}} I_{H3}\mathbf{1} + \left(\frac{\partial \psi_S}{\partial I_{H1}} + \frac{\partial \psi_S}{\partial I_{H2}} I_{H1} \right) \boldsymbol{B}_H - \frac{\partial \psi_S}{\partial I_{H2}} \boldsymbol{B}_H^2 \right] \tag{8}$$

As $\boldsymbol{B}_H^2 = I_{H1}\boldsymbol{B}_H - I_{H2}\mathbf{1} + I_{H3}\boldsymbol{B}_H^{-1}$, Eq. (8) can also be expressed by

$$\tilde{\boldsymbol{\sigma}} = 2\rho_S \left[\left(\frac{\partial \psi_S}{\partial I_{H2}} I_{H2} + \frac{\partial \psi_S}{\partial I_{H3}} I_{H3} \right) \mathbf{1} + \frac{\partial \psi_S}{\partial I_{H1}} \boldsymbol{B}_H - \frac{\partial \psi_S}{\partial I_{H2}} I_{H3}\boldsymbol{B}_H^{-1} \right] \tag{9}$$

In order to distinguish between the volumetric and distorted deformations of porous solid, let $\hat{\boldsymbol{F}}_H = J_H^{-1/3} \boldsymbol{F}_H = J_S^{-1/3} \boldsymbol{F}_S$. So $\hat{\boldsymbol{C}}_H = J_H^{-2/3} \boldsymbol{C}_H = J_S^{-2/3} \boldsymbol{C}_S$, $\hat{\boldsymbol{B}}_H = J_H^{-2/3} \boldsymbol{B}_H = J_S^{-2/3} \boldsymbol{B}_S$. Their corresponding invariants are $\hat{I}_{H1} = J_H^{-2/3} I_{H1}$, $\hat{I}_{H2} = J_H^{-4/3} I_{H2}$, $\hat{I}_{H3} = 1$. Now, $\psi_S(I_{H1}, I_{H2}, I_{H3}, \vartheta_S)$ can be expressed by $\psi_S(\hat{I}_{H1}, \hat{I}_{H2}, J_H, \vartheta_S)$. So $\tilde{\boldsymbol{\sigma}}$ is expressed as:

$$\tilde{\boldsymbol{\sigma}} = 2\rho_S \left[J_H^{-\frac{2}{3}} \frac{\partial \psi_S}{\partial \hat{I}_{H1}} \left(\boldsymbol{B}_H - \frac{1}{3} I_{H1}\mathbf{1} \right) + J_H^{-\frac{4}{3}} \frac{\partial \psi_S}{\partial \hat{I}_{H2}} \left(I_{H1}\boldsymbol{B}_H - \boldsymbol{B}_H^2 - \frac{2}{3} I_{H2}\mathbf{1} \right) + \frac{1}{2} J_H \frac{\partial \psi_S}{\partial J_H} \mathbf{1} \right] \tag{10}$$

As $\text{tr}(\boldsymbol{B}_H - 1/3 I_{H1}\mathbf{1}) = 0$ and $\text{tr}(I_{H1}\boldsymbol{B}_H - \boldsymbol{B}_H^2 - 2/3 I_{H2}\mathbf{1}) = 0$, the first two terms in the square bracket of Eq. (10) determine the deviator effective stress. The end term determines the hydrostatic effective stress resulted from volumetric deformations.

Let us derive the isotropic hyperelastic model in term of $\boldsymbol{B}_S$. The formulations $\boldsymbol{C}_H = J_{RS}^{-2/3} \boldsymbol{C}_S$, $\vartheta_S = \ln(\rho_{RS}/\rho_{RS0})$ and $J_{RS} = \rho_{RS0}/\rho_{RS}$ yield $\boldsymbol{C}_S = \exp(-\frac{2}{3}\vartheta_S)\boldsymbol{C}_H$, so

$$\frac{\partial \boldsymbol{C}_S}{\partial \boldsymbol{C}_H} = J_{RS}^{2/3} = \exp(-\frac{2}{3}\vartheta_S)\mathbf{1}, \quad \frac{\partial \vartheta_S}{\partial \boldsymbol{C}_H} = \mathbf{0}, \quad \frac{\partial \boldsymbol{C}_S}{\partial \vartheta_S} = -\frac{2}{3}\boldsymbol{C}_S, \quad \frac{\partial \vartheta_S}{\partial \vartheta_S} = 1 \tag{11}$$

Using Eq. (11), $\rho_S = n_S \rho_{RS}$, $P_F^* = P_F$, $\tilde{\sigma} = \sigma + P_F \mathbf{1}$, Eq. (4) becomes:

$$\tilde{\sigma} = 2\rho_S \mathbf{F}_S \cdot \frac{\partial \psi_S}{\partial \mathbf{C}_S} \cdot \mathbf{F}_S^T, \quad P_F^* = P_S^* + \frac{\tilde{\sigma}_m}{n_S} = \rho_{RS} \frac{\partial \psi_S}{\partial \vartheta_S} \quad (12)$$

Equation (12) shows that if ψ_S is expressed by $\mathbf{C}_S$ and ϑ_S, P_F^* rather than P_S^* is obtained from Eq. (12).

Similar to the above analysis, when ψ_S is expressed by $\psi_S(I_{S1}, I_{S2}, I_{S3}, \vartheta_S)$, where I_{S1}, I_{S2} and I_{S3} is the three invariants of $\mathbf{C}_S$ and $\mathbf{B}_S$, Eq. (12) yields:

$$\tilde{\sigma} = 2\mathbf{F}_S \cdot \frac{\partial \psi_S}{\partial \mathbf{C}_S} \cdot \mathbf{F}_S^T = 2\rho_S [\frac{\partial \psi_S}{\partial I_{S3}} I_{S3} \mathbf{1} + (\frac{\partial \psi_S}{\partial I_{S1}} + \frac{\partial \psi_S}{\partial I_{S2}} I_{S1}) \mathbf{B}_S - \frac{\partial \psi_S}{\partial I_{S2}} \mathbf{B}_S^2] \quad (13)$$

or $\quad \tilde{\sigma} = 2\rho_S [(\frac{\partial \psi_S}{\partial I_{S2}} I_{S2} + \frac{\partial \psi_S}{\partial I_{S3}} I_{S3}) \mathbf{1} + \frac{\partial \psi_S}{\partial I_{S1}} \mathbf{B}_S - \frac{\partial \psi_S}{\partial I_{S2}} I_{S3} \mathbf{B}_S^{-1}] \quad (14)$

when ψ_S is expressed by $\psi_S(\hat{I}_{S1}, \hat{I}_{S2}, J_S, \vartheta_S)$, where $\hat{I}_{S1} = J_S^{-2/3} I_{S1}$, $\hat{I}_{S2} = J_S^{-4/3} I_{S2}$, the formula $\tilde{\sigma}$ can be expressed by the following equation:

$$\tilde{\sigma} = 2\rho_S [J_S^{-\frac{2}{3}} \frac{\partial \psi_S}{\partial \hat{I}_{S1}} (\mathbf{B}_S - \frac{1}{3} I_{S1} \mathbf{1}) + J_S^{-\frac{4}{3}} \frac{\partial \psi_S}{\partial \hat{I}_{S2}} (I_{S1} \mathbf{B}_S - \mathbf{B}_S^2 - \frac{2}{3} I_{S2} \mathbf{1}) + \frac{1}{2} J_S \frac{\partial \psi_S}{\partial J_S} \mathbf{1}] \quad (15)$$

3 Conclusions

Tensor theory is used to derive the isothermal hyperelastic model of porous media. The isotropic hyperelasticity is formulated in term of the variants of skeleton deformation or porous media deformation.

References

1. Gajo, A.A.: General approach to isothermal hyperelastic modelling of saturated porous media at finite strains with compressible solid constituents. Proc. Roy. Soc. A Math. Phys. Eng. Sci. **466**, 3061–3087 (2010)
2. Hu, Y.Y.: Study on the supper viscoelastic constitutive theory of saturated porous media. Appl. Math. Mech. (Chin. Ed.) **37**(6), 584–598 (2016)
3. Gray, W.G., Schrefler, B.A.: Analysis of the solid phase stress tensor in multiphase porous media. Int. J. Numer. Anal. Meth. Geomech. **31**, 541–581 (2007)
4. Truesdell, C., Noll, W.: The non-linear field theories of mechanics. Spinger-Verlag, Hongkong (1965)

Review of N Models in Simulating Mechanical Behavior of Clays

Jianhong Jiang[1(✉)], Hoe I. Ling[2], and Victor N. Kaliakin[3]

[1] School of Rail Transportation, Soochow University, Suzhou 215131, Jiangsu, China
jianhong.jiang@suda.edu.cn
[2] Department of Civil Engineering and Engineering Mechanics, Columbia University, New York, USA
[3] Department of Civil and Environmental Engineering, University of Delaware, Newark, USA

Abstract. In traditional clay modeling, the form of the yield surface was usually proposed by curve fitting the yield points of the particular clay studied, thus the application of the resulting model could be limited to particular clays. Different from that, through a careful examination of a database of yield surfaces for natural clays, the form of work dissipation based yield surface of Dafalias was adopted. A surface configuration parameter, N, was introduced to the yield surface by emphasizing on the relationship between the shear strength and the shape of yield surfaces for clays. This yield configuration parameter in fact represents a measure of the shear strength over the anisotropic line. Its numerical value can be different from the surface configuration parameter of the plastic potential, M, which also represents a frictional constant associated with the critical state. The resulting models developed with the yield surface configuration parameter, N, would be simply called N models for convenience. Simulations of mechanical behavior of different types of clays under various shearing modes were briefly reviewed here. The review may indicate that the N models have potential wide range of applications to different type of clays, and have the capacity to accurately capture the shear strengths.

Keywords: N models · Clays · Mechanical behavior · Shear strength

1 Review of N Models in Simulating Mechanical Behavior of Clays

The framework of the N models is shown in Fig. 1 and the role of N in capturing different shear strengths with the same M is illustrated in Fig. 2 (Jiang and Ling 2010 in Mech. Res. Commun.). Usually clays behave differently on the compression side and extension side, thus Nc and Ne are adopted for the two sides in the model. A bounding surface N model was developed and was shown to be able to simulate both the strain-hardening Kaolin clay and strain-softening Boston Blue clay, as shown in Figs. 3 and 4, respectively, which usually studied separately in literature (Jiang et al. 2012 in J. Appl. Mech.). In simulating the Kaolin clay behavior under undrained triaxial compression and extension tests with varying overconsolidation ratios, a value of Nc (Ne) smaller than Mc (Me) was adopted to capture the shear strengths. For the triaxial compression tests on

Boston Blue clay, a value of Nc smaller than Mc was adopted to approximately simulate the strain-softening behavior of the normally consolidated specimen. For the triaxial extension tests on Boston Blue clay, which did not show strain-softening behavior, a value of Ne equal to Me was adopted. By focusing on basic features of clays, the bounding surface N model simulated Pisa clay in true triaxial tests as shown in Fig. 5 (Jiang et al. 2013 in J. Appl. Mech.). In the stress-controlled tests, the simulated response of deviatoric strain agreed well with the test data, while the simulated response of volumetric strain provided a guidance for the normal response if the test data did not suddenly change during the test. A time-dependent N model was developed by incorporating the time-dependent effects of clays (Jiang et al. 2017 in Acta Geotech.). With the help of this model, the varying effects of lower-level parameters were discussed on the undrained multistage stress relaxation response for normally consolidated soils which had been ignored in literature. For the Taipei silty clay, the flat yield surface of the clay was simply achieved by adjusting the surface configuration parameters, thus a new heavy formulation specific for this clay was avoided. The resulting simulations for the triaxial tests were generally in a good agreement with the test data as shown in Fig. 6. Some discussions with the help of the basic bounding surface N model were made on approximate simulation of natural structured soft clays (Jiang et al. 2017 in Int. J. Geomech.) and on a damage law for creep rupture of clays (Jiang et al. 2017 in Mech. Res. Commun.). In addition to the above, the N models were validated against test results of several other clays under various shearing modes.

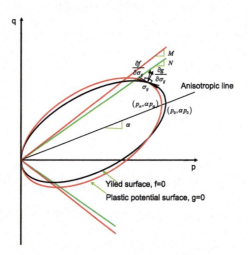

Fig. 1. Framework of N models.

Review of N Models in Simulating Mechanical Behavior of Clays 37

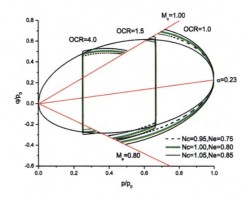

Fig. 2. Role of N.

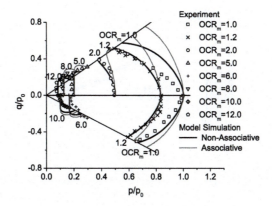

Fig. 3. Simulation of Kaolin clay.

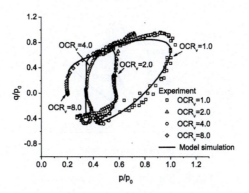

Fig. 4. Simulation of Boston Blue clay.

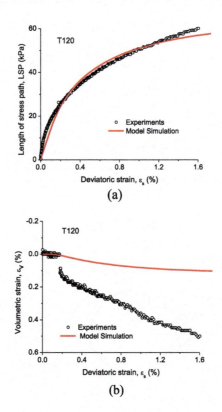

Fig. 5. Simulation of Pisa clay: (a) Deviatoric strain response; (b) Volumetric strain V.S. Deviatoric strain.

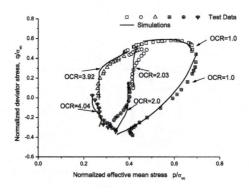

Fig. 6. Simulation of Taipei silty clay.

Strain-Rate and Temperature Dependency for Preconsolidation Pressure of Soft Clay

Ting Li[1], Qi-Yin Zhu[1(✉)], Lian-Fei Kuang[1], and Li Gang[2]

[1] State Key Laboratory for Geomechanics and Deep Underground Engineering, China University of Mining and Technology, Xuzhou 221116, China
qiyin.zhu@gmail.com
[2] Jinan Rail Transit Group Co., Ltd., Jinan 250101, China

Abstract. Both of strain-rate and temperature have significant effects on the one-dimensional (1D) compression behavior of soils. This paper focuses on the time and temperature dependent behavior of the preconsolidation pressure, which is investigated in double logarithm plane. Based on the linear correlation, a strain-rate related parameter and a temperature related parameter are proposed to describe the rate-dependency and temperature dependency, respectively. Finally, a constitutive equation is introduced that expressesing the evolution of the preconsolidation pressure with respect to strain rate and temperature. The evolution of the preconsolidation pressure is the basis of any general elasto-thermo-plastic framework to allow the coupling of the thermoplastic deformation with the modification of the state limit.

Keywords: Structured clay · Strain-Rate · Bonding degradation · Temperature

1 Introduction

Natural soft clays are subjected to action of heat under many different circumstances, for instance, the nuclear waste isolation, heat energy storage, geothermal development, etc. A variety of oedometer tests have been conducted to study the effect of temperature on soft clays [1, 2]. Several constitutive models that describe the influence of temperature on the mechanical behavior have been proposed in the literature [3, 4]. Besides, the compressibility of soft clays is also largely dependent on the strain-rate. After studying 1D compression behavior of soft clays on a variety of tests, [5] concluded that the behavior is controlled by a unique effective stress-strain-strain rate relationship.

However, few of models can capture the combined effect of strain-rate and temperature on the compression behavior at 1D condition of soft clay. In this paper, we focus on the 1D behavior which can bring fundamental features for more general mechanical behavior.

2 Strain-Rate Dependency of Preconsolidation Pressure

In one-dimensional condition, the strain rate-dependency of preconsolidation pressure (σ'_p) has been attracted much attentions since last decade. CRS test results show that larger loading rate can result in larger preconsolidation pressure. Figure 1 plots $\log(\sigma'_p/\sigma'_{v0})$ against $\log(\dot{\varepsilon}_v)$ for seven clays reported in literature (the preconsolidation pressure normalized by in situ vertical stress versus vertical strain rate). It can be seen that the relationship between $\log(\sigma'_p/\sigma'_{v0})$ and $\log(\dot{\varepsilon}_v)$ is essentially linear for each clay. Thus, the effect of strain-rate on σ'_p can be represented by parameter β, which can be estimated from the slope of the best-fit line of the data points. Thereby, preconsolidation pressure σ'_p corresponding to $\dot{\varepsilon}_v$ can be obtained by

$$\frac{\sigma'_p}{\sigma'^{r}_p} = \left(\frac{\dot{\varepsilon}_v}{\dot{\varepsilon}^{r}_v}\right)^{1/\beta} \quad (1)$$

where σ'^{r}_p is the reference preconsolidation pressure corresponding to reference strain rate $\dot{\varepsilon}^{r}_v$.

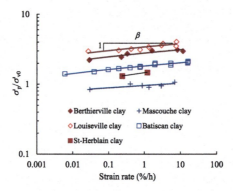

Fig. 1. Effect of strain-rate on the normalized preconsolidation pressure at the same temperature for each clay

3 Temperature Dependency of Preconsolidation Pressure

The influence of temperature on preconsolidation pressure have been studied by oedometer tests and isotropic compression tests with constant strain-rate and variable temperatures. All of the experimental results show that the preconsolidation pressure will decrease with an elevated temperature. Figure 2 presents the normalized preconsolidation pressure with temperature. The regression analyses indicate that it is also reasonable to assume a linear relationship between $\log(\sigma'_p/\sigma'^{r}_p)$ and $\log(T/T^r)$. Thus, the relationship between preconsolidation pressure and temperature can be fitted by the following equation

$$\frac{\sigma'_p}{\sigma'^{r}_p} = \left(\frac{T}{T^r}\right)^{\theta} \tag{2}$$

where θ is a thermal parameter; σ'_p and σ'^{r}_p are the preconsolidation pressures at temperature T and the reference temperature T^r, respectively.

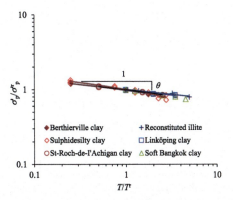

Fig. 2. Effect of temperature on the normalized preconsolidation pressure at the same strain-rate for each clay

4 Combined Expression of Preconsolidation Pressure

As indicated above, the preconsolidation pressure is associating with strain-rate and temperature and can be written as a function of strain rate and temperature. Furthermore, the experiments considering the coupling effect of strain rate and temperature [6, 7] show that β is essentially independent of temperature and, reversely, θ is independent of strain rate. Thus, the preconsolidation pressure can be written as follows:

$$\sigma'_p = f_1(\dot{\varepsilon}_v) f_2(T) \tag{3}$$

Considering Eqs. (1) and (2), the evolution of preconsolidation pressure with respect to strain rate and temperature can be expressed as:

$$\sigma'_{p,T} = \sigma'^{r}_{p,T^r} \left(\frac{\dot{\varepsilon}_v}{\dot{\varepsilon}^r_v}\right)^{1/\beta} \left(\frac{T}{T^r}\right)^{\theta} \tag{4}$$

where $\sigma'_{p,T}$ is the preconsolidation pressure corresponding to the test conducted with strain-rate $\dot{\varepsilon}_v$ and temperature T; σ'^{r}_{p,T^r} is the preconsolidation pressure corresponding to reference strain rate $\dot{\varepsilon}^r_v$ and reference temperature T^r.

5 Conclusion

It is well known that soil behavior is influenced by the rate at which it is strained, as well as by the temperature to which it is subjected to. The strain-rate and temperature dependency of preconsolidation pressure of soft clay has been investigated based on experimental which make the quantification of two effects possible. The evolution model of the preconsolidation pressure could be introduced into any general elasto-thermoplastic framework to allow the coupling of the thermoplastic deformation with the modification of the state limit. The determination of the model parameters is in straightforward.

Acknowledgment. This research project is financially supported by National Natural Science Foundation of China (41502271, 51504245), and 111 Project (B14021).

References

1. Abuel-Naga, H.M., Bergado, D.T., Bouazza, A., Ramana, G.V.: Volume change behaviour of saturated clays under drained heating conditions: experimental results and constitutive modeling. Can. Geotech. J. **44**(8), 942–956 (2007)
2. Jarad, N., Cuisinier, O., Masrouri, F.: Effect of temperature and strain rate on the consolidation behaviour of compacted clayey soils. Eur. J. Environ. Civil Eng. **3**, 1–18 (2017)
3. Leroueil, S., Kabbaj, M., Tavenas, F., Bouchard, R.: Stress–strain–strain rate relation for the compressibility of sensitive natural clays. Géotechnique **35**(2), 159–180 (1985)
4. Tsutsumi, A., Tanaka, H.: Combined effects of strain rate and temperature on consolidation behavior of clayey soils. Soils Found. **52**(2), 207–215 (2012)
5. Yashima, A., Leroueil, S., Oka, F., Guntoro, I.: Modelling temperature and strain rate dependent behavior of clays: one dimensional consolidation. Soils Found. **38**(2), 63–73 (1998)
6. Marques, M.E.S., Leroueil, S.: Viscous behaviour of St-Roch-de-l'Achigan clay, Quebec. Can. Geotech. J. **41**(3), 25–38 (2004)
7. Boudali, M., Leroueil, S., Srinivasa Murthy, B.R.: Viscous behaviour of natural clays. In: Proceedings of the 13th International Conference on Soil Mechanics and Foundation Engineering, pp. 411–416. A.A. Balkema, Rotterdam (1994)

A Nonlinear Elastic Model for Dilatant Coarse-Grained Materials

Si-hong Liu, Yi Sun[✉], and Chao-min Shen

Hohai University, Nanjing 210098, China
sunyihhu@163.com

Abstract. Considering the coupling effect of the mean stress p and the deviatoric stress q on the deformation of coarse-grained materials, a nonlinear elastic K-G-J model that can account for the dilatancy and the intermediate principal stress of coarse-grained materials is established. The dilatancy equation derived from the microstructural changes of granular materials is introduced into the constitutive model to reflect the dilatancy of coarse-grained materials, and the strength nonlinearity of coarse-grained materials is considered by using a logarithmic relationship between the peak internal friction angle and the mean stress. Meanwhile, the SMP criterion is introduced into the model simply by using a transformed stress tensor, so that the model can reflect the influence of the intermediate principal stress. The model prediction of the triaxial compression and the extension tests on Toyoura sands agrees well with the experimental results, illustrating reasonability of the model.

Keywords: Coarse-grained materials · Dilatancy · Intermediate principal stress
Nonlinear elastic · Constitutive model

1 Introduction

As main construction materials of earth/rockfill dams, coarse-grained materials deform mainly due to the slippage between particles and the breakage of particles under loading. Triaxial compression tests have shown that the slippage and the dislocation between particles during the tests result in significant volumetric deformation of coarse-grained materials [1], which are contractive under the low density and high confining pressure, dilatant under the high density and low confining pressure. In addition, the shear strengths of coarse-grained materials are nonlinear and depend on stress paths. The constitutive model of coarse-grained materials should reflect these characteristics reasonably.

In this paper, we establish a nonlinear elastic K-G-J model accounting for the dilatancy and the intermediate principal stress of coarse-grained materials. The model is concise and can be applied to engineering practice.

2 The Nonlinear Elastic Constitutive Model

2.1 The Form of the Model

Considering the coupling effect of the mean stress p and the deviatoric stress q on the deformation of coarse-grained materials, the stress-strain relationship of coarse-grained materials can be expressed as

$$d\varepsilon_v = d\varepsilon_{vp} + d\varepsilon_{vq} = \frac{dp}{K} + \frac{dq}{J_1}$$
$$d\varepsilon_s = d\varepsilon_{sp} + d\varepsilon_{sq} = \frac{dp}{J_2} + \frac{dq}{G} \quad (1)$$

where K, J_1, G, J_2 are the bulk modulus, the dilatancy modulus, the shear modulus and the compression modulus, respectively. In this study, we assume that the dilatancy modulus is equal to the compression modulus, i.e. $J_1 = J_2 = J$, where J is regarded as the coupling modulus.

2.2 Determination of the Bulk Modulus

The experimental results show that the exponential relationship between the volumetric strain and the mean stress exists under isotropic compression [2], which can be expressed as

$$\varepsilon_{vp} = C_t \left[\left(\frac{p}{P_a}\right)^m - \left(\frac{p_0}{P_a}\right)^m \right] \quad (2)$$

The bulk modulus can be determined by differentiating the above formula.

$$K = \frac{dp}{d\varepsilon_{vp}} = \frac{P_a}{C_t m} \left(\frac{p}{P_a}\right)^{1-m} = K_b P_a \left(\frac{p}{P_a}\right)^{n_1} \quad (3)$$

in which K_b, n_1 are the parameters.

2.3 Determination of the Shear Modulus

For coarse-grained materials, the stress-strain relationship under the testing path of the constant p can be expressed as

$$\frac{q}{3} = \frac{\varepsilon_{sq}}{a + b\varepsilon_{sq}} \quad (4)$$

Differentiating the above formula, we can get the following equation for the shear modulus.

$$G = \frac{dq}{d\varepsilon_{sq}} = \frac{3}{a}\left[1 - b\left(\frac{q}{3}\right)\right]^2 = G_i\left(1 - R_f\frac{q}{q_f}\right)^2 \qquad (5)$$

In order to take the intermediate principal stress into account, the SMP criterion [3] proposed by Matsuoka and Nakai is introduced into the Eq. (5) through the transformed stress tensor [4]. Thus, Eq. (5) of the shear modulus is rewritten as

$$G = K_G P_a \left(\frac{\tilde{p}}{P_a}\right)^{n_2} \left(1 - R_f \frac{\tilde{q}}{M_f \tilde{p}}\right)^2 \qquad (6)$$

where $\tilde{p}, \tilde{q}$ are the transformed stress tensors, the expression of which may be found in reference [4].

2.4 Determination of the Coupling Modulus

From Eq. (1), we can get the following relation on the assumption of $J_1 = J_2 = J$

$$\frac{d\varepsilon_v}{d\varepsilon_s}\left(\frac{dp}{J} + \frac{dq}{G}\right) = \frac{dp}{K} + \frac{dq}{J} \qquad (7)$$

Based on the change of the microstructures of granular materials, Liu et al. [5] proposed a stress-dilatancy equation, expressed as

$$D = \frac{d\varepsilon_v}{d\varepsilon_s} = \frac{mM^{m+1} - m\eta^{m+1}}{(m+1)\eta^m} \qquad (8)$$

Where η is the stress ratio q/p and M is the stress ratio corresponding to the zero volumetric strain, names as dilation stress ratio. m is a parameter.

Combining Eqs. (7) and (8), the coupling modulus can be got

$$J = \frac{D - dq/dp}{\frac{1}{K} - D\frac{dq/dp}{G}} \qquad (9)$$

2.5 Strength Nonlinearity of Coarse-Grained Materials

The strength nonlinearity of coarse-grained materials is often considered by using a following logarithmic relationship between the peak internal friction angle φ and the mean stress p in projects of earth-rockfill dams.

$$\varphi = \varphi_0 - \Delta\varphi \lg\left(\frac{p}{p_a}\right) \tag{10}$$

Similarly, the dilation internal friction angle ψ is also assumed to have a logarithmic relationship with the mean stress p.

Peak stress ratio and dilation stress ratio can be obtained by the following formula

$$\left.\begin{aligned} M_f &= \frac{6\sin\varphi}{3-\sin\varphi}, \\ M &= \frac{6\sin\psi}{3-\sin\psi}, \end{aligned}\right\} \tag{11}$$

3 Model Verification

The triaxial compression and extension tests on Toyoura sands are predicted by the proposed model, which are compared with the experimental results as shown in Figs. 1 and 2. The good agreement between them illustrates that the proposed model can not only reflect the nonlinear stress-strain relationship and the dilatancy under the low confining pressure but also consider the influence of the intermediate principal stress.

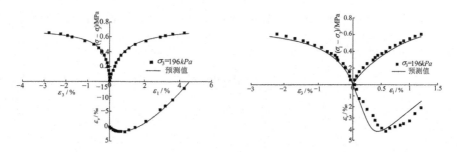

Fig. 1. Triaxial compression test results vs. model predictions (Toyoura sands) (left)

Fig. 2. Triaxial extension test results vs. model predictions (Toyoura sands) (right)

Acknowledgements. This work was supported by the "National Natural Science Foundation of China" (Grant No. U1765205) and the "National Key R&D Program of China" (Grant No. 2017YFC0404800).

References

1. Cheng, Z., Ding, H., Wu, L.: Experimental study on mechanical behavior of granular material. Chin. J. Geotech. Eng. **29**(8), 1151–1158 (2007)
2. Sun, D., Yao, Y.: Practical elastoplastic model for granular materials. Chin. J. Rock Mech. Eng. **21**(8), 1147–1152 (2002)
3. Matsuoka, H., Nakai, T.: Stress - deformation and strength characteristics of soil under three different principal stresses. Doboku Gakkai Ronbunshu **1974**(232), 59–70 (1974)
4. Yao, Y.P., Luo, T., Sun, D.A.: A simple 3-D constitutive model for both clay and sand. Chin. J. Geotech. Eng. **24**(2), 240–246 (2002)
5. Liu, S., Shao, D., Shen, C.: A microstructure-based elastoplastic constitutive model for coarse-grained materials. Chin. J. Geotech. Eng. **39**(5), 777–783 (2017)

Influence of Multidirectional Principal Stress Axes Rotation Sequences on Deformation of Sand

Long Xue, Rui Wang, and Jian-Min Zhang

Department of Hydraulic Engineering, National Engineering Laboratory for Green and Safe Construction Technology in Urban Rail Transit, Tsinghua University, Beijing 100084, China
xuel14@mails.tsinghua.edu.cn, zhangjm@mail.tsinghua.edu.cn

Abstract. Principal stress axes (PSA) rotation tests with different rotation axes are conducted using discrete element modeling to investigate the influence of PSA rotation sequences in various directions on the deformation of sand. Results indicate that the volume strain is dependent on the rotation sequence during the first few cycles of stress rotation, but the ultimate volumetric strain after significant principal stress axes rotation load cycles in all the different directions is independent of the rotation sequence.

Keywords: Multidirectional principal stress axes rotation · Rotation sequence Discrete element method · Volumetric strain

1 Introduction

Soil subjected to cyclic loads such as traffic and wave loads often undergo principal stress axes (PSA) rotation [1–4], and the rotation axes may change during loading. For example, rotation direction of PSA in the seabed varies frequently with the varying wave direction [1]. In this study, discrete element tests are conducted to investigate the deformation behaviors of granular materials under multidirectional PSA rotation, using a method of achieving PSA rotation extended from Tong et al. [5].

2 Numerical Specimens

The numerical tests use spherical specimen composed of clumped particles (see Fig. 1(b), r denotes the radius of each single sphere of the clumped particle) is 'trimmed' from the particles pluviated under gravity, as shown in Fig. 1(a). Two specimens of the particle size distribution representing that of Toyoura sand (see Fig. 1(c)), a dense one (specimen 1) and a loose one (specimen 2), are used with the specimen diameter D = 8.33 mm and specific gravity G_s = 2.6. Each specimen contains around 20000 particles.

A linear force-displacement contact law is chosen in conjunction with the Coulomb's friction criterion in the DEM to describe the interparticle contacts. The normal stiffness k_n and the tangential stiffness k_t are determined from two input parameters E_c and ν_c: $k_n = E_c r_0$, $k_t = \nu_c k_n$, where $r_0 = 2r_1 r_2 /(r_1 + r_2)$ is the common radius of the two particles in contact. In this study, E_c is 500 MPa and ν_c is 0.3. Intersphere friction coefficient μ is 0.5. Normal viscous damping is adopted with coefficient $c = 0.2c_{cr}$, where c_{cr} is the critical viscous damping coefficient of the local vibration system. For numerical stability, numerical damping is also used with coefficient of 0.7.

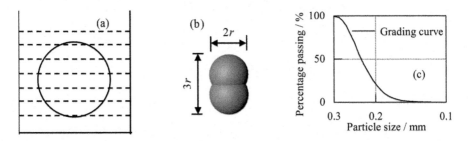

Fig. 1. (a) Sample trimming from the settled particles, (b) a clumped particle, (c) grading curve.

3 PSA Rotation Sequences in Various Directions

The specimens are first consolidated with σ_1, σ_2, σ_3 in the z, x, y directions, respectively. Then, tests with different sequences of 2D pure PSA rotations about the x, y, and z axes are conducted. 2D PSA rotations in various directions are referred to as mode a, b, c, respectively, as shown in Fig. 2. One cycle refers to PSA rotation from 0° to 180°. Stress rotation tests with different intermediate principal stress coefficient b have been conducted. Due to limitation of the paper, only results of $b = 0$ ($\sigma_1 > \sigma_2 = \sigma_3$) are exhibited below.

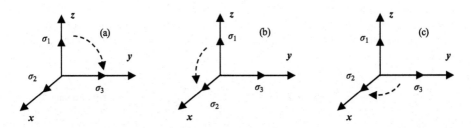

Fig. 2. Three different principal stress rotation modes: (a) mode a, (b) mode b, (c) mode c.

For a transverse isotropic specimen consolidated with $\sigma_2 = \sigma_3$ in the isotropic plane, mode a is equivalent to mode b, while mode c does not induce any change in the specimen, and is hence not analyzed here.

In this study, numerical tests are conducted under stress paths of continuous rotation (C-Rot) under mode a, alternative rotation alternating between mode a and mode b after each cycle (A-Rot), and alternative rotation alternating between mode a and mode b every ten cycles (CA-Rot). The evolution of various strain components of specimen 1 under CA-Rot is exhibited in Fig. 3(a) and volume strains under the three different stress paths are exhibited in Fig. 3(b), with $p = (\sigma_1 + \sigma_2 + \sigma_3)/3 = 100$ kPa, $q = \{[(\sigma_1 - \sigma_2)^2 +$

$(\sigma_2 - \sigma_3)^2 + (\sigma_3 - \sigma_1)^2]/2\}^{1/2} = 60$ kPa. Extensional ε_z and compressive ε_x and ε_y are observed in the overall trend of each cycle, as shown in Fig. 3(a).

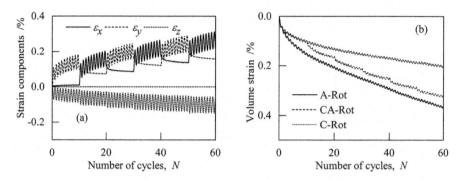

Fig. 3. (a) Evolution of the strain components of specimen 1 under CA-Rot, (b) volume strain of specimen 1 under different stress paths with $b = 0$, $p = 100$ kPa, $q = 60$ kPa.

At the same number of cycles, the volume strain induced by A-Rot or CA-Rot in two orthogonal planes is much larger than that induced by C-Rot in only one plane, as shown in Fig. 3(b). A sudden increase in the volume strain accumulation rate occurs every time the rotation direction changes. The rate of the volume strain accumulation reduces progressively as the rotation continues in the same direction.

Comparisons between the results from A-Rot and CA-Rot shows that the volume strain is rotation-sequence dependent during the first several cycles, but eventually converges after a significant number of cycles. Take the result in Fig. 3(b) as an example, when the number of the cycles reaches twenty, the two specimens under A-Rot and CA-Rot have both undergone stress rotation of ten cycles about x direction and ten cycles about y direction, but distinct difference in volume strains of the two specimens still exists. As the rotation continues, the strains of the two specimens converge progressively.

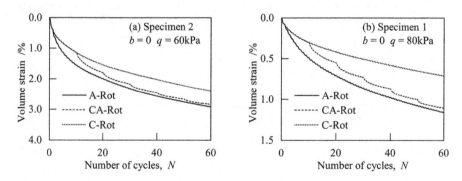

Fig. 4. Volume strain of (a) specimen 2 with $b = 0$, $p = 100$ kPa, $q = 60$ kPa and (b) specimen 1 with $b = 0$, $p = 100$ kPa, $q = 80$ kPa under different stress paths.

Volume strains of specimen 2 under the same three stress paths are demonstrated in Fig. 4(a), while Fig. 4(b) shows the volume strains of specimen 1 under similar stress rotation with larger q ($p = 100$ kPa, $q = 80$ kPa). Similar deformation patterns to those in Fig. 3 are also exhibited in the tests on the loose specimen (specimen 2) and on the dense specimen (specimen 1) under greater q, as shown in Fig. 4.

4 Concluding Remarks

Numerical tests are conducted to investigate the deformation behavior of sand under multidirectional principal stress axes rotation sequences. The main observations are summarized as follows:

(1) The volume strain is rotation-sequence dependent during the first few cycles of stress rotation, but the ultimate volumetric strain after significant principal stress axes rotation load cycles in all the different directions is rotation-sequence independent.
(2) When $b = 0$, the volume strain induced by principal stress axes rotation alternating in both the two orthogonal planes vertical to the bedding plane is much greater than that induced by continuous rotation in only one plane.

References

1. Ishihara, K., Towhata, I.: Sand response to cyclic rotation of principal stress directions as induced by wave loads. Soils Found. **23**(4), 11–26 (1983)
2. Miura, K., Miura, S., Toki, S.: Deformation behavior of anisotropic dense sand under principal stress axes rotation. Soils Found. **26**(1), 36–52 (1986)
3. Tong, Z.X., Zhang, J.M., Yu, Y.L., Zhang, G.: Drained deformation behavior of anisotropic sands during cyclic rotation of principal stress axes. J. Geotech. Geoenviron. Eng. **136**(11), 1509–1518 (2010)
4. Liu, C., Zhang, J.M.: Experimental Research on the deformation of sand under principal stress back-and-forth rotation. China Earthq. Eng. J. **39**(1), 28–31 (2017)
5. Tong, Z.X., Fu, P.C., Dafalias, Y.F., Yao, Y.P.: Discrete element method analysis of non-coaxial flow under rotational shear. Int. J. Numer. Anal. Meth. Geomech. **38**(14), 1519–1540 (2014)

Transversely Isotropic Strength Criterion for Soils Based on the Characteristic Mobilized Plane

Dechun Lu[1](✉), Jingyu Liang[1], Chao Ma[2], and Xiuli Du[1]

[1] Key Laboratory of Urban Security and Disaster Engineering of Ministry of Education, Beijing University of Technology, Beijing, China
dechun@bjut.edu.cn
[2] School of Civil and Transportation Engineering,
Beijing University of Civil Engineering and Architecture, Beijing, China

Abstract. The characteristic stress was proposed to comprehensively reflect the acting stress and the material physical properties. The two methods of establishing a unified strength criterion can be unified under the concept of the characteristic stress, including adjusting the acting stress and adjusting the mobilized plane. The characteristic stress is used here to measure and adjust the position of the mobilized plane which will be called the characteristic mobilized plane (CMP). The proposed CMP can reflect the frictional rules of various geomaterials, and the shear-normal stress ratio on the CMP is used to directly establish a unified strength criterion for geomaterials. The CMP can also be taken as a reference to propose a strength parameter for transversely isotropic soils. By combining the transversely isotropic strength parameter with the CMP criterion, a transversely isotropic CMP (TI-CMP) criterion can be established. The intermediated principal stress effects, as well as the effects of the anisotropy on the peak shear strength can be described by the proposed TI-CMP criterion.

Keywords: Unified strength criterion · Characteristic stress
Characteristic mobilized plane · Transverse isotropy

1 Introduction

The peak shear strength properties of geomaterials can be explained and described by the frictional rules. The difference of various geomaterials can be ascribed to different effects of the intermediate principal stress. The Mohr-Coulomb (M-C) criterion can be interpreted by the frictional rules on the quadrangular plane but with no intermediate principal stress effect considered. The Drucker-Prager (D-P) criterion is interpreted on the bases of the octahedral plane with the intermediate principal stress considered but unsuitable for geomaterials. The Matsuoka-Nakai (M-N) criterion [1], which is more suitable for clays, are proposed based on the frictional rules of the spatial mobilized plane (SMP). These classical single strength theories have clear physical meanings but suits only for a certain type of geomaterials with a certain intermediate principal stress effects. Thus, some unified strength theory [2–9] is tried and proposed to uniformly describe the peak shear strength properties of geomaterials. The characteristic stress concept proposed by Lu et al. [7] is another way to establish a unified strength criterion by directly connecting with the mobilized plane. The concept of the characteristic stress can be used to develop a unified strength criterion by two ways: (*i*) adjust the acting stress; and (*ii*) adjust the mobilized plane. These two ways to establish a unified strength criterion can be unified under the concept of the characteristic stress.

The second way to establish a unified strength theory is used in this paper by acting the general stress on the soil element and then adjusting the mobilized plane measured by the characteristic stress. The characteristic internal friction angle is defined and the CMP is thus proposed. The shear-normal stress ratio on the CMP is used to establish the CMP criterion. The CMP can also be taken as a reference to establish a transversely isotropic strength criterion. By projecting the microstructure tensor onto the direction vector of the CMP, the transversely isotropic strength parameter is proposed and used to extend the CMP criterion into the TI-CMP strength criterion.

2 A Novel Strength Criterion Based on the CMP

Corresponding to the general stress σ_i ($i = 1, 2, 3$), the characteristic stress c_i ($i = 1, 2, 3$) is related to a material parameter which reflects physical properties of geomaterials. The material property reflected in this paper is the intermediated principal stress effects and β is used in the definition.

$$c_i = p_r \left(\frac{\sigma_i}{p_r}\right)^\beta, \quad i = 1, 2, 3 \tag{1}$$

where p_r is the reference stress for dimensionless transforming, and $p_r = 1$ kPa is used in this paper.

2.1 Proposition of the CMP

The characteristic stress is a comprehensive reflection of the acting stress and the material physical properties. The characteristic index β is a constant for a certain type of geomaterials, and varies for different types of geomaterials [7]. The friction rule also varies with different geomaterials, which can be expressed as the characteristic friction angle ψ_{ij}.

$$\sin \psi_{ij} = \frac{c_i - c_j}{c_i + c_j}, \quad i,j = 1,2,3 \ (i<j) \tag{2}$$

Equation (2) can be simplified to determine the position of the CMP as shown in Fig. 1. The CMP can be measured by OA_β, OB_β and OC_β. $OA_\beta = kc_1^{1/2}$, $OB_\beta = kc_2^{1/2}$ and $OC_\beta = kc_3^{1/2}$ can be obtained, in which k is a proportionality coefficient. For $\beta = 0$, $(c_i/c_j)^{1/2} = 1$ and $OA_\beta = OB_\beta = OC_\beta$, then the CMP becomes the octahedral plane.

For $\beta = 1$, $(c_i/c_j)^{1/2} = (\sigma_i/\sigma_j)^{1/2} = \tan(45° + \varphi_{ij}/2)$, the CMP becomes the spatial mobilized plane (SMP). For $0 \leq \beta \leq 1$, $(c_i/c_j)^{1/2} = \tan(45° + \psi_{ij}/2)$, the CMP can continuously change from the upper bound (the octahedral plane) to the lower bound (the SMP).

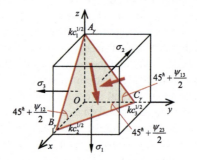

Fig. 1. Characteristic mobilized plane.

2.2 Expression of the Proposed Strength Criterion

Failure is assumed to occur when the ratio of the shear stress τ_β and normal stress σ_β acting on the CMP reaches a critical value. The expression of the proposed strength criterion based on the CMP can be written as follows:

$$\frac{\tau_\beta}{\sigma_\beta} = M_\beta \tag{3}$$

There are a series of continuous, smooth and convex curves on the same deviatoric plane, which can continuously change from the Drucker-Prager strength circle when $\beta = 0$ (upper bound) to the Matsuoka-Nakai strength curved triangle when $\beta = 1$ (lower bound). For a certain type of geomaterials, the characteristic parameter β is a constant, therefore, the corresponding CMP can also be determined.

2.3 Transversely Isotropic Strength Criterion Based on the CMP for Soils

The CMP measured by the characteristic stress can also be taken as a reference to propose a strength parameter for transversely isotropic soils. The fabric of soils measured by a microstructure or fabric tensor **A** is used to reflect the strength properties of transversely isotropic soils. Similar with the proposition of the anisotropic parameter [10], a new 3D transversely isotropic strength parameter χ can be derived by projecting the microstructure tensor **A** onto the direction vector of the CMP. With the transversely isotropic condition $\Omega_1 + 2\Omega_3 = 0$ and $\hat{n}_1^2 + \hat{n}_2^2 + \hat{n}_3^2 = 1$, a 3D transversely isotropic strength parameter can be derived and used to extend the failure stress ratio M_β. The proposed criterion, which is referred as TI-CMP criterion, can be expressed as follows:

$$\frac{\tau_\beta}{\sigma_\beta} = \chi_0 \left[1 + \Omega_3 \left(1 - 3\hat{n}_1^2 \right) \right] \tag{4}$$

where χ_0 reflects the average value of the strength parameter χ. Ω_3 reflects the degree of anisotropy. The proposed transversely isotropic strength criterion can reflects the intermediated principal stress effects and the effects of the anisotropy. The degree of the anisotropy can be measured by the difference between the internal friction angles of the vertical sample (φ_{cz}) and the horizontal sample (φ_{cy}) under the triaxial compression test. Take $p = 200$ kPa, $\beta = 1$, $\varphi_{cz} = 35°$ as a special case, with the increase of φ_{cy}, the change of the strength curves indicates the change of the anisotropy degree, as shown in Fig. 2. The intermediated principal stress effects can also be reflected by the proposed TI-CMP criterion as shown in Fig. 3.

3 Conclusion

The characteristic stress was used to propose the CMP which can describe friction rule of various geomaterials in a unified way. A novel unified strength criterion was thus established by assuming the shear-normal stress ratio acting on the CMP reaches a critical values. It is another way to establish a unified strength criterion based on the characteristic stress. Effects of the intermediated principal stress for various geomaterials can be reflected by the CMP criterion.

Furthermore, the proposed CMP could be taken as a reference to measure the peak shear strength variation rules of the transversely isotropic soils. By projecting the microstructure tensor onto the normal of the CMP, a new transversely isotropic strength parameter was derived and could be used to extend the failure stress ratio. The proposed CMP criterion was thus extended to consider the anisotropy of soils.

References

1. Matsuoka, H., Nakai, T.: Stress-deformation and strength characteristics of soil under three different principal stresses. Proc. Jpn. Soc. Civ. Eng. **232**, 59–70 (1974)
2. Houlsby, G.T.: A general failure criterion for frictional and cohesive materials. Soils Found. **26**(2), 97–101 (1986)
3. Desai, C.S.: Mechanics of materials and interfaces (2001)
4. Yu, M.H., Zan, Y.W., Zhao, J., Yoshimine, M.: A unified strength criterion for rock material. Int. J. Rock Mech. Min. Sci. **39**, 975–989 (2002)
5. Liu, M.D., Carter, J.P.: General strength criterion for geomaterials. Int. J. Geomech. ASCE **3**(2), 253–259 (2003)
6. Yao, Y.P., Lu, D.C., Zhou, A.N., Zou, B.: Generalized non-linear strength theory and transformed stress space. Sci. China, Ser. E Eng. Mater. Sci. **47**(6), 691–709 (2004)
7. Lu, D.C., Ma, C., Du, X.L., Jin, L., Gong, Q.M.: Development of a new nonlinear unified strength theory for geomaterials based on the characteristic stress concept. Int. J. Geomech. ASCE **17**(2), 4016058 (2017)
8. Wu, S., Zhang, S., Guo, C., Xiong, L.: A generalized nonlinear failure criterion for frictional materials. Acta Geotech. **12**(6), 1353–1371 (2017)

9. Yoshimine, M.: 3-D coulomb's failure criterion for various geomaterials. In: Geomechanics II: Testing, Modeling, and Simulation, pp. 71–86 (2005)
10. Pietruszczak, S., Mroz, Z.: Formulation of anisotropic failure criteria incorporating a microstructure tensor. Comput. Geotech. **26**(2), 105–112 (2000)

A Multi-Phase-Field Anisotropic Damage-Plasticity Model for Crystalline Rocks

SeonHong Na and WaiChing Sun

Columbia University, New York, USA
wsun@columbia.edu

Abstract. Many engineering applications, such as geological disposal of nuclear waste, require reliable predictions on the thermo-hydro-mechanical responses of porous media exposed to extreme environments. This presentation will discuss the relevant modeling techniques designed specifically for such environmental conditions. In particular, we will provide an overview of the coupling method of crystal plasticity and multi-phase-field model designed to replicate the thermal- and rate-dependent damage-plasticity of crystalline rock. Special emphasis is placed on capturing the intrinsic anisotropy of salt grain in 3D with respect to damage behavior and plastic flow by incorporating the crystallographic information of salt.

Keywords: Crystal plasticity · Damage-Plasticity · Multi-Phase-Field model · Salt

1 Introduction

The permanent geological disposal of nuclear waste in salt formations is one of the most feasible options to meet the increasing demand for nuclear waste storage. The reason for the popularity of salt formations is attributed to the desirable thermo-hydro-mechanical characteristics of the salt, i.e., the combinations of low permeability (practically impermeable), high thermal conductivity, relatively chemically inactive, high heat resistance, geological stability, and relatively ductile responses compared to other possible host materials (e.g., shale and granite).

In this work, we present a nonlocal damage-plasticity model designed to capture the anisotropic damage-plasticity of salt at the single-grain micromechanical level in a three-dimensional setting. Instead of capturing the anisotropy of the damage-plasticity responses in a purely phenomenological approach, we will introduce a microstructural-aware continuum model for crystalline rock salt. In particular, we will consider salt as a crystalline material with the slip system which dictates the rate-dependent plastic flow. Meanwhile, the anisotropy of fracture behavior will be captured by a multi-phase-field model in which each phase field is assumed to be associated with the orientation of each slip system.

2 Method

In this work, the salt grain is idealized as pure and without any defect or brine inclusion. As such, the work done on the salt grain may lead to increase in the elastic strain energy or can be dissipated via crack growth or plastic flow. Previously, small and finite strain models that couple phase field fracture and cap-plasticity have been proposed to capture the brittle-ductile transition in isotropic materials [1, 2, 6]. Meanwhile, relationship between effective critical energy release rate and the spatial heterogeneity is explored in Na et al. [4]. In addition, the coupling of crystal plasticity and multi-phase-field models for crystalline materials have been reported in Na and Sun [3]. In this work, the model reported in Na and Sun [3] is implemented and simulated in a 3D setting. To couple the nonlocal damage, the phase field model with a crystal plasticity model that is iterative in a local return mapping is used with a non-iterative operator-split algorithm. A very small incremental step is adopted to maintain the equilibrium [1, 3, 5]. The governing equations discretized by a standard Galerkin finite element with equal-order basis functions are the balance of linear momentum, balance of energy and the micro-force balance with the displacement (3-dimensional vector, u), temperature (single scalar, θ) and the multiple phase fields (multiple scalars, $\underline{d}$ denotes a set of d_i where the value of i is consistent with the number of slip systems of salt grain) as the nodal unknown, i.e.,

Balance of Linear Momentum:

$$\nabla \cdot \boldsymbol{\sigma} + b = 0 \text{ with } \boldsymbol{\sigma} = \frac{\partial \hat{\psi}}{\partial \epsilon^e} = g(\underline{d})\hat{\boldsymbol{\sigma}}, \tag{1}$$

where $\hat{\boldsymbol{\sigma}} = \mathbb{C}^e : \epsilon^e - 3\alpha K(\theta - \theta_0)\mathbf{1}$ and $g(\underline{d}) = (1-k)\prod(1-d_i)^2 + k$. Here $\mathbb{C}^e$ denotes the fourth-order elasticity tensor; α is the thermal expansion coefficient; K is the bulk modulus, θ_0 indicates a fixed reference temperature, and $\mathbf{1}$ is the second-order identity tensor. The degradation function $g(\underline{d})$ for the multi-phase-field has been chosen such that $g'(d_i = 1) = 0$. The small parameter $k \ll 1$ is introduced for maintaining the well-posedness of the problem.

Balance of Energy:

$$c_v \dot{\theta} = \underbrace{[D_{mech} - H_\theta]}_{D_{mech} - H_\theta} - \nabla \cdot \boldsymbol{q} + r_\theta$$
$$\text{with } \sum_\alpha (\pi^\alpha \dot{\gamma}^\alpha + g^\alpha \dot{s}^\alpha) + \theta \left(\frac{\partial^2 \hat{\psi}}{\partial \theta \partial \epsilon^e} : \epsilon^e \right). \tag{2}$$

Here c_v denotes the specific heat per unit volume at constant deformation; D_{mech} is the contribution to dissipation due to pure mechanical load and/or thermal flow; H_θ is the non-dissipative (latent) thermoelastic structural heat or cooling. $\boldsymbol{q}$ is the heat flux vector and r_θ indicates the heat source. π^α and g^α are thermodynamic forces power-conjugate to a local plastic slip γ^α and a local scalar measure related to the plastic slip accumulation for each slip system s^α, respectively.

Balance of Micro-forces for Each Crack Phase d_i:

$$2(1-d_i)\mathcal{H}_i + \frac{G_c}{l}d_i + G_c l \nabla \cdot (\boldsymbol{\omega}_i \cdot \nabla d_i) = 0$$

$$\text{with } \mathcal{H}_i = \max_{\tau \in [0,t]} \left\{ \prod_{j \neq i} (1-d_j)^2 [\omega_+^e + \omega^p - \omega_0^p] \right\}, \tag{3}$$

where G_c is the fracture energy or the energy release rate for each slip system. $\mathcal{H}_i$ is the strain-history functional that governs the evolution of the irreversible crack propagation. l indicates a length scale that controls the width of the smooth approximation of the crack, which is assumed to be larger than the mesh size. $\boldsymbol{\omega}_i$ is a second-order structural tensor associated with the slip plane i. ω_+^e denotes the thermoelastic strain, and ω^p is the plastic work coming from the hardening contribution. ω_0^p is introduced to control the plastic deformation in ductile fracture (cf. Na and Sun [3]).

3 Results and Future Directions

We conducted three uniaxial compression test simulations on the domain of the $0.4 \times 0.4 \times 1.0$ mm and compared of the same materials but of different crystallographic orientations. The Euler Angles, in terms of degree, are (0, 0), (30, 10) and (45, 0), respectively. A pair of incremental displacement is applied on the top of the 3D domain to introduce the uniaxial compression. The resultant damage (as illustrated by the combined phase field, $d_{combined} = \sum_i d_i - \sum_{i \neq j} d_i d_j + \sum_{i \neq j \neq k} d_i d_j d_k - \prod_i d_i$) and the plastic strain are shown in Figs. 1 and 2 accordingly.

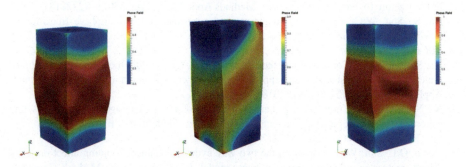

Fig. 1. The resultant damage of uniaxial compression test in 3D domain is described. Fixing the boundary conditions, three sets of different Euler Angles are selected (from the left to right: (0, 0), (30, 10) and (45, 0)) to represent the anisotropic damage behavior.

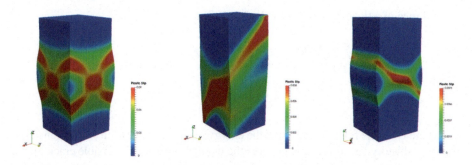

Fig. 2. The resultant plastic behavior of uniaxial compression tests in 3D domain is depicted. As in Fig. 1, three sets of Euler Angles are used from the left to right: (0, 0), (30, 10), and (45, 0), respectively.

These distinct damage and plastic strain clearly indicate that both the damage and the plastic responses are indeed anisotropic. More importantly, the induced anisotropy is replicated simply by incorporating the orientation of the slip system. Currently, the research group is working on a few new directions, including but not limited to incorporating this model for poly-crystalline materials and for field-scale simulations via a recursive homogenization approach as in Wang and Sun [7] or a concurrent coupling approach, and introducing geometrical nonlinearity.

References

1. Choo, J., Sun, W.C.: Coupled phase-field and plasticity modeling of geological materials: from brittle fracture to ductile flow. Comput. Methods Appl. Mech. Eng. **330**, 1–32 (2018)
2. Na, S., Sun, W.C.: Computational thermo-hydro-mechanics for multiphase freezing and thawing porous media in the finite deformation range. Comput. Methods Appl. Mech. Eng. **318**, 667–700 (2017)
3. Na, S., Sun, W.C.: Computational thermomechanics of crystalline rock. Part I: a combined multi-phase-field/crystal plasticity approach for single crystal simulations. Comput. Methods Appl. Mech. Eng. **338**, 657–691 (2018)
4. Na, S., Sun, W.C., Yoon, H., Ingraham, M.: Effects of elastic heterogeneity on the fracture pattern and macroscopic effective toughness of Mancos Shale in Brazilian tests. J. Geophys. Res. Solid Earth **122**(8), 6202–6230 (2017)
5. Ulven, O.I., Sun, W.C.: Capturing the two-way hydro-mechanical coupling effect on fluid-driven fracture in a dual-graph lattice beam model. Int. J. Num. Anal. Methods Geomech. **42**(5), 736–767 (2018)
6. Wang, K., Sun, W.C.: A unified variational Eigen-erosion framework for interacting fractures and compaction bands in brittle porous media. Comput. Methods Appl. Mech. Eng. **318**, 1–32 (2017)
7. Wang, K., Sun, W.C.: A multiscale multi-permeability poroplasticity model linked by recursive homogenizations and deep learning. Comput. Methods Appl. Mech. Eng. **334**(1), 337–380 (2018)

Simulation of the Undrained Behavior of Particulate Assemblies Subjected to Continuous Rotation of Lode Angle

Mohammadjavad Salimi[✉] and Ali Lashkari

Faculty of Civil and Environmental Engineering,
Shiraz University of Technology, Shiraz, Iran
MJ.salimi@sutech.ac.ir

Abstract. Discrete Element Method (DEM) in conjunction with Darcy's law is applied to simulate the undrained behavior of water-saturated particulate assemblies sheared under continuous rotation of Lode angle. DEM simulations show that the undrained rotation of Lode angle leads to liquefaction of particulate assemblies, decrease in coordination number and evolution of contact fabric anisotropy.

Keywords: DEM · Undrained shear · Lode angle · Fabric

1 Introduction

Liquefaction is an important phenomenon intrinsic in the undrained response of granular soils. Liquefaction has been conventionally investigated by means of direct simple shear and axi-symmetric triaxial tests. However, undrained true triaxial tests have revealed that continuous rotation of the Lode angle even under fixed shear stress leads to liquefaction of sands [1, 2]. In true triaxial condition, effective principal stress components are interrelated to the Lode angle parameter (θ) by:

$$\tan \theta = \sqrt{3}\left(\frac{\sigma'_y - \sigma'_x}{2\sigma'_z - \sigma'_y - \sigma'_x}\right) \tag{1}$$

Where σ'_x, σ'_y, and σ'_z are principal effective stresses along perpendicular x, y, and z directions, respectively. In the past, several laboratory investigations have been carried out to investigate the behavior of sands sheared under general undrained conditions. For instance, Yamada and Ishihara [1], Choi and Arduino [2] reported experimental evidence supporting that continuous rotation of the Lode angle under constant shear stress leads to the gradual generation of excess pore-water pressure whose magnitude depends on the applied shear stress and sand void ratio. Using DEM as a versatile tool for simulation of the behavior of granular media, the mechanical response of particulate assemblies subjected to undrained continuous rotation of Lode angle under fixed shear stress is simulated here.

2 Basic Assumptions Behind the DEM Simulations

Liu et al. [3] suggested that the fluid phase in water-saturated particulate assemblies can be considered by participation of Darcy's law in conjunction with the mathematical equations governing the interaction of particles in the granular assemblies. In their numerical scheme, several measurement spheres are defined through the particulate assemblies and the volumetric strain of each measurement sphere represents fluid phase inflow/outflow. Herein, 3D-DEM simulations using the commercial code PFC3D are carried out to simulate the mechanical behavior of water-saturated specimens subjected to continuous rotation of the Lode angle. The selected particle size distribution is similar to that of standard Toyoura sand. Table 1 lists the basic micro-parameters used in DEM simulations.

Table 1. Parameters used in DEM simulations

Particles density	2650 kg/m³	Total number of particles	8438
Maximum diameter of particles	0.54 mm	Minimum diameter of particles	0.14 mm
Damping ratio	0.7	Interparticle friction	0.5
Walls friction	0	Walls stiffness	10^8 kN/m²
Tangential stiffness (k_s)	$k_0 \times r$	Normal stiffness (k_n)	$k_0 \times r$
Permeability (k)	10^{-4} m/s	Density of water (ρ_w)	1000 kg/m³
Standard free fall acceleration (g)	9.81 m/s²	Bulk modules of water (E_w)	10^7 Pa

Contribution of particles in load carrying microstructure of assemblies can be expressed by coordination number (C_N) which is the average contacts per particle:

$$C_N = \frac{1}{N_p} \sum_{N_p} n^c \qquad (2)$$

where n_c and N_p are, respectively, the total number of contact for each particle and the total number of particles. Δ as a measure of anisotropy of the granular assemblies is calculated based on the second invariant of the contact fabric tensor:

$$\Delta = \sqrt{(F_1 - F_2)^2 + (F_3 - F_2)^2 + (F_1 - F_3)^2} \qquad (3)$$

where F_1, F_2, and F_3 are principal values of the contact fabric tensor.

3 Results and Discussions

For a specimen with P'_{in} = 100 kPa and e = 0.684 subjected to continuous rotation of the Lode angle under fixed q = 35.5 kPa, the simulated stress path in the π-plane and the associated generated strains are illustrated in Fig. 1. Inspection of Fig. 1 indicates that the stress path of the specimen resembles a spiral whose radius increases steadily as a direct consequence of the gradual increase in the excess pore-water pressure. Besides, the generated strain path is initially a circle that gradually turns into an inverted triangle at large shear strains.

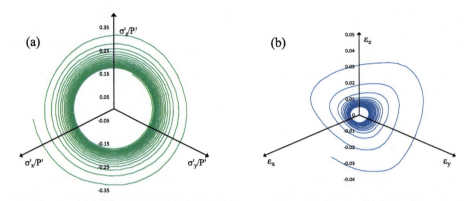

Fig. 1. Response of a particulate specimen with e = 0.684 [-] and P'_{in} = 100 kPa sheared under q = 35.5 kPa: (a) normalized effective stress path in the π-plane; (b) generated strains

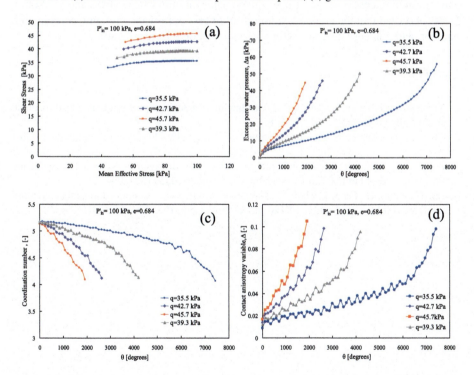

Fig. 2. Response of four particulate specimens with e = 0.684 [-] and P'_{in} = 100 kPa sheared under q = 35.5, 39.3, 42.7, and 45.7 kPa: in: (a) q vs. P' plane; (b) Δu vs. θ; (c) C_N vs. θ; and (d) Δ vs. θ.

For four particulate assemblies with $P'_{in} = 100$ kPa and $e = 0.684$, sheared under $q = 35.5, 39.3, 42.7$, and 45.7 kPa, stress paths in the q vs. P′ plane, accumulated excess pore-water pressure (Δu) vs. Lode angle (θ), and curves for evolution of coordination number (C_N) and contact fabric anisotropy variable (Δ) with θ are, respectively, depicted through parts "a" to "d" of Fig. 2. In Fig. 2(b), the increase in excess pore-water pressure can be attributed to the onset of liquefaction-induced instability and accordingly, it is observed that the samples with higher shear stress become unstable earlier. In Figs. 2(c) and (d), decrease in C_N and increase in Δ with θ become faster subsequent to the onset of instability.

4 Conclusions

The lack of knowledge regarding evolution of the micromechanical parameters in granular soils highlights a need for DEM simulation of their mechanical behavior. In this study, the mechanical behavior of water-saturated particulate assemblies sheared under rotating Lode angle was simulated using DEM method. Simulations confirmed that the generation of excess pore-water pressure depends on initial shear stress and void ratio. It was observed that the gradual evolutions of excess pore-water pressure, coordination number, and contact fabric anisotropy variable become faster concurrently.

References

1. Yamada, Y., Ishihara, K.: Undrained deformation characteristics of sand in multi-directional shear. Soils Found. **23**(1), 61–79 (1983)
2. Choi, C., Arduino, P.: Behavioral characteristics of gravelly soils under general cyclic loading conditions. In: Triantafyllidis, T. (ed.) Cyclic Behaviour of Soils and Liquefaction Phenomena, Proceedings of the International Conference on Cyclic Behavior of Soils and Liquefaction Phenomena, 31 March–02 April, Bochum, Germany, pp. 115–122 (2004)
3. Liu, G., Rong, G., Peng, J., Zhou, C.: Numerical simulation on undrained triaxial behavior of saturated soil by a fluid coupled-DEM model. Eng. Geol. **193**, 256–266 (2015)

On Effective Stress and Effective Stress Equation

Longtan Shao[✉], Xiaoxia Guo, and Boya Zhao

State Key Laboratory of Structural Analysis for Industrial Equipment,
Department of Engineering Mechanics, Dalian University of Technology,
Dalian 116024, China
shaolongtan@126.com

Abstract. The concept of effective stress and the effective stress equation are fundamental for establishing the theory of strength and the relationship of stress and strain in soil mechanics. However, up till now, the physical meaning of effective stress has not been explained clearly, and the theoretical basis of the effective stress equation has not been proposed. Researchers have not yet reached a common understanding of the feasibility of the concept of effective stress and effective stress equation for unsaturated soils. Focusing on these problems, new viewpoints for explicitly elucidating the effective stress and deriving the effective stress equation are given in this paper, including that the effective stress should be defined as the soil skeleton stress due to all the external forces excluding pore fluid pressure, and that the soil skeleton should include a fraction of pore water which can bears and passes the load together with soil particles. The relationship between the effective stress and the shear strength of unsaturated soils is preliminarily verified by experiments and quoting test data from literature.

Keywords: Effective stress · Effective stress equation · Unsaturated soils Soil skeleton stress · Pore fluid pressure

1 Skeleton Water

The soil skeleton may be defined as the structure consisting of the solid phase that can bear and transfer loadings. Due to the strong absorption and the capillary effect to the water on the surface of soil particles, a fraction of pore water is tightly combined with the particle and to bear and transfer load incorporated with the particle. According to the definition of the skeleton, this fraction of pore water should be treated as a constituent of the soil skeleton, which is called skeleton water. The skeleton water may include strong bonding water and may also include a part of weak bonding water and capillary water. The discussion on soil skeleton water illustrates that the pore water in soils is always interconnected, even at low water content. The soils with the water content lower than soil skeleton water content should be called "dry soils" in soil mechanics.

2 Definition of the Effective Stress and Effective Stress Equation

As shown in Fig. 1, taking a saturated soil as example, when analyzing the inner force of an infinitesimal section plane in the soil skeleton, separately considering the effect of pore water pressure and the other external forces, then

$$\sigma_t A = \sigma A + u_w(1-n)A + nu_w A \tag{1}$$

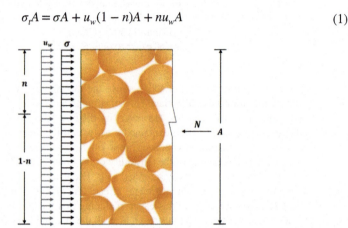

Fig. 1. Force analysis of a saturated soil separately considering the effect of pore water pressure and the other external forces

Where σ_t is the total normal stress, A is cross section area, σ is the normal stress of the skeleton, n is porosity, u_w is the pore water pressure.

From Eq. (1), it can be obtained that

$$\sigma = \sigma_t - u_w \tag{2}$$

This is the Terzaghi's effective stress equation. It indicates that the effective stress is the skeleton stress due to external forces except for pore water pressure.

For unsaturated soils as shown in Fig. 2, according to the equilibrium condition we have

$$N_t = \sigma A + \frac{n_w}{n}(1-n)u_w A + n_w u_w A + \frac{n_a}{n}(1-n)u_a A + n_a u_a A \tag{3}$$

then

$$\sigma = \sigma_t - u_a + S_e(u_a - u_w) \tag{4}$$

in which u_a is air pressure, n_a and n_w are the porosities corresponding to pore air and pore water respectively, $S_e = n_w/n$ is the effective saturation.

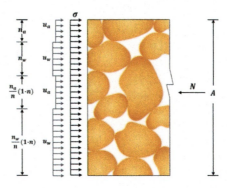

Fig. 2. Force analysis of an unsaturated soil separately considering the effect of pore fluid pressure and the other external forces

Equation (4) is the effective stress equation for soils in both saturated and unsaturated states.

3 Preliminarily Verification of the Effective Stress Equation for Unsaturated Soils

A kind of kaolin is chosen to be tested for verifying that the effective stress governs its shear strength. From Fig. 3, it can be seen that the shear strength of unsaturated kaolin is governed by the effective stress.

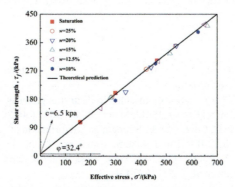

Fig. 3. The test results of the shear strength of the kaolin in saturated and unsaturated states.

Some other test results of the shear strength of unsaturated soils which were found in literatures [1–9] are reanalyzed in the concept of effective stress. It can be found that the shear strength of the soils in both saturated and unsaturated states fits the same formula with the same parameters, as shown in Fig. 4, in which the solid lines show the shear strength of the soils in saturated state. These reanalysis results verify that the effective stress also governs the shear strength of unsaturated soil.

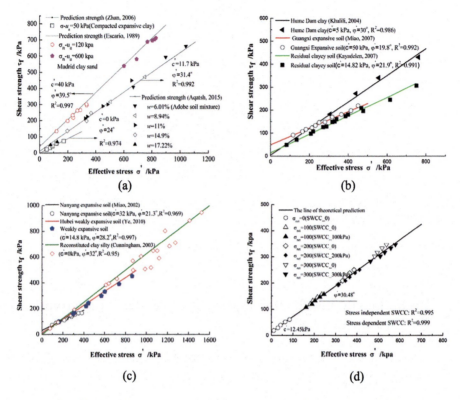

Fig. 4. Reanalysis of the test results of shear strength of unsaturated soils

References

1. Zhan, L.T., Ng, C.W.: Shear strength characteristics of an unsaturated expansive clay. Can. Geotech. J. **43**, 751–763 (2006)
2. Aqtash, U., Bandini, P.: Prediction of unsaturated shear strength of an adobe soil from the soil–water characteristic curve. Constr. Build. Mater. **98**, 892–899 (2015)
3. Escario, V., Juca, J.F.T., Coppe, M.S.: Strength and deformation of partly saturated soils. In: Proceedings of 12th ICSMFE, Rio de Janeiro, Brazil, vol. 11, pp. 43–46 (1989)
4. Khalili, N., Geiser, F., Blight, G.E.: Effective stress in unsaturated soils: review with new evidence. Int. J. Geomech. **4**, 115–126 (2004)
5. Miao, L., Liu, S., Lai, Y.: Research of soil–water characteristics and shear strength features of Nanyang expansive soil. Eng. Geol. **65**(4), 261–267 (2002)
6. Kayadelen, C., Tekinsoy, M.A., Taskiran, T.: Influence of matric suction on shear behavior of a residual clayey soil. Environ. Geol. **53**, 891–901 (2007)
7. Ye, W.M., Zhang, Y., Chen, B., et al.: Shear strength of an unsaturated weakly expansive soil. J. Rock Mech. Geotech. Eng. **2**(2), 155–161 (2010)
8. Cunningham, M.R., Ridley, A.M., Dineen, K., et al.: The mechanical behaviour of a reconstituted unsaturated silty clay. Géotechnique **53**(2), 183–194 (2003)
9. Lee, I.M., Sung, S.G., Cho, G.C.: Effect of stress state on the unsaturated shear strength of a weathered granite. Can. Geotech. J. **42**(2), 624–631 (2005)

Hypoplastic Simulation of Axisymmetric Interface Shear Tests in Granular Media

Hans Henning Stutz[1(✉)] and Alejandro Martinez[2]

[1] Department of Engineering - Geotechnical Engineering, Aarhus University,
Inge Lehmanns Gade 10, 8000 Aarhus C, Denmark
hhs@eng.au.dk
[2] Department of Civil and Environmental Engineering, University of California Davis,
Ghausi Hall, Davis, CA 2001, USA

Abstract. Modelling the soil-structure interface behaviour is an important but often neglected part of a holistic geotechnical modelling approach. The Mohr-Coulomb model, which is often used, is in many cases not a suitable choice to represent the complex phenomena taking place in the localized interface shear zone. Recently developed interface models formulated into the hypoplastic framework have demonstrated their ability to model the barotropy and pyknotropy in the interface shear zone. It is especially important for the models to consider the dependence of the interface shear zone on the initial void ratio to correctly simulate the changes in normal stress around cylindrical inclusions. This paper presents the results of a number of axisymmetric interface shear simulations using a hypoplastic interface and soil model. The simulation results are compared to experimental results obtained from axisymmetric interface shear tests performed between a medium dense sand and surfaces of varying roughness. The surfaces used in these experiments consisted of sheets of sandpaper glued on metal sleeves, and had average surface roughness values between 0.001 and 0.290 mm. The experimental and modelling results are compared considering the stress-deformation responses. Based on this assessment, recommendations are provided regarding the use of the hypoplastic interface shear models.

Keywords: Hypoplasticity · Interface shear behaviour · Sands

1 Introduction

Modelling and testing of granular soil-structure interfaces is an important but often neglected issue for holistic geotechnical predictions and designs. Often the simplified Mohr-Coulomb model is used that is seldom a perfect match for the performance of the geotechnical structures. The Mohr-Coulomb friction model non-state dependent frictional behaviour between the surface and the soil, while the important features of pressure and void ratio dependency are neglected. However, modeling of the volumetric behaviour is of great importance for adequately modelling interface shearing behaviour. While the behavior of interfaces between granular soils and solid materials have been investigated for the last decades, there are still issues that are unresolved. The aim of this paper is to demonstrate the ability of a hypoplastic granular interface model to adequately simulate interface shear behavior performed in an axisymmetric geometry. This model is also capable to address the issue of the parameter calibration in a direct and straight-forward manner.

2 Axisymmetric Shear Test

Results from laboratory tests performed using an axisymmetric device (Fig. 1) are presented herein to compare with the model results. The axisymmetric device consists of a cylindrical steel chamber that allowed controlled application of constant radial pressure to the specimen inside it. The top and bottom caps of the axisymmetric device acted as rigid boundaries, resulting in BC4 type calibration chamber conditions [2]. All the tests were performed on dry sand specimens prepared to "wished in place" conditions. The friction sleeve was placed between two sections of smooth aluminum rods, and displaced upward during testing. A load cell and a Linear Variable Differential Transformer (LVDT) were placed at the top of the testing rod, and were used to monitor the force and displacement. The shear stress on the friction sleeve surface was determined by dividing the force on the sleeve by the sleeve's area, and the stress ratio was defined as the ratio between the shear stress on the sleeve and the radial pressure applied to the specimen. All tests were performed under a radial pressure of 50 kPa on specimens of Ottawa 20–30 sand. Three different surfaces were tested as part of this study, and their surface roughness was quantified in terms of the maximum surface roughness

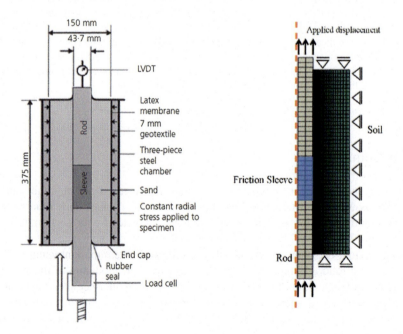

Fig. 1. Axisymmetric device (from Martinez et al. 2015) and numerical boundary value problem.

(R_{max}) and the average roughness (R_a) parameters. One surface consisted of smooth stainless steel sleeve ($R_{max} = 0.01$ mm, $R_a = 0.001$ mm), and the other two surfaces consisted of sleeves wrapped in grit 320 ($R_{max} = 0.08$ mm, $R_a = 0.009$ mm) and grit 36 ($R_{max} = 0.88$ mm, $R_a = 0.171$ mm) sandpaper. Detailed information about the experiments can be found in [5].

3   Hypoplastic Interface Model and Simulations

The hypoplastic interface shear model used was based on the full 3D version of the hypoplastic model for granular materials by [8]. The basic constitutive equation is written as:

$$\dot{\sigma} = f_s \left(L:D + f_d N \| D \| \right) \tag{1}$$

Where N and L are the second and fourth order constitutive tensors, and f_d and f_s are the baro and pyknotropy factors to account for the stiffness dependent pressure and void ratio dependent formulation. D is the Eulerian strechting tensor and σ the stress tensor. More details about the hypoplastic granular model can be found in [8]. This model uses the assumption of simple shear conditions at the interface to develop the enhanced hypoplastic interface model [6].

The reduced strain and stress tensors are used to represent the governing deformation mode at the interface. This model inforporates the assumption that a fully rough surface will induce behaviour identical to soil-soil shearing behaviour. This model uses a surface roughness implementation to model the interface shear behaviour of interfaces without fully rough surfaces. In addition to the hypoplastic continuum soil model parameters described above only two additional constitutive parameters are required. These are the surface roughness $\kappa_r = \varphi_c / \varphi_{c,int}$ (φ_c critical state friction angle; $\varphi_{c,int}$ interface critical state friction angle) and the shear zone thickness d_s^v. Both of these parameters are calibrated according to [1] via back calculation. Different approaches can also be used [1]. Examples for different applications of the hypoplastic model can be found in [3, 7]. The implementation used for the simulation was adapted from [7] for 2D axisymmetric cases and is available from soilmodels.com.

4   Simulation Results Compared with Experiments

The simulation results are compared to experimental results from [4]. In addition, using of the same parameters for the soil and the interface, calibration for only one more parameter is required. The results demonstrate the ability of the model to the behavior observed in the laboratory tests (Fig. 2) with a reasonable accuracy. It is noted that this model was only calibrated for the roughness coefficient parameter. As shown by [5], the thickness of the shear zone is dependent on the roughness of the contacting surface. This highlights the need to necessary to include both parameters to more successfully model the stress-deformation behaviour, which will be the subject of future studies.

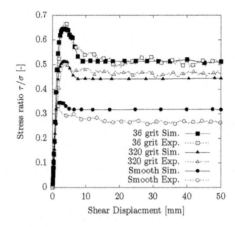

Fig. 2. Comparison of the numerical simulations with the test results from [4]

5 Conclusions and Discussion

The results presented herein highlight the ability of hypoplastic interface models based on 3D continuum models to accurately model interface shear behavior while requiring less parameters than other classical models. The need for many parameters represents a challenge in using such models for holistic simulations and Class A predictions without an exhaustive laboratory testing program. In addition, the hypoplastic interface model is capable of simulating different roughness conditions.

References

1. Arnold, M.: Application of the intergranular strain concept to the hypoplastic modelling of non-adhesive interfaces. In: Proceedings 12th International Conference of International Association for Computer Methods and Advances in Geomechanics (IACMAG), pp. 747–754 (2008)
2. Ghionna, V., Jamiolkowski, M.: A critical appraisal of calibration chamber testing of sands. In: Proceedings of 1st International Symposium on Calibration Chamber Testing (ISOCCT1), Potsdam, N.Y., pp. 13–39 (1991)
3. Gutjahr, S., Hettler, A.: Erdwiderstand bei beliebigem Wanddrehpunkt auf Grundlage der Hypoplastizität. Bautechnik **81**(8), 630–638 (2004)
4. Martinez, A.: Multi-scale studies of particulate-continuum interface systems under axial and torsional loading conditions. Ph.D. dissertation, Georgia Institute of Technology, Atlanta, GA, USA. (2015)
5. Martinez, A., Frost, J.D.: The influence of surface roughness form on the strength of sand – structure interfaces. Geotechnique Lett. (February), 1–8 (2017)
6. Stutz, H., Mašín, D., Wuttke, F.: Enhancement of a hypoplastic model for granular soil-structure interface behaviour. Acta Geotech. **11**(6), 1249–1261 (2016)

7. Stutz, H., Mašín, D., Sattari, A.S., Wuttke, F.: A general approach to model interfaces using existing soil constitutive models application to hypoplasticity. Comput. Geotech. **87**(July), 115–127 (2017)
8. von Wolffersdorff, P.A.: Hypoplastic relation for granular materials with a predefined limit state surface. Mech. Cohesive-Frictional Mater. **1**(3), 251–271 (1996)

A Cam-Clay-Based Fractional Plasticity Model for Granular Soil

Yifei Sun[✉] and Yufeng Gao

Key Laboratory of Ministry of Education for Geomechanics and Embankment Engineering,
Hohai University, Nanjing 210098, China
sunny@hhu.edu.cn

Abstract. A series of fractional plasticity models have been developed to model the stress-strain behaviour of granular soils by using modified Cam-clay yielding function. However, for granular soil, the stress-dilatancy relationship was state-dependent that was better described by the Cam-clay equation. Therefore, an attempt is made in this study to develop a Cam-clay-based fractional plasticity model where numerical solutions of the model are provided for further potential application. To validate the proposed model, a series of triaxial test results on Toyoura sand are simulated. A good agreement between the model predictions and the corresponding test results is observed.

Keywords: Fractional plasticity · Cam-clay equation · Granular soil

1 Introduction

Many constitutive models for granular soils had been proposed in the past [1–4] by following traditional plasticity. Considering the effect of stress history on the stress-dilation behaviour of granular soil, a series of fractional plasticity models within the framework of critical state soil mechanics were proposed [5–7]. In contrast to other models, the fractional order reflected the influence of stress history and material state on the plastic flow of granular soil, which thus had clear physical origins. However, in previous models, the modified Cam-clay model [8] instead of the Cam-clay model [9] was used. But, experimental evidences [10] had shown that the stress-dilation relationship was rather to be linear before reaching the critical state, as described by the Cam-clay equation [9]. As a result, the modified Cam-clay equation [8] is not suitable enough to model the plastic flow of granular soil. Therefore, a Cam-clay based fractional plasticity model for granular soil is proposed in this study.

2 Cam-Clay-Based Fractional Model

The constitutive relation consists of four main parts: the elastic component, loading tensor (**m**), flow tensor (**n**), and hardening modulus (H), which was derived as [4, 5]:

$$\dot{\boldsymbol{\sigma}} = \left[\mathbf{D}^e - \frac{\mathbf{D}^e \mathbf{nm}^T \mathbf{D}^e}{H + \mathbf{m}^T \mathbf{D}^e \mathbf{n}}\right] \dot{\boldsymbol{\varepsilon}} \tag{1}$$

where the superimposed dot indicates increment. $\mathbf{D}^e$ is the elastic stiffness matrix that was defined as shown in Li and Dafalias [4]. Due to the page limitation, expressions of $\mathbf{m}$, H and $\mathbf{D}^e$ are not repeated here and the relevant work of [4] are borrowed for simplicity. However, derivations of $\mathbf{n}$ is presented here by using the fractional derivative which intrinsically accounts for material state [5]. The plastic flow tensor is defined by $\mathbf{n} = [d, 1]^T / \sqrt{d^2 + 1}$, in which d is the dilation ratio that can be obtained as:

$$d = \frac{{}_{p'0}^{C} D_{p'}^{\alpha} f(p', q)}{{}_{0}^{C} D_{q}^{\alpha} f(p', q)} \tag{2}$$

where D mean derivation and C indicates the Caputo's definition of the fractional derivative [7]. The fractional order $\alpha = \exp(\beta \psi)$, where $\psi = e - e_c$, is the state parameter [11]. p' and q are the effective mean principal and deviator stresses, respectively. p'^0 is the initial stress applied on soil sample. f is the Cam-clay function. ${}_{0}^{C} D_{q}^{\alpha} f(p', q)$ can be analytically solved as:

$${}_{0}^{C} D_{q}^{\alpha} f(p', q) = \frac{q^{1-\alpha}}{\Gamma(2 - \alpha)} \tag{3}$$

However, a numerical approximation for solving ${}_{p'0}^{C} D_{p'}^{\alpha} f(p', q)$ is provided. Following Sumelka [12], the loading stress in fractional derivative can be discretized by the following steps: $a = p'^0 < p'^1 < p'^2 < \cdots < p'^k < \cdots < p'^{i-1} < p'^i = p'$, where i is the number of discretized steps, from which the discretized step size (h_p) can be determined by $h_p = (p' - a)/i$. Furthermore, the fractional derivative of f can be approximated by using [12]:

$${}_{p'0}^{C} D_{p'}^{\alpha} f(p', q) \cong \frac{h_p^{n-\alpha}}{\Gamma(n+2-\alpha)} \left\{ \left[(i-1)^{n+1-\alpha} - (i-n-1+\alpha)i^{n-\alpha}\right] f^{(n)}(p'^0) \right. \\ \left. + f^{(n)}(p') + g_i(p'^k) \right\} \tag{4}$$

where the function g_i is expressed as:

$$g_i(p'^k) = \sum_{k=1}^{i-1} [(i - k + 1)^{n-\alpha+1} - 2(i - k)^{n-\alpha+1} + (i - k - 1)^{n-\alpha+1}] f^{(n)}(p'^k) \tag{5}$$

During numerical simulation for step j (= 1, 2, 3, ..., n), input the stress and strain conditions (p'_j, q_j, e_j) and calculate the j-th plastic flow direction is computed by using Eq. (2). However, it is noted that as the sum of series is used in solving ${}_{p'0}^{C} D_{p'}^{\alpha} f(p', q)$, the choice of sub-step size (h_p) influences the calculation accuracy. The decrease of sub-step size would increase the accuracy but result in more computation memory. Therefore, the iteration is suspended once the following is satisfied: $\left|[g_{i-1}(p_j'^k) - g_i(p_j'^k)] / g_i(p_j'^k)\right| \leq 10^{-2}$.

The proposed model has ten parameters, with nine inherited from [4]. There is one additional parameter, β, that determines the fractional order and the associated plastic flow of soil and thus can be obtained by fitting the relationship of $\varepsilon_v \sim \varepsilon_s$. Details for determining other model parameters can be found in [4] and thus not repeated.

To validate the model, both the undrained and drained behaviour of Toyoura sand [10] with a variety of initial material states are simulated, as shown in Figs. 1 and 2. Details of the material properties and test setup can be found in [10]. Values of the model parameters used for validation are: $G_0 = 125$, $v = 0.05$, $M_c = 1.25$, $\lambda = 0.019$, $e_\Gamma = 0.934$, $\xi = 0.7$, $\beta = 0.3$, $k = 1.1$, $h_1 = 3$, $h_2 = 3$. Figure 1 compares the model simulations and the corresponding undrained test results of Toyoura sand with different initial densities, where a good agreement can be found. Figure 2 shows the model simulations of the drained behaviour of Toyoura sand with different initial confining pressures, which are in accordance with the corresponding test results.

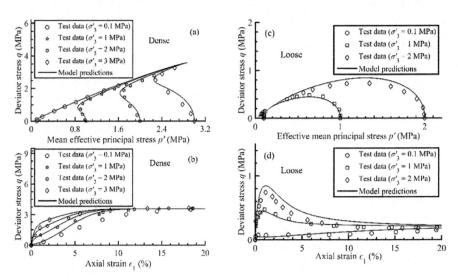

Fig. 1. Model predictions of the undrained behaviour of Toyoura sand

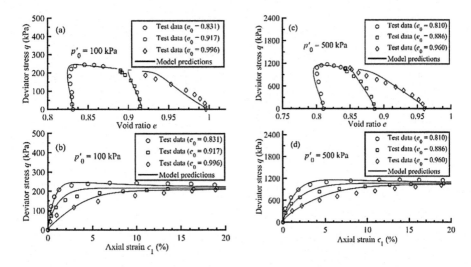

Fig. 2. Model predictions of the drained behaviour of Toyoura sand

3 Conclusions

An attempt was made in this paper to propose a Cam-clay-based state-dependent fractional plasticity model. Model predictions of the test results revealed that the proposed model was capable of predicting the drained and undrained behaviour of granular soil subjected to a wide range of initial states.

References

1. Gajo, A., Wood, D.M.: SevernTrent sand: a kinematic-hardening constitutive model: the $q-p$ formulation. Géotechnique **49**(5), 595–614 (1999)
2. Russell, A.R., Khalili, N.: A bounding surface plasticity model for sands exhibiting particle crushing. Can. Geotech. J. **41**(6), 1179–1192 (2004)
3. Wan, R., Guo, P.: A simple constitutive model for granular soils: modified stress-dilatancy approach. Comput. Geotech. **22**(2), 109–133 (1998)
4. Li, X., Dafalias, Y.: Dilatancy for cohesionless soils. Géotechnique **50**(4), 449–460 (2000)
5. Sun, Y., Xiao, Y.: Fractional order plasticity model for granular soils subjected to monotonic triaxial compression. Int. J. Solids Struct. **118–119**, 224–234 (2017)
6. Sun, Y., Shen, Y.: Constitutive model of granular soils using fractional order plastic flow rule. Int. J. Geomech. **17**(8), 04017025 (2017)
7. Sun, Y., Gao, Y., Zhu, Q.: Fractional order plasticity modelling of state-dependent behaviour of granular soils without using plastic potential. Int. J. Plasticity. **102**, 53–69 (2018)
8. Roscoe, K.H., Burland, J.B.: On the Generalised Stress-Strain Behaviour of 'Wet' Clay Engineering Plasticity. Cambridge University Press, Cambridge (1968)

9. Roscoe, K., Schofield, A., Thurairajah, A.: Yielding of clays in states wetter than critical. Géotechnique **13**(3), 211–240 (1963)
10. Verdugo, R., Ishihara, K.: The steady state of sandy soils. Soils Found. **36**(2), 81–91 (1996)
11. Been, K., Jefferies, M.G.: A state parameter for sands. Géotechnique **22**(6), 99–112 (1985)
12. Sumelka, W., Nowak, M.: On a general numerical scheme for the fractional plastic flow rule. Mech. Mater. **116**, 120–129 (2018)

Modelling Root-reinforced Soils with Nor-Sand

Barbara M. Świtała[1(✉)] and E. James Fern[2]

[1] Institute of Hydro-Engineering, Polish Academy of Sciences, Gdańsk, Poland
b.switala@ibwpan.gda.pl
[2] Department of Civil and Environmental Engineering,
University of California, Berkeley, CA, USA
james.fern@berkeley.edu

Abstract. There has been much effort in the past years to characterise the mechanical behaviour of root-reinforced soils, both in terms of strengths and stiffness. This paper presents a simple extension of bounding surface model called Nor-Sand, which captures the progressive activation of the roots in the frame of continuum mechanics. This is achieved by considering the changes in the yield and the bounding surfaces, which influence the hardening and softening rates as well as the predicted peak strength. The model is implemented in the finite element software Abaqus and used to simulate element tests. The macroscopic modelling approach facilitates the modelling of large-scale problems such as the erosion of dunes, which is of importance for coastal regions.

1 Introduction

Root-reinforced soil form a composite material with a strong soil root interaction. The roots provide tensile strength and the soil provides compressive strength. The reinforcement of soil with roots increases the shear strength and modifies the hydraulic properties of the soil, and has been experimentally investigated by many researchers [4,10–12]. However, less work has been carried out on the constitutive modelling of such composite material. This paper presents a simple extension of an existing model called Nor-Sand [5] for root-reinforced sands.

2 Nor-Sand for Root-reinforced Soils

The Nor-Sand model is a critical state soil model developed by Jefferies [5] and follows the principles of original Cam-Clay [9] with four major differences. (1) It includes a bounding surface, which allows the model to predict the peak strength from dilatancy making the void ratio e a model variable. (2) It sizes the yield surface and the bounding surface with the image pressure p_i and $p_{i,max}$, respectively, which are equivalent expression of the preconsolidation pressure in Cam-Clay. (3) The shape of the yield and bounding surfaces includes a dilatancy parameter N which changes the shape of the surfaces. (4) The hardening rule is driven by the plastic deviatoric strain increment $d\varepsilon_d^p$ rather than the plastic volumetric strain increment $d\varepsilon_v^p$.

It is known that the roots increase the elastic domain and the peak strength of the composite soil [4]. Therefore, additional terms are required to define the initial image pressure and the maximum image pressure. The initial image pressure ($p_{i,ini}$) is defined in Eq. 1

$$p_{i,ini} = p' \left[\frac{1}{1-N} - \left(\frac{N}{1-N}\right) \frac{q}{p' \times M} \right]^{\frac{N-1}{N}} \times \exp\left(m_r^{ini} R_p\right) \quad (1)$$

where m_r^{ini} is a root parameter for the elastic contribution and R_p the root parameter representing the strength of roots.

The peak strength, and thus $p_{i,max}$, is a function of the mobilisation of the tensile strength in the roots and it is driven by both the total volumetric and deviatoric strain ($d\varepsilon_r = d\varepsilon_v + d\varepsilon_d$). The enhanced maximum image pressure is given in Eq. 2.

$$p_{i,max} = p \left(1 + \chi\psi \frac{N}{M}\right)^{\frac{N-1}{N}} \times \exp\left(eR_p d\varepsilon_r\right) \quad (2)$$

where χ is dilatancy coefficient [7], ψ the state parameter [2] and M the critical state stress ratio.

The critical state of the root-reinforced soil is assumed to be the same as for the soil because it is hypothesised that the roots brake at large deformation and cannot influence that state.

3 Numerical Modelling

The extended Nor-Sand model has been implemented in ABAQUS Standard [1] using the formulation of Fern [3], and single element simulation of drained triaxial compression tests were carried. The model parameters of the soil are those for a Erksak sand [6] given in Table 1. First, the element is subjected to an isotropic consolidation up to $p' = 140$ kPa, and then sheared up to an axial strain of $d\varepsilon_1 = 20\%$ at a rate of 0.02% /min.

The results of numerical simulations are shown in Fig. 1. The case without roots ($R_p = 0$) is compared to the experimental results (dashed line) and is in good agreement. Three additional cases are for different root strength parameters and the simulations show an increase in strength, stiffness and dilatancy for different values of R_p.

Table 1. NorSand parameters for Erksak sand (from [6])

Elasticity:	Max. shear modulus at p'_0	G	50 MPa
	Poisson ratio	ν	0.2
Critical state	CS effective stress ratio	M	1.26
	Max. void ratio	$e_{\max}$	0.790
	Min. void ratio	$e_{\min}$	0.676
	CSL fiiting parameter	R	1
	Crushing pressure	Q	10 MPa
Nor-Sand param.	Dilatancy parameter	N	0.37
	Hardening modulus	H	120
	Dilatancy coefficient	χ	4.5
Initial conditions	Initial void ratio	e_0	0.68
	Initial mean effective stress	p'_0	140 kPa

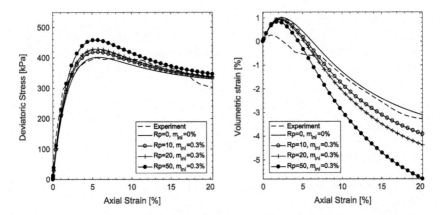

Fig. 1. Results of the numerical modelling of soil-root composite, using modified Nor-Sand model. Sensitivity of the solution to different values of R_p, keeping $m_\mathrm{r}^{\mathrm{ini}} = 0.3\%$.

Figure 2 shows the initial yield surface for each $m_\mathrm{r}^{\mathrm{ini}}$ value and it illustrates the enhancement of the elastic domain.

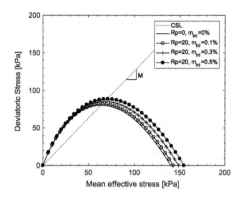

Fig. 2. Sensitivity of the size of the initial yield surface to the changes of $m_\mathrm{r}^\mathrm{ini}$, keeping $R_\mathrm{p}=20$.

4 Conclusions

The extension of the Nor-Sand model is achieved with only two new parameters and permits simulating root-reinforced soil. Although the model is still very simple, fundamental features of the soil-root composites are captured by the model. The progressive activation of the strength, caused by the elongation of the roots with strain, is reflected in the model.

References

1. Abaqus/Standard^{TM} v2016, Dassault Systémes Simulia Corp., Providence, RI (2016)
2. Been, K., Jefferies, M.G.: A state parameter for sands. Géotechnique **35**(2), 99–112 (1985)
3. Fern, E.J. : Constitutive Modelling of Unsaturated Sand and Its Application to Large Deformation Modelling. Ph.D. thesis. University of Cambridge (2016)
4. Ghestem, M., Veylon, G., Bernard, A., Vanel, Q., Stokes, A.: Influence of plant root system morphology and architectural traits on shear resistance. Plant and Soil **377**(12), 43–61 (2014)
5. Jefferies, M.G.: Nor-sand: a simple critical state model for sand. Géotechnique **43**(1), 91–103 (1993)
6. Jefferies, M.G., Been, K.: Soil Liquefaction: A Critical State Approach, 2nd edn. CRC Press, Abingdon (2016)
7. Jefferies, M.G., Shuttle, D.A.: Dilatancy in general Cambridge-type models. Géotechnique **52**(9), 625–38 (2002)
8. Nova, R.: A constitutive model for soil under monotonic and cyclic loading. In: Soil Mechanics – Transient and Cyclic Loading, pp. 343–373, Chichester, Wiley (1982)
9. Roscoe, K.H., Schofield, A.N.: Mechanical behaviour of an idealised wet clay. In: 2nd European Conference on Soil Mechanics and Foundation Engineering, 4754. Wiesbaden (1963)

10. Schwarz, M., Lehmann, P., Or, D.: Quantifying lateral root reinforcement in steep slopes - from a bundle of roots to tree stands. Earth Surf. Proc. Land. **35**, 354–367 (2010)
11. Switala, B.M., Askarinejad, A., Wu, W., Springman, S.M.: Experimental validation of a coupled hydro-mechanical model for vegetated soil. Géotechnique **68**(5), 375–385 (2018). https://doi.org/10.1680/jgeot.16.P.233
12. Wu, T.H.: Investigation of landslides on Prince Wales Island, Alaska. Geotechnical Engineering Report 5, Ohio State University, Department of Civil Engineering (1976)

Experience of Parameter Optimization of the High-Cycle Accumulation Model for Undrained Triaxial Tests on Sand

Yining Teng, Hans Petter Jostad, and Youhu Zhang[✉]

Norwegian Geotechnical Institute, Oslo, Norway
Youhu.Zhang@ngi.no

Abstract. This paper discusses the performance of the high-cycle accumulation model (HCA) for describing the soil response in undrained cyclic triaxial tests of sand, evaluated against a comprehensive experimental database. The parameters for the HCA model calibrated for drained triaxial tests is examined for its applicability in undrained conditions. A modified approach for parameter optimization for undrained conditions is proposed and demonstrated to provide good match to the triaxial test results.

Keywords: Cyclic loading · Triaxial test · Sand · HCA model

1 Introduction

The high-cycle accumulation (HCA) model [1] is an empirical model using mixed implicit-explicit numerical strategy to calculate the accumulation of irreversible strain or pore pressure in the soil under long-term cyclic loading with relatively small cyclic amplitudes (Fig. 1). Strain/pore pressure accumulation in the soil is highly relevant to offshore foundations, such as foundations for supporting offshore wind turbines, subsea mudmats, gravity base structures, etc., which are typically subject to cyclic environmental or operating loading. Prediction of accumulated foundation settlement/displacement is an important design aspect, for which the HCA model may find its application.

The performance of the HCA model and its parameter determination have been published and summarized in [2], mostly for drained condition. But its applicability to describe soil response under undrained conditions is less well understood [3]. This paper will share the experience from a parameter calibration exercise against a database of undrained cyclic triaxial tests on Karlsruhe fine sand [4]. A modified parameter optimization method for undrained conditions is proposed and discussed.

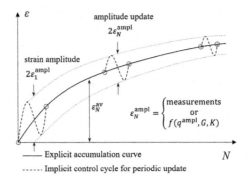

Fig. 1. Schematic of high-cycle accumulation (HCA) model calculation strategy.

2 Basic Formulation

The formulations of HCA model are introduced in [1]. The key formulations to calculate the accumulated volumetric strain for drained and undrained conditions are slightly different as follows:

$$\varepsilon_v^{acc} = f_{ampl} \cdot f_e \cdot f_p \cdot f_Y \cdot f_N \cdot m_v \tag{1}$$

$$\frac{\dot{u}}{K^*} = \dot{\varepsilon}_v^{acc} = f_{ampl}^* \cdot f_e \cdot f_p^* \cdot f_Y^* \cdot \dot{f}_N^* \cdot m_v^* \tag{2}$$

where ε_v^{acc} is accumulated plastic volumetric strain; u is accumulated pore pressure; $(f_{ampl}, f_e, f_p, f_Y, f_N)$ are five influencing functions considering impact of cyclic strain amplitude, void ratio, effective means stress, stress ratio and cyclic loading history respectively; For undrained equations, the * symbol denotes update is required from cycle to cycle; m_v and m_v^* are volumetric portion of total accumulated strain; ε_q^{ampl} and q^{ampl} are deviatoric cyclic strain and cyclic stress; K^* is HCA model bulk modulus [2].

Equation (1) is typical for drained conditions. A modified equation for undrained pore pressure accumulation is given in derivative form (Eq. (2)). For undrained conditions, most functions $\left(K^*, f_{ampl}^*, f_p^*, f_Y^*, \dot{f}_N^*, m_v^*\right)$ require updates due to changing stress conditions.

3 Parameters for Drained and Undrained Conditions

An optimized set of model parameters is presented in [2] based on a series of long-term cyclic drained tests, as listed in Table 1. However, it is reported that optimized parameters for drained long-term cyclic tests may differ from those for undrained cyclic tests, which typically focus on much smaller number of cycles [3]. It is anticipated that the function f_N and $\dot{f}_N$, which accounts for load history may vary from drained to undrained condition due to changing stress states in undrained conditions. A set of isotropically consolidated cyclic undrained tests in [4] are used to examine the function $\dot{f}_N$. In Fig. 2a, the derived $\dot{f}_N$ from the undrained tests, assuming that the remaining functions ($f_{ampl}, f_e, f_p, f_Y, K$ and m_v) and the parameters listed in Table 1 are correct, are plotted against the number of cycles N in a semi-log space. All tests broadly follow a similar trend which indicates that the calibrated functions $f_{ampl}, f_e, f_p, f_Y, K$ and m_v from the drained tests are also reasonable for the undrained tests. However, the fitted $\dot{f}_N$ curve from the parameters for drained condition (dashed line in Fig. 2a) has a different trend than the normalised curves from the undrained tests, which indicates a better fitting is needed.

Table 1. Optimized parameters for drained and undrained cyclic tests.

Drainage condition	C_{ampl}	C_e	C_p	C_Y	$C_{N1}[10^{-4}]$	C_{N2}	$C_{N3}[10^{-3}]$
Drained	1.33	0.60	0.23	1.68	2.95	0.37	2.36
Undrained	1.33	0.60	0.23	1.68	1.60	0.90	0.00

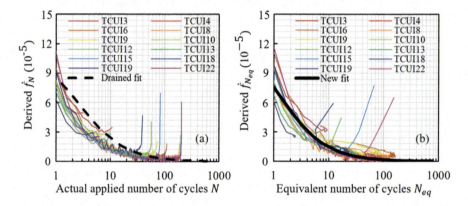

Fig. 2. (a) Calculated $\dot{f}_N$ from undrained tests and fitted $\dot{f}_N$ from the previous optimization; (b) Calculated $\dot{f}_{N_{eq}}$ from undrained tests and fitted $\dot{f}_{N_{eq}}$ from the new fitting method;

In this paper, a different method addressing the current effective stress state and strain accumulation rate using the equivalent number of cycles N_{eq} is proposed. Since the actual applied number of load cycles N may not be appropriate to represent the load history with changing stress condition in undrained triaxial tests, the concept of the equivalent number of cycles N_{eq} is proposed. N_{eq} is defined as the number of cycles to achieve the same amount of accumulated strain in a drained test under the current effective stress state and cyclic strains amplitude (Eq. 3).

$$\dot{f}_{N_{eq}} = \frac{\dot{\varepsilon}^{acc}}{f^*_{ampl} \cdot f_e \cdot f^*_p \cdot f^*_Y} = C_{N1}\left[\ln(1 + C_{N2}N_{eq}) + C_{N3}N_{eq}\right] \quad (3)$$

in this case $\varepsilon^{acc} = \varepsilon_v^{plastic} = u/K^*$ due to negligible development of permanent shear strain for the tests currently looked at (pure two way cyclic loading from an isotopically consolidated initial state).

The equation for $\dot{f}_{N_{eq}}$ remains the same form as $\dot{f}_N$ in Eq. (2), with N replaced by N_{eq} from Eq. (3). The derived $\dot{f}_{N_{eq}}$ from the undrained tests are plotted against the equivalent number of cycles at current state N_{eq} in a semi-log space (Fig. 2b). For new fittings with the method introduced above, the parameters C_{N1} and C_{N2} for $\dot{f}_{N_{eq}}$ have been adjusted to fit better to the trend of the undrained tests, but C_{N3} is set to zero due to minimal impact on the results (Table 1). Curve fittings using the new set of parameters match closely to the test results, and three examples are given in Fig. 3a. Measured strain amplitude is used in Fig. 3a. However, cyclic stress amplitude q^{ampl} could be used as input instead to calculate the strain amplitude ε^{ampl} for each cycle as input to the HCA model. This is demonstrated in Fig. 3b that the fitting to the experimental results is still reasonable without losing much accuracy.

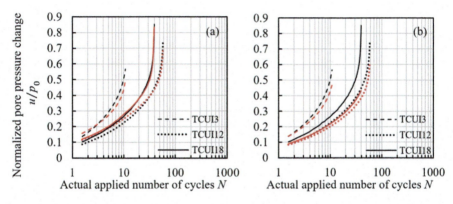

Fig. 3. Test results (black) and fitting (red) of undrained cyclic tests using the new set of parameters: (a) using cyclic strain amplitude as input; (b) using cyclic stress amplitude as input.

4 Summary and Discussion

This paper examines the application of the HCA model to describe soil behaviours under undrained conditions. By considering the effect of changing effective stress state through updating relevant functions from cycle to cycle, it is possible to apply the HCA model to undrained conditions. Best-fit parameters calibrated for drained and undrained cyclic triaxial tests are compared. Parameters calibrated from the drained cyclic tests may not necessarily fit for undrained conditions. It is therefore recommended to calibrate the model parameters based on both drained and undrained tests if the undrained conditions are of interest. A simple modification to the C_N parameters and use of an equivalent number of cycle N_{eq} to represent load history are proposed, and work reasonably well for undrained conditions. It is also demonstrated that it is possible to use the cyclic shear stress instead of cyclic strain amplitude as model input, if the shear modulus G and G/G_{max} is properly defined.

References

1. Niemunis, A., Wichtmann, T., Triantafyllidis, T.: A high-cycle accumulation model for sand. Comput. Geotech. **32**(4), 245–263 (2005)
2. Wichtmann, T.: Soil behaviour under cyclic loading: experimental observations, constitutive description and applications. Doctoral dissertation, Institut für Bodenmechanik und Felsmechanik (2016)
3. Wichtmann, T., Niemunis, A., Triantafyllidis, T.: On the "elastic" stiffness in a high-cycle accumulation model for sand: a comparison of drained and undrained cyclic triaxial tests. Can. Geotech. J. **47**(7), 791–805 (2016)
4. Wichtmann, T., Triantafyllidis, T.: An experimental database for the development, calibration and verification of constitutive models for sand with focus to cyclic loading: part I—tests with monotonic loading and stress cycles. Acta Geotech. **11**(4), 739–761 (2016)
5. Wichtmann, T., Triantafyllidis, T.: Effect of uniformity coefficient on G/G max and damping ratio of uniform to well-graded quartz sands. J. Geotech. Geoenviron. Eng. **139**(1), 59–72 (2012)

Hardening Soil Model - Influence of Plasticity Index on Unloading - Reloading Modulus

Wojciech Tymiński[✉], Tomasz Kiełczewski, and Hubert Daniluk

GEOTEKO Geotechnical Consultants Ltd., 14/16 Wałbrzyska St., 02-739 Warsaw, Poland
wojciech.tyminski@geoteko.com.pl

Abstract. Numerical analyses based on Finite Element Method (FEM) to solve geotechnical problems are carried out using software implementing various soil models. Currently, Hardening Soil Model (HS) is often used in the geotechnical engineering practice. One of the parameters describing the model is the unloading-reloading (E_{ur}) modulus. This paper presents triaxial test results of cohesive soils aimed at determination of unloading-reloading (E_{ur}) modulus. Based on test results, empirical correlations for evaluation of E_{ur} modulus were determined. The correlations determine E_{ur} modulus on the basis of plasticity index and vertical effective stress.

Keywords: Hardening Soil Model · Triaxial tests Unloading-Reloading modulus

1 Introduction

Hardening Soil Model (HS) was formulated in 1999 by Schanz et al. [4]. Its modification, which includes elastic behaviour of soil for the small strain (HSS) was presented by Benz [1]. Both models are among the most popular constitutive models used in the geotechnical practice.

Strain parameters of the model are defined in Fig. 1. The manual [3] and publication [2] listed in the bibliography below, proposed the relationship between E_{ur} and E_{50} as $E_{ur}/E_{50} = 3$. However, Obrzud [2] and Truty [5] noted that E_{ur}/E_{50} ratio should be higher.

Below an empirical equation for E_{ur} modulus calculation based on effective stress and plasticity index (PI) is proposed.

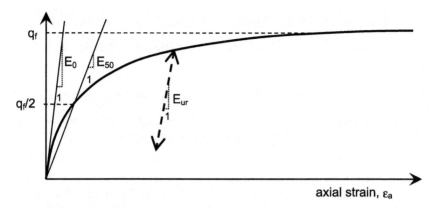

Fig. 1. Evaluation of E_0, E_{ur} and E_{50} based on stress-strain characteristics.

2 Test Methodology and Analyses

2.1 Triaxial Tests

The tests were carried out on undisturbed soil samples in GEOTEKO's laboratory. The triaxial tests included the following stages: back pressure saturation, isotropic consolidation and strain controlled drained shearing along standard stress path i.e. with constant cell pressure and increasing vertical stress.

2.2 Analyses

The analyses were performed using the results of triaxial tests carried out in terms of effective stress of values higher than *in situ* stress. Below, Fig. 2 shows a relationship

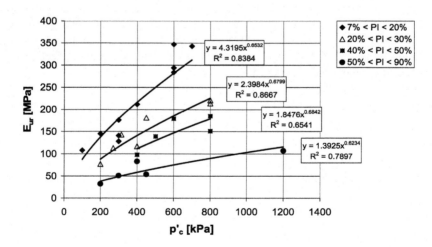

Fig. 2. Eur vs. mean effective stress at the end of consolidation stage.

between E_{ur} modulus and mean effective stress at the end of consolidation stage (p'_c) taking into account plasticity index value.

The relationship between E_{ur} and p'_c can be described by the following general formula:

$$E_{ur} = a \cdot p'^n_c \qquad (1)$$

Parameter (a) in Eq. (1) changes together with plasticity index variations. Figure 3 shows the relationship between parameter (a) and plasticity index.

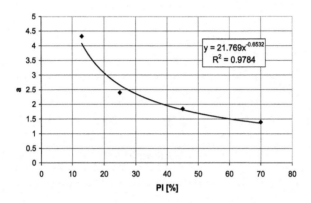

Fig. 3. "a" vs. I_p

Based on the above, E_{ur} moduli values can be calculated from the following formula:

$$E_{ur} = 21.769 \cdot PI^{-0.65} \cdot p'^{0.65}_c \qquad (2)$$

Using the proposed empirical formula, E_{ur} moduli were determined and compared with the relevant values resulting from the laboratory tests (Fig. 4). In conclusion one may declare that a significant part of E_{ur} value determined by empirical formula is within ±30% of E_{ur} value determined in the laboratory tests. Thus it seems to form a better approximation than the ratio $E_{ur}/E_{50} = 3$ which was mentioned earlier.

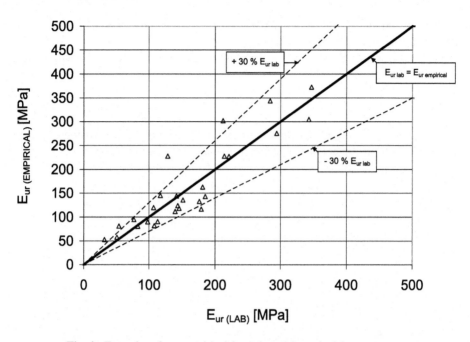

Fig. 4. Eur values from empirical formula and from the laboratory tests

3 Conclusions

The following conclusions can be drawn on the basis of the performed tests and analyses:

- as soil plasticity index increases, the influence of effective stress on E_{ur} value becomes smaller.
- the best correlation between E_{ur} modulus and effective stress was obtained for soils of plasticity index smaller than 30%.
- the proposed empirical relationship allows calculating E_{ur} modulus on the basis of plasticity index (PI) and average effective stress (p'_c) values. A significant part of E_{ur} value determined from the proposed empirical formula is included within the range of ±30% of E_{ur} value determined in the laboratory tests.

The analyses presented in the paper were carried out within R&D activity of Geoteko Geotechnical Consultants Ltd. The a/m works resulted also in the proposal of other empirical formulas for estimation of geotechnical parameters. They will be presented in the next papers to be published.

References

1. Benz, T.: Small-strain stiffness of soils and its numerical consequences. Ph.D., Universitat Stuttgart (2007)
2. Obrzud, R.: On the Use of the Hardening Soil Small Strain Model in Geotechnical Practice. Elmepress International, Lausanne (2010)
3. PLAXIS: Material Models Manual (2017)
4. Schanz, T., Vermeer, P., Bonier, P.: Formulation and verification of the hardening soil model. In: Beyond 2000 in Computational Geotechnics. Balkema, Rotterdam (1999)
5. Truty, A.: Hardening soil model with small strain stiffness. Technical report 080901, Zace Services Ltd., Lausanne (2008)

A Visco-Hypoplastic Constitutive Model and Its Implementation

Guofang Xu[1,2(✉)], Wei Wu[3], Jilin Qi[4], Xiaoliang Yao[1,2], Fan Yu[1,2], and Aiguo Guo[1,2]

[1] SKLGME, Institute of Rock and Soil Mechanics, CAS,
Xiaohongshanstreet 2, Wuhan 430071, China
[2] SKLFSE, Northwest Institute of Eco-Environment and Resources, CAS,
Donggangweststreet 320, Lanzhou 730000, China
gfxu@whrsm.ac.cn
[3] Institute of Geotechnical Engineering,
University of Natural Resources and Life Sciences, Vienna,
Feistmantelstrasse 4, 1180 Vienna, Austria
wei.wu@boku.ac.at
[4] School of Civil and Transportation Engineering,
Beijing University of Civil Engineering and Architecture,
Zhanlanguanstreet 1, Beijing 100044, China
jilinqi@bucea.edu.cn

Abstract. This paper presents a hypoplastic constitutive model for the viscous behaviour of frozen soil. Then the model is implemented into FLAC3D®. The performance of the model is demonstrated by simulating a compression creep test with the tertiary stage and the deformation of an embankment.

1 Introduction

The viscous behavior especially creep of materials is of great importance not only to material science, but also to various engineering problems. In this paper, an integrated constitutive model is proposed for frozen soil based on the extended hypoplastic constitutive model by Xu et al. [1] and the high-order constitutive model by Wu [2]. With the proposed model, not only the common mechanical properties of frozen soil, such as shear dilatancy, path dependence, etc. can be simulated, some viscous behaviors, e.g. creep and rate dependence, can also be described.

2 The Visco-Hypoplastic Constitutive Model

A hypoplastic constitutive model was proposed for the viscous behaviour of frozen soil. The model has the following form:

$$\dot{\boldsymbol{\sigma}} = c_1[\text{tr}(\boldsymbol{\sigma}-\mathbf{s})]\dot{\boldsymbol{\varepsilon}} + c_2\frac{\text{tr}[(\boldsymbol{\sigma}-\mathbf{s})\dot{\boldsymbol{\varepsilon}}]}{\text{tr}(\boldsymbol{\sigma}-\mathbf{s})}(\boldsymbol{\sigma}-\mathbf{s}) + f_\varepsilon \cdot \boldsymbol{f}_{\text{cd}}\left[c_3(\boldsymbol{\sigma}-\mathbf{s})^2 + c_4(\boldsymbol{\sigma}-\mathbf{s})_{\text{d}}^2\right]\frac{\|\dot{\boldsymbol{\varepsilon}}\|}{\text{tr}(\boldsymbol{\sigma}-\mathbf{s})}$$
$$+ \eta_1\sqrt{\eta_2^2 + \text{tr}(\dot{\boldsymbol{\varepsilon}}^2)}\,\ddot{\boldsymbol{\varepsilon}} \tag{1}$$

in which c_i ($i = 1, \cdots, 4$), η_1, and η_2 are parameters; $\mathbf{s}$ is a spherical stress tensor determined by $\mathbf{s} = s \cdot \mathbf{1}$; $\|\dot{\boldsymbol{\varepsilon}}\|$ is the Euclidean norm of $\dot{\boldsymbol{\varepsilon}}$; f_ε is calculated by

$$f_\varepsilon = 2 - \exp(\alpha \cdot l + \beta) \tag{2}$$

in which α and β are parameters, l is the length of the strain trajectory. $\boldsymbol{f}_{\text{cd}}$ is a creep damage function defined as

$$\boldsymbol{f}_{\text{cd}} = \mathbf{1} + \gamma\int_0^t \langle \text{diag}(\text{diag}(\ddot{\boldsymbol{\varepsilon}}(\tau)))\rangle \mathrm{d}\tau \tag{3}$$

in which γ is a parameter; $\langle\ \rangle$ is the Macaulay brackets.

3 Implementation of the Viscous Model

The main function of a constitutive model is to obtain stress increments, given strain increments. For this model, the incremental form can be given as

$$\Delta\boldsymbol{\sigma} = c_1[\text{tr}(\boldsymbol{\sigma}-\mathbf{s})]\Delta\boldsymbol{\varepsilon} + c_2\frac{\text{tr}[(\boldsymbol{\sigma}-\mathbf{s})\Delta\boldsymbol{\varepsilon}]}{\text{tr}(\boldsymbol{\sigma}-\mathbf{s})}(\boldsymbol{\sigma}-\mathbf{s}) + f_\varepsilon \cdot \boldsymbol{f}_{\text{cd}}\left[c_3(\boldsymbol{\sigma}-\mathbf{s})^2 + c_4(\boldsymbol{\sigma}-\mathbf{s})_{\text{d}}^2\right]\frac{\|\Delta\boldsymbol{\varepsilon}\|}{\text{tr}(\boldsymbol{\sigma}-\mathbf{s})}$$
$$+ \eta_1\sqrt{\eta_2^2 + \text{tr}(\dot{\boldsymbol{\varepsilon}}^2)}\,\Delta\dot{\boldsymbol{\varepsilon}} \tag{4}$$

In the implementation of the proposed model, it should be noted that the stresses used for the calculation in the next step are not the stresses directly returned by FLAC3D®, but subtracting the updated stresses by the accumulated stresses resulting from the acceleration term. Hence, when denoting the stress dependent part of the model as $\dot{\hat{\boldsymbol{\sigma}}}$ and the stress independent part as $\dot{\check{\boldsymbol{\sigma}}}$, we have

$$\dot{\hat{\boldsymbol{\sigma}}} = \dot{\boldsymbol{\sigma}} - \dot{\check{\boldsymbol{\sigma}}} \tag{5}$$

On the right side of the incremental form model, the initial stress state $\boldsymbol{\sigma}$ and strain increment $\Delta\boldsymbol{\varepsilon}$ are known condition, the quantities need be determined are f_ε, $\boldsymbol{f}_{\text{cd}}$, $\dot{\boldsymbol{\varepsilon}}$, and $\Delta\dot{\boldsymbol{\varepsilon}}$. $\dot{\boldsymbol{\varepsilon}}$ and $\Delta\dot{\boldsymbol{\varepsilon}}$ can be obtained by Eqs. (6)–(8) as

$$\dot{\boldsymbol{\varepsilon}}_{\text{old}} = \dot{\boldsymbol{\varepsilon}}_{\text{initial}} \tag{6}$$

$$\dot{\boldsymbol{\varepsilon}}_{\text{new}} = 2\Delta\boldsymbol{\varepsilon}/\text{dCrtdel} - \dot{\boldsymbol{\varepsilon}}_{\text{old}} \tag{7}$$

in which dCrtdel is the time step for creep calculation.

$$\Delta\dot{\boldsymbol{\varepsilon}} = \dot{\boldsymbol{\varepsilon}}_{\text{new}} - \dot{\boldsymbol{\varepsilon}}_{\text{old}} \tag{8}$$

Before the calculation of $\boldsymbol{f}_{cd}$, the strain acceleration in the model has to be firstly determined. This is achieved as

$$\ddot{\varepsilon} = (\dot{\varepsilon}_{new} - \dot{\varepsilon}_{old})/\text{dCrtdel} \tag{9}$$

$$\boldsymbol{f}_{cd.old} = \boldsymbol{f}_{cd.initial} \tag{10}$$

$$\boldsymbol{f}_{cd.new} = \boldsymbol{f}_{cd.old} + \boldsymbol{\gamma}\left(|\ddot{\varepsilon}| + \ddot{\varepsilon}\right)/2 \tag{11}$$

in which $\boldsymbol{f}_{cd.initial}$ has the value of **1**. The factor f_ε can be determined by Eqs. (12)–(14) as

$$\Delta l = \sqrt{\Delta\varepsilon_{11}^2 + \Delta\varepsilon_{22}^2 + \Delta\varepsilon_{33}^2} \tag{12}$$

$$l_{new} = l_{old} + \Delta l \tag{13}$$

$$f_\varepsilon = 2 - \exp\left(\boldsymbol{\alpha}\cdot l_{new} + \boldsymbol{\beta}\right) \tag{14}$$

4 Numerical Simulation

In this section, numerical simulations of a compression creep test with the tertiary creep stage and embankment deformation are carried out using the proposed model. The material parameters are determined as $c_1 = -58.99$, $c_2 = -523.26$, $c_3 = -496.05$, and $c_4 = 509.12$, the other parameters are presented in Table 1.

In the creep test, the axial compression stress is 8000 kPa, the radial boundary is let free of stress. The axial strain is presented in Fig. 1. The deformation

Table 1. Model parameters and initial condition for simulating the creep tests

Simulation	α	β	γ	η_1	η_2	$\dot{\varepsilon}_{11.initial}$	$\dot{\varepsilon}_{33.initial}$
Creep test	−0.003	0.005	3.2e1	1.2e12	1.0e−12	1.5e−5	−3.0e−5
Embankment	−0.0001	0.0005	0.0	1.2e11	1.0e−12	0.5e−5	−1.0e−5

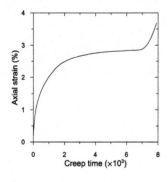

Fig. 1. Axial strain *vs.* time in the creep test

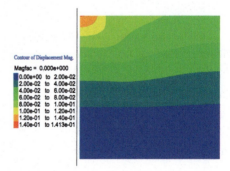

Fig. 2. Deformation of the embankment

of embankment is simulated as a plane strain problem. On the plane, there are 23×23 elements. Normal stress with the magnitude of 9.25×10^4 Pa is applied on the top of the embankment with the X-range of $[0, 3]$. The deformation of the embankment is shown in Fig. 2.

5 Conclusion

In this paper the constitutive model previously proposed by the authors is implemented into FLAC3D®. Then the model is used to simulate a creep test and the deformation of an embankment. It is found that rational results can be given by the model for both the creep test and the embankment deformation.

Acknowledgement. The National Natural Science Foundation of China (grant No. 11702304, 41572268, 41671061), the CAS Pioneer Hundred Talents Program granted to Dr. G. Xu, and the State Key Laboratory of Frozen Soil Engineering (grant No. SKLFSE201714) are acknowledged for their support to this work.

References

1. Xu, G., Wu, W., Qi, J.: Modeling the viscous behavior of frozen soil with hypoplasticity. Int. J. Numer. Anal. Methods Geomech. **40**(15), 2061–2075 (2016)
2. Wu, W.: On high-order hypoplastic models for granular materials. J. Eng. Mech. **56**(1), 23–34 (2006)

Advance in the Constitutive Modelling for Frozen Soils

Guofang Xu[1,2(✉)], Lingwei Kong[1,2], Yiming Liu[3], Cheng Chen[1,2], and Zhiliang Sun[1,2]

[1] SKLGME, Institute of Rock and Soil Mechanics, CAS, Xiaohongshanstreet 2, Wuhan 430071, China
gfxu@whrsm.ac.cn
[2] SKLFSE, Northwest Institute of Eco-Environment and Resources, CAS, Donggangweststreet 320, Lanzhou 730000, China
[3] School of Civil Engineering, Architecture and Environment, Hubei University of Technology, Nanlistreet 28, Wuhan 430068, China

Abstract. This paper presents an intensive review of the constitutive models for frozen soil. The models are classified into five categories, namely, empirically fitted model, classical plastic model, rate process model, element assembly model, and hypo-plastic model. The background and relative merits of each type of the models are elucidated.

Keywords: Constitutive model · Frozen soils

1 Introduction

Engineering activities in permafrost regions are on the rise over the world. For such engineering projects, the strength and deformation characteristics of frozen soil are of great concern. Thereupon, much progress has been made in both experiment and constitutive modeling, to pursue an in-depth understanding of the mechanical and engineering behaviors of frozen soil. This paper aims to give an intensive review of the constitutive models proposed for frozen soils. Relevant models developed by the author are also included in this paper.

2 Constitutive Models

In the past decades, numerous constitutive models have been developed for frozen soil. In this paper, the existing constitutive models are roughly classified into five categories, namely, empirically fitted model, classical plastic model, rate process model, element assembly model, and hypo-plastic model. In this section, the advantage and disadvantage of each type of the constitutive models are elucidated.

2.1 Empirically Fitted Model

Such models are developed by fitting the stress–strain or strain–time curve of frozen soil using common mathematical functions, e.g. linear function, exponential function, power function (particularly hyperbolic function) based on uniaxial or triaxial compression tests, creep tests or dynamic tests. With a specific type of function, different parameters can be obtained for the curves obtained under different test conditions, then the parameters will be formulated with the test conditions to obtain a so called constitutive model, see [1–9]. Such models helped to understand the mechanical behavior of frozen soil in the early days. However, they are obtained by fitting directly or even in a piecewise way the test data, and the fitted functions for the behavior in each axial are generally independent. Moreover, the models cannot take into account the effect of stress path or reverse loading. Therefore the application of the models is rather limited.

2.2 Classical Plastic Model

The plastic constitutive models for frozen soil are generally obtained by modifying one or more of the cornerstones of plasticity theory, i.e. yield criterion, flow rule, and hardening law, to take into account the temperature effect based on the existing plastic models for unfrozen soils. Examples of plastic model are given in [10–23]. In the development of plasticity-based models, one should face up to the following issues. For example, the yield criterion for frozen soil is difficult to determine, which is mainly because the yield point on the stress–strain curve is not as clear as that for metallic materials. Secondly, the determination of a flow rule (an associated one) is generally boiled down to the determination of plastic potential function, which is mainly based on tests or hypotheses. For some materials this potential function even does not exist. In plasticity theory, the hardening law describes the variation of yield criterion with plastic deformation. Again, the ambiguous definition of yielding of frozen soil is encountered. In addition, the plastic models generally have rather complicated mathematical formulation.

2.3 Rate Process Model

Rate process theory (RPT) is a theory derived from statistical mechanics and is based on the assumption that particles participating in a deformation process, termed flow units, are constrained from movement by energy barriers. The displacement of flow unit to new position requires activation through the acquisition of sufficient energy to overcome the energy barrier. Examples of rate process models were given by Andersland and Akili [24] for frozen clay, and Fish and Sayles [25] and Zhu and Carbee [26] for frozen Fairbanks silt. Rate process models can provide some knowledge of the fundamental mechanisms of creep and the nature of interparticle bonding in frozen soils. According to the RPT, the plastic deformation of a material can be described by an Arrhenius equation [27,28] in which the strain rate depends on the differential stress, grain size, and an activation value. However, the Arrhenius law often underestimates the slowdown rate of the motion of the flow units. Besides, more efforts have to be made in evaluating the nature of frequency factor, activation energy, and stress factor in the theory [29].

2.4 Element Assembly Model

Such models are assembled using conceptual physical elements, viz. Hookean spring, Newtonian dashpot, and St. Venant slider. Examples of this type of model have been given by Wang et al. [30] for frozen Huainan soil, Li et al. [31] for frozen clay, Wang et al. [32] for frozen fine sand, and Liao et al. [33] for warm frozen silt. It should be noted that, combined models are originally proposed to describe the rheological behaviors of polymers, rubbers, and numerous metals at a temperature close to their melting point. For modeling frozen soil, owing to the heterogeneity and disordered internal structure, it is difficult to find a suitable combination of the conceptual elements. Therefore, one should proceed with awareness when modeling with this approach: in the simple combined models, one may have difficulties in reflecting the real behavior of frozen soil, while the complex combined models would cause great mathematical difficulties.

2.5 Hypo-plastic Model

Hypo-plasticity is a new theory and has become popular in constitutive modeling of granular materials. In hypo-plasticity, the constitutive model is built by stress rate being a tensor function of stress state and strain rate according to the representation theorem for isotropic tensor function. The hypo-plasticity theory has many advantages in material modeling, such as: (1) the nonlinear stress-strain relationship of granular materials can be well captured by hypo-plastic model; (2) the failure surface and flow rule in hypo-plasticity can be obtained rigorously mathematically; and (3) the distinction between loading and unloading is not necessary during numerical calculations, since the loading–unloading criterion is hidden in hypo-plastic models *per se*. By introducing a spheric stress tensor dependent on the cohesion of frozen soil and a softening parameter related to strain path into the hypo-plastic constitutive model for sand, Xu et al. [34] proposed a simple hypo-plastic constitutive model for frozen sand. Based on this model and by introducing the second time derivative of strain, a visco-hypo-plastic constitutive model is obtained by Xu et al. [35] for the rate dependent behavior of frozen soil. Moreover, the viscous model can describe the three-stage creep process of frozen soil in a unified way. However, some parameters in the model cannot be determined from conventional experiment.

3 Concluding Remarks

Based on this review, it could be concluded that all the constitutive models for frozen soil are simply phenomenological, rather than derived from first principles.

Furthermore, the following rules are recommended for constitutive modeling of frozen soil in the future. First, the constitutive model should be uncomplex and contains less parameters, which should be readily determined from conventional test. Second, the constitutive model should be expedient for practical use.

Acknowledgements. The National Natural Science Foundation of China (Grant 11702304) and the State Key Laboratory of Frozen Soil Engineering (Grant SKLFSE201714) are acknowledged.

References

1. Ladanyi, B.: An engineering theory of creep of frozen soils. Can. Geotech. J. **9**(1), 63–80 (1972)
2. Sayles, F.: Tri-axial constant strain rate tests and tri-axial creep tests on frozen Ottawa sand. In: Proceedings of 2nd International Permafrost Conference, pp. 384–391. National Academy of Sciences, Washington D.C. (1973)
3. Andersland, O., Anderson, D.: Geotechnical Engineering for Cold Regions. McGraw-Hill, New York (1978)
4. Fish, A.: Kinetic nature of the long-term strength of frozen soils. In: Proceedings of 2nd International Symposium on Ground Freezing, pp. 95–108. Balkema, Rotterdam (1980)
5. Finborud, L., Berggren, A.: Deformation properties of frozen soils. Eng. Geol. **18**(S1), 89–96 (1981)
6. Zhu, Y., Zhang, J., Peng, W., Shen, Z., Miao, L.: Constitutive relations of frozen soil in uniaxial compression. J. Glaciol. Geocryology **14**(3), 210–217 (1992)
7. Ma, W., Chang, X.: Comparison of strength and deformation of artificially frozen soil in two testing manners. J. Glaciol. Geocryology **24**(2), 149–154 (2002)
8. Wang, Z., Yuan, S., Chen, T.: Study on the constitutive model of transversely isotropic frozen soil. Chin. J. Geotech. Eng. **29**(8), 1215–1218 (2007)
9. Xu, G., Peng, C., Wu, W., Qi, J.: Combined constitutive model for creep and steady flow rate of frozen soil in an unconfined condition. Can. Geotech. J. **54**(7), 907–914 (2017)
10. Amiri, S., Grimstad, G., Kadivar, M., Nordal, S.: Constitutive model for rate-independent behavior of saturated frozen soils. Can. Geotech. J. **53**(10), 1646–1657 (2016)
11. Li, D., Wang, R., Hu, P., Zhang, Z.: Investigation on viscoelastic constitutive model artificial frozen soil and experimental evaluation. Low Temp. Archit. Technol. **4**, 73–74 (2005)
12. Yang, Y., Lai, Y., Dong, Y., Li, S.: The strength criterion and elastoplastic constitutive model of frozen soil under high confining pressures. Cold Reg. Sci. Technol. **60**(2), 154–160 (2010)
13. Lai, Y., Liao, M., Hu, K.: A constitutive model of frozen saline sandy soil based on energy dissipation theory. Int. J. Plast. **78**, 84–113 (2016)
14. Cai, Z., Zhu, Y., Zhang, C.: Viscoelastoplastic constitutive model of frozen soil and determination of its parameters. J. Glaciol. Geocryology **12**(1), 31–40 (1990)
15. Rong, C., Wang, X., Cheng, H.: An experimental study on finite strain constitutive relations of frozen soil. J. Exp. Mech. **20**(1), 133–138 (2005)
16. Lai, Y., Jin, L., Chang, X.: Yield criterion and elasto-plastic damage constitutive model for frozen sandy soil. Int. J. Plast. **25**, 1177–1205 (2009)

17. Li, Q., Ling, X., Sheng, D.: Elasto-plastic behavior of frozen soil subjected to long-term low-level repeated loading, Part II: constitutive model. Cold Reg. Sci. Technol. **122**, 58–70 (2016)
18. Miao, T., Wei, X., Zhang, C.: Microstructural damage theories of creep of frozen soil. Sci. China (Ser. B) **25**(3), 309–317 (1995)
19. Liu, Z., Zhang, X., Li, H.: A damage constitutive model for frozen soils under uniaxial compression based on CT dynamic distinguishing. Rock Soil Mech. **26**(4), 542–546 (2005)
20. Ning, J., Zhu, Z.: Constitutive model of frozen soil with damage and numerical simulation of the coupled problem. Chin. J. Theor. Appl. Mech. **39**(1), 70–76 (2007)
21. Li, S., Lai, Y., Zhang, S., Liu, D.: An improved statistical damage constitutive model for warm frozen clay based on Mohr-Coulomb criterion. Cold Reg. Sci. Technol. **57**(2–3), 154–159 (2009)
22. He, P., Cheng, G., Zhu, Y.: Constitutive theories on viscoelstoplasticity and damage of frozen soil. Sci. China (Ser. D) **29**(S1), 34–39 (1999)
23. Zhu, Z., Kang, G., Ma, Y., Xie, Q., Zhang, D., Ning, J.: Temperature damage and constitutive model of frozen soil under dynamic loading. Mech. Mater. **102**, 108–116 (2016)
24. Andersland, O., Akili, W.: Stress effect on creep rates of a frozen clay soil. Géotechnique **17**(1), 27–39 (1967)
25. Fish, A., Sayles, F.: Acoustic emissions during creep of frozen soils. In: Acoustic Emissions in Geotechnical Engineering Practice, pp. 194–206. American Society for Testing and Materials, Baltimore (1981)
26. Zhu, Y., Carbee, D.: Creep and strength behaviour of frozen silt in uniaxial compression. Technical report, U.S. Army Cold Regions Research and Engineering Laboratory (1987)
27. Arrhenius, S.: Über die Dissociationswärme und den Einfluss der Temperatur auf den Dissociationsgrad der Elektrolyte. Zeitschrift für Physikalische Chemie **4**, 96–116 (1889)
28. Arrhenius, S.: Über die Reaktionsgeschwindigkeit bei der Inversion von Rohrzucker durch Säuren. Zeitschrift für Physikalische Chemie **4**, 226–248 (1889)
29. Martin, R., Ting, J., Ladd, C.: Creep behavior of frozen sand. Technical report, Department of Civil Engineering, Massachusetts Institute of Technology (1981)
30. Wang, R., Li, D., Wang, X.: Improved Nishihara model and realization in ADINA FEM. Rock Soil Mech. **27**(11), 1954–1958 (2006)
31. Li, D., Fan, J., Wang, R.: Research on visco-elastic-plastic creep model of artificially frozen soil under high confining pressures. Cold Reg. Sci. Technol. **65**(2), 219–225 (2011)
32. Wang, S., Qi, J., Yin, Z., Zhang, J., Ma, W.: A simple rheological element based creep model for frozen soils. Cold Reg. Sci. Technol. **106–107**, 47–54 (2014)
33. Liao, M., Lai, Y., Liu, E., Wan, X.: A fractional order creep constitutive model of warm frozen silt. Acta Geotech. **12**(2), 377–389 (2016)
34. Xu, G., Wu, W., Qi, J.: An extended hypoplastic constitutive model for frozen sand. Soils Found. **56**(4), 704–711 (2016)
35. Xu, G., Wu, W., Qi, J.: Modeling the viscous behavior of frozen soil with hypoplasticity. Int. J. Numer. Anal. Methods Geomech. **40**(15), 2061–2075 (2016)

Towards a Better Understanding of the Mechanics of Soil Liquefaction

Jun Yang[1(✉)], H. Y. Eric Sze[2], Liming Wei[2], and Xiao Wei[1]

[1] The University of Hong Kong, Pokfulam, Hong Kong, China
junyang@hku.hk
[2] Formerly The University of Hong Kong, Pokfulam, Hong Kong, China

Abstract. Liquefaction is essentially a phenomenon in which saturated granular soil loses much of its strength and stiffness under cyclic or monotonic loading. Largely because of the complexity and uncertainty of the phenomenon, soil liquefaction remains an area of great difficulty in soil mechanics and geotechnical engineering, and the worldwide practice in liquefaction evaluation remains to be highly empirical. In this paper, we present some new developments in understanding the mechanics of soil liquefaction, which are firmly based on specifically designed experimental programs involving both clean and silty sands at a broad range of initial states. The key issues examined include the role of initial static shear stress on cyclic behavior and liquefaction resistance of sand, the fabric effects, the role of non-plastic fines in altering the potential for liquefaction, and the role of particle grading. We propose some new perspectives on these issues in a sound theoretical framework, and suggest several important implications for engineering practice.

Keywords: Critical state theory · Fabric · Liquefaction
Particle characteristics

1 Introduction

Soil liquefaction and associated ground deformation during earthquakes may cause huge damage to infrastructure systems that are vital for the safety and economy of our societies. Over the past several decades, a considerable amount of effort has been made worldwide to understand the mechanics of soil liquefaction and to develop methods to evaluate the liquefaction potential of soils [1, 2]. Largely because of the complexity of the problem, the current state-of-the-practice in liquefaction evaluation is highly empirical, involving use of liquefaction charts or correlations developed based on case histories during past earthquakes. However, the empirical approach often contains uncertainties and inconsistent physics, which may lead to geotechnical designs that are unsafe or non-economic. The inconsistencies may also hinder development of understanding in the right direction. This paper presents selected results and findings from a long-term research carried out by the lead author and his co-workers [3–13], with the aim to develop a comprehensive understanding of the mechanics of soil liquefaction (including both seismic and static liquefaction) and to bridge the gap between empiricism and physics.

2 Role of Initial Shear Stress

The role of initial shear stress is a critical issue in liquefaction evaluation involving dams, embankments and slopes, but current literature contains diverse and even contradictory views [1]. Based on systematic data sets for both cyclic and monotonic loading conditions [5, 6], we showed that the presence of initial shear stress is beneficial to the liquefaction resistance of loose sand at low initial shear stress levels but it becomes detrimental at high shear stress levels (Fig. 1). In this connection, we proposed *the concept of threshold α*, where α characterizes the initial shear stress level, along with the use of a no-stress-reversal line to characterize the effect of initial shear stress. Furthermore, within the critical state soil mechanics (CSSM) framework, we showed that the threshold α depends on the state parameter of sand [14]. This proposal leads to a unified and consistent interpretation of the role of initial shear stress.

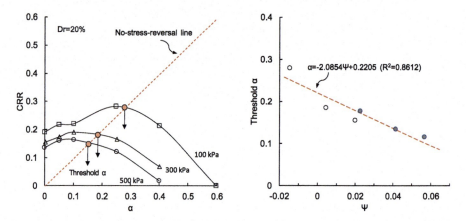

Fig. 1. The concept of threshold α and its dependence on state parameter

3 Fabric

How to take into account the fabric effects of soil in geotechnical analysis remains a difficult problem, particularly in situations involving cyclic loading. We explored in a systematic way the impact of specimen preparation on the cyclic behavior and liquefaction resistance of sand [7], and found that the method used to reconstitute specimens or the soil fabric it forms plays a role that is far more complicated than previously thought. Depending on the combination of relative density, confining stress and degree of stress reversal in cyclic loading, a change of reconstitution method can have a marked or little effect on the nature of the overall response in terms of deformation pattern and failure mechanism (Fig. 2). Under otherwise similar conditions, the fabric formed by dry deposition (DD) can largely reduce the threshold α as compared with the fabric formed by moist tamping (MT). From the microscopic perspective, DD specimens are highly anisotropic whereas MT specimens are more isotropic. Soil fabric should be *regarded as a state index* as important as the conventional ones (e.g. density and confining stress) in describing the overall soil behavior.

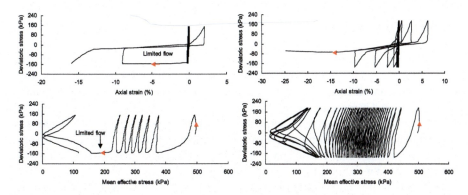

Fig. 2. Undrained responses of medium dense sand specimens under cyclic loading (left: specimen by dry deposition; right: specimen by moist tamping).

4 Role of Fines

Whether the effect of fines is negative or beneficial for liquefaction resistance is a controversial issue in the literature. This is due partly to the oversight of the role of grain characteristics [8, 9] and partly to the use of different density indices as the comparison basis [10]. Based on multiple series of laboratory experiments on sand-fines mixtures, formed by adding non-plastic fines of different shapes into clean sands, we obtained several novel findings on the role of fines, and proposed the *concept of combined roundness* to characterize the coupled effects of grain roundness and fines content [8, 9]. As an example, Fig. 3 shows the undrained responses of specimens of Fujian sand and the

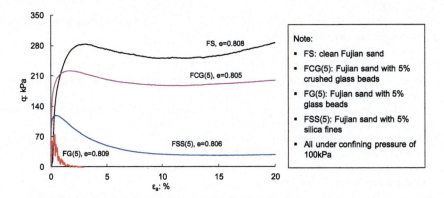

Fig. 3. Undrained shear behavior of sand modified by fines of distinct shape.

sand mixed with three types of fines at the amount of 5%, all sheared from a similar void ratio. Note that the three types of fines all introduced a less stable structure to the base sand. Particularly, the presence of glass beads led to an extremely unstable inter-granular contacts, resulting in stick-slip response. When crushed glass beads were added into the base sand, the fluctuations vanished.

5 Particle Grading

There is a general agreement that well graded sands are less susceptible to liquefaction than uniformly graded sands. However, we observed in recent experiments that under similar post-consolidation conditions in terms of void ratio and confining pressure, a well graded sand is more susceptible to liquefaction than a uniformly graded sand [13]. Whilst this finding appears to be surprising, it is explainable in the CSSM framework. As the coefficient of uniformity increases, the critical state locus (CSL) tends to move downwards in the compression space. This implies that at a given confining pressure and void ratio, a well graded sand would be at a looser state and hence more susceptible to liquefaction. It is noteworthy that the observed movement of the CSL with varying gradation is consistent with the observation on sand-fines mixtures that the CSL tends to move down with increasing fines content [8].

Acknowledgements. The financial support provided by the Research Grants Council of Hong Kong (7191051; 17250316) and by The University of Hong Kong is gratefully acknowledged.

References

1. Youd, T.L., Idriss, I.M.: Liquefaction resistance of soils: summary report from the 1996 NCEER and 1998 NCEER/NSF workshops on evaluation of liquefaction resistance of soils. J. Geotech. Geoenviron. Eng. **127**(10), 817–833 (1997)
2. National Research Council: State of the art and practice in seismically induced soil liquefaction assessment. National Academy Press, Washington, DC (2016)
3. Yang, J.: Non-uniqueness of flow liquefaction line for loose sand. Géotechnique **52**(10), 757–760 (2002)
4. Yang, J., Savidis, S., Roemer, M.: Evaluating liquefaction strength of partially saturated sand. J. Geotech. Geoenviron. Eng. **130**(9), 975–979 (2004)
5. Yang, J., Sze, H.Y.: Cyclic behaviour and resistance of saturated sand under non-symmetrical loading conditions. Géotechnique **61**(1), 59–73 (2011)
6. Yang, J., Sze, H.Y.: Cyclic strength of sand under sustained shear stress. J. Geotech. Geoenviron. Eng. **137**(12), 1275–1285 (2011)
7. Sze, H.Y., Yang, J.: Failure modes of sand in undrained cyclic loading: impact of sample preparation. J. Geotech. Geoenviron. Eng. **140**(1), 152–169 (2014)
8. Yang, J., Wei, L.M.: Collapse of loose sand with the addition of fines: the role of particle shape. Géotechnique **62**(12), 1111–1125 (2012)
9. Wei, L.M., Yang, J.: On the role of grain shape in static liquefaction of sand-fines mixtures. Géotechnique **64**(9), 740–745 (2014)
10. Yang, J., Wei, L.M., Dai, B.B.: State variables for silty sands: global void ratio or skeleton void ratio? Soils Founds. **55**(1), 99–111 (2015)

11. Yang, J., Dai, B.B.: Is the quasi-steady state a real behaviour? A micromechanical perspective. Géotechnique **61**(8), 715–719 (2011)
12. Wei, X., Yang, J.: The effects of initial static shear stress on liquefaction resistance of silty sand. In: Proceedings of 6th International Conference on Earthquake Geotechnical Engineering, Christchurch, New Zealand (2015)
13. Yang, J., Luo, X.D.: The critical state friction angle of granular materials: does it depend on grading? Acta Geotech. **13**, 535–547 (2017). https://doi.org/10.1007/s11440-017-0581-x
14. Been, K., Jefferies, M.: A state parameter for sands. Géotechnique **35**(2), 99–112 (1985)

UH Model for Granular Soils

Yang-ping Yao[✉], Lin Liu, and Ting Luo

School of Transportation Science and Engineering,
Beihang University, Beijing 100191, China
ypyao@buaa.edu.cn

Abstract. A constitutive model is presented to describe the mechanical behaviors of granular soils. A S-shaped normal compression line (NCL) is first expressed by a simple equation. Subsequently, a state parameter (ξ) is defined to quantify the current state of granular soils, and a unified hardening parameter (H) that is a function of the state parameter (ξ) is developed to govern the hardening process of the drop-shaped yield surface. A constitutive model for granular soils is proposed by combining with flow rule. Finally, the comparison between the predictions and the test results of Cambria sand indicates that the model is able to describe the mechanical behaviors of granular soils in a large stress range.

Keywords: Granular soils · Constitutive model · Particle crushing

1 Introduction

The mechanical properties of granular soils are complicated. It is very challenging for researchers to describe these properties well by developing a simple constitutive model. Some constitutive models for granular soils have been proposed [1, 2]. However, these constitutive models can predict negative void ratios at high stresses, so they cannot be used to describe the mechanical behaviors of granular soils at high stresses. This paper focuses on proposing a simple and practical constitutive model for granular soils in a large stress range.

2 Constitutive Model for Granular Soils

In this paper, the NCL can be simply expressed as follow

$$\ln(e - e_\mathrm{L}) = \ln(Z - e_\mathrm{L}) - \lambda \ln\left(\frac{p + p_\mathrm{s}}{1 + p_\mathrm{s}}\right) \tag{1}$$

where Z is the void ratio at $p = 1$ kPa on the NCL; λ is the slope of the NCL in the $\ln(e - e_\mathrm{L}) - \ln(p + p_\mathrm{s})$ space, and is also the slope of the asymptotic line in the $\ln(e - e_\mathrm{L}) - \ln p$ space. e_L is the limit void ratio when $p \to \infty$, as shown in Fig. 1. p_s is the compressive hardening parameter.

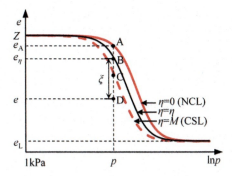

Fig. 1. NCL and CSL in the *e*-ln*p* plane and state parameter

Recently, the authors have proposed a new and simple drop-shaped yield surface [2]. In this paper, the yield surface is modified and is written as follows

$$f = \ln\left[\left(1 + \frac{(1+\chi)q^2}{M^2 p^2 - \chi q^2}\right) p + p_s\right] - \ln(p_{x0} + p_s) - \frac{1}{(\lambda - \kappa)} H = 0 \quad (2)$$

where χ is the critical state parameter; M is the critical state stress ratio; H is the hardening parameter, and is similar to the unified hardening parameter [3, 4]. H is expressed as follows

$$H = \int \frac{M_f^4 - \eta^4}{M_c^4 - \eta^4} \frac{-de^p}{(e - e_L)} \quad (3)$$

where M_f is the potential failure stress ratio, and can be expressed as follows:

$$M_f = 6\left[\sqrt{\frac{12(3-M)}{M^2}} \exp\left(-\frac{\xi}{\lambda - \kappa}\right) + 1 + 1\right]^{-1} \quad (4)$$

M_c is the characteristic state stress ratio, and can be expressed as follows:

$$M_c = M \exp(-m \cdot \xi) \quad (5)$$

where m is a dilatancy parameter. ξ is a state parameter, and can be measured by the distance between point B and point D, as shown in Fig. 1.

The plastic potential function in the paper is the same as that of the UH model for sands, and can be expressed as follows

$$g = \ln\frac{p}{p_y} + \ln\left(1 + \frac{q^2}{M_c^2 p^2}\right) = 0 \quad (6)$$

where p_y is the intersection of the current plastic potential surface with the p-axis. Therefore, the non-associated flow rule is adopted in the present model.

3 Model Validation

In this paper, the experimental results of Cambria sand [5, 6] are adopted to verify the present model for granular soils. All parameters are listed in Table 1. Figure 2 shows the predicted and measured results of Cambria sand at different confining pressures. It is indicated that the proposed model can fit well the test data for the whole testing range from 6.4 MPa to 68.9 MPa, which illustrates that the model is applicable to reflect the stress-strain relations of granular soils at high stresses.

Table 1. Parameters of the simulation data

Parameter	M	λ	k	v	N	x	m	Z	e_L
Cambria sand	1.45	1.2	0.3	0.3	259000	0.7	0.2	0.6	0.07

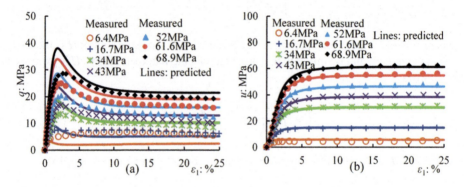

Fig. 2. Triaxial compression test results and model predictions: (a) stress-strain relations; (b) pore pressures.

4 Conclusion

A simple constitutive model is proposed to describe the mechanical behaviors of granular soils 9 parameters are adopted in the proposed model. All the parameters have specific physical meaning and can be measured by laboratory tests. The main features of the model include that (a) the model can describe the S-shaped compression line and critical state line of granular soils in the e-ln p space; (b) the model can predict the mechanical behaviors of granular soils in a large stress range, including the stress-strain relations, positive/negative dilatancy and strain hardening/softening.

Acknowledgement. This work was supported by the National Basic Research Program of China (973 Program, Grant No. 2014CB47006), the National Natural Science Foundation of China (Grant Nos. 51579005 and 11672015).

References

1. Yao, Y.P., Sun, D.A., Luo, T.: A critical state model for sands dependent on stress and density. Int. J. Numer. Anal. Meth. Geomech. **28**, 323–337 (2004)
2. Yao, Y.P., Liu, L., Luo, T.: UH model for sands. Chin. J. Geotech. Eng. **38**, 2147–2153 (2016)
3. Yao, Y.P., Sun, D.A., Matsuoka, H.: A unified constitutive model for both clay and sand with hardening parameter independent on stress path. Comput. Geotech. **35**(2), 210–222 (2008)
4. Yao, Y.P., Hou, W., Zhou, A.N.: UH model: three-dimensional unified hardening model for overconsolidated clays. Géotechnique **59**(5), 451–469 (2009)
5. Lade, P.V., Yamamuro, J.A.: Undrained sand behavior in axisymmetric tests at high pressures. J. Geotech. Eng. **122**, 120–129 (1996)
6. Yamamuro, J.A., Lade, P.V.: Drained sand behavior in axisymmetric tests at high pressures. J. Geotech. Eng. **122**, 120–129 (1996)

An Optimization-Based Parameter Identification Tool for Geotechnical Engineering

Zhen-Yu Yin[1,2(✉)] and Yin-Fu Jin[1,2]

[1] Research Institute of Civil Engineering and Mechanics (GeM), UMR CNRS 6183, Ecole Centrale de Nantes, Nantes, France
zhenyu.yin@gmail.com
[2] Key Laboratory of Geotechnical and Underground Engineering of Ministry of Education, Department of Geotechnical Engineering, Tongji University, Shanghai, China

Abstract. In this paper, ErosOpt, a useful tool for the development, analysis and application of an optimization algorithm, which is devised to solve the problems of parameter identification in geotechnical engineering, is presented. ErosOpt is programmed as an admixture of C#, MATLAB and FORTRAN, which offers a powerful environment for various kinds of optimization tasks in parameter identification. The proposed tool has four important features: (1) it covers a wide range of parameter identification problems; (2) it provides a number of soil models with a user extension interface; (3) it offers various efficient optimization algorithms; (4) it allows for visualization through graphical displays. Furthermore, the entire graphical user interface and usage instructions for ErosOpt are briefly illustrated in simple and practical terms.

Keywords: Parameter identification · Optimization · Geotechnical engineering Constitutive model · Soils · Interface

1 Introduction

An impressive variety of constitutive models have been developed for soils in geotechnical engineering. These range from linear-elastic, perfectly plastic models (such as the Mohr-Coulomb model), to nonlinear models (such as the nonlinear Mohr-Coulomb [1]), to critical-state-based advanced models (such as the critical-state based nonlinear Mohr-Coulomb [1–3]; and the micromechanical models by Yin et al. [4–7]), to hypoplasticity models [8–11]. These have allowed for the garnering of increasingly accurate and reliable descriptions of the mechanical behaviors of soils, which has also resulted in complexities along with additional model parameters. Thus, more parameters are generally required to be determined before the model can be applied to the solving of engineering problems, which poses a considerable challenge for engineers. Therefore, an efficient procedure, in conjunction with a tool for parameter identification, would be extremely useful.

Yin et al. [12] distinguished three approaches: analytical methods, empirical correlations, and optimization methods to determine soil parameters based on experimental data. Among these techniques, the inverse analysis by optimization has been successfully used in the geotechnical area [1–3, 12–16] because it produces a relatively objective determination of the parameters for an adopted soil model, even for those that have no direct physical meaning. Although optimization-based procedures are highly effective for solving parameter identification problems, the writing of a computer program for implementing the sophisticated algorithms according to user needs requires certain programming expertise, not to mention considerable time and effort. Because of the tedious nature of this task, the use of an optimization tool is more advantageous and attractive. To date, various kinds of optimization-based tools have been developed. These offer an object-oriented design for evaluating fitness functions using a variety of optimization algorithms (e.g., Evolvica [17]), MOEAT [18], YALMIP [19]). However, among these tools, the identification of soil parameters has not received any attention. Therefore, a case can be made for the development of a tool offering a powerful environment for various kinds of parameter identification, so that engineers can apply it to solve a range of engineering problems without the need to reproduce its model of operation, which is also a challenge for them.

2 ErosOpt Platform

In this paper, the development of an optimisation-based parameter identification tool (ErosOpt) for geotechnical engineering is described. Figure 1 shows the main interface of ErosOpt, which is divided into three zones: The *Navigation* pane, the *Command* pane and the *Manipulation* pane. The *Navigation* pane features a simple procedure with five modules, *Problem*, *Soil model*, *Algorithm*, *Results* and *Report*, which are aimed at guiding the user to solve problems. The *Command* pane boasts three commands that can help the user to run and exit the ErosOpt or get help on how to use the program.

Figure 2 sets out the basic procedure for using the ErosOpt tool to solve parameter identification problems. This can be divided into six steps:

Step 1: *Select the problem and import the corresponding objective data.*
Step 2: *Select the soil model and determine the model parameters to be optimised. At the same time, the bounds (upper bound, lower bound and step size) of each parameter are also set.*
Step 3: *Select the optimisation algorithm (NMS, NMGA or NMDE) and define the settings.*
Step 4: *Run the tool.*
Step 5: *Export the optimal solutions and the comparisons between the optimal simulations and objectives.*
Step 6: *Generate the report.*

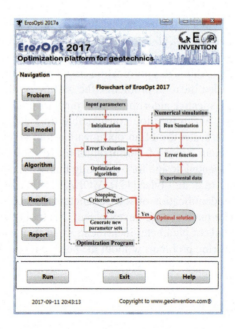

Fig. 1. Main GUI window of start page in ErosOpt

Fig. 2. Basic procedure of main usage instructions for ErosOpt

For future work, the tool can be extended by using other state-of-the-art algorithms such as ACO (Ant Colony Optimization) for further types of tests (biaxial tests, cone penetration tests and so on). The developed software can be freely downloaded from the following URL: http://www.geoinvention.com/en/news.asp?big=14.

Acknowledgments. The financial support for this research came from the National Natural Science Foundation of China (grant numbers 51579179).

References

1. Jin, Y.-F., Yin, Z.-Y., Shen, S.-L., Hicher, P.-Y.: Selection of sand models and identification of parameters using an enhanced genetic algorithm. Int. J. Numer. Anal. Meth. Geomech. **40**(8), 1219–1240 (2016). https://doi.org/10.1002/nag.2487
2. Jin, Y.-F., Wu, Z.-X., Yin, Z.-Y., Shen, J.S.: Estimation of critical state-related formula in advanced constitutive modeling of granular material. Acta Geotech. **12**, 1329–1351 (2017). https://doi.org/10.1007/s11440-017-0586-5

3. Jin, Y.-F., Yin, Z.-Y., Shen, S.-L., Hicher, P.-Y.: Investigation into MOGA for identifying parameters of a critical-state-based sand model and parameters correlation by factor analysis. Acta Geotech. **11**(5), 1131–1145 (2016). https://doi.org/10.1007/s11440-015-0425-5
4. Chang, C.S., Yin, Z.Y.: Micromechanical modeling for inherent anisotropy in granular materials. J. Eng. Mech. **136**(7), 830–839 (2010). https://doi.org/10.1061/(asce)em.1943-7889.0000125
5. Yin, Z.Y., Chang, C.S., Hicher, P.Y.: Micromechanical modelling for effect of inherent anisotropy on cyclic behaviour of sand. Int. J. Solids Struct. **47**(14–15), 1933–1951 (2010). https://doi.org/10.1016/j.ijsolstr.2010.03.028
6. Yin, Z.Y., Chang, C.S.: Stress–dilatancy behavior for sand under loading and unloading conditions. Int. J. Numer. Anal. Meth. Geomech. **37**(8), 855–870 (2013)
7. Yin, Z.-Y., Zhao, J., Hicher, P.-Y.: A micromechanics-based model for sand-silt mixtures. Int. J. Solids Struct. **51**(6), 1350–1363 (2014)
8. Kolymbas, D.: An outline of hypoplasticity. Archive Appl. Mech. **61**(3), 143–151 (1991)
9. Mašín, D.: A hypoplastic constitutive model for clays. Int. J. Numer. Anal. Meth. Geomech. **29**(4), 311–336 (2005)
10. Wu, W., Bauer, E., Kolymbas, D.: Hypoplastic constitutive model with critical state for granular materials. Mech. Mater. **23**(1), 45–69 (1996)
11. Wu, W., Kolymbas, D.: Hypoplasticity then and now. In: Kolymbas, D. (ed.) Constitutive Modelling of Granular Materials, pp. 57–105. Springer, Heidelberg (2000)
12. Yin, Z.-Y., Jin, Y.-F., Shen, J.S., Hicher, P.-Y.: Optimization techniques for identifying soil parameters in geotechnical engineering: comparative study and enhancement. Int. J. Numer. Anal. Meth. Geomech. **42**, 70–94 (2017). https://doi.org/10.1002/nag.2714
13. Jin, Y.-F., Yin, Z.-Y., Riou, Y., Hicher, P.-Y.: Identifying creep and destructuration related soil parameters by optimization methods. KSCE J. Civ. Eng. **21**(4), 1123–1134 (2017). https://doi.org/10.1007/s12205-016-0378-8
14. Jin, Y.-F., Yin, Z.-Y., Shen, S.-L., Zhang, D.-M.: A new hybrid real-coded genetic algorithm and its application to parameters identification of soils. Inverse Probl. Sci. Eng. **25**(9), 1343–1366 (2017). https://doi.org/10.1080/17415977.2016.1259315
15. Ye, L., Jin, Y.-F., Shen, S.-L., Sun, P.-P., Zhou, C.: An efficient parameter identification procedure for soft sensitive clays. J. Zhejiang Univ. SCIENCE A **17**(1), 76–88 (2016). https://doi.org/10.1631/jzus.a1500031
16. Yin, Z.-Y., Jin, Y.-F., Shen, S.-L., Huang, H.-W.: An efficient optimization method for identifying parameters of soft structured clay by an enhanced genetic algorithm and elastic–viscoplastic model. Acta Geotech. **12**, 1–19 (2016). https://doi.org/10.1007/s11440-016-0486-0
17. Rummler, A.: Evolvica: a Java framework for evolutionary algorithms (2007)
18. Sağ, T., Çunkaş, M.: A tool for multiobjective evolutionary algorithms. Adv. Eng. Softw. **40**(9), 902–912 (2009)
19. Lofberg, J.: YALMIP: a toolbox for modeling and optimization in MATLAB. In: 2004 IEEE International Symposium on Computer Aided Control Systems Design, pp 284–289. IEEE (2004)

Constitutive Model for Soaking-Induced Volume Change of Unsaturated Compacted Expansive Soil

Wei-lie Zou[1(✉)], Xie-qun Wang[2], Jun-feng Zhang[3], Zhong Han[1], and Liang-long Wan[1]

[1] School of Civil Engineering, Wuhan University, Wuhan 430072, People's Republic of China
zwilliam@126.com
[2] School of Civil Engineering and Architecture, Wuhan University of Technology, Wuhan, China
[3] Country Garden Real Estate Development Co., Ltd., Foshan 528312, People's Republic of China

Abstract. Several researchers have proposed models extending the independent stress state variables approach to describe the nonlinear nature of the normal consolidation line (NCL) for unsaturated soils. However, there are few models that can describe both the nonlinear nature of the NCL and the non-monotonic variation of collapse potential for unsaturated compacted expansive soils (USCES). In this paper, a constitutive relationship is proposed for the USCES that can be used for interpreting and modeling both the nonlinear NCL and non-monotonic collapse potential by taking account of the influence of macro and microstructure in the term specific volume, v. In this model, effective degree of saturation, S_e is used as one of the state variables, and v is assumed to linearly decrease with increasing net vertical stress, or mean net stress, when S_e is at constant condition. A series of oedometer tests were conducted by soaking the USCES specimens to validate the proposed model. The results of the study suggest that there is a good comparison between the measured data and the prediction curves of NCL during compression using the proposed model. In addition, the model provides excellent predictions of the soaking-induced non-monotonic collapse potential.

Keywords: Constitutive relationships · Compacted expansive soil · Partial saturation · Soaking · Effective degree of saturation

1 Introduction

Settlement and swelling behavior of infrastructure constructed on expansive soils, which are typically in a state of unsaturated condition, is not only significantly affected due to the loads acting on soils but also due to environmental factors associated with precipitation activities, which contribute to soil suction changes. Therefore, the volume change behavior of unsaturated expansive soils is typically interpreted in terms of two independent stress state variables; namely, mean net stress and matric suction [1]. Recent studies suggest the volume change behavior is also related to yield stress-suction and shear strength-suction relationships of unsaturated soils. Furthermore, unsaturated soil properties such as the nominal tensile strength, suction increment yield function and loading-collapse yield function are all related to the volume change behavior [2]. Based on these reasons, it is necessary to propose a constitutive relationship for interpreting unsaturated compacted soils (USCS) behavior that takes into account of volume change (i.e., swell and collapse) behavior.

Several constitutive models have been proposed during the past three decades for the volume change behavior of USCS [1–4]. However, a suitable model is not available in the literature to describe soaking-induced non-monotonic volume change of USCS [5, 6]. In this study, a constitutive model is proposed for both interpreting and modeling the nonlinear normal compression behavior and the soaking-induced non-monotonic collapse potential of unsaturated compacted expansive soils (USCES). In this model, the influence of the macro- and micro-structure (i.e., double-pore structure) associated with the USCES is taken into account in the specific volume, υ. The effective degree of saturation, S_e is used as one of state variables. The proposed constitutive model is successfully validated using a series of oedometer test data conducted on the USCES from Nanyang, China.

2 Constitutive Model

The normally consolidated line (NCL) for saturated soils which is defined as the relationship between the specific volume, υ and natural logarithmic value of mean effective stress, $\ln(p')$, is typically linear in the υ-$\ln(p')$ plane. Equation (1) is conventionally used for expressing this relationship.

$$\upsilon = N - \lambda \ln(p') \tag{1}$$

where, $\upsilon = 1 + e$, e is void ratio; p' is mean effective stress; N is the intercept of the normally consolidated line with υ axis when $\ln(p') = 0$, and λ is compression index of saturated soil, i.e., the slope of normally consolidated line in υ-$\ln(p')$ plane.

For unsaturated soils, λ can be expressed as a function of effective degree of saturation S_e [7]:

$$\lambda = \lambda_0 - (\lambda_0 - \kappa)(1 - S_e)^a \tag{2}$$

where λ_0, λ are the compression indices of a soil at saturated and unsaturated states, respectively; S_e is the effective degree of saturation; κ is the rebound index of the soil, and a is a fitting parameter.

From several shrinkage test results in the literature [8, 9], the relationship between void ratio, e (or specific volume, υ) and gravimetric water content, w under non-loading state is found to be approximately linear in a range from nearly plastic limit to shrinkage limit of soil. In this study, for simplicity, the specific volume, υ is assumed to be a linear function of effective degree of saturation, S_e, as shown in Eq. (3).

$$\upsilon = N(S_e) = N_0 - \beta(1 - S_e) \tag{3}$$

where, $N(S_e)$, N_0 are the specific volumes of specimen at unsaturated and saturated states under the atmospheric pressure (p_a), respectively; and β is the slope of the shrinkage curve of the specimen.

The specific volume, v can be given as Eq. (4) [7].

$$v = N(S_e) - \lambda(S_e) \ln \frac{p' + p_a}{p_a} \qquad (4)$$

Equation (5) can be obtained by substituting Eqs. (3) and (2) into Eq. (4).

$$v = N_0 - \beta(1 - S_e) - [\lambda_0 - \eta(1 - S_e)^a] \ln p' \qquad (5)$$

where $\eta = \lambda_0 - \kappa$.

Considering that compression tests were conducted in the present study under certain constant water content, the relationship between S_e and w (Eq. 6) can be substituted into Eq. (5) to obtain a nonlinear NCL constitutive model Eq. (7).

$$S_e = \frac{wG_s}{v - 1} \qquad (6)$$

$$v = N_0 - \beta(1 - \frac{wG_s}{v-1}) - [\lambda_0 - \eta(1 - \frac{wG_s}{v-1})^a] \ln p' \qquad (7)$$

3 Validation

An expansive soil collected from Nanyang, China, was used to validate the proposed model (Eq. (7)). Oedometers were used for conducting K0-compression and collapsibility tests. A series of USCES specimens with different initial water contents were prepared by static compaction. Vertical stresses were applied on the USCES specimens in the oedometers. The water contents of all the USCES specimens were kept constant during compression. After determining compression behavior under a certain predetermined vertical stress level, all specimens were inundated with distilled water. The measured collapse deformation data from the present study are used to validate the accuracy of the proposed model for predicting the soaking-induced collapse. Figure 1 shows the nonlinear change characteristics of the measured data and the predicted curves of specific volume, v, effective degree of saturation, Se and collapse potential with regard to vertical stress. It can be seen from Figs. 1a, b and c that there are reasonably good comparisons between the measured data and the predicted curves.

Based on this study, a drawn conclusion is that the proposed constitutive model can predict the nonlinear NCL behavior and the non-monotonic collapse potential of unsaturated compacted expansive soils.

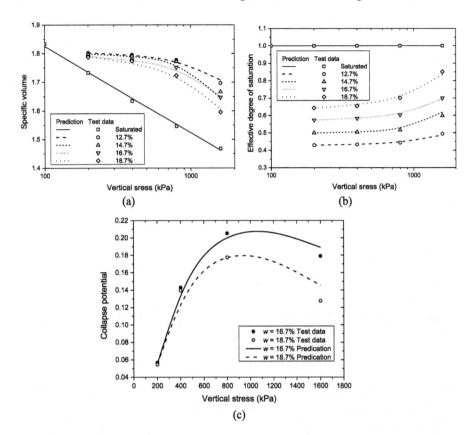

Fig. 1. Comparisons between the measured data and the predicted curves of (a) specific volume, v, (b) effective degree of saturation, S_e and (c) collapse potential with respect to vertical stress

References

1. Fredlund, D.G., Rahardjo, H.: Soil Mechanics for Unsaturated Soils. Wiley, New York (1993)
2. Sheng, D., Gens, A., Fredlund, D.G., Sloan, S.W.: Unsaturated soils: from constitutive modelling to numerical algorithms. Comput. Geotech. **35**(6), 810–824 (2008)
3. Sivakumar, V., Wheeler, S.J.: Influence of compaction procedure on the mechanical behaviour of an unsaturated compacted clay. Part 1: wetting and isotropic compression. Géotechnique **50**(4), 359–368 (2000)
4. Alonso, E.E., Gens, A., Josa, A.: A constitutive model for partially saturated soils. Géotechnique **40**(3), 405–430 (1990)
5. Sun, D., Sheng, D., Sloan, S.W.: Elastoplastic modelling of hydraulic and stress-strain behaviour of unsaturated soils. Mech. Mater. **39**(3), 212–221 (2005)
6. Vilar, O.M., Rodrigues, R.A.: Collapse behavior of soil in a Brazilian region affected by a rising water table. Can. Geotech. J. **48**(2), 226–233 (2011)
7. Zhou, A., Sheng, D., Sloan, S.W., Gens, A.: Interpretation of unsaturated soil behaviour in the stress-saturation space, I: volume change and water retention behaviour. Comput. Geotech. **43**(3), 178–187 (2012)

8. Fredlund, D.G., Zhang, F.: Combination of shrinkage curve and soil-water characteristic curves for soils that undergo volume change as soil suction is increased. In: The 18th International Conference on Soil Mechanics and Geotechnical Engineering, pp. 1109–1112. ISSMGE, London (2013)
9. Zou, W.L., Zhang, J.F., Wang, X.Q.: Volume correction and soil-water characteristics of remodeling expansive soil under dehydration path. Chin. J. Geotech. Eng. **34**(12), 2213–2219 (2012). (in Chinese)

Part II: Micro-Macro Relationship

Linking Microstructural Behavior with Macrostructural Observations on Unsaturated Porous Media

Hiram Arroyo[1(✉)] and Eduardo Rojas[2]

[1] Departamento de Ingeniería Agroindustrial, Universidad de Guanajuato, Campus Celaya-Salvatierra, Av. Javier Barros Sierra 201, 38140 Celaya, GTO, Mexico
hiramarroyo@gmail.com
[2] Facultad de Ingeniería, Universidad Autónoma de Querétaro, Centro Universitario, Cerro de las Campanas, 76160 Querétaro, QRO, Mexico

Abstract. Explaining and predicting the macrostructural behavior of porous media containing air and water phases is of paramount importance. This is not only important for understanding the behavior of unsaturated soils, but it certainly has true impact on engineering applications such as CO2 sequestration, gels and food industry. In this paper we elaborate particularly on the hydraulic relationships between the porous network of a material and the macroscopical consequences that this has on the mechanics of unsaturated soils.

Keywords: Unsaturated soils · Porous model · Effective stress · Pore-size distribution

1 Introduction

A network that is able of simulating the processes of wetting and drying of the pores that form it is presented. The consequences of such processes on the macrostructural behavior of soils are analyzed, particularly on predicting the relationship between the degree of saturation Sr and suction s. A network of pores is proposed, and its behavior is established by imposing restrictions on them to be wet or dry.

2 The Solid-Porous Model

Authors propose that the frequency distribution of pores (sites) can be represented by an exponential function that takes the form:

$$f_1(R) = \frac{1}{\sigma_s \sqrt{2\pi}} \exp\left[-\frac{(\ln R - \mu_s)^2}{2\sigma_s^2}\right] \qquad (1)$$

where σ_s and μ_s are the standard deviation and mean size of sites distribution $f_1(R)$ respectively (see Fig. 1); R represents the radius of those sites that are present in the medium $n = fN$ times (N is the total number of sites that form the porous network) and $v(R) = fNv$ is their total volume; v is the volume of a single site.

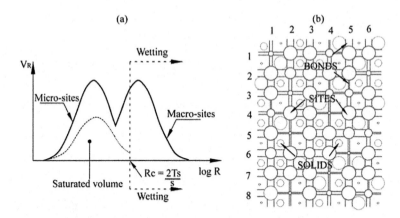

Fig. 1. (a) Modeled pore-size distribution of the material. (b) Porous network.

Pores are given a regular shape. A three-dimensional situation would lead to spherical pores, whereas in this paper we work in a two-dimensional scheme and a circular shape is assigned to the sites; therefore $V(R) = fN\pi R^2$.

The interconnection between sites is achieved with bond elements [1, 2]. These bonds do not contribute to the total volume of the network; however, they are of paramount importance, particularly to correctly predict the different implications that driving a drying process has. These bonds vary in their size as well and have their own frequency distribution.

Soils commonly exhibit two classes of pores that can be visualized through the PSD as two functions with two different maximum peaks [3, 4]. This can be modeled by splitting $f_1(R)$ in two independent functions that will depend on their own σ and μ parameters: f_{1MS} for the larger class of pores and f_{1Ms} for the smaller class of pores. Therefore, the volume of sites will be that of:

$$V_s = N\pi \left(\int_{r\min}^{r\max} FM \cdot f_{1MS} \cdot R^2 + f_{1Ms} \cdot R^2 \right) \quad (2)$$

Where FM represents an model parameter that makes the size of f_{1MS} distribution coherent with f_{1Ms}.

2.1 Conditions for Pores to Wet/Dry

A critical radius will be linked to suction level that the network appertains through the relationship $Rc = 2Ts/s$, where Ts is the air-water surface tension coefficient and s is the suction of the material.

A pore will wet or dry depending on the suction level and the size of its connecting pores (see Fig. 1). Two simultaneous conditions need to be accomplished for a pore to wet: (1) its radius must be smaller than the critical radius and (2) at least one of its connecting bonds must be already water/filled. On the other hand, the conditions that apply for a site to dry are (3) its radius must be larger than the critical radius and at last one of its connecting bonds must be already dry. The degree of saturation of the network is then computed as $Sr = Vw/V_s$; where Vw is the volume of water-filled sites.

2.2 Relationship Between Density and PSD (Pore Shrinking)

Because f_{1MS} and f_{1Ms} are functions of their own σ and μ parameters, V_s will also be. Several authors coincide on the experimental fact that f_{1MS} displaces to smaller sites as the soil suffers void ratio (e) reduction [5–7]. Here, we model soil compression by reducing μ parameter of f_{1MS}. To do so, a reference state characterized by its volume V_0 and void ratio e_0 the needs to be known. Then, the relationship between the volume of any other compressed (or expanded) network and its void ratio can be computed as $Vs = V_0(e/e_0)$.

3 Numerical-Theoretical Result Comparisons

Ng and Pang [8] report experimental results of $Sr - s$ curves as a function of e. Here, we consign model parameters to reproduce such results in Table 1, which were obtained from a fitting procedure taking $e = 0.782$ as the reference state (Fig. 2(a)). Note that two classes of bonds, f_{1MB} and f_{1Mb} are needed as well. The parameters for functions f_{1MS} f_{1MS}, f_{1MB} and f_{1Mb} are contained in Table 1.

Table 1. Model parameters to predict $Sr - s$ relationship as a function of void ratio.

Function	μ	σ	Function	μ	σ
f_{1MS}	0.010	7.000	f_{1MB}	0.500	6.000
f_{1MS}	0.001	4.000	f_{1Mb}	0.001	5.000

Figure 2(b) and (c) show that the model correctly predicts the relationship between soil densification and the $Sr - s$ relationship. Here, Fig. 2(d) shows de evolution of the PSD of the material where the larger class of pores shrink with void ratio increase and slightly displaces to smaller sites, whereas the smaller class of pores do not move and slightly increase its volume.

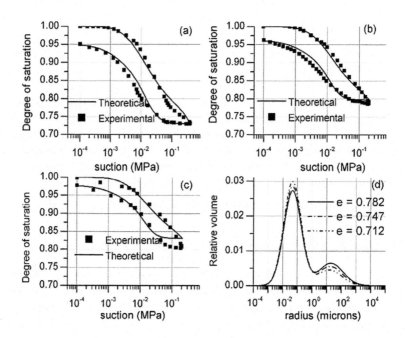

Fig. 2. (a) Modeled pore-size distribution of the material. (b) Porous network.

4 Conclusions

This paper presents a porous model that can predict the consequences of volume strains on the suction state of the material. The behavior of the model is simple but realistic and uses the fact that the larger pores contained within a porous medium are responsible for volume strains. The model is composed of sites that simulate the void space and bonds that interconnect the sites in a rectangular network.

This model gives light to the microstructural behavior of porous media and links it to macrostructural observations such as the relationship of the degree of saturation and suction stress.

References

1. Horta, J., Rojas, E., Pérez-Rea, M.L., López, T., Zaragoza, J.B.: A random solid-porous model to simulate the retention curves of soils. Int. J. Numer. Anal. Meth. Geomech. **37**(8), 932–944 (2013)
2. Rojas, E., Zepeda, A.G., Pérez-Rea, M.L., Leal, J., Gallegos, G.: A four elements porous model to estimate the strength of unsaturated soils. Geotech. Geol. Eng. **29**(2), 193–202 (2009)
3. Futai, M.M., Almeida, M.S.S.: An experimental investigation of the mechanical behaviour of an unsaturated gneiss residual soil. Géotechnique **55**(3), 201–213 (2005)
4. Thom, R., Sivakumar, R., Sivakumar, V., Murray, E.J., Mackinnon, P.: Pore size distribution of unsaturated compacted kaolin: the initial states and final states following saturation. Géotechnique **57**(5), 469–747 (2007)

5. Della Vecchia, G., Dieudonné, A.-C., Jommi, C., Charlier, R.: Accounting for evolving pore size distribution in water retention models for compacted clays. Int. J. Numer. Anal. Meth. Geomech. **39**(7), 702–723 (2014)
6. Hu, R., Chen, Y.-F., Liu, H.-H., Zhou, C.-B.: A water retention curve and unsaturated hydraulic conductivity model for de-formable soils: consideration of the change in pore-size distribution. Géotechnique **63**(16), 1389–1405 (2013)
7. Simms, P.H., Yanful, E.K.: A pore-network model for hydro-mechanical coupling in unsaturated compacted clayey soils. Can. Geotech. J. **42**(2), 499–514 (2005)
8. Ng, C.W.W., Pang, Y.W.: Experimental investigations of the soil-water characteristics of a volcanic soil. Can. Geotech. J. **37**(6), 1252–1264 (2000)

The Effect of Particle Shape Polydispersity on Deformability of Granular Materials

Yuxuan Che[1,2], Fang Liu[1,2(✉)], Gang Deng[3], and Jizhong He[1,2]

[1] State Key Laboratory of Disaster Reduction in Civil Engineering, Tongji University, Shanghai 200092, China
liufang@tongji.edu.cn
[2] Key Laboratory of Geotechnical and Underground Engineering of Ministry of Education, Tongji University, Shanghai 200092, China
[3] State Key Laboratory of Simulation and Regulation of Water Cycle in River Basin, China Institute of Water Resources and Hydropower Research, Beijing 100048, China

Abstract. The shape of the particles that constitute a granular material significantly affects its deformation behavior. This study intends to identify the possible underlying microscopic mechanism of this effect. To this end, a distinct element code is employed to simulate a series of biaxial compression tests on specimens composed of pentagon particles with different levels of roundness and sphericity. The results show that the deformation modulus of the specimens decreases as the particles and the polygons enclosed by force chains turn more regular. Moreover, less isolated contacts are identified with the system getting more regular, contributing to the smaller deformation as well.

Keywords: Shape polydispersity · Deformability · Granular material

1 Introduction and Methodology

Granular materials are commonly used in man-made earthworks such as rockfill dams, of which the deformation has to be reasonably predicted and cautiously controlled so as to ensure safe operations. Particle shape plays an important role in deformation capability of a granular assembly. This paper intends to interpret from microscopic perspective the effect of particle shape on deformability of granular materials.

Following [1–3], this study adopts roundness and sphericity as two shape descriptors and investigates their effects separately. Sphericity here refers to the extent how far the particle is different from a standard pentagon in two dimensions. Referring to [4], the angular position of a vertex k of a standard pentagon can be determined by $\theta_k = \theta_0 + 2\pi k/5$ where θ_0 is the position of the first vertex. Pentagons with different sphericity can then be generated by disturbing randomly the position of vertex k within an angular limit $\pm (1-s)\pi/5$ where s represents sphericity. Roundness describes the scale of major surface features and here can be quantified as the average radius of curvature of each edge of the particle. The particle shape evolves from a standard pentagon to a circle with roundness r varying from 0 to 1.

A discrete element method (DEM) simulator, PPDEM [5], is adopted to simulate a series of biaxial tests on dense specimens composed of 8000 particles with different shape descriptors (Fig. 1). The particle sizes represented by the diameter of the circumscribed circle range from 0.3 mm to 0.9 mm with a random and continuous distribution. Pluviation method with a zero inter-particle friction is employed to generate dense samples. The inter-particle friction angle is switched to 35° during the simulation of biaxial tests. The shear contact stiffness of particles K_n is 2×10^8 MPa and the particle density equals 2650 kg/m^3.

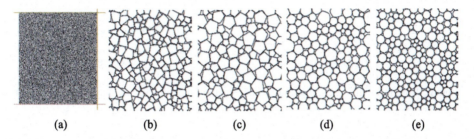

Fig. 1. A biaxial test specimen (a) and particle geometry with different shape descriptors: (b) $s = 0$, $r = 0$; (c) $s = 1$, $r = 0$; (d) $s = 1$, $r = 0.5$; (e) $s = 1$, $r = 1$.

2 Results

For deformation modulus, E_{50} is chosen to reflect the deformation properties, which is defined as the ratio of half of the peak deviator stress to the corresponding axial strain (Fig. 2). Here, E_{50} goes linearly with roundness and sphericity. Void ratios of specimens after consolidation are checked and it is anticipated that particles grow denser as their shapes turn more irregular. Void ratios keep a linear relationship with both shape descriptors (Fig. 3). While coordination number has an evident relation with roundness, it could not correlate with sphericity well (Fig. 3). It is found that irregular particles perform worse in resisting deformation. The reason could be that the pentagon can be acquired by selecting five vertices on the circumscribed circle, thus compared with

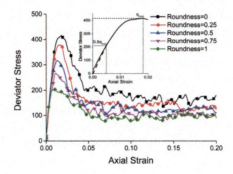

Fig. 2. Stress-strain graph and the determination of E_{50}.

circles with the same area, pentagons could have more space to fit in and lead to larger deformation. Similarly, a regular pentagon accommodates less free deformation space contrary to the situation of a highly irregular one (Fig. 4).

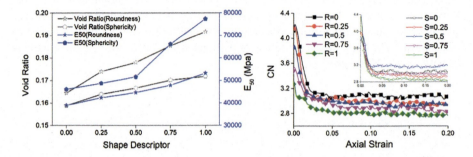

Fig. 3. Void ratio after consolidation and deformation modulus versus shape descriptor (left), and coordination number variation during shearing (right).

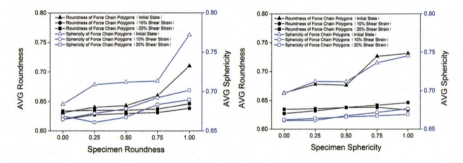

Fig. 4. Average roundness (left) and sphericity (right) of polygons enclosed by force chains versus shape descriptor.

Force chains automatically form closed polygons. Using ImageJ [6], this study also looks into the area, perimeter and shape characteristics of force chain polygons at initial state (after consolidation), 10% shear strain, 20% strain respectively. It is noted that area and perimeter are not sensitive parameters in this study. The number of force chain polygons decreases as the bulk roundness and sphericity go up. As bulk roundness and sphericity increase, average roundness and sphericity of force chain polygons generally rise up accordingly. Then polygons formed by strong force chains are examined and similar conclusions are reached. System with regular shape could be more stable as the anisotropy is not dominant under that situation. Certain contacts between particles are not incorporated in the contact network and do not contribute to force transmission. They are represented by single lines inside closed force chain polygons and we define them as isolated contacts, which are shown in Fig. 5. As particles turn more regular, the number of isolated contacts generally fall down, corresponding to smaller deformation.

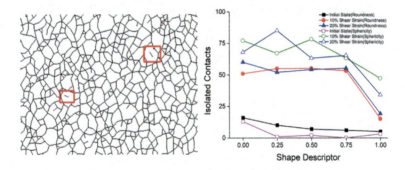

Fig. 5. Isolated contacts versus shape descriptor.

3 Conclusions

Both roundness and sphericity affect deformation of granular material significantly. These two descriptors may be judged as governing variables to control the deformation modulus by linearly affecting specimen void ratio. Coordination number is found to be linked with roundness. Furthermore, the shape feature of force chain polygons resembles that of particles. As specimen sphericity and roundness increase, less isolated contacts are identified and smaller deformation is induced. This study may provide a new insight into interpreting how shape polydispersity affects bulk deformation capability.

Acknowledgement. The work is supported by the National Key Research and Development Program of China (2017YFC0404803), the IWHR Research and Development Support Program (GE0145B562017) and the Chinese National Natural Science Foundation (41572267).

References

1. Barrett, P.J.: The shape of rock particles, a critical review. Sedimentology **27**(3), 291–303 (2010)
2. Krumbein, W.C.: Measurement and geological significance of shape and roundness of sedimentary particles. Radiobiologia Radiotherapia **12**(5), 595–598 (1941)
3. Wadell, H.: Volume, shape, and roundness of rock particles. J. Geol. **40**(5), 443–451 (1932)
4. Nguyen, D.H., Azéma, E., Radjai, F., Sornay, P.: Effect of size polydispersity versus particle shape in dense granular media. Phys. Rev. E Stat. Nonlinear Soft Matter Phys. **90**(1), 12202 (2014)
5. Fu, P., Walton, O.R., Harvey, J.T.: Polyarc discrete element for efficiently simulating arbitrarily shaped 2D particles. Int. J. Numer. Meth. Eng. **89**(5), 599–617 (2012)
6. Abramoff, M.D., Magelhaes, P.J., Ram, S.J.: Image processing with ImageJ. Biophotonics Int. **11**(5–6), 36–42 (2004)

Grain Learning: Bayesian Calibration of DEM Models and Validation Against Elastic Wave Propagation

Hongyang Cheng[1(✉)], Takayuki Shuku[2], Klaus Thoeni[3], Pamela Tempone[4], Stefan Luding[1], and Vanessa Magnanimo[1]

[1] Multi Scale Mechanics (MSM), Faculty of Engineering Technology, MESA+, University of Twente, P.O. Box 217, 7500 AE Enschede, The Netherlands
h.cheng@utwente.nl
[2] Graduate School of Environmental and Life Science, Okayama University, 3-1-1 Tsushima naka, Kita-ku, Okayama 700-8530, Japan
shuku@cc.okayama-u.ac.jp
[3] Centre for Geotechnical and Materials Modelling, The University of Newcastle, Callaghan, NSW 2308, Australia
klaus.thoeni@newcastle.edu.au
[4] Division of Exploration and Production, Eni SpA, Milano, Lombardy, Italy
pamela.tempone@eni.com

1 Introduction

The estimation of micromechanical parameters of discrete element method (DEM) models is a nonlinear history-dependent inverse problem. In order to reproduce the experimental measurements with high accuracy, this work aims to develop a *machine learning*-based calibration toolbox named "*Grain* learning", which can extract grains from X-ray computed tomography (CT) images and perform Bayesian parameter estimation for DEM models of dry granular materials.

2 Bayesian Calibration

We first introduce a feature-based watershed algorithm which performs multi-phase image segmentation and analysis empowered by the WEKA *machine-learning* library [1]. A novel iterative Bayesian filter is developed to estimate the posterior probability distribution of the micromechanical parameters of a DEM model, conditioned to history-dependent experimental data. The iterative application of conventional sequential Bayesian estimation [2,3] allows the virtual granular material to *learn* from all previous experimental measurements of the physical system being modeled in a fast and automated manner.

Bayesian calibration is conducted for DEM modeling of glass beads under cyclic oedometric compression. Using the particle configuration resulting from the CT images, the representative volume of a glass bead packing is reconstructed in DEM simulations. The DEM packing governed by the simplified Hertz-Mindlin contact law and rolling resistance is then calibrated through a iterative Bayesian filtering process, which is able to focus increasingly on highly probable parameter subspaces over iterations. Three iterations are needed to obtain excellent agreement between posterior predictions and experimental data as well as accurate approximation of the posterior probability distribution as shown in Fig. 1a. From the posterior probabilities, micro–macro correlations can be obtained with known uncertainties (see Fig. 1b), which also help understand the uncertainty propagation across various scales.

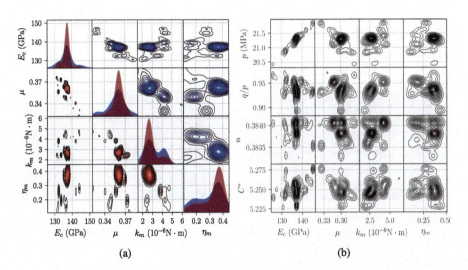

Fig. 1. (a) Posterior PDF estimated at the beginning (blue) and the end (red) of the sequential Bayesian filtering. 2D projections of the posterior PDF in the above- and below-diagonal panels are colored by the posterior probability densities. (b) Approximated posterior distributions for pairs of micromechanical parameters and macroscopic quantities of interest at the maximum stress ratio state.

3 Validation

To demonstrate that the grains are successfully trained by the experimental data, DEM modeling of elastic waves propagating through a long granular column is considered for model validation. The elastic moduli are experimentally measured from ultrasonic traces received along the oedometric compression path. The elastic moduli can be numerically calculated by (1) static probing: load the representative volume with a small strain increment, and (2) dynamic probing: agitate elastic waves through a long granular column constructed with the same representative volume. The wave velocities obtained at different pressures using the two approaches quantitatively agree with those measured in experiments,

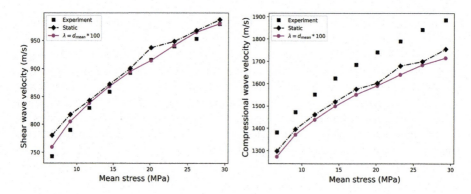

Fig. 2. Comparison of elastic wave velocities predicted by the two probing methods and measured in ultrasonic experiments. A wavelength of 100 times the mean particle diameter is used as the source for agitating elastic waves

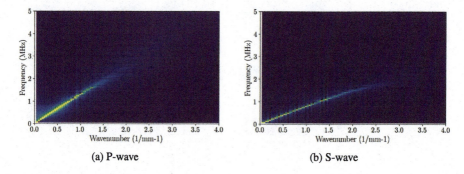

Fig. 3. Dispersion relations obtained by applying two-dimensional fast Fourier transform to layer-averaged particle velocities.

having errors less than 10% for the former and 16% for the latter, as shown in Fig. 2.

In addition to the good agreement between numerical predictions and experimental data, the dispersion relation, namely elastic P- or S-wave velocities as functions of frequency and wavenumber, can be obtained from the DEM simulations as shown in Fig. 3, which is rather difficult in experiments. The initial slopes that correspond to elastic moduli of a continuum agree well with the experimental values, thus validated the robustness of the calibrated DEM model. Although not shown here, a variety of input frequencies and waveforms are considered during the dynamic probing to investigate their effect on the dispersion relations. While the dispersion curves are mostly unaffected by the source, the activated frequency bands show dependency on the characteristics of input signals.

4 Conclusions

The present study show the capability of the *Grain* learning toolbox for calibrating DEM models of granular materials. The new iterative Bayesian filter facilitates a fast and automated search in parameter space from coarse to fine scales. The wave propagation simulations performed with the calibrated DEM model agree well with the ultrasonic experiments conducted during the oedometric loading. It is worth noting that although static and dynamic probing give similar predictions for the elastic moduli of granular materials, the latter generally takes less computational time and provide more useful information than the former.

Acknowledgments. This work was financially supported by Eni S.p.A.

References

1. Arganda-Carreras, I., Kaynig, V., Rueden, C., Eliceiri, K.W., Schindelin, J., Cardona, A., Sebastian Seung, H.: Trainable Weka Segmentation: a machine learning tool for microscopy pixel classification. Bioinformatics **9**, 676–682 (2017)
2. Cheng, H., Shuku, T., Thoeni, K., Yamamoto, H., Radjai, F., Nezamabadi, S., Luding, S., Delenne, J.: Calibration of micromechanical parameters for DEM simulations by using the particle filter. EPJ Web Conf. **140**, 12011 (2017). EDP Sciences
3. Cheng, H., Shuku, T., Thoeni, K., Yamamoto, H.: Probabilistic calibration of discrete element simulations using the sequential quasi-Monte Carlo filter Granul. Matter **20**, 11 (2018)

An Analytical Study of the Bi-rock System Considering Interface Effect

Xing-wei Chen[✉] and Zhong-qi Quentin Yue

Department of Civil Engineering, The University of Hong Kong,
Pokfulam Road, Hong Kong, China
Cxw1110@hku.hk

Abstract. The mechanical behavior of rock mass is largely governed by the joints between the individual intact rocks due to the lower strength of the resulting imperfect interfaces. In order to investigate the influences of the joints, a bi-rock system of laterally infinite extend connected by a planary interface is well employed for mathematical convenience. The objective of this paper is to investigate the influence of imperfect interface on the mechanical behavior of bi-rock system via exact closed-form solution. Four types of idealized interfaces including perfectly bounded, sliding, bilaterally inextensible and unsinkable interface are considered. By comparison with the case of perfectly bounded interface, the contact stresses and displacements at the imperfect interface are well examined to show the influences of different joints. The solution presented in this paper can serve as limiting case bench mark to the problems of bi-rock system with more complicated and general interface, and can as well provide us preliminary prediction of the joint failures.

Keywords: Bi-rock · Imperfect interface · Closed form solution

1 Introduction

The mechanical behavior of rock mass is largely governed by the joints between the individual intact rocks due to the lower strength of the resulting imperfect interfaces. In order to investigate the influences of the joints, a bi-rock system of laterally infinite extend connected by a planary interface (Fig. 1) is well employed for mathematical convenience [1]. Although the real behavior of joined rock system is far more complicated than those that can be modelled as linear elasticity of two solid media, the analytical solutions to the bi-material model with limiting imperfect interface remain equally valuable for investigating the influence of the joint on the overall mechanical behavior of the bi-rock system.

The objective of this paper is to investigate the influence of imperfect interface on the mechanical behavior of bi-rock system via exact closed-form solution. Four types of idealized interfaces including perfectly bounded, sliding, bilaterally inextensible and unsinkable interface are considered. By comparison with the case of perfectly bounded interface, the contact stresses (Fig. 2) and displacements at the imperfect interface are well examined to show the influences of different joints. The solution presented in this paper can serve as limiting case bench mark to the problems of bi-rock system with more complicated and general interface, and can as well provide us preliminary prediction of the joint failures. The works presented in this paper can be regarded as the extension of the solutions by [2, 3] for the point load problems in the bi-material with perfectly bounded and sliding interface respectively.

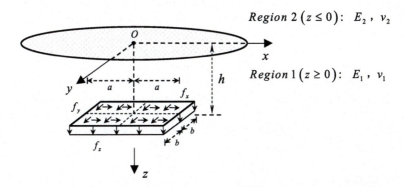

Fig. 1. Bi-rock system joined by planary interface

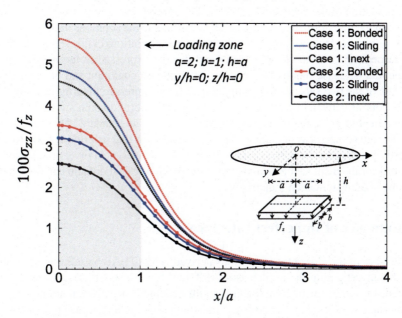

Fig. 2. Normal contact stress σ_{zz} at the interface $(x, 0, 0)$ due to normal loading f_z distributed in a rectangle embedded at depth h

2 Methodology

The solutions presented in this paper are obtained with the aid of an integral transformation methodology developed by [4]. By virtue of this approach, the complete solutions of all the elastic fields induced by 3D uniform load distributed in a horizontal rectangular area embedded in the bi-rock system can be solved systematically, and are expressed as systematical and uniform Green's matrixes with elementary closed form harmonic functions. The main steps of the integral transform based method adopted in this can be summarized as follow:

(1) Transformed field components: With the matrix 2D Fourier's transform, the presented method firstly converts the field components in physical domain to pairs of combined variables in transformed domain. By virtue of this treatment, the integral variables and the complex unit are clearly isolated from the coefficient matrixes of elastic constants, and well incorporated into the combined field variables. Then the original governing PDEs can be consequently reduced to two sets of ODEs with coefficient matrixes of material constants (1).

$$\frac{d}{dz}\mathbf{V}(z) = \rho \mathbf{C}_v \mathbf{V}(z) - \mathbf{G}_v(z) \qquad \frac{d}{dz}\mathbf{U}(z) = \rho \mathbf{C}_u \mathbf{U}(z) - \mathbf{G}_u(z) \qquad (1)$$

(2) Solution of the degenerated ODEs: The special separations and rearrangements applied in the transformed field components can enable the solutions of the degenerated ODEs with respect to z be solved exactly. The general solutions of the degenerated ODEs with respect to z can be solved exactly and then be expressed in terms of couples of exponential decrease functions. Considering boundary condition and body forces, the solutions of the two sets of field components can be expressed in terms of body force.

(3) Fourier's inversion: Once the solutions in transform domain are formulated, the solution in physical domain can be obtained via direct and standard 2D Fourier's inversion. The advantage of systematical solution expressions that in terms of body force can be partly justified by the conveniences to conduct the analytical inversion. Finally, the solutions for the displacements and plane stresses in physical domain can be expressed in terms of compact Green's matrixes.

3 Influence of Imperfect Interface

On the basis of the closed form solutions derived, a MATLAB based program is developed for investigating the influences of different types of interface conditions on the induced stresses by the internal loading. In the ensuing, particular attentions are paid to the interface stresses (Fig. 2) and stress transfer across the interface (Fig. 3).

Figure 2 displays the variations of normal contact stresses, induced by normal pressure, along the horizontal axis $(x, 0, 0)$ at the interface. As it can be seen from the figure, for different material cases, the normal contact stresses at the bilaterally inextensible interface are always the smallest one, and those for perfectly bounded interface are higher than others. The result indicates that the effect of lateral constraint at the interface on the normal contact stress is not monotonic, neither the two limiting cases, bilaterally inextensible nor sliding interface can provide maximum normal contact stress. Figure 3 shows the distribution of all the vertical stresses σ_{zz}, induced by normal pressure, along the vertical axis $(0, 0, z)$ passing through the loading center. As it can be seen that, σ_{zz} for bounded interface and sliding interface are smoothly continuous at the interface, and the σ_{zz} for inextensible interface are non-smoothly continuous. For the unsinkable interface, an jump discontinuity can be observed in the σ_{zz} profile. The result indicates that both perfectly bounded and sliding interface are ideal interfaces for the normal stress transmission.

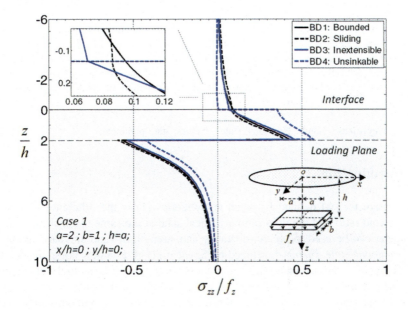

Fig. 3. Veritical stress σ_{zz} along the center axis $(0, 0, z)$ due to normal loading f_z distributed in a rectangle embedded at depth h

References

1. Xiao, H.T., Yue, Z.Q.: Elastic fields in two joined transversely isotropic media of infinite extent as a result of rectangular loading. Int. J. Numer. Anal. Methods Geomech. **37**(3), 247–277 (2013)
2. Rongved, L.: Force interior to one of two joined semi-infinite solids. In: Bogdanoff, J.L. (ed.) Proceedings of 2nd Midwestern Conference on Solid Mechanics, Eng. Exp. Stn., Res. Ser., vol. 129, pp. 1–13. Purdue University (2016)
3. Dundurs, J., Hetényi, M.: Transmission of force between two semi-infinite solids. ASME J. App. Mech. **32**, 671–674 (1965)
4. Yue, Z.Q.: On generalized Kelvin solutions in a multilayered elastic medium. J. Elast. **40**(1), 1–43 (1995)

Natural State Parameter for Sand

Katarzyna Dołżyk-Szypcio[(✉)]

Department of Geotechnics and Structural Mechanics,
Bialystok University of Technology, Wiejska 45E, 15-351 Bialystok, Poland
k.dolzyk@pb.edu.pl

Abstract. Basing on *Frictional State Theory*, the natural state parameter for sand is defined as an extension of state parameter proposed by Been and Jefferies. It is shown that for sand sheared at drained triaxial compression at failure and post-failure, sand behaviour is purely frictional and natural state parameter is $\psi^\circ \approx 0$. This parameter can be used for prediction of sand strength and modeling in the future.

Keywords: Sand · Frictional state · Natural state parameter

1 Introduction

State parameter (ψ) defined by Been and Jefferies [1] is the difference between the current void ratio and the void ratio at critical state at the same mean stress. The state parameter can be used to predict dilatancy and strength of sand and sand modeling.

Frictional State Theory developed by Szypcio [4] gives a new possibility to define more complex state parameter definition named in this paper as natural state parameter (ψ°). The definition of natural state parameter combines the difference between the current void ratio and the void ratio at critical state and the current stress ratio and the stress ratio at critical frictional state.

In this work it is shown that the natural state parameter values at failure are very close to zero for four analyzed sands.

2 Frictional State Theory

The general stress-dilatancy equation has the following form:

$$\eta = Q - AD^p \quad (1)$$

where: $\eta = q/p'$, $Q = M^o - \alpha A^o$, $A = \beta A^o$.

For drained triaxial compression we use the following [5]:
$M^o = 6 \sin \Phi^o / 3 - \sin \Phi^o$, $A^o = 1 - \frac{1}{3}M^o$, $q = \sigma'_1 - \sigma'_3$, $p' = \frac{1}{3}(\sigma'_1 + 2\sigma'_3)$, $D^p = \delta\varepsilon^p_v / \delta\varepsilon^p_q$, $\delta\varepsilon^p_v = \delta\varepsilon^p_1 + 2\delta\varepsilon^p_3$, $\delta\varepsilon^p_q = \frac{2}{3}(\delta\varepsilon^p_1 - \delta\varepsilon^p_3)$, $\delta\varepsilon^p_v = \delta\varepsilon_v - \delta\varepsilon^e_v$, $\delta\varepsilon^p_q = \delta\varepsilon_q - \delta\varepsilon^e_q$.

The elastic parts of strain increments are calculated from equations: $\delta\varepsilon^e_v = \delta p'/K$, $\delta\varepsilon^e_q = \delta q/3G$, $K = \{2(1+v)/3(1-2v)\}G$.

In this paper it is assumed that shear modulus may be expressed by equation:

$$G = G_o^* \frac{(2.973 - e)^2}{1+e} \sqrt{p' p_a} \qquad (2)$$

In *Frictional State Theory* the central role is played by Critical Frictional State and Frictional State [4] illustrated in Fig. 1.

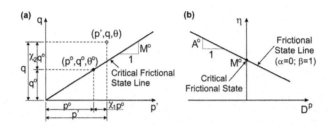

Fig. 1. Definition of Frictional State in planes: (a) q − p'; (b) η−D^p.

The χ_1 and χ_2 can be calculated for drained triaxial compression from equations:

$$\chi_1 = -(D^{pn}/(3+D^{pn})) \qquad (3)$$

$$\chi_2 = (3D^{pn})/(M^o(3+D^{pn})) \qquad (4)$$

where: $D^{pn} = \alpha + \beta D^p$.

3 Natural State Parameter

It is proposed in this paper that the natural state parameter has the following form:

$$\psi^o = \psi_e^o + \psi_\eta^o \qquad (5)$$

where

$$\psi_e^o = e - e_c^o \qquad (6)$$

$$\psi_\eta^o = 1 - \frac{\eta}{M^o} = \frac{\chi_1 - \chi_2}{1 + \chi_1} \qquad (7)$$

are illustrated in Fig. 2.

Therefore ψ_e^o is the difference between the current void ratio and critical void ratio for the reference of mean stress $p^o = p'/(1+\chi_1)$. It is assumed that for sand the critical state line has the power form $e_c = e_\Gamma - \lambda_c (p'/p_a)^\xi$.

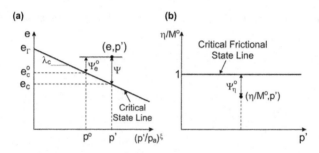

Fig. 2. Natural state parameter components: (a) ψ_e^o; (b) ψ_η^o.

Table 1. Sands and tests parameters and calculated values.

Sand	e_{max}	e_{min}	Φ_{cv}	Φ^o	e_Γ	λ_c	ξ	e_o	e_s	σ_c	G_0^*	ν	D_F^p	ψ_F	ψ_F^o	Φ'_{max}	References
	-	-	[°]	[°]	-	-	-	-	-	kPa	kPa	-	-	-	-	[°]	
Sacramento River	0.906	0.610	32.8	34.9	0.90	0.023	0.7	0.610	0.608	100	97	0.47	-0.903	-0.201	0.010	44.0	Poorooshasb [3]
								0.610	0.605	300	180	0.120	-0.535	-0.172	-0.034	41.2	
								0.610	0.597	1000	190	0.25	-0.230	-0.114	-0.028	38.3	
Toyoura	0.950	0.580	31.0 - 33.4	32.6	0.934	0.019	0.7	0.680	0.670	200	65	0.195	-0.517	-0.181	-0.002	39.7	Yao et al. [7]
								0.680	0.668	500	92	0.155	-0.438	-0.137	-0.002	38.1	
								0.680	0.660	1000	85	0.110	-0.361	-0.095	0.018	37.5	
								0.680	0.620	2000	90	0.080	-0.168	-0.078	-0.008	35.1	
Ottawa	0.790	0.490	30	31.4	0.754	0.017	0.7	0.525	0.525	200	85	0.31	-0.485	-0.149	0.022	38.1	Guo [2]
								0.642	0.642	200	45	0.276	-0.094	-0.053	0.000	33.3	
								0.537	0.537	500	54	0.215	-0.422	-0.104	0.032	37.3	
								0.535	0.535	800	69	0.205	-0.424	-0.075	0.059	36.3	
Portaway	0.790	0.460	29.8	29.8	0.715	0.025	0.7	0.715	0.715	50	16	0.265	0.013	0.036	-0.001	29.6	Wang [6]
								0.538	0.538	50	80	0.060	-0.597	-0.112	0.130	39.9	
								0.537	0.457	300	60	0.020	-0.592	-0.165	0.026	38.2	

e_s – void ratio at the start of shearing; D_F^p, ψ_F, ψ_F^o, Φ'_{max} – values at failure

4 Results

For some data presented in literature the experimental relationships: (σ'_1/σ'_3) – ε_a and $\varepsilon_v - \varepsilon_a$ were sectionally approximated by high degree polynomials and values of $\psi°$, ψ at failure were calculated and shown in Table 1.

For all the analyzed sands tests data the confining pressure is not so high. Therefore, sands dilate during shear and breakage effect may be neglected.

5 Conclusions

Frictional State Theory gives a possibility to define natural state parameter ($\psi°$) as a generalization of classical state parameter. At failure and post-peak failure shearing a sand behaviour is purely frictional and equals to $\psi° \approx 0$. The vanishing of $\psi°$ at failure and post-failure shearing may be used to predict a sand strength and modeling.

Acknowledgements. This work, conducted at Bialystok University of Technology, was supported by Polish Financial Resources on Science under Project No. S/WBiIS/2/2018.

References

1. Been, K., Jefferies, M.: A state parameter for sands. Géotechnique **35**(2), 99–112 (1985)
2. Guo, P.: Modelling granular materials with respect to stress-dilatancy and fabric: a fundamental approach. Ph.D. thesis, University of Calgary (2000)
3. Poorooshasb, H.B.: Description of flow of sand using state parameters. Comput. Geotech. **8**, 195–218 (2009)
4. Szypcio, Z.: Stress-dilatancy for soils. Part I: the frictional state theory. Studia Geotechnica et Mechanica **38**(4), 51–57 (2016)
5. Szypcio, Z.: Stress-dilatancy for soils. Part II: the experimental validation for triaxial tests. Studia Geotechnica et Mechanica **38**(4), 59–65 (2016)
6. Wang, J.: The stress-strain and strength characteristics of Portaway sand. Ph.D. thesis, University of Nottingham (2005)
7. Yao, Y.P., Yamamoto, H., Wang, N.D.: Constitutive model considering sand crushing. Soils Found. **48**(4), 603–608 (2008)

Micro-structure Evolution in Layered Granular Materials During Biaxial Compression

Longlong Fu[1(✉)], Peijun Guo[2], Shunhua Zhou[1], and Yao Shan[1]

[1] Tongji University, Shanghai 201804, China
Longlongfu@tongji.edu.cn
[2] McMaster University, Hamilton L8S4L7, Canada

Abstract. Biaxial compression tests were conducted to investigate the deformation characteristics of layered granular samples by considering the localized deformation characteristics of sublayers. Layered sample refers to a three-layer granular materials stacked in the order hard-soft-hard. The DEM simulation results show that stress and strain development of layered granular samples are positively correlated with the volumetric fraction, which is consistent with the theoretical and FEM results. Directional distribution of contact normal and force chains follows the same pattern: possibilities of near-horizontal and vertical direction increases (decreases) in hard (soft) layer, while possibilities of diagonal direction decreases (increases) in hard (soft) layer.

Keywords: Granular material · Inter-layer-action · Localized strain Contact network · Force chain evolution

1 Introduction

Layered granular soils are commonly found in geotechnical engineering, such as natural stratum in river basin and coastal areas, or subgrade of railway, etc. on basis of the hypothesis of continuity, load induced inter-layer-actions were investigated for decades to homogenize the layered soil [1–3]. However, in granular materials, the spatial heterogeneous distribution of force could result in local failures which may be important but can hardly be obtained via continuous medium analysis methods. In addition, it's invalid to homogenize limited width system with multiple thin layers like subgrade. Therefore, this work focuses on the evolution of macroscopic (stress and strain) and microscopic (contact network and force chain) characteristics of layered cohesionless granular material. Different volumetric fraction was considered.

2 Biaxial Tests of Layered Granular Materials

Numerical model (see Fig. 1) of layered sample was set up by PFC2D [4], where H refers to the relative hard medium while S refers to the relative soft medium, f_H and f_S are the volumetric fraction respectively. Particles forming the sample was in the size of

Gaussian distribution and the linear contact model was employed; the contact bond model was adopted for boundary membrane. Parameters in numerical simulation are shown in Table 1. The axial compression stress σ_1 was applied to the sample by controlling the vertical displacement of the top plate.

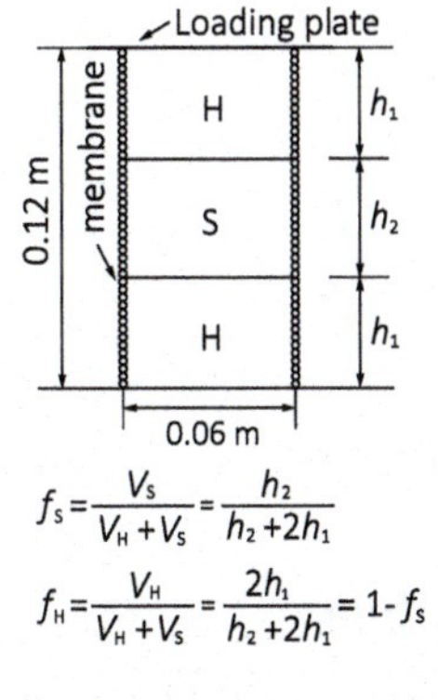

Table 1. Parameters in numerical model.

Parameters	Material		Particles in membrane	Plate
	H	S		
Density (kg/m³)	2600	2000	1800	–
Minimum radius (mm)	0.25	0.25	0.48	–
Maximum radius (mm)	0.65	0.65	0.48	–
Normal stiffness (N/m)	1.5×10^8	7.5×10^7	7.5×10^6	5×10^{12}
Tangential stiffness (N/m)	0.5×10^8	2.5×10^7	2.5×10^6	0
Normal bond (N/m)	–	–	1.0×10^{300}	–
Shear bond (N/m)	–	–	1.0×10^{300}	–
Inter-particle friction	2.0	0.5	0.18	0
Porosity	0.2	0.25	–	–

$$f_S = \frac{V_S}{V_H + V_S} = \frac{h_2}{h_2 + 2h_1}$$

$$f_H = \frac{V_H}{V_H + V_S} = \frac{2h_1}{h_2 + 2h_1} = 1 - f_S$$

Fig. 1. Layered sample

Stress and strain development for various volumetric fraction is depicted in Fig. 2. In order to avoid extreme thin layer in layered sample, f_S was set in the range of 0.2~0.6. It can be found from Fig. 2a that H has obvious strain softening, stress curves of layered samples lie between the stress curves of H and S in the order of f_S, the increase of soft layer make the stress curves close to S. The peak σ_1 decreases nearly linearly as f_S increases from 0 to 0.6 and then various slightly up to $f_S = 1$. Figure 2b shows obvious dilation for H and contraction for S, strain curves of layered samples also lie between the strain curves of H and S in the order of f_S, the increase of soft layer make the strain curves close to S. These indicate that the stress and strain development is positively correlated with f_S.

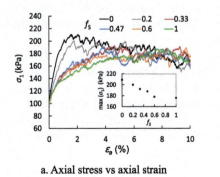

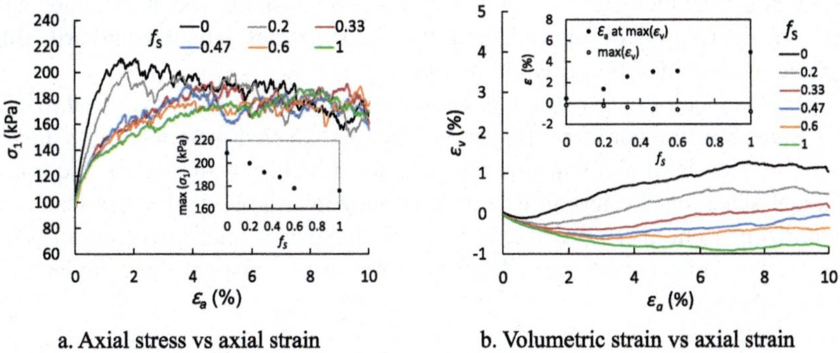

a. Axial stress vs axial strain   b. Volumetric strain vs axial strain

Fig. 2. Stress and strain development during biaxial compression of various volumetric fraction

Figure 3 shows the strain development of whole sample and sublayers, where the localized strains of sublayers were calculated based on Riemann sum. The dilation of hard layers is weakened while the contraction of soft layer is intensified due to the inter-layer-action (see Fig. 4) resulting from the deformation difference. Figures 2, 3 and 4 indicate that the strain characteristics achieved from the DEM simulation are consistent with the theoretical and FEM results [2, 3].

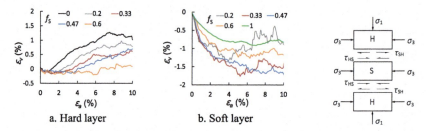

Fig. 3. Strain development of sublayers of various f_S **Fig. 4.** The inter-layer-action

3 Overall and Localized Micro-structure Evolution

The inter-layer-action shown in Fig. 4 is actually obtained based on the continuity hypothesis. In cohesionless granular material, the inter-particle contact behavior between particles in hard medium and soft medium dominate the overall properties. It had been proved that micro-structure of granular materials has significant influence on the macro deformation and strength properties [5–7]. Therefore, the contact network and force chain evolution were analyzed, where force chains in this paper were identified as quasilinear particle strings with particle stress larger than the mean [8, 9].

The directional distributions of contact normal and force chain of both layered sample and each constituent layer were summarized to form the general patterns, as shown in Fig. 5. In Fig. 5a, Contact normal density of near-horizontal and vertical direction increases (decreases) in hard (soft) layer, while contact normal density of diagonal direction decreases (increases) in hard (soft) layer. Consequently, the possibility of force chain in hard (soft) layer oriented in vertical and near-horizontal (diagonal) direction increases, as shown in Fig. 5b. It should be highlighted that the orientation distribution of force chain varies with the strain state, a temporarily different distribution other than Fig. 5 could occur.

The difference of micro-structure between hard and soft layer may lead to different break patterns of the constituent layer and thus the layered material. If the inter-layer-actions between hard and soft layer is considered as Fig. 4 shows, i.e. the horizontal additional stress on the interface, then this horizontal additional stress results in not only horizontal but also vertical and diagonal changes of micro-structure in both hard and soft layers, this demonstrates the high nonlinear responses of the granular materials induced by loading.

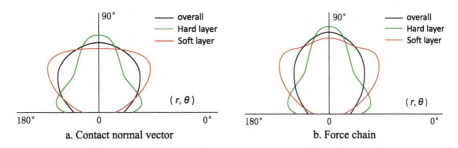

Fig. 5. Typical pattern of directional distribution of micro-structures

4 Discussion

DEM-achieved stress and strain development of layered granular samples are positively correlated with the volumetric fraction, which is consistent with the theoretical and FEM results. The micro-structures of hard and soft layers are quite different due to inter-layer-action, which may result in some deformation or strength peculiarities not revealed yet in layered granular material. Rolling resistance ignored in this work could be included in the future either by introducing a theoretical rolling resistance factor or considering polygons to make the force chain evolution, especially bulking, closer to the practical condition.

Acknowledgement. The NSFC (National Natural Science Foundation of China) Program, Grant NO.51708423, is greatly appreciated for providing financial support for this research.

References

1. Postma, G.W.: Wave propagation in a stratified medium. Geophysics **20**, 780–806 (1955)
2. Lydzba, D., Pietruszczak, S., Shao, J.F.: On anisotropic of stratified rocks: homogenization and fabric tensor approach. Comput. Geotech. **30**, 289–302 (2003)
3. Guo, P., Stolle, D.: Lower and upper limits of layered-soil strength. Can. Geotech. J. **46**, 665–678 (2009)
4. Jiang, M.J., Konrad, J.M., Leroueil, S.: An efficient technique for generating homogeneous specimens for DEM studies. Comput. Geotech. **30**(7), 579–597 (2003)
5. Radjaï, F., Wolf, D.E., Jean, M.: Bimodal character of stress transmission in granular packings. Phys. Rev. Lett. **80**(1), 61–64 (1998)
6. Campbell, C.S.: A problem related to the stability of force chains. Granul. Matter **5**, 129–134 (2003)
7. Behringer, R.P., Daniels, K.E., Majmudar, T.S.: Fluctuations, correlations and transitions in granular materials: statistical mechanics for a non-conventional system. Philos. Trans. R. Soc. A **366**, 493–504 (2008)
8. Howell, D.W., Behringer, R.P.: Fluctuations in granular media. Chaos **9**(3), 559–572 (1999)
9. Peters, J.F., Muthuswamy, M., Wibowo, J.: Characterization of force chains in granular material. Phys. Rev. E **72**(041307), 1–8 (2005)

Dilatancy Phenomenon Study in Remolded Clays – A Micro-Macro Investigation

Qian-Feng Gao[1(✉)], Mohamad Jrad[1], Lamine Ighil Ameu[1], Mahdia Hattab[1], and Jean-Marie Fleureau[2]

[1] Laboratoire d'Etude des Microstructures et de Mécanique des Matériaux, CNRS, UMR 7239, Université de Lorraine, Metz, France
qianfeng.gao@univ-lorraine.fr
[2] Laboratoire de Mécanique des Sols, Structures et Matériaux, CNRS, UMR 8579, Université Paris-Saclay, CentraleSupélec, Gif-sur-Yvette, France

Abstract. The aim of this study is to analyze the influence of stress path on the dilatancy behavior of clays. Triaxial tests were conducted on saturated, remolded clay specimens. During triaxial loading, two different stress paths were considered to bring the stress level to a given point in the $(p' - q)$ plane. After triaxial testing, scanning electron microscopy (SEM) observations and X-ray microtomography (XR-μCT) scans were performed on subsamples. The obtained images were processed using different software and thus the geometric characteristics of pores and cracks were identified. The microstructural characteristics were linked to the dilatancy phenomenon of specimens. The results indicate that dilatancy is influenced by stress path and may be attributed to the evolution of pore geometry and the presence of local open cracks in highly over-consolidated (OC) clays.

Keywords: Clays · Dilatancy · Stress path · Pore orientation Cracks

1 Introduction

Dilatancy is a typical property of clay during shearing. Normally consolidated (NC) or slightly over-consolidated (OC) clay exhibits contractancy; highly OC clay shows dilatancy. Previous work has shown that the dilatancy behavior of clay depends on the overconsolidation ratio (OCR) value, stress ratio ($\eta = q/p'$) and Lode angle [1–4]. Meanwhile, the influence of the stress path on dilatancy should be emphasized. In addition, many experimental studies have revealed that the macroscopic behavior of clay is related to the evolution of its microstructure [5–8]. This makes it possible to give new insight into the dilatancy phenomenon by microscopic studies.

The aim of this research is to examine if stress path has an influence on the dilatancy phenomenon of clays. The dilatancy is approached at two different scales, the macroscale and the microscale. The experimental procedure consists in performing triaxial tests on clay specimens, then carrying out microstructural observations using scanning electron microscopy (SEM) and X-ray microtomography (XR-μCT). The results were analyzed to link the dilatancy behavior with the microscopic phenomena.

2 Materials and Methodology

The studied material is an industrial kaolin clay named Kaolin K13. Its basic properties are: specific gravity $G_s = 2.63$, liquid limit $w_L = 42\%$, plastic limit $w_P = 20\%$ and specific surface area $s = 27$ m^2/g.

Four consolidated drained (CD) triaxial tests were performed on saturated remolded clay specimens (Table 1). Two tests concern the NC clay and the others concern the OC clay (OCR = p'_0/p'_1). Two different stress paths, i.e., the constant σ'_3 stress path (S) and the constant p' stress path (P) were considered for the triaxial shearing stage. All triaxial stress paths were stopped when stress levels approached the fixed P2 point ($p'_2 = 300$ kPa, $q_2 = 150$ kPa) in the (p', q) plan (Fig. 1).

Table 1. Triaxial tests on saturated remolded clay.

Specimen	OCR	Stress path	p'_0 (kPa)	p'_1 (kPa)
NC_S250_P2	1.0	S	250	–
OCR4.0_S250_P2	4.0	S	1000	250
NC_P300_P2	1.0	P	300	–
OCR3.3_P300_P2	3.3	P	1000	300

Notes: p'_0 and p'_1 are the isotropic effective stresses for NC clay and OC clay, respectively.

After triaxial loading, specimens were recovered and then cut into subsamples for SEM observations and XR-μCT scans. ImageJ and Avizo software were then used to process and analyze the clay microstructure (i.e., pores and cracks).

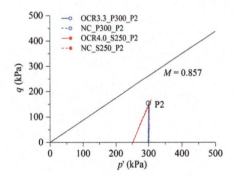

Fig. 1. Two different stress paths.

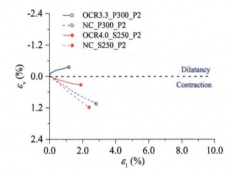

Fig. 2. Evolution of volumetric strain.

3 Results and Discussion

3.1 Macroscopic Behavior

Figure 2 shows the volumetric behavior of the specimens under different loading conditions. One can note that the two NC specimens are contractive regardless of stress paths. Nevertheless, the two OC specimens exhibit two different volumetric behaviors. The specimen (OCR4.0_S250_P2) under constant σ'_3 stress path shows contractive behavior, while the specimen (OCR3.3_P300_P2) under the constant p' stress path shows dilatant behavior. These results indicate that the dilatancy phenomenon is affected by stress paths.

3.2 Pore Shape and Pore Orientation

The pore roundness (R_s = ratio of the minor axis to the major axis) and the pore orientation (θ = angle of the major axis with respect to the horizontal plane) were analyzed.

The pore roundness curves in Fig. 3a are characterized by a unimodal distribution for all specimens under triaxial loading. Moreover, compared to the contractive specimens, the curve of the dilatant specimen (OCR3.3_P300_P2) presents a wider peak shifted to the right, which corresponds to more open pores.

The pore orientation curves in Fig. 3b are away from the depolarization (D) line (that represents a perfectly random orientation) and show preferential orientation of pores. Furthermore, the curve of the dilatant specimen (OCR3.3_P300_P2) is relatively close to the D line, which means that the pores tend toward a random orientation.

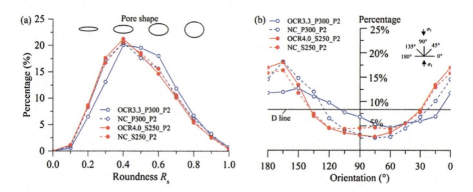

Fig. 3. Micropore structure: (a) pore roundness; (b) pore orientation.

3.3 Micro- and Meso-Cracks

From SEM images (Fig. 4a, b), we can notice the appearance of micro-cracks between weakly-linked and parallel-oriented particle groups in both NC and OC specimens. These micro-cracks have lengths varying from around 5 μm to 15 μm and small thicknesses of 0.5–1.0 μm. However, several open meso-cracks (whose lengths are larger than 100 μm) with tortuous surfaces are found only in specimen OCR3.3_P300_P2 (Fig. 4c). It seems that the activation of open meso-cracks is due to the propagation of micro-cracks and closely related to the dilatancy phenomenon. Notice that meso-cracks may also exist in contractive specimens, but they are probably closed and cannot be detected with XR-μCT.

From the above analyses, one can note the strong relationship observed between the dilatancy phenomenon and the microstructure of clay. The results show that in highly OC clay, dilatancy develops owing to the increasing opening of pores associated with a random pore orientation. During loading, along the stress path, these pores lead to open microcracks that can propagate up to the mesoscopic level.

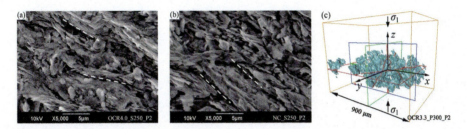

Fig. 4. Micro-cracks in (a) OCR4.0_S250_P2 and (b) NC_S250_P2; and meso-cracks in (c) OCR3.3_P300_P2.

4 Conclusions

The following conclusions can be drawn from this study: (i) for a given stress level, the dilatancy phenomenon is clearly affected by stress paths; (ii) the dilatancy development may be attributed to the evolution of pore geometry, and the formation of local open cracks.

Acknowledgement. We would like to acknowledge the scholarship offered by China Scholarship Council (CSC) for the first author's Ph.D study.

References

1. Ye, G.L., Ye, B., Zhang, F.: Strength and dilatancy of overconsolidated clays in drained true triaxial tests. J. Geotech. Geoenviron. Eng. **140**(4), 06013006 (2014)
2. Hattab, M., Hicher, P.Y.: Dilating behaviour of overconsolidated clay. Soils Found. **44**(4), 27–40 (2004)
3. Shimizu, M.: Effect of overconsolidation on dilatancy of a cohesive soil. Soils Found. **22**(4), 121–135 (1982)
4. Matsuoka, H.: Dilatancy characteristics of soil. Soils Found. **14**(3), 13–24 (1974)

5. Hattab, M., Hammad, T., Fleureau, J.M.: Internal friction angle variation in a kaolin/montmorillonite clay mix and microstructural identification. Géotechnique **65**(1), 1–11 (2015)
6. Hattab, M., Fleureau, J.M.: Experimental analysis of kaolinite particle orientation during triaxial path. Int. J. Numer. Anal. Methods Geomech. **35**(8), 947–968 (2011)
7. Hattab, M., Fleureau, J.M.: Experimental study of kaolin particle orientation mechanism. Géotechnique **60**(5), 323–331 (2010)
8. Li, X., Zhang, L.M.: Characterization of dual-structure pore-size distribution of soil. Can. Geotech. J. **46**(2), 129–141 (2009)

DEM Simulation of Wave Propagation in Anisotropic Granular Soil

Xiaoqiang Gu[1,2(✉)] and Shuocheng Yang[1,2]

[1] Department of Geotechnical Engineering, Tongji University, Shanghai 200092, China
guxiaoqiang@tongji.edu.cn
[2] Key Laboratory of Geotechnical and Underground Engineering of the Ministry of Education, Tongji University, Shanghai 200092, China

Abstract. The anisotropy of elasticity in soil can be evaluated using the velocities of compression waves and shear waves propagated in different manner. In this paper, discrete element simulations were performed to determine the wave velocities in Toyoura sand where isotropic confining stress and anisotropic consolidations in vertical direction were conducted. Based on the discussion of the launch conditions of elastic wave and travelling time interpretation, satisfying output wave signals were presented. The elastic constants determined by wave velocities were compared with those obtained by drained triaxial tests and simple shear tests at very small strain levels. It was found that the elastic constants determined by these two methods are consistent with each other, which stands as a solid proof to apply the continuum assumption to wave propagation in granular soils. Since affected by the anisotropy of fabric and stress state, the wave velocities experience nonlinear change. At anisotropic stress states, there is a threshold stress ratio reflecting dramatic velocity decreasing and fabric rearrangement. Moreover, the ratio of stress-normalized wave velocities can capture the evolution of fabric and reflect the granular nature of stress-dependency, which can also be served as a macroscopic index of the soil fabric.

Keywords: Discrete element method · Granular soil · Anisotropy Wave velocity

1 Introduction

The elastic constants of soil are considered as significant parameters concerning with small strain conditions. Usually by capturing the velocities of compression wave (V_p) and shear wave (V_S) using bender elements, elastic constrained modulus (M_0) and initial shear modulus (G_0) of isotropic medium are evaluated, previous researchers [1] proposed a general relation to link the elastic modulus with material fabric and the three principal stresses: the stress in the direction of wave propagation (σ_{pro}), the stress in the direction of particles oscillation (σ_{osc}) and the stress in the direction perpendicular to the plane determined by the former two (σ_{out}).

$$M_0(G_0) = \rho V_P^2(V_S^2) = A_p(A_s) \times f(e) \times [(\sigma_{pro}/p_a)^{n_{prop}}(\sigma_{osc}/p_a)^{n_{osc}}(\sigma_{out}/p_a)^{n_{out}}]^2 \qquad (1)$$

where A is a constant of grain properties and fabric, $F(e)$ is a void ratio function and n_{prop}, n_{osc} and n_{out} are exponents indicating the disparate influence of three principal stress normalized by atmospheric pressure p_a. Learn from this equation, the varied distribution of fabric and stress state differ the elastic stiffness in directions as well as wave velocities. For anisotropic soils, experiment and simulation results [1, 2] shows the velocities of shear wave (S-wave) and compression wave (P-wave) depend on σ_{pro} and σ_{osc} meanwhile σ_{out} makes few contributions. However, the lack of effective measurement of fabric in laboratory prevents further discussion on the respective contributions made by fabric and stress. In this study, DEM investigation in Toyoura sand is aimed to divide the influence of fabric and stress state on the mechanics of wave propagation in granular soils.

2 DEM Modeling

In DEM simulation, 62073 balls in a cubic space were generated with no anisotropy in the initial place and their sizes fitted in the particle size distribution of Toyoura sand. The non-linear Hertz-Mindlin contact model was considered. More basic parameters are consistent with previous Toyoura sand simulation [3]. After isotropic consolidation, we increased or decreased the vertical stress in the consolidation process to introduce stress anisotropy, while keeping the horizontal stress at rest. The samples with stress ratio (SR) from 0.45 to 2.3 were obtained. In wave propagation tests, the movement of particles near the wall were fixed as the single period sine function, meanwhile two sets of particles on the opposite were monitored as receiver 1 (R1) and receiver 2 (R2). We launched five elastic waves directly associated with certain stiffness: (i) velocity of P-wave propagated horizontally: V_{Ph}; (ii) velocity of P-wave propagated vertically: V_{Pv}; (iii) velocity of S-wave propagated vertically (horizontally) and oscillated horizontally (vertically): V_{Svh} (V_{Shv}); (iv) velocity of S-wave propagated and oscillated in the horizontal direction: V_{Shh}. Meanwhile probe tests like triaxial compression tests and simple shear tests are performed to provide control values of velocities interpretation.

3 Discussions and Conclusions

Figure 1 shows the output signals in time domain. In our tests, elastic wave suffers serve distortion and energy dissipation because of the granular nature and boundary reflections. Referring to previous researches [1, 2], our signals capture the same properties. Comparing to the velocities obtained by probe tests, travelling time determined by peak-peak method and cross correlation method is satisfied as the deviations stay under 5% which also guarantee to apply continuum assumption to wave theory in granular materials. Then, investigation on the anisotropic conditions indicates that wave velocities are not in the same relationship with stress state as Eq. 1. For better illustration, we present the velocities normalized by σ_{pro} and σ_{osc} with exponent 1/3 in Hertz-Mindlin contact law and plot them as a function of SR in Fig. 2. It seems that there are some threshold

SR values in S-wave and P-wave. When SR rises to 1.9 or drops to 0, normalized wave velocities begin to fall dramatically. Thus the threshold indicates that the soil start to mainly adjust the fabric to resist the external loads.

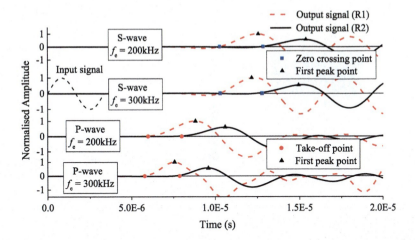

Fig. 1. Wave propagation signals with different excitation frequencies (f_e) at isotropic 100 kPa

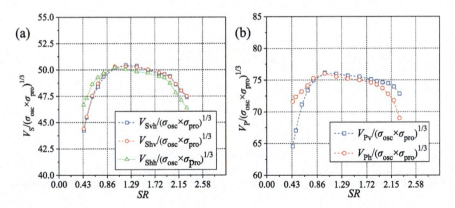

Fig. 2. Effect of SR on the normalized wave velocity: (a) shear wave; (b) compressive wave

As for microscopic information, Fig. 3a shows the rose diagram of coordination number (CN) reflecting contact rearrangement. Moreover in Fig. 3b the fabric anisotropy factor a_r generally shows the fabric evolution of contact vectors where the SR thresholds in a_r coincide with those in wave velocity. So to link the macroscopic velocities with microscopic information, contact normal function $E(\theta)$ in propagation direction (θ_{pro}) and oscillation direction (θ_{pro}) are obtained. In Fig. 4, the ratio of two type of normalized wave velocity (V_1, V_2) are plotted against the ratio of contact normal function: $\left(E\left(\theta_{1_{pro}}\right)+E\left(\theta_{1_{osc}}\right)\right)\Big/\left(E\left(\theta_{2_{pro}}\right)+E\left(\theta_{2_{osc}}\right)\right)$. It's interesting that the normalized velocities ratio is linearly proportional to the ratio of fabric. The data from Gu et al. [3] simulation on Toyoura sand is also fitted. The sole relation indicates that the fabric contribution on the wave velocity is exclusive and not concerned about the propagation manner. Besides the way stress state influences wave velocities is same as the exponential relation in Eq. 1.

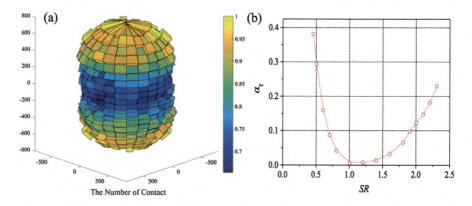

Fig. 3. Fabric evolution: (a) rose diagram of contact number at $SR = 2.2$; (b) relation between anisotropy factor and SR.

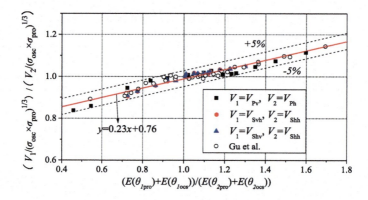

Fig. 4. Effect of the ratio of contact normal function on the ratio of normalized wave velocity

Acknowledgements. The work presented in this paper is supported by National Key Research and Development Program (Grant no. 2016YFC0800200) and National Natural Science Foundation of China (Grant no. 41772283).

References

1. Fioravante, V., Giretti, D., Jamiolkowski, M.: Small strain stiffness of carbonate Kenya Sand. Eng. Geol. **161**(14), 65–80 (2013)
2. O'Sullivan, C., Marketos, G., Muir Wood, D.: Anisotropic stress and shear wave velocity: DEM studies of a crystalline granular material. Géotech. Lett. **5**, 224–230 (2015)
3. Gu, X., Hu, J., Huang, M.: Anisotropy of elasticity and fabric of granular soils. Granul. Matter **19**(2), 33 (2017)

On Collapse of 2D Granular Columns: A Grain-Scale Investigation

Xuzhen He[1](✉), Wei Wu[1], Dichuan Zhang[2], and Jong Kim[2]

[1] Institut für Geotechnik, Universität für Bodenkultur, 1180 Wien, Austria
xuzhen.he@boku.ac.at
[2] Nazarbayev University, Astana 010000, Kazakhstan

Abstract. This study uses the Discrete Element Method (DEM) to investigate the grain-scale mechanisms that give rise to the diverse flow phenomena of granular material, particularly the collapse of granular columns. The small-scale 2D experiments conducted with aluminium rods are used as benchmarks. It is found that the stiffness or the viscous dissipation at the contacts are not important factors to influence the kinetic, but the apparent friction angle is the dominant one, which is contributed by several sources.

Keywords: Granular columns · DEM

1 Introduction

Landslides are catastrophic geophysical phenomena that may cause heavy fatality and property losses. Hence, it is of vital importance to understand their mechanisms and evaluate their travel distance so that appropriate measures can be taken to mitigate their risk.

Recently, a number of studies [1–3] have been conducted on the transient granular flows formed by the sudden release of vertical columns of grains onto horizontal planes (Fig. 1), which include experiments [1] and continuum-based numerical analyses [2,3]. In the present study, Discrete Element Method (DEM) is used to study the grain-scale mechanism and the small-scale experiments conducted by [1] are used as benchmarks.

2 The DEM Model

Nguyen et al. [1] conducted a series of two-dimensional (2D) collapse experiment with aluminium rods. The diameters of the rods were 16 or 30 mm, and they were mixed in the ratio of 5.27:1 by number. In the 2D DEM model, circular disks with the same configuration are used. The linear rolling resistance model embedded in PFC^{3D} is used for the contacts and parameters are listed in Table 1, where μ is the friction coefficient and μ_r is the rolling resistance coefficient at contacts.

Table 1. DEM simulation parameters

D (mm)	ρ_{grain} (kg/m^3)	E (MPa)	κ	β_n	β_s	μ and μ_r
1.6 and 3 at the ratio of 5.27:1	2700	45.5	2	0.2	0.2	Varies

It is found that in this collapsing problem, ρ_{grain}, E, κ, β_n and β_s all have very limited influence on the kinetic, therefore, the fixed values listed in Table 1 are used all the time. Nevertheless, different combinations of friction coefficient and rolling resistance coefficient leads to distinct flow behaviour.

Figure 1 presents the simulated collapse of a shallow column ($\mu = 0.3$, $\mu_r = 0.157$). In comparison with experiments, the time evolution of the free surface, especially the deposition profile, is predicted very well by the DEM model.

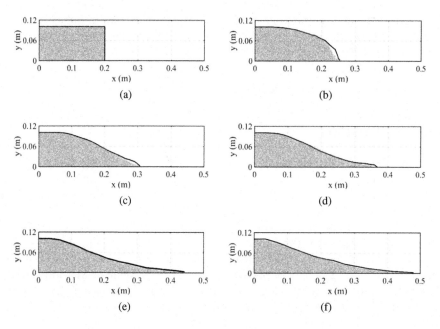

Fig. 1. Profiles from Experiments (solid lines by [1]) and from DEM simulations (grey circles, $\mu = 0.3$, $\mu_r = 0.157$). (a) 0 s. (b) 0.11 s. (c) 0.17 s. (d) 0.24 s. (e) 0.34 s. (f) 0.6 s.

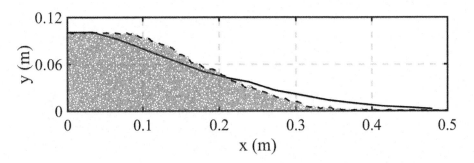

Fig. 2. Final profile from Experiments (solid lines by [1]) and from DEM simulations (grey circles, $\mu = 0.5$, $\mu_r = 0.5$).

3 Discussions

Figure 2 presents the final profile of another simulation ($\mu = 0.5$, $\mu_r = 0.5$). The deposited heap has a steeper slope than the aluminium rods. If the experimental profile and the DEM-simulated profile are mathematically represented by $f^{exp}(x,t)$ and $f^{DEM}(x,t)$, respectively, an error can be defined as follow to measure the difference between them.

$$E(t) = \frac{\int_{x=0}^{L_\infty} |f^{exp}(x,t) - f^{DEM}(x,t)| dx}{\int_{x=0}^{L_\infty} f^{exp}(x,t) dx} \quad (1)$$

A series of DEM simulations are conducted with different combinations of μ (ranges from 0.2 to 0.5 with a step of 0.025) and μ_r (0 to 0.5 with a step of 0.05), the results regarding $E(\infty)$ are presented in Fig. 3, which suggests that when μ and μ_r are on the solid line, the DEM-predicted profiles agree well with experimental ones ($E(\infty) < 5\%$).

The finding above suggests that, in the collapsing flows, the stiffness or the viscous dissipation at the contacts are not important factors to influence the kinetic, but an apparent (or bulk) friction angle ϕ is dominant. The apparent friction angle ϕ is contributed by several sources such as (a) the friction between grains ϕ_μ, (b) the rolling resistance between grains ϕ_{μ_r}, (c) the resistance mobilised by dilation or particle climbing ϕ_d, (d) resistance mobilised by particle rearrangement and damage ϕ_p, etc.

$$\phi = \phi_\mu + \phi_{\mu_r} + \phi_d + \phi_p + \ldots \quad (2)$$

Figure 3 shows that by varying the contribution from ϕ_μ and ϕ_{μ_r}, the same apparent friction angle ϕ can be obtained and the kinetic of the flow is eventually the same despite the grain-scale difference.

It must also be addressed that the kinetic during the whole process is the same, not just the final profile. Eleven simulations are conducted with μ and μ_r chosen on the solid line (squares in Fig. 3), an average error $E = \sum_{t=t_1}^{t_5} E(t)/5$ calculated at five different time snaps (0.11 s, 0.17 s, 0.24 s, 0.34 s and 0.6 s) are presented in Fig. 4, which suggests that the DEM-predicted profiles agree well with experimental ones during the whole process ($E(\infty) < 5\%$) and the apparent friction angle is the same for all tests.

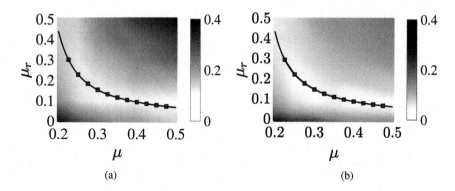

Fig. 3. Error of the predicted final profile $E(\infty)$ with various model parameters. (a) $L_0 = 0.1$ m, $H_0 = 0.1$ m. (b): $L_0 = 0.2$ m, $H_0 = 0.1$ m

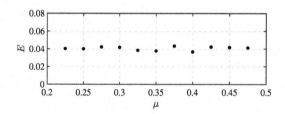

Fig. 4. Average error of the predicted profile E with various model parameters.

Acknowledgements. The first author acknowledges the financial support from the Otto Pregl Foundation for Geotechnical Fundamental Research, Vienna, Austria and Jiangsu Province natural sciences fund subsidisation project (BK20170677). The research is also partly funded by the project "GEORAMP" within the RISE programme of Horizon 2020 under grant number 645665.

References

1. Nguyen, C.T., Nguyen, C.T., Bui, H.H., Nguyen, G.D., Fukagawa, R.: A new SPH-based approach to simulation of granular flows using viscous damping and stress regularisation. Landslides **14**(1), 69–81 (2017)
2. He, X., Liang, D., Bolton, M.D.: Run-out of cut-slope landslides: mesh-free simulations. Géotechnique **68**(1), 50–63 (2018)
3. He, X., Liang, D., Wu, W., Cai, G., Zhao, C., Wang, S.: Study of the interaction between dry granular flows and rigid barriers with an SPH model. Int. J. Numer. Anal. Meth. Geomech. https://doi.org/10.1002/nag.2782

DEM Simulation of the Behavior of Rockfill Sheared with Different Intermediate Principle Stress Ratios

Juntian Hong, Ming Xu[✉], and Erxiang Song

Department of Civil Engineering, Tsinghua University, Beijing 100084, China
Mingxu@mail.tsinghua.edu.cn

Abstract. This study investigates the influence of intermediate principle stress ratio on the mechanical behavior of crushable rockfill materials using discrete element method. Rockfill particles are modeled as breakable agglomerates, and reasonable consistency is found between the predicted and experimental results. The evolution of particle breakage is also obtained. Consistent relationships between the macro and micro parameters are established.

Keywords: Discrete element method · Intermediate principal stress ratio Rockfill · Particle breakage

1 Numerical Simulation

1.1 Numerical Model

A discrete element code PFC3D was used for this investigation. Xu et al. (2017) developed an agglomerate model that allows particle breakage. Each agglomerate was created by bonding 13 spheres together into a regular assembly in hexagonal close packing. The sizes of the agglomerates ranged from 21.6 mm to 36 mm. The statistical variability of particle strength is achieved by assuming normal distribution of the normal and shear bond strengths b_n and b_s. Single grain crushing tests were carried out and the survival probability curve was found to match the Weibull distribution. A total of 944 different sized and randomly orientated agglomerates were created to form a sample.

Agglomerates parameters were calibrated following a systematic methodology based on the results of triaxial tests with different stress paths on limestone rockfills performed by Xu et al. (2012). Figure 1 shows the simulation results and the corresponding experimental results. The parameters are summarized in Table 1. Detailed description about the DEM model and the calibration procedure can be found in Xu et al. (2017).

1.2 Simulation Procedure

The assembly of randomly orientated agglomerates was first isotropically compressed to a confining pressure of 10 kPa to simulate the preparation procedure in the

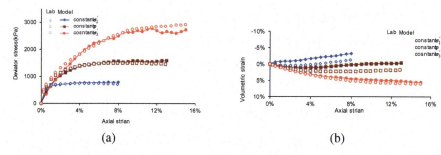

Fig. 1. DEM simulation results and corresponding laboratory results of triaxial tests with different stress paths: (a) deviator stress against axial strain; (b) volumetric strain against axial strain

Table 1. Properties of agglomerate

Input parameter	Numerical value
Normal bond strength (b_n): N	Obey normal distribution $\mu = 1600$, $\sigma = 1600$
Shear bond strength (b_s): N	Obey normal distribution $\mu = 1200$, $\sigma = 1200$
Normal stiffness of each sphere (k_n): N/m	8×10^6
Shear stiffness of each sphere (k_s): N/m	10^9
Frictional coefficient of sphere	0.6

laboratory test. The inter-particle friction coefficient was set to a very low value during the generation stage to achieve a dense packing of the particles.

After sample generation, the inter-particle friction coefficient was set to its final values and the samples were isotropically compressed to different confining pressures (600 kPa, 1000 kPa and 2000 kPa), with the size of the cubic samples becoming about 235 mm.

Subsequently, the samples were subjected to true triaxial shearing with different stress paths, while maintaining a constant minor principal stress. The intermediate principal stress ratio b, defined as $b = (\sigma_1 - \sigma_3)/(\sigma_2 - \sigma_3)$, was kept constant in each test. The values of b vary among 0 (triaxial compression), 0.5 and 1 (triaxial extension).

2 Simulation Results

2.1 Macro-behavior

A series of true triaxial tests were simulated with different minor principal stresses (600 kPa, 1000 kPa and 2000 kPa) and different intermediate principal stress ratios ($b = 0$, $b = 0.5$, $b = 1$).

Typical simulated stress strain behaviors and volume changes at a constant minor principal stress of 2000 kPa are shown in Fig. 2. Figure 2a shows the plot of stress ratio, defined as the ratio of the deviator stress q to the mean effective stress p', versus axial strain. The stress ratio decreases with an increase of b value, which illustrates that the stress-strain behaviors of rockfill materials were dependent on the intermediate principal stress ratio. Similar observations were made by Yang et al. (2016), who performed true triaxial tests on dense rockfill materials with different intermediate principal stress ratios. Figure 2b shows the evolution of volumetric strain against axial strain. It reveals that, with constant minor principal stress, the volumetric contraction increases with b value, except that the simulation result overestimates the dilation with $b = 1$ at large axial strain.

The mobilized friction angle can be defined as

$$\varphi = (\sigma_1 - \sigma_3)/(\sigma_1 + \sigma_3) \tag{1}$$

Figure 3 shows the peak friction angles with different minor principal stresses σ_3 and intermediate principal stress ratios b. The peak friction angle at a given b decreases with σ_3, which indicates the stress-dependence of strength. While at a given σ_3, the peak friction angle increases to a peak value and then decreases with the increase of b value. These observations are also in agreement with results reported by Yang et al. (2016) in the laboratory tests.

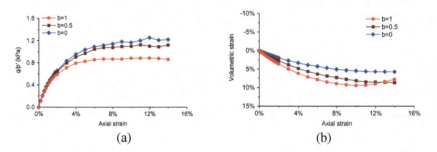

Fig. 2. DEM results of true triaxial tests at a constant minor principal stress of 2000kPa: (a) deviator stress against axial strain; (b) volumetric strain against axial strain

2.2 Particle Breakage

The breakage of each individual agglomerates was monitored and counted during the true triaxial tests. Figure 4 shows the evolution of agglomerate breakage number at a constant minor principal stress of 2000 kPa. The agglomerates continue to break during the tests and the number of breakage increases with the axial strain. With the same axial strain, the number of agglomerate breakage is larger for the test with a high intermediate principal stress ratio.

The energy input per unit volume is calculated during the true triaxial tests. And the number of agglomerate breakage during the tests is plotted against it in Fig. 5. It appears that a consistent relationship exists for all these tests with different minor principal stresses and intermediate principal stress ratios. And this relationship can be applied to the constitutive modeling in further research.

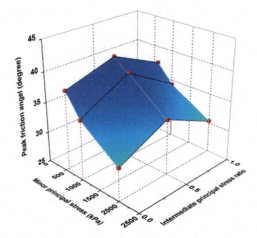

Fig. 3. Peak friction angles with different minor principal stresses and intermediate principal stress ratios

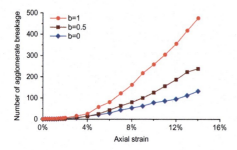

Fig. 4. The number of agglomerate breakage during true triaxial tests at a constant minor principal stress of 2000 kPa

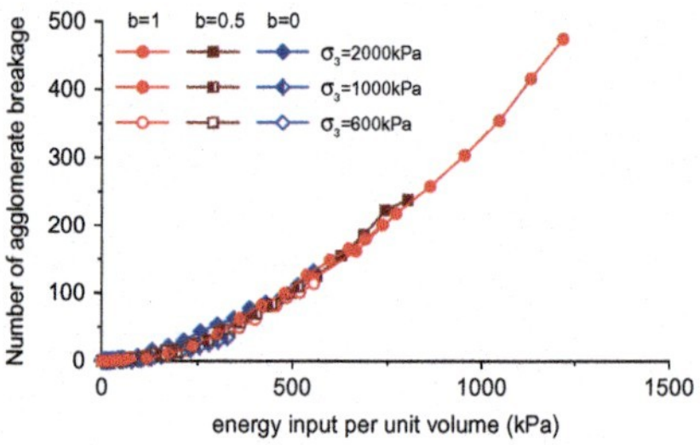

Fig. 5. Number of agglomerate breakage against the energy input per unit volume during true triaxial tests

Acknowledgments. The authors are grateful for the research support received from the China 973 Program (2014CB047003) and the National Natural Science Foundation of China (41272280).

References

Yang, X., Liu, H., Liu, H., Chen, Y., Zhang, W.: Strength and dilatancy behaviors of dense modeled rockfill material in general stress space. Int. J. Geomech. **16**(5), 04016015 (2016)

Xu, M., Song, E., Chen, J.: A large triaxial investigation of the stress-path-dependent behavior of compacted rockfill. Acta Geotech. **7**(3), 167–175 (2012)

Xu, M., Hong, J., Song, E.: DEM study on the effect of particle breakage on the macro- and micro-behavior of rockfill sheared along different stress paths. Comput. Geotech. **89**, 113–127 (2017)

DEM Simulation of Sand Liquefaction Under Partially Drained Conditions

Qian-Qian Hu[✉], Rui Wang, and Jian-Min Zhang

Department of Hydraulic Engineering, National Engineering Laboratory for Green & Safe Construction Technology in Urban Rail Transit, Tsinghua University, Beijing 100084, China
huqq15@mails.tsinghua.edu.cn

Abstract. The drainage condition of sand under cyclic and seismic loading often deviates from the ideal undrained and drained conditions, being partially drained. Drainage conditions can significantly influence the liquefaction behavior of sand. In order to study the liquefaction of sand under partially drained conditions, 2D discrete element method (DEM) cyclic biaxial simulations are conducted on granular materials with partial drainage. Volumetric strain is controlled in the simulations to achieve designated partially drained conditions. The simulation results show that drainage condition plays an important role in both the liquefaction susceptibility and the liquefaction induced deformation of sand.

Keywords: DEM · Liquefaction · Partially drained condition

1 Introduction

Drainage condition can be influential to the liquefaction susceptibility and deformation of sand under seismic and other cyclic loading. Soil is often considered as undrained during seismic events, due to the short duration of loading. However, this interpretation has been suggested to be unrealistic [1], especially in cases of layered soil with strong contrast in inter-layer permeability and prolonged cyclic loading (e.g. wave loading), where the drainage condition of the soil is often partially drained, between fully undrained and fully drained. Some existing laboratory tests and numerical simulations have assessed importance of drainage conditions on soil behavior [1–3]. In this study, 2D DEM is used to analyze the influence of drainage condition on soil liquefaction susceptibility and deformation.

2 DEM Simulations

In this study, a 2D DEM simulation package *PPDEM* is employed to conduct cyclic biaxial tests with various drainage conditions on polydisperse circular particle granular material. The details of the material used can be found in Wang et al. [4]. Rectangular shaped samples are used in the simulations, with two pairs of walls in the x and y directions used to apply stress boundary conditions.

The ratio ($d\varepsilon_v/d\gamma$) between the incremental volumetric strain and the incremental shear strain is controlled in the simulations to achieve the desired drainage conditions: When $d\varepsilon_v/d\gamma = 0$, fully undrained condition is achieved. When $d\varepsilon_v/d\gamma > 0$, the sample experiences forced contraction, where "pore water" is drained from the sample. When $d\varepsilon_v/d\gamma < 0$, the sample experiences forced expansion, where "pore water" is absorbed into the sample. The latter two conditions are both partially drained conditions. Compressive volumetric strain is considered positive.

26 cyclic biaxial simulations are conducted under initial mean effective stress of 100 kPa and deviatoric stress amplitude of 25 kPa, on samples with two different initial void ratios, as listed in Table 1.

Table 1. DEM simulations list

e	ID[a]	$d\varepsilon_v/d\gamma$	ID[a]	$d\varepsilon_v/d\gamma$	e	ID[a]	$d\varepsilon_v/d\gamma$	ID[a]	$d\varepsilon_v/d\gamma$
0.2125	e21C0	0			0.1946	e19C0	0		
	e21C1	0.001	e21C9	0.009		e19C1	0.001	e19D1	−0.001
	e21C2	0.002	e21D1	−0.001		e19C2	0.002	e19D2	−0.002
	e21C3	0.003	e21D1.5	−0.0015		e19C2.5	0.0025	e19D2.5	−0.0025
	e21C4	0.004	e21D2	−0.002		e19C3	0.003	e19D3	−0.003
	e21C5	0.005	e21D2.5	−0.0025		e19C4	0.004	e19D3.5	−0.0035
	e21C7	0.007	e21D5	−0.005		e19C5	0.005	e19D4	−0.004

[a]ID consists of prescribed void ratio, mode of partial drainage and the value of $d\varepsilon_v/d\gamma$. The letter "C" indicates that the sample is forced to contract and "D" stands for forced expansion. For example, e21C3 means the test had a prescribed void ratio of approximately 0.21, a partially drained mode of forced contraction and the value of $d\varepsilon_v/d\gamma$ of 0.003

Figure 1 plots the stress path and deviatoric stress-shear strain relationship in three simulations with different drainage conditions on samples with the same void ratio, including e21C4, e21C0 and e21D5, with $d\varepsilon_v/d\gamma = 0.004$, $d\varepsilon_v/d\gamma = 0$ and $d\varepsilon_v/d\gamma = -0.005$, respectively. The stress path and stress-strain curves before initial liquefaction (pre-liquefaction) are plotted with dashed lines, while those after initial liquefaction (post-liquefaction) are plotted with solid lines. Simulation e21C0 achieves initial liquefaction after 2 load cycles, while simulation e21C4 requires 2.5 load cycles to reach initial liquefaction, whereas e21D5 only requires 1.5 load cycles. This shows that drainage condition has a significant influence on the liquefaction susceptibility of soil under cyclic loading. When the material is allowed or forced to contract, it becomes less susceptible to liquefaction. In contrast, when the material is allowed or forced to dilate, it becomes more susceptible to liquefaction. The drainage condition is also observed to affect the deformation of soil after initial liquefaction. The double amplitude shear strain in simulation e21C0 in each load cycle increases initially and eventually reaches a stable value. Figure 1(a) shows that after a number of post-liquefaction cycles, forced contraction of the sample in simulation e21C4 eventually causes the effective stress to increase and the sample no longer liquefies, the shear strain generated in each load cycle also becomes smaller (dotted line). The maximum shear strain in simulation e21C4 is smaller than that in simulation e21C0. In simulation e21D5, as volumetric expansion is enforced, the shear strain is much greater, eventually cause the sample to experience uncontrolled flow slide.

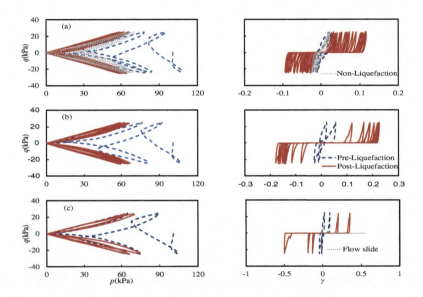

Fig. 1. Stress path and stress-strain relationship in simulations: (a) *e21C4*, (b) *e21C0*, (c) *e21D5*

The post-liquefaction shear strain γ_0, the amplitude of the post-liquefaction shear strain at zero effective stress [4], is plotted in Fig. 2 against the number of load cycles and the void ratio of the samples in simulations *e21C3*, *e21C2*, *e21C1*, *e21C0*, *e21D1*, and *e21D2*, with the same initial void ratio but different drainage conditions. When $d\varepsilon_v/d\gamma = 0$, γ_0 reaches a stable value after a certain number of load cycles and e remains unchanged. When $d\varepsilon_v/d\gamma < 0$, γ_0 is greater than that of the undrained condition, and both γ_0 and e continue to increase with increasing load cycle N. When $d\varepsilon_v/d\gamma > 0$, as load cycle N increases, γ_0 initially increases to a peak value, then γ_0 decrease. Greater $d\varepsilon_v/d\gamma$ leads to smaller γ_0.

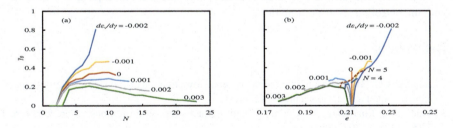

Fig. 2. (a) γ_0 development in reference to load cycles N under different $d\varepsilon_v/d\gamma$, (b) γ_0 development in reference to void ratio e under different $d\varepsilon_v/d\gamma$

The maximum post-liquefaction shear strain γ_{0max} achieved in all the simulations are plotted against $d\varepsilon_v/d\gamma$ in Fig. 3. When flow slide occurs, γ_{0max} tends to infinity. These results suggest that liquefaction with bounded post-liquefaction shear strain can only occur when the drainage of the material is within a certain range. Take the sample with initial void ratio of 0.1946 for example, when the value of $d\varepsilon_v/d\gamma$ exceeds 0.003, the sample does not reach liquefaction under the applied cyclic biaxial loading, and when the value of $d\varepsilon_v/d\gamma$ becomes less than −0.004, the sample experiences flow slide with uncontrolled shear deformation.

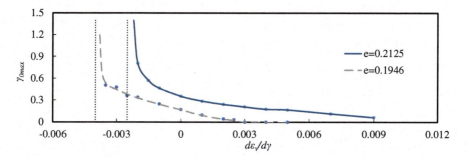

Fig. 3. γ_{0max} development in reference to $d\varepsilon_v/d\gamma$ when e equals to 0.2125 and 0.1946 respectively

3 Conclusion

The drainage condition significantly influences the liquefaction behavior of soils. When $d\varepsilon_v/d\gamma < 0$, where "pore water" is absorbed into the sample, liquefaction susceptibility and shear deformation of sand become greater than those under undrained condition. In contrast, when $d\varepsilon_v/d\gamma > 0$, where "pore water" is drained from the sample, liquefaction susceptibility and shear deformation of sand become smaller.

Liquefaction with bounded post-liquefaction shear strain can only occur when the drainage of the material is within a certain range. Under a certain cyclic load, when $d\varepsilon_v/d\gamma$ exceeds a certain value, the material is unable to reach zero effective stress, and when $d\varepsilon_v/d\gamma$ becomes smaller than a certain value, the material experiences flow slide with uncontrolled shear deformation.

References

1. Adamidis, O., Madabhushi, S.P.G.: Experimental investigation of drainage during earthquake-induced liquefaction. Géotechnique (in press)
2. Vaid, Y.P., Eliadorani, A.: Undrained and drained stress-strain response. Can. Geotech. J. **37**(5), 1126–1130 (2000)
3. Kamai, R., Boulanger, R.W.: Single-element simulation of partial-drainage effects under monotonic and cyclic loading. Soil Dynam. Earthq. Eng. **35**, 29–40 (2012)
4. Wang, R., Fu, P., Zhang, J.M., DEM Dafalias, Y.F.: Study of fabric features governing undrained post-liquefaction shear deformation of sand. Acta Geotech. **11**(6), 1321–1337 (2016)

A Hard-Sphere Model for Wet Granular Dynamics

Kai Huang[✉]

Experimentalphysik V, Universität Bayreuth, 95440 Bayreuth, Germany
kai.huang@uni-bayreuth.de

Abstract. Due to the large number of particles in geo-physical or engineering applications, simplified physical models for liquid mediated particle-particle interactions are desirable for predicting wet granular dynamics with numerical simulations. Based on an analysis of various energy dissipation mechanisms, I introduce a hard-sphere model for predicting wet granular dynamics.

1 Introduction

As large agglomerations of macroscopic particles with liquid mediated particle-particle interactions, wet granular materials are ubiquitous in nature (e.g. soil), industries (e.g. mining and construction materials) and our daily lives (e.g. ground coffee beans) [5]. Depending on the amount of liquid added, different wetting regimes with distinct particle-particle interactions arise: From the cohesionless dry to the completely immersed slurry state [4,8]. According to a previous experiment, tens of atomic layers liquid coverage is sufficient to enhance the rigidity of a granular material dramatically [7]. Therefore, one should always keep the influence of liquid in mind for applications on our blue planet largely covered with water.

From the 'microscopic' perspective of particle-particle interactions, the energy loss during wet particle impact can be influenced by more factors than capillary interactions [1,3]. Based on an analysis of various energy loss terms, an analytical prediction of the coefficient of restitution for wet particle impacts was introduced recently [9]. Using this model, it is possible to use event driven (ED) molecular dynamics (MD) simulations to predict 'macroscopic' collective behavior in wet granular materials such as wave propagation [6] and pattern formation [2].

2 Wet Particle Impact Model

As sketched in Fig. 1(a), the normal COR is defined as the relative rebound over the impact velocities $e_\mathrm{n} = v_\mathrm{o}/v_\mathrm{i}$. The sources of energy loss include inelastic deformation of the rigid bodies themselves and the wetting liquid. As shown in Fig. 1(b), the capillary force F_{cap} arises as soon as the sphere touches the liquid film (i.e., $s = \delta$ with liquid film thickness δ) at the end of regime I along with the formation of a liquid bridge. As the particle penetrates further into the liquid film till the minimum separation distance $s = s_1$ is reached at the end of regime II, the cohesive force increases until its maximum $\pi d \gamma \cos \phi$ [10] is reached, where γ and ϕ correspond to the surface tension and contact angle, respectively. However, the qualitative trend of how F_{cap} changes with s stays the same. For thin liquid films, an integration to the rupture distance δ_r leads to an estimation of the rupture energy of the liquid bridge [3]

$$\Delta E_{\text{cap}} \approx \pi \gamma \sqrt{2 V_b d} = \pi \gamma d^2 \sqrt{\frac{2\pi W}{3 \rho_0 N_c}}, \qquad (1)$$

In an agglomeration of wet particles, the bridge volume V_b can be estimated with $\pi d^3 W / (3 \rho_0 N_c)$, where the liquid content W (i.e., volume of the liquid added to that occupied by the wet granular material), mean packing density ρ_0 and coordination number N_c can be determined experimentally. For the saturated case, the liquid content W equals the porosity $(1 - \rho_0)$ of the sample. Note that this estimation is based on experimental and numerical investigations in the quasi-static limit.

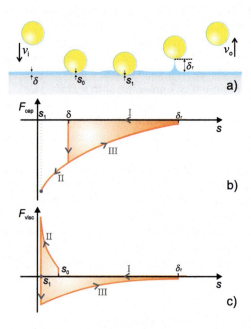

Fig. 1. (a) A sketch showing the impact of a wet sphere on a wet rigid plane with definitions of the associated length scales. (b) and (c) are schematics of capillary and viscous drag forces as a function of separation distance s, respectively. Negative force denotes attractive interactions. For both types of force, there exist three distinct regimes during the impact. The shaded hysteresis regions illustrate the energy loss during the impact.

For the dynamic case, we need to consider additionally the viscous drag force applied to the particle. In the regime where lubrication theory applies $s < s_0$, the viscous drag force reads $F_{\text{visc}} = 3\pi\eta d^2 v_i/(2s)$ with dynamic viscosity of the liquid η and s_0 the length scale at which lubrication theory starts to apply [3]. It contributes to both approaching (II) and departure (III) regimes during impact. As F_{visc} diverges for infinitely small s, it is intuitive to assume that most of the energy loss takes place between $s_1 < s < s_0$. Consequently, the corresponding energy dissipation can be estimated with [9]

$$\Delta E_{\text{visc}} \approx 3\pi\eta d^2 v_i \ln(s_0/s_1). \tag{2}$$

Moreover, the kinetic energy of the liquid being pushed aside also contribute to the energy loss. For spheres, this part of energy dissipation reads [9]

$$\Delta E_{\text{int}} \approx \frac{\pi}{24}\rho_l v_i^2 d^3 \left(\frac{W}{\rho_0} - \frac{W^2}{3\rho_0^2} + \frac{W^3}{27\rho_0^3}\right) \tag{3}$$

with liquid density ρ_l.

The above analysis suggests that except for the influence of capillary force, the other two energy dissipation factors are both impact velocity dependent and hence contribute to the total energy loss for the dynamic case. As the next step, I check the contributions of various energy loss terms with respect to the initial kinetic energy $E_i = \pi\rho_p d^3 v_i^2/12$ with particle density ρ_p.

As shown in Fig. 2, both $\Delta E_{\text{cap}}/E_i$ and $\Delta E_{\text{visc}}/E_i$ scale with d^{-1} with a constant ratio $\Delta E_{\text{visc}}/\Delta E_{\text{cap}}$ arising from the capillary number $\text{Ca} = v_i\eta/\gamma$ and W. For the dynamic case, the contribution from the viscous drag force is comparable to that from capillary interactions, thus its influence cannot be ignored.

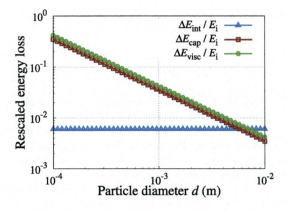

Fig. 2. Kinetic energy loss due to inertial ΔE_{int}, capillary ΔE_{cap} and viscous drag ΔE_{visc} forces rescaled by the initial kinetic energy E_i as a function of particle diameter for the impact of a water wetting glass bead with $v_i = 1\,\text{m/s}$, liquid content $W = 2\%$, packing density $\rho_0 = 0.64$, and the factor $\ln(s_0/s_1) \approx 3$ [9]. Other parameters are $d = 0.8\,\text{mm}$, $\rho_p = 2580\,\text{kg m}^{-3}$, $\rho_l = 1000\,\text{kg m}^{-3}$, $\eta = 1.0\,\text{mPa s}$, $\gamma = 0.072\,\text{Nm}^{-1}$ and $\phi = 0$.

However, the influence from liquid inertia can be ignored as $\Delta E_\text{int}/E_\text{i}$ stays constantly low at ~0.6%, a value dependent on $\rho_\text{l}/\rho_\text{p}$ and W in the system. For sand grains with $d \approx 0.3\,\text{mm}$, ΔE_visc may contribute to more than 10% of the initial kinetic energy. For centimeter sized particles, the inertia of the particle starts to dominate and hence the relative energy loss due to the wetting liquid decays to be less than 1%. This dependence on d indicates that a consideration of F_visc for millimeter sized or smaller particles is necessary in modeling wet granular dynamics.

In practice, we can implement the impact velocity dependent wet COR,

$$e_\text{wet} = e_\text{inf}[1 - 9\eta \text{St}_\text{c}/(\rho_\text{p} d v_\text{i})] \tag{4}$$

in an ED algorithm for predicting the collective behavior [6]. Note that the two control parameters e_inf and St_c can be predicted analytically [9]. This prediction includes all aforementioned energy dissipation mechanisms, and the implementation can be more efficient in comparison to DEM simulations because of the assumption of instantaneous collisions.

3 Conclusion and Outlook

To summarize, a wet particle impact model is briefly reviewed with a focus on the dependence of the liquid induced energy dissipation on particle size. It shows that viscous drag force should be considered in addition to capillary interactions for modeling wet granular dynamics. This hard-sphere model provides an alternative way of predicting wet granular dynamics with tuning velocity dependent COR based on 'microscopic' particle-particle interactions.

Acknowledgements. The author thanks Ingo Rehberg and Simeon Völkel for helpful discussions. This work is partly supported by the DFG through Grant No. HU1939/4-1.

References

1. Barnocky, G., Davis, R.H.: Elastohydrodynamic collision and rebound of spheres: experimental verification. Phys. Fluids **31**(6), 1324 (1988)
2. Butzhammer, L., Völkel, S., Rehberg, I., Huang, K.: Pattern formation in vertically agitated wet granular matter. Phys. Rev. E **92**, 012202 (2015)
3. Gollwitzer, F., Rehberg, I., Kruelle, C.A., Huang, K.: Coefficient of restitution for wet particles. Phys. Rev. E **86**(1), 011303 (2012)
4. Herminghaus, S.: Wet Granular Matter: A Truly Complex Fluid. World Scientific, Singapore (2013)
5. Huang, K.: Wet granular dynamics: From single particle bouncing to collective motion. Habilitation thesis (2014)
6. Huang, K.: Internal and surface waves in vibrofluidized granular materials: role of cohesion. Phys. Rev. E **97**, 052905 (2018)
7. Huang, K., Sohaili, M., Schröter, M., Herminghaus, S.: Fluidization of granular media wetted by liquid he4. Phys. Rev. E **79**(1), 010,301 (2009)

8. Mitarai, N., Nori, F.: Wet granular materials. Adv. Phys. **55**(1–2), 1–45 (2006)
9. Müller, T., Huang, K.: Influence of liquid film thickness on the coefficient of restitution for wet particles. Phys. Rev. E **93**, 042904 (2016)
10. Willett, C.D., Adams, M.J., Johnson, S.A., Seville, J.P.K.: Capillary bridges between two spherical bodies. Langmuir **16**(24), 9396–9405 (2000)

Distinct Element Modelling of Constant Stress Ratio Compression Tests on Pore-Filling Type Methane Hydrate Bearing Sediments

Mingjing Jiang[1,2,3(✉)], Wenkai Zhu[1,2,3], and Jie He[4]

[1] Department of Geotechnical Engineering, College of Civil Engineering, Tongji University, Shanghai 200092, China
mingjing.jiang@tongji.edu.cn
[2] Key Laboratory of Geotechnical and Underground Engineering of Ministry of Education, Tongji University, Shanghai 200092, China
[3] State Key Laboratory of Disaster Reduction in Civil Engineering, Tongji University, Shanghai 200092, China
[4] Shanghai Jianke Engineering Consulting Co., Ltd., Shanghai 200092, China

Abstract. The mechanical behavior of methane hydrate bearing sediments (MHBS) subjected to various stress-path loading is relevant to seabed subsidence or hydrate sediment strata deformation and well stability during the exploration and methane exploitation from methane hydrate (MH) reservoir. This study presents a numerical investigation into the mechanical behavior of pore-filling type MHBS along constant stress ratio path using the distinct element method (DEM). In the simulation, methane hydrate was modelled as agglomerates of sphere cemented together in the pores of soils. The numerical results indicate that the compressibility of MHBS is low before yielding but higher after. Bond breakage in MH agglomerates occurs during loading and the bond contact ratio decreases when the major principal stress increases. As for the fabric tensors, the minor principal fabric slightly increases to a constant value while the major principal fabric decreases to a constant value.

Keywords: Methane hydrate bearing sediments · Distinct element method · Pore-filling · Constant stress ratio compression test · Bond breakage

1 Introduction

Methane hydrate (MH) is a crystalline solid in which methane molecules are captured by cages formed by water molecules framework under high pressure and low temperature [1]. Methane hydrate bearing sediments (MHBS) is a soil deposit containing MH in its pores. The exploration of MH has attracted worldwide attention for its high energy density. However, MHs only exist stably under certain conditions and changes of pore pressure or temperature caused by gas exploitation may lead to MH dissociation, which may change the mechanical properties of MHBS and trigger a series of and geo-hazards. Therefore, a deep investigation on the mechanical behaviors of MHBS is necessary.

The mechanical behavior of MHBS is influenced by many factors, among which the formation of MHs in the pore of MHBS plays an important role. According to previous research, three formation habits of MHs have been summarized: pore-filling, cementation and load-bearing [2]. In the last decades, considerable laboratory tests have been conducted to study the mechanical properties of natural and artificial MHBS, such as triaxial compression tests [3, 4] and direct shear tests [5]. However, the mechanical behavior of MHBS is still insufficient, especially the microscopic mechanism related with the macroscopic behavior. The distinct element method (DEM) [6] is efficient to capture the evolution of microscopic particle interactions and thus was employed. In this study, a pore-filling types of MHBS was generated by a novel technique. The constant stress ratio (CSR) compression tests with different stress ratios are performed to analyze its volumetric strain and contact fabric.

2 Numerical Simulation

A new technique proposed by He and Jiang [7] was used to generate the pore-filling type MHBS samples. At first, a cubic sample with an initial void ratio of 0.84 was generated by multi-layer under-compaction method (UCM) [8]. The size of the sample is 5.445 × 5.445 × 5.445 mm^3. The grain size distribution is similar to that of Toyoura sand.

After the generation, N particles with radius of R_a was randomly created in the pores of soil particles according to the MH saturation S_{mh} which is defined by the ratio of MH volume V_{mh} to the total pore volume V_v. Subsequently, the particle with radius of R_a was replaced by a near-spheroidal cluster, which was formed by bonding n MH particles with radius r together. The void ratio of the cluster is e_a. The number of MH clusters N can be calculated by:

$$N = \frac{V_{mh}}{\frac{4}{3}n\pi r^3 \times (1+e_a)} \quad (1)$$

In this study, the void ratio of the MH cluster e_a is 0.82, the radius of the MH particle r is 0.024 mm and the number of MH particles in a cluster n is set to 60. When S_{mh} = 10%, the number of MH clusters N is 1166. The parallel bond model (BPM) is employed in the cluster and the material parameters used in this paper is the same as in [9].

3 Numerical Results

In CSR tests, the intermediate principal stress σ_2 and the minor principal stress σ_3 were kept equal and the stress ratio η was maintained constant, i.e., $\sigma_2 = \sigma_3 = \eta\sigma_1$. The load was applied step by step. In the first step, σ_1 was set to 1.0 MPa and the latter σ_1 was 1.15 times the former σ_1 during the subsequent loading. To investigate the influence of stress ratio, three different stress ratio, i.e., 0.5, 0.7, 0.9, were used.

Figure 1 shows relationships between void ratio and major principal stress of MHBS with S_{mh} = 10% under different stress ratios. The void ratio decreases when the major principal stress increases. And the compressibility of pore-filling type MHBS is low before yielding but higher after, though the yield point is not obvious. Besides, the void ratio is lower when stress ratio is higher, which indicates that the yield of the MHBS sample is related to stress path.

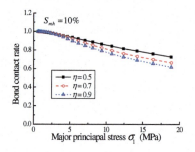
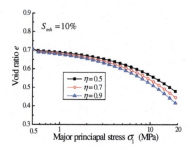

Fig. 1. Relationships between void ratio and major principal stress of MHBS with $S_{mh} = 10\%$ under different stress ratios.

Fig. 2. Relationships between bond contact rate and major principal stress of MHBS with $S_{mh} = 10\%$ under different stress ratios.

Relationships between bond contact ratio and major principal stress of MHBS with $S_{mh} = 10\%$ under different η is illustrate in Fig. 2. The bond contact ratio is defined as the ratio of bond contact number to initial bond contact number. The bond contact ratio decreases when the major principal stress increases.

The fabric tensor is defined based on Satake [10] to describe the soil anisotropy:

$$F_{ij} = \frac{1}{N_c} \sum_{k=1}^{N_c} n_i^k n_j^k, i = j = 1, 3 \qquad (2)$$

where N_c is the contact number, n_i^k is the component of the unit vector n^k at a contact.

The fabric tensor for the strong contacts is described by [11]:

$$F_{ij}^s = \frac{1}{N_c^s} \sum_{s=1}^{N_c^s} n_i^s n_j^s, i = j = 1, 3 \qquad (3)$$

where N_c^s is the number of the strong contacts and n_i^s is the component of unit vector n^s at a strong contact.

The development of the eigenvalues of contacts (F_{11}, F_{22}, F_{33}) for MHBS samples with $S_{mh} = 10\%$ when $\eta = 0.7$ during the test is presented in Fig. 3(a). It can be seen that the minor principal fabric F_{33} slightly increases to a constant value while the major principal fabric F_{11} slightly decreases to a constant value. As for the eigenvalues of the strong contacts ($F_{11}^s, F_{22}^s, F_{33}^s$), shown in Fig. 3(b), F_{11}^s reaches a peak value then decreases, while F_{22}^s and F_{33}^s decrease to a minimum value then increase.

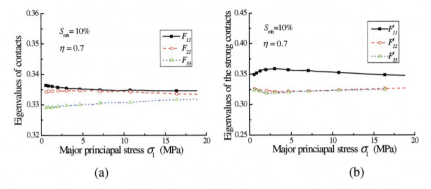

Fig. 3. The development of the eigenvalues of (a) contacts and (b) the strong contacts for MHBS samples with $S_{mh} = 10\%$ when $\eta = 0.7$.

References

1. Max, M.D., Lowrie, A.: Oceanic methane hydrates: a "frontier" gas resource. J. Pet. Geol. **19**(1), 41–56 (1996)
2. Waite, W.F., Santamarina, J.C., Cortes, D.D., et al.: Physical properties of hydrate-bearing sediments. Rev. Geophys. **47**, 38 (2009)
3. Hyodo, M., Nakata, Y., Yoshimoto, N., et al.: Basic research on the mechanical behavior of methane hydrate-sediments mixture. J. Jpn. Geotech. Soc. Soils Found. **45**(1), 75–85 (2005)
4. Miyazaki, K., Tenma, N., Aoki, K., et al.: A nonlinear elastic model for triaxial compressive properties of artificial methane-hydrate-bearing sediment samples. Energies **5**(12), 4057–4075 (2012)
5. Liu, Z., Dai, S., Ning, F., et al.: Strength estimation for hydrate-bearing sediments from direct shear tests of hydrate-bearing sand and silt. Geophys. Res. Lett. **45**(2), 715–723 (2018)
6. Cundall, P.A., Strack, O.D.L.: A discrete numerical model for granular assemblies. Géotechnique **30**(30), 331–336 (1979)
7. He, J., Jiang, M.J.: Three-dimensional distinct element novel sample-preparing method and mechanical behavior for pore-filling type of methane hydrate-bearing soil. J. Tongji Univ. (Nat. Sci.) **44**(5), 709–717 (2016). (in Chinese)
8. Jiang, M.J., Konrad, J.M., Leroueil, S.: An efficient technique for generating homogeneous specimens for DEM studies. Comput. Geotech. **30**(7), 579–597 (2003)
9. He, J., Jiang, M.J., Liu, J.: Effect of different temperatures and pore pressures on geomechanical properties of pore-filling type of methane hydrate soils based on the DEM simulations. In: Li, X., et al. (eds.) Proceeding of the 7th International Conference on Discrete Element Methods 2017, SPPHY, vol. 188, pp. 827–835. Springer, Singapore (2017)
10. Satake, M.: Fabric tensor in granular materials. In: Vermeer, P.A., Luger, H.J. (eds.) Proceedings of IUTAM Symposium on Deformation and Failure of Granular Materials 1982, pp. 63–68. Balkema, Netherlands (1982)
11. Kuhn, M.R., OVAL and OVALPLOT: Programs for analyzing dense particle assemblies with the discrete element method. http://faculty.up.edu/kuhn/oval/doc/oval_0618.pdf. Accessed 28 Feb 2018

Investigating the Mechanical Behavior of Pore-Filling Type Methane Hydrate Bearing Sediments by the Discrete Element Method

Mingjing Jiang[1,2,3(✉)], Jun Liu[1,2,3], and Jie He[4]

[1] Department of Geotechnical Engineering College of Civil Engineering, Tongji University, Shanghai 200092, China
mingjing.jiang@tongji.edu.cn
[2] Key Laboratory of Geotechnical and Underground Engineering of Ministry of Education, Tongji University, Shanghai 200092, China
[3] State Key Laboratory of Disaster Reduction in Civil Engineering, Tongji University, Shanghai 200092, China
[4] Shanghai Jianke Engineering Consulting Co., Ltd., Shanghai, China

Abstract. This study presents a numerical investigation into the mechanical behavior of pore-filling type methane hydrate bearing sediments (MHBS) using the discrete element method. Based on the available experimental data of methane hydrate (MH) samples, MH is modeled as strong bonded clusters. A series of drained triaxial compression tests are simulated and the results show that for the pore-filling type MHBS with MH saturation from 5% to 20%, a higher MH saturation induces a slightly higher shear strength and slightly larger dilatancy. The inclination of the critical state line on the $q - p'$ plane increases when MH saturation increases. In addition, the evolution rule of deviator fabrics of the strong contact is similar to that of stress ratio q/p'-axial strain.

Keywords: Methane hydrate bearing sediments · Discrete element method Pore-filling · Drained triaxial tests

1 Introduction

Methane hydrate (MH) is regarded as a new energy resource to alleviate current and future energy needs [1, 2]. MHs usually form in deep seabeds and permafrost regions where MHs can remain stable and can enhance the strength of the host sediments [3]. However, human interventions may give rise to dissociation of MHs, which may further lead to large marine landslide and other hazards such as instability of submarine pipeline and drilling platform. Therefore, it is crucial to study the mechanical properties of methane hydrate bearing sediments (MHBS) for submarine disaster prevention and MH exploitation in the future.

Some pioneering works have been carried out to investigate the mechanical properties of MHBS through a number of experimental and numerical studies [4–7]. It has been revealed that the shear strength, dilation and stiffness of MHBS are influenced by a series of factors such as MH saturation, confining pressure and temperature etc. Typically, the morphology of MH in sandy MHBS controls the mechanical properties of MHBS from microscopic origin and can be categorized into four types, i.e. pore-filling, load-bearing, cementation, and grain-coating [7]. To explore the isolation effect of pore-filling MH on the mechanical properties of MHBS, the discrete element method (DEM) [8] is used in this study to carry out a series of drained triaxial compression tests on MHBS.

2 Numerical Simulation

To generate pore-filling type MHBS, a homogeneous DEM sample with an initial void ratio of 0.84 was generated. The size of the DEM sample is 4.322 × 4.322 × 8.644 mm^3. After that, MH particles are generated within the scope of a sphere with radius of R_a, and then cemented together as a whole, forming a cluster of MH. A cluster consists $n = 60$ small MH particles with radius of $r = 0.024$ mm. The total number of cluster N in the MHBS sample with hydrate saturation S_{MH} is decided by:

$$N = \frac{V_{MH}}{\frac{4}{3}\pi r^3 \times n \times (1+e_a)} \quad (1)$$

$$S_{MH} = \frac{V_{MH}}{V} \quad (2)$$

where V is the volume of the MHBS sample and $e_a = 0.82$ is the void ratio of MH cluster. Note that N sphere particles with radius R_a were firstly generated in the pore of soil sample, subsequently, 60 small particles were generated within the scope of large sphere, which was deleted after small particles were generated. The small particles within each scope were bonded together by the parallel bond model to form a cluster.

After the MHBS sample was generated, drained triaxial compression tests were performed with different effective confining pressures and hydrate saturations.

3 Results

Figure 1 illustrates the stress-strain and volumetric responses of MHBS samples under effective confining pressure of 2 MPa. Figure 1(a) shows that the deviator stress and elastic modulus of MHBS gradually increase with S_{MH}. Figure 1(b) shows that a larger S_{MH} induces a slightly higher dilatancy.

Figure 2 illustrates the critical state lines of MHBS with different S_{MH} on the $q - p'$ plane. It can be observed that the inclination of the critical state line slightly increases with the increase of S_{MH}.

Figure 3 presents relations between the stress ratio q/p', deviator fabrics of the strong contact $F_{11}^s - F_{33}^s$ and axial strain. The unit normal vector for the strong contacts is described by Kuhn [9]:

$$F_{ij}^{s} = \frac{1}{N_c} \sum_{s=1}^{N_c^s} n_i^s n_j^s, \quad i=j=1,d \tag{3}$$

where N_c is the number of contact, N_c^s is the number of the strong contacts, n_i^s is the component of the unit vector at a strong contact, and d is equal to 3 for 3D.

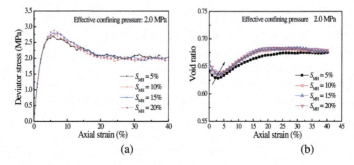

Fig. 1. Macroscopic behavior of MHBS under triaxial compression: (a) stress-strain relationships and (b) volumetric responses.

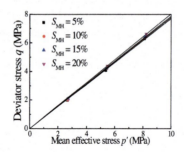

Fig. 2. Critical state lines on $q - p'$ plane.

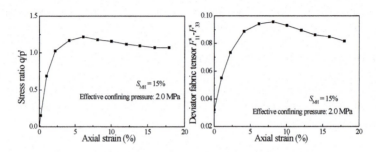

Fig. 3. Relations between stress ratio q/p', deviator fabrics of the strong contact $F_{11}^s - F_{33}^s$ and axial strain.

Figure 3 shows that the evolution rule of deviator fabrics of the strong contact $F_{11}^s - F_{33}^s$ is similar to that of stress ratio q/p'-axial strain. The macro behavior changes along with the micro response.

4 Conclusions

The DEM simulation can effectively capture the macro-mechanical response of stress–strain and volumetric response of pore-filling type MHBS with different hydrate saturations. As the hydrate saturation increases from 5% to 20%, the strength, elastic modulus and dilatancy of MHBS gradually increase. The evolution rule of deviator fabrics of the strong contact $F_{11}^s - F_{33}^s$ is similar to that of stress ratio q/p'-axial strain.

Acknowledgements. This work was funded by the National Natural Science Foundation of China with Nos. 51639008 and 51579178, Major Project of Chinese National Programs for Fundamental Research and Development (973 Program) with No. 2014CB046901. All of these supports are greatly appreciated.

References

1. Collett, T.S., Kuuskraa, V.A.: Hydrates contain vast store of world gas resources. Oil Gas J. **96**(19), 90–95 (1998)
2. Kvenvolden, K.A.: Methane hydrate—a major reservoir of carbon in the shallow geosphere? Chem. Geol. **71**(1–3), 41–51 (1988)
3. Waite, W.F., et al.: Physical properties of hydrate-bearing sediments. Rev. Geophys. **47**, 465–484 (2009)
4. Masui, A., Haneda, H., Ogata, Y., Aoki, K.: Effects on methane hydrate formation on shear strength of synthetic methane hydrate sediments. In: Proceedings of the 15th international offshore and polar engineering conference, pp. 364–369. International Society of Offshore and Polar Engineers, Seoul (2005)
5. Hyodo, M., Yoneda, J., Yoshimoto, N., Nakata, Y.: Mechanical and dissociation properties of methane hydrate-bearing sand in deep seabed. Soils Found. **53**, 299–314 (2013)
6. Yu, Y., Cheng, Y.P., Xu, X., Soga, K.: Discrete element modelling of methane hydrate soil sediments using elongated soil particles. Comput. Geotech. **80**, 397–409 (2016)
7. Shen, Z.F., Jiang, M.J.: DEM simulation of bonded granular material. Part II: extension to grain-coating type methane hydrate bearing sand. Comput. Geotech. **75**, 225–243 (2016)
8. Cundall, P.A., Strack, O.D.L.: A discrete numerical model for granular assemblies. Géotechnique **29**(1), 47–65 (1979)
9. Kuhn, M.R., OVAL and OVALPLOT: Programs for analyzing dense particle assemblies with the discrete element method. http://faculty.up.edu/kuhn/oval/doc/oval_0618.pdf. Accessed 28 Feb 2018

A New Bottom-Up Strategy for Multiscale Studying of Clay Under High Stress

Lianfei Kuang[1,2(✉)], Guoqing Zhou[1], and Yazhou Zou[2]

[1] State Key Lab for Geomechanics and Deep Underground Engineering,
China University of Mining and Technology, Xuzhou 221116, Jiangsu, China
klfcumt@126.com
[2] Institute of Soil Mechanics and Foundation Engineering,
University of Bunderswehr Munich, 85577 Munich, Germany

Abstract. Greater demands are being placed on studying the mechanical properties of clay under high stress (>1 MPa) by growing construction of deep underground engineering (>500 m). Moreover, as a kind of porous media, the clay particle properties span a wider range. Consequently, in this work, we present a new bottom-up strategy for multiscale studying of clay. At microscale, the structure of interlayer species in hydrated clay minerals were systematically studied by molecular dynamics simulation (MD). Further by mapping the interaction parameters from MD results, larger hydrated clay mineral systems were simulated by dissipative particle dynamics (DPD) at mesoscale. The morphology and the structure of system at microscale and mesoscale were finally used as evidences to reveal the basic mechanisms of mechanical response for clay under high stress by theoretical analysis at the macroscale. As a message-passing approach, the force filed (FF) parameters of this bottom-up multiscale method are rationalistic and have definite physical meaning. The strategy suggested here provides a new though for the multiscale study of geotechnical materials.

Keywords: Multiscale · High stress · Clay · Molecular simulation

1 Introduction

At present, large-scale deep mine shafts are being constructed in China [1], it needs to research the basic composition of deep clay, the mechanical response characteristics under high overlying pressure and the internal control mechanism of its strength and deformation. On the other hand, the worldwide deep underground nuclear waste storage projects are carried out like a raging fire, it is also urgent to study the high pressure physical and mechanical properties of bentonite which is an ideal candidate buffer material and the diffusion of radioactive nuclides under actual conditions.

Under these backgrounds, some scholars [2, 3] have carried out high pressure compression and shear tests on deep clay and bentonite. The experiments show that the mechanical response of clay under high pressure will be significantly different from normal pressure conditions, for example, the $e \sim \log P$ curves of deep clay and bentonite all show obvious bilinear characteristics, of which 0.5–2.0 MPa is the turning range. And the strength envelopes of deep clay shear are also characteristic of two fold lines, which converge near 1.6–1.8 MPa. It is preliminarily considered that this bilinear representation has relations with the change of soil water characteristics and clay microstructure under different load level, and the key nodes of this change close to 2.0 MPa. Under high pressure, the clay crystal lamellar will be compressed into a few nanometers, and the layer number of interlaminar water molecules is very limited (≤ 4). This microstructure and the interaction between water and mineral surface cannot be described in detail by the recently popular discrete element method (DEM) at the mesoscale. Consequently, only microscopic description of the essential physical and mechanical responses of clay mineral can further reveal its mesoscale and macroscale mechanical characteristics from bottom to top, and the bottom-up approach will be more rationalistic.

2 Multiscale Molecular Modelling Strategy

In geotechnical engineering, the traditional finite element method (FEM) which is based on continuum theory has encountered great difficulties in dealing with discontinuous interfaces. Thus, discontinuous numerical method has been developed, for example, the DEM which is typically represented by the program of particle flow code (PFC). Contact model is the core of DEM calculation, However, due to the top-down development history, the parameters of mesoscopic discrete element contact model are mostly determined by regression fitting of macroscopic tests. This fitting method has a certain phenomenology, and the physical meaning of some parameters is not explicit. Therefore, the bottom-up multiscale approach which is based on accurate first principles quantum scale, and passing information into molecular scales and eventually to macroscopic scales is a new though for the study of rock-soil mass [4]. The multiscale simulation protocol is diagrammed in Fig. 1.

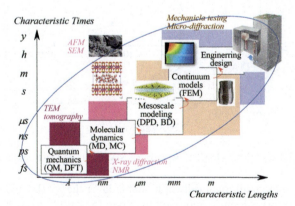

Fig. 1. General schema of a unified message-passing multiscale approach

At the lower quantum scale ($\sim 10^{-10}$ m and $\sim 10^{-12}$ s), methods classified as quantum mechanics (QM) can be employed to treat electrons explicitly and accurately, and to optimize value for the FF parameters at the atomic level. At the atomistic scale (~10^{-9} m and ~$10^{-9} - 10^{-6}$ s), MD and Monte Carlo (MC) simulations are usually performed employing FF to study microstructure. And the binding energies among components as the DPD input parameters can be further mapped from MD simulations results, at the mesoscale (~10^{-6} m and ~$10^{-6} - 10^{-3}$ s), various simulation methods have been proposed to study the mesoscale structures and mechanical properties, the most common being DPD and Brownian Dynamics (BD). Eventually, the mesoscopic structure is transferred to FEM tools to calculate properties for the systems of interest at the upper macroscale (~10^{-3} m and ~$10^0 - 10^1$ s). The ultimate goal of the multiscale modelling is, hence, to predict the macroscopic behavior of material for engineering design. In this protocol, it is important to be able to compare the simulation results with experimental evidences at each scale, fortunately, these methods available now allows the comparison along the entire multiscale procedure, as also shown in Fig. 1.

3 Results and Discussion

The application of the step-by step message passing multiscale modelling on the study of the basic mechanisms that govern the high stress compression mechanical properties of saturated montmorillonite (MMT) is presented and discussed as followed. The results of Yong's test are selected for comparison discussion, the maximum pressure of these test is up to 400 MPa, which covers the whole process of high pressure compression of clay [5], as shown in Fig. 2.

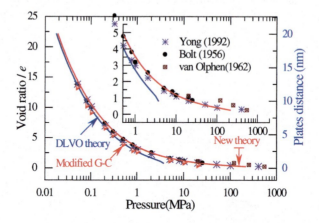

Fig. 2. Comparison of the $e \sim \log P$ curves for Na-MMT between tests and theory calculation

According to the traditional diffusion double layer theory (DLVO theory) and the modified G-C theory which is improved by Yong [5] based on Stern model, the low pressure test data can be described better by the theory calculation. But as the van der Waals (VDW) playing a role, the total force turns from repulsion to attraction, and the force direction show reversal. These theory curves continue to deviate from test data.

According to our new theory, the theoretical compression curve can well spread across the entire range of measured results. When the layer number of the interlayer water is as large as 4, its characteristics are still show as adsorbed water according to the microscale MD simulation. And the mesoscale DPD simulation result shows the interlayer water at content of 78.4% is close to the free water, where the layer number is about 7. The calculated pressure is 1.652–11.5 MPa that corresponding the spacing layer number from 7 to 4 layers. And this pressure range is consistent with the bilinear turning pressure of about 2 MPa. That is to say, with the increase of pressure, the water that extruded from the clay transforms from free water to adsorbed water, thus, the compression curve is characterized by a bilinear feature. When there are only 2 layers and 3 layers of water molecules between the particle crystal lamellar, the corresponding pressure is 52.46 MPa and 21.61 MPa respectively. These quantitative pressures indicate that under the maximum pressure of 40 MPa which is corresponding to the backgrounds of maximum depth of 2 km mine shaft construction and nuclear waste storage projects, the interlayer water molecules cannot be removed completely, but remain be extruded partially, that is, the partially dehydrated of bound water govern the basic mechanism of high stress compression for saturated clay.

4 Conclusions

In this study, a new bottom-up strategy for multiscale studying of clay is proposed. At microscale, the microstructure and nano-mechanical properties of clay minerals are systematic investigated by MD simulation. By message-passing approach, the morphology and structure are further simulated by DPD at the mesoscale. And these microscale and mesoscale results were finally used as evidences to reveal the basic mechanisms of mechanical response for clay under high stress by theoretical analysis at macroscale.

Acknowledgements. The study presented in this article are supported by the National Natural Science Foundation of China (51504245, 41502271, 41772338, 51408595), and Jiangsu Provence Postdoctoral Science Foundation (1402009B). These supports are gratefully acknowledged.

References

1. Zhou, G.Q., Kuang, L.F., Ma, J.R., et al.: Status quo and prospects of the deep soil mechanical properties. J. Chn. Univ. Min. Technol. **45**(2), 195–204 (2016)
2. Shang, X.Y., Zhou, G.Q., Kuang, L.F., et al.: Compressibility of deep clay in East China subjected to a wide range of consolidation stresses. Can. Geotech. J. **52**(2), 244–250 (2015)
3. Marcial, D., Delage, P., Cui, Y.J.: On the high stress compression of bentonites. Can. Geotech. J. **39**(4), 812–820 (2002)
4. Fermeglia, M., Posocco, P., Pricl, S.: Nano tools for macro problems: multiscale molecular modeling of nanostructured polymer systems. Compos. Interfaces. **20**(6), 379–394 (2013)
5. Yong, R.N., Mohamed, A.-M.O.: A study of particle interaction energies in wetting of unsaturated expansive clays. Can. Geotech. J. **29**(6), 1060–1070 (1992)

Evaluation of Grain Shape Parameters of Fujian Standard Sand

Shuai Li, Wan-Huan Zhou[✉], Xue-Ying Jing, and Hua-Xiang Zhu

Department of Civil and Environmental Engineering,
University of Macau, Macau, China
hannahzhou@umac.mo

Abstract. Grain shape is one of the most significant factors that influence the mechanical properties of sand material, e.g. shear strength, compactness. In this study, five groups of Fujian standard sand samples with different ranges of particle sizes were characterized respectively. The digital photos of samples on the sensitive plate have been processed by using the image recognition technology and the two-dimensional (2D) contour of each sand particle has been extracted. The geometries of the sands have been characterized based on the statistical analysis of the geometric parameters of a large number of grains. Finally, the sand grain shape is evaluated in terms of five common grain shape parameters, including the length-width ratio, flatness, roundness, practical sphericity and sphericity.

Keywords: Grain shape parameters · Fujian standard sand

1 Introduction

Sand material is commonly used in the geotechnical engineering. In the framework of conventional continuum mechanics, soil material is treated as a continuum material. However, the soil in nature is discrete. Thus, in the particle-scale perspective, the properties of individual soil particle, i.e. particle size and particle shape significant affect the behavior of soil assembly [1, 2]. To correlate the shape parameter to mechanical property of soil assembly, it is necessary to characterize the grain shape. Nevertheless, it is a difficult research problem due to three main reasons. Firstly, the shape of particle is mainly determined by environment condition, such as mother rock composition, rock weathering, handling, deposition and so on, which makes it difficult to sum up shape variation rules [3]. The second reason is that different research unities have different opinions on the description of sand shape parameters [4, 5]. So far, there is not a unified evaluation standard. The last reason is lacking efficient and accurate methods to extract a large number of sand shape parameters because of small sizes and diversity of sand material. As Fujian standard sand is widely used in practical engineering and laboratory experiments, this paper will characterize and analyze diverse grain shape parameters of this sand. The python image recognition program library named opencv (Open Source Computer Vision) is used to extract shape parameters of samples with different grain sizes from digital photos. Then five grain shape parameters are quantitatively analyzed and evaluated.

2 Description and Selection of Shape Parameters

The basic geometric parameters have been summarized in Table 1, which illustrate the diagrammatic sketch of relevant geometric dimensions [6, 7].

Table 1. Basic geometric parameters of sand grains

Parameter	Description
A=Area	Particle image projection area
L=Length	The maximum distance between outer contour points
Width	$B = \dfrac{4A}{\pi L}$
P=Perimeter	Particle image projection perimeter
Equivalent circle radius	$R = \sqrt{A/\pi L}$
Maximum/ Minimum Feret diameter	Fm/Fn
Maximum inscribed circle radius	Rm
Minimum circumcircle radius	Rn

Image analysis and identification method were used to digitize 3D solid grain into 2D images. Five kinds of shape parameters are selected for analyzing, including length-width ratio, flatness, roundness, practical sphericity and sphericity. Table 2 lists these five concerned parameters and their corresponding formulas [7, 8].

Table 2. Particle shape evaluation parameters

Parameters	The length-width ratio	Flatness	Roundness	Practical sphericity	Sphericity
Calculation formula	$\alpha = \dfrac{L}{B}$	$\pounds = \dfrac{F_m}{F_n}$	$\beta = \dfrac{2\sqrt{\pi A}}{P}$	$sp = \dfrac{R}{R_n}$	$s = \dfrac{R_m}{R_n}$

3 Acquisition and Processing of Sample Grain Image

Five groups number one to five of Fujian standard sand were obtained by vibratory sieving. The grain size for each group corresponds to 0.85–1, 1–1.18, 1.18–1.45, 1.45–1.66 and 1.66–2 (mm) respectively. The image recognition technology allows processing a large number of grains each time, about 400 grains.

The high-quality photos of sand grains on shadowless LED flat lamp were taken by 24.3-million pixel cameras. Python and opencv are used to identify pixel point from digital images. We set a threshold value to divide the image data into two parts, black and white, called binarization processing. Then, we can get coordinates by identifying the reference ruler and converting the external contour into a polygon consisting of a set of arrays. After field segmentation, target separation and other operations, the coordinates of all grains in each photo can be extracted.

4 Assessment of Shape Parameters

Concentrated interval means the shape parameters in this range accounted for more than 95% of the total number of grains. As shown in Fig. 1, there is a common phenomenon that shape parameter values, for all the five groups, concentrate on a specific interval. Whereas, their average values do not at center of the interval, which means that they belong to the skewed distribution. The length of concentrated interval (LCI) of different shape parameters can stand for sensitivity degree to particle shapes description.

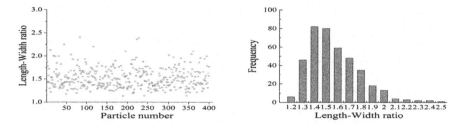

Fig. 1. The distribution of length-width ratio from size 1.45 mm–1.66 mm

The Coefficient of variation (COV) is the ratio of variance to average value, which is frequently used to reflect dispersion degree of a set of data. Thus, the shape parameters sensibility degree can be reflected by COV. As shown in Fig. 2 the sensitivity degrees of roundness and practical sphericity are relatively lower and the

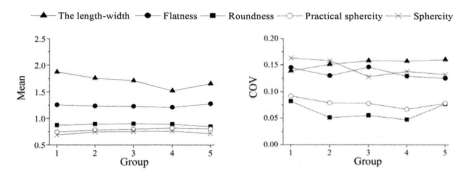

Fig. 2. Mean of shape parameters and COV varying with the grain size

Table 3. Mean length of concentrated interval of each shape parameter

Shape parameters	The length-width ratio	Flatness	Roundness	Practical sphericity	Sphericity
Mean of LCI	0.93	0.57	0.228	0.24	0.348
Mean of COV	0.153	0.135	0.0624	0.0788	0.1438

length-width ratio has a high sensitivity degree to particle shapes description. As shown in Table 3, the mean of COV of the length-width ratio, flatness and sphericity are higher than those of the roundness and sphericity, which mean that their sensibility to description of particle shapes is different. Moreover, the two parameters' extent of reaction to deviating from spheroidal particles is also worse than that of other three parameters.

5 Conclusions

According to the shape parameter analysis, two main conclusions can be drawn: (1) the five parameters obey skewed distribution. When the grain diameter ranges between [0.85 mm–1.66 mm], particle shapes close to standard ball with increase of grain diameters. In contrast, when grain diameter ranges between [1.66 mm–2 mm], the irregularity of particle shapes increases, and (2) sensibility of flatness and roundness are at intermediate level by analyzing CI length and COV. Roundness and practical sphericity have the weakest sensibility to particle shapes descriptions.

Acknowledgments. The authors wish to thank the financial support from the Macau Science and Technology Development Fund (FDCT) (125/2014/A3), the University of Macau Research Fund (MYRG2017-00198-FST, MYRG2015-00112-FST).

References

1. Jing, X.Y., Zhou, W.H., Zhu, H.X., Yin, Z.Y., Li, Y.: Analysis of soil-structural interface behavior using three-dimensional DEM simulations. Int. J. Numer. Anal. Methods Geomech. (2018)
2. Zhu, H., Zhou, W.H., Yin, Z.Y.: Deformation mechanism of strain localization in 2D numerical interface tests. Acta Geotech., 1–17 (2017)
3. Folk, R.L.: Petrology of Sedimentary Rocks. Hemphill Publishing Company, Austin (1980)
4. Ehrlich, R., Weinberg, B.: An exact method for characterization of grain shape. J. Sediment. Res. **40**(1) (1970)
5. Cho, A.G., Dodds, J., Santamarina, J.C.: Particle shape effects on packing density, stiffness, and strength: natural and crushed sands. J. Geotech. Geoenviron. Eng. **132**(5), 591–602 (2006)

6. Hentschel, M.L., Page, N.W.: Selection of descriptors for particle shape characterization. Part. Part. Syst. Charact. **20**(1), 25–38 (2003)
7. Chen, H.Y., Wang, R., Li, J.G., Zhang, J.M.: Grain shape analysis of calcareous soil. Rock Soil Mech. **26**(9), 1389 (2005)
8. Liu, Q., Xiang, W., Budhu, M.: Study of particle shape quantification and effect on mechanical property of sand. Rock soil Mech. **32**(1), 191–197 (2011)

DEM Simulation of Oedometer Tests with Grain Crushing Effects

Jia Lin[1], Erich Bauer[2], and Wei Wu[1]

[1] Institution of Geotechnical Engineering, University of Natural Resources and Life Sciences, Vienna, Austria
jia.lin@boku.ac.at
[2] Institution of Geotechnical Engineering, Graz University of Technology, Graz, Austria

> **Abstract.** The grain crushing effects in oedometer tests are studied in this paper with discrete element method (DEM) in both 2D and 3D. The crushing of grain is modeled with a novel method which combines the replacement method and agglomerate method together. The results of oedometer tests with and without grain crushing are compared.
>
> **Keywords:** Grain crushing · DEM · Oedometer test

1 Combined Method to Model Grain Crushing in DEM

There are generally two kinds of methods to model grain crushing with DEM, replacement methods and agglomerate methods. The replacement methods use a predefined failure criterion to check whether particles of the grain ensemble break under the given arrangement of contact forces. If the failure criterion is met, the grain is deleted and subsequently replaced by smaller grains. Replacement methods have been used for instance in [2, 4, 9, 10]. On the other hand, the agglomerates methods employ crushable grains built up from a large number of smaller grains, which are connected by breakable bonds. The failure criterion is checked implicitly, through the strength-check of the bonds. Examples of using agglomerates methods are given in [3, 5–8, 11, 12].

However, the replacement methods cannot model the actual crushing process of grains accurately and the agglomerates methods are very slow. Hence, we proposed a combined method to model grain crushing with DEM. As the name implies, this method combines the replacement methods and the agglomerates methods together. The basic idea of the combined method is to replace the high-stressed particles with agglomerates before numerical simulation of their breakage. In this way, the crushing of grains can be accurately modeled by breakage of bonds and crushing of agglomerates. At the same state, particles with low stress will not be replaced, which saves computational resources and time. An own subroutine is written to realize this method in particle flow code (PFC3D and PFC2D) (Fig. 1).

The subroutine checks the stress of each particle at a certain time interval. We use the second stress invariant as a measurement of the stress state of each particle. If J2 of a particle is greater than the prescribed value of the breakage condition, the particle will be replaced by a corresponding agglomerate. An agglomerate is a set of smaller particles bonded together with so-called parallel bonds. The size of the small particles in the agglomerate can be chosen freely.

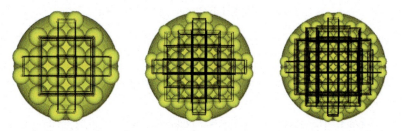

Fig. 1. Agglomerates with different size of small particles.

No additional rule needs to be defined for the breaking process of the agglomerates, since the breakage depends on the forces acting on the agglomerate and the strength of the bonds. Even if the strengths of bonds are the same for all particles forming an agglomerate, the stress states of each agglomerate is different and is changing during the simulations, which will lead to different breaking mechanisms for each agglomerate.

2 Oedometer Tests in 2D and 3D

The combined method is applied to DEM simulations of oedometer tests in both 3D and 2D. For 3D simulation, a cylindrical shaped wall with a height of 40 mm and diameter of 20 mm are generated, which is filled with 13199 particles with diameters from 0:8 to 1:2 mm. For 2D simulations, a rectangular box with a width of 20 mm and height of 40 mm is filled with 20000 particles (diameters from 0.16 mm to 0.24 mm). The stress-strain curves of oedometer test with and without grain crushing for 3D and 2D simulations are plotted in Figs. 2 and 3.

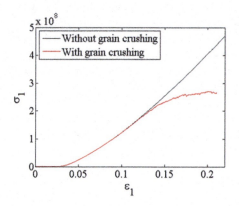

Fig. 2. Stress strain curves of the DEM simulations of oedometer tests in 3D.

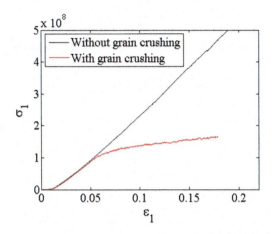

Fig. 3. Stress strain curves of the DEM simulations of oedometer tests in 2D.

It can be seen that for both cases, similar results were obtained. If the model do not have grain crushing effects, the vertical stress keep increasing with loading in the vertical direction. The relationship between stress and strain is close to the exponential function. If grain crushing effects are considered, the increasing rate of the vertical stress becomes much smaller after a certain point, which corresponds to the crushing of grains. Since there is no obvious shear bands in the oedometer tests, it is not easy to determine where the grain crushing should take place. The combined method goes through each particle and checks the stress state. Therefore, the method can determine whether a particle should be replaced or crushed.

3 Conclusion

A combined method to model grain crushing in DEM is presented. The performance of this method is shown by simulations of oedometer tests in both 3D and 2D. The method can be further applied to other DEM simulations to study grain crushing in different loading conditions and used to improve continuum models with grain crushing effects [1].

References

1. Bauer, E.: Calibration of a comprehensive hypoplastic model for granular materials. Soils Found. **36**(1), 13–26 (1996)
2. Bolton, M.D., Nakata, Y., Cheng, Y.P.: Micro- and macro-mechanical behaviour of DEM crushable materials. Geotechnique **58**(6), 471–480 (2008)
3. Cheng, Y.P., Bolton, M.D., Nakata, Y.: Crushing and plastic deformation of soils simulated using DEM. Geotechnique **54**(2), 131–141 (2004)
4. Ciantia, M.O., Arroyo, M., Calvetti, F., Gens, A.: An approach to enhance efficiency of DEM modelling of soils with crushable grains. Geotechnique **65**(2), 91–110 (2015)

5. De Bono, J.P., McDowell, G.R.: DEM of triaxial tests on crushable sand. Granular Matter **16**(4), 551–562 (2014)
6. Laufer, I.: Grain crushing and high-pressure oedometer tests simulated with the discrete element method. Granular Matter **17**(3), 389–412 (2015)
7. Liu, J., Liu, H., Zou, D., Kong, X.: Particle breakage and the critical state of sand. Soils Found. **54**(3), 451–461 (2015). Ghafghazi, M., Shuttle, D.A., DeJong, J.T.: Soils Found. **55**(1), 220–222(2015)
8. Lobo-Guerrero, S., Vallejo, L.E.: Discrete element method analysis of railtrack ballast degradation during cyclic loading. Granular Matter **8**(3–4), 195–204 (2006)
9. Marketos, G., Bolton, M.D.: Compaction bands simulated in Discrete Element Models. J. Struct. Geol. **31**(5), 479–490 (2009)
10. McDowell, G.R., de Bono, J.P.: On the micro mechanics of one-dimensional normal compression. Geotechnique **63**(11), 895–908 (2013)
11. Refahi, A., Aghazadeh Mohandesi, J., Rezai, B.: Discrete element modeling for predicting breakage behavior and fracture energy of a single particle in a jaw crusher. Int. J. Miner. Process. **94**(1–2), 83–91 (2010)
12. Whittles, D.N., Kingman, S., Lowndes, I., Jackson, K.: Laboratory and numerical investigation into the characteristics of rock fragmentation. Miner. Eng. **19**(14), 1418–1429 (2006)

Mechanical Behaviour of Meso-Scale Structure at Critical State for Granular Soils

Yang Liu(✉), Xiaoxiao Wang, and Shunchuan Wu

University of Science and Technology Beijing, Beijing, China
yangliu@ustb.edu.cn

Abstract. The macro- and meso-numerical simulations of biaxial tests for ideal granular samples with different initial densities were carried out using discrete element method. Loop structures, which are represented by voronoi polygons, were taken as the basic meso-mechanical unit of granular material by mesh-subdivision. The meso-mechanical characteristic at critical state for granular media was especially analyzed. The contact forces of side particles and inner sliding rates of same kinds of loops in granular samples with different initial densities have respectively reached critical state. From the mesoscopic point of view, critical state of granular media is the consequence of mutual transformation for higher-order loops and lower-order loops, the combined average and external manifestation of all loops.

Keywords: Critical state · Meso-scale structure · Discrete element method

1 Introduction

The critical state theory proposed by Roscoe [1] is widely accepted. The critical state of granular materials is defined as a state at which the materials are still continuously deforming in shear with constant stresses and volume. However, granular material is multi-scale materials, so multi-scale investigations [2, 3] are needed to analyze its mechanical response, and usually this can be done on three levels, i.e., micro-scale (contacts) level, meso-scale (particle loop or force-chain) level and macro-scale (particle assembly) level. Two kinds of structure elements at meso-scale are often used in granular materials, including force-chains [4–6] and meso-loops (Satake [7], Kruyt [8], Kuhn [9], Nguyen [10]). Meso-loops including mutual contacting particles and voids are nonoverlapping sub-domains decomposed by contact network of granular assembly.

2 DEM Simulation and Definition of the Meso-Loops

The two-dimensional Particle Flow Code (PFC2D) [11] was used to implement the numerical tests. Two kinds of samples with different initial density were prepared. The radius of particle uniformly falls into a range of 0.3–0.6 mm, with $d_{50} = 0.9$ mm.

Stress-strain curves are showed in Fig. 1. Loose sample shows a strain-hardening phenomenon, while dense sample shows an increasing deviatoric stress before reaching peak stress. Four characteristic state points, defined as A, B, C, and D, mean zero volumetric strain rate, zero volumetric strain, largest volumetric strain rate, zero volumetric strain rate, respectively. Point D can be considered as critical state of the simulated granular samples.

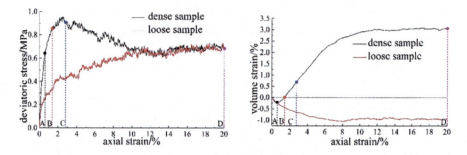

Fig. 1. Deviatoric stress-axial strain and volume strain-axial strain curves.

Meso-loop is defined as a closed region which is formed by mutual contact particles. Figure 2 illustrates the definition of loop element in Voronoi-Delaunay mesh. Note that, particles, which have only one contact or no contact with the surrounding particles, are not included in the loop. Loop structure is a self-organization structure by which particle assembles can maintain mechanical stability under external load. Tordesillas suggests that the loops can be classified into different groups according to the number of edges. In other words, if the loop has n edge particles, it can be defined as "n-cycle", where n is called valence of loop element.

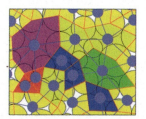

Fig. 2. Definition of loop in Voronoi-Delaunay mesh

3 Mechanical Characters of the Meso-Mechanical Structure

Figure 3 shows the evolution of mean contact force, where the black solid line and dash line respectively represent the average contact force of all loops in both samples. The average contact force of two samples achieves the same value which is close to the value of "5-cycle". Compared to average contact force, the higher the orders, the larger the force. Meanwhile, contact forces of the same loop category in both samples are almost equal, which reveals that the contact force of particles achieve stability for the same types of loops, so-called reaching at their own critical state of contact force.

Mechanical Behaviour of Meso-Scale Structure at Critical State 197

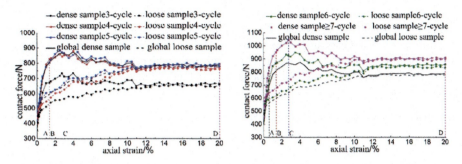

Fig. 3. Evolution of contact force for different loops

Figure 4 shows the evolution of mean contact force ratio for different loops. One can see from Fig. 4 that the ratio of the same types of loops achieves accordance at the critical state which is close to the value of "5-cycle". The ratio of "3-cycle" and "4-cycle" is higher than "5-cycle", "6-cycle" and "≥7-cycle" no matter in loose or dense sample. Hence, the development of friction resistance in lower orders is higher than that in higher orders, which indicates contacts between adjacent particles in lower orders are more likely to slide.

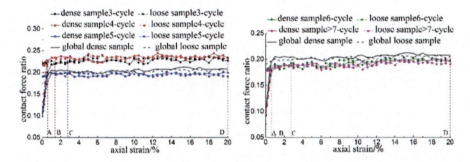

Fig. 4. Evolution of contact force ratio for different loops.

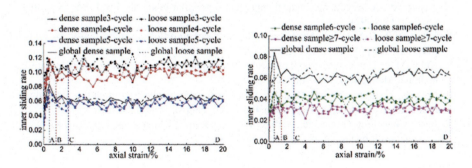

Fig. 5. Evolution of inner sliding rate for different loops.

We define the inner sliding ratio as the ratio of the sliding contact number to total contact number for a specific category of loops. Figure 5 shows the inner sliding ratios for different loops. One can see from Fig. 5 that the ratios of each same loop category in both samples turn to a same value at critical state, which indicates that particles in loops are still in motion state, but achieve the dynamic balance.

4 Summaries and Conclusions

In summary, loop structures, which are represented by Voronoi polygons, were taken as the basic meso-mechanical unit of granular material by mesh-subdivision. The stability against sliding of higher orders is stronger than that of lower orders due to the forming of strong force chain and weak force chain in higher and lower orders, respectively. At the critical state, inter-particle contact force state for the same types of loops reaches to their critical states and of which the inner sliding ratios are approximately equal. From mesoscopic point of view, critical state of granular media is the consequence of mutual transformation of higher-order loops and lower-order loops, the combined average and external manifestation of all loops.

References

1. Roscoe, K.H., Schofield, A., Wroth, C.P.: On the yielding of soils. Geotechnique **8**(1), 22–53 (1958)
2. Drescher, A., De Jong, G.D.J.: Photoelastic verification of a mechanical model for the flow of a granular material. J. Mech. Phys. Solids **20**(5), 337–340 (1972)
3. Sun, Q.C., Jin, F., Wang, G.Q., Zhang, G.H.: Force chains in a uniaxially compressed static granular matter in 2D. Acta Phys. Sinica **59**(1), 30–37 (2010)
4. Peters, J.F., Muthuswamy, M., Wibowo, J., Tordesillas, A.: Characterization of force chains in granular material. Phys. Rev. E **72**(4), 041307 (2005)
5. Tordesillas, A., Muthuswamy, M.: On the modeling of confined buckling of force chains. J. Mech. Phys. Solids **57**(4), 706–727 (2009)
6. Tordesillas, A., Shi, J.Y., Tshaikiwsky, T.: Stress-dilatancy and force chain evolution. Int. J. Numer. Anal. Meth. Geomech. **35**(2), 264–292 (2011)
7. Satake, M.: A discrete-mechanical approach to granular materials. Int. J. Eng. Sci. **30**(10), 1525–1533 (1992)
8. Kruyt, N.P., Rothenburg, L.: Micromechanical definition of the strain tensor for granular materials. J. Appl. Mech. **63**(3), 706–711 (1996)
9. Kuhn, M.R.: Heterogeneity and patterning in the quasi-static behavior of granular materials. Granular Matter **4**(4), 155–166 (2003)
10. Nguyen, N., Magoariec, H., Cambou, B.: Local stress analysis in granular materials at a mesoscale. Int. J. Numer. Anal. Meth. Geomech. **36**(14), 1609–1635 (2012)
11. Itasca Consulting Group Inc.: Particle Flow Code in 2 Dimensions, PFC2D Version. 4.0. Itasca Consulting Group Inc, Minneapolis (2008)

Calibrating the Strength Ratio and the Failure Envelope for DEM Modeling of Quasi-Brittle Materials

Haiying Huang[✉] and Yifei Ma

Georgia Institute of Technology, Atlanta, GA, USA
haiying.huang@ce.gatech.edu, yifei85.ma@gmail.com

Abstract. A displacement-softening contact model is formulated for discrete element modeling of quasi-brittle materials such as rocks and concretes. This local contact law modifies from the parallel bond option in the DEM code PFC2D/3D. It is shown that the displacement-softening contact model can yield realistic compressive over tensile strength ratios as large as ∼30. The high nonlinearity in the failure envelope over the confined extension range, a typical characteristic of rocks, can also be simulated. Calibrations against two well studied rocks, Lac du Bonnet granite and Berea sandstone, are given as examples.

1 Introduction

A common issue in numerical modeling with the discrete element method (DEM) based on bonded spherical particles is that the largest compressive over tensile strength ratio (UCS/UTS) a particle assembly can attain is only about 3–5 [1,2], if the interactions between particles are limited to short-range, elasto-perfectly brittle and frictional. In comparison, the strength ratio for quasi-brittle materials such as rocks and concretes is in the range of 10–30 [3]. Furthermore, the failure envelope for the particle assembly generally fails to capture the high nonlinearity in the confined extension range [1]. If failure in the problem of interest involves both plastic flow and tensile fracturing, the strength ratio, which can be viewed as a measure of material brittleness, then affects the transition between the two failure modes. In order to quantitatively model complex rock behaviors, it is critical that the strength ratio as well as the failure envelope of a DEM model are properly calibrated.

Several numerical strategies such as clumping/clustering particles, increasing the particle interaction range or using multiscale representation of the grain structure and rock fabric have been proposed in the literature to address the issue of the low UCS/UTS ratio (see ref. in [4]). Nevertheless, DEM modeling with bonded spherical particles having only short-range interactions still has the appeal in its computational efficiency.

In this work, we show that a high strength ratio and the associated nonlinear failure envelope can in fact be achieved by implementing a displacement-softening contact law within the conventional framework of soft particle DEM. Formulation of the displacement-softening contact law is first introduced. Calibration against two well studied rocks, Lac du Bonnet granite and Berea sandstone, is then given as examples to illustrate the capability of the contact model.

2 Displacement-Softening Contact Model

The displacement-softening contact model modifies from the parallel bond option in the DEM code PFC2D/3D [5]. Our modification is restricted to the normal bond component. The shear bond strength $\bar{\tau}_c$ is set to be much larger than the normal bond strength $\bar{\sigma}_c$ to ensure that the bond failure is dictated by the condition in the normal component only. For a bond in stretch, onset of softening occurs if the normal bond force reaches its maximum $\bar{\sigma}_c A$, where A is the cross sectional area of the bond. The softening path is defined by a softening coefficient, β, a ratio between the normal stiffnesses along the softening and elastic loading paths. The parallel bond model is recovered if $\beta \to \infty$. If we make an analogy between the bond and a beam cross section, the bond failure criterion is essentially equivalent to state that a bond breaks when the stretch at the outer edge of the bond reaches a critical value $\bar{\delta}_c = \bar{\sigma}_c A \left(1 + \beta\right)/\bar{k}_\ell \beta$, where $\bar{k}_\ell$ is the normal bond loading stiffness.

Softening coefficient β is the only additional parameter introduced in this model. For quasi-static loading, it follows from dimensional analysis that if $\bar{\tau}_c \gg \bar{\sigma}_c$, β is the primary parameter affecting the strength ratio. As long as the particle assembly is finely discretized, effect of β on UCS/UTS is independent of the mean particle size. Numerical results from a series of uniaxial simulations in both 2D and 3D show that UCS/UTS remains nearly constant ~ 4 if $\beta \geqslant 3.5$, but increases to UCS/UTS ~ 30 at $\beta = 0.015$ as β decreases from 3.5 to 0.015 [4] (see Fig. 1a). The reason that the softening model is ineffective when $\beta \geqslant 3.5$ can be explained by the fact that unloading of the system, consisting of the bond plus the particles around it, is basically controlled by the softer one between the two. As β decreases from 3.5 to 0.015, the macro-scale failure mechanisms, i.e., a mode I tensile crack in direct tension and shear localization in uniaxial compression, remain unchanged. Nevertheless, there is substantial increase in the number of broken bonds prior to the peak in the compression test as β decreases. It seems that while UTS is primarily affected by $\bar{\sigma}_c$, UCS depends strongly on $N\bar{\sigma}_c\bar{\delta}_c/2\overline{R}$, which is a nominal measure of the energy loss due to breakage of N number of bonds.

Confined extension and compression tests in 2D and 3D performed with $\beta = 0.1$ offer an interesting insight. While the overall failure envelopes can be very well fitted by the Hoek-Brown criteria, the tensile strength remains nearly constant if the magnitude of the compressive principal stress is limited to a few times of uniaxial tensile strength (see Fig. 1b). This means that the use of a tension cutoff in junction with a failure criterion for shear is indeed justified.

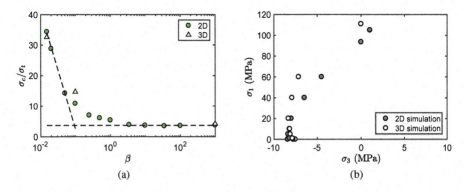

Fig. 1. Variation of the strength ratio with the softening coefficient β (a); a close-up view of the failure envelopes from the confined extension tests with $\beta = 0.1$ (b).

3 Calibration Against Lac du Bonnet Granite and Berea Sandstone

Key contact parameters for modeling Lac du Bonnet (LdB) granite and Berea sandstone are chosen as follows:

- LdB granite (2D) $\beta = 0.03$, $\overline{\sigma}_c = 13 \pm 1.3$ MPa, point contact and parallel bond modulus $E_c = \overline{E}_c = 67.59$ GPa;
- LdB granite (3D) $\beta = 0.045$, $\overline{\sigma}_c = 15 \pm 1.5$ MPa, $E_c = \overline{E}_c = 78.79$ GPa;
- Berea sandstone (3D) $\beta = 0.15$, $\overline{\sigma}_c = 15 \pm 1.5$ MPa, $E_c = \overline{E}_c = 20.0$ GPa.

The numerically obtained macro-scale properties are: LdB granite (2D) - $E = 67.52$ GPa, UCS = 204.73 MPa and UTS = 9.05 MPa; LdB granite (3D) - $E = 69.08$ GPa, UCS = 207.94 MPa and UTS = 11.3 MPa; Berea sandstone (3D) - $E = 17.98$ GPa, UCS = 108.35 MPa and UTS = 6.80 MPa.

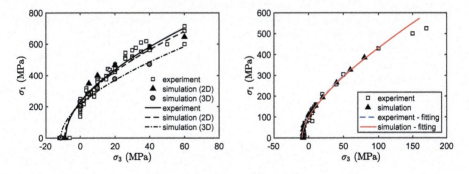

Fig. 2. Comparison of the failure envelopes between the simulations and the experiments; (a) Lac du Bonnet granite [6], (b) Berea sandstone [7]; both are fitted by the Hoek-Brown criteria.

Comparisons between the numerical and experimental results from the confined tests are shown in Fig. 2. It can be seen that for LdB granite, the 2D results match with the experiments better than those from 3D; and excellent agreement is obtained in the results for Berea sandstone up to $\sigma_3 \sim 100$ MPa.

4 Conclusions

It is shown that the displacement-softening contact model can yield realistic compressive over tensile strength ratios as large as $\sim$30. The high nonlinearity in the failure envelope over the confined extension range, a typical characteristic of rocks, can be very well simulated. In this work, displacement-softening is only implemented in the normal bond component. A general softening contact model taking into account the effect of the shear bond component needs to be explored in future work.

Acknowledgements. Partial financial support from the National Science Foundation through grant NSF/CMMI-1055882 and the Sand Control Client Advisory Board of Schlumberger is gratefully acknowledged.

References

1. Huang, H.: Discrete element modeling of tool-rock interaction. Ph.D. thesis, University of Minnesota (1999)
2. Potyondy, D.O., Cundall, P.A.: A bonded-particle model for rock. Int. J. Rock Mech. Min. Sci. **41**(8), 1329–1364 (2004)
3. Hoek, E., Martin, C.D.: Fracture initiation and propagation in intact rock-a review. J. Rock Mech. Geotech. Engng. **6**(4), 287–300 (2014)
4. Ma, Y., Huang, H.: A displacement-softening contact model for discrete element modeling of quasi-brittle materials. Int. J. Rock Mech. Min. Sci. **104**, 9–19 (2018)
5. Itasca Consulting Group Inc. PFC 5.0 manual, Minneapolis, MN (2015)
6. Martin, C.D.: The strength of massive Lac du Bonnet granite around underground openings. Ph.D. thesis, University of Manitoba (1993)
7. Bobich, J.K.: Experimental analysis of the extension to shear fracture transition in Berea sandstone. Master's thesis, Texas A&M University (2005)

Investigation on Stress-Fabric Relation for Anisotropic Granular Material

Weiyi Li, Jiangu Qian[✉], and Xiaoqiang Gu

Department of Geotechnical Engineering, Tongji University, Shanghai 200092, China
qianjiangu@tongji.edu.cn

Abstract. In order to build a proper micro-mechanics-based constitutive model, the stress-fabric should be investigated in detail. A micromechanics-based constitutive framework is presented in this work first to predict the mechanics characteristic of anisotropic granular material. From the micromechanical aspect, the origin of deviatoric stress is decomposed into two components: contact force anisotropy and contact normal anisotropy. Based on past research, the rate of the reduced stress tensor is related with the rate of the fabric tensor. In this work, the relationship between them is verified by the DEM method. After derivation, the back stress is also related to the fabric tensor. Then the back stress is deduced from the stress-fabric-force relationship and determined with reference to the deviation of the principal directions between the reduced stress and its rate. The DEM simulation in this paper refers to Oda's experiment to calibrate parameters. A series of three-dimensional simple shear tests simulated by DEM is first analyzed to study the stress-fabric relationship. The influence of fabric anisotropy on granular materials is specifically investigated.

Keywords: Micromechanics-based model · Fabric anisotropy · Non-coaxiality DEM

1 Introduction

Establishing a proper constitutive equation is an important problem in geotechnical research. Over the past decades, extensive constitutive equations have been proposed to predict various mechanical properties of assorted soils within the framework of continuum mechanics [1, 5]. However, for the great dispersion and diversity of geomaterials, the existing constitutive equations based on continuum mechanics still lack rationality at the micromechanics level.

Based on past researches, the existing two-dimensional models [6] have been attempted to expand to three dimensions, leading to more generalized framework. In order to have a better understanding of the relationship between micro and macro and to establish a more precise constitutive model, the parameters ω_c, which connect the reduced stress tensor and the fabric tensor, is carefully investigated in this analysis.

2 Micromechanics-Based Constitutive Framework

A micromechanics-based constitutive framework was developed by Qian et al. [6]. Some brief introduction for this framework is given below. First, from the micromechanical point of view, macro deviatoric stress tensor s_{ij} origins from two distinct micromechanisms, namely contact force anisotropy and contact normal anisotropy.

$$s_{ij} = \sigma_{ij} - \sigma_{ij}^0 = \sigma_{ij}^r + \sigma_{ij}^f \tag{1}$$

This formula refers to Ouadfel and Rothenburg's theory [4], ignoring the term of multiplication of contact force anisotropy and contact normal anisotropy and assuming the particle is spherical. σ_{ij}^r and σ_{ij}^f denote stresses induced by contact normal anisotropy and contact force anisotropy respectively. Then the stress-fabric relationship is

$$\sigma_{ij}^r = \frac{2}{5} a_{ij}^r \tag{2}$$

a_{ij}^r denotes a second-order anisotropy tensor relating contact normal anisotropy. According to Oda's experimental observations [3]

$$\sigma_{ij}^r = -\omega_c r_{ij} \tag{3}$$

where $r_{ij} = t_{ij}/p$, and t_{ij} denotes the reduced stress. More explanations can be found in Qian et al. [6]. After derivation, the yield equation, plastic potential equation and elastoplastic matrix have the same form as the conventional form proposed by Qian et al. [6], but a parameter ω_c linking macro stress to micro fabric need be given.

3 DEM Investigation

Since the evolution of fabric has been studied in Konishi's laboratory tests [2] of photoelastic materials. The DEM simulation in this paper refer to Oda's experiment to determining parameters. A series of numerical tests including simple shear test, triaxial test are conducted to investigate the relationship between fabric tensor and stress tensor. As an example, the relationship varying with the initial fabric, calibrated in simple shear tests,

An increase of deviatoric stress versus development of induced anisotropy for 4 different inherent anisotropic samples is plotted in Fig. 1. J_2, I_1 denote the second deviatoric stress invariant and the first invariants of the stress tensor. As to J_2^F, I_1^F, it refers to the corresponding invariants of fabric tensor. From Fig. 1, we can see that the rate of stress changing with fabric varies with the initial anisotropic direction θ. For comparison, this trend is shown in Fig. 2. From these results, it can be concluded that initial anisotropy direction significantly affects the stress-fabric relationship.

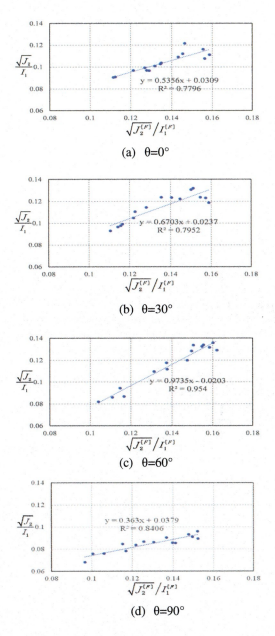

Fig. 1. An increase of deviatoric stress versus development of induced anisotropy for 4 different inherent anisotropic samples.

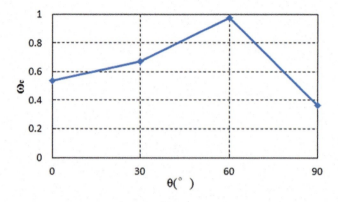

Fig. 2. ω_c varies with bedding angle.

References

1. Hashiguchi, K., Tsutsumi, S.: Shear band formation analysis in soils by the subloading surface model with tangential stress rate effect. Int. J. Plast. **19**, 1651–1677 (2003)
2. Konishi, J., Oda, M., Nemat-Nasser, S.: Induced anisotropy in assemblies of oval cross-sectional rods in biaxial compression. Stud. Appl. Mech., 31–39 (1983)
3. Oda, M.: Inherent and induced anisotropy in plasticity theory of granular soils. Mech. Mater. **16**, 35–45 (1993)
4. Ouadfel, H., Rothenburg, L.: Stress-force-fabric relationship for assemblies of ellipsoids. Mech. Mater. **33**(4), 201–221 (2001)
5. Qian, J.G., Yang, J., Huang, M.S.: Three-dimensional non-coaxial plasticity modeling of shear band formation in geomaterials. J. Eng. Mech. ASCE **34**(4), 322–329 (2008)
6. Qian, J., You, Z., Huang, M., et al.: A micromechanics-based model for estimating localized failure with effects of fabric anisotropy. Comput. Geotech. **50**(3), 90–100 (2013)

Micromechanical Analysis of the At-Rest Lateral Pressure Coefficient of Granular Materials

Chaomin Shen and Sihong Liu[✉]

College of Water Conservancy and Hydropower Engineering,
Hohai University, Nanjing 210098, China
sihongliu@hhu.edu.cn

Abstract. The resistance of the lateral earth pressure of rigid retaining structures is related to a series of important geotechnical problems. This paper reports a micromechanical analysis of the at-rest lateral pressure coefficient K_0 of granular materials. The influence of the spring stiffnesses and the fabric anisotropy is micromechanically investigated. It is revealed that the at-rest lateral pressure coefficient increases with the increasing stiffness ratio as well as the increasing fabric anisotropy.

Keywords: Earth pressure · Granular materials
One-dimensional compression · Spring stiffness ratio
Fabric anisotropy

1 Introduction

In engineering, the lateral pressure is considered in the design of lots of geotechnical structures such as the retaining walls, foundations, tunnels and basements. Various approaches have been adopted to understand and to determine the earth pressure [1, 2]. To evaluate the lateral pressure, the at-rest lateral pressure coefficient, represented as K_0, is defined as the ratio of the horizontal effective stress to the vertical effective stress under the condition of zero lateral strain.

Apart from the laboratory tests, K_0 can be evaluated by being related to other material properties. For example, it was found that for normally consolidated soil, K_0 can be related to the internal friction angle [4]

$$K_0 = 1 - \sin \phi \qquad (1)$$

In the theory of elasticity, the at-rest earth pressure can be also related to the Poisson's ratio, given by

$$K_0 = \frac{v}{1 - v} \qquad (2)$$

In this study, we formulate a micromechanically-based equation for the estimation of K_0. Emphasis of this investigation will be placed on revealing the mechanism of at-rest earth pressure of granular soils.

2 Micromechanical Formulation

2.1 Outline of the Previous Micromechanical Work

In our previous work [5], we have demonstrated that the oedometric modulus, designated as E_s, of 2D granular materials can be quantitatively related to the microscopic quantities by the micromechanical constitutive relationship [6], written as

$$E_s = \frac{C_v}{\pi(1+e)} \left(K^n \left(\frac{3}{8} - \frac{1}{4}\beta \right) + \frac{1}{8} K^t \right) \tag{3}$$

where C_v is the averaged coordination number of the granular sample; K_n and K_t are the normal and tangential contact stiffnesses of particles; β is the fabric anisotropy ranging from zero to 1.

Similarly, micromechanical formulation of the bulk modulus for isotropic compression can also be obtained. The relationship is found as follows

$$K_{iso} = \frac{C_v}{\pi(1+e)} \left(\frac{3}{16} K^n + \frac{1}{16} K^t \right) \tag{4}$$

2.2 Formulation of K_0

Considering the causes for volumetric strain ε_v during one-dimensional compression, it is agreed that the volumetric strain of soils is mainly caused by the hydrostatic compaction (designated as ε_{v-c}) and the dilatancy (designated as ε_{v-d}) due to the deviatoric stress. This can be expressed in the following form

$$d\varepsilon_v = d\varepsilon_{v-c} + d\varepsilon_{v-d} \tag{5}$$

We remark that because of the existence of deviatoric stress, the compressibility of granular materials during one-dimensional compression might be different from that during isotropic compression. To be more specific, if we introduce the definition of the bulk modulus of one-dimensional compression K_{one} and of isotropic compression K_{iso} into Eq. (5), respectively. Equation (5) can be transformed into

$$K_{one} = \frac{K_{iso}}{1+\xi} \tag{6}$$

where $\xi = d\varepsilon_{v-d}/d\varepsilon_{v-c}$ is defined as a parameter of the dilatancy with respect to hydrostatic compaction.

Equation (6) relates the bulk modulus during one-dimensional compression to the bulk modulus during isotropic compression and the dilatancy. Nevertheless, the dilatancy for one-dimensional compression is hard to be detected in the laboratory due to the following reasons: As is agreed by most researchers, the dilatancy of soil is caused by the interparticle rolling and sliding. Despite the existence of shearing stress in one-dimensional compression, the confining pressure increases almost proportionally, which ensures the interlocking among the particles. The dilatancy during one-dimensional compression is therefore relatively small vis-à-vis the compressive deformation.

Neglecting the dilatancy, Eq. (6) becomes

$$K_{one} = K_{iso} \tag{7}$$

Both moduli above can be expressed as functions of microscopic parameters: The bulk modulus for one-dimensional compression can be derived from the oedometric modulus based on the following relationship

$$K_{one} = \frac{1}{2}(1 + K_0)E_s \tag{8}$$

with the form of E_s shown in Eq. (3); The bulk modulus for isotropic compression is given by Eq. (4).

Combing Eqs. (3), (4) and (8), one can obtain

$$K_0 = \frac{2\beta\eta}{(3 - 2\beta)\eta + 1} \tag{9}$$

where η is the stiffness ratio defined in the previous section, which is related to the Poisson's coefficient of particles.

3 Discussion

Equation (9) shows that the at-rest lateral pressure coefficient is related to both the microscopic material property (contact stiffness ratio) and the microscopic structure (anisotropy). Figure 1 shows the evolution of the at-rest lateral pressure as a function of the contact stiffness ratio and the anisotropy. It can be seen that K_0 increases with the increase of the contact stiffness ratio and anisotropy.

However, when η tends to infinity, which corresponds to incompressible particles [7], K_0 becomes

$$\lim_{\eta \to \infty} K_0 = \frac{2\beta}{3 - 2\beta} \tag{10}$$

Since K_0 can be related to the Poisson's coefficient according to (2), Eq. (11) indicates the granular material can be compressible, even single particles, of which the granular material is composed, are incompressible. To conclude, the compressibility of granular materials depends not only of the material properties of the particles, but also on the microscopic structure.

As is discussed in the introduction, the at-rest lateral pressure coefficient can be related to other macroscopic properties. Substituting also the contact stiffness ratio by the Poisson's coefficient of the particles on the basis of Eq. (2), the macroscopic properties of granular materials can be therefore written as functions of microscopic parameters, as shown in Table 1.

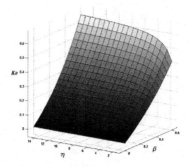

Fig. 1. The effect of contact stiffness ratio and anisotropy on K_0

Table 1. Microscopic description of macroscopic material properties

At-rest lateral pressure coefficient K_0	Poisson's ratio ν	Internal friction angle ϕ
$\dfrac{4\beta(1-v_g)}{4(2-\beta)(1-v_g)-1}$	$\dfrac{4\beta(1-v_g)}{7-8v_g}$	$arcsin\left(\dfrac{8(1-\beta)(1-v_g)-1}{4(2-\beta)(1-v_g)-1}\right)$

Acknowledgement. This work was supported by the "National Natural Science Foundation of China" (Grant No. U1765205) and the "National Key R&D Program of China" (Grant No. 2017YFC0404800).

References

1. Mesri, G., Hayat, T.: The coefficient of earth pressure at rest. Can. Geotech. J. **30**(4), 647–666 (1993)
2. Michalowski, R.L.: Coefficient of earth pressure at rest. J. Geotech. Geoenviron. Eng. **131**(11), 1429–1433 (2005)
3. Jaky, J.: Pressure in silos. In: Proceedings of the 2nd International Conference on Soil Mechanics and Foundation Engineering, Rotterdam, vol. 1, pp. 103–107 (1948)
4. Shen, C.M., Liu, S.H., Wang, Y.S.: Microscopic interpretation of one-dimensional compressibility of granular materials. Comput. Geotech. **91**, 161–168 (2017)
5. Luding, S.: Micro–macro transition for anisotropic, frictional granular packings. Int. J. Solids Struct. **41**(21), 5821–5836 (2004)
6. Xin, J., Yu, L., Zhang, R., Wang, Y.: Determination of calculating parameters and solution methods of discrete element method. Chin. J. Comput. Mech. **01**, 47–51 (1999)

Shear Modulus Degradation and Its Association with Internal Damage in Cemented Granular Material

Zhifu Shen[1(✉)], Mingjing Jiang[2], Zhihua Wang[1], and Shengnian Wang[1]

[1] Institute of Geotechnical Engineering, Nanjing Tech University, Nanjing 210009, China
zhifu.shen@njtech.edu.cn
[2] Department of Geotechnical Engineering, College of Civil Engineering, Tongji University, Shanghai 200092, China

Abstract. Cemented granular material is featured by the presence of inter-particle bonds. In mechanical loading, bond breakage results in a transition from an intact state to a fully disturbed state. However, it is impossible to directly track and measure the breakage of each individual bond among the millions of bonds in a small sample. This study applied discrete element method simulation to associate the evolution of a micro-scale-based damage variable with the degradation of measurable shear modulus in cemented sand. The simulations show that bond breakage is highly non-uniform in a sample. As a result, the sample is deteriorated by the detachment of particles from the main skeleton and the detached particles are found to take no load in the initial stage. This explains the observed sharp degradation of shear modulus. With further bond breakage, the skeleton is disassembled into assembly of particle-clusters, but the shear modulus only varies slightly. The implication for establishment of a physically robust constitutive model for cemented sand is discussed.

Keywords: Cemented sand · Structural damage · Shear modulus degradation

1 Introduction

Cemented granular materials are common in geotechnical field, e.g. cemented sand [1], gas hydrate bearing sediment [2] and weak rock [3]. Constantly increasing knowledge in micro-structure facilitates the development of constitutive model with consideration of particle-scale mechanism [4]. However, the lack of understanding in the association of microscopic quantities and macroscopic measurable quantities hinders reliable determination of some model parameters that has been embedded with microscopic meaning. This study focuses on the association of particle-scale bond breakage and degradation of macroscopic shear modulus.

2 DEM Simulation of Cemented Sand

Cemented sand was idealized as an assembly of bonded spheres in this study. The bonding material is assumed to sparsely distribute at distinct contacts. The bond contact model, particle size distribution and the basic set of contact model parameters in [5] were used. Cubic samples with 40,000 spherical particles were simulated. Four samples were modelled to study the effects of sample density and bond content: S1 (initial void ratio $e_0 = 0.80$, bond content $c = 1.0\%$ which was defined as the ratio of total bond material volume over total particle volume), S2 ($e_0 = 0.80$, $c = 3.0\%$), S3 ($e_0 = 0.92$, $c = 1.0\%$), and S4 ($e_0 = 0.92$, $c = 3.0\%$). Constant-p triaxial shear tests were simulated with $p = 400$ kPa.

Figure 1 shows that increase in sample density (decrease in void ratio) and bond content can increase the peak strength and dilation. The softening behavior of sample S3 is accompanied with contraction, which is a feature of loose weakly-cemented soil.

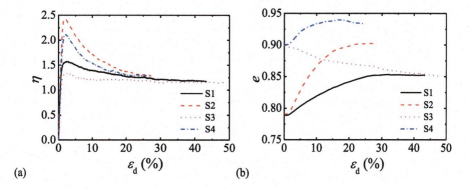

Fig. 1. Effects of sample density and bond content on mechanical behavior ($\eta = q/p$, ε_d = deviator strain): (a) stress ratio versus deviator strain, (b) void ratio versus deviator strain.

3 Bond Breakage and Shear Modulus Degradation

A damage variable θ is defined as the fraction of unbonded contacts in an assembly to describe the degree of bond breakage. Shear modulus G was measured at various points along the constant-p loading path by applying one unloading-loading loop with a variation in deviator strain of 0.2%. The loop is realized by reversing the velocity of the loading plate perpendicular to the major principal stress direction. The shear modulus was determined from the average slope of the unloading-loading path in the deviator strain- deviator stress plot. Figure 2 presents the degradation of shear modulus (normalized by the initial value G_0) with the development of bond breakage. Figure 2 shows that the shear modulus drops sharply when θ is less than 0.1, which is followed by a much slower degradation to a minimum around $\theta = 0.2$. After that, little variation in shear modulus is observed. The degradation of shear modulus keeps sound pace with the decrease in coordinate number Z as shown in Fig. 2.

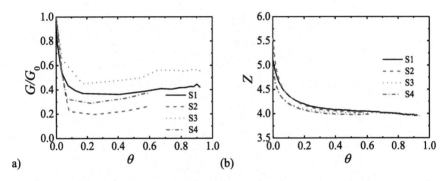

Fig. 2. Degradation of shear modulus: (a) degradation of shear modulus, (b) evolution of coordinate number.

The microscopic structure change is examined here to explain the numerically observed nonlinear degradation of shear modulus with bond breakage. Our DEM simulation results show that the heterogeneous nature of granular material leads to spatially non-uniform bond breakage events. This allows the formation of bonded particle clusters, which are groups of particles connected through bonded contacts. The presence of clusters had been confirmed in experiments [6]. With the aid of DEM simulation, Fig. 3 visualizes the spatial configuration of particle clusters, which are distinguished by colors (see the online version of this paper). The initial intact cemented assembly can be viewed as a particle skeleton. Bond breakage mainly leads to detachment of small clusters (usually consisting of several particles) from the skeleton when θ is less than 0.2. The skeleton bears external load and it is gradually deteriorated due to the detachment of particles. With further bond breakage, the skeleton is disassembled into assembly of particle-clusters. To quantitatively examine this feature, the average stress σ_{ij} of a granular assembly is expressed as

$$\sigma_{ij} = \frac{1}{V}\sum_{k=1}^{N_p}\sum_{m=1}^{N_{c,k}} f_j l_i = \underbrace{\frac{1}{V}\sum_{k=1}^{N_s}\sum_{m=1}^{N_{c,k}} f_j l_i}_{\sigma_{ij,s}} + \underbrace{\frac{1}{V}\sum_{k=1}^{N_p-N_s}\sum_{m=1}^{N_{c,k}} f_j l_i}_{\sigma_{ij,d}} = \sigma_{ij,s} + \sigma_{ij,d} \tag{1}$$

where V is the sample volume, f_j is the contact force, l_i is a vector pointing from particle centroid to contact center, N_p is total particle number, N_s is particle number in the skeleton and $N_{c,k}$ is the number of contacts around particle k. The stress can be partitioned into skeleton stress ($\sigma_{ij,s}$) and detached particle stress ($\sigma_{ij,d}$), respectively.

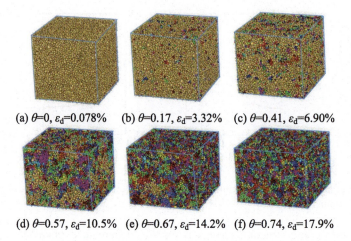

(a) $\theta=0$, $\varepsilon_d=0.078\%$ (b) $\theta=0.17$, $\varepsilon_d=3.32\%$ (c) $\theta=0.41$, $\varepsilon_d=6.90\%$

(d) $\theta=0.57$, $\varepsilon_d=10.5\%$ (e) $\theta=0.67$, $\varepsilon_d=14.2\%$ (f) $\theta=0.74$, $\varepsilon_d=17.9\%$

Fig. 3. Spatial configurations of particle clusters

Figure 4 presents the contribution of detached particles to the overall stress. When θ is less than 0.2, the very little contribution of $\sigma_{ij,d}$ (less than 1.0%) suggests that detached particles actually float in the skeleton voids. That is, the detached particles take no load and act as voids in the sample. The degradation in shear modulus therefore is a result of the deterioration of the initially intact skeleton.

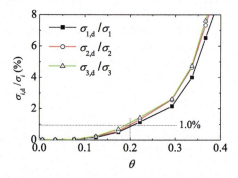

Fig. 4. Contribution of detached particles to the overall stress

The damage variable θ can be inserted into a constitutive model for cemented soil while its association with the measurable shear modulus allows the determination of the damage evolution law. This issue deserves further study.

Acknowledgement. The research is funded by National Natural Science Foundation of China with Grant Nos. 51678300 and 51639008 (key program) and the Youth Program of Natural Science Foundation of Jiangsu Province of China with Grant No. bk20171006, which are sincerely appreciated.

References

1. Collins, B.D., Sitar, N.: Stability of steep slopes in cemented sands. J. Geotech. Geoenviron. Eng. **137**(1), 43–51 (2011)
2. Hyodo, M., Yoneda, J., Yoshimoto, N., Nakata, Y.: Mechanical and dissociation properties of methane hydrate-bearing sand in deep seabed. Soils Found. **53**(2), 299–314 (2013)
3. Ciantia, M.O., Castellanza, R., Crosta, G.B., Hueckel, T.: Effects of mineral suspension and dissolution on strength and compressibility of soft carbonate rocks. Eng. Geol. **184**, 1–18 (2015)
4. Jiang, M.J., Zhang, F.G., Sun, Y.G.: An evaluation on the degradation evolutions in three constitutive models for bonded geomaterials by DEM analyses. Comput. Geotech. **57**, 1–16 (2014)
5. Shen, Z.F., Jiang, M.J., Thornton, C.: DEM simulation of bonded granular material. Part I: contact model and application to cemented sand. Comput. Geotech. **75**, 192–209 (2016)
6. Lo, S.R., Wardani, S.P.: Strength and dilatancy of a silt stabilized by a cement and fly ash mixture. Can. Geotech. J. **39**(1), 77–89 (2002)

Induced Crack Network Evolution in Geomaterials: μct Examination and Mathematical Morphology Based Analysis

Maciej Sobótka[1(✉)], Michał Pachnicz[1], and Magdalena Rajczakowska[2]

[1] Faculty of Civil Engineering, Wrocław University of Science and Technology,
27 Wybrzeże Wyspiańskiego, 50-370 Wrocław, Poland
`maciej.sobotka@pwr.edu.pl`
[2] Department of Civil, Environmental and Natural Resources Engineering,
Luleå University of Technology, Luleå, Sweden

Abstract. Fracture parameters of the rock material are known to change due to applied loading, having effect on, e.g. the rock permeability coefficient which can be estimated based on the geometrical descriptors of the crack space. In this paper, a method of crack network evolution analysis is proposed for the rock samples subjected to uniaxial compression. Fracture development is investigated using ex-situ time-lapse micro-computed tomography (μCT). The sets of images are acquired for each specimen at three damage levels: before loading, at approximately 50% of compressive strength and, finally, after reaching micro-dilatancy threshold (or compressive strength). The reconstructed and segmented 3D crack network is examined at each loading stage. The analysis consists of image processing and determination of the selected morphological parameters, i.e. volume fraction of the crack, spatial distribution of the fracture aperture, tortuosity as well as the structure model index (SMI).

Keywords: Tortuosity · Porosity · Morphology · Microstructure

1 Introduction

The quality of the image obtained from the X-ray micro-computed tomography depends on the contrast in the absorption of X-ray radiation between the individual constituents of the tested object. Therefore, this technique has been successfully applied to verify the porosity of soils [3] and rocks [1, 4], due to the high contrast between the air or water filling the voids, and the mineral skeleton. In addition, a number of publications on the characterization of the failure mechanism of rocks can be found, e.g. evaluation of the effect of anisotropy (e.g. [2]), mechanical behavior under unloading [5], freeze-thaw cycles [6] or different high temperature treatments [7]. In this study, the ex-situ 4D imaging of dynamic processes in rocks is conducted. The X-ray micro-computed tomography is applied followed by a morphological analysis of the acquired images in order to describe the evolution of the crack network in the rock material subjected to different load levels.

2 Basics of Microtomography

Computed microtomography is a modern technique that enables the visualization of the internal structure of the object under study. It consists in a mathematical reconstruction of the 3D microstructure of the examined material from a series of X-ray radiographs made in high resolution. Registration of the images is based on the detection of the intensity of X-rays emitted by the source, weakened during the passage through the sample. The X-ray microtomography examination includes scanning and reconstruction of a three-dimensional image of a sample followed by the analysis of the obtained results using image processing techniques. In order to perform a quantitative evaluation of the material microstructure, the segmentation procedure is applied, i.e. the individual constituents of the microstructure are segregated based on their gray scale values and represented by the sets of binary images. Then, based on the obtained binary images selected morphological measures of the microstructure can be determined.

3 Materials and Methods

The study was carried out to identify the microstructural changes in the rock medium subjected to three different compression load levels (stages): Stage 1 - before loading (reference microstructure image), Stage 2 - approximately 50% of the compressive strength, and Stage 3, after reaching the strength limit.

After each loading of the specimen, a series of ex-situ scans was performed with the use of X-ray micro-computed tomography. The Bruker Skyscan 1172 with a 11 Mp resolution camera was applied. NRecon software based on the Feldkamp algorithm was utilized for the reconstruction of the microstructure of the rock material. Next, images corresponding to different load stages were adjusted in terms of their spatial orientation in order to analyze the evolution of the cracks in a unified coordinate system. Orientation was determined "manually" in DataViewer software using 3D Registration procedure. To segment the crack network, the images were binarised, i.e. grayscale histogram based thresholding was adopted. Obtained binary images were then limited to the cubic area selected from the inside of the sample (VOI - volume of interest).

The sets of slices, i.e. cross-sections of the reconstructed volume of the material, were used for further quantitative analysis. For each loading stage, the following morphological measures of the microstructure were determined: void fraction (porosity, volume fraction of crack space), fracture aperture, tortuosity, crack surface area, SMI index (structure model index). All measures were calculated using CTAnalyser software and author's own Wolfram Mathematica code.

4 Results

The reconstructed 3D VOI regions with segmented fracture networks corresponding to each damage level are presented in Fig. 1. For each case quantitative analysis was performed and selected morphological measures were determined according to definitions given in Chapter 3. Summary of the calculated values is presented in Table 1.

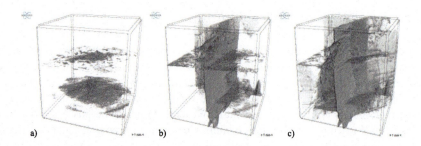

Fig. 1. Segmentation of crack network – 3D rendering: (a) Stage 1, (b) Stage 2, (c) Stage 3

Table 1. Summary of analysis results

Stage	Crack volume fraction [-]	Medium crack aperture [mm]	Medium tortuosity in direction			Crack surface area [1/mm]	SMI [-]
			X [-]	Y [-]	Z [-]		
1	0.57	0.124	1.44	1.45	–	0.150	1.48
2	2.24	0.189	1.27	2.31	1.35	0.506	1.56
3	3.77	0.205	1.24	1.62	1.24	0.731	1.15

The exemplary spatial distributions of the crack aperture as well as the tortuosity in the Z direction, determined for the load level3 are presented in Fig. 2.

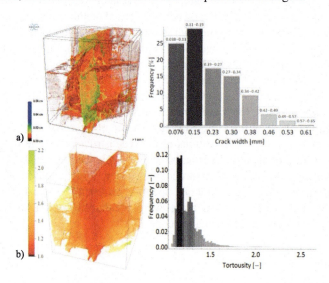

Fig. 2. Stage 3: (a) spatial distribution and histogram of crack aperture, (b) spatial distribution and histogram of tortuosity in Z direction 0

5 Final Remarks

The paper presented a method for evaluation of the crack network evolution based on the changes in the morphological properties of rock fracture. A 4D quantitative and qualitative analysis of the crack development under three different loading levels was performed with the use of X-ray micro-computed tomography. For each damage level, the selected morphological measures of the fracture network were determined e.g.: volume fraction of the crack, crack width, tortuosity, surface area. Based on the presented results, it is visible that as the applied load increases, the cracks volume fraction as well as their surface area increase due to micro-incrustations. On the other hand, this trend is not exibited by the average crack width changes. New cracks emerge mainly in the vertical direction: the mean tortuosity in the Z (vertical) direction reaches an average value close to 1. The aforementioned morphological parameters give the possibility to estimate the permeability of the pore network based on, e.g. Kozeny–Carman equation (e.g. [8]).

References

1. Tetsuro, H., Manabu, T., Satoru, N.: In situ visualization of fluid flow image within deformed rock by X-ray CT. Eng. Geol. **70**(1), 37–46 (2003)
2. Nasseri, M.H.B., Young, R.P., Rezanezhad, F., Cho, S.H.: Application of 3D X-ray CT scanning techniques to evaluate fracture damage zone in anisotropic granitic rock. In: 3rd US-Canada Rock Mechanic Symposium, Toronto. Canada, pp. 55–56 (2009)
3. Peyton, R.L., Haeffner, B.A., Anderson, S.H., Gantzer, C.J.: Applying X-ray CT to measure macropore diameters in undisturbed soil cores. Geoderma **53**(3–4), 329–340 (1992)
4. de Argandoña Vicente, R., Rey, Á.R., Celorio, C., del Río, L.S., Calleja, L., Llavona, J.: Characterization by computed X-ray tomography of the evolution of the pore structure of a dolomite rock during freeze-thaw cyclic tests. Phys. Chem. Earth Part A: Solid Earth Geodesy **24**(7), 633–637 (1999)
5. Zhou, X.P., Zhang, Y.X., Ha, Q.L.: Real-time computerized tomography (CT) experiments on limestone damage evolution during unloading. Theor. Appl. Fract. Mech. **50**(1), 49–56 (2008)
6. De Kock, T., Boone, M.A., De Schryver, T., Van Stappen, J., Derluyn, H., Masschaele, B., Cnudde, V.: A pore-scale study of fracture dynamics in rock using X-ray micro-CT under ambient freeze–thaw cycling. Environ. Sci. Technol. **49**(5), 2867–2874 (2015)
7. Yang, S.Q., Ranjith, P.G., Jing, H.W., Tian, W.L., Ju, Y.: An experimental investigation on thermal damage and failure mechanical behavior of granite after exposure to different high temperature treatments. Geothermics **65**, 180–197 (2017)
8. Costa, A.: Permeability-porosity relationship: a reexamination of the Kozeny-Carman equation based on a fractal pore-space geometry assumption. Geophys. Res. Lett. **33**(2) (2006)

Kinematic Nature of Sand Strength

Zenon Szypcio[✉]

Department of Geotechnics and Structural Mechanics,
Bialystok University of Technology, Wiejska 45E, 15-351 Bialystok, Poland
z.szypcio@pb.edu.pl

Abstract. Sand strength has kinematic nature and can be expressed as a sum of strength at the critical state and an additional, kinematic component, which is a function of the strain increment at failure. The strain increments at failure are dependent on the initial density, stress level and mode of deformation. The sand strength obtained according to our theory is well corroborated by test data from literature.

Keywords: Sand strength · Dilatancy · Critical state · Kinematic constraint

1 Introduction

The influence of dilatancy on sand strength is considered by many researchers, e.g. Taylor [12], Cam-clay model [6], Rowe [7, 8] and Horne [3] and Bolton [1]. The general stress-dilatancy equation according to the *Frictional State Theory* [9] has the following form:

$$\eta = Q - AD^p \tag{1}$$

where: $\eta = q/p'$ with q the deviatoric stress and p' the mean stress; $Q = M^o - \alpha A^o$, $A = \beta A^o$; $M^o = 6 \sin \Phi^o / 3 - \sin \Phi^o$ (Φ^o is the critical friction angle), $A^o = 1 - \frac{1}{3}M^o$, $q = \sigma'_1 - \sigma'_3$, $p' = \frac{1}{3}(\sigma'_1 + 2\sigma'_3)$, $D^p = \delta\varepsilon^p_v / \delta\varepsilon^p_q$, $\delta\varepsilon^p_v = \delta\varepsilon^p_1 + 2\delta\varepsilon^p_3$, $\delta\varepsilon^p_q = \frac{2}{3}(\delta\varepsilon^p_1 - \delta\varepsilon^p_3)$, $\delta\varepsilon^p_v = \delta\varepsilon_v - \delta\varepsilon^e_v$, $\delta\varepsilon^p_q = \delta\varepsilon_q - \delta\varepsilon^e_q$. In this work, the dilatancy is treated as internal kinematic. Therefore, we can say that a sand strength has kinematic nature.

2 Strength of Sand

The strength of sand can by expressed by the maximum value of secant friction angle

$$\Phi'_{max} = 2 \tan^{-1} \sqrt{\sigma'_1 / \sigma'_3} - \pi/2 \tag{2}$$

The stress ratio in the above equation can be expressed as

$$\sigma'_1 / \sigma'_3 = \left\{ 3 - 2\eta \sin\left(\theta - \tfrac{2}{3}\pi\right) \right\} / \left\{ 3 - 2\eta \sin\left(\theta + \tfrac{2}{3}\pi\right) \right\} \tag{3}$$

The stress ratio η can be obtained from Eq. (1). For many sands the critical friction angle can be found in our previous work [10, 11]. The Lode angle for the stress and strain increments are: $\tan\theta = (1-2b)/\sqrt{3}$, $\tan\theta_\varepsilon = (1-2b_\varepsilon)/\sqrt{3}$, $b = (\sigma'_2-\sigma'_3)/(\sigma'_1-\sigma'_3)$, $b_\varepsilon = (\delta\varepsilon^p_2 - \delta\varepsilon^p_3)/(\delta\varepsilon^p_1 - \delta\varepsilon^p_3)$. At failure, $\delta p' = \delta q = 0$ therefore, $\delta\varepsilon_{ij} = \delta\varepsilon^p_{ij}$ and index p is omitted later without inducing confusion. It is assumed that at failure we have $(\delta\varepsilon_v/\delta\varepsilon_a)_e = -0.5(\delta\varepsilon_v/\delta\varepsilon_a)_c$, $(\delta\varepsilon_v/\delta\varepsilon_a)_b = (\delta\varepsilon_v/\delta\varepsilon_a)_c$ with $\theta = 15°$ [11].

The parameters α and β for sand at failure can be found in [10, 11] with $\alpha = 0$, $\beta = 1$ for drained triaxial compression and extension, $\alpha = 0$, $\beta = 1.4$ for drained biaxial compression. Indexes c, e and b mean triaxial compression, extension and biaxial compression respectively. It is also assumed that

$$(\delta\varepsilon_v/\delta\varepsilon_a)_c = (\delta\varepsilon_v/\delta\varepsilon_a)_r I^o_R \tag{4}$$

where the natural dilatancy ratio $I^o_R = 1 - R^o \ln\left(p'_f/p'_r\right)^\xi$ and $(\delta\varepsilon_v/\delta\varepsilon_a)_r$ is the reference strain increment ratio for triaxial compression. The variable p'_f is the mean pressure at failure, p'_r is the reference mean pressure at $(\delta\varepsilon_v/\delta\varepsilon_a)_c = 0$ for $I_D = 0$ and $(\delta\varepsilon_v/\delta\varepsilon_a)_c = -1$ for $I_D = 1$ according to the theory by Rowe and Horne [3, 7, 8].

Let us consider Brasted sand as a representative for silica sand, we can write out the following relationship [2]:

$$(\delta\varepsilon_v/\delta\varepsilon_a)_r = (\delta\varepsilon_v/\delta\varepsilon_1)_r = I_D(0.36 + 0.64 I_D) \tag{5}$$

The critical state line in e-p' has the following power form:

$$e_c = e_\Gamma - \lambda_c \left(p'/p_a\right)^\xi \tag{6}$$

where $p_a = 101$ kPa (atmospheric pressure). For $p'_f = p'_r$ the $(\delta\varepsilon_v/\delta\varepsilon_a)_c = (\delta\varepsilon_v/\delta\varepsilon_a)_r$. For $p'_f = p'_c$ (critical state) $(\delta\varepsilon_v/\delta\varepsilon_a)_c = 0$ (Fig. 1).

This condition gives rise to $e_r = e_\Gamma - \lambda_c(p'_r/p_a)^\xi$, $R^o = 1/\ln\{(e_\Gamma - e)/(e_\Gamma - e_r)\}$, $I_{Dr} = (e_{max} - e)/(e_{max} - e_{min})$, $p'_{fmax} = p_a\{(e_\Gamma - e_{min})/\lambda_c\}^{1/\xi}$.

It is further assumed that at failure:

$$(\delta\varepsilon_v/\delta\varepsilon_1) = (\delta\varepsilon_v/\delta\varepsilon_1)_c \text{ for } 0 \leq b < b_o \tag{7a}$$

$$(\delta\varepsilon_v/\delta\varepsilon_1) = (\delta\varepsilon_v/\delta\varepsilon_1)_c\{1 - (\delta\varepsilon_v/\delta\varepsilon_1)_c[(b-b_o)/(1-b_o)]/[2+(\delta\varepsilon_v/\delta\varepsilon_1)_c]\}, \text{ for } b_o \leq b \leq 1 \tag{7b}$$

and

$$(\delta\varepsilon_2/\delta\varepsilon_1) = -\frac{1}{2}(1-b/b_o)\{1-(\delta\varepsilon_v/\delta\varepsilon_1)\} \text{ for } 0 \leq b \leq b_o \tag{8a}$$

$$(\delta\varepsilon_2/\delta\varepsilon_1) = (b-b_o)/(1-b_o) \text{ for } b_o \leq b < 1 \tag{8b}$$

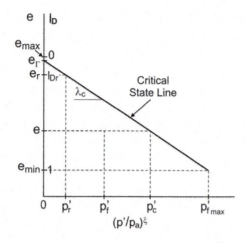

Fig. 1. Characteristic values in e (I_D) – $(p'/p_a)^\xi$ plane.

where b_o is a value of b for the plane strain condition. We assume $\alpha = 0$ at failure.

$$\beta = 1 + 0.4(b/b_o) \text{ for } 0 \leq b < b_o \quad (9a)$$

$$\beta = 1.4 - 0.4(b - b_o)/(1 - b_o) \text{ for } b_o \leq b < 1 \quad (9b)$$

Now we are able to obtain Φ'_{max} from Eq. (2) with use of Eqs. (3)–(9a and 9b).

3 Results

The strength of Sacramento River sand was calculated by the proposed method. The following parameters are used: $e_{max} = 0.906$, $e_{min} = 0.610$, $p'_r = 190$ kPa, $\Phi° = 34.9°$, $e_\Gamma = 0.900$, $\lambda_c = 0.023$, $\xi = 0.7$. The following characteristic values are known: $e_r = 0.864$, $I_{Dr} = 0.14$, $p'_{fmax} = 3773$ kPa. The influence of mean pressure on strength of dense Sacramento River sand ($I_D = 1.0$) are shown in Fig. 2.

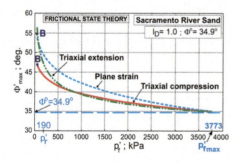

Fig. 2. The Φ'_{max}-p'_f relationships.

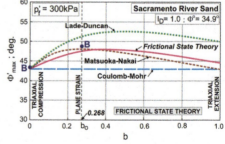

Fig. 3. Influence of intermediate stress on sand strength.

The proposed method shows that for Sacramento River sand and $p'_f < 700$ kPa, Φ'_{max} for drained triaxial extension is higher than for triaxial compression. The difference grows with decreasing mean pressure at failure. For $p'_f = 50$ kPa the difference is about 10°. For plane strain Φ'_{max} is higher than for triaxial compression. The difference decreases with increasing mean pressure at failure. The influence of intermediate stress on Φ'_{max} of dense ($I_D = 1.0$) Sacramento River sand at $p'_f = 300$ kPa is shown in Fig. 3. The strength Φ'_{max} obtained from Mohr-Coulomb, Lade-Duncan [4] and Matsuoka-Nakai [5] are also shown in Fig. 3. The maximum values of Φ'_{max} calculated according to Bolton [1] for triaxial compression and plane strain are also shown in Figs. 2 and 3 as points B. It is evident that the proposed method and Bolton's proposal give similar strength for dense silica sand at a low stress level. The proposed method correctly describes the influence of intermediate stress on the sand strength.

4 Conclusions

Sand strength has kinematic nature and can be expressed as the sum of a critical state strength and a kinematic component generated by strain increments. This component can be computed by our *Frictional State Theory*. The strain increments at failure are dependent on the initial density and mean pressure.

Acknowledgements. This work, conducted at Bialystok University of Technology, was supported by Polish Financial Resources on Science under Project No. S/WBiIS/2/2018.

References

1. Bolton, M.D.: The strength and dilatancy of sand. Géotechnique **36**(1), 65–78 (1986)
2. Cornforth, D.H.: Some Experiments on the Influence of Strain conditions on the Strength of Sand. Géotechnique **14**(2), 143–167 (1964)
3. Horne, M.R.: The Behaviour of an Assembly of Rotund, Rigid, Cohesionless Particles. Proc. Roy. Soc. of London, Series A. Math. Phys. Sci. **286**(1404), 62–97 (1965)
4. Lade, P.V., Duncan, J.M.: Elasto-plastic stress-strain theory for cohesionless soil. J. Geotech. Eng. ASCE **101**(10), 1037–1053 (1975)
5. Matsuoka, H., Nakai, T.: Stress-deformation and strength characteristics of soil under three different principal stresses. Proc. JSCE **232**, 59–74 (1974)
6. Roscoe, K.H., Schofield, M.A., Thurairajah, A.: Yielding of Clays in States Wetter than Critical. Géotechnique **13**(3), 211–240 (1963)
7. Rowe, P.W.: The stress-dilatancy relation for static equilibrium of an assembly of particles in contact. Proc. Roy. Soc. of London, Series A. Math. Phys. Sci. **269**, 500–527 (1962)
8. Rowe, P.W.: The relation between the shear strength of sands in triaxial compression plane strain direct shear. Géotechnique **19**(1), 75–86 (1969)
9. Szypcio, Z.: Stress-dilatancy for soils. Part I: the frictional state theory. Studia Geotechnica et Mechanica. **38**(4) (2016)

10. Szypcio, Z.: Stress-dilatancy for soils. Part II. Experimental validation for triaxial tests. Studia Geotechnica et Mechanica **38**(4), 59–65 (2016)
11. Szypcio, Z.: Stress-dilatancy for soils. Part III. Experimental validation for the biaxial condition. Studia Geotechnica et Mechanica **39**(1), 73–90 (2017)
12. Taylor, D.W.: Fundamentals of Soil Mechanics. Wiley, New York (1948)

Shortcomings of Existing Scan Line Void Fabric Tensors

A. I. Theocharis[1(✉)], E. Vairaktaris[1], and Y. F. Dafalias[1,2]

[1] School of Applied Mathematical and Physical Sciences,
Department of Mechanics, National Technical University of Athens,
Athens, Greece
atheomec@mail.ntua.gr
[2] Department of Civil and Environmental Engineering,
University of California Davis, Davis, CA, USA

Abstract. Fabric of granular materials is related to the statistical distribution of orientation of different microstructural vector-like entities based on the solid or the void phase. Due to difficulties associated with the accurate representation of the solid structure and in particular contact normal vectors, void fabric appears today as simpler to quantify within the current laboratory techniques for granular media, such as X-ray CT, and thus is of particular interest. Void fabric tensors can be determined by image-based quantification methods of voids, which are well de-fined and easy to apply to both physical and numerical experiments. Such a promising void fabric characterization approach is based on the scan line method originally proposed by Oda et al. [1]. In this work, existing scan line void fabric anisotropy tensors definitions are proven analytically to inherit serious shortcomings and, thus, should be modified for future use; such modifications are proposed and verified in Theocharis et al. [2] and briefly stated herein.

Keywords: Void fabric · Anisotropy · Scan line

1 Introduction

Fabric of granular materials is commonly related to three main material elements: the primary orientation of the solid grains, the contact normal vectors, and the voids. Void fabric appears of particular importance since it seems to be the most easily accessible through the knowledge of the structure of the material obtained with existing laboratory methods, such as X-Ray CT scans. Oda et al. [1] underlined the importance of the voids and void fabric anisotropy for granular materials and provided the initial scan line concept and definition to quantify them.

The scan line method is based on the concept of scanning the sample of a granular material with parallel lines at fixed distance d, at an inclination θ; the scanning is repeated for small intervals of θ between limits covering the whole sample. The scanning lines intersect the solid grains and as a result the line segments defined between grain boundaries can be characterized as voids or solids (Fig. 1). This method

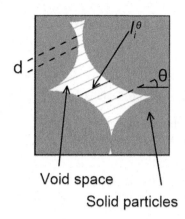

Fig. 1. Scan line parameters.

can be directly applied in 2D and 3D samples and provides a direct way to quantify void fabric.

Oda et al. [1] introduced the method by applying it to snapshots of small-scale physical experiments while Ghedia and O'Sullivan [3] applied it on images using advanced imaging tools. In the present work, it will be proved analytically that the scan line void fabric tensor definition by [1, 3] have an inherent shortcoming, and thus should be modified for future use.

2 Scan Line Void Fabric Tensor Definitions

Initially, some basic definitions and notations are introduced, where the notations may differ from those of the original contributions, in order to unify the presentation: l_i^θ is the length of an *ith* void scan line segment along angle θ (Fig. 1); N^θ is the total number of scan line segments at an angle θ; $L^\theta = \sum_{i=1}^{N^\theta} l_i^\theta$ is the sum of all void scan line segments l_i^θ along θ; $l^\theta = L^\theta/N^\theta$ is the mean length of scan line segments for each angle θ; $L = \sum_{\theta=-90°}^{\theta=90°} l^\theta$ is the sum of all mean lengths l^θ over all θ, where the summation is executed in predefined intervals of θ within the angular domain $[-90° : 90°]$; and N^v denotes the total number of voids. A summation over θ, in the limit of very small intervals of θ, is equivalent to integration over θ in the chosen angular domain since the original scan line void definitions were given in terms of integrals.

Based on the scan line approach as described above, void fabric is quantified using a corresponding scan line void fabric tensor created from all void segments in all directions. The original definition of this void fabric tensor was proposed in 2D by Oda et al. [1] and reads:

$$\mathbf{V} = 4\left(\frac{1}{L} \sum_{\theta=-90°}^{\theta=90°} l^\theta \mathbf{n}^\theta \otimes \mathbf{n}^\theta - \frac{1}{4}\mathbf{I}\right) \quad (1)$$

where $\mathbf{n}^\theta$ is the unit vector along all scan line void vectors inclined at a specific angle θ, I is the identity tensor and $\mathbf{n}^\theta \otimes \mathbf{n}^\theta$ represents the tensor product of $\mathbf{n}^\theta$ by itself. Same Eq. (1) with ½ substituting for ¼ in front of the identity tensor was used in Kuo et al. [4].

Ghedia and O'Sullivan [3] modified this definition and employed only the first part in parentheses of Eq. (1) to define a scan line void fabric tensor by

$$\mathbf{V}_2 = \frac{1}{L} \sum_{\theta=-90°}^{\theta=90°} \bar{l}^\theta \mathbf{n}^\theta \otimes \mathbf{n}^\theta \qquad (2)$$

where $\bar{l}^\theta = L^\theta/N^v$ is the mean total length of voids along θ per void and acts as the weighting factor for the tensors $\mathbf{n}^\theta \otimes \mathbf{n}^\theta$ (compare this with $l^\theta = L^\theta/N^\theta$ entering Eq. (1)).

3 Analytical Proof of Shortcomings of Existing Scan Line Void Fabric Tensors

The definition of Eq. (1) gives rise to an important implication that leads to an inherent shortcoming regarding the accurate quantification of void fabric anisotropy; an even greater shortcoming applies to the definition by Eq. (2). For the following analysis 2D conditions are assumed, but the conclusion applies as well to 3D.

When the assumed fixed distance d between pairs of parallel scan lines multiplies a scan line segment l_i^θ, it yields the differential void area element $\delta A_{vi} = l_i^\theta d$ of the total void surface area A_v in 2D. Based on the previous definitions, it follows by summation over all i's for a given θ, and in the limit case as $d \to 0$ and $N^\theta \to \infty$ that:

$$A_v = \sum_{i=1}^{N^\theta} \delta A_{vi} = \sum_{i=1}^{N^\theta} l_i^\theta d = L^\theta d = N^\theta l^\theta d = \text{constant} \qquad (3)$$

It follows that $Ld = \sum_{\theta=-90°}^{\theta=-90°} l^\theta d = A_v \sum_{\theta=-90°}^{\theta=-90°} (1/N^\theta)$. Thus, multiplication of numerator and denominator of the first part in parenthesis of Eq. (1) by d, and use of the foregoing expressions for $l^\theta d$ and Ld, eliminates the common factor A_v yielding the expression:

$$\mathbf{V} = 4\left(\left(\sum_{\theta=-90°}^{\theta=-90°} (1/N^\theta) \right)^{-1} \sum_{\theta=-90°}^{\theta=-90°} (1/N^\theta) \mathbf{n}^\theta \otimes \mathbf{n}^\theta - \frac{1}{4}\mathbf{I} \right) \qquad (4)$$

Surprisingly any notion of void length disappears from the form of Eq. (4), despite the fact the void fabric tensor was defined in Eq. (1) on the basis of void length segments. Instead, the weighting factors of the tensors $\mathbf{n}^\theta \otimes \mathbf{n}^\theta$ are limited to the inverse $1/N^\theta$ of the total number of the void scan line segments for each angle θ, and do not depend on the lengths of the void segments or the mean void length. This shortcoming has even more serious implications when a very large number N^θ of very small void lengths can overwhelm the effect of larger voids and induce erroneously an almost isotropic fabric tensor $\mathbf{V}$.

By similarly analyzing the definition of Eq. (2), an even more serious shortcoming arises: recalling that $\bar{l}^\theta = L^\theta/N^v$, $l^\theta = L^\theta/N^\theta$ and $l^\theta d = A_v/N^\theta$, it follows that $\bar{l}^\theta d = (N^\theta/N^v)l^\theta d = A_v/N^v$. Multiplication now of numerator and denominator of Eq. (2) by d and use of the above-mentioned expressions leads to the elimination of the common factor A_v and yields the expression $\mathbf{V}_2 = \left(N^v \sum_{\theta=-90°}^{\theta=90°} \frac{1}{N^\theta} \right)^{-1} \frac{1}{2}\mathbf{I}$. The result implies that $\mathbf{V}_2$ is always an isotropic tensor, a serious drawback of Eq. (2) that is intended to measure anisotropy.

In Theocharis et al. [2] two remedies to the foregoing shortcomings were proposed and verified. The first consist of using a cut-off threshold on the void length l^θ that will eliminate the influence of very small such lengths; it applies to both Eqs. (1) and (2). The second, which applies to Eq. (1), is more natural and simply uses the void vector $\mathbf{v}^\theta = l^\theta \mathbf{n}^\theta$ as the basis for defining $\mathbf{v}^\theta \otimes \mathbf{v}^\theta = l^{\theta^2} \mathbf{n}^\theta \otimes \mathbf{n}^\theta$ as the main element of void fabric tensor; the appearing square length l^{θ^2} as the weighting factor of $\mathbf{n}^\theta \otimes \mathbf{n}^\theta$ instead of l^θ of Eq. (1), prohibits its elimination because it has no connection to void surface area if multiplied by d.

4 Conclusions

In this work the existing scan line void fabric definitions were analyzed and sever shortcomings were analytically proved in 2D, but same conclusions apply to 3D. This deficiency of the definitions limits their applicability and usefulness, and requires modifications, as per those proposed in Theocharis et al. [2].

Acknowledgements. The research leading to these results has received funding from the European Research Council under the European Union's Seventh Framework Program (FP7/2007-2013)/ERC IDEAS Grant Agreement n 290963 (SOMEF).

References

1. Oda, M., Nemat-Nasser, S., Konishi, J.: Stress-induced anisotropy in granular masses. Soils Found. **25**(3), 85–97 (1985)
2. Theocharis, A.I., Vairaktaris, E., Dafalias, Y.F.: Scan line void fabric anisotropy tensors of granular media. Granular Matter **19**(4), 1–12 (2017)
3. Ghedia, R., O'Sullivan, C.: Quantifying void fabric using a scan-line approach. Comput. Geotech. **41**, 1–12 (2012)
4. Kuo, C.-Y., Frost, J.D., Chameau, J.-L.: Image analysis determination of stereology based fabric tensors. Geotechnique **48**(4), 515–525 (1998)

DEM Simulation of Coarse-Grained Soil in Slip Zone

Shun Wang[1(✉)], Wei Wu[1], Wei Xiang[2], and Deshan Cui[3]

[1] Institute of Geotechnical Engineering,
University of Natural Resources and Life Sciences, Vienna,
Feistmantelstraße 4, 1180 Vienna, Austria
{shun.wang,wei.wu}@boku.ac.at
[2] Three Gorges Research Center for Geo-hazard,
Ministry of Education, Wuhan 430074, Hubei, China
xiang.wei@cug.edu.cn
[3] Faculty of Engineering, China University of Geosciences,
Wu Han 430074, People's Republic of China
cuideshan@cug.edu.cn

Abstract. Most landslides in Three Gorges Reservoir (TGR) area (China) contain coarse-graded soil forming the sliding zone. Conventional laboratory tests with small size samples are not suitable to study the mechanical behavior while large-scale triaxial tests and field tests are often too costly and time-consuming. In the paper, a DEM approach is described for modeling coarse-graded soils. DEM samples are generated based on 2D thin-section of X-ray CT images of intact coarse-graded samples. A series of DEM biaxial compression tests are then conducted to investigate the mechanical behaviors of a coarse-graded sample.

1 Introduction

Most landslides in TGR region are characterized by preexisting surfaces. The large deformation within the sliding zones usually gives rise to clastic soil formation. This clastic soil usually contains coarse-grained particles crushed from underlying bedrocks or floating rocks from overlying colluvium and varying proportions of fine-grained soils composed of sand, silt, and clay. However, the existing of gravels in the soil limits the experimental study of coarse-graded soil, which constitutes a strong motivation for the present study. In this paper, we introduce a DEM approach for modeling the mechanical behaviour of intact coarse-graded soil from slip zone.

2 Digital Image Processing Based Modeling

The proportion of coarse particles in soil is a crucial factor that affects the overall mechanical behavior of coarse-gained soils. To investigated the coarse-gained soil in sliding zones, a fresh intact soil sample is taken from the slip zone of the Huangtupo landslide in TGR area. The soil sample is then CT scanned to obtain the spatial distribution of the coarse fractions.

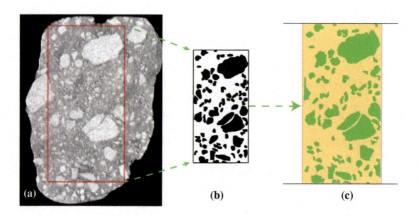

Fig. 1. Steps for the transformation from X-ray CT image to DEM sample: (a) CT image [4], (b) Digital image and (c) DEM sample

The method used in this paper provides a straightforward way to generate the DEM sample. Figure 1(a) presents a 2D thin-section of X-ray CT image. The image shows that the fine and coarse particles can be easily distinguished by binarization processing of the CT image. The CT images are 100 mm in width and 200 mm in height, in which the spatial distribution of the gravels can be recognized by geometry vectorization. The vectorial images are then transformed into discrete pixels format, e.g. digital image. We assume each digital image contains 100 × 200 pixels and each pixel represents a soil particle, either fine or coarse particle. Therefore, a soil sample can be expressed as a discrete function in the i and j Cartesian coordinate system [1].

$$Digital\ image = f_{ij} = \begin{bmatrix} f_{11} & f_{12} & \cdots & f_{1H} \\ f_{21} & f_{22} & \cdots & f_{2H} \\ \vdots & \vdots & \ddots & \vdots \\ f_{W1} & f_{W2} & \cdots & f_{WH} \end{bmatrix}$$

where i from 1 to 100, and j from 1 to 200. By Making use of the corresponding digital image, the particles representing gravels can be determined. A step for the transformation of X-ray CT image to DEM sample is shown in Fig. 1.

3 DEM Biaxial Compression Tests

A series of DEM biaxial compression test was conduced to investigate the mechanical behaviour of the intact coarse-graded soil. During the biaxial compression, the confining pressures remain unchanged. The stress and strain were monitored by measurement circles. It is worth noting that flexible lateral boundaries consisting of highly bonded particles is applied for the biaxial compression test, which mimic the soft rubber membrane used in triaxial tests as originally proposed by Kuhn [2] and widely used by Jiang [3]. The flexible membranes allows capturing the deformation of numerical sample in good coincidence with physical experimental results.

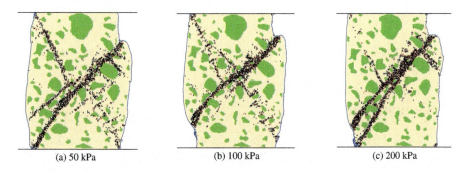

(a) 50 kPa (b) 100 kPa (c) 200 kPa

Fig. 2. The shear band and bonds breakage for coarse-graded soil samples under different confining pressures of 50 kPa, 100 kPa and 200 kPa

The deformation patterns in a numerical sample observed under different confining pressures are shown in Fig. 2. These figures show that the shear bands develop from the top-right to the bottom-right corner of the sample and exhibit similar inclination to the horizontal. It is noted that a bifurcation is observed in the sample under 50 kPa of confining pressure, while the bands become centralized with the confining pressure. For the sample under 200 kPa of confining pressure, a shear band developed from the top-right corner of the sample decomposes into two shear bands which surround some coarse particles. This phenomenon is known as "bypass road effect". During the deformation process, the shear bands search for the weakest path and formed in the fine-grained subdomain, as observed in FEM simulation [4].

The results obtained from biaxial compression test under different confining pressures of 50 kPa, 100 kPa and 200 kPa are presented in Fig. 3. It shows that all samples are characterized by strain softening and shear dilatency. Likewise, the volumetric responses show more pronounced dilative with lower confining pressure. The internal friction angle and cohesion of pure fine-grained sample are 20.4° and 18.8 kPa, while the coarse-graded samples give rise to a internal friction angle of 21.8° and cohesion of 2.8 kPa. The increase in frictional angle and decrease in cohesion is due to the existing of coarse particles. This results coincide with results obtained from some field tests and FEM simulations of coarse-graded soils [5,6].

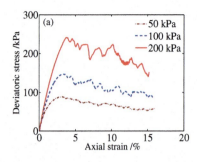 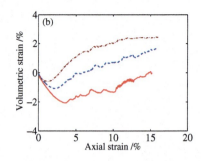

Fig. 3. (a) Stress-strain and (b) Volumetric responses of coarse-graded soil in DEM simulations

4 Conclusions

We described a DEM approach modeling of intact coarse-graded soil. X-ray CT images of the intact coarse-graded soil are used to generate the DEM samples. DEM biaxial compression tests were conducted to investigate the mechanical behaviors of a coarse-graded sample. The numerical simulation can give reasonable mechanical responses of the intact coarse-graded soil.

Acknowledgements. This work was funded by the National Natural Science Foundation of China (No. 41772304) and the project "GEORAMP" within the RISE programme of Horizon 2020 under grant number 645665.

References

1. Yue, Z., Chen, S., Tham, L.: Finite element modeling of geomaterials using digital image processing. Comput. Geotech. **30**(5), 375–397 (2003)
2. Kuhn, M.R.: A flexible boundary for three-dimensional DEM particle assemblies. Eng. Comput. MCB UP Ltd. **12**(2), 175–183 (1995)
3. Jiang, M., Zhu, H., Li, X.: Strain localization analyses of idealized sands in biaxial tests by distinct element method. Front. Archit. Civil Eng. China **4**(2), 208–222 (2010)
4. Jiang, J.W., Xiang, W., Rohn, J., Schleier, M., Pan, J., Zhang, W.: Research on mechanical parameters of coarse-grained sliding soil based on CT scanning and numerical tests. Landslides **13**(5), 1261–1272 (2016)
5. Xu, W.J., Hu, R., Tan, R.: Some geomechanical properties of soil-rock mixtures in the Hutiao Gorge area China. Géotechnique **57**(3), 255–264 (2007)
6. Xu, W.J., Yue, Z.Q., Hu, R.: Study on the mesostructure and mesomechanical characteristics of the soil-rock mixture using digital image processing based finite element method. Int. J. Rock Mech. Min. Sci. **45**(5), 749–762 (2008)

Discrete Element Experiment and Simulation

Zhaofeng Li[1], Jun Kang Chow[1], Yu-Hsing Wang[1(✉)], Quan Yuan[2], Xia Li[3], and Yan Gao[4]

[1] Hong Kong University of Science and Technology, HKSAR,
Clear Water Bay, Hong Kong, China
ceyhwang@ust.hk
[2] Guangzhou Metro Design and Research Institute Co., Ltd., Guangzhou, China
[3] Southeast University, No. 2 Sipailou, Nanjing 210096, Jiangsu, China
[4] Sun Yat-sen University, Guangzhou, China

Abstract. The advancement of sensing and manufacturing technologies allows us to carry out the discrete element experiment and simulation in parallel; this possibility grants us a precious opportunity to reassess the validity of discrete element method (DEM). In this study, the biaxial test of randomly packed printed elliptical rods and the corresponding DEM simulation are carried out in parallel. We first present the specially designed biaxial testing system, for which the 3D printing technique is applied to ease the manufacturing process. A packing of elliptical rods, also produced by the 3D printer, is used as the testing sample. The particle motion during shearing is traced using the particle image velocimetry (PIV) and close-range photogrammetry (CRP) techniques. In the DEM simulations, the contact model is derived and then validated by finite element analysis; other associated parameters are calibrated by different tailor-made experiments. Overall, the DEM simulation is found effective to reproduce the experimental findings of the macroscopic stress-strain-volumetric response and the microscopic strain localization.

Keywords: 3D printing · Biaxial test · DEM simulation

1 Introduction

Technology breakthroughs on data sensing and 3D printing are revolutionizing our world, including the geotechnical research; in this field, the 3D printing technique has been adopted in geomaterial testing [1] and device producing [2]. Inspired by these works, this study aims to leverage these advanced techniques to study the granules by experimentation. Considering that the experiment and simulation, the two common approaches for particulate study, are usually performed separately without comparisons, the experiment is specially designed herein so that the representative simulation using discrete element method (DEM) could be conducted in parallel. Besides, the validity of DEM could be reassessed. This paper begins with the details of a biaxial system made with the 3D printing technique. Then, the details of conducting the DEM simulation, in parallel to the experiment, are described. Finally, the physical and numerical results are presented and compared to examine the validity of DEM.

2 Biaxial Shearing System

A biaxial testing system is fabricated with the aid of 3D printing technique. As shown in Fig. 1a, the system is composed of (1) a loading frame and a biaxial cell, (2) a volume-measuring unit, and (3) a control unit. Figure 1b displays the biaxial cell, which exhibits a complicated geometry and is produced using the 3D printer, Objet Connex350™ (Stratasys, Inc. Minn. USA). The highest printing resolution of 16 µm is adopted to ensure the production quality. The raw printing material, i.e., VeroWhitePlus Full-Cure835, is selected, because of its high moduli to prevent failure occurring on the cell. Compared with the traditional methods (i.e., working with lathes, drills, stamping presses and moulding machines), 3D printing not only provides high-quality manufacturing, but also significantly saves time and costs; in this production, the material cost is only ~2,000 US dollars and the producing time is <2 days. Testing elliptical particles (rods), as shown in Fig. 1c, are also 3D printed, with two object dots printed at the center and the edge of the cross section. Two raw materials are used for the printing, with VeroWhitePlus FullCure 835 and TangoBlackPlus FullCure 930 for the white and black part, respectively.

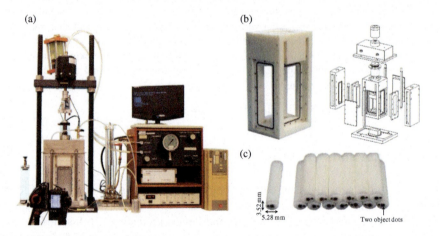

Fig. 1. The biaxial testing system: (a) the picture; (b) the middle cell which is manufactured by 3D printing technique; and (c) the testing elliptical rod, also 3D printed.

An experiment of shearing an assembly of randomly packed rods is carried out with this biaxial system. The initial rod packing before shearing is shown in Fig. 2a. A confining pressure σ_c of 50 kPa is used for consolidation. Then, the vertical loading is applied at a speed of 0.08%/min under constant lateral pressure. The axial stress σ_a, axial strain ε_a and volumetric strain ε_v are monitored continuously during the test. To capture high-resolution images of the particle arrangement, close-range photogrammetry (CRP) is used, where a digital camera (Canon EOS 7D) equipped with a timer controller (Canon TC-80N3) is used. The sample images are automatically taken at intervals of 1 min. Additionally, as two object dots are designed on the rods, Particle Image Velocimetry (PIV) technique is adopted to trace the change in rod positions and orientations during shearing based on the images taken.

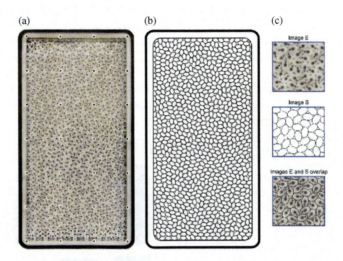

Fig. 2. Reconstruction of initial packing: (a) the packing from experiment; (b) the reconstructed packing in the DEM simulation; and (c) one-to-one mapping between the two packings.

3 Details of DEM Simulation

Based on the 3D printing, CRP and PIV techniques, the experiment described in Sect. 2 allows conducting the DEM simulation in parallel. The initial packing from the experiment could be reconstructed in the simulation, as shown in Fig. 2b. Starting from the contact level, the contact behavior between rods and the associated model is proposed and validated against the finite element analysis (FEA). Details of the contact model could be referred to Li et al. [3]. Young's modulus and Poisson's ratio of the printing material, as needed in the contact model, are calibrated as 1.58 GPa and 0.36, respectively. Indentation test of spherical particle printed with the same material is carried out for this calibration, and the parameters are obtained from indentation-force response. Interparticle friction angle is measured by stacking two piles of testing rods on an inclined table with an adjustable angle. The slope angle of the table is gradually increased until the rod pile slides, and this slope angle corresponds to the friction angle, which is 27.5°. Finally, using the calibrated contact model, the position and orientation of each rod, i.e., the initial packing is reconstructed in the DEM simulation. Figure 2c shows that between the physical and numerical samples, particles are one-to-one mapping in terms of shape, position and orientation. Upon this reconstruction, the biaxial shearing test could be reproduced in the DEM simulation; essentially, the experiment and simulation are carried out in a parallel and meaningful way.

4 Result Comparison between Experimental and Simulation

Figure 3a shows the stress-strain-volumetric responses from the experiment and simulation. Both curves show a good agreement between each other. At small strain $\varepsilon_a <$ 0.4%, the samples exhibits similar stiffness, suggesting the effectiveness of the calibrated contact model. As shearing continues, the initial structure is changed and discrepancies are gradually enlarged. However, at large deformation $\varepsilon_a < 10.0\%$, as expected, both responses become similar again since they reach critical state. Similar phenomena could be found at the microscale, as shown Fig. 3b. Initially, in both experiment and simulation, the local strain is distributed randomly. As shearing proceeds, small differences emerge but ultimately two crossing shear bands can be observed in both samples. All these have demonstrated the effectiveness of the biaxial system and the validity of DEM in reproducing the behavior of granules.

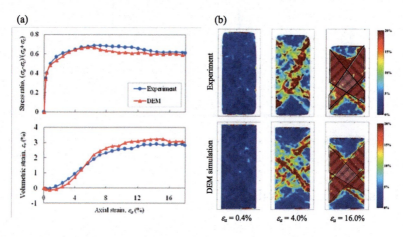

Fig. 3. Result comparison between experiment and DEM simulation: (a) the stress-strain-volumetric response; and (b) the distribution of local strain.

5 Summary and Conclusions

In this study, the 3D printing technique is used to manufacture a tailor-made biaxial shearing system, including the testing cell and rods. Together with the CRP and PIV techniques, these techniques enable users to carry out the DEM simulation in parallel. In the simulation, the contact model is derived and validated by FEA; the associated parameters are calibrated by different experiments. In general, the results in terms of stress-strain-volumetric response and local strain distribution from the experiment and simulation are similar, which demonstrates the effectiveness of the shearing system and the validity of DEM in reproducing the behavior of granules.

References

1. Tian, W., Han, N.: Preliminary research on mechanical properties of 3D printed rock structures. Geotech. Test. J. **40**(3), 483–493 (2017)
2. Shen, H., Haegeman, W., Peiffer, H.: 3D printing of an instrumented DMT: design, development, and initial testing. Geotech. Test. J. **39**(3), 492–499 (2016)
3. Li, Z., Wang, Y.H., Li, X., Yuan, Q.: Validation of discrete element method by simulating a 2D assembly of randomly packed elliptical rods. Acta Geotech. **12**(3), 541–557 (2017)

Analytical Data Based Time-Dependent Mechanical Responses of an Intergranular Bond Considering Various Bond Sizes

H. N. Wang[1(✉)], N. Che[1], H. Gong[1], and M. J. Jiang[2]

[1] School of Aerospace Engineering and Applied Mechanics,
Tongji University, Shanghai, China
wanghn@tongji.edu.cn
[2] Department of Geotechnical Engineering, College of Civil Engineering,
Tongji University, Shanghai, China

Abstract. Geo-materials usually exhibit time-dependent mechanical behavior sand deform gradually with time even under an applied constant strain. In this study, the time-dependent stiffness and strength of an intergranular bond in discrete element method (DEM) are proposed based on the modified Dvorkin's analytical solutions. The proposed stiffness can provide an alternative approach to determine the reasonable micro parameters in DEM according to actual bond size and bond materials. Meanwhile, through introducing the long-term strength of materials into the unified strength theory, the time-dependent bond strength can be formulated by fitting the analytical data. It is believed that our achievements can give valuable reference to the determination of bond stiffness and bond strength at different time points for various bond sizes in DEM.

Keywords: DEM · Time-Dependence · Intergranular bond

1 Introduction

Geo-materials such as soil and rock usually exhibit time-dependent behaviors [1], and deform gradually with time even though a constant strain is applied. It is significant to appropriately capture various time-dependent behaviors of different materials in numerical simulations. Discrete element method (DEM) provides a powerful mean to reveal the contribution of mechanically evolving intergranular bonds in cemented materials. The core of DEM for cemented materials is the establishment of contact models for intergranular bonds. Many studies have explored various contact models to simulate the time-dependent deformation of granular materials via DEM [2–5]. However, to the best of our knowledge, the bond contact model that can capture the size dependence of bonds is still in urgent need for some particular geo-materials such as methane hydrate bearing soils (MHBSs) and sedimentary rocks. In this study, considering the bond shape and size, the time-dependent stiffness and strength of an intergranular bond are proposed based on an analytical model.

In two-dimensional (2D) DEM, the particle is usually simplified as a disc (see Fig. 1). Based on an analytical solution for stress field of a generic bond proposed by

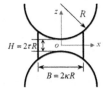
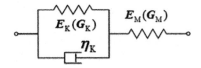

Fig. 1. Geometric model of disc particle bond

Fig. 2. Generalized Kelvin viscoelastic model

Dvorkin [6], the expressions of normal, tangential and bending bond stiffness have been deduced in Ref. [7]. According to the stiffness expression and the corresponding elastic-viscoelastic principle [8], the bond stiffness with respect to time can be obtained by substitution of elastic parameters with viscoelastic ones in elastic stiffness and subsequent Laplace and inverse transformations.

Since the expressions of stiffness are fairly complex after Laplace and inverse transformations, it is necessary to seek simple expressions by fitting the exact solutions. In the following analysis, the generalized Kelvin viscoelastic model (see Fig. 2) is employed to simulate the viscoelastic properties of bond materials under lower stresses, where G_M and G_K are the shear moduli of Hook and Kelvin parts of the model respectively, and η_K is the coefficient of viscosity. For the sake of generality, the stiffness are normalized as follows in subsequent analysis:

$$\overline{K}_n^{HK}=K_n^{HK}/(G_K B), \quad \overline{K}_s^{HK}=K_s^{HK}/(G_K B), \quad \overline{K}_m^{HK}=K_m^{HK}/(R^2 B G_K B) \quad (1)$$

where K_n^{HK}, K_s^{HK} and K_m^{HK} represent the time-dependent normal, shear and bending stiffness, respectively; B and R are the bond width and radius of the disc-shaped particle, respectively; $T_K = \eta_K/G_K$ is the relaxation time of the generalized Kelvin model. Thus, the normalized stiffness is only related to the normalized material parameter (G_M/G_K) and time (t/T_K). Fig. 3 presents the variation of normalized stiffness (only the component containing material parameters) versus time for different G_M/G_K values. According to massive data fitting, the simplified expressions of the normalized stiffness are obtained as follows:

$$\begin{aligned}
\overline{K}_n^{HK} &= (-1.57e^{(-0.439\frac{G_M}{G_K})} + 1.58)(4.14e^{(-2.475\frac{t}{T_K})} + 2.63)\frac{\kappa}{\tau} \\
\overline{K}_s^{HK} &= (-1.03e^{(-0.46\frac{G_M}{G_K})} + 1.08)(3.21e^{(-4.54\frac{t}{T_K})} + 1.04)(\frac{\kappa}{\tau})^{0.84} \\
\overline{K}_m^{HK} &= (-1.03e^{(-0.44\frac{G_M}{G_K})} + 1.06)(2.03e^{(-3.0\frac{t}{T_K})} + 1.05)\kappa^{2.34}\tau^{-0.66}
\end{aligned} \quad (2)$$

where κ and τ are the width and thickness coefficients, respectively.

1.1 Time-Dependent Bond Loading Resistance

Assume that the compressive strength of bond material is a function of time, i.e., $\sigma_c = \sigma_{c\infty} + (\sigma_{c0} - \sigma_{c\infty})e^{-at}$, where $\sigma_{c\infty}$ and σ_{c0} are the long-term and initial strengths

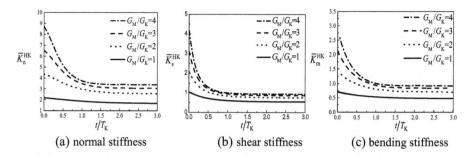

Fig. 3. Normalized stiffness versus time.

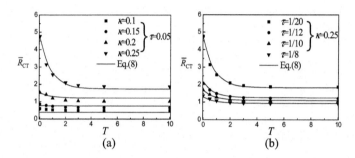

Fig. 4. Bond compression resistance versus normalized time for (a) various bond thicknesses with bond width $\tau = 0.05$, and (b) various bond widths with bond thickness $\kappa = 0.25$.

respectively and a reflects the strength decreasing rate. To investigate the bond properties of a certain brittle material such as rock, the ratio of tensile strength to compressive strength is set as 0.1 and that of long term strength to initial strength is 0.7. The normalization of bond loading resistance can be found in Ref. [7]. Figure 4 presents the normalized bond compression resistance ($\bar{R}_{CT}$) against the normalized time ($T = at$) for various bond shapes. The data can be fitted by the following equation:

$$\bar{R}_{CT} = 5088\kappa^{10.4}\tau^{-2.3}e^{-1.3T} + \bar{R}_{C\infty} \qquad (3)$$

where $\bar{R}_{C\infty}$ is the final bond compressive resistance and $\bar{R}_{C\infty} = 2.2\kappa^{1.6}\tau^{-0..67}$.

In order to analyze the bond shear resistance under combined tension/compression shear load, Fig. 5 plots the shear resistance at different time points versus compression load for various bond sizes. By data fitting, the following equation is proposed for describing the relationship between the normalized shear ($\bar{R}_{ST}$) and compression ($\bar{R}_c$) resistance:

$$\left(\frac{\bar{R}_c - \xi}{\delta}\right)^2 + \left(\frac{\bar{R}_{ST}}{\bar{R}_{CS}}\right)^2 = 1 \qquad (4)$$

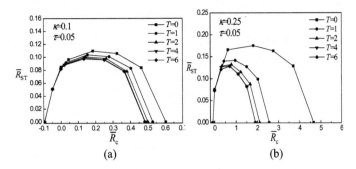

Fig. 5. Shear resistance at different times versus compression load for various bond size: (a) $\kappa = 0.1$, $\tau = 0.05$; (b) $\kappa = 0.25$, $\tau = 0.05$

where $\delta = \frac{\bar{R}_{lt} + \bar{R}_{CT}}{2}$ and $\xi = \delta - \bar{R}_{lt}$; $\bar{R}_{lt}$ is the bond tension resistance and $\bar{R}_{lt} = 3.82\kappa^{0.3}\tau^{0.2}$; $\bar{R}_{CS} = 0.28\kappa^{2.96}\tau^{-0.78}e^{-1.1T} + 0.066\kappa^{0.25}\tau^{-0.33}$.

2 Conclusions

Considering bond shape coefficients, a series of time-dependent expressions for bond stiffness is proposed in this study. Meanwhile, through introducing the long-term strength of materials into the unified strength theory, the time-dependent bond loading resistance can be formulated by fitting the analytical data. Our achievements can give valuable reference to the determination of bond stiffness and bond strength at different time points for various bond sizes in DEM.

References

1. Li, X.Z., Shao, Z.S.: Micro–macro modeling of brittle creep and progressive failure subjected to compressive loading in rock. Environ. Earth Sci. **75**(7), 583–593 (2016)
2. Silvani, C., Désoyer, T., Bonelli, S.: Discrete modelling of time-dependent rockfill behaviour. Int. J. Numer. Anal. Meth. Geomech. **33**(5), 665–685 (2009)
3. Zhou, M.J., Song, E.X.: A random virtual crack DEM model for creep behavior of rockfill based on the subcritical crack propagation theory. Acta Geotech. **11**(4), 827–847 (2016)
4. Zhao, Z., Song, E.X.: Particle mechanics modeling of creep behavior of rockfill materials under dry and wet conditions. Comput. Geotech. **68**, 137–146 (2015)
5. Jiang, M.J., Liao, Z.W., Zhang, N., et al.: Discrete element analysis of chemical weathering on rock. Eur. J. Environ. Civ. Eng. **19**(sup1), 15–28 (2015)
6. Dvorkin, J., Mavko, G., Nur, A.: The effect of cementation on the elastic properties of granular material. Mech. Mater. **12**(3–4), 207–217 (1991)
7. Wang, H.N., Gong, H., Liu, F., Jiang, M.J.: Size-dependent mechanical behavior of an intergranular bond revealed by an analytical model. Comp. Geot. **89**, 153–167 (2017)
8. Kim, Y.R., Lee, Y.C., Lee, H.J.: Correspondence principle for characterization of asphalt concrete. J. Mater. Civ. Eng. **7**(1), 59–68 (1995)

Visualization of Geogrid Reinforcing Effects Under Plane Strain Conditions Using DEM

Zhijie Wang[1(✉)], Martin Ziegler[2], and Guangqing Yang[1]

[1] School of Civil Engineering, Shijiazhuang Tiedao University, Shijiazhuang 050043, China
zwang@stdu.edu.cn
[2] Institute of Geotechnical Engineering, RWTH Aachen University, 52074 Aachen, Germany

Abstract. In order to visualize the geogrid reinforcing effects under plane strain conditions, numerical biaxial compression tests of unreinforced and geogrid reinforced specimens were carried out using the discrete element method (DEM). Both the total and the horizontal particle displacement were obtained as well as the contact force development in the specimens and the tensile force development along the geogrids at different loading stages. The visualization results in this study provide researchers improved understanding of geogrid reinforcing effects under plane strain conditions.

Keywords: Geosynthetics · Plane strain · Discrete element method

1 Introduction

Geogrids have been widely used in practice to increase the bearing capacity and the stability of various reinforced soil structures, such as reinforced embankments and reinforced retaining walls, etc. [1, 2]. Geogrid reinforcement mechanisms are performed via the interaction between soil and geogrid. Hence, the geogrid-soil interaction is the key issue to investigate the geogrid reinforcement mechanism.

In order to describe the interaction between soil and geogrid, plenty of laboratory tests have been carried out, e.g. Ezzein and Bathurst [3]. However, in real geogrid reinforced soil structures, e.g. reinforced embankments and/or reinforced retaining walls, the reinforced soil is generally under the plane strain conditions. Therefore, it is of great importance to investigate the geogrid reinforcing effects under plane strain conditions.

The discrete element method (DEM), which has particular advantages of getting detailed insights into the geogrid-soil interaction at a microscopic scale, has been commonly used to visualize the geogrid reinforcing effects, such as Stahl et al. [4]. Therefore, numerical investigations and analyses using the DEM are carried out in this study based on unreinforced biaxial specimens and reinforced biaxial specimens with two layers of geogrids. The numerically obtained particle displacement distributions are also compared with the corresponding experimental observations, which were conducted at RWTH Aachen University.

2 Force Development Under Plane Strain Conditions

Similar to previous work [5], the geogrid reinforcing effects were investigated under plane strain conditions in this study. It should be noted that only half of the numerical specimens were simulated so as to save the computational costs. The micro input parameters of the DEM investigations were calibrated by the experimental results of large-scale biaxial compression tests.

Figure 1 shows the qualitative contact force development in the unreinforced specimen at different vertical strains. The black lines represent the contact forces in the specimen and the thickness of the lines is proportional to the magnitude of the contact forces.

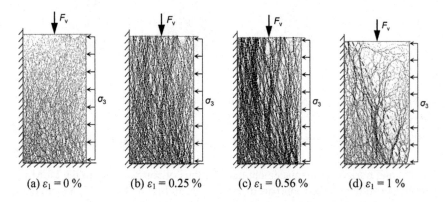

(a) $\varepsilon_1 = 0\%$ (b) $\varepsilon_1 = 0.25\%$ (c) $\varepsilon_1 = 0.56\%$ (d) $\varepsilon_1 = 1\%$

Fig. 1. Contact force development in unreinforced specimen.

At the vertical strain of $\varepsilon_1 = 0\%$ (Fig. 1a), the contact forces between soil particles in the specimen are relatively small. With increasing vertical load, the contact forces increase in the specimen and the orientations of the contact forces are mainly vertical, especially at the vertical strain of $\varepsilon_1 = 0.56\%$ (Fig. 1c), which corresponds to the maximum vertical load. Due to further development of shear zones in the post-peak phase, failure occurs in the specimen and the contact force decreases in the specimen and the orientations of the contact forces in the shear zone become diagonal. At the vertical strain of $\varepsilon_1 = 1\%$ (Fig. 1d), the contact forces between the top plate and the specimen are quite small, which visualizes the complete failure state of the numerical unreinforced specimen.

Figure 2 presents the qualitative contact force distributions in the specimen and the qualitative tensile force distributions along the geogrid at different vertical strains. The red lines indicate the tensile forces within the geogrid.

At the vertical strain of $\varepsilon_1 = 0\%$ (Fig. 2a), the contact forces between soil particles in the reinforced specimen are relatively small as well as the tensile forces along the geogrids. With increasing vertical strain, the contact forces in the specimen increase and the compression load applied on the top plate is transferred gradually from soil to the geogrids, which activates the geogrid tensile forces (see Fig. 2b). Due to the geogrid frictional and bearing resistance, soil particles in the vicinity of both geogrid layers are restrained laterally and soil arching is developed in the specimen between two geogrid layers as well as in the specimen between the geogrid and the top/bottom plate, as shown in Fig. 2c. At the vertical strain of $\varepsilon_1 = 2.2\%$ (Fig. 2d), sliding failure occurs at the upper geogrid layer, which leads to the failure of the numerical reinforced specimen.

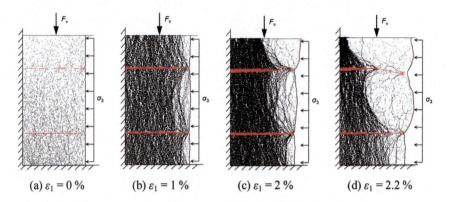

(a) $\varepsilon_1 = 0\%$ (b) $\varepsilon_1 = 1\%$ (c) $\varepsilon_1 = 2\%$ (d) $\varepsilon_1 = 2.2\%$

Fig. 2. Contact force and tensile force development in geogrid reinforced specimen.

3 Particle Displacement Under Plane Strain Conditions

The total and horizontal displacement distributions of soil particles in the DEM investigations are shown in the right part of each figure in Fig. 3. As a comparison, the total and horizontal displacement distributions of soil particles obtained in the laboratory tests using the Digital Image Correlation (DIC) method are also presented in Fig. 3 (the left part of each figure). The total and horizontal particle displacement distributions in the numerical unreinforced specimen are quite similar to those observed in the experimental studies, i.e. sliding wedges develop from the middle top to the lateral bottoms of the specimen.

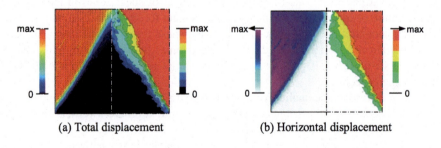

(a) Total displacement   (b) Horizontal displacement

Fig. 3. Particle displacement distributions of unreinforced specimens.

The total and horizontal displacement distributions of soil particles in the numerical reinforced specimens are shown in the right part of each figure in Fig. 4. For the purpose of qualitative comparison, the total and horizontal displacement distributions of soil particles in the physical reinforced specimens are also presented in Fig. 4 (the left part of each figure). The total and horizontal particle displacement distributions of the numerical reinforced specimens show similar kinematic behavior to the experimental observations, i.e. the shear zones do not develop from the middle top to the lateral bottoms of the specimen directly. The particle displacements, especially the horizontal particle displacements, have been greatly restricted by the two layers of geogrid reinforcements.

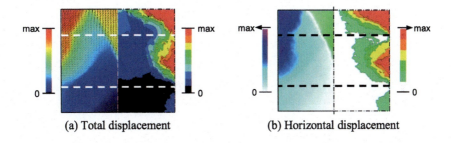

(a) Total displacement   (b) Horizontal displacement

Fig. 4. Particle displacement distributions of reinforced specimens.

4 Conclusion

The presented results in this study is based on the discrete element modeling of unreinforced and geogrid reinforced specimens under biaxial compression loads. The contact force development in the specimens and the tensile force development along the geogrids at different loading stages visualized the geogrid reinforcing effects as well as the total and horizontal particle displacement distributions in unreinforced and reinforced specimens. The DEM investigation results in this study provide researchers improved understanding of geogrid reinforcement mechanisms under plane strain conditions.

Acknowledgements. This work was supported by the Youth Foundation of Hebei Education Department (No. QN2017134) and the Natural Science Foundation of Hebei Province (No. E2017210148).

References

1. Han, J., Thakur, J.K.: Sustainable roadway construction using recycled aggregates with geosynthetics. Sustain. Cities Soc. **14**, 342–350 (2015)
2. Yang, G., Liu, H., Zhou, Y., Xiong, B.: Post-construction performance of a two-tiered geogrid reinforced soil wall backfilled with soil-rock mixture. Geotext. Geomembr. **42**(2), 91–97 (2014)

3. Ezzein, F.M., Bathurst, R.J.: A new approach to evaluate soil-geosynthetic interaction using a novel pullout test apparatus and transparent granular soil. Geotext. Geomembr. **44**(3), 246–255 (2014)
4. Stahl, M., Konietzky, H., te Kamp, L., Jas, H.: Discrete element simulation of geogrid-stabilised soil. Acta Geotech. **9**(6), 1073–1084 (2014)
5. Wang, Z., Jacobs, F., Ziegler, M.: Experimental and DEM investigation of geogrid–soil interaction under pullout loads. Geotext. Geomembr. **44**(3), 230–246 (2016)

DEM Investigation of Fracture Characteristic of Calcareous Sand Particles Under Dynamic Compression

Lei Wang, Xiang Jiang, Hanlong Liu, Zhichao Zhang, and Yang Xiao(✉)

Chongqing University, Shabei Street. 83, Chongqing 400045, China
hhuxyanson@163.com

Abstract. Calcareous sand as one kind of fragile geotechnical materials, however it is often under dynamic loading in engineering. In this study, discrete elements method (DEM) simulation was used to investigate the fracture characteristics of single and contacting spherical calcareous sand particles under dynamic compression. Calcareous Sand particle was simulated by cluster made up of elemental ball. The DEM models included two kinds of particle arrangements with one and five particles bounded in a cylindrical wall, and the dynamic compression loading were applied by a drop hammer made of clumps. The numerical simulation results of single particle under dynamic compression included stress-strain relation and Weibull distribution were coincide with the experimental results. For five particles numerical model, grain size distribution was obtained for all five cluster to differentiating the degree of breakage of particles in different positions. The transfer of force between particles was also exposed through the observation of the force chain evolution during compression.

Keywords: Calcareous sands · DEM simulations · Dynamic compression
Weibull distribution · Grain size distribution

1 Brief Introduction

Unique causes and geographical distribution of calcareous sand make it more crushable than quartz sand and often subjected to dynamic load [1]. Particle breakage significantly affects the strength of materials for crushable geotechnical materials such as carbonate sand [2]. Among a number of factors that affect the degree of particle breakage, it has been reported that the strength of particles is the most significant one [3]. Taking into account the strength of brittle materials with great dispersion, Weibull distribution has been proposed and proved to be effective to describe the distribution of particle strength for geotechnical brittle material such as quartz sand and carbonate sand [4]. In view of the limitations of the experimental conditions, discrete element method (DEM) has been widely used to simulate particle breakage behavior since the clustering of balls method first introduced by Thomas and Bray [5]. The most fundamental and important issue for DEM simulation is to find the suitable bond strength between elemental particles [6]. Besides, the effects of particle-particle contact condition have rarely been considered during the process of particle breakage which might have a great influence on the sequence and the degree of particle breakage [7].

Thus, the main purpose of this paper is to carry out DEM simulation for single and contacting spherical calcareous sand particles under dynamic compression. Weibull distribution and stress-strain relationship are presented to verify the reliability of calcareous sand clusters sample for single particle numerical model. Then the five particles numerical model is established to differentiating the degree of breakage of particles in different positions by grain size distribution for all five cluster and the transfer of force between particles was also exposed through the observation of the force chain evolution during compression.

2 Method and Key Results

Calcareous sand particle was simulated by cluster made up of elemental ball. The size of cluster ranged from 0.8 to 1.0 mm and the size of elemental ball was from 0.075 to 0.5 mm. Elemental balls were bonded to a whole by parallel bond. The tensile strength of parallel bond followed a Gaussian distribution with average equal to 24 MPa and standard deviation equal to 8 MPa. Dynamic load was applied by a hummer simulated by clump and the strain rate used in experiment was 5 m/s. For single particle numerical model, the simulation was repeat 30 times with different random factors each time to replace 30 similar calcareous sand particles. The breakage stress was recorded after each simulation for further analysis and typical stress-strain relation was also obtained. For five particles numerical model, clusters were created from the middle to both sides and naming followed the order of generation. Figure 1 represents schematic of the numerical model for five calcareous sand particles dynamic compression simulation. During the process of compression, the grain size distribution of all five clusters were obtained and the force chain between clusters was also investigated.

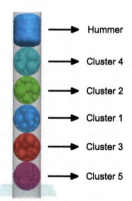

Fig. 1. Schematic of the numerical model for five calcareous sand particles dynamic compression simulation.

Figure 2(a) and (b) show the Weibull distribution of calcareous sands for numerical results and experimental results respectively. The compression experiment was carried out by the Deben MICROTEST compression apparatus on 30 calcareous sand samples with the same size to numerical samples. Weibull modulus for numerical samples is 2.0873 and the value for calcareous sand samples is 2.0926. The characteristic stress are 40.9 MPa and 41.9 MPa for numerical and experimental results respectively. The comparison between numerical and experimental results was described in Fig. 2(c) which is also consistent with each other. Hence the strength distribution of numerical sample is coincided with the calcareous sand sample. Figure 2(d) describes that the breakage of calcareous sand follows avalanche destruction.

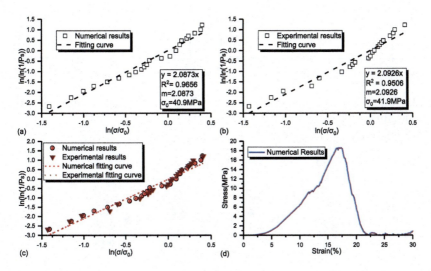

Fig. 2. Weibull distribution of calcareous sands for (a) numerical results, (b) experimental results, (c) comparison between numerical and experimental results, and (d) typical stress-strain relationship during compression.

Figure 3 presents PSD evolution of five calcareous sand clusters during dynamic compression. Figure 4 shows force chain evolution during dynamic compression. From the results we can find that although cluster 2 and cluster 4 under the same contacting condition but cluster 2 crushed first and cluster 4 crushed last which may be caused by the distance to hummer. The other three one crushed between this two cluster. Figure 4 indicated that force chain expanded with the crack growth.

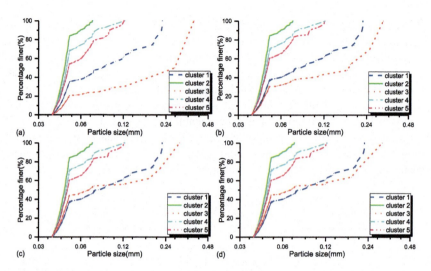

Fig. 3. PSD evolution of five calcareous sand clusters during dynamic compression for (a) t = 100 μs, (b) t = 200 μs, (c) t = 300 μs, (d) t = 400 μs.

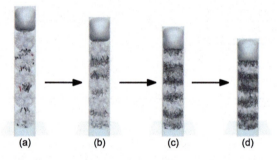

Fig. 4. Force chain evolution of five calcareous sand clusters during dynamic compression for (a) t = 100 μs, (b) t = 200 μs, (c) t = 300 μs, (d) t = 400 μs.

3 Conclusions

Firstly, a suitable model for calcareous sand particle breakage under dynamic compression was established whose strength distribution is coincide with the calcareous sand. Secondly, the sequence and degree of particle breakage can't be determined by the contacting condition. Finally, force chain spread along with crack propagation.

References

1. Xiao, Y., Liu, H., Xiao, P., Xiang, J.: Fractal crushing of carbonate sands under impact loading. Géotech. Lett. **6**(3), 199–204 (2016)
2. Einav, I.: Breakage mechanics—Part I: theory. J. Mech. Phys. Solids **55**(6), 1274–1297 (2007)

3. Yamamuro, J.A., Lade, P.V.: Drained sand behavior in axisymmetric tests at high pressures. J. Geotech. Eng. **122**(2), 120–129 (1996)
4. Mcdowell, G.R., Amon, A.: The application of Weibull statistics to the fracture of soil particles. J. Jpn. Geotech. Soc. Soils Found. **40**(5), 133–141 (2000)
5. Thomas, P.A., Bray, J.D.: Capturing nonspherical shape of granular media with disk clusters. J. Geotech. Geoenvironmental Eng. **125**(3), 169–178 (1999)
6. Lobo-Guerrero, S., Vallejo, L.E.: Discrete element method evaluation of granular crushing under direct shear test conditions. J. Geotech. Geoenvironmental Eng. **131**(10), 1295–1300 (2005)
7. Parab, N.D., Guo, Z., Hudspeth, M.C., Claus, B.J., Fezzaa, K., Sun, T., Chen, W.W.: Fracture mechanisms of glass particles under dynamic compression. Int. J. Impact Eng. **106**, 146–154 (2017)

Validation of Synthetic Images for Contact Fabric Generated by DEM

Max Wiebicke[1,3(✉)], Václav Šmilauer[2], Ivo Herle[1], Edward Andò[3], and Gioacchino Viggiani[3]

[1] Institute of Geotechnical Engineering, Technische Universität Dresden, 01062 Dresden, Germany
max.wiebicke@tu-dresden.de
[2] woodem.eu, Prague, Czech Republic
[3] Univ. Grenoble Alpes, CNRS, Grenoble INP, 3SR, 38000 Grenoble, France

Abstract. X-ray tomography can be used to acquire three-dimensional images of granular materials. In order to check the performance of image analysis tools to extract information from such images, a benchmarking strategy is presented.

Keywords: Synthetic images · Contact fabric · DEM

1 Introduction

When measurements of any kind are acquired, the accuracy of the tools that are used for the acquisition is of utmost importance. No experimentalist would run an experiment without calibrating sensors and checking their accuracy. In the measurement of the fabric in granular materials from tomographies, however, the question of accuracy is rarely posed, as a lack of publications in this area shows. The accuracy of the tools that are commonly used for determining the fabric of granular materials from x-ray tomographies was investigated in [3]. This contribution worked on individual contacts, *i.e.*, only two particles in contact. It was found that common image analysis tools produce inaccurate results to a point that these should not be used for any quantitative analysis. Improvements and advanced algorithms were proposed and shown to yield more accurate results on images of individual contacts.

This contribution aims to relate the findings and to validate the proposed methods of the study on individual contacts to assemblies of thousands of particles. Therefore, synthetic images are created from numerical reference samples that are the result of simulations with the discrete element method (DEM).

1.1 Measuring Contact Fabric

In order to determine contact fabric, the following basic steps have to be applied to the initial grey-scale images [3]: binarization of the grey-scale image;

segmentation and labelling of the binary image; contact detection and finally the determination of contact orientations

In [3] chosen common tools to apply these steps in image analysis were evaluated regarding their accuracy. It was found that the main shortcomings to detect contacts and determine their orientation were the binarisation and the segmentation: (1) Contacts are systematically over-detected using a standard global thresholding approach for the binarisation. (2) Contact orientations are strongly biased when applying simple, topological watershed for the segmentation. In order to overcome problem (1), a local thresholding was proposed and validated on individual contacts. This local threshold is applied to the globally detected contacts in order to exclude false contacts. An advanced watershed approach, the random walker, was proposed to determine contact orientations and validated on images of individual contacts.

2 Creation of Synthetic Images

In order to directly relate the accuracy of individual measurements of contact fabric on measurements of granular assemblies, a known reference fabric has to be created. The discrete element method (DEM) allows numerical assemblies of particles to create and loaded. Realistic images of these numerical samples can then be created using Kalisphera [1], a tool to create images of spheres considering the partial volume effect.

We choose the DEM software WooDEM [2] to create an initially (almost) isotropic sample of spherical particles with periodic boundary conditions. In order to mimic a triaxial compression test, this sample is initially compressed to a certain isotropic pressure and then sheared by a displacement condition in the main loading direction at constant radial stress. This triaxial compression is then followed by a final unloading step until an isotropic stress state is reached. The macroscopic response of the numerical assembly is plotted in Fig. 1.

The fabric that is exported at the chosen stages of the loading by the DEM simulation serves as the reference fabric for the image analysis. Grey-scale images

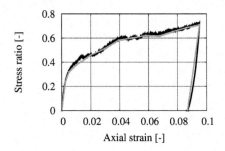

Fig. 1. Macroscopic response of the DEM simulation. Black line – continuous stress-strain curve. Grey line with points – chosen stages of the loading at which the fabric is saved.

of the assembly at these stages are created by Kalisphera. To create realistic images, that can be compared to real x-ray tomographies, certain defects have to be applied on these otherwise perfect images. Following the analysis of noise and blur in real tomographies in [3], blur and noise are applied on the images.

3 Image Analysis

The basic image analysis steps, as described in Sect. 1, are performed on the synthetic images. A local threshold, as found in [3], is applied on the identified contacts to refine the list of contacts. As recommended in [3], the random walker is used to segment particles in contact and determine their contact normal orientation.

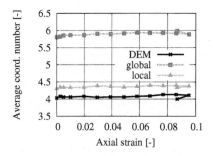
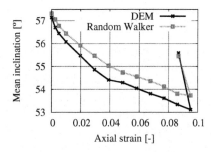

Fig. 2. Evolution of the average coordination number with the macroscopic loading. DEM as the reference fabric, global and local as in the respective contact detection approaches.

Fig. 3. Evolution of the mean inclination with the macroscopic loading. DEM as the reference fabric *vs* the Random Walker.

The average coordination number can be used to check the results of the contact detection. Figure 2 shows the evolution of the average coordination number with macroscopic loading for the reference, the common global and the refined local approach. This result validates the finding of [3] that the global approach systematically over-detects contacts of spherical particles. A local refinement has to be applied for spherical shapes in order to detect contacts more accurately.

The contact normal orientations can be expressed in spherical coordinates: inclination – the angle with the main loading direction, azimuth – the angle in the minor loading direction. As this experiment is axi-symmetric, it is sufficient to consider the inclination. Figure 3 shows the mean inclination of all contact orientations with the loading. The mean inclination tends towards the principal stress direction ($0°$ being the principal stress direction). The contact fabric determined with the random walker agrees qualitatively as well as quantitatively with the reference fabric.

4 Conclusions

In this contribution, synthetic images are used as benchmarks to evaluate the performance of the recommendations in [3]. DEM simulations are used to create numerical samples of particles that can be loaded. At chosen stages of the loading, the fabric is saved and synthetic images of the samples are created using Kalisphera [1].

The propositions suggested in [3] are found to work on these synthetic images: (1) The contact detection is improved when using the local refinement compared to the global standard procedure. (2) Contact orientations and their evolution with the macroscopic loading determined using the random walker segmentation agree with the reference fabric from the numerical simulations.

The proposed and investigated methods can be applied with confidence to real tomographies of granular assemblies. Care must be taken, however, when angular particles are tested, as these yield more problems in the image analysis [3].

Acknowledgements. The research leading to these results has received funding from the Deutsche Forschungsgemeinschaft (DFG, German Research Foundation) – 254872581 and from the European Research Council under the European Union's Seventh Framework Program FP7-ERC-IDEAS Advanced Grant Agreement No. 290963 (SOMEF). Laboratoire 3SR is part of the LabEx Tec 21 (Investissements d'sAvenir-Grant Agreement No. ANR-11-LABX-0030). We thank the Center for Information Services and High Performance Computing (ZIH) at TU Dresden for generous allocations of computational resources.

References

1. Tengattini, A., Andò, E.: Kalisphera: an analytical tool to reproduce the partial volume effect of spheres imaged in 3D. Measur. Sci. Technol. **26**(9) (2015)
2. Šmilauer, V.: Woo Documentation (2016). https://www.woodem.org
3. Wiebicke, M., Andò, E., Herle, I., Viggiani, G.: On the metrology of interparticle contacts in sand from x-ray tomography images. Measur. Sci. Technol. **28**(12), 124007 (2017)

Particle-Scale Observations of the Pressure Dip Under the Sand Pile

Qiong Xiao[1,2] and Xia Li[1(✉)]

[1] The University of Nottingham, Nottingham, NG7 2RD, UK
[2] School of Engineering, Southeast University, Nanjing 210096, People's Republic of China
xia.li@seu.edu.cn

Abstract. The pressure dip phenomenon remains an interesting question to interpret. This paper presents a numerical investigation on the pressure dip under sand piles. Discrete element simulations have been carried out to reproduce the pressure dip observed underneath the conical sand pile formed by point-deposition. Compared with the average scheme method, the Voronoi tessellation could better interpret the base plane pressure at the representative radius distances. The normal pressure profile shows the peak pressure appearing at the radius distance of approximately 0.3 R, where R denotes the radius distance of the base plane. The stress tensor was employed to study the spatial distribution of stress state and principal stress direction. The results indicated more arches are formed at the higher height than the base plane in the center of sand pile to further transmit the stress to the outer edges.

Keywords: Sand pile · Pressure dip · Stress distribution · DEM

1 Introduction

The sand pile problem is well known as the pressure dip phenomenon, referring to the depression in the normal pressure below the apex of a sand pile. It has been studied for around a century in the pressure distribution. Knowledge in this is important for practical applications such as the stability of earth embankments, dams, and silos [4]. The majority of studies in this area were inspired by the experimental results of [5, 9]. Some good reviews are available for this sand pile problem [8, 12]. To understand the formation of dip phenomenon, it has inspired attention with the stress/force propagation with analytical solutions [11] and internal structure of sand pile in terms of the distribution of particle orientation, contact orientation and contact force orientation [1, 3]. It implies that the pressure dip phenomenon relating to the formation of arches. However, the understanding of arches formation inside the heap is still limited and it gives rise to the aims of this paper.

In the recent decades, with the aid of fast growing technology, the microscopic material behaviour could give a fundamental understanding of the macroscopic behaviour. Cundall and Strack [2] developed the discrete element method (DEM) to simulate the discrete particles as rigid assemblies. Treating the particle interactions as only point contact, the macro-stress tensor (σ_{ij}) could be evaluated from the tensor of contact force (f_i) and contact vector (l_j) as Eq. (1) [6, 7]. It could estimate the stress state of each particle, using the particle volume (V_p) and the average void ratio of the heap (e), to obtain the spatial distribution of stress behavior inside sand pile for interpreting the arches formation.

$$\sigma_{ij} = \sum f_i l_j / (1+e) V_p \tag{1}$$

2 Numerical Model Set-up

Tests are performed with spherical particles under the Hertz contact model using the 3D DEM package LIGGGHTS (LAMMPS Improved General Granular and Granular-heat Transfer Simulations). Table 1 presents the material properties for the simulation, which are set with a reference of [10]. The hopper boundary is adopted to control the flow rate, with the outlet length as ten average particle sizes. To save the computational effort, particles are continuously generated within the shadow region portrayed in Fig. 1 and deposited to the base plane. A baffle boundary is generated first at the edge of base plane to avoid excess rotation of particles. It is generated by the uniformly distributed particles with the radius of 0.8 mm, set at the radius distance of ($R + 0.4$) with z-value of 0.

Table 1. Particles properties for the simulation.

Diameter	Young's modulus	Poisson's ratio	Restitution coefficient	Friction coefficient
1.2 ± 0.4 mm	70 GPa	0.35	0.5	1.0

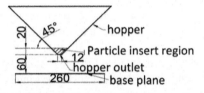

Fig. 1. The numerical model for simulation.

3 Observations of the Pressure Dip Phenomenon

The pressure distribution is one of the major aspects regarding the sand pile problem. The normal pressure is commonly obtained by the averaging values of the concentric annulus regions at representative radius distances. In this paper, the voronoi tessellation method was employed to measure the influential area for particles interacted with the base plane. For each representative radius distance, it is measured as the average value of the tessellation cells lay on it. Figure 2(a) gives an example of the voronoi tessellation cells on the base plane and cells near the boundary is reconstructed by the edge of the base plane. Figure 2(b) compares the results of pressure distribution using the two methods. The initial radius distance and the interval value are both set as 2.5 mm. The two methods are both observed a dip phenomenon with the peak pressure appearing at 40 mm. Due to the small number of particle in the center, the pressure is increased using the average scheme. It shows the voronoi tessellation method could better reflect the actual condition at specified radius distances. A smaller relative pressure dip is found compared with the results of [10]. This may be related to the lower value of repose angle using the spherical shape.

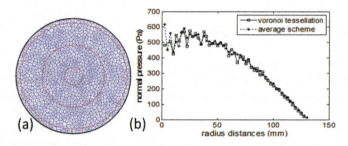

Fig. 2. The base plane result: (a) voronoi tessellation formation; (b) pressure profile

4 Observations of Local Material Behaviour

To understand the arches formation inside sand pile, it requires obtaining the local the stress tensor inside the heap. Due to the symmetrical condition, the results are analyzed employing the cylindrical coordinate system. Four different radius distances are selected to observe the stress behavior, with the initial value of 5 mm and the interval value of 12 mm as ten average particle sizes. To eliminate the random fluctuations results, the result is evaluated with the average value of particles located in the span of $[x, x + 1.2]$ mm, which are determined by the central coordinates, for each representative radius distance (x).

The stress behavior of σ_{zz} is presented in Fig. 3(a). At $h = 0$ mm, the peak magnitudes is observed at $r = 29$ mm, with a dip phenomenon appearing in the center. From the base plane to upwards, the center of sand pile also shows a lower magnitude of σ_{zz}. Figure 3(b) shows the result of principal stress direction, estimated by $\theta = \arccos(n_z)$, where n_z is the z value of unit principal stress direction vector. After the sand pile constructed, the base plane contains a significant rotation of principal stress direction at $r = 5$ mm. Inside the heap, the larger principal stress direction appears at a higher height than the results at $h = 0$ mm. It suggests that the arches are not only formed at the center of base plane but also inside the heap during construction.

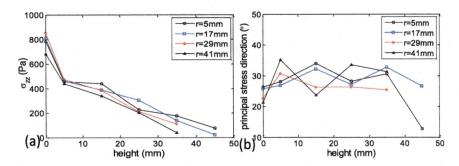

Fig. 3. The spatial distribution of stress behavior: (a) σ_{zz}; (b) principal stress direction.

5 Conclusion

This paper studies the pressure dip phenomenon in terms of the local behavior of stress state and principal stress direction after the pile construction. The results imply the arches formation in the center of sand pile with respect to the rotation of principal stress direction, which results in the decrease of pressure. The pressure profile reveals that the voronoi tessellation method could better interpret the stress state on the base plane. The 3D tessellation method and the impact of the construction history are studied in the on-going work to further understand the formation of the pressure dip phenomenon.

References

1. Ai, J., Chen, J.F., Rotter, J.M., Ooi, J.Y.: Numerical and experimental studies of the base pressures beneath stockpiles. Granular Matter **13**, 133–141 (2011)
2. Cundall, P.A., Strack, O.D.L.: A discrete numerical model for granular assemblies. Géotechnique **29**(1), 47–65 (1979)
3. Geng, J.F., Longhi, E., Behringer, R.P., Howell, D.W.: Memory in two-dimensional heap experiments. Phys. Rev. E **64**(6), 060301–060304 (2001)
4. Hummel, F.H., Finnan, E.J.: The distribution of pressure on surfaces supporting a mass of granular material. Proc. Instn. Civ. Eng. **212**, 369–392 (1920)
5. Jotaki, T., Moriyama, R.: On the bottom pressure distribution of the bulk materials piled with the angle of repose. J. Soc. Powder Technol. Jpn. **16**(4), 184–191 (1979)
6. Li, X., Yu, H.-S.: Application of stress-force-fabric relationship for non-proportional loading. Comput. Struct. **89**(11–12), 1094–1102 (2011)
7. Rothenburg, L., Bathurst, R.J.: Analytical study of induced anisotropy in idealized granular materials. Géotechnique **39**(4), 601 (1989)
8. Savage, S.B.: Modeling and granular material boundary value problems. In: Physics of Dry Granular Media, pp. 25–96 (1998)
9. Smid, J., Novosad, J.: Pressure distribution under heaped bulk solids. In: Proceedings of 1981 Powtech Conference, Industrial Chemical Engineering Symposium, vol. 63, pp. D3/V/1–D3/V/12 (1981)
10. Vanel, L., Howell, D., Clark, D., Behringer, R.P., Clement, E.: Memories in sand: experimental tests of construction history on stress distributions under sandpiles. Phys. Rev. E **60**(5), 5040–5043 (1999)

11. Wittmer, J.P., Claudin, P., Cates, M.E., Bouchaud, J.P.: An explanation for the central stress minimum in sand piles. Nature **382**(25), 336–338 (1996)
12. Wittmer, J.P., Cates, M.E., Claudin, P.: Stress propagation and arching in static sandpiles. J. Phys. I France **7**, 39–80 (1997)

Experimental Study on Clay Pore Microstructure Based on SEM and IPP

Ri-qing Xu[1,2(✉)], Li-yang Xu[1,2], and Xiao-ran Yan[1,2]

[1] Research Center of Coastal and Urban Geotechnical Engineering,
Zhejiang University, Hangzhou 310058, China
Xurq@zju.edu.cn, Xuliyang1991@126.com
[2] Engineering Research Center of Urban Underground Development of Zhejiang Province, Zhejiang University, Hangzhou 310058, China

Abstract. The soil macro-mechanical properties depend on soil microstructure. In order to build the quantitative relation between the macro porosity and pore microstructure, Zhejiang clay was taken as an example in the paper. The pore microstructure was observed by scanning electron microscope (SEM), and was quantitatively studied by Image Pro-Plus (IPP) software. The microcosmic indexes that describe the morphological feature and permutation characteristic were mainly measured. Especially, the pore grading curve was obtained to describe the permutation characteristic, and the nonuniformity coefficient of the pore grading curve was proposed, as well as the curvature coefficient. The regression analysis methods were used to build the quantitative relation. The results show that the structure and size of the clay pore are varied. The macro porosity is significantly related to the two microcosmic indexes of the roundness and the pore grading curve. Finally, the function relationship between macro porosity and pore microcosmic indexes was established.

Keywords: Microstructure · SEM · IPP · Porosity

1 Introduction

The pore structure of soil has important effects on its strength, permeability resistance, deformation, and many other macro performances. Previous study found that there is a big difference between measured soil deformation and the calculated deformation in actual engineering. Mainly because that the pore structure has changed when the soil is compressed. So the study of pore structure distribution, the size of pore diameter, and pore channels have become important research contents [1–3].

The quantitative relationship between microstructure and macro properties hasn't been established in the previous studies. It's useful to explore the relation between the porosity and pore structure. In this paper, Zhejiang clay was taken as an example. The pore microstructure was observed by scanning electron microscope (SEM), and was quantitatively studied by Image Pro-Plus (IPP) software. Especially, the pore grading curve was proposed to describe the permutation characteristic, and the nonuniformity coefficient of the pore grading curve was proposed,as well as the curvature coefficient. The regression analysis methods were used to build the quantitative relation.

2 Materials and Methods

The raw clay in this experiment was adopted from Zhejiang University, then the raw clay was made into 6 saturated soil specimens of different porosity, the porosities are 0.368, 0.384, 0.392, 0.413, 0.426 and 0.458. Then the SEM specimens were prepared. Images of the soil microstructure were taken by the GSED probe in the SEM. Then the qualitative and quantitative analysis was made. The pore microcosmic indexes include pore configuration parameters and pore arrangement parameters [4]. Pore configuration parameters include the roundness R, the abundance C, the average shape factor F and the average diameter D. Pore arrangement parameters include the rate of anisotropic I_n, the directional probability entropy H_m, the nonuniformity coefficient C_u and the curvature coefficient C_c.

3 Results

3.1 Soft Clay Pore Microstructure Characteristics

Soft clay pore microstructure characteristics include elevated pore, inter-granular pore and intra-granular pore (Fig. 1). The Fig. 1(a) shows elevated pore, which can be found in the soil skeleton. The Fig. 1(b) shows inter-granular pore, which exist among contacted soil particles in the plane. For this inter-granular pore, long axis is 6.7 μm, and the short axis is 1.9 μm, and the area is 14.7 μm^2; The Fig. 1(c) shows intra-granular pore, which is small pore in the bigger particles. The aperture of this intra-granular pore is 2.1 μm, and the area is 14.7 μm^2.

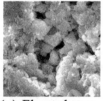
(a) Elevated pore.

(b) Inter-granular pore.

(c) Intra-granular pore.

Fig. 1. The characteristic of soft clay pore.

3.2 Pore Microcosmic Indexes

The Table 1 below shows the results of pore microcosmic indexes. It shows that with different porosity, the roundness R almost remains unchanged. Other microstructure parameters fluctuate within a certain range. Especially, with the increase of porosity, the trend of the abundance C and the average shape factor F are similar. The trend of directional probability entropy H_m and rate of anisotropic I_n are also similar.

According to the average particle size, the pore grading curve of the 6 soil samples (Fig. 2) were drawn, and the nonuniformity coefficient C_u and the curvature coefficient C_c

Table 1. Pore microcosmic indexes results.

Porosity	R	C	F	D(pixel)	I_n	H_m	C_u	C_c
0.368	0.286406	0.847764	0.812483	2.78544	0.152236	0.21175	1.5625	0.9025
0.384	0.282693	0.825185	0.80281	3.830614	0.174349	0.269633	1.8303	0.8055
0.392	0.288169	0.854628	0.815599	2.812079	0.145372	0.181961	1.5178	0.8876
0.413	0.286324	0.821303	0.808423	3.708502	0.178697	0.231789	2.0353	0.8314
0.426	0.293577	0.87296	0.841417	2.034932	0.12704	0.145536	1.4234	0.9488
0.458	0.293246	0.874478	0.859163	2.082353	0.125522	0.129885	1.4594	0.9398

(Table 1) were calculated. Here $C_u = \frac{d_{80}}{d_{20}}$, $C_c = \frac{d_{60}^2}{d_{20} \cdot d_{80}}$, and d_{20}, d_{60}, d_{80} respectively represents the pore size when the cumulative percentage of pore whose size are less than a certain aperture equals to 20%, 60% and 80%.

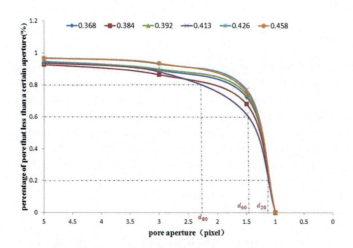

Fig. 2. The soft clay pore grading curve (It shows d_{20}, d_{60} and d_{80} of the 0.413 porosity sample).

3.3 Single Factor Regression Analysis

Single factor regression analysis was made between the pore configuration parameters and the porosity. The square of correlation coefficient (R-Square) between porosity and R, C, F, D is 0.9998, 0.9253, 0.8465, 0.4096 respectively. So the porosity is significantly influenced by the roundness R.

Single factor regression analysis was also made between the pore arrangement parameters and the porosity. The R-Square between porosity and I_n, H_m, C_u, C_c is 0.4229, 0.5591, 0.5451, 0.4049 respectively. It is difficult to determine which parameter is most significantly related to porosity.

3.4 Multiple Factors Regression Analysis

Multiple regression analysis was made in order to find a better relationship between the pore microcosmic indexes and the porosity. Both the pore configuration parameters and arrangement parameters were taken into consideration. Among the configuration parameters, the roundness R was chosen because of the largest R-Square. Among the arrangement parameters, all the 4 parameters were tried in the regression analysis respectively, because their R-Square were approximated. Finally, the curvature coefficient C_c of pore grading curve was found to be the best. The function relationship between macro porosity n and pore microcosmic indexes R, C_c is:

$$n = 10.1R^2 - 6.3C_c^2 + 19.5RC_c - 23.2935R$$

The regression effect is good: with 95% confidence level, R-Square is 0.999418. The P-Value of each coefficient is 0.02141, 0.029963, 0.026999 and 0.023335.

4 Conclusion

- The pore structure of Zhejiang clay includes elevated pore, inter-granular pore and intra-granular pore, and their apertures are varied.
- The pore arrangement characteristic can be described by the pore grading curve, the nonuniformity coefficient C_u and the curvature coefficient C_c.
- The results of single factor regression show that among the configuration parameters, the roundness R has the most significant effect on porosity. Among the arrangement parameters, the 4 parameters have similar significant effect on porosity.
- Multiple regression analysis was made with the roundness R, the curvature coefficient C_c and porosity n. The function relationship between macro porosity and pore microcosmic indexes was fitted to be:

$$n = 10.1R^2 - 6.3C_c^2 + 19.5RC_c - 23.2935R$$

References

1. Liu, S.Y., Zhang, J.W.: Fractal approach to measuring soil porosity. J. Southeast Univ. **27**(3), 127–130 (1997)
2. Zhang, X.W., Kong, L.W., et al.: Evolution of microscopic pore of structured clay in compression process based on SEM and MIP test. Chin. J. Rock Mech. Eng. **31**(2), 406–412 (2012)
3. Jiang, M.J., Wu, D., et al.: Connected inner pore analysis of calcareous sands using SEM. Chin. J. Geotech. Engi. **39**(supp. 1), 1–5 (2017)
4. Zhao, A.P.: Microscopic Mechanism Research on Frost Heave of Subgrade Soil in Season Frozen Area. Heilongjiang University Press, Heilongjiang (2010). 80–85

Investigating the Effect of Intermediate Principal Stress on Granular Materials via DEM

Dunshun Yang[1], Xia Li[2(✉)], and Hai-Sui Yu[3]

[1] Arup, 13 Fitzroy Street, London W1T 4BQ, UK
[2] School of Civil Engineering, Southeast University,
Nanjing, Jiangsu 211196, People's Republic of China
xia.li@seu.edu.cn
[3] School of Civil Engineering, Faculty of Engineering,
University of Leeds, Leeds LS2 9JT, UK

Abstract. The intermediate principal stress σ_2 has a signifant effect on the strength and deformation characteristics of granular materials. This paper reports a numerical investigation into this issue via Discrete element simulations. A series of true triaxial numerical experiments have been conducted on the initially isotropic samples to reproduce qualitatively the material behaviour. Lower peak and critical stress ratios have been observed at a greater intermediate principal stress ratio $b = (\sigma_2 - \sigma_3)(\sigma_2 - \sigma_3)$. And the intermediate ratio for strain increment is consistently higher than b. With the particle-scale information provided by DEM, observations on the internal structure and contact force have been presented. Findings from this study provides particle-scale insight into the failure criterion of granular materials.

Keywords: Intermediate principal stress · Strength
Deformation · Fabric · DEM

1 Introduction

Strength and deformation of granular materials are related to all three stress invariant. With Mohr-Column criterion initially defining material failure in terms of the major and minor principal stresses, a non-dimensional parameter, i.e., the intermediate principal stress ratio $b = (\sigma_2 - \sigma_3)(\sigma_2 - \sigma_3)$, could be been introduced as an additional variable into the formulation of yield surface to quantify the effect of the intermediate principal stresses. Difference in stress-strain behaviour of triaxial compression, triaxial extension and plane strain provides clear evidence of the importance of intermediate principal stress, in particular, true triaxial apparatus developed in 1960s [2,4,5,11,12] and Hollow cylindrical apparatus in 1970s [9,10,13] enabled independent control of three principal stresses, and provided many experimental observations with b varying from 0 to 1, which however are often mixed with the influence of cross-anisotropy, shear band occurrence, boundary effect and et al. A commonly accepted failure criterion is still missing [7].

Challenges in continuum-scale phenomenologically-based constitutive modelling promotes interests in multi-sale investigations. Discrete element method has become a popular tool in studying the stress-strain behaviour of granular materials [1,14–16]. DEM simulation results are in qualitative agreement with those reported in laboratory experiments, and can be well captured by [6]'s failure criteria. With three dimensional Stress-Force-Fabric relationship, the macroscale stress can be expressed analytically in terms of the fabric tensor and the force tensor [6], providing additional insight into the strength and deformation of granular materials from the particle scale.

2 Numerical Model Set-Up and Simulation Results

A series of numerical experiments have been conducted using the commercial DEM software, Particle Flow Code in Three Dimensional (PFC3D) [3]. The samples are of polyhedral shape with $n = 8$ and $R = 0.01$ m as detailed in [8]. Particles are clumps formed by two overlapping spheres with thee distance between their centres of 1.4 times the sphere radius. The equivalent particle radius is uniformly distributed between [0.3 mm, 0.5 mm]. The sample is prepared isotropically using the radius expansion approach. The particle frictional coefficient during the preparation stage is set to be $\mu_g = 0.1$ and $\mu_g = 0.5$ in order to prepare dense and loose samples, respectively. The constant stiffness contact model has been used in the simulations with $k_n = k_s = 1.0 \times 10^5$ N/m, and the particle frictional coefficient has been set as 0.5. No cohesion or particle breakage are considered.

Numerical simulations have been carried out following the loading path with constant p, b and fixed loading direction. Deviatoric strain has been increased continuously. Triaxial shearing results with different b values are plotted in Fig. 1.

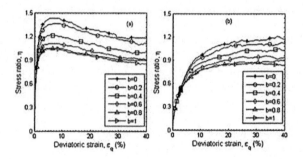

Fig. 1. Observations from numerical experiments

3 Particle-Scale Observations

The Stress-Force-Fabric relationship is an analytical expression of the macro-stress tensor in terms of the particle-scale fabric tensor and the contact force tensor as:

$$\sigma_{ij} = \frac{\omega N f_0 v_0 \zeta}{3V}(\delta_{ij} + G^f_{ij} + C^c_{ij} + \frac{2}{5}D_{ij}) \quad (1)$$

where ω is the coordination number; N is the total particle number; V is the sample volume; f_0 is the average contact force; v_0 is the average contact vector; ζ is the scalar reflecting the statistical dependence between contact forces and contact vectors; δ_{ij} is the Kronecker delta; G^f_{ij}, C_{ij} and D_{ij} characterize the anisotropies in contact force, contact vector and contact normal density, respectively.

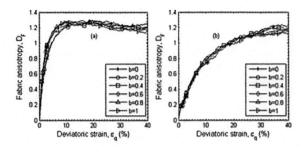

Fig. 2. Contact normal anisotropy

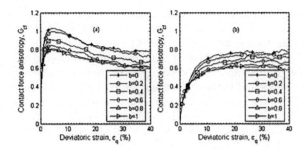

Fig. 3. Anisotropy in contact forces

4 Discussions

The anisotropic tensors could be equivalently expressed in terms of their anisotropy indices and b values. Those of contact normals and contact forces are given in Figs. 2 and 3, respectively. These figures show that the fabric anisotropy index for different b values are close but that of contact forces vary significantly. A higher force anisotropy is observed for lower b values, and is the main contributor to a higher stress ratio η. The b value of contact force has been almost identical to that of stress tensor but that of fabric shows a non-linear relationship with that of stress ratio. A strong correlation between the b values of fabric anisotropy and strain increment has been seen evident.

References

1. Barreto, D., O'Sullivan, C.: The influence of inter-particle friction and the intermediate stress ratio in soil response under generalised stress conditions. Granular Matter **14**, 505–521 (2012)
2. Hambly, E.: A new true triaxial apparatus. Géotechnique **19**(2), 307–309 (1969)
3. Itasca Consulting Group Inc.: PFC3D (Particle Flow Code in Three Dimensions). Minneapolis, ICG (2004)
4. Ko, H.-Y., Scott, R.F.: Deformation of sand in hydrostatic compression. J. Soil Mech. Found. Div. **93**, 137–156 (1967)
5. Lade, P.V., Duncan, J.M.: Cubical triaxial tests on cohesionless soil. J. Soil Mech. Found. **99**, 793–812 (1973)
6. Lade, P.V.: Elasto-Plastic stress-strain theory for cohesionless soil with curved yield surfaces
7. Lade, P.V.: Assessment of test data for selection of 3-D failure criterion for sand. Int. J. Numer. Anal. Meth. Geomech. **30**, 307–333 (2005)
8. Li, X., Yang, D.S., Yu, H.-S.: Macro deformation and micro structure of 3D granular assemblies subjected to rotation of principal stress axes. Granular Matter **18**(3), 1–20 (2016)
9. Miura, K., Miura, S., Toki, S.: Deformation behaviour of anisotropic dense sand under principal stress axes rotation. Soils Found. **26**(1), 36–52 (1986)
10. Nakata, Y., Hyodo, M., Murata, H., Yasufuku, N.: Flow deformation of sands subjected to principal stress rotation. Soils Found. **38**, 115–128 (1998)
11. Reades, D.W., Green, G.E.: Independent stress control and triaxial extension tests on sand. Géotechnique **26**, 551–576 (1976)
12. Sutherland, H.B., Mesdary, M.S.: The influence of the intermediate principal stress on the strength of sand. In: Proceedings of the Seventh International Conference on Soil Mechanics and Foundation Engineering, Mexico city, pp. 391–399 (1969)
13. Symes, M.J.P.R., Gens, A., Hight, D.W.: Undrained anisotropy and principal stress rotation in saturated sand. Géotechnique **34**(1), 11–27 (1984)
14. Thornton, C.: Numerical simulation of deviatoric shear deformation of granular media. Géotechnique **50**, 43–53 (2000)
15. Thornton, C., Zhang, L.: On the evolution of stress and microstructure during general 3D deviatoric straining of granular media. Géotechnique **60**, 333–341 (2010)
16. Zhao, J., Guo, N.: Unique critical state characteristics in granular media considering fabric anisotropy. Géotechnique **63**(8), 695–704 (2013)

DEM Analysis of Cone Penetration Test in Sand

Fuguang Zhang[✉]

Shijiazhuang Tiedao University, Shijiazhuang 050043, China
fuguang.zhang@stdu.edu.cn

Abstract. The cone penetration test has proved to be a very effective method for predicting pile capacity, evaluating soil properties, and delineating between layers of different soil types and states. However, few distinct element method (DEM) researches have been conducted to understand the effect of soil layering on penetrometer response. This paper presents a numerical investigation into the cone penetration test in two-layered sand using two-dimensional DEM. The effect of soil layering on the variation of penetration resistance with penetration depth are studied and compared with previous experimental data. In addition, the final displacement paths and mean stresses during penetration are also examined at the microscopic scale.

Keywords: Cone penetration test · Layering · Distinct element method

1 Introduction

The cone penetration test (CPT) has been widely used to evaluate soil properties and to delineate soil stratigraphy. Natural soils generally consist of layers with different thickness and mechanical behavior, leading to difficulty in interpreting penetrometer readings. This is because the CPT measurements are influenced by the soil near the cone tip and the soil layers beneath and above it.

Few analytical solutions [1], experimental tests [2], and Finite element modeling [3] have been carried out to study the effect of soil layering on CPT measurements. However, no particular method has received universal acceptance. An alternative tool is the distinct element method (DEM), developed by Cundall and Strack [4], which treats soil as an assembly of discrete grains, and can provide macro- and micro-mechanical responses of granular system. Some researchers have analyzed the penetration mechanism of uniform sand deposits using two-dimensional (2D) or 3D DEM simulations. Nevertheless, the DEM data on the CPT in layered sand is scarce.

In this paper, the effect of soil layering on the penetration resistance, final displacement path and mean stress during penetration was studied using 2D DEM simulations, where loose, dense and loose-over-dense sand ground were generated.

2 CPT Simulations

The DEM material contains about 300,000 discs with diameter ranging from 6 to 9 mm, which is often employed in the previous DEM simulations [5]. Two uniform sand specimens with void ratio of 0.20 and 0.27 were generated. The consolidation under an amplified gravity field of 10 g and the CPT simulations can be found in Reference [5]. To generate the loose-over-dense sand specimen, a dense specimen with 150,000 particles of void ratio of 0.20 was first generated, and then a rigid wall was set above the dense specimen. Afterwards, a loose specimen with 150,000 particles of void ratio of 0.27 was generated above the rigid wall. The middle rigid wall was then deleted followed by cycling until static equilibrium was reached. For the dense DEM material, the peak (residual) internal friction angle is 22° (19°). Figure 1 presents a schematic illustration of the CPT DEM model. Table 1 gives a summary of all parameters.

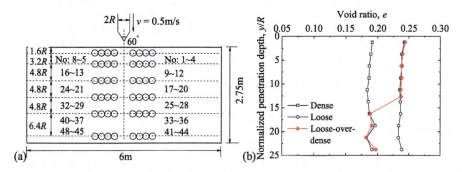

Fig. 1. The CPT DEM model: (a) layout of measurement circles; (b) void ratio

Table 1. Parameters used in the DEM tests.

Parameter	Value
Particle density/(kg.m^{-3})	2.6×10^3
Inter-particle frictional coefficient during loading	0.5
Penetrometer-soil frictional coefficient	0.0
Normal stiffness of particle/(N.m^{-1})	2.0×10^8
Tangential stiffness of particle/(N.m^{-1})	1.33×10^8
Normal stiffness of wall/(N.m^{-1})	1.5×10^9
Tangential stiffness of wall/(N.m^{-1})	1.0×10^9
Local damping coefficient	0.7

As shown in Fig. 2(a), the generated sand ground has a height of 2.75 m and a width of 6.0 m, where a total of 44 measurement circles were distributed near the symmetric line. Figure 2(b) illustrates that the dense and loose sand ground are uniform along the vertical direction. For the two-layered sand ground, the void ratio of the top half part is identical to that of loose sand, while the void ratio of the other half part is the same as that of dense sand.

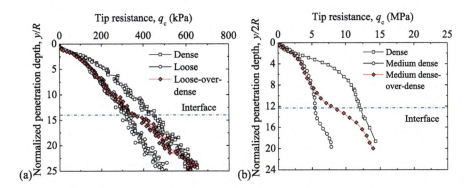

Fig. 2. Cone tip resistance against normalized penetration depth: (a) DEM results; (b) Experimental data

3   DEM Results

Figure 2 provides the cone tip resistance during penetration. For comparison, experimental data is included. Figure 2(a) shows that the tip resistance of the dense sand is always larger than that of the loose sand. For the loose-over-dense sand, the penetration curve first moves along that of the loose sand, and then gradually tends to that of the dense sand when the cone tip passes through the interface, which agrees with the experimental results qualitatively as shown in Fig. 2(b).

Figure 3 presents the final displacements at different depths. Figure 3(a) shows that almost all the particles at $y/R = 1.6$ move horizontally at first and then vertically upward, with the magnitude of displacement decreasing with x/R. As the depth increases, the displacement value decreases. Compared with dense sand, the magnitude of displacement of loose sand is larger. For the loose-over-dense sand, the displacement path of particles at depth above the interface is almost the same as that of loose sand, and the displacement path of particles beneath the interface is the same as dense sand.

Figure 4 provides the mean stresses during penetration. As shown in Fig. 4, the maximum mean stress is reached when the cone tip passes through the location of the measurement circles. For the loose-over-dense sand, the variations of mean stresses at $y/R = 4.8$ and 9.6 are the same as those of loose sand, and the changes of mean stresses at other depths follow those of dense sand.

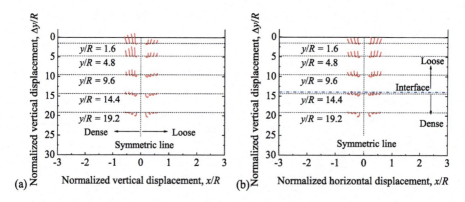

Fig. 3. The final displacements: (a) loose and dense sand; (b) loose-over-dense sand

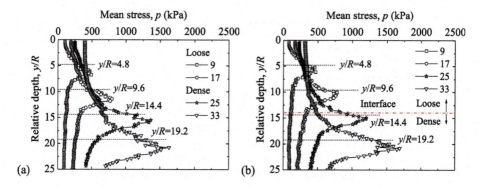

Fig. 4. Mean stresses during penetration: (a) loose and dense sand; (b) loose-over-dense sand

4 Conclusions

The effect of layering on the penetration resistance can be captured qualitatively using 2-D DEM. In the case of loose-over-dense sand, the displacement paths and the variations of mean stresses above the interface is almost the same as those of loose sand, while those beneath the interface tend to follow dense sand.

References

1. Mo, P., Marshall, A., Yu, H.: Interpretation of cone penetration test data in layered soils using cavity expansion analysis. J. Geotech. Geoenviron. **143**(1), 04016084 (2016)
2. Tehrani, F., Arshad, M., Prezzi, M.: Physical modeling of cone penetration in layered sand. J. Geotech. Geoenviron. **144**(1), 04017101 (2018)
3. Ma, H., Zhou, M., Hu, Y.: Interpretation of layer boundaries and shear strengths for stiff-soft-stiff clays using cone penetration test: LDFE analyses. Int. J. Geomech. **17**(9), 06017011 (2017)

4. Cundall, P., Strack, O.: The distinct numerical model for granular assemblies. Geotechnique **29**(1), 47–65 (1979)
5. Jiang, M., Yu, H., Harris, D.: Discrete element modelling of deep penetration in granular soils. Int. J. Numer. Anal. Meth. Geomech. **30**(4), 335–361 (2006)

DEM Investigation of Particle Crushing Effects on Static and Dynamic Penetration Tests

Ningning Zhang[1(✉)], Marcos Arroyo[1], Matteo Ciantia[2], and Antonio Gens[1]

[1] Department of Geotechnical Engineering and Geosciences,
Polytechnic University of Catalonia (UPC), Barcelona, Spain
ningning.zhang@upc.edu
[2] School of Science and Engineering, University of Dundee, Dundee, UK

Abstract. A 3-dimensional discrete element method model has been developed to simulate static and dynamic rod penetration test in a calibration chamber. The chamber has been filled with a scaled analogue of Fontainebleau sand. Crushing effects on penetration results have been examined. It has been found that particle crushing reduces penetration resistance for both static and dynamic tests. Microscale observation of crushed particles has been conducted. It is shown that dynamic impact causes more crushing events. The crushed particles are distributed within 2–3 radius from the rod for both tests.

Keywords: Dynamic penetration test · Static penetration test · Crushing

1 Introduction

In geotechnical engineering, penetration tests have been widely applied to characterize soils, such as estimating soil strength and stiffness and to evaluate soil liquefaction potential. Considering different penetration principles, penetrometers can mainly be divided into two groups: static and dynamic. In static penetration test, a tip is pushed into soil at a constant speed. In contrast, dynamic penetration involves driving a device formed by rods or rigid tips into the soil by striking it with a weight. When a rod is driven into a granular material, grain can be broken due to stress concentration around the rod. Discrete Element Method (DEM) has been applied as a powerful tool to address particle crushing in penetration tests (Ciantia et al. 2016a; Lobo-Guerrero and Vallejo 2007).

It is interesting to investigate the influence of crushing on penetration results under the two different penetration principles. In this paper, after briefly introducing the employed methodology for modelling particle crushing, simulations of a rod driven statically and dynamically into a 3-dimensional sample of Fontainebleau sand are presented. Particular attention has been paid to the effect of particle crushing on evolution of rod tip resistance. Microscale observation of crushed particles is also presented.

2 Methodology

2.1 Particle Crushing Model

The modelling approach for particle crushing developed by Ciantia et al. (2015) was adopted. For completeness, a general description of the model is given here. Coulomb friction and the simplified Hertz-Mindlin model were used for contacts. The crushing limit is based on the formulation proposed by Russell et al. (2009). Breakage of a given particle is activated if the maximum contact force F reaches the limit condition:

$$F \leq \sigma_{\lim} A_F \Rightarrow$$

$$F \leq \left\{ \sigma_{\lim 0} f(\text{var}) \left(\frac{d}{Nd_0}\right)^{-3/m} \pi \left[\frac{3}{4}\left(\frac{1-v_1^2}{E_1} + \frac{1-v_2^2}{E_2}\right)\left(\frac{1}{r_1} + \frac{1}{r_2}\right)^{-1}\right]^{2/3} \right\}^3 \quad (1)$$

Where $\sigma_{\lim}$ is the limit strength of the material and A_F is the contact area. $f(\text{var})$ is a function used to incorporate the natural material variability into the model. The limit strength, $\sigma_{\lim}$, is assumed to be normally distributed for a given sphere size. The coefficient of variation of the distribution, var, is taken to be a material parameter. The mean strength value $\sigma_{\lim 0}$ depends on the particle diameter (d) where m is a material constant, d_0 is the reference diameter and N is the scaling factor. r_i, E_i and v_i ($i = 1, 2$) are the radii, Young's Moduli and Poisson's ratio of the contacting spheres, respectively. Once the limit condition is reached, the spherical particle will split into 14-ball inscribed tangent spheres. Mass loss issues related to the choice of using such replacement technique are clearly addressed by Ciantia et al. (2016b). The crushed fragments assume the velocity and material parameters of the original particle.

2.2 Model Construction and Simulation

The DEM code PFC3D (Itasca 2016) has been used to construct the numerical models. The construction of a 3-dimensional calibration chamber model was carried out by adapting the procedure described by Arroyo et al. (2011) (Fig. 1a). The chamber was filled with an analogue of Fontainebleau dense sand, which was previously tested by Ciantia et al. (2015, 2018). A scaling factor of 79 was applied in order to achieve a manageable number of particles. Isotropic compression to 100 kPa was performed by using wall servo-control.

The flat-ended rod was then created by using frictional rigid walls (Fig. 1a). Particle rotation was inhibited to roughly mimic the effect of non-spherical particle shapes. Table 1 shows the model parameters of Fontainebleau sand. A local non-viscous damping of 0.05 was employed. In static tests, the rod was driven at a rate of 40 cm/s. As described by Zhang et al. (2018), dynamic penetration was simulated by prescribing a time-dependent force on the rod. In both tests, particle crushing effects were examined.

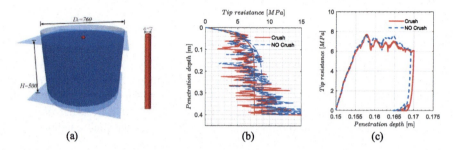

Fig. 1. (a) View of DEM model of calibration chamber and rod. Crushing effects on penetrograms (b) static test (raw and adjusted curves), (c) dynamic test

Table 1. Model parameters for Fontainebleau sand.

μ/-	E/GPa	v/-	$\sigma_{lim,0}$/GPa	d_0/mm	m/-	var/-
0.275	9	0.2	1.9	2	10	0.36

3 Outcomes

3.1 Penetration Curves

Penetrograms of static tests for crushable and uncrushable sand exhibit large oscillations, as shown in Fig. 1b, due to particle size effect. The approach which has been successfully used by Arroyo et al. (2011) to filter out the noise is adopted. The adjusted curve of penetration resistance is also shown in Fig. 1b. It is noted that crushing causes a resistance reduction due to particle rearrangement and finer particle generation. Additionally, one single impact was conducted at the depth of 15 cm (Fig. 1c). It shows that during the rising phase of dynamic resistance, particles haven't reached limit condition. Afterwards, crushing takes place at the steady-stage of penetration curve, leading to a reduction of resistance value and an increase of penetration depth.

3.2 Microscale Observations

The rod was driven from 15 cm to 30 cm, then a direct comparison can be made on crushed particle numbers (Fig. 2a). It is clear that the dynamic impact breaks more particles. DEM-based model can provide microscale observation at particle or contact level to provide meaningful insights. It is interesting to visualize the spatial distribution of crushing events that occur in both static and dynamic penetration tests. Figure 2b, c show the location of crushed particles in both tests, where it can be clearly observed that crushing events are distributed within 2–3 radius from the cone.

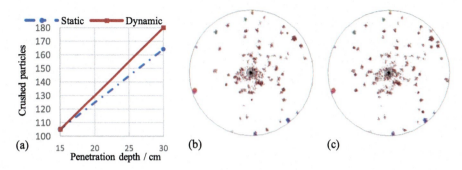

Fig. 2. Penetration tests from the depth of 15–30 cm (a) crushed particle number, (b) crushed particle distribution in static test and (c) crushed particle distribution in dynamic test

4 Conclusions

A 3-dimensional DEM model has been created to simulate static and dynamic rod penetration into calibration chambers filled with a scaled analogue of Fontainebleau sand. Crushing effects on penetration results has been examined. Particle crushing causes a reduction of penetration resistance for both static and dynamic tests. Dynamic impact causes more crushing events. The crushed particles are distributed within 2–3 radius from the rod. In forthcoming research, a more comprehensive study will be conducted to investigate other penetration test factors, such as sample relative density and confining pressure.

References

Arroyo, M., Butlanska, J., Gens, A., Calvetti, F., Jamiolkowski, M.: Cone penetration tests in a virtual calibration chamber. Géotechnique **61**(6), 525–531 (2011)

Ciantia, M.O., Arroyo, M., Calvetti, F., Gens, A.: An approach to enhance efficiency of DEM modelling of soils with crushable grains. Géotechnique **65**(2), 91–110 (2015)

Ciantia, M.O., Arroyo, M., O'Sullivan, C., Gens, A., Liu, T.: Grading evolution and critical state in a discrete numerical model of Fontainebleau sand. Géotechnique (2018)

Ciantia, M.O., Marcos, A., Butlanska, J., Gens, A.: DEM modelling of cone penetration tests in a double-porosity crushable granular material. Comput. Geotech. **73**, 109–127 (2016a)

Ciantia, M.O., Arroyo, M., Calvetti, F., Gens, A.: A numerical investigation of the incremental behavior of crushable granular soils. Int. J. Numer. Anal. Meth. Geomech. **40**(13), 1773–1798 (2016b)

Itasca Consulting Group. PFC3D: Itasca Consulting Group, Inc. (2014) PFC—Particle Flow Code, Ver. 5.0 Minneapolis, USA (2016)

Lobo-Guerrero, S., Vallejo, L.E.: Influence of pile shape and pile interaction on the crushable behavior of granular materials around driven piles: DEM analyses. Granular Matter **9**(3–4), 241–250 (2007)

Russell, A.R., Muir Wood, D., Kikumoto, M.: Crushing of particles in idealised granular assemblies. J. Mech. Phys. Solids **57**(8), 1293–1313 (2009)

Zhang, N., Arroyo, M., Gens, A., Ciantia, M.: DEM modelling of dynamic penetration in granular material. In: 9th European Conference on Numerical Methods in Geotechnical Engineering, Porto (2018)

Simulation of One-Dimensional Compression of Sand Considering Irregular Grain Shapes and Grain Breakage

Fan Zhu and Jidong Zhao[✉]

The Hong Kong University of Science and Technology, Hong Kong, China
jzhao@ust.hk

Abstract. Natural sand grains possess complex morphologies which evolve significantly through breakage of grains. The numerical modeling of irregular grain shapes and grain breakage remains far from mature. Traditional DEM based approach often suffers from high computational cost. In this paper we introduce an impulse-based dynamics (IBD) approach to simulate irregular shape sand grains with consideration of continuous grain breakage. Simulation of a 1-D compression is presented which showed reasonable results with respect to fractal particle size distribution and normal compression. The presented approach allows more efficient simulation of sand grains in irregular shapes and enables more in-depth study on micro-structural behaviors of sand.

Keywords: Impulse-based dynamics · Irregular shape · Particle breakage 1-D compression

1 Introduction and Methodology

Natural sand grains often possess complex morphologies with varying aspect ratio, roundness and convexity. The morphologies of sand grains may evolve significantly through breakage of particles under shear, compression or impact. Particle shape and breakage are known to be important factors underpinning many mechanical behaviors of sand such as packing, deformation and failure [1, 2]. Despite of great efforts made by multi-disciplinary research communities, the modeling of irregular shape particles and particle breakage remains far from mature. A numerical method that can incorporate irregular particle shapes while considering continuous particle breakage is highly desirable but unavailable. In current practice, the irregular particle shapes are often modeled using discrete element method (DEM) with "glued" or overlapped elementary disks or spheres [3, 4], or using other DEM based approaches such as polyhedral DEM [5] and level set DEM [6]. Particle breakage is commonly handled by adopting over-simplified, yet diversified criteria [7]. The traditional DEM based approach is in general computationally intensive when modeling complex particle morphologies. If particle breakage is to be considered, dramatic increase in particle number and evolution of grain size/mass would further impose stringent requirements on the use of much smaller time step, leading to overwhelming magnification of computational cost which may become quickly unaffordable to most researchers. There is an urgent demand for alternative, efficient numerical methods.

In this paper we introduce a novel numerical approach based on impulse-based dynamics (IBD) [8] to simulate irregular shape sand particles with consideration of particle breakage which is modeled by peridynamics [9, 10]. The IBD method, featured by its high computational efficiency and convenient integration of irregular particle shapes, has been commonly employed in physical simulation of robots and in the field of virtual reality, animation, games, and is gaining increasing attentions from engineering society. Some pioneer studies on granular materials using IBD [11–13] have showed promising results. The IBD method is different from traditional DEM in many aspects. The basic equations in IBD are formulated on the level of velocity and impulse. A symplectic Euler time integration scheme is adopted and the scheme is known to provide excellent numerical stability [14]. Therefore, large time steps are often used in IBD simulations, offering high computational efficiency. Many open-source libraries have been written based on the IBD method. In this study we utilized Bullet physics library [15] for simulation. The particles are modeled as convex and non-convex polyhedrons. It will be shown from a 1-D compression simulation that the presented approach is able to produce reasonable results with respect to fractal particle size distribution and normal compression.

2 Simulation of One-Dimensional Compression

A 1-D compression of a small sand sample has been simulated. The initial sample consists of a total of 216 particles with diameters ranging from 1.2 to 1.5 mm. Weibull distribution of particle strength has been considered with a Weibull modulus of 3.3. Particles are placed in a cubic rigid box of 8 mm long on each side and smooth surfaces. The initial void ratio of the sample is 0.84, representing a loose packing. Vertical pressure is applied incrementally on the top platen to a maximum level of 20 MPa. A summary of modeling parameters is provided in Table 1 below. Figure 1 shows snapshots of the sample at different stress levels in the simulation.

Table 1. Parameters used in the simulation of 1-D compression.

Parameter	Adopted value
Young's modulus (GPa)	100
Poisson's ratio	0.15
Friction coef. (particle-particle)	0.5
Restitution	0
Time step (s)	1.5×10^{-5}
Total simulated time (s)	2.7

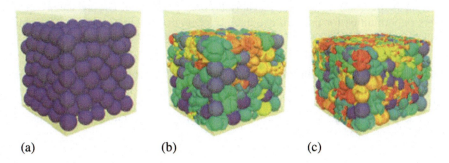

Fig. 1. Snapshots of the sample at different vertical stresses: (a) initial condition; (b) at 10 MPa; (c) at 20 MPa. Loading platen at the top of sample is not shown for clarity.

The simulation results are first reviewed with respect to fractal particle size distribution, which defines the number of particles greater than a certain size has a power law relation to that size, as expressed by:

$$N_d(d > d_0) = A\, d_0^{-D} \tag{1}$$

where A is a constant and D is regarded as fractal dimension. The simulations in Fig. 2 showed that D approaches a level near 2.3 and exhibits a stabilizing trend with increasing load (Fig. 2). The obtained D sits within typical range for rock and sand (i.e., 2.0 to 3.0) [16]. Also shown in Fig. 2 is the obtained normal compression line (NCL). The slope of NCL is found to be approximately 0.38 which is a reasonable value according to past studies [17].

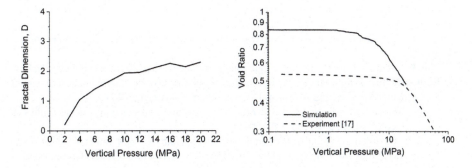

Fig. 2. Obtained fractal dimension (left) and normal compression line (right) from the 1-D compression simulation using IBD.

3 Conclusions

This paper presents an IBD based approach to simulate granular materials with consideration of irregular particle shapes as well as particle breakage. A simulation of 1-D compression of sand has been performed and reasonable results are obtained with respect to fractal particle size distribution and normal compression line. The presented approach offers opportunities for more in-depth study of micro-structural behaviors of sand in future.

Acknowledgments. The author acknowledges financial supports from National Science Foundation of China under Project No. 51679207, Research Grants Council of Hong Kong under GRF Project No. 16210017 and a Collaborative Research Fund Project No. C6012-15G. The first author acknowledges financial support from Hong Kong Ph.D. Fellowship Scheme funded by Research Grants Council of Hong Kong.

References

1. Mirghasemi, A.A., Rothenburg, L., Matyas, E.L.: Influence of particle shape on engineering properties of assemblies of two-dimensional polygon-shaped particles. Géotechnique **52**(3), 209–217 (2002)
2. Hardin, B.O.: Crushing of soil particles. J. Geotech. Eng. **111**(10), 1177–1192 (1985)
3. Cheng, Y.P., Nakata, Y., Bolton, M.D.: Discrete element simulation of crushable soil. Géotechnique **53**(7), 633–641 (2003)
4. Price, M., Murariu, V., Morrison, G.: Sphere clump generation and trajectory comparison for real particles. In: 2007 Conference on Discrete Element Methods, Brisbane, Australia (2007)
5. Lee, S.J., Hashash, Y.M.A., Nezami, E.G.: Simulation of triaxial compression tests with polyhedral discrete elements. Comput. Geotech. **43**, 92–100 (2012)
6. Kawamoto, R., Andò, E., Viggiani, G., Andrade, J.E.: Level set discrete element method for three-dimensional computations with triaxial case study. J. Mech. Phys. Solids **91**, 1–13 (2016)
7. de Bono, J., McDowell, G.R.: Particle breakage criteria in discrete-element modelling. Géotechnique **66**(12), 1014–1027 (2016)
8. Mirtich, B.V.: Impulse-Based Dynamic Simulation of Rigid Body Systems. Ph.D. Dissertation, University of California, Berkeley (1996)
9. Zhu, F., Zhao, J.: A peridynamic investigation on crushing of sand particles. Géotechnique, accepted
10. Zhu, F., Zhao, J.: Investigation on single sand particle crushing criterion using peridynamic method. In: Proceedings of the 15th International Conference of the International Association for Computer Methods and Advances in Geomechanics, Wuhan (2017)
11. Pytlos, M., Gilbert, M., Smith, C.C.: Modelling granular soil behaviour using a physics engine. Géotech. Lett. **5**(4), 243–249 (2015)
12. Izadi, E., Bezuijen, A.: Simulation of granular soil behaviour using the Bullet physics library. In: Geomechanics from Micro to Macro, pp. 1565–1570. Taylor & Francis Group, London (2015)
13. Toson, P., Khinast, J.G.: Impulse-based dynamics for studying quasi-static granular flows: application to hopper emptying of non-spherical particles. Powder Technol. **313**, 353–360 (2017)
14. Holmes, M.H.: Introduction to Numerical Methods in Differential Equations. Springer, New York (2007)
15. Coumans, E.: Bullet Physics Library (version 2.85) (2016). http://bulletphysics.org
16. Turcotte, D.L.: Fractals and fragmentation. J. Geophys. Res. **91**(B2), 1921–1926 (1986)
17. McDowell, G.R.: On the yielding and plastic compression of sand. Soils Found. **42**(1), 139–145 (2002)

Size Effects in Cone Penetration Tests in Sand

Pei-Zhi Zhuang[✉] and Hai-Sui Yu

School of Civil Engineering, Faculty of Engineering, University of Leeds, Leeds, LS2 9JT, UK
P.Zhuang@leeds.ac.uk

Abstract. Size effects in miniature cone penetration tests (CPTs) are examined by performing a series of 1 g laboratory tests using three penetrometers of 3, 6, and 12 mm in diameter (D) in two grades of dry Leighton Buzzard sand respectively. It is found that the size effects primarily depend on three non-dimensional geometrical parameters, including relative penetration depth (H/D), normalised surface roughness of the cone (R_a/d_{50}), and normalized cone size (D/d_{50}). Test results showed that: (1) H/D is a major size factor influencing the cone resistance at relatively shallow depths, and its influence may disappear while the localized failure mechanism dominates. (2) the cone resistance may increase with a decreasing value of D/d_{50} in some circumstances, and this effect attenuates in loose sand; (3) the cone resistance is positively related to R_a/d_{50}, especially for cones with an intermediate rough interface. These size dependent behaviour is attributed to the dependency of the failure pattern and sand properties on the stress level, strain level, and non-local interactions of underlying microstructures and the dependency of the shearing resistance of sand-cone interface on R_a/d_{50}.

Keywords: Size effect · Cone penetration tests · Experimental analysis

1 Introduction

As one of the most versatile and reliable in-situ devices, cone penetrometers have been extensively used for soil explorations (e.g., soil classification and characterization) in both geotechnical practices and agricultural investigations. In addition to recommended standard penetrometers (e.g., 12.83 mm, 20.27 mm, 35.7 mm, 43.7 mm in diameter [1, 2]), miniature and needle cone penetrometers also attracted growing interests due to their inherent advantages such as: (1) energy-saving as smaller downward thrust needed; (2) higher mobility and site accessibility; (3) higher sensitivity to variations of the soil stratigraphy; (4) lower requirement on the sample size to get rid of the possible boundary effect, which can effectively reduce labour, time, and cost in the sample preparation and make it easier to be integrated with other techniques (e.g., centrifuge, nondestructive testing methods like X-ray CT); and (5) geometrical similarity in modelling the interaction between soil and plant root-tips or earthworms. Considering the wide applications of various small sized cone penetrometers, possible size effects associated with the cone size and the sand particle size in 1 g shallow CPTs are experimentally investigated and briefly discussed in this paper.

2 Test Description

Two grades of the Leighton Buzzard sand (see Table 1) were used in the tests. Reconstituted dry sand samples of different relative densities (Dense: $D_r = 90\%$; Medium dense: $D_r = 65\%$; Loose: $D_r = 40\%$) were prepared by using the air pluviation technique. The repeatability of penetration resistance data has been verified in duplicated tests.

Table 1. Basic sand properties

Sand type	Median grain size d_{50} / mm	Special gravity G_s	Maximum void ratio e_{max}	Minimum void ratio e_{min}	Critical state friction angle φ_{cs} / °
Fraction C (FC)	0.51	2.65[a]	0.805	0.55	32[a]
Fraction E (FE)	0.12	2.65[b]	1.014[b]	0.613[b]	32[b]

Note: [a]after Lee [3]. [b]after Tan [4].

Three sized cone penetrometers of 3 mm, 6 mm, and 12 mm in diameter were designed and used, which all have the same apex angle of 60° and are made from stainless steel with polished surfaces. The penetrometers have an arithmetical mean surface roughness (R_a) of 0.607 um in average measured by the Contour GT-I 3D optical microscope. The cone resistance (q_c) and the average sleeve friction can be measured separately by the 12 mm sized penetrometer, equipped with strain gauges in two sections. In tests with the 6 mm and 3 mm sized probes, only the total load has been measured with a load cell mounted on the top of the probe. The soil resistance during both penetrations and extractions were recorded in all tests.

To minimise the potential lateral boundary effect and the bottom boundary effect [3, 5], a rigid walled cylinder of 490 mm in diameter and 500 mm in height was used. The smallest diameter ratio of the container over the penetrometer (B/D) is 40.8, and the sand samples were deposited into heights within the range of 315 mm–415 mm.

The cone penetrometers were driven by a SKF linear actuator (CAHB-10, rated load: 500 N, stroke: 300 mm) with a speed of approximately 1.5 mm/s. The possible penetration rate effect has been evaluated with moving speeds in the range of 1–4 mm/s, which showed negligible influences in tests with the dry sand samples.

3 Test Results and Discussion

Measured by the 12 mm sized penetrometer, the ratio of the maximum tensile shaft capacity over the maximum shaft capacity during penetrations (Q_{ft}/Q_{fc}) lies in a range of 0.62–0.76 in the tests. The shaft friction experienced by the 6 mm and 3 mm probes during penetrations are estimated by dividing the Q_{ft} measured by them and the corresponding value of Q_{ft}/Q_{fc}, and thus the cone resistance experienced by them can be obtained by extracting the shaft frictional resistance from the total load.

Dimensional analysis is used to interpret the 1 g CPT results. Factors affecting the normalised cone resistance (N_{CPT}) are summarised in Eq. (1).

$$N_{CPT} = \frac{q_c}{\sigma'_v} = \left\{ \frac{\sigma'_v}{p'_c}, \frac{H}{D}, \frac{R_a}{D}, \frac{D}{d_{50}}, \frac{B}{D}, \frac{S}{D}, D_r, \varphi_{cs}, \text{etc.} \right\} \quad (1)$$

where σ'_v is the effective vertical stress. p'_c is an index of aggregate crushing strength [6]. S is the distance from the nearest wall boundary to the location of penetration.

All test results of N_{CPT} are presented in Fig. 1. The tests are identified with codes comprising the sand fraction, cone diameter and sand density classification.

Due to the strong dependency of the friction and dilation angles of sand on the stress level, p'_c, D_r, and φ_{cs} [6], no doubt that N_{CPT} may significantly vary with these factors (e.g., Fig. 1). As the maximum q_c is less than 3 MPa in the present 1 g CPTs, it is anticipated that no significant particle crushing takes place in the tested silica sand [7]. The lateral boundary effect is avoided with $B/D \geq 40.8$. Results of multiple penetrations in the same dense FC sand samples showed a negligible difference between tests with $S/D = 20.5$ and $S/D = 10$. Preserving the above factors in the comparison analysis of data in Fig. 1, it is observed that: (1) the tip resistance measured in samples of the FE sand are generally higher than that in the FC sand; (2) higher tip resistances are experienced by smaller probes at the same penetration depth, and this phenomenon is more significant in the FC sand and attenuates with an increasing penetration depth. It is found that the size-dependent behaviour of these are mainly due to the combined size effect associated with the normalized geometry sizes of R_a/d_{50}, H/D, and D/d_{50} as

(1) The shearing resistance of sand-structure interface may increase with increasing values of R_a/d_{50}, especially for interfaces lie in the intermediate rough zone (approximately, $0.001 \leq R_a/d_{50} \leq 0.1$) [8, 9]. As a result, R_a/d_{50} may influence the cone resistance. According to the correlated relationship between the critical state interface friction angle (δ_{cs}) and R_a/d_{50}, δ_{cs} between the cone surface and the FE sand and the FC sand equals 19.2° and 14.3° respectively [9]. This partly explains the difference between tests in the FC sand and the FE sand while preserving other influencing factors.

(2) Apart from its influence to the sand-cone interaction, the particle size effect is also strongly dependent on the value of D/d_{50} [5, 10], for example, CPTs performed by the 'modelling of model' method suggested $D/d_{50} > 20$ to get rid of the particle size effect. The D/d_{50} effect may primarily stem from the dependency of the failure pattern (e.g., progressive failure along the shear band) and sand properties on the stress level, strain level and strain gradients (non-local contribution of underlying microstructures to the macroscopic response), and this effect may still apply in deep penetrations.

For tests in the same sand, the observed size-dependent behaviour is mainly attributed to the effects of H/D and D/d_{50}. It is observed that the values of N_{CPT} measured by different sized penetrometers are in good agreement while plotted against H/D in similar sand samples. It indicates that H/D is a major factor causing the size variation of q_c or N_{CPT} at the same initial stress level in the shallow CPTs at 1 g, and this effect attenuates with increasing values of H/D. The critical depth from shallow failure mode to deep failure mode is found around 30–40D in the 1 g tests, varying with the sand relative density. It is deeper than that found in CPTs on centrifuge (e.g., 20D [5]) probably due to the dependency of sand properties on the stress level.

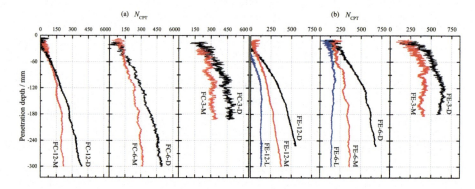

Fig. 1. Normalised cone resistance. (a) Fraction C sand; (b) Fraction E sand

4 Conclusion

Evident size effects have been observed in the shallow 1 g CPTs with miniature cone penetrometers. The dependency of the failure pattern and sand properties on the stress level, strain level, and non-local contributions of underlying microstructures and the dependency of the shearing resistance along the sand-cone interface on R_a/d_{50} were concluded as the main reasons leading to the size-dependent behaviour in CPTs. These characteristics are closely related to the dimensionless size parameters of H/D, R_a/d_{50}, and D/d_{50}. Hence, particular care should be taken to these size parameters while scaling them in the model test design and interpretations of tests with different scales.

References

1. American Society of Agricultural Engineers (ASAE), Standard for Soil Cone Penetrometer (S313. 3 FEB04). St. Joseph, Michigan, USA (2004)
2. Lunne, T., Robertson, P.K., Powell, J.J.M.: Cone Penetration Testing in Geotechnical Practice. Blackie Academic & Professional, London (1997)
3. Lee, S.Y.: Centrifuge modelling of cone penetration testing in cohesionless soils. Ph.D. thesis, University of Cambridge, UK (1990)
4. Tan, F.S.C.: Centrifuge and theoretical modelling of conical footings on sand. Ph.D. thesis, University of Cambridge, UK (1990)
5. Bolton, M.D., Gui, M.W.: The Study of Relative Density and Boundary Effects for Cone Penetration Tests in Centrifuge (Report CUED/D-SOILS/TR256). University of Cambridge, UK (1993)
6. Bolton, M.D.: The strength and dilatancy of sands. Géotechnique **36**(1), 65–78 (1986)
7. McDowell, G.R.: On the yielding and plastic compression of sand. Soils Found. **42**(1), 139–145 (2002)

8. Dietz, M.S.: Developing an holistic understanding of interface friction using sand with direct shear apparatus. Ph.D. thesis, University of Bristol, UK (2000)
9. Zhuang, P.: Cavity expansion analysis with applications to cone penetration test and root-soil interaction. Ph.D. thesis, University of Nottingham, UK (2017)
10. Wu, W., Ladjal, S.: Scale effect of cone penetration in sand. In: 3rd International Symposium on Cone Penetration Testing, Las Vegas, Nevada, USA, pp. 459–465 (2014)

Part III: Numerical Simulation

SPH-FDM Boundary Method for the Heat Conduction of Geotechnical Materials Considering Phase Transition

Bing Bai, Dengyu Rao, Nan Wu, and Tao Xu

School of Civil Engineering, Beijing Jiaotong University, Beijing 100044, China
bbai@bjtu.edu.cn

Abstract. A coupled SPH-FDM boundary is proposed for the analysis of thermal process in geotechnical materials, in which the smoothed particle hydrodynamics (SPH) method is used in the inner computational domain; and the finite difference method (FDM) is used as the function approximation near the boundary. The proposed coupled SPH-FDM boundary is applicable to the analysis of heat conduction in geotechnical materials, including the problems with discontinuous interface in the computational domain and the solidification of porous materials with a moving phase transition boundary. The effect of latent heat released in the phase transition is considered in the SPH method, and a weighted mean is used to determine the thermal parameters at the phase transition interface, which can be easily used to analyze the evolution of the moving phase transition interface over time. In general, there is a good agreement between the SPH results obtained using the three boundary treatments (coupled SPH-FDM boundary, no virtual particles, and virtual particles) and the analytical solution. The maximum error decreases gradually over time. Besides, it shows that as the thermal diffusion coefficient increases after the phase transition, the velocity of the moving phase transition interface increases, and the reduction rate in temperature in the non-solidified region decreases.

Keywords: Smoothed-particle hydrodynamics · Function approximation · Coupled SPH-FDM boundary · Porous medium

1 Introduction

In this study, a coupled SPH-FDM boundary is proposed for the analysis of thermal process in porous media, in which SPH is used in the inner computational domain and finite difference method (FDM) is used in the function approximation near the boundary. This approach allows for the achievement of second-order accuracy in the entire solution domain, and overcomes the difficulties in dealing with the second-type (Neumann) and the third-type (Robin) boundary conditions. At last, this boundary is applied to solve heat conduction problems in porous media, including the problems with discontinuous interface in the computational domain and the solidification of porous materials with moving phase transition boundary.

2 The SPH Method and a New Boundary Treatment

Detailed introduction to the conventional SPH method is available in Liu's work [1]. Thus, the new boundary treatment is introduced in detail here. For an one-dimensional heat conduction problem containing a heat source, the heat balance equation can be expressed as [2, 3]:

$$m_c \frac{\partial T}{\partial t} = K \frac{\partial^2 T}{\partial x^2} + q \qquad (1)$$

where T is the temperature, t is the time, K is the heat conduction coefficient, $m_c = \rho \cdot c$ is the internal heat capacity with ρ being the density and c being the specific heat capacity, and q is the heat source, respectively.

The first-, second-, and third-type boundary conditions (i.e., Dirichlet, Neumann and Robin) are: $T = T_0$, $K \cdot \partial T/\partial x = q_0$ and $K \cdot \partial T/\partial x = \lambda(T_0 - T)$, respectively [3].

Under the first-type boundary condition, $T = T_0$, the temperature of particles at the boundary should be maintained at a constant temperature in the SPH method. For near-boundary correction, the central difference scheme is adopted: $(dT/dx)_i = (T_{i+1} - T_{i-1})/2\Delta x$, $(d^2T/dx^2)_i = (T_{i+1} - 2T_i + T_{i-1})/(\Delta x)^2$. It follows from Eq. (1) that the temperature of particles near the boundary is calculated by

$$\left(\frac{dT}{dt}\right)_i = \frac{K}{m_c}\left[\frac{T_{i+1} - 2T_i + T_{i-1}}{(\Delta x)^2}\right] \qquad (2)$$

It is assumed that there is a particle i (e.g., $i = 1, 2$) at the boundary (Fig. 1) with only two neighboring particles, particle $i + 1$ and particle $i + 2$. A second-type boundary condition, $\partial T/\partial x = q_0(t)$, is imposed at particle i, where $q_0(t)$ is a known function. The temperature for particle i is as follows based on the forward difference of the three particles:

$$T_i = (-2\Delta x q_0 + 4T_{i+1} - T_{i+2})/3 \qquad (3)$$

Similarly, under the third-type boundary conditions $\partial T/\partial x = (\lambda/K) \cdot (T_0 - T)$, the temperature for particle i at the boundary is

$$T_i = (-2H\Delta x T_0 + 4T_{i+1} - T_{i+2})/[3 - 2H\Delta x] \qquad (4)$$

where $H = \lambda/K$.

The temperature of each particle i ($i = 1, 2, 3,\ldots$) in the computational domain at the first time step is calculated using conventional SPH method under the given initial and boundary conditions, and then the temperature of boundary particles i (e.g., $i = 1, 2$) is substituted by Eq. (3) or Eq. (4). As such, the temperature of each particle i at the next time step can be consecutively obtained using the coupled SPH-FDM method.

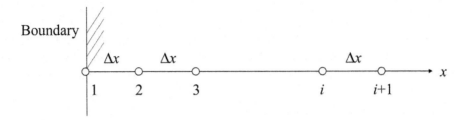

Fig. 1. Arrangement of particles near the boundary of a porous medium

3 Heat Conduction Problems Considering Phase Transition

The physical properties of porous media are as follows: $\rho_s = \rho_l = 2000$ kg/m^3, $c_s = c_l = 2000$ J/(kg·K), $L_t = 10000$ J/kg, and $K_s = 2$ J/(m·s·K). The solidification phase-transition temperature of the porous media is assumed to be $T_m = -10$ °C. The latent heat is released when the temperature is reduced below the phase-transition temperature, and moisture migration due to temperature gradient is neglected. The initial temperature of porous media is set as $T_i = 20$ °C, and a temperature of $T_0 = -30$ °C is applied at the boundary (first-type boundary). The calculation length is set to 10 m. Particles are uniformly distributed in the computational domain at a distance of $\Delta x/L = 0.01$ (taking $L = 1$ m). The smoothing length of the kernel function is 1.2 times the particle distance Δx, and the time step is $\Delta t = 10$ s. It is difficult to determine the thermal parameters at the phase transition interface due to the release of latent heat in the phase transition. The weighted mean of the thermal diffusion coefficients before and after phase transition is adopted (Fig. 2(a)) in this study, and thus $\kappa_p = (1-\alpha)\kappa_s + \alpha\kappa_l$, where $\alpha = \delta_2/\Delta x$.

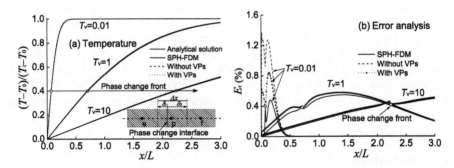

Fig. 2. The evolution of phase transition under different boundary treatments: (a) Temperature distribution; and (b) Error analysis

Figure 2 shows the evolution of the temperature distribution and the moving phase transition interface of porous media at different times (i.e., $T_v = 0.01$, 1, and 10) and the calculation errors when the thermal diffusion coefficient remains unchanged (i.e., $\kappa_s/\kappa_l = 1$) after phase transition. In general, there is a good agreement between the SPH results obtained using the three boundary treatments (coupled SPH-FDM boundary, no virtual particles, and virtual particles) and the analytical solution (Fig. 3(a)). The maximum error decreases gradually over time ($T_v = 0.01 \rightarrow 1 \rightarrow 10$). The minimum error is observed for the coupled SPH-FDM boundary ($E_r = 0.48\%$ when $T_v = 0.01$, Fig. 3(b)), whereas the maximum error is observed in the case of no virtual particles ($E_r = 1.27\%$).

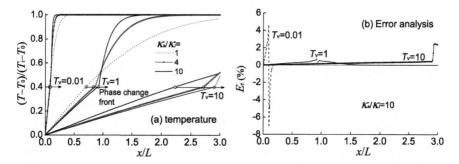

Fig. 3. The effect of the change of diffusion coefficient induced by phase transition on calculation results: (a) Temperature distribution; and (b) Error analysis.

Figure 3 shows the results considering the change of thermal diffusion coefficient of porous media after phase transition (assuming $\kappa_s/\kappa_l = 1$, 4 and 10). It shows that as the thermal diffusion coefficient increases after the phase transition, the velocity of the moving phase transition interface increases (Fig. 3(a), e.g., $\kappa_s/\kappa_l = 10$), and the reduction rate in temperature in the non-solidified region decreases. In addition, the calculation error at the phase transition interface increases significantly with increasing thermal diffusion coefficient ratio κ_s/κ_l. When $\kappa_s/\kappa_l = 10$ and $T_v = 0.01$, the calculation error ranges from $E_r = -7.04\% \sim +4.57\%$ (Fig. 3(b)), and decreases with the increase of time (e.g., $T_v = 1$), but increases slightly as the accumulated calculation error increases (e.g., $T_v = 10$). Besides, as the inter particle distance decreases, e.g., $\Delta x/L = 0.01 \rightarrow 0.005 \rightarrow 0.001$ ($\kappa_s/\kappa_l = 10$ and $T_v = 0.01$), the calculation error can be significantly reduced from $E_r = -7.04\% \sim +4.57\%$ to $E_r = 0 \sim +0.60\%$. The number of particles corresponding to $x/L = 0.01$, 0.005 and 0.001 is 1001, 2001, and 10001, respectively. Thus, the SPH results could become more accurate simply by increasing the number of particles.

4 Conclusions

A coupled SPH-FDM boundary is proposed for the analysis of thermal process in porous media in this study, in which SPH is used in the inner domain and FDM is used as the function approximation near the boundary. This coupled SPH-FDM method can not only improve the calculation accuracy under the first-type boundary conditions, but also convert heat conduction problems under the second- and the third-type boundary conditions into that under the first-type boundary conditions. As a result, a second-order accuracy can be achieved in the entire solution domain.

The effect of latent heat released in the phase transition is considered in the SPH method, and a weighted mean is used to determine the thermal parameters at the phase transition interface, which can be easily used to analyze the evolution of the moving phase transition interface over time. It is critical to determine the phase transition conditions and the corresponding physical parameters.

References

1. Liu, M.B., Liu, G.R.: Smoothed Particle Hydrodynamics (SPH): an overview and recent developments. Arch. Comput. Methods Eng. **17**(1), 25–76 (2010)
2. Jeong, J.H., Jhon, M.S., Halow, J.S., van Osdol, J.: Smoothed particle hydrodynamics: applications to heat conduction. Comput. Phys. Commun. **153**(1), 71–84 (2003)
3. Sikarudi, M.A.E., Nikseresht, A.H.: Neumann and Robin boundary conditions for heat conduction modeling using smoothed particle hydrodynamics. Comput. Phys. Commun. **198**, 1–11 (2016)

MPM Simulations of the Impact of Fast Landslides on Retaining Dams

Francesca Ceccato[✉] [iD], Paolo Simonini, and Veronica Girardi

DICEA, University of Padua, Via Ognissanti 39, 35129 Padua, Italy
francesca.ceccato@dicea.unipd.it

Abstract. Possible protection systems against flow-like landslides are earth dams built to stop or deviate the flow. The evaluation of impact forces on the structures is still based on oversimplified empirical approaches, which may lead to a very conservative design, with high costs and environmental impact. Numerical methods able to capture the essential features of the phenomenon can offer a valuable tool to support the design of protection measures. This paper shows the potentialities of the Material Point Method (MPM) in this field. A dry granular flow, modelled with the Mohr-Coulomb model is considered. The landslide is placed in front of the barrier with a prescribed velocity and the impact forces on the slanted face is monitored with time.

Keywords: MPM · Fast landslides · Protection dams

1 Introduction

Landslides of the flow-type, such as rock or debris avalanches, and debris flows, are among the most dangerous natural hazards worldwide. To reduce the associated risk, defense structures such as earth dam can be used to stop, deviate or slow down the flow. The design of these structures requires the knowledge of the impact forces, which are customarily estimated with empirical approaches. These simplified methods have a high level of uncertainties that sometimes lead to very conservative design of the structures with a significant environmental impact. Numerical models can support the design of these structures, thus leading to a more efficient implementation of the risk mitigation measures. In this paper, we analyze the potentialities of the Material Point Method (MPM) in this field.

MPM is a continuum particle-based method specifically developed for large deformations of history dependent materials. It simulates large displacements by Lagrangian points moving through an Eulerian grid. A 3D MPM code (Anura3D) featuring a specific algorithm to model soil-structure interaction and frictional sliding is applied in this study. The software has been recently validated in a number of geomechanical problems see e.g. [2, 3] and references therein.

In this paper, the propagation phase of the landslide is disregarded and we focus on the impact forces generated on a slated rigid structure by placing the dry soil mass, with an initial prescribed velocity, in front of the barrier.

2 Numerical Model

The mass is initially positioned in front of the barrier with a prescribed velocity v_0 varying between 4 and 32 m/s. The flow is 3.0 m thick and 15.0 m long. The model is 0.2 m wide. The numerical results are normalized with respect to the model width. The obstacle is 6 m-high and all the boundaries, as well as the structure face, are assumed to be smooth. Different inclinations of the barrier face (β) are considered. The mesh is refined in the proximity of the obstacle. 20 MPs are initially placed inside each active element (yellow color in Fig. 1). The structure is assumed to be rigid, i.e. the MPs do not move and the nodal velocity is zero (thus the MPs do not move and the nodal velocities are set to zero).

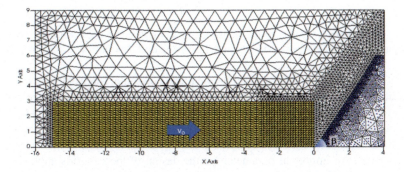

Fig. 1. Geometry and discretization of the numerical model

The behavior of the soil is modelled with a linear elastic perfectly plastic model with Mohr-Coulomb failure criterion. The reference parameters are summarized in Table 1.

Table 1. Material constitutive parameters

Parameter	Symbol	Value
Solid density [kg/m^3]	ρ_s	2650
Porosity [-]	n	0.45
Friction angle [°]	φ	33
Young modulus [kPa]	E	58000
Poisson ratio [-]	ν	0.2

3 Results

During the impact, the material climbs the wall decelerating and it exerts a pressure on the surface, which has a horizontal and a vertical component (Fig. 2a). To properly design these structures, both these components are important and they are considered separately in the following. Both the impact force components (F_x, F_y) increase with time up to a peak value, at slightly different instants (Fig. 2b).

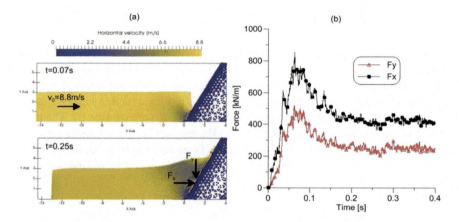

Fig. 2. (a) Horizontal velocity of the landslide impacting the structure at different instants. (b) horizontal and vertical forces on the structure face along time. ($\beta = 60°$, $v_0 = 8.8$ m/s)

Decreasing the structure face inclination β, the horizontal force decreases, while the reduction of maximum vertical force is less significant. The horizontal force is significantly lower than the case of $\beta = 90°$, moreover F_{max} varies non-linearly with v_0 (Fig. 3).

Slanted structures can be built with compacted granular materials or with the use of geosynthetics, to improve soil strength and reduce the dimensions of the dam by using steeper faces. An important parameter for the design of the geogrids applied in earth reinforced walls (ERW) is the horizontal force at the level of a potential sliding plane between the soil and the geogrid. Moreover, the global stability of the dam must be considered. Several procedures have been proposed for the design of these structures in static conditions [1], but their design under impact loading is not yet supported by a comprehensive scientific literature.

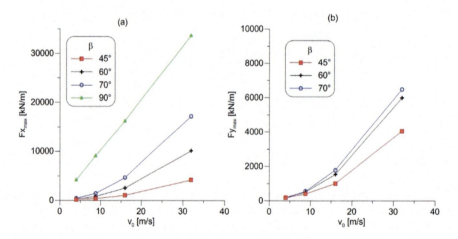

Fig. 3. Maximum horizontal (a) and vertical (b) force on the structure for different impact velocities and face inclinations.

The structure face is divided in 6 parts and the evolution of the force exerted by the landslide at the level of each grid is monitored (Fig. 4a). This may be of interest for the definition of the reinforcement length in ERW. The reinforcements that are closer to the base of the wall experiences the highest impact forces (grid #1 in Fig. 4).

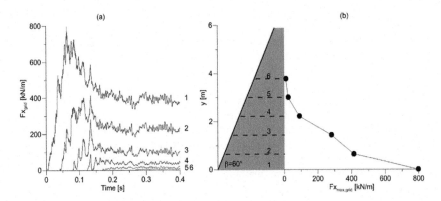

Fig. 4. (a) Evolution of the horizontal force at the level of each grid with time. (b) Maximum horizontal force at the level of each grid.

The impact force reaches its maximum at a different time for each considered grid, thus the pressure distribution assumed for the global stability of the structure could be inaccurate to study the internal stability of the wall. The maximum impact forces for each strip is shown in Fig. 4b where it should be noted that the maximum force increases non-linearly with depth.

4 Conclusions

This paper presents some preliminary analyses that show the potentialities of the MPM in simulating the impact of flow-like landslides on ERW. This is relevant for the global and internal stability of these structures, for which a comprehensive scientific literature of design under dynamic loading is still missing.

Future developments of the research will investigate the internal stability of these barriers and the impact of soil-water mixtures.

References

1. Berg, R.: Design and construction of mechanically stabilized earth walls and reinforced soil slopes–Volume I. Federal Highway Administration, November 2009
2. Ceccato, F.: Study of flow landslide impact forces on protection structures with the Material Point Method. In: Landslides and Engineered Slopes. Experience, Theory and Practice Proceedings of the 12th International Symposium on Landslides, pp. 615–620 (2016)
3. Soga, K.: Trends in large-deformation analysis of landslide mass movements with particular emphasis on the material point method. Géotechnique **66**(3), 248–273 (2016)

Extensions to the Limit Equilibrium and Limit Analysis in Geomechanics

Zuyu Chen

China Institute of Water Resources and Hydropower Research, Beijing, China
chenzy@tsinghua.edu.cn

Abstract. This article is an extended abstract that summarizes the author's work on the methods of slices to render practicable approaches to slope stability, earth pressure and bearing capacity analysis. They are derived based on the upper-bound and lower-bound theory of plasticity and extended to the 3D-area.

Keywords: Slope stability · Earth pressure · Bearing capacity 3D analysis

1 Introduction

The limit equilibrium method is a major analytical approach to tangle various geotechnical stability problems. In theory, it falls in the framework of lower bound of Plasticity. On the other hand, the upper-bound theorems, as illustrated by Chen [1], have more rigorous theoretical background. However, this upper bound work is limited to analytical approaches that can hardly be extended to practical problems. This lecture reviews the author's work mainly based on the methods of slices to render practicable approaches to performing these methods.

2 The Generalized Method of Slices: Lower-Bound Approaches (GMS)

The method proposed by Morgenstern and Price [2], as well as by others (Bishop [3]; Spencer [4]), imply a lower bound approach. Chen and Morgenstern [5] provided the following governing equations satisfying the force and moment equilibrium conditions[1] (Fig. 1):

$$G_n(F, l) = \int_a^b p(x)s(x)dx = 0 \qquad (1)$$

$$M_n(F, \lambda) = \int_a^b p(x)s(x)t(x)dx - M_e = 0 \qquad (2)$$

---

[1] The meanings of all symbols in this text can be found in the relevant references being cited.

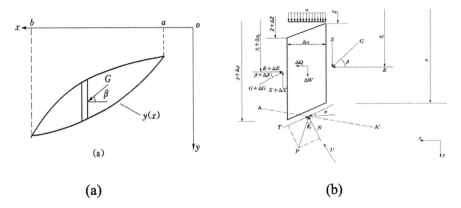

Fig. 1. The generalized method of slices. (a) The sliding mass. (b) Forces applied on a slice.

For Spencer method [4], which assumes parallel inter-slice forces, Eqs. (1) and (2) can be further simplified as [6]

$$\int_a^b p(x)\sec(\varphi_e - \alpha + \beta)dx = 0 \qquad (3)$$

$$\int_a^b p(x)\sec(\varphi_e - \alpha + \beta)[\sin\beta(x - x_a) - \cos\beta(y - y_a)]dx = M_e \qquad (4)$$

3 The Energy Method of Inclined Slices: Upper-Bound Approaches (EMU)

A simplified upper-bound approach can be formulated based on the method of inclined slices to solve various practical problems of geotechnical engineering [7]. This method is based on the associated flow law and Mohr-Coulomb failure criterion. By equating the work done by the external load to the internal energy dissipation, the governing equation for calculating the factor of safety or the ultimate load can be found by the following equation (Fig. 2):

$$\int_{x0}^{x_n}\left[(c_e\cos\varphi_e - u\sin\varphi_e)\sec\alpha - (\frac{dW}{dx} + \frac{\eta dT_y}{dx})\sin(\alpha - \varphi_e)\right]E(x)dx$$
$$-\int_{x0}^{x_n}\left[(c_e^j\cos\varphi_e^j - u^j\sin\varphi_e^j)L\csc(\alpha - \varphi_e - \theta_j)\right]\frac{d\alpha}{dx}E(x)dx + K_i = 0 \qquad (5)$$

Figure 3 shows a uniform slope without weight but subjected to a vertical surface load. Sokolovski [8] gave a closed-form solution. With parameters c = 98 kPa, φ = 30°, the ultimate load T is 111.44 kPa. The initial guessed failure mode with 4 blocks as shown in Fig. 3a gave a value of factor of safety F = 1.047 by Eq. (5).

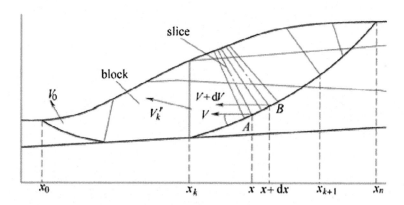

Fig. 2. The upper-bound method as applied for slope stability analysis.

The optimization further gave the minimum value of F as shown in Fig. 3b with a solution Fm = 1.013. If the failure mass is divided into 16 slices, a failure mode almost identical to the one suggested by the closed-form solution was obtained as shown in Fig. 3c, associated with Fm = 1.006.

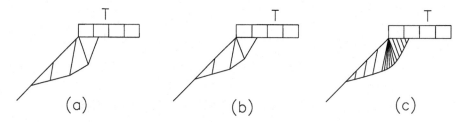

Fig. 3. An example describing the upper bound approach. (a) A four-slice failure mode, initial estimate, $F = 1.047$; (b) Results of the optimization search, $F_m = 1.013$; (c) Result of the optimization search using 16 slices, $F_m = 1.006$.

4 From Slope Stability to Earth Pressure Analyses

The Coulomb's theory for active earth pressure considers a straight-line slip surface and force equilibrium condition, which implies that the point of earth pressure application is located at the lower one-third of the wall. Terzaghi and Peck [9] pointed out that for flexible wall, the deformation conditions could lead to a center of earth pressure anywhere between lower and upper third points. By considering the requirement for moment equilibrium condition, they offered the empirical equations for evaluating the earth pressure of deep excavation.

GMS as formulated by Eqs. (1) and (2) has been extended by Chen and Li [10] to determine the active earth pressure associated with different input of point of application.

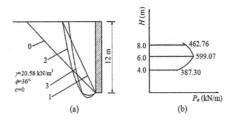

Fig. 4. An illustrative example for earth pressure.

Figure 4 shows a 12 m high retaining structure. The parameters for the backfills are: $\varphi = 36°$, $c = 0$, and $\gamma = 20.58$ kN/m^3. Figure 4a shows the critical slip surfaces associated with different locations of points of application.

Starting from the same initial slip surface numbered 0. 1, 2, 3 correspond to the cases k = 1/3, 1/2, 2/3 associated with Pa = 387.3, 599.1 and 462.8 kN respectively. The active pressure obtained for the case k = 1/3 was very close to the classical solution.

5   From Slope Stability to Bearing Capacity Analyses

The traditional bearing capacity analysis method adopts Prantdl solution that is based on the slip-line field method. Since this method employs purely analytical approaches, it is necessary to introduce various empirical coefficients to handle practical problems that normally have complicated geometry and material properties. The upper-bound method EMU is able to produce calculation results identical to those provided by slip-line field theory. Therefore, various empirical approaches are no longer necessary.

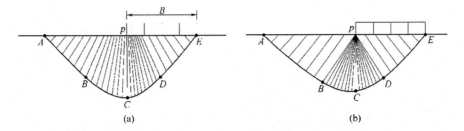

Fig. 5. Calculating the bearing capacity for a uniform foundation. (a) The initial failure mode; (b) The critical failure mode.

Figure 5 shows an example that offers the closed-form solution by the numerical approach by EMU.

Wang et al. [11] investigated the credit of various empirical equations for the coefficient Nγ accounting for the soil weight involved in the conventional bearing capacity analysis.

6 From 2D to 3D Analyses

Both GMS and EMU have been extended to the area of three-dimensional analysis by the 'method of columns'. As shown in Fig. 6a.

6.1 The Lower-Bound Approach

This method is an extension of Spencer method, which assumes that the inter-column forces G as shown in Fig. 6b are inclined at a same angle β [12]. By projecting all of the forces on a column in a direction perpendicular to G (refer to Fig. 6b), a straight forward equation calculating the normal force Ni, applied on the column base can be obtained,

$$N_i = \frac{W_i \cos \beta + (uA_i \tan \varphi'_e - c'_e A_i)(m_x \sin \beta + m_y \cos \beta)}{n_x \sin \beta + n_y \cos \beta + \tan \varphi'_e (m_x \sin \beta + m_y \cos \beta)} \quad (6)$$

The factor of safety F, in combination with other two variables, can to be solved by the following force and moment equilibrium equations.

$$S = \sum [N_i(n_x \cos \beta + n_y \sin \beta)_i + T_i(m_x \cos \beta + m_y \sin \beta)_i - W_i \sin \beta] = 0 \quad (7)$$

$$Z = \sum (N_i \cdot n_z + T_i \cdot m_z) = 0 \quad (8)$$

$$M = \sum [-W_i x - N_i \cdot n_x \cdot y + N_i \cdot n_y \cdot x - T_i \cdot m_x y + T_i \cdot m_y \cdot x] = 0 \quad (9)$$

This method allows satisfaction of complete overall force equilibrium condition and the moment equilibrium requirement about the main axis of rotation. The computational procedure is simple because it involves only three unknowns.

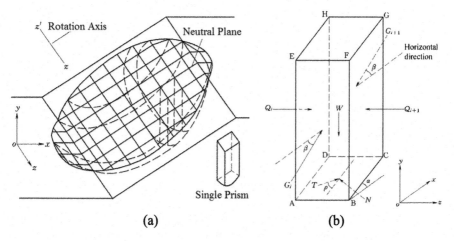

Fig. 6. The discretization pattern for a three-dimensional failure mass. (a) The method of columns. (b) Projecting all of the forces on a column.

6.2 The 3D Upper Bound Analysis

Chen et al. proposed a 3D extension to the 2D upper-bound method [13]. The sliding mass is divided into a number of columns with an axis representing the 'Neutral plane' on which the velocity is parallel to it, as shown in Fig. 7a.

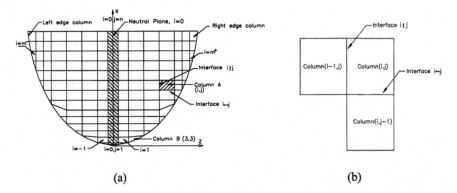

Fig. 7. The 3D upper bound analysis. (a) Discretization of the failure mass (b) Calculating the velocity field.

Using the Mohr-Coulomb associative flow law the plastic velocity is determined by

$$\Phi(\mathbf{V},\mathbf{N}) = \cos(\frac{\pi}{2} - \varphi'_e) = \sin \varphi'_e \qquad (10)$$

Where the symbol Φ means a dot product of the two unit vectors involved in the parenthesis (here are $\mathbf{V}$ and $\mathbf{N}$). $\mathbf{N}$ is normal of the slip surface where $\mathbf{V}$ is developed.

Symbol $\updownarrow$ is used to represent the interfaces between two adjacent columns and $\leftrightarrow$, between two adjacent rows of prisms. The requirement for kinematical compatibility as expressed by the following equations enables the calculation of the velocity field, refer to Fig. 7b,

$$V_{i \leftrightarrow j} = V_{i,j} - V_{i,j-1} \qquad (11)$$

$$V_{i \updownarrow j} = V_{i,j} - V_{i-1,j} \qquad (12)$$

The calculations start from the 'neutral plane' [Fig. 7b], which represents the main direction of sliding movement and is assumed to move without lateral component, using the two-dimensional approach [7]. Equations (11) and (12) allow a successive determination of $V_{i,j}$. The 3D work-energy balance equation is given as Eq. (13) from which the factor of safety can be calculated.

$$\sum D^*_{i \leftrightarrow j,e} + \sum D^*_{i \updownarrow j,e} + \sum D^*_{i,j,e} = \mathbf{W}\mathbf{V}^* + \mathbf{T}^o\mathbf{V}^* \qquad (13)$$

7 Conclusions

The solution of a slope stability analysis problem can be bracketed by its upper and lower bounds in Plasticity. The generalized method of slices GMS implies a lower bound while the Energy Method EMU falls in the framework of upper bound but formulated by inclined slices.

GMS has been successfully extended to the area of active earth pressure analysis for flexible retaining structure whose point of application is near the center of the wall.

EMU has been successfully extended to the area of bearing capacity analysis in which various empirical coefficients are no longer necessary.

The three-dimensional upper and lower bound methods based on the method of columns have been formulated with successful practical applications.

References

1. Chen, W.F.: Limit Analysis and Soil Plasticity. Elsevier, Amsterdam (1975)
2. Morgenstern, N.R., Price, V.: The analysis of the stability of general slip surface. Geotechnique 15(1), 79–93 (1965)
3. Bishop, A.W.: The use of the slip circle in the stability (1955)
4. Spencer, E.: A method of analysis of embankments assuming parallel inter-slice forces. Geotechnique 17, 11–26 (1967)
5. Chen, Z., Morgenstern, N.R.: Extensions to the generalized method of slices for stability analysis. Can. Geotech. J. 20(1), 104–119 (1983)
6. Chen, Z., Ugai, K.: Limit equilibrium and finite element analysis – a perspective of recent advances. In: Proceedings of the 10th International Symposium on Landside and Engineered Slopes, vol. 1, pp. 25–38 (2008)
7. Donald, I.B., Chen, Z.Y.: Slope stability analysis by an upper bound plasticity method. Can. Geotech. J. 34(6), 853–862 (1997). https://doi.org/10.1139/cgj-34-6-853
8. Sokolovski, V.V.: Statics of Soil Media. Butterworth, London (1960). Translated by Jones, D.H., Scholfield, A.N.
9. Peck, R.: Fifty years of lateral earth pressure. In: Proceedings of Conference on Design and Performance of Earth Retaining Structures, Cornel University, Ithaca, New York. ASCE Geotechnical Special Publication, No. 25 (1990)
10. Chen, Z.Y., Li, S.M.: Evaluation of active earth pressure by the generalized method of slices. Can. Geotech. J. 35(4), 591–599 (1998). https://doi.org/10.1139/cgj-35-4-591
11. Wang, Y.J., Yin, J.H., Chen, Z.Y.: Calculation of bearing capacity of a strip footing using an upper bound method. Int. J. Numer. Anal. Meth. Geomech. 25(8), 841–851 (2001). https://doi.org/10.1002/nag.151
12. Chen, Z., Mi, H., Zhang, F., Wang, X.: A simplified method for 3D slope stability analysis. Can. Geotech. J. 40(3), 675–683 (2003). https://doi.org/10.1139/t03-002
13. Chen, Z., Wang, X., Haberfield, C., Yin, J., Wang, Y.A.: Three-dimensional slope stability analysis method using the upper bound theorem, Part I: Theory and methods. Int. J. Rock Mech. Min. Sci. 38(3), 369–378 (2001). https://doi.org/10.1016/s1365-1609(01)00012-0

Extrusion Flow Modelling of Concentrated Mineral Suspensions

Chuan Chen[1], Damien Rangeard[1], and Arnaud Perrot[2(✉)]

[1] Laboratoire de Génie Civil et Génie Mécanique, INSA, Rennes, France
[2] Institut de Recherche Dupuy De Lôme, UBS, Lorient, France
arnaud.perrot@univ-ubs.fr

Abstract. This study deals with the stability of the flow of cement-based materials induced by the extrusion forming process. Recently, such process has gained much interest because of its involvement in 3D printing process. The occurrence of the relative filtration of the fluid through the solid part of the material is governed by a competition between the permeability of the granular skeleton and the velocity of the extrusion process. Based on extrusion tests, two types of plastic behavior are proposed to describe the extrusion properties. The extrusion flow are tested under different conditions, such as sand proportion, ram velocity, die diameter, etc. The moisture contents on extrudates are calculated to describe the fluid filtration and the development of solid volume fraction. The experimental results are to be compared with the model prediction of the extrusion tests.

Keywords: Extrusion test · Fluid filtration · Friction

1 Introduction

Extrusion is a highly efficient forming process used in a wide range of industries such as food engineering, ceramics, building materials, etc. For concentrated suspensions, the load distribution induces a differential repartition of the stresses between the solid part and the fluid part of the material. The fluid pressure gradient induced by the solicitation leads to a fluid filtration through the granular skeleton, leading to local variations of the porosity of the material along the extruder. The evolution of porosity changes the behavior of the material from an entirely plastic to frictional behavior [1–3]. Besides, the velocity of extrusion, geometry of die, and formulation of material are found to be amongst the most important variables during extrusion [4]. After a complete experimental hydro-mechanical characterization [5], extrusions of two mineral suspensions, a cement based material and a kaolin paste, are performed at different velocities in order to obtain the total range of behavior of the tested materials.

2 Materials and Extrusion Tests

2.1 Mixture of Kaolin and Sand

In this study, a mixture of kaolin clay powder and fin sable regarded as concentrated mineral suspensions is performed. Firstly, the water/kaolin mass ratio is 40%. The mixture 40% represents 40% of fin sand and 60% of kaolin clay (with 0.4 water/kaolin mass ratio), while mixture 20% represents 20% of fin sand in weight. Through vane test, the initial yield stress is measured (from 15 kPa to 22 kPa) to ensure the material shape for extrusion. For each test, the mass of extruded material is equal to 700 g and the billet length around 180 mm.

2.2 Extrusion Device and Conditional Control

The extrusion device consisted of a vertical Steel barrel of 300 mm in length with 50.0 mm diameter cylinder (denoted D) [2]. The ram is fitted with a load transducer and the device placed in a universal compression machine in order to measure the total extrusion force F. Firstly, We propose two drainage conditions: drained condition, medium-drained condition. Then, the tests are performed for criterion validation at different ram velocities (respectively 1 mm/s, 0.1 mm/s, 0.033 mm/s) and different exit die diameters (8 mm, 16 mm, 24 mm) during the extrusion process.

3 Theoretical Framework

3.1 Perfect Plastic Behavior

When extrusion velocity is sufficiently high, no drainage occurs and the material in the extruder remains homogeneous and behaves as a pure plastic model. Figure 1 presents a typical experimental extrusion force on the mixture materials. In this case, stress distribution, internal and wall friction remain constant and the process corresponds to a

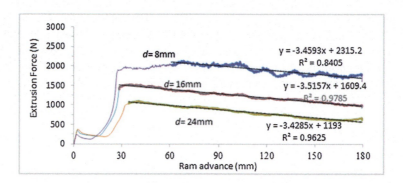

Fig. 1. Extrusion force of mixture 40% in mid-drained condition under three different die diameters with 1 mm/s ram velocity

relatively proper extrusion. The extrusion force decreases linearly with the ram advance due to the decrease of friction length until the ram reaches the die land.

3.2 Frictional Plastic Behavior

When extrusion velocity is sufficiently slow, the material undergoes drainage in the barrel. Thus, the water flow becomes heterogeneous, and the material sample becomes drier with a consequent increase in the plastic yield stress. Leakage could occur at both ends of the extruder. In the case of heterogeneous flow, a frictional plastic behavior is proposed to analysis those properties of concentrated mineral suspensions. The behavior could be based on Drucker-Prager plastic criterion and Coulomb friction law.

4 Experimental Results

The evolutions of moisture content along the extruder are tested to quantify liquid filtration and distribution. Figure 2 proves that fluid filtration occurs towards the extrudates and creates heterogeneities in the billet during the whole extrusion process.

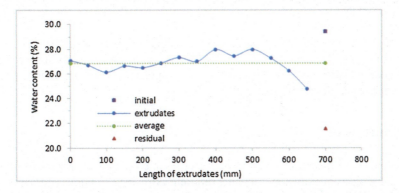

Fig. 2. Typical results of water content along the extrudates and extreme points (Mixture 40% with 8 mm die diameter and 0.033 mm/s ram velocity under drained condition)

It is clearly observed that the moisture content intends to decrease slightly at the early stage due to compression force. Then, filtration happens under the lowest ram velocity, the moisture is transferred into two extreme exits of extruder through the granular skeleton. The extrudates start to contain more liquid and reach a peak value when fluid paste is fully filtrated. In the last stage, the solid volume fraction increases gradually, inducing a radical decrease of moisture content as well as an exponential increase of extrusion load.

To compare the evolution of moisture contents under different conditions, we define Δw as the subtraction between the average water content along the extrudates and the that of remaining material in the barrel:

$$\Delta w = w_{average} - w_{residual} \tag{1}$$

In this test, the Δw is equal to 7.20%, which shows a relatively high fluid filtration and an increase of solid volume fraction during extrusion. In the same way, the different evolutions of moisture content are calculated under different extrusion conditions, such as ram velocities, die diameters and drainage conditions. The results are listed below (Fig. 3):

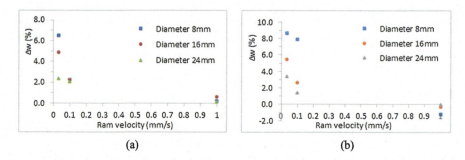

Fig. 3. Evolution of Δw in mid-drained condition under different ram velocities (a: mixture 20%; b: mixture 40%)

The Δw increases with the reduction of die diameter under speed 0.033 mm/s. For the mixture with a higher proportion of fin sand (mixture 40%), the Δw starts to present an obvious variation under the middle speed and a higher fluid leakage, indicating a high fluid filtration in the extrusion barrel.

5 Conclusions

In this paper, we focus on the extrusion properties of concentrated mineral suspensions. Through a series of extrusion test under different extrusion conditions (fin sand proportion, extrusion velocity, die geometry and drainage condition), two types of plastic behavior are described. For a high ram velocity, no fluid filtration occurs and the material shows perfect plastic behavior. For a relatively lower speed, the material shows frictional plastic behavior because of leakage. Besides, the evolution of moisture contents on the extrudates is calculated to describe the fluid filtration in the barrel. For a certain condition under a lower speed, the moisture is transferred into two extreme exits of extruder, the material becomes drier and results in higher extrusion load.

References

1. Perrot, A., Rangeard, D., Melinge, Y.: Prediction of the ram extrusion force of cement-based materials. Appl. Rheol. **24**(5) (2014). 53320 (7 p.)
2. Khelifi, H., Perrot, A., Lecompte, T., Rangeard, D., Ausias, G.: Prediction of extrusion load and liquid phase filtration during ram extrusion of high solid volume fraction pastes. Powder Technol. **249**, 258–268 (2013)

3. Perrot, A., Mélinge, Y., Rangeard, D., Micaelli, F., Estellé, P., Lanos, C.: Use of ram extruder as a combined rheo-tribometer to study the behaviour of high yield stress fluids at low strain rate. Rheol. Acta **51**, 743–754 (2012)
4. Yu, A.B., Bridgwater, J., Burbidge, A.S., Saracevic, Z.: Liquid maldistribution in particulate paste extrusion. Powder Technol. **103**, 103–109 (1999)
5. Rangeard, D., Perrot, A., Picandet, V., Mélinge, Y., Estellé, P.: Determination of the consolidation coefficient of low compressibility materials: application to fresh cement-based materials. Mater. Struct. **48**(5), 1475–1483 (2015)

Enriched Galerkin Finite Element Method for Locally Mass Conservative Simulation of Coupled Hydromechanical Problems

Jinhyun Choo[1(✉)] and Sanghyun Lee[2]

[1] Department of Civil Engineering, The University of Hong Kong,
Pokfulam, Hong Kong
jchoo@hku.hk

[2] Department of Mathematics, Florida State University,
Tallahassee, FL 32306, USA
lee@math.fsu.edu

Abstract. Numerical simulation of coupled hydromechanical processes is crucial to address many problems in geotechnical engineering. Conventionally, the seepage flow equation in a hydromechanical model is solved by the continuous Galerkin (CG) finite element method. However, the CG method does not ensure mass conservation in a local (element-wise) manner, which may be a critical drawback in a number of situations. Here we introduce an enriched Galerkin (EG) finite element method that allows locally mass conservative simulation of coupled hydromechanical problems. The major advantage of the EG method is that its computational cost is appreciably less than that of the discontinuous Galerkin (DG) method. We describe the proposed method's key idea and demonstrate its performance and significance through an example simulation of wetting collapse of a heterogeneous unsaturated soil.

1 Introduction

Numerical simulation of hydromechanical processes—coupled seepage flow and soil deformation—plays key roles in a wide range of geotechnical engineering problems from consolidation settlements to rainfall-induced landslides. The fidelity of coupled hydromechanical simulation hinges upon the accuracy of the numerical method employed for the sub-problem of seepage flow. In geotechnical engineering, the seepage flow problem has conventionally been solved by the continuous Galerkin (CG) finite element method. However, the CG method may be sub-optimal for modeling flow in porous media, especially when the media exhibit heterogeneous permeability fields and/or the fluid flow is associated with transport phenomena. The major reason is that the CG method is not conservative in a local (element-wise) manner.

To overcome this limitation of the CG method for coupled hydromechanical problems, we have recently proposed an enriched Galerkin (EG) finite element method that efficiently allows for local mass conservative in hydromechanical simulation [2]. In what follows, we describe the key idea of the EG method, and demonstrate the method's performance and significance through a numerical example that involves wetting collapse of a heterogeneous unsaturated soil. It will be shown that the use of a locally mass conservative method can be critical to accurately predict not only the unsaturated flow pattern but also the timing of wetting collapse.

2 Numerical Method and Example

Using mixture theory we conceptualize an unsaturated soil as a three-phase continuum comprised of solid, water, and air. Without much loss of generality, it is assumed that the solid deformation is infinitesimal, the water is incompressible, and the air pressure is atmospheric. Also, the mechanical behavior of the unsaturated soil is postulated to be governed by the effective stress $\boldsymbol{\sigma}'$, which is related to the total stress $\boldsymbol{\sigma}$ as $\boldsymbol{\sigma} = \boldsymbol{\sigma}' - S_r p \mathbf{1}$ with S_r denoting the water saturation, p the pore water pressure, and $\mathbf{1}$ the second-order identity tensor. The hydromechanical problem at hand is described by a coupled system of the following two equations [1]:

$$\text{Solid deformation:} \quad \nabla \cdot (\boldsymbol{\sigma}' - S_r p \mathbf{1}) + \rho \boldsymbol{g} = \mathbf{0}, \tag{1}$$

$$\text{Fluid flow:} \quad \phi \dot{S}_r + S_r \nabla \cdot \dot{\boldsymbol{u}} + \nabla \cdot \boldsymbol{q} = 0, \tag{2}$$

where ρ is the overall mass density, $\boldsymbol{g}$ is the gravitational force, ϕ is the porosity, $\boldsymbol{u}$ is the solid displacement, and $\boldsymbol{q}$ is the seepage (Darcy) flow vector. The overdot denotes the material time derivative following the motion of the solid matrix.

The proposed formulation uses the EG method to discretize Eq. (2). The EG finite element space is constructed by augmenting piecewise constant functions to the CG finite element space with k-th order polynomials. Figure 1 illustrates the EG degrees of freedom for $k = 1$ (linear), in comparison with its CG and DG counterparts. It is noted that the EG method provides local mass conservation with a substantially fewer unknowns than the DG method. The variational (weak) form of the EG method can be developed in the same way as the DG method. Denoting the jump operator by $[\![\cdot]\!]$ and the average operator by $\{\cdot\}$, the semi-discrete variational form of the fluid flow equation may be written as (for zero flux boundary conditions)

$$\sum_{T \in \mathcal{T}_h} \int_T \psi (\phi \dot{S}_r + S_r \nabla \cdot \dot{\boldsymbol{u}}) \, \mathrm{d}V - \sum_{T \in \mathcal{T}_h} \int_T \nabla \psi \cdot \boldsymbol{q} \, \mathrm{d}V + \sum_{e \in \mathcal{E}_h^\circ \cup \mathcal{E}_h^{\partial p}} \int_e [\![\psi]\!] \cdot \{\boldsymbol{q}\} \, \mathrm{d}A$$

$$+ \sum_{e \in \mathcal{E}_h^\circ \cup \mathcal{E}_h^{\partial p}} \frac{\alpha}{h^e} \int_e \kappa_e [\![\psi]\!] \cdot [\![p_h]\!] \, \mathrm{d}A - \sum_{e \in \mathcal{E}_h^{\partial p}} \frac{\alpha}{h^e} \int_e \kappa_e [\![\psi]\!] \cdot (\hat{p} \boldsymbol{n}) \, \mathrm{d}A = 0, \tag{3}$$

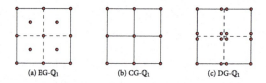

Fig. 1. Degrees of freedom for the EG, CG, and DG methods.

for all ψ in the EG finite element space. The last two terms are penalty terms. Details of the formulation are explained in Choo and Lee [2].

An example problem is conducted to examine the performance of the EG method and the significance of local mass conservation for hydromechanical simulation. The problem involves infiltration of water into an unsaturated soil deposit manifesting a heterogeneous permeability field. Figure 2 illustrates the problem setup, along with the distribution of the absolute permeability. The mechanical response of the soil is modeled by linear elasticity and Drucker–Prager plasticity. The van Genuchten equation is used for modeling water retention behavior. Using the EG and CG methods, we simulate this unsaturated flow problem until the onset of wetting collapse.

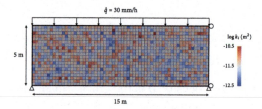

Fig. 2. Setup of the numerical example with the absolute permeability field.

Figure 3 comparatively shows the EG and CG results of pore water pressure fields and mass balance errors in individual elements after 1000 min of water infiltration. It is seen that the fluid flow is highly heterogeneous throughout, and the EG and CG results manifest non-trivial differences in the pore pressure fields and infiltration fronts. Observe that the EG result has virtually no mass balance errors, whereas the CG result shows orders of magnitude larger errors in element-wise mass balances throughout the domain.

Figure 4 presents the plastic strain and pore pressure fields after 1800 min, at which the unsaturated soil begins to collapse in the EG simulation. In the EG result, the pore pressure has built up significantly around the left bottom conner. On the other hand, the CG result does not predict collapse at this point, as the pore pressure buildup is not significant yet. This result shows that the differences in flow patterns (see Fig. 3) can have a marked impact on the deformation and failure responses.

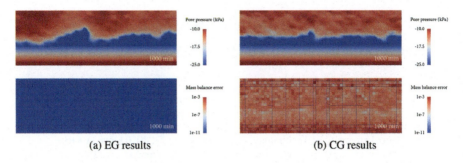

Fig. 3. EG and CG results of pore pressures and mass balance errors at 1000 min.

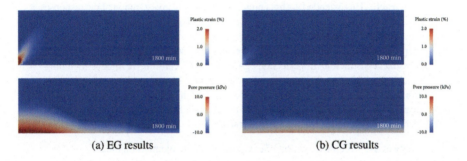

Fig. 4. EG and CG results of plastic strains and pore pressures at 1800 min.

3 Closing Remarks

An EG finite element method has been introduced for locally mass conservative simulation of coupled hydromechanical problems. It has also been shown that local mass conservation can be critical not only for fluid flow but also for deformation and failure. The EG method will be even more useful for accurate simulation of chemohydromechanical problems (*e.g.*, [3]), since local mass conservation is essential whenever fluid flow is coupled with transport.

References

1. Choo, J., White, J.A., Borja, R.I.: Hydromechanical modeling of unsaturated flow in double porosity media. Int. J. Geomech. **16**(6), D4016002 (2016)
2. Choo, J. and Lee, S.: Enriched Galerkin finite elements for coupled poromechanics with local mass conservation. Comput. Methods Appl. Mech. Engrg. (2018). https://doi.org/10.1016/j.cma.2018.06.022
3. Choo, J., Sun, W.C.: Cracking and damage from crystallization in pores: coupled chemo-hydro-mechanics and phase-field modeling. Comput. Methods Appl. Mech. Engrg. **335**, 347–379 (2018)

Influence of Individual Strut Failure on Performance of Deep Excavation in Soft Soil

Kamchai Choosrithong[✉] and Helmut F. Schweiger

Institute of Soil Mechanics, Foundation Engineering and Computational Geotechnics,
Graz University of Technology, Rechbauerstraße 12, 8010 Graz, Austria
{kamchai.choosrithong,helmut.schweiger}@tugraz.at

Abstract. This extended abstract presents a numerical investigation of the response of individual strut failure by analysing a 30 m deep excavation in marine clay supported by a diaphragm wall and ten levels of struts. Insufficient embedment depth of the wall into the stiff soil layer has also been accounted for, simulating possible construction imperfections. An advanced constitutive model including double hardening plasticity and small strain stiffness is employed for the analysis. It could be shown that provided a robust design is put in place in the first place significant stress redistribution capacity is available avoiding catastrophic failure of the excavation when individual support elements fail.

Keywords: Deep excavation · Strut failure · Finite element method

1 Introduction

Safety and robustness of structural support systems have to be ensured in construction of underground structures. Temporary structures, e.g. struts, are usually the critical elements and overall failure may occur, e.g. either due to inadequate capacity of the struts or due to poor strut/waler connection, in particular if a complete strutting level fails. However, failure of individual struts should not lead to an overall stability problem duet to redundancies in the supporting system and 3D effects allowing for stress redistributions. An additional factor is the embedment depth of the wall.

2 Sequential Failure of Individual Struts

Construction of a deep excavation has been modelled using the finite element code Plaxis [1]. Figure 1a shows the schematic sketch of the assumed soil profile and excavation sequence. Low permeability clays were treated as undrained material and the analyses are performed in terms of effective stresses employing an advanced constitutive model [2]. The potential failure of structural members was considered assuming elastoplastic behaviour, i.e. a maximum compressive and bending capacity is specified for struts and wall respectively. The mesh used is depicted in Fig. 1b.

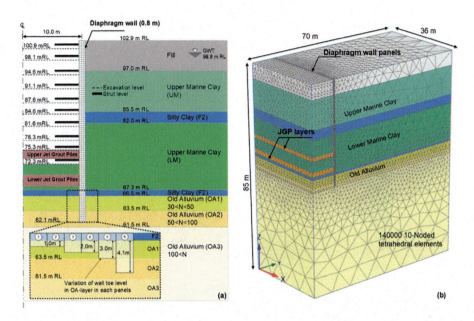

Fig. 1. (a) Schematic sketch of soil profile and excavation sequence; (b) Finite element for 3D.

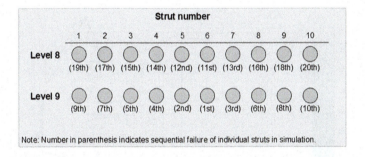

Fig. 2. Schematic diagram for simulation failure sequence of individual struts.

The load of failed struts is generally transferred vertically and horizontally to neighbouring struts. In order to simulate a consecutive failure of individual struts the scheme of sequential failure of struts as indicated in Fig. 2 was assumed. The analysis was performed as follows: excavation down to excavation level 10 was performed in steps, including groundwater lowering inside the excavation and activating the prestressed struts. At this stage individual struts were removed according to Fig. 2. This was performed for different layouts of embedment depths of neighbouring diaphragm wall panels. Only two extreme cases are compared here, namely "wall type I", which is the reference configuration, and "wall type II" as indicated in the insert of Fig. 5. The calculated lateral wall deflections and bending moments for the reference case ("wall type I") and "wall type II" are shown in Figs. 3 and 4 respectively, and comparison is made for two constitutive models, namely the Hardening Soil model (HS) and the Hardening Soil Small model (HSS) as implemented in the finite element code Plaxis [1, 2].

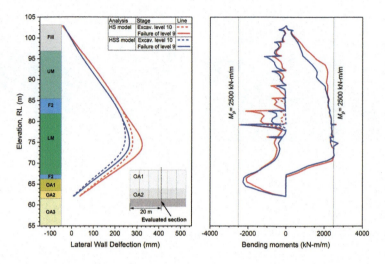

Fig. 3. Lateral wall deflections and bending moments for reference embedment depth.

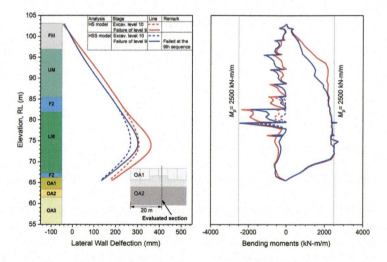

Fig. 4. Lateral wall deflections and bending moments for embedment depth "wall type II".

The following can be observed: as expected, slightly smaller wall movements are calculated with HSS model and wall movements increase when the 9th level of struts is removed. However, equilibrium can still be achieved with the reference embedment depth. However, for "wall type II", i.e. a reduced embedment depth of panels, the HSS model predicts failure when removing the 9th strut of level 9 whereas the HS model still achieves equilibrium. This is a consequence of the HSS model because in addition to small strain stiffness effects the flow rule is also different as compared to the HS model leading to a lower undrained shear strength in effective stress analysis of undrained conditions. As indicated in Fig. 4 plastic hinges occur on both sides of the wall in this case.

As an example for load redistribution Fig. 5 depicts the normalized loads (as ratio to maximum capacity) on struts at levels 8 and 7 when the first 3 struts of level 9 have been assumed to fail. It is obvious that struts at level 8 quickly reach their bearing capacity and therefore the overall factor of safety, which can be evaluated by means of a so-called strength-reduction-technique drops significantly, eventually leading to failure as indicated in Fig. 4.

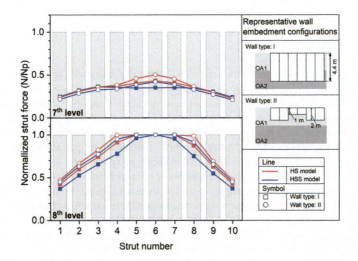

Fig. 5. Load distribution after the third failure sequence of individual struts.

Acknowledgements. The authors would like to acknowledge the financial support with Ernst Mach Grant, ASEA-UNINET provided by the Austrian Federal Ministry of Science, Research and Economy (BMWFW) for doctoral degree study.

References

1. Brinkgreve, R.B.J., Swolfs, W.M., Engin, E.: PLAXIS 3D 2017-User Manual. The Netherlands, Delft, Plaxis bv (2017)
2. Benz, T.: Small-strain stiffness of soil and its numerical consequences. Ph.D. thesis, University of Stuttgart (2007)

Implementation of Advanced Constitutive Models for the Prediction of Surface Subsidence After Underground Mineral Extraction

Yury Derbin[1(✉)], James Walker[1], Dariusz Wanatowski[2], and Alec M. Marshall[3]

[1] Nottingham University Ningbo China, 199 Taikang East Road, Ningbo 315100, China
yury.derbin@nottingham.edu.cn
[2] University of Leeds, Leeds LS2 9JT, UK
[3] Nottingham University, Nottingham NG7 2RD, UK

Abstract. Surface subsidence is a typical problem for any underground mineral extraction. In order to choose an appropriate method of extraction, predict damage to infrastructure and other unwanted consequences of surface subsidence, it is necessary to model the surface settlement. Previous research has shown that the conventional constitutive models, which are built into commercial software, encounter considerable difficulties when attempting to predict subsidence troughs. The troughs should be modelled deeper and narrower.

To improve the predictions, the authors explore the possibilities of the implementation of advanced constitutive models by programming the Clay And Sand Model or CASM [1] into the commercial finite-difference software FLAC3D by Itasca Consulting Group, Inc. CASM is a Critical State model that differs from the popular modified Cam-clay model, which is embedded within FLAC3D, by only two new parameters. Therefore, the implementation of CASM could be considered as a first step towards programming more advanced constitutive models in FLAC3D. After programming and validation of CASM, the model was used to simulate surface subsidence after a fictitious mine collapse. The simulations showed that CASM did not improve the prediction of the surface subsidence, and a different model may be more suitable.

Keywords: FLAC · CASM · Surface subsidence

1 Introduction

In order to minimize mining cost and subsidence damage by means of improvements in planning, a reliable tool is needed to simulate surface subsidence after a mine collapse. According to Kratzsch [2], subsidence engineering is of mounting importance to minimise costs occasioned by mining damage. The forecast of the surface subsidence can help to choose the appropriate method of mining, i.e. caving, longwall, room-and-pillar or mining with stowing.

Surface subsidence can be somewhat mitigated by stowing the empty space left after coal mining. According to Brady and Brown [3], stowing can reduce subsidence; however, it increases mining costs and the efficiency of mineral extraction. Some tradeoff should be found between costs and consequences of surface subsidence during designing a mine.

Numerical modelling can be a useful tool for mine planning. In this work, FLAC3D by Itasca Consulting Group, Inc was implemented to model surface subsidence after a mine collapse. This software has already been employed to model surface settlement due to longwall mining with the earliest examples dating back to the years of 1997 [4] or 1999 [5]. However, exact predictions from numerical simulations are challenging because of the complex rock-soil behavior [6]. It was showed that FLAC's embedded constitutive models could not predict the measured surface subsidence trough [7]. The modelled subsidence trough was too wide and too shallow if compared with field measurements.

In attempt to improve the predictions of the surface subsidence, the Critical State constitutive model CASM is implemented in this work to study surface subsidence after a mine collapse. CASM is a relative simple model developed by Yu in 1995 [1]. It can forecast the behaviour of both sand and clay, which usually both present in the overburden of the shallow mines (about 50 m depth), under loads. At the same time, the popular Critical State model, the modified Cam-clay model, catches mostly clay responses to the loads. Therefore, the modified Cam-clay model should be expended to CASM to model the behavior of sand, which can constitute the overburden.

After programming, the model was validated for clay and sand properties by numerically imitating triaxial tests. These properties were obtained from the work of Khong [8]. Both sets of properties were used to run a model of a collapse of a fictitious shallow mine (at a depth of 45 m). The results were compared between each other and with the results of the modified Cam-clay model.

2 Modelling Results

After the programming and successful validation of CASM, the same properties, which had been used in the validation, were implemented to model surface subsidence. Figure 1 presents the modelling results.

In Fig. 1, the y-axis indicates the depths of the subsidence trough. This axis is the line of symmetry of the excavation. The x-axis is the horizontal distance from the centre of the goaf. Three curves, namely CASM clay, CASM sand and modified Cam-clay, correspond to the modelled troughs by CASM and the modified Cam-clay model for clay and sand. Figure 1 shows that the results of both constitutive models with clay properties agreed well, and the sand properties did not result into narrower and deeper subsidence trough either. This means that implementation of CASM did not improve the predictions of the surface subsidence.

Figure 1 also gives a clue that the properties did not influence the trough size because the results obtained for sand and clay agreed well. This is due to the distribution of the plastic deformations. Earlier it was showed that this distribution played an important role in the modelling of surface subsidence [7]. In the case of an equal area of plastic deformations above and under the goaf, the plastic zone acts as a large goaf. In the case at hand, the plastic deformation occurs above the rib of the goaf and the seam in the overburden. It is likely that this distribution of the plastic deformations gave priority to the influence of the elastic properties on surface subsidence; therefore, the Cam-clay properties did not affect the subsidence trough.

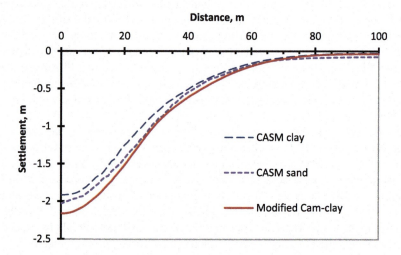

Fig. 1. Half-Subsidence Profiles obtained with modified Cam-clay model and CASM.

3 Conclusions

The work examined the capabilities of the authors' programmed CASM constitutive model to predict rock-soil behaviour during surface subsidence. The work showed that CASM could not improve the prediction of the surface subsidence for both sand and clay overburden. Therefore, a different constitutive model is required.

The work also showed that the trough size was not influenced by the properties of the overburden material once the goaf height of the simulation was adjusted to the mutual value of the goaf height. Summarizing, two basic difficulties were met: finding an advanced constitutive model and finding suitable material properties.

Acknowledgements. The authors greatly acknowledge the open fund SKLGDUEK1512 of the China University of Mining and Technology.

References

1. Yu, H.S.: A unified critical state model for clay and sand. Civil Engineering Research Report No 112.08.1995. University of Newcastle, NSW 2308, Australia (1995)
2. Kratzsch, H.: Mining Subsidence Engineering. Springer, Berlin, New York, 543 (1983)

3. Brady, B.H.G., Brown, E.T.: Rock Mechanics for Underground Mining. Luwer Academic Publishers, New York, p. 628 (2005)
4. Lloyd, P., Mohammad, N., Reddish, D.: Surface subsidence prediction techniques for UK coalfields-an innovative numerical modelling approach. In: Proceeding of the 15th Mining Congress of Turkey, Ankara, pp. 6–9 (1997)
5. Alejano, L.R., Alonso, E.: Subsidence prediction with FLAC. In: International FLAC Symposium on Numerical Modeling in Geomechanics, pp. 225–232 (1999)
6. Xu, N., Kulatilake, P.H.S.W., Tian, H., Wu, X., Nan, Y., Wei, T.: Surface subsidence prediction for the WUTONG mine using a 3-D finite difference method. Comput. Geotech. **48**, 134–145 (2013)
7. Derbin, Y., Walker, J., Wanatowski, D., Marshall, A.: Numerical simulation of surface subsidence after the collapse of a mine (in press)
8. Khong, C.D.: Development and numerical evaluation of unified critical state models. University of Nottingham, p. 190 (2004)

Effect of Travelling Waves on Tunnels in Soft Soil

Stefania Fabozzi[1,2(✉)], Emilio Bilotta[1], Haitao Yu[3,4], and Yong Yuan[5]

[1] Department of Civil, Architectural and Environmental Engineering,
University of Naples Federico II, Naples, Italy
stefania.fabozzi@unina.it
[2] Institute of Environmental Geology and Geo-Engineering,
Italian National Research Council, Rome, Italy
[3] Key Laboratory of Geotechnical and Underground Engineering of Ministry of Education,
Tongji University, Shanghai, China
[4] State Key Laboratory of Geo-Mechanics and Deep Underground Engineering,
China University of Mining and Technology, Beijing, China
[5] State Key Laboratory for Disaster Reduction in Civil Engineering,
Tongji University, Shanghai, China

Abstract. The paper proposes a three-dimensional numerical model able to catch the main deformation patterns of the soil and of an underground tunnel structure subjected to non-uniform seismic shaking. The proposed 3D FE model for instance, investigate the effect of the non-uniform seismic loading condition due to the wave passage along the longitudinal axis of a tunnel and a comparison between the uniform and non-uniform case is proposed both in terms of free field and soil-structure interaction analysis. The numerical results show the non-negligible effect of wave passage on the tunnel structure compared with the uniform case. The dynamic increment of the transversal component of the internal tunnel forces is higher if the asynchronism is considered, in function of the shear waves velocity of propagation in the bedrock. In addition, longitudinal components of the internal forces arise on tunnel, i.e. longitudinal normal force, horizontal bending and shear force acting in the horizontal plane along the structure.

Keywords: Travelling waves · Numerical FE analysis · Tunnel

1 Introduction

Most of the seismic damages on tunnels in soft ground occurs for ground shaking caused by the propagation of shear waves [1]. This loading condition determines on long underground structures (i.e. tunnels, pipelines, subways) axial deformation, longitudinal bending and ovaling/racking [2]. The longitudinal deformation component in particular, is caused by the spatially varying ground motion that is not always considered in the seismic design of underground long structures engineering although it produces such modification of the seismic soil response with respect to the case in which the spatial variability is not considered. As consequence, because the behavior of the underground structure is governed by the surrounding soil deformation during seismic shaking, the different seismic shaking conditions affect the seismic response of the tunnel structure in different ways. During the propagation of the earthquake signal in fact, the arrival time of the seismic wave along the structure is different producing an overlapping effect of the waves in the direction of the length of the structure.

Current research on the topic distinguishes (a) analytical solutions and (b) numerical methods to approach the problem. Many analytical solutions are available in the technical literature: (1) Free field deformation approach [3]; (2) Seismic deformation method [4]; (3) Beam springs model [5]; (4) Mass-beam-springs model [6]; (5) Multi masses beam springs model [7]. Numerical solutions, instead, are less common since they are highly time consuming and need very large computer memory: (1) Pseudo-static 3D numerical analysis [8]; (2) 3D multi-scale method [9]; (3) 3D full dynamic analysis [10].

This paper proposes a three-dimensional numerical model able to catch the main deformation patterns of the soil and of an underground tunnel structure subjected to non-uniform seismic shaking. The paper focuses on the effect of the travelling wave passage along the tunnel axis on the seismic response of the structure via numerical analysis. A three-dimensional numerical model able to catch the main deformation patterns of the soil subjected to multi-directional seismic motion has been developed in the FE code Plaxis 3D.

2 Numerical Analyses

The effect of travelling waves along the longitudinal tunnel direction is dealt assuming the input wave propagating vertically upward from the rigid bedrock up to the ground surface while the asynchronism of the seismic motion is induced by propagating the seismic waves longitudinally at the same time.

Under these hypotheses, a three-dimensional FE model has been developed in Plaxis 3D to investigate the free-field soil response under such seismic condition, considering a soil domain with a depth H = 60 m, a width B = 200 m and a length Y = 400 m.

The 3D full dynamic analyses have been carried out considering the simplified hypothesis that the soil behaves elastically (γ_{soil} = 20 kN/m³, V_{soil} = 250 m/s, ν_{soil} = 0.3). The velocity of the bedrock is assumed equal to 800 m/s and it has been used only to determine the time delay (or time-lag, TL) of the wave along the longitudinal tunnel direction, that is a harmonic single-oscillation wave with an amplitude of 0.1 g and a frequency equal to 4 Hz. The input signal is assigned as a time history of acceleration to the base surface. The asynchronism of the seismic motion has been simulated by dividing the base in 16 sub-surfaces each with a width y_i equal to 25 m and the same input wave has been assigned to each of them, with a time lag TL_{yi} to each other, calculated as Y_i/V_b and equal to 0.03 s. A simplified scheme assumed for the numerical model is shown in Fig. 1a while the numerical mesh is represented in Fig. 1b.

The same numerical model has been used also for the soil-structure interaction analysis for which a tunnel structure with a constant diameter equal to 6 m and a 30 m deep axis has been considered. The tunnel lining has been modelled with a continuous structural plate, with 'no slip' interface with the soil assuming its elastic behaviour. The input signal used for this analysis is similar to that adopted in the free-field condition: a harmonic wave with an amplitude of 0.1 g and a frequency of 4 Hz, repeated in five cycles to reproduce the effect of overlapping waves during a real earthquake.

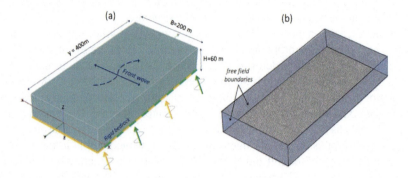

Fig. 1. (a) Scheme of the model adopted in Plaxis 3D to simulate the wave passage along y-axis, (b) numerical mesh of the model.

The free-field analysis demonstrates that there is a time delay between the input wave and the arrival time of the wave in correspondence of each point of the model along the signal propagation direction. This is the time that the input needs to arrive at the considered points, depending on the travelled distance and the wave velocity; in general, decreasing the bedrock velocity, thus increasing the time lag, the effect of the asynchronism increases. The effect of wave passage becomes more important and not negligible when it involves the tunnel structure. With regard of the soil-tunnel interaction analysis indeed, the non-uniform seismic load increases the dynamic increment of internal forces in the lining in its transversal plane, hoop force and bending moment with respect to the case of uniform seismic load (Time Lag = 0 s). Moreover, the non-uniform seismic load induces axial force along the tunnel axis, bending in the horizontal plane and shear force perpendicular to the tunnel axis. These components of forces are null in the case of uniform seismic load.

Another interesting issue is the effect of the bedrock velocity that affects the time-lag during the travelling of the wave. A set of parametrical analysis for instance, has been carried out varying longitudinal time-lag. To this end, the following velocities have been considered in the analyses: $V_{bedrock,1} = 800$ m/s → Time-lag = 0.030 s, $V_{bedrock,2} = 1000$ m/s → Time-lag = 0.025 s, $V_{bedrock,3} = 1250$ m/s → Time-lag = 0.020 s. The obtained results show that increasing the bedrock velocity, thus decreasing the time-lag along the tunnel axis, there is a decreasing of the dynamic increment of the internal forces in the tunnel lining and vice versa.

3 Conclusions

The paper proposes a three-dimensional numerical FE model to investigate the effect of the non-uniform seismic loading condition due to the wave passage along the longitudinal axis of a tunnel. The results of the 3D full dynamic analyses, carried out for the case of a continuous tunnel lining, show the non-negligible effect of non-uniform seismic loading compared with the uniform case, both in terms of dynamic increment of the transversal normal forces and bending moment. Their magnitude is higher than in the case of uniform seismic loading, depending on the shear waves velocity of propagation in the bedrock. It is found that increasing the bedrock velocity, thus decreasing the time-lag along the tunnel axis, there is a decreasing of the dynamic increment of the internal forces in the tunnel lining. For the investigated range of bedrock velocity in fact, the difference between the uniform and non-uniform seismic loading can achieve a maximum value of up to 50% in terms of hoop force and bending moment. This is interesting when thinking to the possible longitudinal joints opening of segmental tunnel lining under non uniform seismic shaking, which could potentially lead to the loss of water-tightness under high level of ground shaking. The asynchronous ground motion produces also longitudinal forces, horizontal bending and transversal shear along the tunnel axis and their magnitude is not negligible. The dynamic increment of these components of internal tunnel forces could lead the possible opening of circumferential joints when they are present.

References

1. Hashash, Y.M.A., Hook, J.J., Schmidt, B., Yao, J.I.: Seismic design and analysis of underground structures. Tunnel. Undergr. Space Technol. **18**, 377–393 (2001)
2. Owen, G.N., Scholl, R.E.: Earthquake engineering of large underground structures, Report no. FHWA/RD-80/195. Federal Highway Administration and National Science Foundation (1981)
3. John, C.M.St., Zahrah, T.F.: Aseismic design of underground structures. Tunn. Undergr. Space Technol. **2**(2), 165–197 (1987)
4. Kawashima, K.: Seismic design of underground structures in soft ground: a review. In: Kusakabe, O., Fujita, K., Miyazaki, Y. (eds.) Geotechnical Aspects of Underground Construction in Soft Ground. Balkema, Rotterdam (1999). ISBN 90 5809 1 066
5. Yu, H.T., Cai, C., Guan, X.F., Yuan, Y.: Analytical solution for long lined tunnels subjected to travelling loads. Tunn. Undergr. Space Technol. **58**, 209–215 (2016)
6. Kiyomiya, O.: Earthquake-resistant design features of immersed tunnels in Japan. Tunn. Undergr. Space Technol. **10**(4), 463–475 (1995)
7. Li, C., Yuan, J.Y., Yu, H.T., Yuan, Y.: Mode-based equivalent multi-degree-of-freedom system for one-dimensional viscoelastic response analysis of layered soil deposit. Earthq. Eng. Eng. Vibr. **17**, 103–124 (2017)
8. Park, D., Sagong, M., Kwak, D.Y., Jeong, C.G.: Simulation of tunnel response under spatially varying ground motion. Soil Dyn. Earthq. Eng. **29**, 1417–1424 (2009)
9. Yu, H.T., Yuan, Y., Qiao, Z., Gu, Y., Yang, Z., Li, X.: Seismic analysis of a long tunnel based on multi scale method. Eng. Struct. **49**, 572–587 (2013)
10. Li, P., Song, E.: Three-dimensional numerical analysis for the longitudinal seismic response of tunnels under an asynchronous wave input. Comput. Geotech. **63**, 229–243 (2015)

Numerical Simulation of a Post-grouted Anchor and Validation with In-Situ Measurements

Carla Fabris[1(✉)], Václav Račanský[2], Boštjan Pulko[3], and Helmut F. Schweiger[1]

[1] Graz University of Technology, Rechbauerstraße 12, 8010 Graz, Austria
cfabris@tugraz.at
[2] Keller Grundbau GmbH, Guglgasse 15, 1110 Vienna, Austria
[3] University of Ljubljana, Jamova cesta 2, 1000 Ljubljana, Slovenia

Abstract. Prior to the pull-out test of a post-grouted ground anchor, numerical simulations were performed in order to predict the load-displacement behaviour of the structure. Advanced constitutive models were employed to simulate the pull-out test. The soil parameters were determined based on laboratory and in-situ investigations. The stress distribution along the soil was evaluated and the numerical model employed for the grout was capable of reproducing the crack pattern in the post-grouted body. The strains throughout the tendon were strongly influenced by the grout conditions. The measured and the predicted load-displacement curves agreed very well, suggesting that the assumptions for the soil and the grout were reasonable.

Keywords: Ground anchor · Pull-out test · Numerical simulation

1 Introduction

Modelling an anchor pull-out test numerically provides a very useful tool not only to predict the ultimate load of the anchor but also to have a better insight into the interactions between the tendon, the grout and the soil. In this respect, numerical simulations of a ground anchor were performed prior to an in-situ anchor pull-out test. The results are discussed and compared with the in-situ measurements of loads and displacements.

2 Pull-Out Test and Soil Description

The pull-out test was performed by Keller Grundbau on the 6th December 2017 in St. Kanzian, Austria. The anchor was vertically installed and was post-grouted. The anchor was 12 m at the free length and 8 m at the fixed length and 6 strands were employed. The borehole diameter was 178 mm and, due to the post-grouting, the diameter at the bond length was increased approximately to 280 mm.

Laboratory testing of soil samples extracted about 80 m from the test area classified the soil as clayey, sandy silt of low plasticity and the grain size distribution was approximately 70% silt, 15% clay and 15% sand. Direct shear tests and oedometer test were performed in samples located at 17 m depth and 24 m depth, respectively. After the anchor installation, an additional in-situ seismic dilatometer test (sDMT) was conducted in the area. The soil is referred herein as "seeton".

3 Numerical Simulation

The numerical simulation was performed using the finite element software Plaxis 2D 2016 (Brinkgreve et al. 2016). The model was axisymmetric and only half of the anchor geometry was modelled. The model geometry is presented in Fig. 1. The material parameters are shown in Tables 1 and 2.

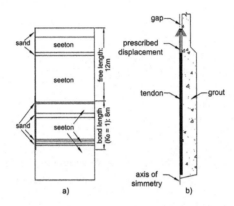

Fig. 1. Model geometry – (a) overview and (b) anchor detail

Table 1. Soil properties

Parameter	Description	Unit	Seeton	Sand
$E_{50,ref}$	Primary loading stiffness at ref. pressure	kPa	6625	24000
$E_{oed,ref}$	Oedometric stiffness at ref. pressure	kPa	5300	24000
$E_{ur,ref}$	Un/reloading stiffness at ref. pressure	kPa	48000	72000
G_{0ref}	Small strain shear modulus	kPa	120000	120000
$\gamma_{0.7}$	Shear strain at 70% G_{0ref}	–	0.15E–3	0.15E–3
c'	Effective cohesion	kPa	10	5
φ'	Effective friction angle	°	29	35

The tendon was considered only in the fixed length and the grout was applied in the free and fixed length. The load was applied by means of vertical prescribed displacements on top of the tendon.

The pressure grouted body was considered by increasing the radial stresses along the entire section comprising the fixed length and, for this purpose, the earth coefficient at rest (K_0) was set equal to one. The soil was composed of two different materials: seeton and sand layers and the soil properties were calibrated according to the laboratory and in-situ investigations. The material model for the soil was the Hardening soil model with small strain stiffness and was set as drained. The tendon was modelled as a linear elastic material and the grout with the Shotcrete model. The Shotcrete model is a nonlinear constitutive model which allows post-peak softening in compression and tension and is therefore able to capture the development of cracking in the grout. For more details on the model the reader is referred to Schädlich and Schweiger (2014).

Table 2. Tendon and grout properties

Parameter	Description	Unit	Tendon	Grout
E	Young's Modulus	kPa	195000000	16260000
$f_{c,28}$	Uniaxial compressive strength	kPa	–	32120
$f_{t,28}$	Uniaxial tensile strength	kPa	–	2000
$G_{c,28}$	Compressive fracture energy	kN/m	–	50
$G_{t,28}$	Tensile fracture energy	kN/m	–	0.15
f_{tun}	Ratio Residual/Peak tensile strength	–	–	0.05
φ'	Maximum friction angle	°	–	40

4 Results

Figure 2a shows the load-displacement curves obtained numerically and after the in-situ measurements and the shear stress distribution along the interface grout-soil is presented in Fig. 2b.

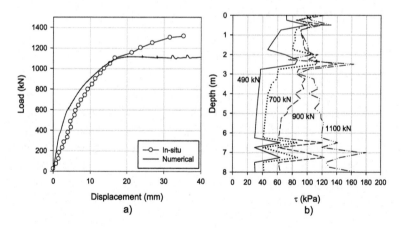

Fig. 2. Numerical results – (a) load-displacement curve and (b) shear stress distribution

In Fig. 2a, both curves showed good agreement and after 1100 kN the in-situ anchor had already reached the failure criteria, which was a creep value larger than 2 mm. The stresses in Fig. 2b are higher within the stiffer sand layers. The ultimate pull-out load of about 1100 kN is achieved when the maximum shear stress is mobilised along the entire fixed length.

The contribution of the grout on the strain distribution along the tendon can be evaluated by the tension softening parameter H_t, an output of the Shotcrete model. If H_t exceeds zero softening in tension starts and the cracks start to develop, leading to an increase of the strains along the tendon. If H_t is higher than 1, the tensile stress decreases to its residual value and the strains oscillate along the tendon. The strains distribution and the variation of H_t in the grout column are shown in Figs. 3a and b.

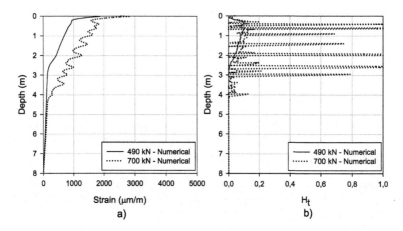

Fig. 3. Numerical results – (a) strains along the tendon and (b) H_t distribution in the grout

5 Conclusions

The predicted load-displacement curve ("class-A" prediction) and the one obtained in-situ showed very good agreement. The ultimate pull-out load is achieved when the shear strength is mobilised along the entire interface grout-soil. The numerical simulation indicated that the strains in the tendon are highly influenced by the development of cracks in the grout of the fixed length.

References

Brinkgreve, R.B.J., Kumarswamy, S., Swolfs, W.M.: Finite element code for soil and rock analyses, User Manual, PLAXIS 2016. Plaxis bv, Delft, The Netherlands (2016)

Schädlich, B., Schweiger, H.F.: A new constitutive model for shotcrete. In: Hicks, M.A., Brinkgreve, R.B.J., Rohe, A. (eds.) Proceedings Numerical Methods in Geotechnical Engineering, pp. 103–108. Taylor & Francis Group, London (2014)

Numerical Analysis of Monopile Behavior Under Short-Term Cyclic Loading

Anfeng Hu[1,2], Peng Fu[1,2(✉)], and Kanghe Xie[1,2]

[1] Research Center of Coastal and Urban Geotechnical Engineering,
Zhejiang University, Hangzhou, China
fupeng@zju.edu.cn
[2] MOE Key Laboratory of Soft Soils and Geoenvironmental Engineering,
Zhejiang University, Hangzhou, China

Abstract. A numerical model to analyze the behavior of monopile foundation under short-term cyclic lateral loading is presented here. On the basis of subroutine UMAT provided by the finite element program ABAQUS, the generalized theory of plasticity proposed by Pastor-Zienkiewicz is implemented in this model. Using the numerical model, the behavior of monopile under short-term cyclic loading is presented in this study. Numerical results show that the deflection of monopile and the excess pore water pressure within the soil around the pile increase with the number of cycles.

Keywords: Monopile foundation · Pastor-Zienkiewicz model
Short-term cyclic loading · Lateral displacement · Pore pressure

1 Introduction

Large diameter monopile is commonly used to support lateral loaded structure in recent years. The monopile foundation will be subjected to the cyclic loading from wind, earthquake and other impact load over its lifetime. The short term cyclic loading will lead to permanent lateral displacement of monopile foundation. The standard analysis method for an offshore monopile under lateral loading is the p-y curve method which was adopted in the design guidelines of American Petroleum Institute [1]. However, many studies have shown that p-y curve is not suitable for analyzing the behavior of large diameter monopile foundation under cyclic loading [2, 3].

The theory of generalized plasticity proposed by Pastor and Zienkiewicz [4, 5] is capable of describing the feature of sands under cyclic loading [6, 7]. Therefore, the Pastor-Zienkiewicz model is used in this study to analyze the behavior of monopile foundation under short term cyclic loading. On the basis of this constitutive model, the monopile behavior under short term cyclic loading is simulated cycle by cycle. Besides, the accumulation of excess pore pressure within the soil with the number of cycles is obtained.

2 Pastor–Zienkiewicz Model for Sands

In general, the constitutive models of soils play an important role to the analysis of the behavior of soils. The generalized theory of plasticity proposed by Pastor and Zienkiewicz [4, 5] can display the elastic-plastic property of cohesionless soil subject to cyclic loading. In its basic form, it relates the increments of stress $d\sigma$ and strain $d\varepsilon$ in a material as follows

$$d\varepsilon = C_{EP} \cdot d\sigma = C_E \cdot d\sigma + \frac{n_{gL/U} n^T \cdot d\sigma}{H_{L/U}} \quad (1)$$

The physical meaning of parameters in Eq. 1 has been described by Pastor and Zienkiewicz [4, 5].

3 Numerical Results and Discussion

3.1 Numerical Model of Monopile Foundation

The monopile foundation with diameter of 5 m, embedment length of 20 m and pile wall thickness of 0.05 m is considered here. In application, the finite element program ABAQUS is used herein. The three-dimensional numerical model of a monopile foundation is illustrated in Fig. 1. The monopile is considered as linear elastic material (200 GPa of elastic modulus, 0.3 of Poisson's ratio, density of 7.9×10^3 kg/m^3). The adopted values for the soil parameters of the Pastor-Zienkiewicz model used in this study are summarized in Table 1 [8]. The horizontal load is in the form of sine wave and the frequency of the load is 1 Hz, the amplitude of the horizontal load is 5MN.

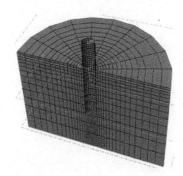

Fig. 1. Three dimensional finite element model of monopile foundation.

Table 1. Parameters of the Pastor-Zienkiewicz model.

K_0^e (kPa)	G_0^e(kPa)	M_g	α_g	M_f	α_f	H_0 (kPa)	β_0	β_1	γ	$H_{U,0}$ (kPa)	γ_U	p_0 (kPa)
5500	10000	1.18	0.45	0.77	0.45	65	17.2	0.43	1	250	5	100

3.2 Monopile Behavior Under Short Term Cyclic Loading

The deflection lines of the monopile are presented in Fig. 2. As can be seen from the figure, monopile exhibits some rigid response, which the pile rotates around the zero deflection point. The simulation results with the Pastor-Zienkiewicz model show an accumulation of lateral deformation under short term cyclic lateral loading. The deflection increases with the number of cycles, and the zero deflection point almost remain stationary.

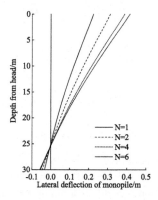

Fig. 2. Lateral deflection of monopile under cyclic loading.

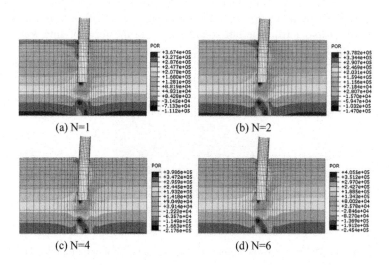

Fig. 3. Nephogram of pore water pressure (unit: Pa).

Figure 3 shows the variation of pore water pressure within the soil with the number of load cycles. The results show that the excess pore water pressure within the soil around the pile increases with the number of load cycles. There is negative excess pore pressure within the soil below the pile. Figure 4 shows that the excess pore water pressure within the soil near the pile tip increases with the number of load cycles. The accumulation of pore water pressure means that the effective stresses within the soil decrease and softening of the soil with every cycle. Therefore, the pile displacement increases with the number of cycles.

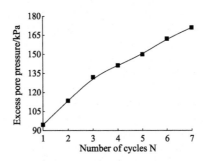

Fig. 4. Accumulation of excess pore pressure at pile toe.

References

1. API: Recommended practice for planning, designing and constructing fixed offshore platforms–working stress design. RP2A-WSD, 21st edn. American Petroleum Institute, Washington, DC (2000)
2. Long, J.H., Vanneste, G.: Effects of cyclic lateral loads on piles in sand. J. Geotech. Eng. **120**(1), 225–244 (1994)
3. LeBlanc, C., Houlsby, G.T., Byrne, B.W.: Response of stiff piles in sand to long-term cyclic lateral loading. Geotechnique **60**(2), 79–90 (2009)
4. Pastor, M., Zienkiewicz O C.: A generalized plasticity, hierarchical model for sand under monotonic and cyclic loading. In: Proceedings of 2nd International Symposium on Numerical Models in Geomechanic, pp. 131–149. Ghent (1986)
5. Pastor, M., Zienkiewicz, O.C., Chan, A.H.C.: Generalized plasticity and the modelling of soil behavior. Int. J. Numer. Anal. Met. Geomech. **14**(3), 151–190 (1990)
6. Chai, H.Y., Cui, Y.J., Lu, Y.F.: Simulation of loess behaviours under cyclic loading. Chin. J. Rock Mech. Eng. **24**(23), 4272–4281 (2005). (in Chinese)
7. Manzanal, D., Fernández Merodo, J.A., Pastor, M.: Generalized plasticity state parameter-based model for saturated and unsaturated soils. Part 1: Saturated state. Int. J. Numer. Anal. Met. Geomech. **35**(12), 1347–1362 (2011)
8. Cuéllar, P., Mira, P., Pastor, M., et al.: A numerical model for the transient analysis of offshore foundations under cyclic loading. Comput. Geotech. **59**, 75–86 (2014)

An Improved Beam-Spring Model for Excavation Support

Biao Huang[1], Mingguang Li[1(✉)], Jinjian Chen[1], Yongmao Hou[2], and Jianhua Wang[1]

[1] Department of Civil Engineering,
Shanghai Jiao Tong University, Shanghai, China
lmg20066028@sjtu.edu.cn
[2] Shanghai Tunnel Engineering Co., Ltd., Shanghai, China

Abstract. The servo struts are widely used in recent years to control deformations of retaining walls. However, the steel support servo system lacks theoretical guidance for the control thresholds of axial forces. This paper proposes an improved beam-spring method for modeling servo system. The revised method uses a combined model of prestress and springs to simulate servo struts. Considering the couple relationship between earth pressure and wall deformation, the non-limit earth pressure is used as applied load and the equilibrium equation can be solved by iteration method. Moreover, the comparison is carried out to analyze bending moment and deformations of retaining walls with or without servo struts. Results indicate that servo struts can effectively control wall deformations, while bending moment of walls on unexcavated sides increases significantly, which may cause structure failure during construction.

Keywords: Servo strut · Wall deformation control · Beam-spring method
Non-limit earth pressure

1 Introduction

The steel servo struts are popularized in recent years for its high efficiency on controlling the deformations of retaining walls and mitigating impacts on surroundings. In servo support system, two jacks are installed between the both ends of the steel struts and retaining walls. The axial forces of struts can be artificially changed by adjusting oil pressure in jacks to ensure wall deformations limited within requirements.

Currently, most researchers analyze field data based on relative cases to study the deformation behaviors of retaining walls under the control of servo struts, such as the foundation pit of Suzhou's Shin Kong Mitsukoshi project studied by Cai [1], the pit of Shanghai Concord City World Plaza project studied by Zhang [2] or other cases. However, no design method is available at present to determine the relationship between thresholds of axial forces and allowable wall deformations. Therefore, based on beam-spring method, the paper proposes a new method to model steel support servo system. Furthermore, the stress and deformation behaviors of retaining walls using servo struts is studied by comparing with traditional support system.

2 Simplified Model

Based on beam-spring method, the basic simplified model of servo support system is shown as Fig. 1. A modified model by combining prestress and spring is adopted to simulate servo struts. Considering the coupling relationships between earth pressure and lateral displacements of retaining walls in control stages, the non-limit earth pressure model is used as applied load in the equilibrium equations.

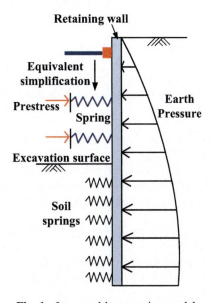

Fig. 1. Improved beam-spring model.

3 Improved Beam-Spring Method for Servo Struts

Solving static equilibrium of the soil slice elements in horizontal and vertical directions [3, 4], the active earth pressure, which is dependent on the active lateral stress ratio k, and the angle of the sliding plane α, can be obtained:

$$p_z = \frac{-k\gamma}{kC_1+2}(H-z) + C_2 k(H-z)^{-kC_1-1} \tag{1}$$

where $C_1 = -\frac{\cos(\alpha-\varphi-\delta)\tan\alpha}{\sin(\alpha-\varphi)\cos\delta}$, $C_2 = \frac{\gamma}{kC_1+2}(H-h)^{kC_1+2}$.

Two friction angles, relevant to the non-limit earth pressure on retaining walls, are the friction angle of the soil-wall interface, denoted as δ, and the internal friction angle of soil, denoted as φ. Previous studies by Liu [5] proposed the following empirical model for mobilization of friction angles based on extensive field observations.

$$\delta_m = \begin{cases} \frac{4\arctan(s/s_c)}{\pi}\delta, & s<s_c \\ \delta, & s \geq s_c \end{cases}, \quad \varphi_m = \begin{cases} \varphi_0 + \frac{4\arctan(s/s_a)}{\pi}(\varphi - \varphi_0), & s<s_a \\ \varphi, & s \geq s_a \end{cases} \quad (2)$$

The new model by combining prestress and springs to describe the mechanism of servo struts can be written as

$$F_s = F_{pre} + EA \cdot (\delta_{n+1} - \delta_n) \quad (3)$$

Where F_{pre} = prestress of per strut, δ_n = the deformations of struts in per stages.

When the servo struts are controlled to adjust axial force, the deformations of struts can be determined by:

$$\delta = \Delta L - \Delta F/(EA) \quad (4)$$

where ΔL = the deformations of jacks, ΔF = changed axial force.

Considering construction of struts lags behind the wall deformations in per stage, the wall deformations can be determined by solving the equilibrium equation:

$$[F_{e,n}] = ([K] + [K_m] + [K_s])[\Delta_n] - [K_s][\Delta_{s,n-1}] \quad (5)$$

where $[F_e]$ = matrix of earth pressure, $[K]$ = stiffness matrix of retaining wall, $[K_m]$ = stiffness matrix of soil springs, $[K_s]$ = stiffness matrix of struts, $[\Delta]$ = displacement matrix of retaining wall.

Since the deformations of jacks is unknown, the improved method modified stiffness matrix of struts during servo control to solve the problem of deformation compatibility between struts and walls in two stages.

Stage1: When the servo struts adjust axial force, the model just uses a prestress to simulate the controlling process:

$$[F_{e,n}] = ([K] + [K_m] + [K'_s])[\Delta_n] - [K'_s][\Delta_{s,n-1}] - [F_{s,pre}] \quad (6)$$

where $[K'_s] = [K_s] - [K_{s,i}]$, $[K_{s,i}]$ = stiffness matrix of the adjusting struts, $[F_{s,pre}]$ = controlling thresholds of servo struts.

Stage2: When the servo struts stop adjusting axial force during excavation, the equilibrium equation of next-stage excavation can be determined by

$$[F_{e,n+1}] = ([K] + [K_m] + [K_s])[\Delta_n] - [K_s][\Delta_{s,n-1}] - [F_{s,pre}] \quad (7)$$

where the stiffness matrix of struts transformed into $[K_s]$ again.

4 Comparison Analysis

The proposed method is used to calculate deformations and bending moment of retaining walls. In the case, the first supporting level is concrete supports and other two levels are steel servo struts. Figure 2 show the comparisons of stress and deformation behaviors of retaining walls with or without servo struts. The servo struts have high efficiency on limiting wall deformations. However, the maximum bending moment of retaining walls increase significantly on unexcavated side during servo control, which may cause structure failure by using conventional method for reinforcement.

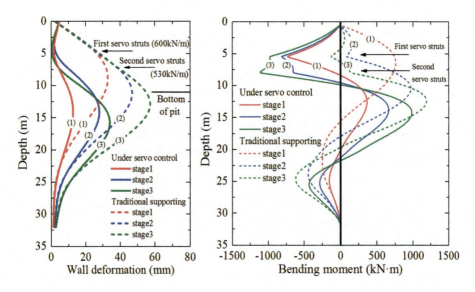

Fig. 2. Deformations and bending moment of retaining walls with or without servo struts.

References

1. Cai, J.T.: Design practice of deep foundation pit excavation adjacent to metro in soft soil area. Geotech. Eng. Tech. **29**(4), 163–168 (2015)
2. Zhang, D.B., Fei, W., Wang, C.Y., Wen, S.J.: Application of servo system of axial stress for steel support in deep foundation excavation next to the subway. Constr. Technol. **40**(341), 67–70 (2011)
3. Wang, Y.Z.: Distribution of earth pressure on a retaining wall. Géotechnique **52**(3), 231 (2000)
4. Ying, H.W., Huang, D., Xie, Y.X.: Study of active earth pressure on retaining wall subject to translation mode considering lateral pressure on adjacent existing basement exterior wall. Rock Mech. Rock Eng. **30**(S1), 2970–2978 (2010)
5. Liu, F.Q.: Lateral earth pressures acting on circular retaining walls. Int. J. Geomech. **14**(3), 04014002 (2014)

Numerical Modelling of a Monopile for Estimating the Natural Frequency of an Offshore Wind Turbine

David Igoe[1(✉)], Luke J. Prendergast[2], Breiffni Fitzgerald[1], and Saptarshi Sarkar[1]

[1] Trinity College Dublin, Dublin, Ireland
igoed@tcd.ie
[2] Delft University of Technology, Delft, Netherlands

Abstract. Monopiles are the most common foundation system for supporting Offshore Wind Turbines (OWT's), accounting for more than 80% of all OWT substructures installed in Europe to date [1]. Significant reductions in the cost of developing OWTs have been realized over the past few years, to the point where offshore wind can now be developed subsidy free in favorable locations. Optimizing the engineering design of these structures has played a key role in ensuring these cost reductions are possible. The largest uncertainty with respect to modelling the dynamic response of an OWT often relates to the geotechnical design. This paper examines the influence of soil-structure interaction on the dynamic response of an OWT structure. The below ground pile-soil behavior was modelled using (i) a conventional DNV (De Norske Veritas) '*p-y*' approach and (ii) an advanced in-situ calibrated 3D FE geotechnical design approach. The results for the soil-structure interaction were inputted into a separate dynamic wind turbine model and the dynamic response using the two separate SSI approaches were compared.

Keywords: Monopiles · Numerical modelling · Dynamics

1 Introduction

Monopile foundations are single large diameter open-ended tubular steel piles, usually driven into the sea bed, which rely on the stiffness and strength of the surrounding soil to provide resistance against large environmental loads. In most practical situations, considerations of the dynamic response and natural frequency of the OWT structure will govern the monopile diameter.

The industry standard approach for the geotechnical design of monopiles in Europe are those recommended by DNV-GL [2], which uses a decoupled Winkler beam approach [3] where the lateral soil reaction is described by non-linear '*p-y*' springs. The methods were calibrated using a limited number of pile tests performed on slender jacket piles with diameters less than 1 m, and are now recognized as being unsuitable for predicting the response of large diameter monopiles [2]. Recent research [4–7] has attempted to address some of the shortfalls of the DNV approach and there is significant research effort in this area currently ongoing. This paper compares the dynamic response of an OWT structure where the below ground pile-soil behavior was modelled using (i) the conventional DNV 'p-y' approach and (ii) an advanced in-situ calibrated 3D FE geotechnical design approach.

2 Modelling

2.1 Site Location, Pile Geometry and Loading Conditions

In order to model real-world offshore wind conditions, a soil profile from the Holland Kust (zuid) offshore wind farm zone in the Netherlands was chosen due to the public availability of the soil profile information and geotechnical testing reports. After a careful review, the HKZ2_BH03 location was selected as it was deemed to be representative of a typical North Sea medium dense to dense sand site. All soil parameters required for both the DNV approach and the 3D FE modelling were derived from the CPT profiles following the method defined by [4]. Based on the mean sea level at the location a water depth of 30 m was selected for analysis.

For the purposes of modelling the below mudline pile response, an initial approximate monopile geometry for the site was estimated as 7 m in diameter, 28 m embedment with a constant 70 mm wall thickness below mudline. Above the mudline, the monopile and wind-turbine geometry were developed based on a 5 MW reference turbine as defined by NREL [8]. For this initial study, a fatigue damage equivalent horizontal load, H, of 4000 kN was assumed to act at an eccentricity, e, of 40 m above mudline.

2.2 Monopile Modelling

The 3D FE modelling was undertaken using a static analysis in Plaxis 3D 2017 software. The pile was modelled in half space to reduce computation time. After a sensitivity analysis to optimize the model geometry, the boundaries were set at ± 60 m in the y-axis (direction of loading), 0 to +30 m in the x-axis and +40 m to −50 m in the z-axis (with mudline defined at +0 m). The soil elements were modelled as ten-node tetrahedral elements. The pile wall was modelled using six-node plate elements with interface elements added to allow a reduction in interface shear strength by a factor, R_{inter}, of 0.7. The pile plates were modelled as linear elastic elements with a Young's Modulus, E of 210 GPa, Poisson's ratio, v of 0.3 and a unit weight, γ of 77 kN/m^3. The Hardening Soil (HS) model as defined by Schanz [9] was used to define the soil response. All the required soil parameters for this model were derived from the CPT profile following the procedure defined in [4], which has been validated against a suite of monopile field tests. Once the 3D FE analysis was successfully completed, the soil reactions were extracted using the procedure described in [4] and defined in terms of linearized springs for use in the dynamic turbine model using a Matlab code.

For comparison with the 3D FE, the DNV 'p-y' approach was also considered (for brevity, details of the formulation of DNV approach will not be discussed in this paper). To summarize the approach taken, the DNV cyclic p-y curves were applied in a 1D FE Winkler beam model using a static solver developed in Matlab (similar to the commercial LPile software). Under the applied fatigue loading (H = 4000 kN), the secant stiffness of each non-linear p-y soil spring was then output from the monopile model for use as a linearized spring stiffness in a full dynamic coupled OWT model.

The dynamic wind turbine model used was similar to that described in detail in [10, 11]. The monopile and tower elements were formulated using four-degree-of-freedom (4-DOF) Euler-Bernoulli beam elements. Each element was 1 m in length and the individual elements are assembled to create global matrices for the full system. The soil was modelled by means of linear Winkler springs [12] attached to each embedded node of the monopile. The soil springs were linearized from the static monopile analyses as described previously. The nacelle was incorporated by means of a lumped mass attached do the final node of the tower-top beam. The undamped natural frequencies were extracted by means of solution to the Eigenproblem [13], obtained by solving for the eigenvalues of a system matrix, [D], defined as $[M]^{-1}[K]$.

3 Results/Conclusions

The displacement and linearized stiffness profiles, output from the 3D FE and DNV p-y static loading analyses are shown in Fig. 1. It is evident that the 3D FE provides a stiffer pile response, especially near the pile toe, albeit with a similar linear stiffness over the top 10 m. Calculations of the natural frequency of the OWT system from the dynamic turbine model suggest a 1.03% increase in first natural frequency using linearized stiffness values from the 3D FE approach when compared with the DNV p-y approach. Further research is ongoing to assess the accuracy of the dynamic modelling approaches and to assess effects of using non-linear springs on the first natural frequency of the OWT system.

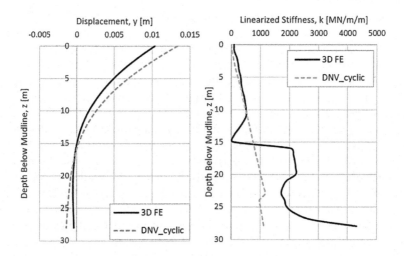

Fig. 1. (a) Displacement and (b) Linearized Stiffness Profiles from Static Monopile Models

References

1. Wind Europe: Offshore wind in Europe - Key Statistics and Trends 2017. Wind Europe (2017). https://windeurope.org/wp-content/uploads/files/about-wind/statistics/WindEurope-Annual-Offshore-Statistics-2017.pdf
2. DNV.GL: DNVGL-ST-0126: Support structures for wind turbines, Dnv Gl As, no, April 2016
3. Winkler, E.: Theory of Elasticity and Strength. Dominicus, Prague (1867)
4. Murphy, G., Igoe, D., Doherty, P., Gavin, K.G.: 3D FEM approach for laterally loaded monopile design, Comput. Geotech. **100**, 76–83 (2018)
5. Byrne, B.W., et al.: New design methods for large diameter piles under lateral loading for offshore wind applications. In: Frontiers in Offshore Geotechnics III - Proceedings of the 3rd International Symposium on Frontiers in Offshore Geotechnics, ISFOG 2015 (2015)
6. Byrne, B.W., et al.: Field testing of large diameter piles under lateral loading for offshore wind applications. In: Geotechnical Engineering for Infrastructure and Development - Proceedings of the XVI European Conference on Soil Mechanics and Geotechnical Engineering, ECSMGE, vol. 3 (2015)
7. Byrne, B., et al.: PISA: new design methods for offshore wind turbine monopiles (Keynote). In: International Conference on Offshore Site Investigations and Geotechnics (2017)
8. Jonkman, J., Butterfield, S., Musial, W., Scott, G.: Definition of a 5-MW Reference Wind Turbine for Offshore System Development, no. February 2009
9. Schanz, T., Vermeer, A., Bonnier, P.: The hardening soil model: formulation and verification. In: Beyond 2000 Computers Geotechnics 10 years PLAXIS – Proceedings of International Symposium beyond 2000 Computers Geotechnics, Amsterdam, Netherlands 1820 March 1999, p. 281 (1999)
10. Prendergast, L.J., Gavin, K., Doherty, P.: An investigation into the effect of scour on the natural frequency of an offshore wind turbine. Ocean Eng. **101**, 1–11 (2015)
11. Prendergast, L.J., Reale, C., Gavin, K.: Probabilistic examination of the change in eigenfrequencies of an offshore wind turbine under progressive scour incorporating soil spatial variability. Mar. Struct. **57**, 87–104 (2018)
12. Dutta, S.C., Roy, R.: A critical review on idealization and modeling for interaction among soil–foundation–structure system. Comput. Struct. **80**, 1579–1594 (2002)
13. Clough, R.W., Penzien, J.: Dynamics of structures (1993)

Modelling of Deep Excavation Collapse Using Hypoplastic Model for Soft Clays

Jan Jerman(✉) and David Mašín

Faculty of Science, Charles University, Albertov 3, 128 43 Prague, Czech Republic
jermanj@natur.cuni.cz

Abstract. The extended abstract presents an application of the newly developed hypoplastic model for soft clays. The model is applied to simulation of the well-documented geotechnical failure of a deep excavation occurring at the Nicoll Highway in Singapore. The marine clays occurring at the site are typical example of natural normally consolidated soft soils, which are difficult to characterize and model as they typically suffer from large deformations under loading and their stress-strain behavior is complex. In the new hypoplastic model, anisotropic asymptotic state boundary surface (ASBS) has been implemented to account for soil strength anisotropy. The apparent rotation of the ASBS is implemented by skewing the stress space. Additionally, the tensor L was made bilinear in D to more realistically predict the stress path of natural K_0 consolidated soils. The model has been implemented in the form of a user-defined subroutine for Abaqus and Plaxis finite element software. Firstly, it is demonstrated that the model predictions compare well with experimental data on element tests, demonstrating the effect of various features of the model on predictions. Secondly, numerical simulations of Nicoll highway excavation focusing on specific cross-section within the collapse area are presented, demanding good fit between measurement and simulations up to the final stages immediately preceding the excavation collapse.

Keywords: Constitutive modelling · Hypoplasticity · Anisotropy · Excavation

1 Introduction

Nicoll Highway collapse presents a prominent case of a geotechnical failure and has been excessively documented, see [2, 3] and many others. The disaster occurred during very deep excavations for the cut-and-cover tunnels for the Mass Rapid Transit (MRT) Circle Line in Singapore. The key reason leading to the collapse was under-design of the diaphragm walls due to procedures used in finite element analyses to design the excavation and under-design of excavation bracing. The calculations used effective stress strength parameters (c', φ') to represent behavior of soft clays under undrained conditions, which lead to significant overestimation of the undrained shear strength of the soil. Thus, the diaphragm wall deflections and bending moments were underestimated. Following the collapse, extensive investigation was concluded. Most of the experts used the same constitutive model to investigate the failure process, however, total stress strength parameters were used. Corral and Whittle [2] used advanced elastoplastic model to simulate the collapse. All the above-mentioned models either lacked the complexity to be able to simulate the behavior in element tests or were using very high number of parameters [2]. By using hypoplastic model for soft clays all these disadvantages could be overcome by the model – hypoplasticity belongs to group of incrementally non-linear models, thus providing non-linearity, however, the model uses only 8 parameters, which are obtainable from 3 element test types. This fact means, that the procedure described below could be easily used by practicing engineers and not only for academical purposes.

2 Hypoplastic Model for Soft Clays

The presented hypoplastic model for soft clays is based on hypoplastic model for clays by Mašín [6]. This approach to hypoplasticity was developed at the University of Karlsruhe and is represented by the work of Kolymbas [5].

The motivation for development of the hypoplastic model for soft clays was driven by a need for realistic prediction of experimental results of undrained triaxial tests on normally consolidated or slightly overconsolidated soft clays, whilst the original hypoplastic model for clays [7] has displayed an inability to predict undrained triaxial tests (in extension in particular) realistically. Two distinct improvements to the original hypoplastic model [7] aiming to enhance model performance were implemented:

- Strength anisotropy - apparent rotation of ASBS has been done by skewing the stress space similarly to Gajo and Muir Wood [3].
- Stress path correction for K_0 consolidated samples (LD approach, abbreviated as LD hereafter).

The new model with both enhancements combined is entitled soft clay model hereafter. By setting the additional soft clay model parameters to reference values the model reduces to the original hypoplastic model for clays by Mašín [7]. For detailed mathematical formulation of soft clay model the readers are referred to [4].

2.1 Model Calibration

The hypoplastic soft clay model adds three parameters to the original model by Mašín [7]:

- β – Defines inclination of the ASBS with respect to p-axis in q-p plane.
- γ – Defines skewing of the stress space for calculation of all variables denoted with dl for use in LD approach.
- ζ – Parameter for control of the stress path shape in DdepL approach.

For calibration of parameter β, extension triaxial test is necessary to evaluate strength difference in compression and extension. For parameters γ and ζ anisotropically (ideally K_0) consolidated triaxial tests are required in compression and extension to find the best agreement between experiments and stress paths predicted by the model. Table 1 shows calibration results using hypoplastic soft clay model to simulate element tests of Upper and Lower marine clays.

Table 1. Material parameters for two clays used in calibrations – Upper (UMC) a Lower (LMC) marine clays

Parameter	Symbol	UMC	LMC
CS friction angle	φ	32	27
NCL slope	λ	0.165	0.165
Slope of URL	κ	0.005	0.01
NCL for p = 1 kPa	N	1.76	1.77
Shear modulus parameter	ν	0.21	0.21
Stress path factor	ζ	1	1
Stress space rotation for DdepL	γ	31	30
ASBS rotation	β	8	2

Model predictions (abbreviated "soft clay model" for soft clay model and "original model" for model by Mašín [7]) are shown in Figs. 1 and 2. Experiments on Upper and Lower marine clays, respectively, are abbreviated "experiment". The soft clay model shows significant improvement compared to the hypoplastic model by Mašín [7]. Lower ultimate strength in extension is predicted and the stress paths show much better representation of experiments. Mohr-Coulomb model simulations are shown as well for comparison.

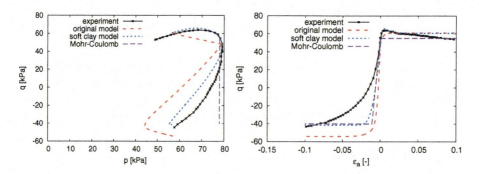

Fig. 1. Lower marine clay. Soft clay, original model and Mohr-Coulomb model predictions compared to experimental results, compression and extension triaxial tests.

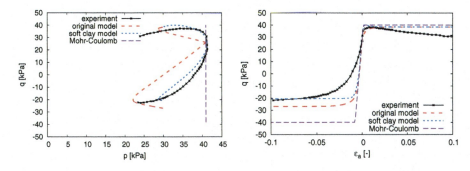

Fig. 2. Upper marine clay. Soft clay, original model and Mohr-Coulomb model predictions compared to experimental results, compression and extension triaxial tests.

3 Finite Element Model

Plaxis 2D software has been used for the back-analysis of the Nicoll highway collapse. Lateral wall deflections are compared for 3 constitutive models: elasto-plastic Mohr-Coulomb model using effective parameters as in the original design as well as total parameters as used in back-analyses, hypoplastic model for clays by Mašín [7] and soft clay hypoplastic model. Additionally, soft clay model with $\tan(\varphi_c)$ reduced by factor of safety 1.25, which would have been presumably used by practicing engineers analyzing the excavation, are shown. For clarity, four stages of excavation are shown, however, 20 stages were calculated in the FEM model to simulate all the steps of the excavation construction. Lateral wall deflections are compared in Fig. 3.

The collapse occurred in the excavation at the level RL 72.3. During the construction of the diaphragm wall except the last phases before the collapse (levels 75.3 and 72.3) the soft clay model predicts the lateral wall deflection better compared to the original hypoplastic model, which agrees with improved predictions of element tests. Mohr-Coulomb model with total parameters significantly overpredicts deformation of the wall deflection for all excavation phases, on the other hand, Mohr-Coulomb model using effective parameters as in the original design significantly underestimates deformations in all phases. Hypoplasticity underpredicts the excess wall deflections for the last two phases before the collapse, however, while applying factor of safety 1.25 to the soft clay model, the prediction follows monitoring data even in the phase immediately preceding the collapse.

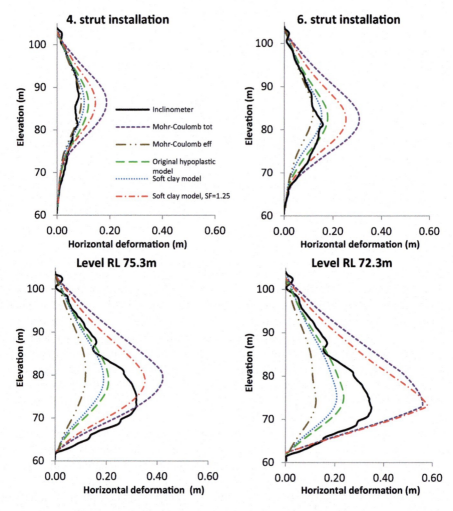

Fig. 3. Comparison of computed and measured lateral wall deflections. Comparison of the simulations using original hypoplastic model, soft clay model and Mohr-Coulomb model with inclinometer data. Four stages of excavation are shown, the stages RL 75.3 and RL 72.3 are showing excessive deformations before the collapse.

4 Summary and Conclusions

Two modifications of a hypoplastic model for clays [6] are presented in the paper, first aimed at improving the performance of the model for K_0 consolidated soils and the second one defining anisotropic state boundary surface. Fundamental features of the model formulation were defined, followed by evaluation of performance of the model in element tests and the retrospective simulations of a deep excavation collapse occurring at the Nicoll Highway in Singapore. The model adds three additional parameters, which can be obtained from K_0 normally consolidated triaxial tests.

Model predictions compare well with element test experimental data as well as lateral wall deflections where it clearly provides better predictions than the reference Mohr-Coulomb analyses.

Acknowledgement. The first author has also been supported by grant No. 1075516 of the Charles University Grant Agency. Institutional support by Center for Geosphere Dynamics (UNCE/SCI/006) is greatly appreciated.

References

1. Corral, G., Whittle, A.J.: Re-analysis of deep excavation collapse using a generalized effective stress soil model. In: Earth Retention Conference, vol. 3, pp. 720–731 (2010)
2. Davies, R.V., Fok, P., Norrish, A., Poh, S.T.: Nicoll highway collapse: field measurements and observations. In: Proceedings of International Conference on Deep Excavations, Singapore, June 2006
3. Gajo, A., Muir Wood, D.: A new approach to anisotropic, bounding surface plasticity: general formulation and simulations of natural and reconstituted clay behaviour. Int. J. Numer. Anal. Meth. Geomech. **25**(3), 207–241 (2001)
4. Jerman, J., Mašín, D.: A hypoplastic model for soft clays incorporating strength anisotropy. In: Proceedings of 9th NUMGE Conference on Numerical Methods in Geotechnical Engineering in Porto, Portugal, pp. 25–27 (2018)
5. Kolymbas, D.: An outline of hypoplasticity. Arch. Appl. Mech. **61**(3), 143–151 (1991)
6. Mašín, D.: A hypoplastic constitutive model for clays. Int. J. Numer. Anal. Meth. Geomech. **29**(4), 311–336 (2005)
7. Mašín, D.: Clay hypoplasticity model including stiffness anisotropy. Géotechnique **64**(3), 232–238 (2014)

Simulation of Caisson Foundation in Sand Using a Critical State Based Soil Model

Zhuang Jin[1,3], Zhen-Yu Yin[1,2,3(✉)], Panagiotis Kotronis[1,3], and Ze-Xiang Wu[1,3]

[1] Ecole Centrale de Nantes, Université de Nantes, CNRS, Institut de Recherche en Génie Civil et Mécanique (GeM), 1 rue de la Nöe, 44321 Nantes, France
zhenyu.yin@gmail.com
[2] Department of Geotechnical Engineering, College of Civil Engineering, Tongji University, Shanghai, China
[3] Key Laboratory of Geotechnical and Underground Engineering of Ministry of Education, Tongji University, Shanghai 200092, People's Republic of China

1 Introduction

A caisson is a closed-top steel tube, which is first lowered to the seafloor allowing to penetrate under its own weight, and then push to full depth with suction force producing by pumping water out of its interior. Recently, caissons have been widely used for different types of constructions, such as gravity platform jackets, jack-ups, offshore wind turbines, subsea systems and seabed protection structures. For an optimum design, a better understanding of the performance of caisson foundations is therefore necessary.

In recent years, with the development of the finite element method (FEM) and the rising of commercial FEM codes, a number of 2D and 3D numerical studies have been performed to research the caisson bearing capacity under different loading combinations and drainage conditions [1–4]. However, this type of numerical analysis is computationally expensive as calculations may need days or weeks to finish. In this sense, more simple, fast and robust numerical methods are needed.

In this article, numerical simulations are presented of a caisson foundation in sand. A critical state based soil model is used for the soil. The numerical results are compared with laboratory experiments. Different horizontal and rotation loading combinations under different vertical loading levels are considered. The different loading paths are adopted to find the failure surface. The bearing strength surface of combined horizontal and rotation loadings under different level of vertical loading was found and analyzed.

2 SIMSAND Model

2.1 Mathematical Description

In order to simulate the installation process of caisson foundation, a recently developed critical state based model, the SIMSAND model is adopted hereafter. The simple sand model (SIMSAND) was developed based on the Mohr-Coulomb model by implementing the critical state concept [5–7] with non-linear elasticity, non-linear plastic hardening, and a simplified three-dimensional strength criterion. The density-dependent peak strength and stress-dilatancy(contraction or dilation) are well captured by the SIMSAND model. The model parameters with their definitions are summarized in Table 1. The calibration of the model parameters can be carried out using a straight-forward way [7] or using optimization methods [6, 8, 9].

Table 1. Model parameters for the Baskarp No. 15 sand.

Groups	Elastic			CSL				Shear-sliding	
Symbols	K_0 (kPa)	n	v	e_{ref}	λ	ζ	ϕ_u (°)	k_p	A_d
Values	344	0.58	0.25	1.25	0.038	0.11	35.1	0.0034	0.45

The SIMSAND model parameters are divided in three groups: (1) elastic parameters (K_0, n and v), (2) critical state line related parameters (e_{ref}, λ, ζ and ϕ_u), and (3) plastic shear-sliding parameters (k_p and A_d). The calibration of the model parameters can be carried out using optimization methods [5, 9–11].

According to the three drained triaxial tests on Baskarp No. 15 sand [12], the model parameters were identified as follows.

3 Numerical Modelling

3.1 Finite Element Model

In this study, a three-dimensional finite element model of a caisson foundation was developed and utilized in this study. The ABAQUS software was adopted for the simulations. Figure 1 shows a three-dimensional view of the numerical model with geometrical properties. As shown, 5 times the bucket diameter were chosen as the length and the width of the model. In addition, a depth of 3 times the bucket skirt was considered as the depth of soil under the bucket tip. Displacements in all directions of the model boundaries (the bottom, all directions of periphery) were fixed.

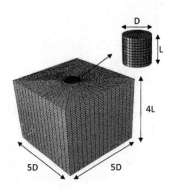

Fig. 1. Finite element mesh and boundary extensions of suction bucket foundation.

Since the main goal of this paper was to evaluate the bearing capacity of the bucket foundation, the wind turbine and superstructure were beyond the scope of this study and were not evaluated. To simulate the applied loads from the superstructure and wind energy converter system, point load was used for simulation of the vertical dead load. The displacement-controlled method was also employed to model the horizontal and rotational loads.

3.2 Model Validated and Simulations

One field test study for suction bucket foundation at Frederikshavn [13] was analyzed by the above model to validate the numerical modeling adopted in this study.

To verify the capability of the bucket foundation used for a 3 MW Vestas wind turbine, large-scale experiments were conducted at Frederikshavn, Denmark. A lateral load at a fixed eccentricity under constant vertical load was applied in the experiments. In one of the experimental tests, a caisson with diameter D of 2 m, skirt length L of 2 m and skirt thickness of 12 mm was installed in a shallow pond near the sea for modeling of the bucket foundation. The load eccentricity h and vertical load V were 17.4 m and 37.3 kN respectively. The caisson was installed in very dense fine sand with the unit weight 19.5 kN/m^3, relative density, 90% and specific gravity, 26.5 kN/m^3. As mentioned before, the soil parameters are presented in Table 1.

Experiment results of moment-rotation curves and simulation results are shown in Fig. 2. The results obtained from the numerical simulation properly agree with those achieved from field test. Consequently, the capability of the finite element modeling for assessment of bearing capacity of suction caisson is validated.

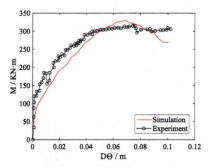

Fig. 2. Comparison of numerical modeling and field test results for the caisson foundation.

In order to find the bearing strength surface under different loading combinations, a series of loading paths were adopted. Figure 3 presents the normalized bearing strength surface with different level of vertical loading. It is pointed that the non-perpendicular intersection of the envelope with the load axes indicates the cross-coupling between the rotational and horizontal degrees of freedom. Perhaps expectedly, mainly due to the nonlinear nature of the caisson-soil interface, the level of vertical loading affects the shape of the surface in a non-straightforward manner. More specifically, an expansion of the apex is observed as the vertical load increases until the value increases until reaches to 40%~60% of vertical loading bearing capacity, followed by a contraction at higher loading levels up to bearing failure.

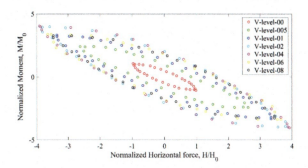

Fig. 3. Normalized bearing strength surfaces with different level of vertical loading.

References

1. Sukumaran, B., McCarron, W., Jeanjean, P., Abouseeda, H.: Efficient finite element techniques for limit analysis of suction caissons under lateral loads. Comput. Geotech. **24**, 89–107 (1999)
2. El-Gharbawy, S., Olson, R.: Modeling of suction caisson foundations. In: Proceedings of the Tenth International Offshore and Polar Engineering Conference, pp. 670–677. International Society of Offshore and Polar Engineers (Year)
3. Liu, M., Yang, M., Wang, H.: Bearing behavior of wide-shallow bucket foundation for offshore wind turbines in drained silty sand. Ocean Eng. **82**, 169–179 (2014)
4. Li, D., Zhang, Y., Feng, L., Gao, Y.: Capacity of modified suction caissons in marine sand under static horizontal loading. Ocean Eng. **102**, 1–16 (2015)
5. Jin, Y.-F., Yin, Z.-Y., Shen, S.-L., Hicher, P.-Y.: Selection of sand models and identification of parameters using an enhanced genetic algorithm. Int. J. Numer. Anal. Met. Geomech. **40**, 1219–1240 (2016)
6. Jin, Y.-F., Yin, Z.-Y., Shen, S.-L., Hicher, P.-Y.: Investigation into MOGA for identifying parameters of a critical-state-based sand model and parameters correlation by factor analysis. Acta Geotech. **11**, 1131–1145 (2016)
7. Wu, Z.-X., Yin, Z.-Y., Jin, Y.-F., Geng, X.-Y.: A straightforward procedure of parameters determination for sand: a bridge from critical state based constitutive modelling to finite element analysis. Eur. J. Environ. Civ. Eng., 1–23 (2017)
8. Jin, Y.-F., Wu, Z.-X., Yin, Z.-Y., Shen, J.S.: Estimation of critical state-related formula in advanced constitutive modeling of granular material. Acta Geotechnica, 1–23 (2017)
9. Yin, Z.-Y., Jin, Y.-F., Shen, S.-L., Huang, H.-W.: An efficient optimization method for identifying parameters of soft structured clay by an enhanced genetic algorithm and elastic–viscoplastic model. Acta Geotech. **12**, 849–867 (2017)
10. Jin, Y.-F., Yin, Z.-Y., Shen, S.-L., Zhang, D.-M.: A new hybrid real-coded genetic algorithm and its application to parameters identification of soils. Inverse Prob. Sci. Eng. **25**, 1343–1366 (2017)
11. Yin, Z.-Y., Jin, Y.-F., Huang, H.-W., Shen, S.-L.: Evolutionary polynomial regression based modelling of clay compressibility using an enhanced hybrid real-coded genetic algorithm. Eng. Geol. **210**, 158–167 (2016)

12. Foglia, A., Ibsen, L.B.: Laboratory experiments of bucket foundations under cyclic loading. Department of Civil Engineering, Aalborg University (2014)
13. Houlsby, G.T., Ibsen, L.B., Byrne, B.W.: Suction caissons for wind turbines. Frontiers in Offshore Geotechnics: ISFOG, Perth, WA, Australia, pp. 75–93 (2005)

Analytical Solution for the Transient Response of Cylindrical Cavity in Unsaturated Porous Medium

Wei-Hua Li[✉], Feng-Cui Feng, and Zhao Xu

Key Laboratory of Urban Underground Engineering of Ministry of Education,
Beijing Jiaotong University, Beijing 100044, China
whli@bjtu.edu.cn

Abstract. The transient response of a pressurized long cylindrical cavity in an infinite unsaturated porous medium is considered for the first time. Solutions are presented for three different prescribed transient radial pressure types, namely step load, gradually applied step load, and triangular pulse load, acting on the surface of a permeable as well as an impermeable cavity surface. Time domain solutions are obtained by inverting Laplace domain solutions using a reliable numerical scheme. The validity of the solution is confirmed by comparing with the one in which the infinite medium is assumed as a saturated porous medium. The time histories of radial displacements, hoop stresses and pore fluid pressures corresponding to a cavity in a representative unsaturated porous material are presented. A detailed parametric study is described to illustrate the saturation influence on the responses of the radial displacements, hoop stresses, and pore fluid pressures at the cavity surface.

Keywords: Unsaturated porous medium · Transient response
Pressurized cylindrical cavity

1 Introduction

The problem of the transient response of a pressurized cavity in an infinite medium is of considerable significance, owing to its manifold applications. In most previous studies, the infinite medium was treated only as ideal elastic or saturated porous medium. In reality, unsaturated porous media are encountered in various areas, such as near the earth surface, and notorious geomaterials known as unsaturated soils can be grouped into a certain family of unsaturated porous media. Therefore, a study of dynamic responses in unsaturated porous media should be of considerable interest. In the current state of the knowledge, the study of the dynamic behavior of unsaturated porous media constitutes a relatively new area. No analysis has been conducted regarding the transient response of a pressurized cavity in an unsaturated porous medium. Based on the dynamic equations for an unsaturated porous medium formulated by Wei and Muraleetharan [1], The solutions for the transient response of a pressurized long cylindrical cavity in an infinite unsaturated porous medium are established in this paper for the first time.

2 Theoretical Formulation

A cross-section of the two-dimensional model to be analyzed is illustrated in Fig. 1.

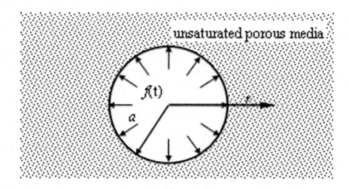

Fig. 1. Deep buried cavity model in unsaturated soil

An infinite-length cylindrical cavity is embedded in full space, which consists of unsaturated porous media. To solve this problem, the cylindrical coordinates are only related to the r coordinates; therefore, the dynamic equations of unsaturated porous media established by Wei and Muraleetharan [1] can be reduced to the following form

$$
\begin{aligned}
n_0^S \rho_0^S \frac{\partial^2 u_r^S}{\partial t^2} &= (M_{SS} + n_0^S \mu_S) \frac{\partial}{\partial r}\left(\frac{\partial u_r^S}{\partial r} + \frac{u_r^S}{r}\right) + n_0^S \mu_S \left(\nabla^2 - \frac{1}{r^2}\right) u_r^S \\
&+ M_{SW} \frac{\partial}{\partial r}\left(\frac{\partial u_r^W}{\partial r} + \frac{u_r^W}{r}\right) + M_{SN} \frac{\partial}{\partial r}\left(\frac{\partial u_r^N}{\partial r} + \frac{u_r^N}{r}\right) \\
&+ \hat{\mu}^W \frac{\partial}{\partial t}\left(u_r^W - u_r^S\right) + \hat{\mu}^N \frac{\partial}{\partial t}\left(u_r^N - u_r^S\right)
\end{aligned}
\quad (1a)
$$

$$
\begin{aligned}
n_0^W \rho_0^W \frac{\partial^2 u_r^W}{\partial t^2} &= M_{SW} \frac{\partial}{\partial r}\left(\frac{\partial u_r^S}{\partial r} + \frac{u_r^S}{r}\right) + M_{WW} \frac{\partial}{\partial r}\left(\frac{\partial u_r^W}{\partial r} + \frac{u_r^W}{r}\right) \\
&+ M_{WN} \frac{\partial}{\partial r}\left(\frac{\partial u_r^N}{\partial r} + \frac{u_r^N}{r}\right) - \hat{\mu}^W \frac{\partial}{\partial t}\left(u_r^W - u_r^S\right)
\end{aligned}
\quad (1b)
$$

$$
\begin{aligned}
n_0^N \rho_0^N \frac{\partial^2 u_r^N}{\partial t^2} &= M_{SN} \frac{\partial}{\partial r}\left(\frac{\partial u_r^S}{\partial r} + \frac{u_r^S}{r}\right) + M_{WN} \frac{\partial}{\partial r}\left(\frac{\partial u_r^W}{\partial r} + \frac{u_r^W}{r}\right) \\
&+ M_{NN} \frac{\partial}{\partial r}\left(\frac{\partial u_r^N}{\partial r} + \frac{u_r^N}{r}\right) - \hat{\mu}^N \frac{\partial}{\partial t}\left(u_r^N - u_r^S\right)
\end{aligned}
\quad (1c)
$$

where the superscript S denotes the solid component; W and N denote the wetting and non-wetting fluids, respectively; n_0^α and ρ_0^α (α = S, W, N) are the initial volume fraction and initial density of individual components, respectively; M_{SS}, M_{WW}, M_{NN}, M_{SW}, M_{SN}, and M_{WN} are the elastic constants related to the degree of saturation; $\hat{\mu}^f$ (F = W, N) can be related to the permeability k and the dynamic viscosity of an F-fluid.

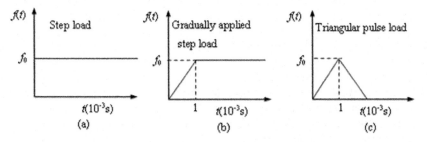

Fig. 2. Transient loadings considered in the analysis

By introducing scalar potentials φ^α such that $u_r^\alpha = \frac{\partial \varphi^\alpha}{\partial r}$ (α = S, W, N) and using the Laplace transform about time t, Eq. (1) can be solved as

$$\left(\nabla^2 - k_j^2\right)\bar{\varphi}_j^S = 0 \quad (j=1,2,3), \quad \bar{\varphi}^W = \sum_{j=1}^{3} \delta_j^W \bar{\varphi}_j^S, \quad \bar{\varphi}^N = \sum_{j=1}^{3} \delta_j^N \bar{\varphi}_j^S \quad (2)$$

It can be seen that there are three types of compression waves. Since the domain in Fig. 1 is unbounded, the admissible solutions for $\bar{\varphi}_j^S$ are

$$\bar{\varphi}_j^S = A_j K_0(k_j r) \quad (j = 1, 2, 3) \quad (3)$$

where A_j is the arbitrary constant, and K_n is the modified Bessel functions of the second kind of order n. Then the displacements of unsaturated porous medium in the cylindrical coordinate can be expressed as

$$\bar{u}^S = -\sum_{i=1}^{3} A_i k_i K_1(k_i r), \quad \bar{u}^W = -\sum_{i=1}^{3} \delta_i^W A_i k_i K_1(k_i r), \quad \bar{u}^N = -\sum_{i=1}^{3} \delta_i^N A_i k_i K_1(k_i r) \quad (4)$$

Based on the stress-strain relationship of the unsaturated porous media, the stress tensors in unsaturated porous media can be expressed as

$$\bar{\sigma}_{rr}^S = \sum_{i=1}^{3} \left[(M_{SS} + M_{SW}\delta_i^W + M_{SN}\delta_i^N + n_0^S\mu_S) A_i k_i^2 K_0(k_i r) + n_0^S\mu_S A_i k_i^2 K_2(k_i r) \right] \quad (5a)$$

$$\bar{\sigma}_{\theta\theta}^S = \sum_{i=1}^{3} \left[(M_{SS} + M_{SW}\delta_i^W + M_{SN}\delta_i^N + n_0^S\mu_S) A_i k_i^2 K_0(k_i r) - n_0^S\mu_S A_i k_i^2 K_2(k_i r) \right] \quad (5b)$$

$$\bar{\sigma}^W = -\sum_{i=1}^{3}(M_{SW} + M_{WW}\delta_i^w + M_{WN}\delta_i^N)A_i k_i^2 K_0(k_i r) \quad (5c)$$

$$\bar{\sigma}^N = -\sum_{i=1}^{3}(M_{SN} + M_{WN}\delta_i^w + M_{NN}\delta_i^N)A_i k_i^2 K_0(k_i r) \quad (5d)$$

In this paper, the response of a deeply buried cylindrical cavity in unsaturated porous media, under the action of an axial symmetrical radial force, is discussed. Three different prescribed transient radial pressure types (step load, gradually applied step load, and triangular pulse load) are illustrated in Fig. 2, and two different hydraulic boundary conditions (permeable and impermeable cavity surface) are considered.

Considering the permeable boundary, the boundary conditions corresponding to the loading case can be expressed as

$$\sigma_{rr}^S = -\bar{f}(s), \; \sigma^W = \sigma^N = 0 \text{ at } r = a \quad (6)$$

where $\bar{f}(s)$ is the Laplace transform of the applied load as shown in Fig. 2.

The undetermined parameters in Eq. (4) can be solved as follows by the above boundary conditions. The analytical solution for the transient response of the pressurized cylindrical cavity in the unsaturated porous medium is then established.

When the pores in the unsaturated porous medium are filled with the same liquid phase, the medium is completely saturated. At this time, ignoring the items related with the gas phase, there will be two kinds of compression waves in the medium. By comprising with the solutions established by Senjuntichai [2] for the saturated condition, the presented solutions are thus validated.

3 Influence of Saturation on Dynamic Response

To study the response of the cavity with time, the Laplace domain solutions are converted into the time domain using the numerical inversion method proposed by Stehfest [3]. The time histories of radial displacements and hoop stresses around the cavity in unsaturated sandstone are analyzed. The main conclusions obtained are as follows: The maximum and stable radial displacement values decrease with increasing saturation, and the maximum values accordingly appear earlier for the three transient radial load types. The stable radial displacement values decrease with increasing saturation for the step load and gradually applied step load cases, while for the triangular pulse load case, the change is not obvious. Compared to the step load and gradually applied step load cases, the maximum radial displacement values for the triangular pulse load case are smaller and the maximum values arrive earlier. The maximum hoop stress values decrease with increasing saturation, and the maximum values accordingly appear earlier for the step load and gradually applied step load cases, while for the triangular pulse load case, the change is not obvious. The initial and stable hoop stress values do not change with saturation.

References

1. Wei, C.F., Muraleetharan, K.K.: A continuum theory of porous media saturated by multiple immiscible fluids. I. Linear poroelasticity. Int. J. Eng. Sci. **40**, 1807–1833 (2002)
2. Senjuntichai, T., Rajapakse, R.K.N.D.: Transient response of a circular cavity in a poroelastic medium. Int. J. Numer. Anal. Geomech. **17**, 357–383 (1993)
3. Stehfest, H.: Numerical inversion of Laplace transforms. Commun. ACM. **13**, 47–49 (1970)

Seepage and Deformation Analyses in Unsaturated Soils Using ANSYS

Yangyang Li[1(✉)], Harianto Rahardjo[2], and K. N. Irvine[3]

[1] Interdisciplinary Graduate School,
Nanyang Technological University, Singapore 639798, Singapore
liya0030@e.ntu.edu.sg
[2] School of Civil and Environmental Engineering,
Nanyang Technological University, Singapore, Singapore
[3] School of Humanities and Social Studies Education,
National Institute of Education, Nanyang Technological University,
Singapore 637616, Singapore

Abstract. Unsaturated soil mechanics has become the solution to many geotechnical challenges, including slope stability and tree stability. Compared to saturated soil, the additional cohesion from the matric suction strengthens the unsaturated soil. Slope stability problems could be analyzed using two-dimensional software such as GeoStudio by assuming soil under plane strain or plane stress condition. However, two-dimensional modelling may not be appropriate for complex analyses such as tree stability problems that involve soil-structure interaction. Such assumptions may lead to model inaccuracy since structural behaviour is three-dimensional. Unfortunately, software for three-dimensional analyses that could handle unsaturated soils is not readily available. In this study, a method for three-dimensional seepage and deformation analyses in unsaturated soils was done using ANSYS, incorporating the unsaturated soil shear strength, the permeability and the elastic modulus as functions of the net normal stress and the matric function. The results from two-dimensional GeoStudio and ANSYS analyses were compared to validate the method.

Keywords: Numerical modelling · Unsaturated soil
Three-dimensional analyses

1 Introduction

Singapore has been identified as the city with most trees being planted in the world [1], and as such, tree stability becomes particularly important since failures could endanger public safety and cause infrastructure damage. Tree failures could be triggered when the strength and the stiffness of the supporting soil would reduce as rainwater infiltrates into the unsaturated soil.

Two-dimensional (2D) modelling could be carried out in GeoStudio to analyse the behaviour of unsaturated soil under rainfall and wind load. However, in plane strain conditions, the thickness of soil and tree in the direction perpendicular to the plane are required to be much larger than the size within the plane so that the strain in the perpendicular plane could be assumed zero. Trees usually grow in a radial direction and therefore, three-dimensional (3D) analyses are necessary for tree stability problems.

This paper aims to provide a method that could help in analysing tree stability or other 3D structure-soil interaction problems in ANSYS with the incorporation of unsaturated soil properties.

2 Seepage Analyses

Transient or steady state thermal analyses in ANSYS can be used to conduct transient or steady state seepage analyses in unsaturated soils as the heat transfer equation and the seepage flow equation are very similar.

Equation (1) presents the heat equation [2]:

$$\frac{\partial}{\partial x}\left(k_x \frac{\partial T}{\partial x}\right) + \frac{\partial}{\partial y}\left(k_y \frac{\partial T}{\partial y}\right) + \frac{\partial}{\partial z}\left(k_z \frac{\partial T}{\partial z}\right) + q = \rho c \frac{\partial T}{\partial t} \qquad (1)$$

where k_x, k_y, and k_z are the thermal conductivities in the x, y and z direction respectively, T is the temperature, q is the heat flux, ρ is the density, c is the specific heat and t is the time.

Equation (2) presents the seepage equation [3]:

$$\frac{\partial}{\partial x}\left(k_{wx} \frac{\partial h}{\partial x}\right) + \frac{\partial}{\partial y}\left(k_{wy} \frac{\partial h}{\partial y}\right) + \frac{\partial}{\partial z}\left(k_{wz} \frac{\partial h}{\partial z}\right) + q_w = m_2^w g \rho \frac{\partial h}{\partial t} \qquad (2)$$

where k_{wx} and k_{wy} are the major and minor permeability functions for unsaturated soils, h is the total head in the water phase (i.e. elevation head plus pressure head), q_w is the water flux, m_2^w is the water storage function and g is gravitational acceleration.

In this research, T and c in the thermal equation were used to represent h and $m_2^w g$ in the seepage equation. The permeability function for unsaturated soil is shown in Eq. (3) [4] as a function of matric suction $(u_a - u_w)$:

$$k = \frac{k_s}{\left\{\ln\left[e + \left(\frac{u_a - u_w}{a}\right)^n\right]\right\}^{mp}} \qquad (3)$$

where a, n and m are fitted parameters from the best fit of the soil–water characteristic curve (SWCC), p is a fitted parameter determined by curve fitting the permeability data, k_s is the saturated permeability, u_a and u_w are pore-air pressure and pore-water pressure respectively.

Since $(u_a - u_w)$ is equal to $(-u_w)$ or $(z - h)g$ when u_a is 0 kPa, the permeability function could be represented by the thermal conductivity as a function of T in Eq. (4). Therefore, the calculated temperature T is equivalent to the total head h in the seepage analyses.

$$k = \frac{k_s}{\left\{\ln\left[e + \left(\frac{(z-T)*g}{a}\right)^n\right]\right\}^{mp}} \tag{4}$$

Seepage analyses were carried out in both ANSYS and GeoStudio with a 2D geometry. The seepage process when rainwater infiltrates into the soil was analyzed. The geometry and the boundary conditions are illustrated in Fig. 1. The initial ground water table (GWT) is at 2 m below the ground.

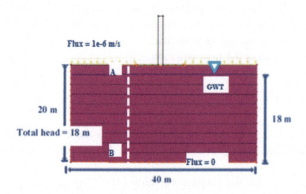

Fig. 1. Illustration of geometry and boundary conditions during seepage analyses.

The fitted parameters a, n, m and p are 23 kPa, 1.2, 1.9 and 5.26 respectively. The saturated permeability is 1E-5 m/s. Figure 2 shows the change in total head with elevation head along line A-B (Fig. 1) after 300 s. The results in ANSYS and GeoStudio are similar.

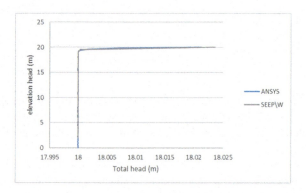

Fig. 2. Comparison of total heads in ANSYS Fluent and GeoStudio SEEP/W.

3 Deformation Analyses

The deformation analyses were carried out in both ANSYS static structural and GeoStudio SIGMA/W.

Equation 5 presents the Young's Modulus as a function of matric suction [5]. Because GeoStudio could not incorporate Young's Modulus as a function of matric suction, the soils were layered with different Young's Modulus values assigned according to the stress at the centroid position in both GeoStudio and ANSYS. The soil unit weight is 19.6 kN/m^3 and the soil shear strength parameters c', ϕ', ϕ^b are 10 kPa, 30° and 20°. A lateral force was applied at the top of the tree (10 m) and Fig. 3 shows the relationship between the maximum displacements (at the tree top) and the lateral force obtained from ANSYS and GeoStudio. Results for 2D analyses in ANSYS and GeoStudio are similar. However, compared to 2D analyses, 3D analyses lead to higher displacements and the 3D analysis results are more reliable without assumption of no stress or strain in the direction perpendicular to the plane. Moreover, the Young's Modulus could be represented as a function of matric suction in ANSYS and hence the change in stiffness due to flux boundary changes can be considered in the analyses.

$$E = 90.5 P_a \left(\frac{\sigma_n - u_a}{P_a}\right)^{0.69} + 92.8 P_a \left(\frac{u_a - u_w}{P_a}\right)^{0.18} + 18.9 P_a \left(\frac{(\sigma_n - u_a)(u_a - u_w)}{P_a^2}\right)^{0.99} \quad (5)$$

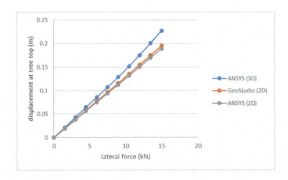

Fig. 3. Comparison of maximum displacements in ANSYS and GeoStudio SIGMA/W.

4 Conclusion

Seepage analyses and deformation analyses in ANSYS were verified against the results from GeoStudio. The comparison shows that ANSYS can be used to solve 3D problems including structure-soil interaction with the incorporation of unsaturated soil properties.

References

1. Li, X., Zhang, C., Li, W., Ricard, R., Meng, Q., Zhang, W.: Assessing street-level urban greenery using Google Street View and a modified green view index. Urban For. Urban Greening **14**(3), 675–685 (2015)
2. Fluent, A.: 12.0 User's guide. Ansys Inc. (2009)
3. Fredlund, D.G., Rahardjo, H.: Soil Mechanics for Unsaturated Soils. Wiley, New York (1993)
4. Leong, E.C., Rahardjo, H.: Permeability function for unsaturated soils. J. Geotech. Geoenviron. Eng. **123**(12), 1118–1126 (1997)
5. Rahardjo, H., Melinda, F., Leong, E.C., Rezaur, R.B.: Stiffness of a compacted residual soil. Eng. Geol. **120**(1–4), 60–67 (2011)

Implicit and Explicit Integration Schemes of a Double Structure Hydro-Mechanical Coupling Model for Unsaturated Expansive Clays

Jian Li[✉], Pengyue Wang, Lu Hai, Yanhua Zhu, and Yan Liu

Key Laboratory of Urban Underground Engineering of Ministry of Education,
Beijing Jiaotong University, Beijing 100044, China
jianli@bjtu.edu.cn

Abstract. Comparative studies on the performances of implicit and explicit integration schemes are conducted for a double structure hydro-mechanical coupling model of unsaturated expansive clays at an integration point level. The published constitutive model is built on the basis of two pore levels in expansive clay. In this paper, firstly, the implicit and explicit schemes based on the Euler backward and forward integration algorithms are proposed. These schemes can deal with multi-plastic mechanisms defined in the model and distinguish various cases in relation to the number of plastic mechanism active. Secondly, the performances of the integration schemes are demonstrated by comparisons of calculation results of a control-strain test with different strain increment sizes.

Keywords: Unsaturated expansive soils · Double structure
Hydro-mechanical coupling · Implicit integration algorithm
Explicit integration algorithm

1 Introduction

Establishment of constitutive models that accurately depict mechanical even water retention behaviors simultaneous of geo-materials has always been one of focuses for researchers. It is advised to combine bounding surface theory, multiple yield surface and so on to improve predictive abilities of models, which leads to complicate the models inevitably. In order to deal with engineering problems by these complex models, proposing a robust numerical implement method is necessary. According to solution strategies, the integration algorithm can be divided into implicit and explicit schemes [1, 2]. The comparisons of performances by the implicit and explicit schemes for a complex plastic models is carried out in this paper, which is seldom carried out in published literatures.

2 Stress Integration Algorithms

A double structure hydro-mechanical coupling model for unsaturated expansive clays was proposed by the author [3]. The model is established on the basis of bounding surface theory. It considers interactions between macro- and microstructure, distinguishes differences between capillary and adsorbed water retention behaviors, and combines a rotational hardening law to reflect an anisotropic structure of soil. The model is composed of two parts: mechanical and water retention models. For the former one, the stress variables are the Bishop's stress σ' and the microstructural effective stress $\hat{p}$, and the strain increments are the macro- and microstructural strain increments, i.e., $d\varepsilon^M$ and $d\varepsilon^m$. Besides, for the latter one, the modified suction s^* and the increment of macrostructural degree of saturation dS_r^M are chosen as a couple of constitutive variables. For a finite element method, the total strain and suction increments, i.e., $d\varepsilon$ and ds are known variables except for the initial values of stresses $(\sigma', S^*, \hat{p})$, hardening parameters $(p_{c0}^b, \alpha, s_\alpha^{*b}, \hat{p}_\alpha^b)$, and state variables (e_M, e_m, S_r^M).

A conventional implicit integration scheme based on the closest point projection method [1] is modified for the above complex model. In an implicit scheme, the solving process can be split into two stages, i.e., elastic trial and plastic corrector. The multiple plastic mechanism is adopted in the model, and two different cases can be distinguished in relation to the number of irreversible mechanisms active. For case I, the plastic mechanical mechanism associated with the coupling between two structural levels and the plastic water retention mechanism are active. For case II, the above two plastic mechanisms and the plastic mechanical mechanism associated with the yielding of the macrostructure is active.

Limited by space, the flow of implicit scheme is directly stated as follows. (1) Set $k = 0$, $s_{n+1} = s_n + \Delta s_{n+1}$, $\varepsilon_{n+1} = \varepsilon_n + \Delta \varepsilon_{n+1}$. (2) According to the case type, determine expressions of plastic multipliers γ_i where i indicates the active plastic mechanism. (3) Calculate residual values of constitutive equations r and check each element r_i< Tolerance. If it is yes, go to step 6. (4) Calculate increments of each variables $\delta U'^{(k+1)}$ by Newton Raphson method with $U' = \{\sigma', p_{c0}^d, p_{c0}^b, \alpha_d, \hat{p}, \hat{p}_\alpha^b, s^*, S_r^M, s_\alpha^{*b}\}$. (5) Update variables U' by $U'^{(k+1)}_{n+1} = U'^{(k)}_{n+1} + \delta U'^{(k)}$, set $k = k+1$ and go to step 3. (6) Check γ_i< -Tolerance and whether the updated stress fall outside the current loading surface which plastic mechanism is assumed not active. If it is yes, go to the next case and retune to step 2. (7) Get $U_{n+1} = U'_{n+1}$, and the calculation is finished.

The flow of explicit scheme combined with an automatic sub-stepping technique [2] is stated as follows. (1) Set $k = 0$, $s_{n+1} = s_n + \Delta s_{n+1}$, $\varepsilon_{n+1} = \varepsilon_n + \Delta \varepsilon_{n+1}$. (2) Check the pseudo time $T < 1$. If it is yes, go to step 3, otherwise the calculation is finished. (3) According to the case type, determine expressions of plastic multiples γ_i and set $j = 1$. (4) Calculate variables, i.e., stresses, state variables and hardening parameters by the initial stress state for $j = 1$ and by the updated stress state for $j = 2$, respectively. (5) if $j = 1$ go to step 6, otherwise, go to step 7. (6) Check γ_i < -Tolerance and whether the stress fall outside the current loading surface which plastic mechanism is assumed not active. If it is yes, go to the next case and retune to step 3; otherwise, go back to step 4 and define $j = 2$. (7) Calculate average values of the variables and store. (8) Determine a maximum relative error R_m of variables calculated based on the current stress state and the update stress sate. (9) Check $R_m >$ *Tolerance*. If it is yes, it means that the calculation fails, and a smaller strain step $\Delta T^{(k)}$ is desired by $\Delta T^{(k)} = q\Delta T^{(k)}$ with $q = \max \left(0.9(Tolerance/R_m)^{1/2}, 0.1\right)$. Return to step 2. (10) Update of the variables by the stored values. (11) Check the absolute value of current loading surfaces $|f_i^{(k)}| >$ *Tolerance*. If it is yes, go to step 12, otherwise go to step 13. (12) Correct the variables to drift of the current loading surfaces and return to step 11. (13) Calculate the next strain increment size $\Delta T^{(k+1)} = q\Delta T^{(k)}$ with $q = \min\left(0.9(Tolerance/R_m)^{1/2}, 1.1\right)$, and $\Delta T^{(k+1)} = \min(\Delta T^{(k+1)}, 1-T)$ to ensure $T \leq 1$. (14) Update $T = T + \Delta T^{(k)}$ and set $k = k+1$.

3 Performances of Integration Schemes

The performances of the implicit and explicit integration schemes, i.e., accuracy and convergence, are assessed through a strain controlled triaxial test. ε_a and ε_r increase from 0.0 to 0.01 during the consolidation stage with 10^5 strain increments. Then ε_a increases from 0.01 to 0.11 and ε_r decreases from 0.01 to -0.035 during the shear stage with different strain increments. The model parameters used in the calculation are as follows: $\lambda(0) = 0.104$, $r_\lambda = 1.4$, $\beta_\lambda = 0.005$, $N(0) = 1.937$, $r_N = 1.25$, $\beta_N = 0.005$, $\kappa = 0.01$, $M = 0.86$, $\omega = 10.0$, $\omega_d = 0.318$, $\lambda_m = 0.005$, $d_I = d_D = 5.0$, $\lambda_w = 0.115$, $\kappa_w = 0.02$, $k_{ws} = 1.5$, $\gamma_{LC} = \gamma_{MID} = \gamma_{SID} = 8.0$. The initial values of the state variables are as follows: $p_{net} = 20$ kPa, $s = 540$ kPa, $e_0 = 0.716$, $e_{m0} = 0.286$, $S_{r0} = 0.68$, $p_{c0}^b = 622.0$ kPa, $s_I^{*b} = 170.0$ kPa, $s_D^{*b} = 50.0$ kPa, $\hat{p}_I^b = 1170.0$ kPa, $\hat{p}_D^b = 390.0$ kPa.

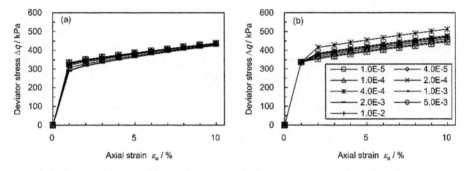

Fig. 1. Calculation results for shear tests at different strain increment sizes (a) implicit scheme (b) explicit scheme

Figure 1 shows calculation results by the implicit and explicit schemes. As shown in Fig. 1(a), the stress-strain curves of implicit scheme move downward with the increasing of strain increment size in the $\varepsilon_a - \Delta q$ plot. Noted that it doesn't converge when the strain increment size is greater than and equal to 5.0×10^{-3}. Besides,

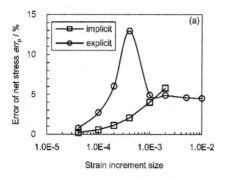
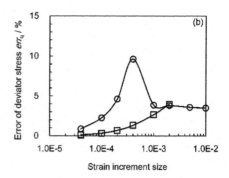

Fig. 2. Calculation errors for shear tests at different strain increment sizes (a) mean net stress (b) deviator stress

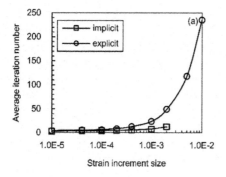
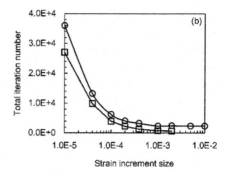

Fig. 3. Iteration numbers for shear tests at different strain increment sizes (a) mean iteration number (b) total iteration number

as shown in Fig. 1(b), the stress-strain curves of explicit scheme move upward with the increasing of step size when the strain increment size is lower than and equal to 4.0×10^{-4}. Figure 2 shows calculation errors by the two schemes. The error of implicit scheme increases with the increasing of strain increment size, and it is lower than that of the explicit one. The error of explicit scheme suddenly drops to a steady value when the automatic sub-stepping technique is active. Figure 3 shows iteration numbers by the two schemes. As shown in Fig. 3(a), the average iteration numbers of both schemes increase with the increasing of strain increment size. Besides, as shown in Fig. 3(b), the total iteration numbers decrease significantly with the increasing of strain increment size and tend to a steady value. Overall, the accuracy and the convergence of explicit scheme combined with the automatic sub-stepping technique are well at small and large strain increment sizes. However, the convergence of implicit scheme is weakened at the large strain increment size.

References

1. Ortiz, M., Simo, J.C.: An analysis of a new class of integration algorithms for elastoplastic constitutive relations. Int. J. Numer. Met. Eng. **23**(3), 353–366 (1986)
2. Sloan, S.W.: Substepping schemes for the numerical integration of elastoplastic stress-strain relations. Int. J. Numer. Met. Eng. **24**(5), 893–911 (1987)
3. Li, J., Yin, Z.Y., Cui, Y.J., et al.: Work input analysis for soils with double porosity and application to the hydro-mechanical modeling of unsaturated expansive clays. Can. Geotech. J. **54**(2), 173–187 (2017)

Influence of Different Soil Constitutive Models on SSI Effect on a Liquefiable Site

Zheng Li[✉]

Key Laboratory of Urban Security and Disaster Engineering of Ministry
of Education, Beijing University of Technology, Beijing, China
zheng.li@bjut.edu.cn

Abstract. With the rapid development of computer science and computational techniques, numerical simulations are regarded as powerful tools to solve complex geotechnical problems such as soil-structure interaction effects on liquefiable sites. For the simulation of soil liquefaction, soil constitutive models play important role in the correctly reproducing and interpreting the behavior liquefied soil and so as to the consequent soil structure interaction (SSI) problem. In this study, a comparative work was carried out to investigate the influence of different soil constitutive models on SSI effect on a liquefiable site. Two well know soil constitutive models are exploited: SANISAND model (Dafalias et al. 2004) and Hypoplastic model (von Wolffersdorff 1996). These two numerical models were first numerically implemented by semi-explicit 3rd order Runge-Kutta integration scheme with adapting sub-steps. The numerical implementation is done by UMAT subroutines in Abaqus. The model parameters of these two constitutive models were calibrated using triaxial test data of Fontainebleau sand (NE34). The performance of these two constitutive models is investigated numerically by comparing the seismic responses of a shallow foundation on the same liquefiable site. Results show that although these two constitutive models can be calibrated successfully against oridinary triaxial test data, their performance in terms of a real case SSI are different. The possible reasons are also disccussed in this study.

1 Background and Main Conclusions

Liquefaction is an important, complex and destructive phenomena which is usually produced by earthquake. It can induce enormous damage. In decades, many researchers worked on this topic, trying to model the liquefaction of soil. However, this is really a challenging task, there are still many issuers involved should be clarified and studied. With the rapid development of computer science and computational techniques, numerical simulations are regarded as powerful tools to solve complex geotechnical problems such as soil-structure interaction effects on liquefiable sites. For the simulation of soil liquefaction, soil constitutive models play important role in the correctly reproducing and interpreting the behavior liquefied soil and so as to the consequent SSI problem. Among the constitutive models, the two popular and most used ones could be the SANISAND elasoplastic model and Hypoplastic model. Dafalias et al. [1–8] successfully developed series constitutive models in a family – SANISAND, for sands. For the family of hypoplasticity theory [9–11], the von Wolffersdorff model with 'intergranular strain' is often recommended for the simulation of sand under cyclic loadings.

This paper presents a preliminary study on the influence of different soil constitutive models on SSI effect on a liquefiable site. Numerical comparative work was carried out by using commercial code ABAQUS. In terms of undrained condition, the changing of the volumetric strain of the soil should be zero i.e. $\Delta\varepsilon_v = 0$. In abaqus, in the dynamic modulus 'implicit' there is no $u-p$ element to simulate the undrained condition of soil, however this can be done alternatively in abaqus UMAT. In abaqus UMAT, in order to maintain the volumetric strain to be approximately 0, the bulk modulus $bulk_w$ of water (a relatively large number) was added to DDSDDE, by which abaqus calculated the strain increment for the next loading step. Figure 1 shows the calculated liquefaction 'butterfly' curves by using the two constitutive models.

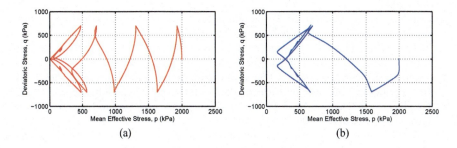

Fig. 1. Response of a single element under sinusoidal excitation (a) SANISAND model (b) Hypoplastic model

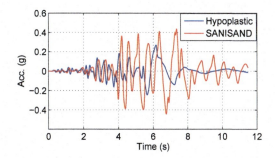

Fig. 2. Acceleration responses of the top of the building

Then, these two different constitutive models are compared based on the response of a shallow foundation on a liquefiable site. Although these two models are well calibrated against the triaxial test data, the final performances of these two models are significantly different, for example the acceleration response of the structure Fig. 2. Due to the lack of experimental data, it is difficult to evaluate which model gives better results. The practical application of advanced constitutive model on the modelling of liquefaction of soil still a challenging task in the field of geotechnical engineering.

References

1. Dafalias, Y.F.: Bounding surface plasticity I: mathematical foundation and hypoplasticity. J. Eng. Mech. **112**(9), 966–987 (1986)
2. Dafalias, Y.F., Herrmann, L.R.: Bounding surface plasticity II: application to anisotropic cohesive soils. J. Eng. Mech. **112**(12), 1292–1318 (1986)
3. Manzarit, M.T., Dafalias, Y.F.: A critical state two-surface plasticity model for sands. Géotechnique **47**(2), 255–272 (1997)
4. Li, X.S., Dafalias, Y.F.: Dilatancy for cohesionless soils. Géotechnique **50**(4), 449–460 (2000)
5. Dafalias, Y.F., Manzarit, M.T.: Simple plasticity sand model accounting for fabric change effects. J. Eng. Mech. **130**(6), 622–634 (2004)
6. Taiebat, M., Dafalias, Y.F.: SANISAND: simple anisotropic sand plasticity model. Int. J. Num. Anal. Methods Geomech. **32**(8), 915–948 (2008)
7. Taiebat, M., Dafalias, Y.F.: A true zero elastic range sand plasticity model. In: 6th International Conference on Earchquake Geotechnical Engineering, pp. 1–4 (2015)
8. Dafalias, Y.F., Taiebat, M.: SANISAND-Z: zero elastic range sand plasticity model. Géotechnique **66**(12), 999–1013 (2016)
9. Wu, W., Bauer, E.: A simple hypoplastic constitutive model for sand. Int. J. Num. Anal. Methods Geomech. **18**(12), 833–862 (1994)
10. Bauer, E.: Calibration of a comprehensive hypoplastic model for granular materials. Soils Found. **36**(1), 13–26 (1996)
11. Niemunis, A., Herle, I.: Hypoplastic model for cohesionless soils with elastic strain range. Mech. Cohesive-Frictional Mater. **2**, 279–299 (1997)

Performance of Composite Buttress and Cross Walls to Control Deformations Induced by Excavation

Aswin Lim[1,2(✉)] and Chang-Yu Ou[2]

[1] Universitas Katolik Parahyangan, 40141 Bandung, Indonesia
aswin_lim@yahoo.co.id
[2] National Taiwan University of Science and Technology, Taipei 10672, Taiwan

Abstract. A series of three-dimensional finite element analyses were conducted to investigate the performance of the composite buttress and cross walls (U-shape wall) including the depth of the cross wall, the length of the buttress wall, and the spacing of U-shape wall. For yielding an optimum performance, the normalized cross wall depth to excavation depth ratio (D_{cw}/H_e) could be designed around 0.15–0.2. Moreover, the design of buttress wall length depends on the target value of reduction. The 8 m interval of U-shape wall was recommended for producing a uniform reduction of the wall deflection. For 8 m interval of the U-shape wall, the 5 m length of the buttress wall could reduce the maximum wall deflection and the ground surface settlement around 60%–80%, respectively, from those without U-shape wall installation.

Keywords: Cross wall · Buttress wall · Wall deflection · Ground settlement Deep excavation

1 Introduction

In a recent decade, buttress walls and cross walls were widely used as an alternative measure to control excessive movements induced by deep excavation in some Asia countries, such as Taiwan and Singapore. The buttress wall is a concrete wall perpendicular to the diaphragm wall with a limited length. If the buttress wall is connected to the opposite diaphragm wall, then it is called the cross wall. Some researchers reported the successful excavation case histories with the buttress wall solely, or combinations with the cross wall to control movements induced by deep excavation [1, 2]. Due to the larger size of a cross wall compared to a buttress wall, cross walls might not be economical, yet the application of cross walls can efficiently and significantly reduce wall deflections because the cross-wall functions as a lateral strut and exists before excavation.

Indeed, buttress walls and cross walls could be integrated as a wall system, so called the U-shape wall (see Fig. 1), but no study has been carried out in discovering the performance and deformation control mechanism yet. In this article, a series of three-dimensional finite element analyses were conducted to investigate the performance of the U-shape wall.

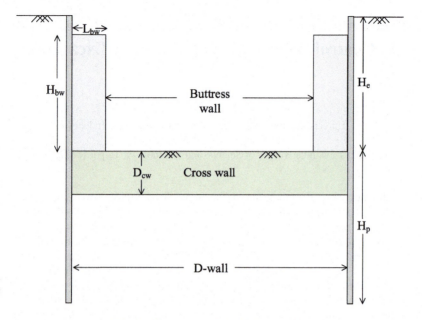

Fig. 1. Illustration of the U-shape wall.

2 Numerical Modelling

Three-dimensional numerical analyses were performed using PLAXIS 3D [3]. The Hardening Soil model [4] was used to model the soil behavior. Figure 2 demonstrates a three-dimension finite element mesh used to conduct the analyses. The excavation length (L) and the excavation width (B) were assumed to be 56 m and 80 m, respectively. Also,

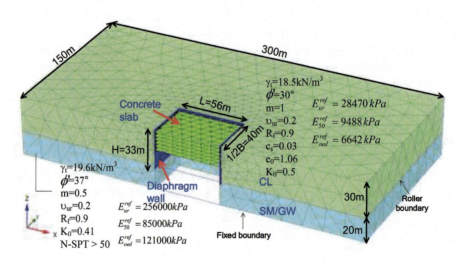

Fig. 2. Finite element mesh and parameters used for analyses.

the final excavation depth (He) and the height of diaphragm wall (H) were assumed to be 20 m and 33 m, respectively. The toe of diaphragm walls was embedded 3 m in the SM/GW (referred as the hard-stratum) layer. Under such conditions, no stability issues were expected to happen. Only a half of the excavation geometry was adopted in the analyses due to symmetry. The excavation geometry was assumed based on a consideration that the generation of the unbalanced force was large enough to cause excessive movements induced by deep excavation. Due to the space limitation, the detail of parameter selection and finite element modeling could refer to Lim [5].

2.1 Numerical Analysis Program

Table 1 lists the simulation program of the U-shape. The length of the buttress wall (L_{bw}) and the depth of the cross wall (D_{cw}) was varied from 1 m to 5 m. Also, for each combination, the number (n) of U-shape wall was changed from 1, 3, and 7, and they were uniformly distributed along the longitudinal side of the diaphragm wall.

Table 1. Finite element analysis program for the U-shape wall

H_{ve} (m)	L_{ve} (m)	D_{cw} (m)	n	Objectives
20	1, 2, 3, 4, 5	1, 2, 3, 4, 5	1, 3, 7	To investigate the effect of the interval, the cross wall height, and the buttress wall length of the U-shape wall

Moreover, the maximum deflection ratio (MDR) and the maximum settlement ratio (MSR) are introduced to quantify the performance of each analyses The MDR and the MSR are according to the maximum value of excavation induced-movements. The MDR, and MSR are defined as:

$$MDR = \frac{(\delta_{h\max-0} - \delta_{h\max-i})}{\delta_{h\max-0}} \times 100\% \tag{1}$$

$$MSR = \frac{(\delta_{v\max-0} - \delta_{v\max-i})}{\delta_{v\max-0}} \times 100\% \tag{2}$$

Where $\delta_{h\max-0}$ is the maximum lateral wall deflection without U-shape walls, $\delta_{h\max-i}$ is the maximum lateral wall deflection with U-shape walls, $\delta_{v\max-0}$ is the maximum ground surface settlement without U-shape walls, and $\delta_{v\max-i}$ is the maximum ground surface settlement with U-shape walls.

3 Concluding Remarks

The U-shape wall could be formed by combining the buttress wall and the cross wall. For the U-shape wall, the top of the cross wall was located at the final excavation level until a certain depth which can be beneficial to save the construction cost. It was suggested that the normalized cross wall depth ratio (D_{cw}/H_e) was 0.15–0.2 to yield an optimum performance (see Fig. 3). Also, the design of the buttress wall length depends on the target value of the reduction. For 8 m intervals of the U-shape wall, 5 m length of the buttress wall could yield MDR and MSR around 60%–80%. Moreover, the U-shape was recommended to be installed with the 8 m interval (7 U-shape walls) for producing the better U-shape performance along the diaphragm wall. The larger spacing among the U-shape wall would localize the effect of the U-shape wall.

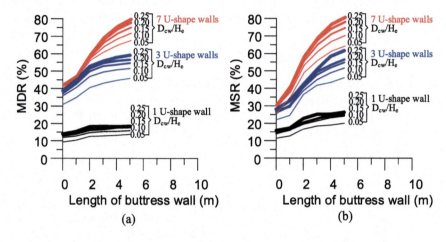

Fig. 3. Reduction ratio of U-shape wall. (a) MDR, (b) MSR.

References

1. Hsieh, P.G., Ou, C.-Y., Hsieh, W.H.: Efficiency of excavations with buttress walls in reducing the deflection of the diaphragm wall. Acta Geotech. **11**, 1087–1102 (2016)
2. Lim, A., Ou, C.-Y., Hsieh, P.G.: Investigation of the integrated retaining system to limit deformations induced by deep excavation. Acta Geotech. (2017). https://doi.org/10.1007/s11440-017-0613-6
3. Brinkgreve, R.B.J., Engin, E., Swolfs, W.M.: PLAXIS 3D Manual. PLAXIS, Delft (2013)
4. Schanz, T., Vermeer, P.A., Bonnier, P.G.: Formulation and verification of the hardening-soil model. In: Brinkgreve, R.B.J. (ed.) Beyond 2000 in Computational Geotechnics, pp. 281–290. Balkema, Rotterdam (1999)
5. Lim, A.: Investigation of integrated buttress and cross walls to reduce movements induced by excavation. Dissertation, Taiwan Tech. (2018)

Study on the Envelope of Stress Path During Deep Excavation

Li Liu(✉), Hong-ru Zhang, and Jian-kun Liu

School of Civil Engineering, Beijing Jiaotong University, Beijing 100044, China
liu.li@bjtu.edu.cn

Abstract. In the pit excavation process, soil deformation and earth pressure is highly depending on the loading process of soil. It is essential to characterize soil loading process, as in the form of stress path, during pit excavation process. However, there are two problems existed: Firstly, the stress path of the test is single. There are two stress paths commonly applied to experimental research, i.e., lateral stress reduction with the constant of axial stress and axial stress reduction with the constant of lateral stress. Therefore, the whole process of excavation is calculated by using the software of numerical analysis in this paper. The calculation results are shown as follows. (1) In the process of excavation, the stress paths of soil in different parts are different. (2) In the process of excavation, the stress paths of the soil are not all the stress reduction or increase in the single direction. (3) The main characteristic of soil stress path is the decrease of stress in a certain direction or multi direction in the process of excavation. According to the calculation conclusion, the concept of stress path envelope of foundation pit excavation is proposed. The probability of soil deformation around the foundation pit is large when the stress path outside the envelope of soil is experienced. The soil deformation probability around the foundation pit is small when the soil is subjected to the stress path in the envelope.

Keywords: The stress path · Excavation · The envelope of soil
The foundation pit

1 Introduction

The deformation of the surrounding soil during the excavation of the foundation pit is related to the excavation process [1], external load and support structure. Excavation is actually an unloading process [2], which is completely different from the application of external load and the arrangement of the support structure [3, 4]. Therefore, in the process of foundation pit excavation, the soil stress path is not the same as the stress path in the process of loading. The stress path experienced by soil during excavation and unloading is further studied, which helps to predict the deformation characteristics of soil around foundation pit and reduce the occurrence of foundation pit accidents. It is necessary to put forward a constitutive model considering the variation of stress path. Before the establishment of the constitutive model, it is necessary to understand the typical stress path of the soil during the excavation. In this paper, numerical simulation is used to analyze typical stress paths during excavation, and the stress path envelope is used to describe the possibility of soil deformation around foundation pit.

2 Numerical Simulation

The PLAXIS software is used in this numerical calculation and the soil around the foundation pit is all of the same clay. The Hardening Soil Model is selected as the constitutive model of soil. The position of soil in the study is shown in Fig. 1. The two groups of soil layer parameters are selected for comparison study to see Table 1.

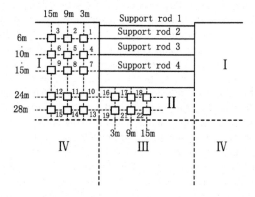

Fig. 1. Influence regions of foundation pit excavation.

Table 1. The main soil physical and mechanical indicators

	γ (kN·m^{-3})	c (kN/m^2)	φ (°)	E_{50}^{ref} (kPa)	E_{oed}^{ref} (kPa)	E_{ur}^{ref} (kPa)	m
1	18.2	5	25	8000	8000	24000	0.8
2	18.2	5	25	6000	6000	18000	0.8

Among them, γ is heavy of soil; c is effective cohesion; φ is friction angle; E_{50}^{ref} is three axis drainage test 50% secant modulus under reference confining pressure; E_{oed}^{ref} is loading modulus of main consolidation instrument; E_{ur}^{ref} is unloading/reloading stiffness; m is stiffness and stress level related parameters.

3 Calculation Results

The 11 test points in Fig. 1 are selected for research. Figure 2 is the depth of 10 m stress-path of test points 4, 5, 6 from the side of foundation pit 3 m, 9 m, 15 m. Figure 3 is the depth of 28 m stress-path of test points 13, 14, 15 from the side of foundation pit 3 m, 9 m, 15 m. Figure 4 is the depth of 24 m stress path of test points 19, 20, 21 from the side of foundation pit 3 m, 9 m, 15 m.

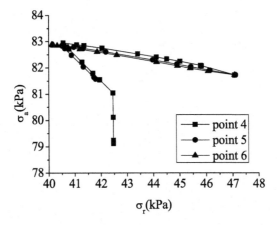

Fig. 2. The stress-path of test points 4, 5, 6 from the side of foundation pit 3 m, 9 m, 15 m

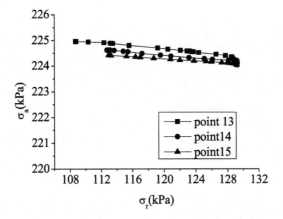

Fig. 3. The stress-path of test points 13, 14, 15 from the side of foundation pit 3 m, 9 m, 15 m

According to the analysis of soil stress path on the bottom side of foundation pit, the results have been found: during the excavation of foundation pit, the soil stress path around the foundation pit is not a single axis (side). In the process of excavation, the stress has different degrees of increase and reduction, which is not only related to the construction process but also to the influence of the depth of the excavation. In the first area of foundation pit, there are two main stress paths: the first lateral stress decreases, the axial stress decreases, the second one is the lateral stress decreases, and the axial stress basically remains unchanged.

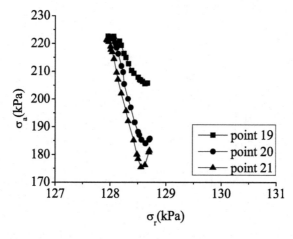

Fig. 4. The stress-path of test points 19, 20, 21 from the side of foundation pit 3 m, 9 m, 15 m

According to the soil stress path analysis at the bottom of the foundation pit, it can be seen that the stress path of the soil after the wall can also be considered as two kinds: The first axial stress decreases and the lateral stress is slightly increased; the second is the decrease in axial stress and the constant lateral stress. The increase of the first lateral stress is about 1 kPa. The path of the soil after the wall is similar to that of the soil after the wall, and it is not the stress increase or decrease in the single direction.

References

1. Zhang, H.R., Hou, X.Y.: Prediction of influence of deep excavation on working facilities. J. North. Jiao Tong Univ. **20**(2), 205–209 (1996)
2. Hou, X.Y., Chen, Y.F.: Calculation of soil subsidence in the surrounding ground caused by deep foundation pit excavation. Geotech. Eng. **1**(1), 1–13 (1989)
3. Song, E.X., Lou, P., Lu, X.Z., et al.: Nonlinear three-dimensional finite element analysis of a special deep foundation pit support. Geotech. Mech. **4**(25), 538–543 (2004)
4. Liu, W.N., Zhang, M., Cheng, H.: Influence of excavation on the engineering properties of soil layer around foundation pit. J. Rock Mech. Eng. **1**(21), 60–64 (2002)

New Bond Model in Disk-Based DDA for Rock Failure Simulation

Feng Liu, Kaiyu Zhang, and Kaiwen Xia[✉]

State Key Laboratory of Hydraulic Engineering Simulation and Safety,
School of Civil Engineering, Tianjin University, Tianjin 300072, China
kaiwen@tju.edu.cn

Abstract. Discontinuous Deformation Analysis (DDA) is an implicit discrete method and has its unique advantages in the simulation of block systems. In order to simulate the process of rock failure, a bonding algorithm between blocks is necessary to describe the deformation characteristics of continuous medium. In the framework of 2D particle-based DDA, a new bonding model is established in this paper, which is composed of a normal, a tangential and a rotational spring. The new bonding model can resist not only normal penetration and tangential slip as has been done in the literature, but also relative rotation. Several numerical examples are calculated to demonstrate the effectiveness of the proposed model.

Keywords: Discontinuous deformation analysis · Disk-based DDA
Rock failure simulation · Bond model

1 Introduction

Rock mass contains a large number of discontinuities, such as faults and joints. Compared to continuum methods, the discrete methods are more suitable to model the discontinuities and large deformation of the rock mass than continuous numerical methods. Discontinuous deformation analysis (DDA) [1] is one of the discrete methods. Although the original DDA has already made great achievements, it has some drawbacks, such as the complexity of contact algorithm and the efficiency of the open-close iteration. To overcome these drawbacks, particle-based DDA [2] are proposed. Compared to the success of particle-based discrete element method, intensive studies are needed for particle-based DDA. The improvement of particle-based DDA is focused on following aspects, such as contact model of particles [3], shape or attribute of particles [4] and three dimensional particle-based DDA [5]. Jiao et al. [6] developed a three-dimensional spherical DDA (SDDARF3D) for simulating the whole process of rock failure using a bonding and cracking algorithm. However, the bonding algorithm cannot resist relative rotation. So, this study will focus on the bonding algorithm of the particle-based DDA.

2 Improvement of Particle DDA

In this paper, each discrete block is viewed as a 2D rigid disk. The deformation variable of each disk can be written as:

$$D_i = [u_0 \ v_0 \ \bar{w}_0]^T \tag{1}$$

where u_0, v_0 and $\bar{w}_0$ are three rigid body translations and rotation angle. Here, w_0 is replaced by $\bar{w}_0$, where $\bar{w}_0 = w_0 R_i$, R_i the radius of disk i. $\bar{w}_0$ represents the displacement components due to rigid body rotation. Using $\bar{w}_0$ instead of w_0 leads to relatively small condition number and better numerical properties of the stiffness matrix, especially when the radius of disk is too large or too small.

In order to simulate the failure of rock properly, it's necessary to develop an effective bond model. The simplest contact model between disks is point-point contact consisting of a normal and tangential spring. The most straightforward way to bond disk i and j is to stick them together at the contact point. However, this bond model cannot resist relative rotation. To overcome this issue, as shown in Fig. 1, we add an additional rotational bond into the model. Since one disk has three DOFs, three bond springs can exactly restrain the relative movement of these two disks. The formulas of the rotational bond are given as follow.

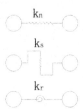

Fig. 1. Three types of bonding model between disks

The relative rotation between contact disk i and disk j:

$$\Delta\theta = \theta_{i0} + \theta_i - (\theta_{j0} + \theta_j) \tag{2}$$

where θ_{i0} and θ_{j0} are rotations of disk i and disk j respectively at end of the last step. θ_i and θ_j are rotation of disk i and disk j at the current step.

The deformation energy of a rotational bonding π_r is given by:

$$\pi_r = \frac{1}{2} k_r R_i R_j [\theta_{i0} + \theta_i - (\theta_{j0} + \theta_j)]^2 \tag{3}$$

where R_i and R_j represent radius of disk i and disk j, respectively. k_r is the stiffness of rotational spring.

By expanding and minimizing the potential energy π_r, we have:

$$k_r \frac{R_i R_j}{R_i^2} YY^T \to K_{ii}, \quad -k_r YY^T \to K_{ij} \tag{4}$$

$$-k_r YY^T \to K_{ij}, \quad k_r \frac{R_i R_j}{R_j^2} YY^T \to K_{jj} \tag{5}$$

$$-\delta_0 R_j Y \to F_i, \quad \delta_0 R_i Y \to F_j \tag{6}$$

where $Y^T = [0, 0, 1]$ and $\delta_0 = \theta_{i0} - \theta_{j0}$.

Before the bond breaks, a linear elastic response is provided by the linear elastic springs between the disks. Once the normal force exceeds the maximum acceptable tensile force, the bond breaks. These two disks may be in contact in the subsequent steps. Once the shear force exceeds the maximum shear force, the bond breaks and contact occurs.

In the next part, the bond model is tested by two examples.

3 Numerical Example

The effectiveness of the new bond model will first be certified by uniaxial compression test. The assembly consists of 1263 disks as shown in Fig. 2(a). The axial stress-strain curves in Fig. 2(b) represents the typical response of brittle material. The failure mode is shown in Fig. 2(c). From the developing curve of crack numbers shown in Fig. 2(d), it can be seen that tension cracks are the main failure mode all the time. The computed

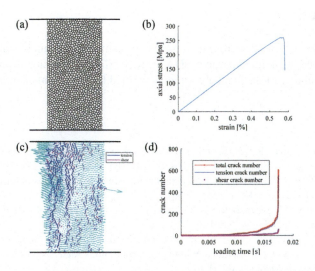

Fig. 2. Uniaxial compression test, (a) particle assembly; (b) axial stress-strain curve; (c) failure mode; (d) developing curve of crack numbers.

Young's modulus and uniaxial compressive strength are 47 Gpa and 261 Mpa respectively, which agree with material properties 46 GPa and 259 MPa.

The second example is a tunnel impacted by a missile. The particle assembly is shown in Fig. 3. After the missile impact on the tunnel, tension cracks and shear cracks occur and extend to the surrounding. The maximum deformation of the left tunnel appears at the vault. The failure process along with stress wave propagation can be observed clearly, proving that the proposed method is correct and feasible.

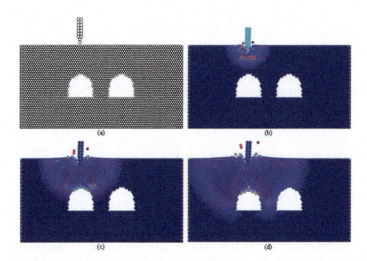

Fig. 3. The failure process of a tunnel impacted by a missile, (a) particle assembly (t = 0 s); (b) t = 0.005 s; (c) t = 0.010 s; (d) t = 0.015 s. (The color of disks represents their displacement)

4 Conclusions

By introducing a new bond algorithm which consisting of a normal, a tangential and a rotational bond, a modified particle DDA for rock failure simulation is proposed in this paper. The validity of the bond algorithm is verified by numerical examples.

References

1. Shi, G.H., Goodman, R.E.: Discontinuous deformation analysis - a new method for computing stress, strain and sliding of block systems. Intern. Med. J. **44**(3), 309–311 (1988)
2. Ke, T.C., Bray, J.: Modeling of particulate media using discontinuous deformation analysis. J. Eng. Mech. **121**(11), 1234–1243 (1995)
3. Beyabanaki, S.A.R., Bagtzoglou, A.C.: Non-rigid disk-based DDA with a new contact model. Comput. Geotech. **49**, 25–35 (2013)
4. Ohnishi, Y., Miki, S.: Development of circular and elliptic disc elements for DDA. In: Proceedings of the 1st International Forum on Discontinuous Analysis (DDA) and Simulations of Discontinuous Media. TSI (1996)

5. Beyabanaki, S.A.R., Jafari, A., Yeung, M.R.: High-order three-dimensional discontinuous deformation analysis (3-D DDA). Int. J. Numer. Methods Biomed. Eng. **26**(12), 1522–1547 (2010)
6. Jiao, Y.Y., Huang, G.H., Zhao, Z.Y.: An improved three-dimensional spherical DDA model for simulating rock failure. Sci. China Technol. Sci. **58**(9), 1533–1541 (2015)

On the Pseudo-Coupled Winkler Spring Approach for Soil-Mat Foundation Interaction Analysis

Dimitrios Loukidis[✉] and Georgios-Pantelis Tamiolakis

Department of Civil and Environmental Engineering, University of Cyprus,
Kallipoleos 75, 1678 Nicosia, Cyprus
loukidis@ucy.ac.cy

Abstract. Analysis of the interaction between mat foundations and the supporting soil is often performed with the soil being represented by vertical translational springs. This study sheds light on what must be the distribution of spring stiffness coefficients under a mat in order to get the same mat deflections and bending moment diagrams as in the case of a mat resting on a three-dimensional linear elastic continuum. For this purpose, the spring stiffness coefficients were directly back-calculated at each mat node from the results of finite element analyses that treat the soil as a 3D continuum. Parametric runs are performed to investigate the influence of the soil elastic properties, slab geometrical characteristics and column load configurations on the back-calculated spring stiffness spatial distributions. The computational results of the present study indicate that spring stiffness distributions often assumed currently in practice may lead to a significant underestimation of the mat bending moments.

Keywords: Mat foundations · Winkler springs · Finite element analysis

1 Introduction

The structural design of flexural foundation elements, such as mats and piles, requires performing static analysis of the interaction of these elements with the soil. In the case of mat foundations, the methods currently available are (1) finite element analysis (FEA) that treats the soil as a three-dimensional continuum, (2) the coupled analysis approach and (3) the pseudo-coupled analysis approach. Three-dimensional FEA is the most versatile and accurate method, but at the same time is the most computationally expensive and, consequently, its penetration in routine mat design is still very limited. In the coupled approach (e.g. [1, 2]), the soil is represented by vertical (Winkler) springs and other mechanical components, such as shear and flexural layers and membranes that account for the effects of soil shearing. The coupled approach is accurate but rather difficult to be applied in common structural analysis software because of the mechanical components other than the Winkler springs. The pseudo-coupled analysis approach is a simpler, more approximate alternative, in which the soil is represented solely by Winkler springs (Fig. 1), but with the springs near the mat edge having substantially larger stiffness than those near the mat center [3] in order to indirectly take into account the effects of soil shearing. The present study aims at

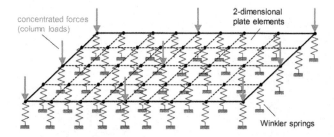

Fig. 1. Static analogue of a mat resting on a bed of Winkler springs.

establishing the distribution of spring stiffness coefficients underneath a mat that is needed to obtain mat analysis results as accurate as those produced by 3D FEA.

2 Numerical Methodology and Results

The appropriate spring stiffness coefficient K_s was back-calculated at each mat node using directly the results of FEA (Abaqus) modeling the soil as a linear elastic continuum (cFEA) on which a rectangular concrete ($E_{mat} = 32$ GPa, $v_{mat} = 0.25$) foundation slab (modeled using shell elements) rests (Fig. 2a). The back-calculation is achieved by using the mat deflections resulting from cFEA as input in a static analysis analogue of the mat resting on springs (Fig. 1) and solving with respect to the unknown K_s values at each node. Several parametric runs were performed, varying the soil elastic properties (E_s, v_s), the mat thickness d, the mat aspect ratio, the column load relative magnitude and spacing, and the thickness H (Fig. 2a) of the deformable soil layer.

Figure 2b shows an example of back-calculated K_s distribution. It can be seen that K_s remains relatively constant inside a large, central portion of the mat and increases sharply ($\times 3.5$) toward the edge, especially the corners ($\times 6$). This cup-shaped distribution was observed in all analyses performed, and appears to be relatively insensitive to E_s (Fig. 3a), v_s, d, B/L, and column spacing, except H. It was found that the thinner the deformable layer is, the flatter the K_s distribution becomes. Moreover, the distribution appears rotated in cases of unsymmetrical column load arrangement (Fig. 3b).

The observed K_s distributions can be fitted adequately by the following equation:

$$K_s = K_r(0.55 + C_{H1})\left\{1 + 2C_{H2}\left[\left(\frac{x+0.1e_x}{L/2}\right)^6 + \left(\frac{y+0.1e_y}{B/2}\right)^6\right] + 4\left[\frac{e_x}{L}\left(\frac{x}{L/2}\right) + \frac{e_y}{B}\left(\frac{y}{B/2}\right)\right]\right\} \quad (1)$$

where e_x and e_y are the eccentricities of the resultant of all column loads with respect to the mat center in the x and y directions, respectively, and K_r is a reference stiffness coefficient corresponding to a perfectly rigid foundation. K_r can be estimated using the following equation established based on Fraser and Wardle [4] and Chow [5]:

$$K_r = \frac{E_s}{C_f(1-v_s^2)B}A_{inner} \quad (2)$$

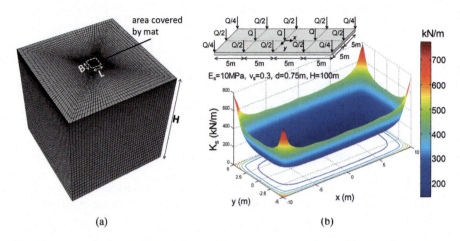

Fig. 2. (a) Typical mesh used in cFEA and (b) example of back-calculated K_s distribution.

where A_{inner} is the tributary area of a given node calculated as if it is an inner node (by doubling and quadrupling the actual tributary area of edge and corner nodes, respectively). The factors C_{H1}, C_{H2} and C_f introduce the effect of H on K_s as follows:

$$C_{H1} = 0.45 \exp(-2.2H/B), \quad C_{H2} = \exp(-0.4B/H) \quad (3a, 3b)$$

$$C_f = \frac{0.85(L/B)^{0.45}}{[1 + 0.1(2 + L/B)(B/H)]^{1+\exp(5v_s^3)}} \quad (4)$$

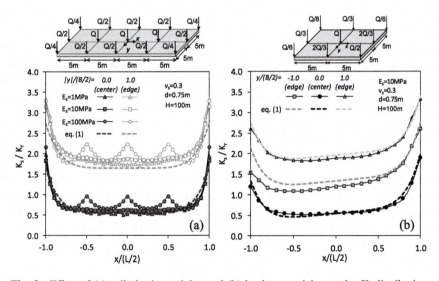

Fig. 3. Effect of (a) soil elastic modulus and (b) load eccentricity on the K_s distribution.

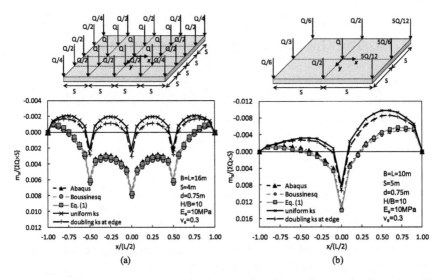

Fig. 4. Comparison of bending moment diagrams predicted by various mat analysis approaches.

Figure 4 compares the bending moment diagrams along the mat centerline (x-axis) predicted by mat-on-spring analysis using the K_s values yielded by Eq. (1) and those produced by the two most common approaches employed in practice, namely (i) assuming a constant value of modulus of subgrade reaction k_s ($K_s = k_s A_{tributary}$) and (ii) doubling the k_s for the nodes at the mat perimeter. In addition, Fig. 4 plots the moment diagrams predicted using the deflections yielded by cFEA (Abaqus) and a semi-analytical iterative procedure proposed by Ulrich [6] and based on the Boussinesq solution. It can be seen that the use of the proposed K_s distribution results in bending moment diagrams that are almost identical to those from cFEA and the semi-analytical method. Contrarily, the k_s approaches frequently adopted in practice may largely underpredict (by a factor of 1.5 to 2.5) the positive bending moments (bottom fiber in tension), leading to an unsafe design. Finally, it was observed that the shear force diagrams are practically unaffected by the choice of the K_s distribution.

References

1. Kerr, A.D.: A study of a new foundation model. Acta Mech. **1**(2), 135–147 (1965)
2. Horvath, J.S.: New subgrade model applied to mat foundations. J. Geotech. Eng. **109**(12), 1567–1587 (1983)
3. Bowles, J.E.: Foundation Analysis and Design, 5th edn. McGraw-Hill, New York (1996)
4. Fraser, R.A., Wardle, L.J.: Numerical analysis of rectangular rafts on layered foundations. Géotechnique **26**(4), 613–630 (1976)
5. Chow, Y.K.: Vertical deformation of rigid foundations of arbitrary shape on layered soil media. Int. J. Numer. Anal. Methods Geomech. **11**(1), 1–15 (1987)
6. Ulrich Jr., E.J.: Subgrade reaction in mat foundation design. Concr. Int. **13**(4), 41–50 (1991)

Numerical Analysis of Soil Ploughing Using the Particle Finite Element Method

Lluís Monforte[1], Marcos Arroyo[1(✉)], Maxat Mamirov[2], and Jong R. Kim[2]

[1] Universitat Politècnica de Catalunya – BarcelonaTech, Barcelona, Spain
marcos.arroyo@upc.edu
[2] Nazarbayev University, Astana, Kazakhstan

Abstract. This contribution illustrates some results from a parametric analysis of the factors affecting soil ploughing. The numerical simulations rely on the Particle Finite Element method, a method known for its capabilities to tackle large deformations and rapid changing boundaries at large strains. A total stress analysis – assuming a quasi-incompressible elastic model along with a Tresca plastic model - is used to simulate a clayey soil behavior. As a first step, a two-dimensional geometry is used and the effect of contact roughness in the horizontal force along the chip formation is assessed.

Keywords: Particle Finite Element Method · Plough · Large strain

1 Introduction

The optimization of the geometry of the tillage tool has been undertaken by several means including both experiments and simulations employing a variety of techniques [1]. Whereas traditional mathematical simulations of the soil-tool interaction relied on the passive earth pressure theory, nowadays, numerical models are predominant.

The Finite Element method allows to accurately simulate the non-linear behavior of clayey soils and to include the interaction between various deformable or rigid bodies. However, advanced methods in order to overcome mesh tangling and distortion are required.

In this contribution, the Particle Finite Element method is employed to simulate the ploughing process in clayey soil. The work is structured as follows: first, the numerical method is outlined; afterwards, preliminary results of a parametric analysis of soil ploughing are presented and, finally, some conclusions are drawn.

2 Numerical Model

The Particle Finite Element Method is characterized by a particle discretization of the domain: every time step a finite element mesh –whose nodes are the particles– is constructed using a Delaunay's tessellation and the solution is evaluated using this mesh with well-shaped, low-order elements. The continuum is modeled using an Updated Lagrangian formulation [2, 3].

Numerical simulations have been carried out by means of the numerical code G-PFEM [4], specially developed for the analysis of large strain contact problems in geomechanics. The code is able to accurately simulate the interaction between fluid-saturated porous media and rigid structures.

Ploughing in clayey soils occurs at a relatively high velocity compared with the hydraulic properties of clay; as such, it is modeled with a total stress approach. Then, the material is assumed to satisfy a Tresca yield criterion in conjunction with a linear elastic model. A Poisson's ratio of 0.49 is used. Therefore, techniques to alleviate volumetric locking are required: in this work, a mixed stabilized formulation [5].

The plough is considered completely rigid; this hypothesis is approximate enough due to the high between the plough Young's modulus and that of the soil. The contact constraints are imposed to the solution by the Penalty Method. As customary in total stress analyzes, the maximum tangential contact stress is a proportion of the soil undrained shear strength.

3 Representative Numerical Simulations

The geometry of the problem is depicted in Fig. 1. At the left and right boundaries, the vertical displacement is restricted whereas all the components of the displacement vector are null at the lower boundary. As a first step, the problem is simulated by assuming plane strain conditions. The soil is characterized by an undrained shear strength of 10 kPa and a Shear Modulus of $G = 1000$ kPa; thus, representing a rigidity index of $Ir = G/Su = 100$. The Poisson's ratio is equal to 0.49. The soil self-weight is also considered.

At the initial state, the plough is placed at a depth of 0.5 m; during all the simulation the plough remains at a 45° with respect to the vertical and moves horizontally. The tip of the plough is rounded. Two different interface behaviors are assumed: on one simulation a completely smooth interface is used whereas in the other the maximum tangential contact stress is 0.3 times the soil undrained shear strength.

Figure 1(a) presents the total horizontal stress after 0.5 m of advancement of the plough. High horizontal stresses are found in the vicinity of the tool. Meanwhile, in the upper part of the chip, the soil experience horizontal tension.

In terms of the failure mechanism, Fig. 1(b) depicts the incremental plastic shear strain. Plastic strains just appear in a very narrow zone: there is a straight localization plane that departs from the tip of the plough and goes up to the surface; this plane has an angle of 45° with respect to the horizontal. It should be noted that the angle of the localization plane and the plough is coincident. The rest of the soil remains in elastic regime.

These previous observations refer to the smooth case. However, the same plastic failure mechanism is observed for the rough case.

Figure 2 presents the evolution of the horizontal force in terms of the plough displacement. The horizontal force is slightly larger for the rough case. In both cases, the horizontal force continuously increases and a stationary state is not reached.

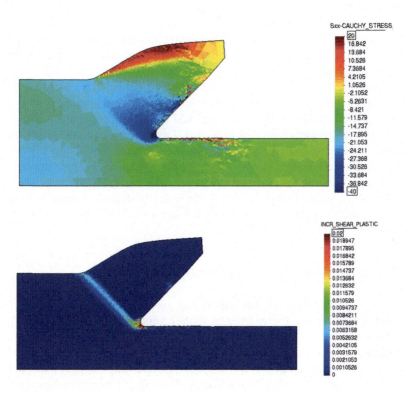

Fig. 1. Horizontal component of the Cauchy stress tensor, (on top), and incremental plastic shear strain, (on the bottom).

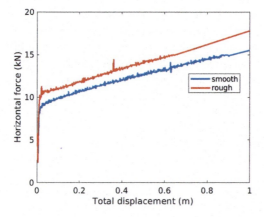

Fig. 2. Evolution of the horizontal force acting on the plough in terms of the horizontal displacement assuming a smooth and a rough interface behavior.

4 Conclusions

This contribution has outlined preliminary results of the numerical simulation of ploughing process by means of the Particle Finite Element method. In particular, the effect of contact roughness on the interaction of a round-tipped tool has been analyzed. It has been shown that the failure mechanism is independent of the contact roughness; however, the horizontal force acting on the tool increases when contact roughness increases. In this contribution, a total stress approach has been used; the extension to a fully-coupled hydro-mechanical model is the topic of ongoing research.

The developed numerical scheme appears to be a promising tool for the simulation of tool-soil interaction.

Acknowledgment. This work has been supported by the REA of the European Union, through grant 645665—GEO-RAMP—H2020-MSCA-RISE-2014.

References

1. Ibrahmi, A., Bentaher, H., Hbaieb, M., Maalej, A., Mouazen, M.A.: Study the effect of tool geometry and operational conditions on mouldboard plough forces and energy requirement: part 1. Finite element simulation. Comput. Electron. Agric. **117**, 258–267 (2015)
2. Oñate, E., Idelsohn, S.R., Del Pin, F., Aubry, R.: The particle finite element method – an overview. Int. J. Comput. Methods **1**(2), 267–307 (2004)
3. Rodriguez, J.M., Carbonell, J.M., Cante, J.C., Oliver, J.: The particle finite element method (PFEM) in thermo-mechanical problems. Int. J. Numer. Methods Eng. **107**(9), 733–785 (2016)
4. Monforte, L., Arroyo, M., Carbonell, J.M., Gens, A.: Numerical simulation of undrained insertion problems in geotechnical engineering with the particle finite element method (PFEM). Comput. Geotech. **82**, 144–156 (2017)
5. Monforte, L., Carbonell, J.M., Arroyo, M., Gens, A.: Performance of mixed formulations for the particle finite element method in soil mechanics problems. Comput. Part. Mech. **4**(3), 269–284 (2017)

Numerical Analyses with an Equivalent Continuum Constitutive Model for Reinforced Soils with Angled Bar Components

Seyed Mehdi Nasrollahi[1(✉)] and Ehsan Seyedi Hosseininia[2]

[1] Civil Engineering Department, Ferdowsi University of Mashhad, Mashhad, Iran
s.m.nasrollahi@mail.um.ac.ir
[2] Civil Engineering Department, Faculty of Engineering,
Ferdowsi University of Mashhad, Mashhad, Iran

Abstract. The discontinuity and hardness variation in the reinforced soil, makes modeling difficult, so use of simple methods that are developed to estimate the heterogeneous field behavior has been found. In this paper, an equivalent continuum constitutive model for three-dimensional analysis reinforced soil in FLAC^{3D}, was developed. The soil characteristics of deformation and strength are considered containing up two sets of reinforcements with arbitrary spatial setting and equivalent stiffness tensor of soil are conducted based on superposition of all components. By new constitutive relationships, anisotropy of reinforced soil can be implemented; however plastic strains can develop in pure soil mass. The model C++ code is written and implemented in FLAC3D and used for modeling of a triaxial compressive test on a soil specimen with one set of reinforcement. The stiffness values calculated by the model are compared with the results from an analytical solution. The results derived are in good agreement with those obtained from analytical method.

Keywords: Reinforced soil · Anisotropy · Equivalent continuum model

1 Introduction

The behavior analysis of reinforced soil directly by means of a classical finite element or finite difference numerical tool, considering to strong heterogeneity, leading almost inevitably to computational difficulties since such reinforced structures have been formed of different elements with different characteristics and behavior. Of course, the small cross-section and very large number of the inclusions (up to several hundreds) that requires a very locally fine meshing, leading to oversized numerical model. Moreover, due to the geometry of the composite reinforced soil, a three-dimensional analysis is needed as well as elastoplastic constitutive models have to be utilized. For all these reasons, the computational cost may eventually be prohibitively large.

In order to overcome such difficulties, the classical homogenization technique appears to be a good way as an alternative approach to direct numerical simulations, since the Non-homogeneous media could be replaced by an equivalent anisotropic homogeneous one [1]. Using the advanced techniques of this tool, in finite element (or finite difference) code (performed by Hooper [2]), helps to the problem is simplified and the time of analysis is extremely small [3, 4].

In the present paper, a new extension method of homogenization called the multiphase method (proposed by Sudret and de Buhan [5, 6]) was used and developed for multi unidirectional inclusions. According to this model the reinforced soil is described as the superposition of two mutually interacting continua called phase, which represent the soil and the network of inclusions, respectively. This paper starts from formulation of methods, and then the model is numerically implemented in a specifically three-dimensional version of standard FDM code. The extension model validated by comparison to analytical solutions. The results derived from such a simulation appear to be in good agreement. The aim of the paper is to present and validate a new equivalent medium for reinforced material.

2 Background of Theory and the Model's Implementation

Consider a continuous medium reinforced by a network of uniformly distributed inclusions oriented along arbitrary direction. This composite material model might be viewed as the superposition of two mutually interacting continua, matrix and reinforcement.

As regards the statics of such material, under the action of an external force, the total elastic stress increment tensor $\dot{\overline{\Sigma}}$ in the equivalent model is the sum of the elastic stress in each component of the system, i.e. matrix and reinforcement. In a global coordinate system this can be written as follow:

$$\dot{\Sigma}_{ij} = \dot{\sigma}_{ij}^{m} + \dot{\sigma}_{ij}^{r1} + \dot{\sigma}_{ij}^{r2} \tag{1}$$

where $\dot{\sigma}_{ij}^{m}$ is the stress increment in the matrix continua and $\dot{\sigma}_{ij}^{r1}$, $\dot{\sigma}_{ij}^{r2}$. are the stress increment in the reinforcement sets. In local coordinate system, $\bar{\sigma}^{r}$ is not the stress in the reinforcement material. It is a macroscopic stress variable calculated as the axial force F_{inc} developed in the inclusion per unit transverse area of REV (Eq. 2).

$$\bar{\sigma}^{r} = \frac{F_{inc}}{A_{REV}} = \frac{F_{inc}}{s^{2}} \tag{2}$$

which s represents the spacing between two neighboring inclusions. The local coordinate system of bars is oriented such that 3-axis points to inclined direction of inclusion and 1&2-axis is defined perpendicular to bar direction, that together form a right-handed coordinate system. To obtain the total stress (Eq. 1) $\bar{\sigma}^{r}$ must be converted from the local to the global coordinate system, so it is:

$$\sigma_{ij}^{r} = \bar{\sigma}^{r}\left(n_{i}n_{j}\right) \tag{3}$$

In global coordinate system by summing of stiffness matrix of all components, the equivalent elastic matrix of reinforced soil can be determined and implemented.

The user can also determine the number of inclusion sets involved in the analysis that can vary from zero to two sets. An equivalent reinforced soil model with zero inclusion set works as the built-in Elastic model in FLAC3D.

The parameters required for the model are those defining the spatial geometry of the inclusion, i.e. unit vector of inclusions direction and spacing, and the elastic and failure parameters of the soil and inclusions. The bulk and shear moduli of the soil are the only parameters required for elastic analysis.

3 Deformation and Anisotropy Validation

The deformation of a block of soil in a triaxial loading condition is analyzed by two methods: the equivalent soil model implemented in FLAC^{3D} and the analytical method. The model consists of a $1 \times 1 \times 1$ m soil block that contains one set of inclusion in yz-plane at different angles ranging $\beta = 0$ to $90°$ measured from z-axis. Inclusions are spaced at $S = 0.2$ m. In all models plastic deformations are prevented. Table 1 shows the mechanical properties that are used for these models.

Table 1. Mechanical and strength parameters for triaxial test.

Parameter	Soil	Inclusion
Elastic modulus	1500	150000
Poison's ratio	0.25	—
Cohesion	0	—
Friction angle	25	—
Inclusion area (mm^2)	—	625
Inclusion spacing	—	0.25

Figure 1 shows the variation of the axial stiffness versus inclusion inclination. In comparison with the analytical solution, the equivalent soil model tends to slightly overestimate the deformations for β values in the elementary of the study range. For the assumed mechanical parameters, the difference reaches its maximum of 10% at $\beta = 25°$ (Fig. 2).

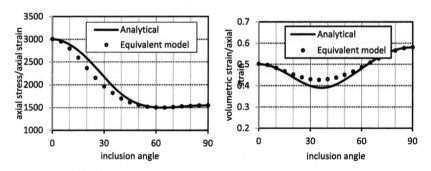

Fig. 1. Axial behavior of reinforced soil in triaxial test

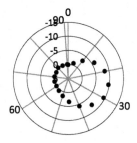

Fig. 2. Difference between calculated axial stress strain ratio by the equivalent model and analytical solution for different inclusion angle

4 Conclusions

A three-dimensional equivalent continuum constitutive model was formulated for reinforced soil masses containing up to two persistent inclusion sets. The model was developed to efficiently handle the deformation anisotropy and stress dependency in reinforced soils. The model was implemented in FLAC3D and the results compared to analytical solutions. The results show that the equivalent soil model provides an efficient tool for deformation analysis of reinforced soil in which the general anisotropy of a soil mass is of importance to the analysis. The analytical method tends to slightly underestimate the axial deformations over the applied range of inclusion angle except for the two cases of $\beta = 0°$ and $\beta = 90°$.

References

1. Poulos, H.G.: Piled raft foundations: design and applications. Geotechnique **51**, 95–113 (2001)
2. Hooper, J.: Observations on the behaviour of a pile-raft foundation on London Clay. Presented at the proceedings of the institution of civil engineers (ICE), London (1973)
3. de Buhan, P., Sudret, B.: Micropolar multiphase model for materials reinforced by linear inclusions. Eur. J. Mech. **19**, 669–687 (2000)
4. Herrmann, L.R., Welch, K.R., Berg, R.R.: Composite FEM analysis for layered systems. J. Eng. Mech. ASCE **110**, 1284–1302 (1984)
5. de Buhan, P., Sudret, B.: A two-phase elastoplastic model for unidirectionally-reinforced materials. Eur. J. Mech. A. Solids **18**, 995–1012 (1999)
6. Sudret, B.: Multiphase model for inclusion-reinforced structures. Ph.D., Ecole Nationale des Ponts et Chauss'ees Paris, (1999)

A GPU-Accelerated Three-Dimensional SPH Solver for Geotechnical Applications

Chong Peng[1(✉)], Wei Wu[2], and Hai-sui Yu[3]

[1] ESS Engineering Software Steyr GmbH, Berggasse 35, 4400 Steyr, Austria
pengchong@boku.ac.at
[2] Institute of Geotechnical Engineering,
University of Natural Resources and Life Sciences, Vienna,
Feistmantelstraße 4, 1180 Vienna, Austria
wei.wu@boku.ac.at
[3] School of Civil Engineering, Faculty of Engineering,
University of Leeds, Leeds LS2 9JT, UK
h.yu@leeds.uk.at

Abstract. A three-dimensional GPU-accelerated SPH solver is developed and ready to be released as an open-source tool. In this paper, the SPH formulation, geomechanical aspects of SPH modeling and the GPU parallelization are presented. A numerical case of granular flow passing a cylindrical obstacle is simulated to show the performance of the solver.

1 Introduction

Smoothed particle hydrodynamics (SPH) is a Lagrangian particle-based meshless methods with wide applications in astrophysics and computational fluid dynamics (CFD). It has several advantages over conventional grid-based methods: (1) Its Lagrangian and meshless nature makes it convenient in solving problems involving free-surface flow and large deformation; (2) The generation of computational model is easier compared with the mesh preparation in grid-based methods; (3) SPH has a structure very suitable for parallel computing.

In recent years, SPH attracts many attentions from the geotechnics community. Examples of its application include large deformation analysis, granular flow modeling, fluid-soil mixture coupling and debris flow simulation, among others. However, there are several factors which discourage researchers and engineers to apply SPH in their work. First, in single-cpu mode SPH is slow, mainly due to the large amount of particle interactions and the small time step. Second, although SPH has a long history, only in recent years it is applied to geotechnical problems. Available resources are inadequate. Particularly, there is no commercial software or open-source code allowing geotechnical simulations with SPH. These two factors make the development of an efficient open-source geotechnical SPH solver essential and urgent.

In this work, we developed a three-dimensional SPH solver for geotechnical simulations. The GPU computing is chosen as the acceleration technique, because it is widely available, and more efficient and much cheaper than CPU-based parallelization. The SPH fundamentals and the SPH adoption in geotechnical simulation are introduced. The efficient GPU implementation of SPH is then detailed. A numerical example is given to demonstrate the validity, accuracy and efficiency of the solver.

2   SPH for Geotechnical Simulation

In SPH, a set of scattered particles are used to discretize the problem domain. These particles carry field variables such as mass, velocity, density and stress, and move with the material. A field function $f(\boldsymbol{x})$ and its gradients can be approximately by

$$f(\boldsymbol{x}_i) = \sum f(\boldsymbol{x}_j) W_{ij} m_j / \rho_j, \; \nabla f(\boldsymbol{x}_i) = \sum f(\boldsymbol{x}_j) \nabla_i W_{ij} m_j / \rho_j \qquad (1)$$

where $\boldsymbol{x}_i$ and $\boldsymbol{x}_j$ are the coordinates of particle i and j, respectively; W_{ij} is a smoothing function dependent on r_{ij}, the distance between the particles; m_j and ρ_j denotes the mass and density at particle j, $\nabla_i W_{ij}$ indicates the gradient of the kernel W_{ij} evaluated at particle i.

Using the above SPH approximations, the conventional continuity equation and momentum equation can be written in the following form

$$\frac{\mathrm{d}\rho_i}{\mathrm{d}t} = \sum m_j (\boldsymbol{u}_i - \boldsymbol{u}_j) \cdot \nabla_i W_{ij} \qquad (2)$$

$$\frac{\mathrm{d}\boldsymbol{u}_i}{\mathrm{d}t} = \sum m_j \left(\frac{\boldsymbol{\sigma}_i}{\rho_i^2} + \frac{\boldsymbol{\sigma}_j}{\rho_j^2} - \Pi_{ij} \boldsymbol{I} + \boldsymbol{S}_{ij} \right) \nabla_i W_{ij} + \boldsymbol{g}_i \qquad (3)$$

where $\boldsymbol{u}$ is the velocity, $\boldsymbol{\sigma}$ is the Cauchy stress tensor, $\boldsymbol{I}$ is a 3×3 identity matrix, $\boldsymbol{g}$ is the body forces. Π_{ij} and $\boldsymbol{S}_{ij}$ denotes the contribution from artificial viscosity and artificial stress, used to enhance the stability of SPH computation. For details of the derivation of formulations, please refer to [1,2].

Constitutive models should be included in SPH to correctly capture the mechanical behaviors of geomaterials. In the SPH solver, three constitutive models are implemented: the advanced $\mu(I)$ rheological model for granular media, and the Drucker-Prager elastoplastic (D-P) model and hypoplastic model for general simulation. The hypoelastic approach is used to extend the infinitesimal strain-based D-P model and hypoplastic model to large deformation analysis. The SPH solution are obtained by numerical integrating Eqs. (3), (4) and the constitutive equation using explicit integration schemes. In the solver, the second-order Predictor-Corrector scheme is employed.

3 SPH Implementation on GPU

In three-dimensional simulations, large number of SPH particles are needed. Parallel computing technique is necessary for fast SPH simulation. The GPU based parallel computing becomes increasingly popular in recent years in large-scale computing for both research and application. Compared with CPU-based parallelization, GPU is much cheaper and easily available, and is extremely fast because usually a main stream graphic card has more than one thousand computing cores.

The solver is implemented using the CUDA programming language, an extension to the standard C. All the time-expensive parts of SPH computation, including neighbor search, particle interaction and numerical integration, are performed using graphic car. CPU-based code is used to control the whole computation flow. The efficiency of SPH solver is highly dependent on the neighbor search algorithm. In the solver, the efficient cell based searching scheme is adopted. First, the computational domain is partitioned into regular three-dimensional cells and each cell has a cell number according to its location, the size of the cell is $2h$. All particles are assigned a cell index in which cell it locates. Then all the particles are sorted according to their cell index, and the beginning particle and ending particle of each cell are computed. With the constructed neighbor list, a particle only has to search interacting particles in its own and the neighboring cells. This neighbor search scheme fully makes use of the GPU parallel architecture and is highly efficient. In our simulations, usually less than 5% of the total simulation time is spent on neighbor search.

4 Numerical Example

A simple yet representative example is numerically simulated using the GPU-based SPH solver. The example is based on a test performed by Cui and Gray [3] where granular flow passes a cylinder obstacle. This test case is of practical interest in industry and engineering. A body of granular material of diameter 0.002 m is at rest in a container initially at the top of a inclined chute. Once the gate of the container opens, the granular material flows down the inclined chute and passes a cylinder. A three dimensional simulation is carried out using the presented SPH solver. The chute is set as 1 m wide and 1.5 m long. The cylinder has a diameter of 0.2 m and is placed 0.3 m downside of the gate. The open of the gate is 8 cm. Two simulations with different resolutions, 0.004 m and 0.002 m, are performed. In the two simulations, the particle numbers are 3.3 million and 22.8 million, respectively. The simulated physical time is 3 s. With GTX 1080Ti graphic card, the simulation times for the two cases are approximately 7 h and 47 h, respectively. Although a huge number of particles are simulated, the solver can deliver results in reasonable time. The results at several time instances are shown in Fig. 1. Salient features of the granular flow are captured by the SPH simulation.

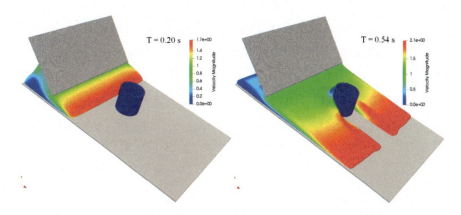

Fig. 1. Snapshots of the granular flow simulation with resolution 0.004 m.

5 Summary

We developed a three-dimensional GPU-accelerated SPH solver for geotechnical applications. State-of-the-art SPH formulations for geotechnical problems are implemented. The GPU parallelization is efficient and the solver is capable of performing large-scale simulation using conventional desktops and laptops. Preliminary validation are performed and it is shown that the solver is applicable to geotechnical problems. We hope the solver can promote the application of SPH in geotechnical engineering once it is mature and released to public.

References

1. Peng, C., Wu, W., Yu, H.S., Wang, C.: A SPH approach for large deformation analysis with hypoplastic model. Acta Geotech. **10**(6), 703–717 (2015)
2. Peng, C., Guo, X.G., Wu, W., Wang, Y.Q.: Unified modeling of granular media with smoothed particle hydrodynamics. Acta Geotech. **11**(6), 1231–1247 (2016)
3. Cui, X.J., Gray, J.M.N.T.: Gravity-driven granular free-surface flow around a circular cylinder. J. Fluid Mech. **720**, 312–337 (2013)

Numerical Analysis of Three Adjacent Horseshoe Galleries Subjected to Seismic Load

Sid Ali Rafa[✉], Allaoua Bouaicha, Idriss Rouaz, and Taous Kamel

National Centre for Studies and Integrated Researches of Building "CNERIB",
Cité Nouvelle El Mokrani Souidania, Algiers, Algeria
rafa.sidali@gmail.com

Abstract. Tunnels are widely constructed all over the world. They help to solve many infrastructures problems especially in large cities. However, their construction is a great challenge for engineers and constructers since they have to deal with a lot of uncertainties, such as, soil starts, water table, adjacent structures and earthquakes. In order to resist to seismic effect, engineers developed many calculation methods to design a tunnel under the effect of seismic load. Unfortunately, these methods are limited to specific shapes of tunnel sections' and do not take into account the interaction that occurs between the tunnel and the soil during an earthquake. To check the stability of three adjacent horseshoe galleries subjected to lateral seismic loads. A numerical model using the finite element method software PLAXIS 3D is developed where the soil and the galleries were modeled by mean of plan deformation. The soil behavior was modelled with Mohr-Coulomb model while three behavior models were used to model the structure of the galleries, namely, linear elastic model, Mohr-Coulomb model (MC) and Shotcrete Model (SCM). The results of these analysis show that the behavior model of concrete has an important effect on tunnel's response to seismic load.

Keywords: Tunnel · Seismic loads · PLAXIS 3D · Shotcrete Model

1 Introduction

During the last decades, Algeria has constructed many tunnels in the north of the country, which are either parts of the east-west highway project or Algiers subway. Since Algeria is considered as high seismic area, the regulation codes impose to engineers the consideration of the seismic effect in their design. Several methods can be used to design a tunnel under seismic loads such as Forced-Based Methods and Displacement-Based Methods. However, these two methods were developed only for circular or rectangular tunnels and cannot be used for other shapes like a horseshoe gallery. To analyze these types of galleries, Pitilakis et al. [1] recommended the use of numerical methods based either on finite element method or finite difference method. In this article, the transversal behavior of three adjacent horseshoe galleries dug into sandstone formation was analyzed under the effect of lateral seismic loads using the finite element method implemented in the finite element software PLAXIS 3D.

2 Numerical Analysis

2.1 Numerical Model

Commonly, when a FE model is used, the structure is modelled with beam elements and the surrounding soil with plane strain elements. Nevertheless, more sophisticated models can also be adopted where the structure and the surrounding soil are modelled with plane strain elements.

In order to simulate these three adjacent horseshoes, a developed model was used where the structure of the galleries and the surrounding soil were modelled with plane strain elements (Fig. 1). The soil behavior was modelled with Mohr coulomb model while three behavior models were used to model the structure of the galleries, namely, linear elastic model, Mohr-Coulomb model (MC) and Shotcrete Model (SCM) that was initially developed for shotcrete concrete [2] and extended for reinforced concrete [3].

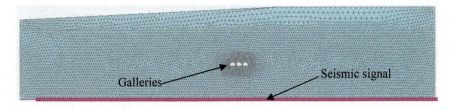

Fig. 1. Numerical model of the galleries

2.2 Seismic Loads

In order to analyze the behavior of these galleries and rule about their stability. Six recorded accelerograms during the major earthquake of Boumerdes City in 2003 are calibrated to a return period of 1000 years, which lead to a peak ground acceleration "PGA" of 0.63 g. The choice of these accelerograms is by the fact that they are local, and their frequency content contains the fundamental frequencies of the soil. Table 1 gives the predominant frequency of each accelerograms calculated after AFPS method [4].

Table 1. Predominant frequency of accelerograms.

Accelerograms	Fréquence prédominante des mouvements sismiques [Hz]
	AFPS method
Kouba E-W	12.22
Keddara E-W	11.67
Boumerdes E-W	14.33
Hussein Dey E-W	8.89
Re 1	8.08
Re 2	6.42

3 Results and Discussion

The numerical analysis results show that when the galleries' structure is modelled with a linear elastic model, a low plastic zone appear around the galleries. However, when the galleries' structure is modelled with MC model and SCM model the plastic zone is much important than the former model. Moreover, the linear elastic model shows no failure in the galleries' structure while the MC model shows a little failure of the concrete mainly in the junction between the raft and the wall of the galleries' structure, whereas the SCM model shows that many parts of the galleries' structure failed, namely, the junction between the raft and the wall, the wall and the vault. These results match with the findings by Asakura et al. [5], who numerically studied the effect of an earthquake on mountain tunnels.

References

1. Pitilakis, K., Tsinidis, G.: Performance and seismic design of underground structures. In: Earthquake Geotechnical Engineering Design, vol. 28, pp. 279–340. Springer, Cham (2014)
2. Schädlich, B., Schweiger, H.F.: A new constitutive model for shotcrete. In: Hicks, M.A., Brinkgreve, R.B.J., Rohe, A. (eds.) Numerical Methods in Geotechnical Engineering, pp. 103–108 (2014)
3. Maatkamp, T.W.P.: The capabilities of the Plaxis Shotcrete material model for designing laterally loaded reinforced concrete structures in the subsurface. MSc thesis, Faculty of Civil Engineering and Geosciences Section Geo-Engineering, Delft University (2016)
4. AFPS: Guide AFPS, Conception et Protection des Ouvrages Souterrains (2011)
5. Asakura, T., Shiba, Y., Matsuoka, S., Oya, T., Yashiro, K.: Damage to mountain tunnels by earthquakes and its mechanism. J. Jpn. Soc. Civil Eng. (659/III-52), 27–38 (2000)

Dynamic Analysis by Lattice Element Method Simulation

Zarghaam Haider Rizvi, Frank Wuttke[(⊠)], and Amir Shorian Sattari

Geomechanics and Geotechnics, Kiel University, Ludewig-Meyn-Street 10,
24118 Kiel, Germany
frank.wuttke@ifg.uni-kiel.de

Abstract. Quantifying the fracture in materials and structures is daunting task and to add to this misery there is no unique approach or model to wrestle with this problem. The challenges to model this complexity that arise with failure are plentiful, beginning with mesh dependency for softening to various numerical hitches and instabilities, tracking procedures, multiple cracking with crack interactions etc. Formation of cracks in the dynamic framework come with additional challenges bringing inertial effects, crack branching etc. In this paper, lattice model for dynamic failure is presented. The concluding objective is to mimic crack origination and propagation in 2D brittle and quasi-brittle material prone to dynamic loading. The advantage of the lattice element models is in their effective demonstration of failure mechanisms. The presented model is established on Delaunay triangulated lattice of Bernoulli beams which act as interconnected links among the Voronoi cells used to calculate beam cross sections. The removal of elements exceeding the failure criteria serve for depiction of failure mechanisms in modes I, whereas the mass and the inertial properties are encompassed into lattice network. Representative numerical simulation is implemented comparing the outcomes for static and dynamic analysis.

Keywords: Lattice element method · Dynamics analysis · Fracture mechanics
Numerical method

1 Introduction

Among the most important, challenging and exciting recent steps forward in geo-engineering is the development of coupled numerical method including dynamic problem. As most of the geomechanical interactions are at meso scale a numerical method based on this scale is better suited to capture the effects which are included in ad hoc manner based upon fitting parameters in continuum-based models such as FEM, BEM and FDM. These meso scale models represent the fundamental physics of geo-engineering processes which can include the effects of heat, water, mechanics. The complex interactions of these physics could easily be captured with meso scale numerical methods to capture the interactions. The method is best suited to estimate, evaluate, optimize and develop and new knowledge in this field based on rigorous mathematical and physical sense. The lattice element models, as a class of discrete models, in which the structural solid is represented as an assembly of one-dimensional elements. This idea allows one to provide robust models for propagation of discontinuities, multiple cracks interaction or cracks coalescence. Many procedures for computation of lattice element parameters for representing linear elastic continuum have been developed. Special attention is dedicated to presenting the ability of this kind of models to consider heat transfer, fluid flow, biological and chemical reactions, material disorder, heterogeneities and multi-phase materials, which makes lattice models attractive for meso- or micro-scale simulations of failure phenomena in quasi-brittle materials, such as rocks, cemented sand and low consistent clay for dynamic loading. Common difficulties encountered in material failure and a way of dealing with them in the lattice models framework are easy to incorporate and are given in [1]. Namely, the size of the localized fracture process zone around the propagating crack plays a key role in failure mechanism, which is observed in various models of linear elastic fracture mechanics, multi-scale theories, homogenization techniques, finite element models, and molecular dynamics. An efficient way of dealing with this kind of phenomena is by introducing the embedded strong discontinuity into lattice elements, resulting with mesh-independent computations of failure response. In this work we present the extension of lattice-based method to dynamic problems considering Euler-Bernoulli beams as cohesive elements. Failure of these beams are calculated based on linear elastic fracture mechanics (LEFM) [2] and the elements are assigned a lower stiffness based on bilinear softening model as these elements reach the threshold.

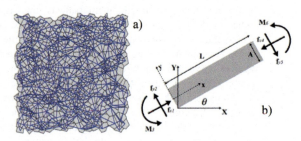

Fig. 1. (a) Generation of lattice elements (blue) with Voronoi and Delaunay triangulation. (b) 2D Euler-Bernoulli beam element which acts as connections among the Voronoi cells [3].

2 Mathematical Formulation

In the lattice element models, lattice nodes which can be considered as the centers of the unit cells, are connected by beams that can carry normal force, shear force and bending moment (Fig. 1a, b). As the strain energy stored in each element exceed the threshold the element is either removed or assigned a lower stiffness value. The method is based on minimizing the stored energy of the system. The total energy is comprised of two components, namely, the strain energy (ξ_{Strain}) and the external applied loading ($\xi_{Loading}$). The displacement and movement of lattice elements are calculated due to application of external loads. For 2D representation of elements with Euler-Bernoulli beam with 6 DoF, the corresponding axial (f_x), shear (f_y) and moment (M) are computed due to corresponding local displacements in translational (u_x, u_y) and rotational (ϕ) directions. The stiffness of beam element (k_{Local}) is, computed as

$$k_{local} = \begin{bmatrix} k_1 & 0 & 0 & -k_1 & 0 & 0 \\ 0 & k_2 & k_3 & 0 & k_2 & k_3 \\ 0 & k_3 & 2k_4 & 0 & -k_3 & k_4 \\ -k_1 & 0 & 0 & k_1 & 0 & 0 \\ 0 & -k_2 & -k_3 & 0 & k_2 & -k_3 \\ 0 & k_3 & k_4 & 0 & -k_3 & 2k_4 \end{bmatrix} ; k_1 = \frac{E_b A}{L} ; k_2 = \frac{12 E_b I}{L^3} ; k_2$$
$$= \frac{6 E_b I}{L^2} ; k_4 = \frac{2 E_b I}{L}$$

where, L is length, A is cross-section area, I is moment of inertia and E_b is Young's modulus of beam elements. Therefore, the global stiffness (k_{global}) is, $[k_{global}] = [T_m]^T [k_{local}] [T_m]$ where, $[T_m]$ is a transformation matrix. Minimizing the total potential energy, the displacements in global system are obtained.

2.1 Failure Model

Lattice based modelling is best suited for brittle and quasi brittle materials. For modelling a brittle material, elements are removed once the failure limit is reached. The reduction in Young's modulus of the element is achieved by removing the elements as they reach the critical strain (ε_p) corresponding to the failure stress ε_p. The critical strain is computed based on strain energy release rate from linear elastic fracture mechanics failure criteria [4] (Fig. 2).

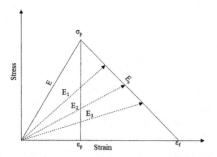

Fig. 2. An element with bi-linear softening failure [4].

2.2 Dynamic Implementation

Consider no damping in the system, the equation of motion of one element is written as $M_e a^t_{e+1} + K_e x^t = F^{applied}_{e+1} - F^{t-1}_{e+1}$ where M is the mass and K is the stiffness matrix. Following the steps defined in [5, 7] the acceleration is written in terms of velocity and displacement with applying trapezoidal rule. The velocity and acceleration is approximated as $v_{e+1} = -v_e + 2h^{-1}(x_{e+1} - x_e)$ and $a_{e+1} = -a_e - 4h^{-1} v_e + 4h^{-2}(x_{e+1} - x_e)$. Applying the Newton's method and rearranging the equations we get,

$$\left[\frac{4}{h^2} M + K_e \right] x^t_{e+1} = F^{applied}_{e+1} - F^{t-1}_{e+1} - M \left[a_e - \frac{4}{h} v_e + \frac{4}{h^2} (x^{t-1}_{e+1} - x_e) \right]$$

A detail description of the implementation is given in [5, 7].

3 Results and Discussion

We generated a three-phase material following a segregated scheme with two phases of smaller size having weak material properties and these elements are allowed to break.

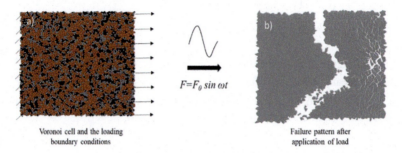

Fig. 3. (a) Generated Poisson random lattice with three phase segregated distribution [6]. (b) Failure pattern obtained after application of a dynamic sin wave load.

The generated assembly is completely broken as shown in the Fig. 3b due to dynamic loading. There are also cracking near the force application side. The pattern of cracks are showing many nucleation centers and propagation in separate directions. The meso scale response of the cracked system is capture with relative ease (Fig. 4).

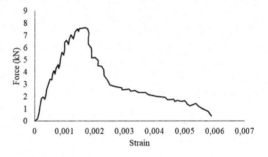

Fig. 4. Force strain graph of the heterogeneous system. The response of the dynamic system is visible with kinks in the graph.

Acknowledgement. The authors wish to thank Federal Ministry of Education and Research (BMBF), Germany for providing grant for GeoMInt (03G0866B).

References

1. Nikolić, M., Karavelić, E., Ibrahimbegovic, A., Miščević, P.: Lattice element models and their peculiarities. Arch. Comput. Methods Eng. **25**(3), 753–784 (2017). https://doi.org/10.1007/s11831-017-9210-y
2. Wuttke, F., Sattari, A.S., Rizvi, Z.H., Motra, H.B.: Advanced meso-scale modelling to study the effective thermo-mechanical parameter in solid geomaterial. In: Advances in Laboratory Testing and Modelling of Soils and Shales (ATMSS). Springer Series in Geomechanics and Geoengineering (2017)

3. Sattari, A.S., Rizvi, Z.H., Motra, H.B., Wuttke, F.: Meso-scale modeling of heat transport in a heterogeneous cemented geomaterial by lattice element method. Granular Matter. **19**, 66 (2017)
4. Bazant, Z.P.: Fracture in concrete and reinforced concrete. In: Bazant, Z.P. (ed.) Mechanics of Geomaterials, pp. 259–303. John Wiley (1985)
5. Nikolić, M., Cesic, J., Ibrahimbegovic, A., Nikolić, Z.: Lattice Model for failure based on Embedded strong discontinuities in dynamic framework. In: XIV International Conference on Computational Plasticity, Fundamentals and Application (COMPLAS XIV) (2017)
6. Rizvi, Z.H., Shrestha, D., Sattari, A.S., Wuttke, F.: Numerical modelling of effective thermal conductivity for modified geomaterial using Lattice Element Method. Heat Mass Transf. **54** (2), 483–499 (2018)
7. Nikolic, M., Do, X.N., Ibrahimbegovic, A., Nikolic, Z.: Crack propagation in dynamics by embedded strong discontinuity approach: enhanced solid versus discrete lattice model. Comput. Methods Appl. Mech. Eng. **340**, 480–499 (2018)

Numerical Simulation of Plane Wave Propagation in a Semi-infinite Media with a Linear Hardening Plastic Constitutive Model

Erxiang Song, Abbas Haider, and Peng Li

Department of Civil Engineering, Tsinghua University, Beijing 100084, China
lipengcivil@mail.tsinghua.edu.cn

Abstract. The theory of linear elastic waves is one of the pioneer works in the development of stress waves in soil mechanics which is involved in many practical problems in geophysics, earthquake engineering and geomechanics. However, geotechnical materials under normal engineering conditions present a plastic mechanical behavior, and the development of the theory of plastic waves was developed a century later. Till now the numerical modelling of wave propagation in a semi-infinite media with nonlinear material constitutive equation has seldom been addressed. This paper presents a systematic research on this topic. A linear hardening plastic stress-strain relationship that is a typical and representative model is chosen. A rectangular impulse load is considered to conduct the numerical and analytical studies. Numerical results are verified by the analytical solutions that is derived based on character line method. Results have shown the phenomenon of wave chase as well as the interaction of the loading slow wave and unloading fast wave. The conclusions and the proposed numerical scheme will facilitate the further investigation in many practical earthquake engineering problems.

Keywords: Wave propagation · Linear hardening plastic model · Plastic wave · Finite element modeling

1 Introduction

The rate-independent theory of plastic waves was founded a 100 years after the development of theory of elastic waves in 1920's by Poisson, Stokes *et al.* and Rayleigh *et al.* Incorporating the concepts of nonlinear stress-strain relationship and the interaction between the loading and unloading waves, this theory was fully developed by Von Karman [1], Taylor and Rakhmatulin independently.

As geotechnical materials such as soils show plastic behaviors, a number of constitutive relationships have been developed following this theory. Since stress-strain curves of these constitutive models turn sharply at yielding, the stress strain curve has been idealized as bilinear by many researchers to simplify the analysis of interaction of elastic and plastic waves. Analytical solutions for classical problem of an semi-infinite bar loaded uniformly by pressure load have been proposed in [2, 3] using the method of characteristics in the t-x plane but the numerical modeling of this problem has not caught much attention in the scientific society.

An effort to numerically validate the analytical solution has been made in this study. The case where the secondary impact velocity exceeds the initial yield velocity of the bar is considered. The problem is simulated using FORTRAN to validate the derived analytical solution where a semi-infinite bar of linear hardening material is impacted by a rectangular impulse. It has been demonstrated that the numerical scheme used is strong enough to replicate the phenomenon of interaction of elastic and plastic waves as well as the propagation of unloading waves.

2 Analytical Solution

Propagation of waves in a semi-infinite media is governed by certain kinematic and dynamic equations and the stress-strain relationship of the material. These conditions are listed in Ref. [3], based on which Wang used method of character line to derive analytical solution where he considered a semi-infinite bar of linear hardening material which is impacted by a sudden rectangular impulse at $t_0 = 0$. The load is suddenly released to 0 at time $t_1 = 1.50$ s. As the unloading wave travels faster than the precursor plastic loading wave, it will catch up and interact with the plastic loading wave. In this study the results of this derivation are extended to the case where the initial impact velocity is greater than two times the yield velocity of the material (i.e. $v^* > 2v_y$). The material properties chosen are homogeneous and without material damping (Figs. 1 and 2):

$M_0 = 121153846.2\text{Pa}, M_1 = 13461538.46\text{Pa}, \sigma_y = 300000\text{Pa}, \rho = 2070\,\text{kg}/\text{m}^3,$
$\mu = 0.3.$

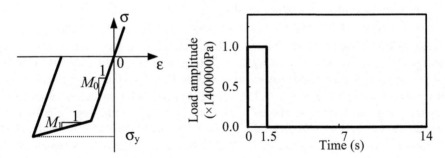

Fig. 1. Linear hardening plastic model

Fig. 2. Applied rectangular load impulse

As the unloading wave catches up and interacts with the precursor plastic loading wave, it produces secondary plastic wave of reduced stress propagating rightward and a secondary elastic wave of same stress propagating leftward in the bar $(t = t_1')$. This leftward secondary elastic wave reaches the impact end where the stress is again unloaded to 0 and an unloading wave travels towards the fore running rightward secondary plastic loading wave. This cycle is repeated until $(t = t'_3)$ when the secondary waves produced are no more plastic in nature and the stress waves are unloaded to 0 throughout the bar after the last reflection from the impact end.

3 Numerical Model

To solve this problem numerically, finite element formulation of the momentum balance equation has been carried out by using FORTRAN. The momentum balance equation is a partial differential equation which is discretized in space and integrated by parts resulting in a second order differential equation. The link between the successive values is provided by the GN$_{22}$ truncated series expansion [4]. Then the final numerical scheme is

$$\left[\underline{M} + \frac{1}{2}\underline{K}\beta_2 \Delta t^2\right] \Delta \ddot{u}_i = F_{i+1} \tag{1}$$

where $F_{i+1} = f_{i+1} - \underline{K}u_i - \underline{K}\dot{u}_i \Delta t - \frac{1}{2}\underline{K}\ddot{u}_i \Delta t^2 - \underline{M}\ddot{u}_i$, $\underline{M} = \int N^T \rho N dV$, $\underline{K} = \int B^T MB dV$.

As previously stated the problem considered is one dimensional only, therefore constrained modulus is used in calculating strains and stresses. Time step and element length for the example are chosen as 0.001 s and 1 m respectively.

4 Results and Discussion

The rectangular impulse is applied on the left end of the bar at $t = 0$ and unloaded at $t = t_1$. The time (t'_1) and location (l'_1) of the interaction of slow unloading wave $C_0 = \sqrt{M_0/\rho}$ with the precursor fast plastic loading wave $C_1 = \sqrt{M_1/\rho}$ can be determined by $t'_1 = t_1/(1 - C_1/C_0)$ and $l'_1 = l_1/(1 - C_1/C_0)$, where C_0 and C_1 are the velocities of propagation of elastic and plastic stress waves. The unloading wave interacts with the plastic loading wave at $t'_1 = 2.25$ s, $l'_1 = 181$ m producing two stress waves travelling in opposite directions. The leftward elastic wave after reflecting again from the impact end interacts with the fore running secondary plastic loading wave and unloads it further. This process repeats itself till $t'_3 = 9.00$ s, $l'_3 = 726$ m after which all the plastic stress is unloaded and only elastic waves propagate in the bar. Results from the two solutions are plotted and compared in Fig. 3.

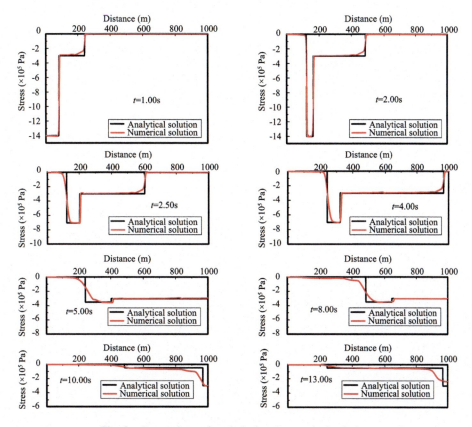

Fig. 3. Comparison of analytical and numerical solutions

5 Conclusion

The classical theory of linear elastic waves and rate-independent plastic waves lay the foundation of stress waves propagation in soil mechanics. The former has been extensively used to solve different problems in geotechnical engineering but the later remained neglected. In this study, the phenomenon of plastic wave propagation in a semi-infinite bar of linear hardening plastic material has been simulated numerically by using a simple linear hardening plastic model. The numerical solution has duplicated the complex phenomenon of propagation of stress waves of varying velocities and their interaction when they pursue each other. The proposed numerical scheme can be used to investigate the plastic wave propagation in complex plastic materials.

Acknowledgement. This work is financially supported by the National Key Fundamental Research and Development Program of China (Project no. 2014CB047003), the National Natural Science Foundation of China (Project No. 51408331) and the Science and Technology Research and Development Program of China Railway Corporation (Project No. 2014G004-C).

References

1. Von Kármán, T., Bohnenblust, H.F., Hyers, D.H.: The Propagation of Plastic Waves in Tension Specimens of Finite Length: Theory and Methods of Integration. National Defense Research Committee (1942)
2. Clifton, R.J., Bodner, S.R.: An analysis of longitudinal elastic-plastic pulse propagation. J. Appl. Mech. **33**(2), 248–255 (1966)
3. Wang, L.: Foundations of Stress Waves. Elsevier, Amsterdam (2011)
4. Zienkiewicz, O.C., et al.: Computational Geomechanics. Wiley, Chichester (1999)

Construction Simulation and Sensitivity Analysis of Underground Caverns in Fault Region

Chao Su and Yijia Dong[✉]

College of Water Conservancy and Hydropower Engineering, Hohai University,
Nanjing 210098, Jiangsu, China
yijiadonghhu@gmail.com

Abstract. The underground caverns of hydropower stations have a large scale and its stability depends on the complicated condition of the rock quality, the ground stress distribution, the shape and size of caverns, the excavation sequence, the support methods. To ensure the safety of the underground structures, it is necessary to study the influencing factors based on the realistic construction methods. During the exploration survey of the rock mass and geological conditions, the mechanical behaviors are usually obtained with uncertainty. Among these factors, the mechanical properties of rock and faults are important to the displacement and stress of underground caverns. In this study, the displacements and stresses during the excavation procedure of underground caverns of a hydropower station in fault region were analyzed using plastic finite element method. Subsequently, a sensitivity analysis was performed to quantitatively analyze the influence of uncertainty of the exploration survey to the safety of the underground caverns.

Keywords: Construction simulation · Sensitivity analysis · Deep excavation

1 Introduction

Underground power house scheme is beneficial to the layout design of power station, the construction of diversion structure, and speed of the dam construction [1]. Because of the complicated geological conditions of surrounding rocks, the excavation of cavern groups is a complex project especially in regions of multiple faults [2]. In general, the influencing factors of the safety of underground caverns include the quality of surrounding rocks, layout and mechanical properties of geological structures, the distribution of geostresses, the depth and scale of cavern groups, the influence of underground water and so on [1, 3]. Among these factors, the mechanical properties of rock and faults are important to the displacement and stress of underground caverns. During the exploration survey of the rock mass and geological conditions, the mechanical behaviors are usually obtained with uncertainty. Sensitive analysis is required to analyze the influence of the uncertainty to the safety of underground caverns. In this study, the procedure of the construction of underground caverns is simulated with commercial finite element software ABAQUS. Initial geostresses and construction methods such as supports and construction sequence were considered. Finally, to determine the most critical factor to the safety of the underground caverns, a sensitivity analysis was performed by changing mechanical properties of surrounding rocks and faults.

2   Numerical Modeling and Project

The hydropower station is in the south of China. The underground structures are composed of the main power house, the transformer house, four busbar chambers diversion tunnels and draft tubes. To construct the underground structures, excavation was performed about 200 m under the surface of the mountain. The main power house is 180 m long, 29.5 m wide and 74 m high. The transformer house is 178 m long, 20 m wide and 21 m high. There are four faults near the cavern. The position relationship between underground structures and faults are given in Fig. 1(a). The excavation of the caverns is divided into nine stages and the supports include steel fiber reinforced shotcrete, system anchors and prestressed anchor cable. Three-dimensional finite element method was adopted to simulate the excavation process. Rocks and faults were meshed with tetrahedron elements and supports were meshed with two-dimensional bar elements. There were 393,741 elements in the model. The meshes of the excavated rocks are given in Fig. 1(b).

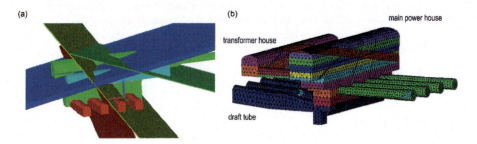

Fig. 1. Location of faults (a) and meshes of underground caverns (b)

With the proceeding of excavation, the stresses will redistribute after achieving the yield criteria. In this study, Mohr-Coulomb criteria were adopted to simulate the plastic behaviors of rock and faults [4]. When the material reaches its yield condition, the normal stress σ and the shear stress τ satisfy Eq. (1):

$$\tau = c - \sigma \cdot \tan \varphi \tag{1}$$

in which c and φ are internal friction and friction angle separately.

Sensitivity analysis can determine help engineers estimate the influence of the uncertainty of the exploring survey on the stability of underground caverns. Therefore, the reliability of the project will be improved. To study the sensitivity of the mechanical properties of rock and faults, their elastic modulus and values of c, ϕ were changed to 80%, 90%, 110%, 120% of the original values to perform numerical simulations. In each simulation, only one parameter was different from its original value.

3 Results and Discussion

3.1 Simulation of the Excavation Procedure

Most of the rock mass around the caverns is fresh rock. According to Chinese specification for design of hydraulic tunnel, the percentage of rock of grade III_1, III_2, II, IV are 55%, 30%, 10% and 5%. The grade of faults is IV.

The distribution of maximum normal stress and minimum normal stress of the section at the center of the third generator sets is given in Fig. 2. The maximum normal stress and compressive stress are 1.2 MPa and 5.9 MPa separately. They both appear at the middle height of the main power house. Because of sufficient supports, the stress does not exceed the bearing capacity of the rock mass.

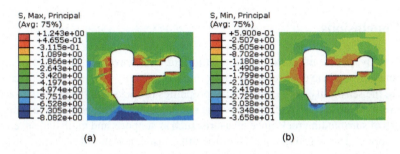

Fig. 2. Maximum Normal Stress (left) and Minimum Normal Stress (right) distribution of the middle section of the third Generator Sets (MPa)

3.2 Sensitivity Analysis Mechanical Properties of Rocks and Faults

Based on the stress and plastic strain distribution, five key points were selected for the sensitivity analysis. The location of the key points is given in Fig. 3.

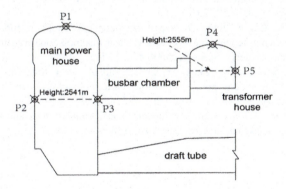

Fig. 3. Location of key points

From the results of the simulation after changing the mechanical properties of rock. It can be found that the displacement decreases with the increase of elastic modulus. When elastic modulus of rock increased by 20%, the displacement of the roof of the main power house decreased by 9.37%, and the displacement of downstream wall of the power main house decreased 16%. Compared with the influence on the displacements, the stresses at the key points does not change much when elastic modulus of rock changes. The influence of strength of rocks show obvious nonlinearity since the change of strength causes the redistribution of stresses. As the strength increases, the rocks are harder to yield. Therefore, the displacements decrease because of the increment of stiffness. For the influence of mechanical properties of faults, they show a similar tendency with that of rock mass. But the extent of their influence is less. The reason may be the volume of faults is much less than the volume of the rock mass.

4 Conclusions

In this study, plastic finite element simulation for the excavation procedure of a large-scale underground cavern group was performed. Sensitivity analysis was carried out to analyze the sensitivity of the mechanical properties of rocks and faults on the safety of the caverns. The simulated results show the stresses of the surrounding rocks satisfy the strength of surrounding rock since measures such as supports and multistage excavation were adopted. The sensitivity analysis shows the elastic modulus of rock mass influence the displacements of caverns most. The stress distribution is most sensitive to the strength of rock mass. The influence of mechanical properties of faults is less because its volume is smaller than rock mass.

References

1. Wu, A., Wang, J., Zhou, Z., Huang, S., Ding, X., Dong, Z., et al.: Engineering rock mechanics practices in the underground powerhouse at Jinping I hydropower station. J. Rock Mech. Geotech. Eng. **8**(5), 640–650 (2016)
2. Xu, D., Feng, X., Cui, Y., Jiang, Q.: Use of the equivalent continuum approach to model the behavior of a rock mass containing an interlayer shear weakness zone in an underground cavern excavation. Tunn. Undergr. Space Technol. **47**(2), 35–51 (2015)
3. Feng, X., Pei, S., Jiang, Q., Zhou, Y., Li, S., Yao, Z.: Deep fracturing of the hard rock surrounding a large underground cavern subjected to high geostress. in situ observation and mechanism analysis. Rock Mech. Rock Eng. **5**, 2155–2175 (2017)
4. Zhang, W.G., Goh, A.T.C.: Regression models for estimating ultimate and serviceability limit states of underground rock caverns. Eng. Geol. **188**, 68–76 (2015)

A Stress Correction Algorithm for a Simple Hypoplastic Model

Shun Wang[1(✉)], Wei Wu[1], Xuzhen He[1], Dichuan Zhang[2], and Jong Ryeol Kim[2]

[1] Institute of Geotechnical Engineering, University of Natural Resources and Life Sciences, Vienna, Feistmantelstraße 4, 1180 Vienna, Austria
{shun.wang,wei.wu,xuzhen.he}@boku.ac.at
[2] Department of Civil Engineering, Nazarbayev University, Astana, Kazakhstan
{dichuan.zhang,jong.kim}@nu.edu.kz

Abstract. In this paper, we consider the numerical integration of a simple hypoplastic constitutive equation. The stress drift away from the failure surface is corrected with a predictor-corrector scheme, which is verified by a boundary value problems, i.e., failure process of a homogeneous slope.

1 Introduction

Hypoplastic models are characterized by simple formulation and few material parameters, and have been used to simulate soil behavior in element tests and to solve boundary value problems with finite element method (FEM). While the existing integration schemes perform well prior to reaching the failure surface, none of properly handle the stress drift away from the failure surface. Consequently, the error resulted from stress drift away from the failure surface can accumulate in numerical computations and eventually lead to unphysical behaviour. For numerical calculations, it makes sense not to allow stress to wander outside the failure surface. A solution to this problem is to use the return mapping method originally proposed for elastoplastic models [1]. In this work, we will introduce a stress correction scheme and the effects of the stress correction scheme on FEM simulation of a slope failure process. In doing so, we consider a fairly general and simple hypoplastic model so that our approach can be easily adopted to handle more sophisticated hypoplastic models.

2 A Simple Hypoplastic Constitutive Model

A simple hypoplastic constitutive equation for granular materials is used in this paper, which is an improvement of an early version by Wu and Bauer [2].

$$\overset{\circ}{\boldsymbol{\sigma}} = C_1(\mathrm{tr}\boldsymbol{\sigma})\dot{\boldsymbol{\varepsilon}} + C_2(\mathrm{tr}\dot{\boldsymbol{\varepsilon}})\boldsymbol{\sigma} + C_3 \frac{\mathrm{tr}(\boldsymbol{\sigma}\cdot\dot{\boldsymbol{\varepsilon}})}{\mathrm{tr}\boldsymbol{\sigma}}\boldsymbol{\sigma} + C_4(\boldsymbol{\sigma}+\boldsymbol{\sigma}^*)\|\dot{\boldsymbol{\varepsilon}}\|I_e, \qquad (1)$$

in which C_i ($i = 1, 2, 3, 4$) are dimensionless parameters. The deviatoric stress tensor $\boldsymbol{\sigma}^*$ in Eq. (1) is defined by $\boldsymbol{\sigma}^* = \boldsymbol{\sigma} - 1/3(\mathrm{tr}\boldsymbol{\sigma})\boldsymbol{\delta}_{ij}$, with $\boldsymbol{\delta}_{ij}$ being the Kronecker delta. I_e is adopted as the critical state function.

$$I_e = \left(\frac{e}{e_{crt}}\right)^\alpha, \quad \text{and} \quad e_{crt} = e_{co}\exp\left[-\lambda\left(\frac{-\mathrm{tr}\boldsymbol{\sigma}}{p_r}\right)^\xi\right] \tag{2}$$

where e and e_{crt} are the current void ratio and critical state void ratio, respectively. α, e_{co}, λ, and ξ are parametric constants. p_r is a reference pressure. The effect of cohesion is realized by using a translated stress tensor:

$$\boldsymbol{\sigma}_c = \boldsymbol{\sigma} - p_t \boldsymbol{\delta}_{ij} \tag{3}$$

where $p_t = c/\tan\phi$ with c and ϕ being respectively the cohesion and friction angle of soil.

The failure criterion is defined by vanishing stress rate for a non-vanishing strain rate, and can be readily derived through:

$$f(\boldsymbol{\sigma}) = \boldsymbol{N}^\mathrm{T} : (\boldsymbol{\mathscr{L}}^\mathrm{T})^{-1} : \boldsymbol{\mathscr{L}}^{-1} : \boldsymbol{N}(I_e)^2 - 1 = 0 \tag{4}$$

The explicit formula of the failure surface can be obtained using the symbolic computational program *Mathematica*, which gives rise to a Drucker-Praguer type of failure surface.

3 Stress Correction Algorithm

This hypoplastic is implemented in FEM code using an adaptive explicit method (the RKFsec method [3,4]) combined with a stress correction algorithm.

Let us consider a stress state that is lying inside the failure surface, e.g., stress $\boldsymbol{\sigma}_n$ at the $(n)^{th}$ step of analysis. The hypoplastic model allows some stress state lying outside the failure surface. For the strain path shown in Fig. 1, no matter how accurate the integration method, the stress defined by $\boldsymbol{\sigma}_{n+1}^{trial}$ at the $(n+1)^{th}$ step of analysis will violate the consistent condition, so that

$$f(\boldsymbol{\sigma}_{n+1}^{trial}) = \sqrt{J_2(\boldsymbol{\sigma}_{n+1}^{*trial})} + \varsigma(C_i, I_e)I_1 > FTOL, \tag{5}$$

In order to return the stress state to the failure surface, it is desirable that the total strain increment, $\Delta\boldsymbol{\varepsilon}$, remains unchanged, since this is consistent with the philosophy of the displacement finite element procedure. The corrected stress state in Eq. (5) satisfies the consistency condition. Using Eq. (5), together with the assumption that departures from the failure surface are sufficiently small that only one return step is required, the consistent condition is expressed as follows:

$$f = \eta\sqrt{J_2(\boldsymbol{\sigma}_{n+1}^{*trial})} + \varsigma(C_i, I_e)I_1 = 0, \tag{6}$$

which yields the unknown multiplier η.

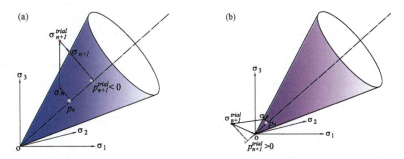

Fig. 1. Sketch of return mapping scheme, (a) return to the cone, and (b) return to the apex, the direct of σ is negative

On the other hand, if the mean stress $p_{n+1}^{trial} > 0$, as shown in Fig. 1(b), the stress is corrected to the apex of the failure surface

$$\sigma_{n+1} = 0, \quad p_{n+1} = 0, \tag{7}$$

After solving the above equation, dependent on the position of the stress state, we correct the violated stress either to the cone or to the apex of the failure surface via Eqs. (6) and (7).

4 Numerical Simulation of a Homogeneous Slope

The developed stress correction treatment is validated by evaluating the safety factor of a homogeneous slope and simulating the subsequent failure process. The slope is assumed to consist of cohesive soil with the material parameters $\phi = 20°$ and $c = 12$ kPa, including an initial void ratio of the soil of $e_i = 0.88$. The friction angle ϕ and cohesion c are the two shear strength parameters subjected to strength reduction.

The effect of the stress correction scheme is interpreted in Fig. 2, which shows contour plots of the shear surface obtained using an the RKF23sec method with and without stress correction. Analysis of this figure reveals that whereas a

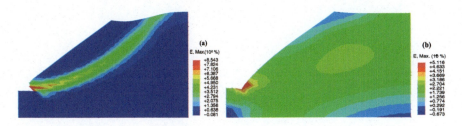

Fig. 2. The shear surface at the final increment obtained using the RKF23sec method, (a) with and (b) without stress correction

failure surface, as depicted by the equivalent strain in Fig. 2(a), can be observed in simulation using a stress correction scheme, no failure surface is generated in the computation undertaken without a stress correction scheme, as shown in Fig. 2(b). Therefore, the slope failure process can be captured in simulations carried out using the hypoplastic constitutive model with a combined stress correction scheme.

5 Conclusions

This paper introduces a stress correction treatment used in the FEM implementation of a simple hypoplastic model. Numerical analysis of a homogeneous slope reveals that by using the developed stress correction algorithm, the slope failure process can be successfully captured.

Acknowledgement. The research carried out in this paper is partly funded by the project "GEORAMP" within the RISE programme of Horizon 2020 under grant number 645665. The first author is grateful to the financial support from the Otto Pregl Foundation for Fundamental Geotechnical Research in Vienna.

References

1. Krieg, R.D., Krieg, D.B.: Accuracies of numerical solution methods for the elastic-perfectly plastic model. J. Pre. Vessel Tech. **99**(4), 510–515 (1997)
2. Wu, W., Bauer, E.: A simple hypoplastic constitutive model for sand. Int. J. Numer. Anal. Meth. Geomech. **18**(12), 833–862 (1994)
3. Tamagnini, C., Viggiani, G., Chambon, R., et al.: Evaluation of different strategies for the integration of hypoplastic constitutive equations. Application to the CLoE model. Mech. Cohesive-Frictional Mater **4**(5), 263–289 (2000)
4. Ding, Y.T., Huang, W.X., Sheng, D.C., et al.: Numerical study on finite element implementation of hypoplastic models. Comput. Geotech. **68**, 78–90 (2015)

Experimental and Numerical Study of the Mechanical Properties of Granite After High Temperature Exposure

Su-ran Wang(✉) and You-liang Chen

Department of Civil Engineering, University of Shanghai for Science and Technology,
516 Jungong Road, Shanghai 200093, China
evanwangsuran@foxmail.com

Abstract. The macroscopic mechanical properties of granite collected from the Fujian Province, China were measured using uniaxial compression and three-point bending after the granite was exposed to high temperatures. The stress-strain relationship was measured and mechanical properties were calculated. The microstructure and content of minerals are detected. The failure mechanisms and the mechanical characteristics of the granite are explained in terms of the microstructure and the minerals present. Where after the Universal Distinct Element Code (UDEC) was used to numerically simulate the mechanical behavior of granite specimens under uniaxial compressive loading after the granite was exposed to different high temperatures. The model parameters were calculated based on previous experiments. The effects of heat treatment on granite was investigated numerically. The model predictions were compared with experimental results and an acceptable agreement was achieved. A modified failure criterion was developed and a corresponding fish function was written to investigate the development of shear and tensile cracking during loading. It was found that initial cracking is in the form of shear cracks which increase rapidly in number until the model reached the unstable fracture developing stage. The number of tensile cracks increases slowly as the axial stress increases from the crack initiation stress (CIS) to the unstable fracture developing stage. Once the specimen enters into unstable fracture developing stage, there is a transition between the type of failure recorded in the model where the shear cracks change to tensile cracks.

Keywords: UDEC · Heat treatment · Mechanical properties
Numerical simulation

1 Methodology and Results

1.1 Methodology

Experimental uniaxial compression tests were conducted on 30 granite specimens which were divided into 6 groups based on the temperature of heat treatment. The Young's modulus and longitudinal wave velocity for each of the granite specimens were measured before heating. The specimens were then heat treated at different temperature and atmospheric pressure. Samples were heated at a rate of 5 °C/min until the nominal temperature was reached. The temperature was then maintained for 6 h before the specimen was cooled naturally in the furnace. After cooling, the mass, longitudinal wave velocity and Young's modulus of each of the specimens were measured. Uniaxial compression tests were performed under load control at a rate of 0.5 kN/s until the point of failure.

After uniaxial compression tests, the mineral composition and microstructure of the samples were obtained using XRD analysis and SEM in order to understand the failure mechanisms under uniaxial compression. Three-point bending tests were also carried out on 24 specimens in 6 groups prepared in the same way as the samples for uniaxial compression tests. Three-point bending test was carried out under load control at a rate of 50 N/s until the point of failure. The cracks and micro-cracks were observed by stereomicroscope.

UDEC software were used for building and simulating uniaxial compression test after high temperature.

1.2 Results

Comparing the experimental and simulated curves shows the results are quite different. For example, in Fig. 1 there are 3 distinct stages to the "200°C-experimental" curve and the post failure stage was not recorded in the laboratory testing. In contrast, the UDEC curves mainly correspond to elastic loading. After the elastic stage, a short elastic-plastic stage occurs before unstable fracture develops. As the unstable fracture occurs, the UDEC stress-strain curves fluctuate unstably and many macro-cracks appear in the model before final failure.

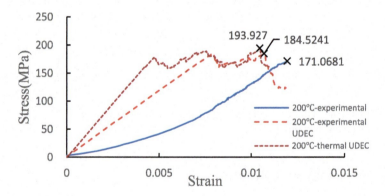

Fig. 1. Comparison of experimental and predicted stress-strain relationships for rocks subjected to heat treatment at 200 °C.

The experimental value for UCS was 171.0681 MPa, and the predicted values were 184.5241 MPa and 193.927 MPa for the "200 °C-experimental UDEC" and "200 °C-thermal UDEC" models, respectively. The experimental value was therefore the lowest, and the value from the "200 °C-thermal UDEC" model was the highest. Although the magnitude of the error is small, these results should be noted. The measured value of UCS is lower than the predicted value because the choice of input parameters can affect the results. As there are many parameters to consider, this increases the chance of a greater error.

Thermal damage causes thermal stress in a rock-type material. The 1000 °C heating cycle will be used as an example for comparing the high temperature and post-cooling behavior of rocks here. Figure 2 shows the predicted stress-strain relationships for granite at 1000 °C and for granite which has been cooled from 1000 °C to room temperature. These curves have a distinct difference. In particular, the elastic modulus is lower and the strain at peak stress is higher after cooling to room temperature. The strain and the UCS after cool-down are around 134.1% and 63.7% of the strain and UCS at 1000 °C, respectively. After heat treating at 1000 °C, the thermal stress in the specimen affects its mechanical properties.

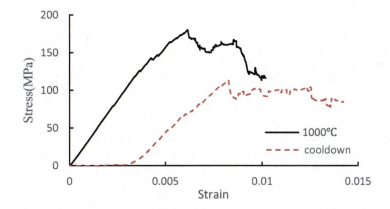

Fig. 2. Stress-strain relationship for rock at 1000 °C and post cool-down.

2 Conclusions

(1) The UCS of the granite specimens increases slightly for samples heat treated at 200 °C but then decreases linearly for heat treatment temperatures between 200 °C and 800 °C. The Young's modulus decreases with increasing heat treatment temperature up to 800 °C and then remains constant;

(2) The fracture toughness of granite decreases with increasing heat treatment temperature. It decreases linearly with increasing heat treatment temperature up to 600 °C and drops considerably above 600 °C due to the structural and chemical changes of the constituent minerals;

(3) The included angle between the loading direction and the micro- structural surfaces and micro-cracks affects the bearing capacity and other mechanical properties of

the granite. Cracks appear on the boundaries of the mineral crystals after heat treatment at temperatures below 800 °C, and the development of intracrystallines can be observed at temperatures above 800 °C;
(4) The modelling of the porosity which exists in rock materials affects the accuracy of the simulation results. If the material under study has a high porosity which cannot be neglected, a suitable value of porosity must be included in the numerical model;
(5) The onset of the unstable fracture development stage occurs when the number of failed shear contacts peaks and starts to decrease. At this point, shear cracks are subjected to tensile loads and turn into tensile cracks thus the number of tensile cracks increases and the specimen enters the unstable fracture development stage. The number of contacts under a tensile load then continues to increase until the point of failure.

References

1. Aliha, M.R.M., Bahman, A., Akhondi, S.H.: A novel test specimen for investigating the mixed mode I + III fracture toughness of hot mix asphalt composites-experimental and theoretical study. Int. J. Solids Struct. **90**, 167–177 (2016)
2. Balme, M.R., Roochi, V., Jones, C., Sammonds, P.R., Meredith, P.G., Boon, S.: Fracture toughness measurements on igneous rocks using a high-pressure, high-temperature rock mechanics cell. J. Volcanol. Geotherm. Res. **132**, 159–172 (2004)
3. Bauer, S.J., Handin, J.: Thermal expansion and cracking of three confined water-saturated igneous rocks to 800 °C. Rock Mech. Rock Eng. **16**, 181–198 (1983)
4. Cao, P., Liu, T.Y., Pu, C.Z., Lin, H.: Crack propagation and coalescence of brittle rock-like specimens with pre-existing cracks in compression. Eng. Geol. **187**, 113–121 (2015)
5. Chen, M., Zhang, G.Q.: Laboratory measurement and interpretation of the fracture toughness of formation rocks at great depth. J. Pet. Sci. Eng. **41**, 221–231 (2004)
6. Chen, Y.L., Ni, J., Shao, W., Azzam, R.: Experimental study on the influence of temperature on the mechanical properties of granite under uni-axial compression and fatigue loading. Int. J. Rock Mech. Min. Sci. **56**, 62–66 (2012)
7. Chen, Y.L., Ni, J., Jiang, L.H., Liu, M.L., Wang, P., Azzam, R.: Experimental study on mechanical properties of granite after freeze-thaw cycling. Environ. Earth Sci. **71**, 3349–3354 (2014)
8. Funatsu, T., Seto, M., Shimada, H., Matsui, K., Kuruppu, M.: Combined effects of increasing temperature and confining pressure on the fracture toughness of clay bearing rocks. Int. J. Rock Mech. Min. Sci. **41**, 927–938 (2004)

Conformable Derivative Modeling of Pressure Behavior for Transport in Porous Media

R. Wang, H. W. Zhou[✉], S. Yang, and Z. Zhuo

School of Mechanics & Civil Engineering,
China University of Mining and Technology, Xuzhou, China
zhw@cumtb.edu.cn

Abstract. A mathematical model is developed for studying the pressure behavior in porous media which is a modification of the diffusivity equation in concept of conformable derivative. The solution of the new model is semi-analytically obtained for constant-rate flow condition. Type curves of the dimensionless pressure are plotted by the Stehfest algorithm. The effects of conformable derivative order parameters on the pressure responses are discussed in detail.

Keywords: Pressure behavior · Conformable derivative · Anomalous diffusion

1 Introduction

Fractional calculus is widely used in describing anomalous diffusion phenomenon and non-linear mechanical problems in last decades [1–5]. By replacing a generalized time derivative order ($0 < \alpha \leq 1$), fractional diffusion equation involves the memory effects on the basis of continuous time random walk [6–8]. Transport in reservoirs fellows the similar diffusion process in which the pressure response is the major concern [9–11].

More recently, conformable derivative is introduced by Khalil [12], the new definition shows its advantages relative to the classical fractional derivative [13, 14]. Zhou [15] proposed a modified diffusion equation in terms of conformable derivative. The analytical solutions of the new model showed the non-linear relationship between mean square displacement and time which can capture the anomalous diffusion. In this study, we further expand the conformable diffusion equation to radial flow with constant-rate flow condition.

2 Definition and Property of Conformable Derivative

Fractional derivative has been defined in many ways to meet theoretical and engineering demands. The most widely known fractional derivatives are Riemann-Liouville, Caputo fractional derivative. Although the two fractional derivatives mentioned above have shown their advantage in describing non-linear phenomenon, the integral form definition seems complicated and lacks some of the basic properties that typical first derivative has. Conformable derivative is a local fractional derivative:

Definition 2.1. The conformable derivative of a function $f(t) : [0, \infty) \to \mathbb{R}$ and $t > 0$, $0 < \alpha \leq 1$ was defined by [12]

$$T_\alpha f(t) = \lim_{\varepsilon \to 0} \frac{f(t + \varepsilon t^{1-\alpha}) - f(t)}{\varepsilon} \tag{1}$$

Definition 2.2. Let $t_0 \in \mathbb{R}$ $0 < \alpha \leq 1$ and $f(t) : [0, \infty) \to \mathbb{R}$, the conformable Laplace transform of order α starting t_0 from of $f(t)$ is defined by [13]

$$L_\alpha^{t_0}\{f(t)\}(s) = \int_{t_0}^{\infty} e^{-s\frac{(t-t_0)^\alpha}{\alpha}} f(t)(t - t_0)^{\alpha-1} dt = F_\alpha^{t_0}(s) \tag{2}$$

Property 2.1. Let $t_0 \in \mathbb{R}$ $0 < \alpha \leq 1$ and $f(t) : [0, \infty) \to \mathbb{R}$ is differentiable real valued function. Then

$$L_\alpha\{T_\alpha(f(t))\}(s) = sF_\alpha(s) - f(a) \tag{3}$$

Property 2.2. Let $f(t) : [0, \infty) \to \mathbb{R}$, $t_0 \in \mathbb{R}$ and $0 < \alpha \leq 1$ and $L_\alpha^{t_0}\{f(t)\}(s) = F_\alpha(s)$ exsits. Then

$$F_\alpha^{t_0}(s) = L\{f(t_0 + (\alpha t)^{1/\alpha})\}(s) \tag{4}$$

where $L\{f(t)\}(s) = \int_0^\infty e^{-st} f(t) dt$.

3 Mathematical Model

Radial flow model in fractured rock with wellbore storage and skin effect is initially proposed by barker [16]. We take the same assumptions in this paper except for considering anomalous diffusion phenomenon. The time derivative is replaced by terms of conformable derivative.

All the parameters are defined in the Notation section. The governing equations are rewritten by the dimensionless variables:

$$\frac{\partial^\alpha p_D}{\partial t_D^\alpha} = \frac{1}{r_D} \frac{\partial}{\partial r_D}\left(r_D \frac{\partial p_D}{\partial r_D}\right) \quad 1 < r_D < \infty, \ t_D = 0 \tag{5}$$

Then we take the conformable Laplace transform with respect to t_D which gives:

$$s\widehat{p}_D = \frac{1}{r_D} \frac{\partial}{\partial r_D}\left(r_D \frac{\partial \widehat{p}_D}{\partial r_D}\right) \tag{6}$$

Where $\hat{p}_D$ is the conformable Laplace transform of $\bar{p}_D$ defined in. The general solution of Eq. (6) is

$$\hat{p}_D(r_D, s) = AI_0(r_D\sqrt{s}) + BK_0(r_D\sqrt{s}) \tag{7}$$

Where A and B are functions to be determined, I_0 and K_0 are modified Bessel function. With the zero-boundary condition, $A = 0$, and

$$B = \frac{1}{s} \cdot \frac{1}{C_D s K_0(\sqrt{s}) + (1 + C_D S s\sqrt{s})K_1(\sqrt{s})} \tag{8}$$

Finally, we obtain the pressure distribution function:

$$\hat{p}_D(r_D, s) = \frac{1}{s} \cdot \frac{K_0(r_D\sqrt{s})}{C_D s K_0(\sqrt{s}) + (1 + C_D S s)\sqrt{s}K_1(\sqrt{s})} \tag{9}$$

The wellbore pressure distribution function at bottom is:

$$\hat{p}_{wD}(s) = \frac{1}{s} \cdot \frac{K_0(\sqrt{s}) + S\sqrt{s}K_1(\sqrt{s})}{\sqrt{s}K_1(\sqrt{s}) + C_D s(K_0(\sqrt{s}) + S\sqrt{s}K_1(\sqrt{s}))} \tag{10}$$

The solutions in real space can be obtained by using the Stehfest numerical inversion algorithm [17]. It is worth noting that with numerical inversion procedure we don't get the pressure at time t. According to Property 2.2, it is the pressure at time t^α/α.

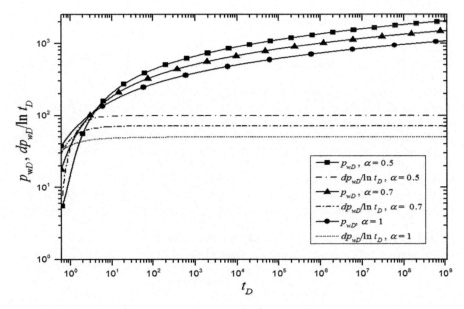

Fig. 1. Transient pressure curves at different conformable derivative order with $S = 5 \times 10^{-3}$ and $C_D = 5 \times 10^{-5}$.

Therefore, when the well is pumped at a constant rate, we can obtain the standard log-log type dimensionless curves of p_{wD} and $dp_{wd}/d\ln t_D$ in Fig. 1.

Figure 1 shows a quantitative picture of the pressure response and its logarithmic derivative curves. Both of them vary with the increase of conformable derivative order α. The higher α value leads to an earlier pressure response but a smaller one in long term.

4 Conclusion

In this paper, we have introduced a mathematical model with conformable derivative to describe pressure behavior in porous media. Based on conformable Laplace transform and Stehfest algorithm, a semi-analytical solution is obtained for constant-rate flow condition. A detailed analysis of conformable derivative order α shows the pressure responses is a power law function of time. The negative correlation between anomalous diffusion factor and Log derivative of pressure is also inferred.

Acknowledgments. The present work is supported by the National Natural Science Foundation of China (51674266), and the State Key Research Development Program of China (2016YFC 0600704). The financial supports are gratefully acknowledged.

References

1. Metzler, R., Glöckle, W.G., Nonnenmacher, T.F.: Phys. A Stat. Mech. Appl. **211**, 13–24 (1994)
2. Fomin, S., Chugunov, V., Hashida, T.: Transp. Porous Media **81**, 187–205 (2009)
3. Mendes, G.A., Lenzi, E.K., Mendes, R.S., da Silva, L.R.: Phys. A Stat. Mech. Appl. **34**, 6271–6283 (2005)
4. Zhou, H., Wang, C., Mishnaevsky Jr., L., Duan, Z., Ding, J.: Mech. Time Depend. Mater. **17**, 413–425 (2013)
5. Wu, W., Bauer, E., Kolymbas, D.: Mech. Mater. **23**, 45–69 (1996)
6. Zhang, Y., Benson, D.A., Meerschaert, M.M., LaBolle, E.M., Scheffler, H.P.: Phys. Rev. E Stat. Nonlin. Soft Matter Phys. **74**, 026706 (2006)
7. Magdziarz, M., Weron, A., Burnecki, K., Klafter, J.: Phys. Rev. Lett. **103**, 180602 (2009)
8. Gorenflo, R., Mainardi, F., Vivoli, A.: Chaos Solitons Fractals **34**, 87–103 (2007)
9. Park, H.W., Choe, J., Kang, J.M.: Energy Sources **22**, 881–890 (2000)
10. Raghavan, R.: J. Petrol. Sci. Eng. **80**, 7–13 (2011)
11. Razminia, K., Razminia, A., Torres, D.F.: Appl. Math. Comput. **257**, 374–380 (2015)
12. Khalil, R., Al Horani, M., Yousef, A., Sababheh, M.: J. Comput. Appl. Math. **264**, 65–70 (2014)
13. Abdeljawad, T.: J. Comput. Appl. Math. **279**, 57–66 (2015)
14. Atangana, A., Baleanu, D., Alsaedi, A.: Open Math. **13**, 889–898 (2015)
15. Zhou, H.W., Yang, S., Zhang, S.Q.: Phys. A Stat. Mech. Appl. **491**, 1001–1013 (2018)
16. Barker, J.A.: Water Resour. Res. **24**, 1796–1804 (1988)
17. Stehfest, H.: Commun. ACM **13**, 47–49 (1970)

Boundary Element Analysis of Geomechanical Problems

Sha Xiao[✉] and Zhongqi Yue

Department of Civil Engineering, The University of Hong Kong,
Hong Kong, People's Republic of China
xiao812@connect.hku.hk

Abstract. This paper presents a new boundary element analysis to solve elastic problems in layered halfspace whose material interface is non-horizontal to the horizontal boundary surface. Generalized Kelvin solutions (or Yue's solutions) are applied as the fundamental solutions. Infinite boundary elements and Kutt's numerical quadrature are applied to solve the influence of the far-field areas and strongly singular integrals, respectively. Finally, numerical examples show the influence of material interface for the distributions of displacements and stresses.

Keywords: Boundary element method (BEM) · Geomechanical problems
Generalized Kelvin solutions · Non-horizontally layered halfspace

1 Introduction

Geomechanical problems deals with the boundary value problems of geomaterials such as soil and rocks under various internal or external loadings made by human beings or nature. The complexity of geomechanical problems has led to the development of various numerical methods for solution and analysis of geoscientific and/or geotechnological interests. The boundary element method (BEM), as one of the efficient and accurate numerical methods, has been developed and used in analyzing the stress and displacement fields of geomechanical problems since 1970s. The most important advantage of the BEM as compared with other domain-based numerical methods is the reduction of the dimensions of the problems to be solved. The second advantage is its accuracy and its third advantage the capability of accurately modeling geomechanical problems of fullspace or/and halfspace.

On the other hand, the geomaterials of halfspace extent can be formed with different soils and rocks in layered formation. Such layered halfspaces are common material domain for geomechanical problems in geotechnical engineering. They can be further divided at least two cases according to the orientation of material interface in layered media. One is that the material interfaces are parallel to the boundary surface and the other is that the material interfaces are inclined to the boundary surface. BEM has been widely used to examine the distributions of displacements and stresses in layered halfspace under the action of various loads. The layered halfspace is normally modeled as a linear elastic, isotropic and horizontally layered halfspace [1]. There are a few publications on the elastic analysis of non-horizontally layered halfspace [2, 3]. In these existing BEM analysis, both the horizontal boundary surface and the material interface planes are needed to be discretized together, which can reduce the efficiency and accuracy of the numerical results.

2 Numerical Method

This paper adopts the generalized Kelvin solution (or Yue's solution) [4–6] as the fundamental singular solution in the boundary element analysis. Since the generalized Kelvin solution automatically satisfies the material interface conditions, the new BEM only needs to discretize the horizontal boundary surface. It does not need to discretize material interfaces, which is an advantage. Hence, its calculation can be more efficiency and accuracy. Moreover, the infinite boundary elements are employed to consider the influence of far-field area [7, 8] and Kutt's numerical quadrature [9] is used to calculate the strongly singular integral in the discretized boundary integral equation.

3 Numerical Examples

Numerical examples of the inclined layered halfspace are calculated to analyze the displacement and stress components of internal pointes. As shown in Fig. 1, an elastic halfspace has two different materials that are fully bonded at their non-horizontal interface. The dip angle of the non-horizontal interface plane is θ. The strike line of the inclined material interface plane with the horizontal boundary surface is located at the line of $x = 1.5$ m, $-\infty < y < +\infty$ and $z = 0$ m. A square area (1 m × 1 m) of its horizontal boundary surface is subject to a uniformly distributed footing pressure of 100 MPa.

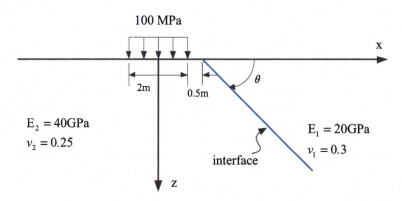

Fig. 1. Square footing on bi-material halfspace with a non-horizontally placed interface

The boundary surface is discretized into the mesh with 1192 nodes for 284 finite continuous elements, 36 finite discontinuous elements, 60 infinite continuous elements and 4 infinite discontinuous elements.

Using the proposed numerical method, all the components of the elastic fields at any point can be obtained. Here, only the displacement u_z and the stress σ_{xx} for different θ are presented.

In Fig. 2, it can be found that the value of u_z in the bi-material halfspace for each θ value is almost bonded by the two corresponding u_z values in the two homogeneous elastic halfspaces.

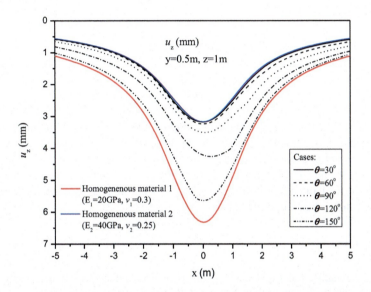

Fig. 2. Variation of the displacement u_z with the horizontal distance

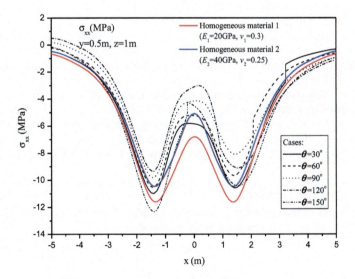

Fig. 3. Variation of the stress σ_{xx} with the horizontal distance

Figure 3 shows the variation of the stress σ_{xx} with the horizontal distance. Effect of the bi-materials to the stress is evident comparing to these of the two homogeneous elastic halfspaces. In particular, σ_{xx} is discontinuous across the material interface plane for θ = 30°, 60°, 120° or 150° and is non-smoothly continuous for θ = 90° across the material interface plane.

4 Conclusions

This paper used the new BEM to examine various geomechanical problems in layered elastic halfspace. In particular, the numerical examples are given for the displacements and stresses in a non-horizontally layered halfspace under distributed tractions over the horizontal boundary surface. These numerical results can quantify the influence of non-horizontally layered materials to the elastic fields induced by traction.

The new BEM is further applied to simulate elastic fields of underground tunnels and foundations in either horizontally or non-horizontally layered halfspace. Moreover, crack problems and contact problems in the layered halfspace can be analyzed. In conclusion, the new BEM can be used to examine many geomechanical problems in the layered elastic halfspace. More quantitative results and knowledge of geotechnical interests can be obtained accordingly.

References

1. Pan, E., Yang, B., Cai, G., Yuan, F.G.: Stress analyses around hoels in composite laminates using boundary element method. Eng. Anal. Bound. Elem. **25**, 31–40 (2001)
2. Pereira, O.J.B.A., Parreira, P.: Direct evaluation of Cauchy-principal-value integrals in boundary elements for infinite and semi-infinite three-dimensional domains. Eng. Anal. Bound Elem. **13**, 313–320 (1994)
3. Moser, W., Duenser, Ch., Beer, G.: Mapped infinite elements for 3D multi-region boundary element analysis. Int. J. Numer. Meth. Eng. **61**, 317–328 (2004)
4. Yue, Z.Q.: On generalized Kelvin solutions in a multilayered elastic medium. J. Elast. **40**, 1–43 (1995)
5. Yue, Z.Q.: Yue's solution of classical elasticity in n-layered solids: part 1, mathematical formulation. Front. Struct. Civ. Eng. **9**(3), 215–249 (2015)
6. Yue, Z.Q.: Yue's solution of classical elasticity in n-layered solids: part 2, mathematical verification. Front. Struct. Civ. Eng. **9**(3), 250–285 (2015)
7. Beer, G., Watson, J.O.: Infinite boundary elements. Int. J. Numer. Meth. Eng. **28**, 1233–1247 (1989)
8. Davies, T.G., Bu, S.: Infinite boundary elements for the analysis of the halfspace. Comput. Geomech. **19**, 137–151 (1996)
9. Kutt, H.R.: Quadrature formulae for finite-part integrals. Special Report WISK 178, National Research Institute for Mathematical Sciences, Pretoria (1975)

Numerical Simulation of the Bearing Capacity of a Quadrate Footing on Landfill

Dawei Xue[1,2] and Xilin Lü[1,2(✉)]

[1] Department of Geotechnical Engineering, Tongji University,
Shanghai 200092, China
xilinlu@tongji.edu.cn
[2] Key Laboratory of Geotechnical and Underground Engineering of Ministry
of Education, Tongji University, Shanghai 200092, China

Abstract. A deviatoric hardening plasticity model for municipal solid waste (MSW) is proposed to study the ultimate bearing capacity and failure mode of a quadrate footing on landfill. Taking into account the nonlinear characteristics of MSW strength, the power form shear strength criterion is adopted. The implicit integration scheme was used to update the stress increment, state parameters and Jacobian matrix in the finite element analysis. According to the triaxial test results, the material parameters were calibrated; and parameters were used to simulate the bearing behavior of landfill. Compared to the ideal elasto-plasticity based on the nonlinear failure criterion, the numerical simulation based on the deviatoric hardening plasticity gives out a smaller failure area but a slightly higher ultimate load.

Keywords: Municipal solid waste · Implicit integration · Hardening plasticity Bearing capacity

1 Introduction

Landfill is a common method for the disposal of municipal solid waste. Stringent regulations and public rejection of building landfills make it difficult to find adequate new sites for new landfills, and the existing solid waste landfill sites are often required to increase its capacity so as to store more MSW disposal [1].

Since the physical and mechanical properties of MSW are extremely complicated, a reasonable constitutive model must be established for the bearing capacity analysis of landfill sites [2]. Terzaghi's formula and finite element method is usually used to solve the bearing capacity of foundations [3]. However, the simulation mainly adopts ideal elasto-plasticity constitutive model and the failure criterion of often based on the Mohr-Coulomb criterion. Previous studies have shown the nonlinearity of shear strength and strain hardening characteristics of MSW [4, 5]. Obviously, the analysis of bearing capacity of foundation without considering these two characteristics is unreasonable. Existing models which can reflect this behavior of MSW have been proposed, but there is still lack of applications to practical engineering.

This paper proposes a hardening plasticity constitutive model [6] with nonlinear strength criterion. The landfill bearing capacity and failure modes were analyzed by the proposed model; the results were compared to those results obtained by an ideal elasto-plasticity model.

2 Hardening Plasticity for Municipal Solid Waste

The strength criterion F and the plastic potential function Q are as follows.

$$\begin{cases} F = q + Mp^{\xi} = 0 \\ Q = q - \frac{A_d M_c p}{1-A_d} \left[1 - \left(\frac{p}{p_0}\right)^{-(1-A_d)} \right] = 0 \end{cases} \quad (1)$$

The hardening rule is described by assuming that the incremental equivalent plastic shear strain obeys

$$d\varepsilon_s^p = \frac{-pM}{h_s G(M_f - M)} dM \quad (2)$$

where G is the elastic shear modulus; M_f defines the condition at peak state; hs is a fitting parameter.

3 Finite Element Analysis of Bearing Capacity of Landfill

3.1 Material Parameters Calibration

A medium age artificial MSW was adopted. According to tests data by Lü et al. [6], material parameters were calibrated. The initial shear modulus G_0 is 150, Poisson ratio υ is 0.2, void ratios e and e^* are 2.55 and 2.97, the strength parameters M_f and n are 1.6 and 0.41. M_c is 1.6, and A_d is 0.55. The hardening parameter h_s is 4.0. The comparison of the simulated results against experimental data is shown in Fig. 1.

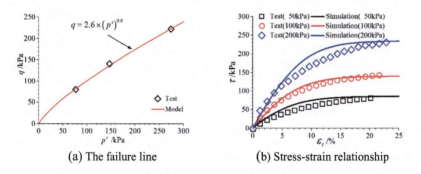

(a) The failure line (b) Stress-strain relationship

Fig. 1. Calibration of the material parameters (experimental data after [6])

3.2 Bearing Characteristics S by Different Models

The adopted finite element mesh for a square foundation is shown in Fig. 2, the length of the model is 10 m, the height is 8 m, and the width is 5 m. The size of the square footing is 2 m × 1 m. The bottom is fixed and lateral sides are horizontally constrained.

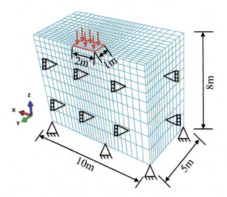

Fig. 2. Finite element model

The results are shown in Fig. 3. Figure 3(a) shows the results of the ideal elasto-plastic constitutive model built-in ABAQUS. The MSW around the bearing zone entered into plastic state, and the model shows a failure mode similar to the Terzaghi's mode, the results are close to the results of conventional soil. Figure 3(b) shows the results of the model in this paper. The plastic area of the model is limited to the bearing area and did not extend to the vicinity of the area. No slip surface was formed in the MSW below the bearing area. The failure mode of the model looks like a local one. Figure 4 shows the load-displacement curves calculated by different constitutive models. Comparing to the results by ideal elasto-plasticity, the deformation of the ground is smaller when a same force is imposed, and a higher ultimate load force can be obtained.

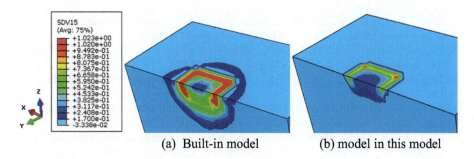

(a) Built-in model   (b) model in this model

Fig. 3. The equivalent plastic strain distribution by different models

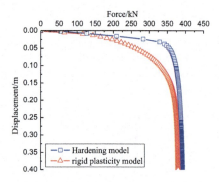

Fig. 4. Load-displacement curve of model MSW

4 Conclusions

This paper presented the formulation of a deviatoric hardening plasticity model for the analysis of bearing capacity and failure mode of landfill. The backward Euler implicit integration algorithm is employed.

The bearing behavior and failure mode are numerically simulated by both ideal elasto-plasticity and deviatoric hardening plasticity. The results show that the failure mode calculated by the ideal plastic model gives out a global failure mode, while the deviatoric hardening plasticity gives out a local one. The ultimate bearing capacity obtained from hardening plasticity is slightly higher than that of ideal elasto-plasticity model. It can be concluded that the nonlinear shear strength and hardening behavior of MSW present a different ultimate state of a landfill.

Acknowledgements. The financial supports by the National Natural Science Foundation of China (grant No. 41672270) and fundamental Research Funds for the Central Universities are gratefully acknowledged.

References

1. Zhan, T.T., Chen, Y.M., Ling, W.A.: Shear strength characterization of municipal solid waste at the Suzhou landfill. Chin. Eng. Geol. **97**(3–4), 97–111 (2008)
2. Machado, S.L., Vilar, O.M., Carvalho, M.F.: Constitutive model for long term municipal solid waste mechanical behavior. Comput. Geotech. **35**(5), 775–790 (2008)
3. Terzaghi, K., Peck, R.B., Mesri, G.: Soil Mechanics in Engineering Practice, 3rd edn. Wiley, New York (1996)
4. Zekkos, D., Bray, J.D., Riemer, M.F.: Drained response of municipal solid waste in large-scale triaxial shear testing. Waste Manag. **32**, 1873–1885 (2012)
5. Lü, X., Lai, H., Huang, M.: Nonlinear strength criterion for municipal solid waste. In: Oka, F., Murakami, A., Uzuoka, R. (eds.) Computer Methods and Recent Advances in Geomechanics, Kyoto, Japan, pp. 279–284. Taylor & Francis Group, London (2014)
6. Lü, X., Zhai, X., Huang, M.: Characterization of the constitutive behavior of municipal solid waste considering particle compressibility. Waste Manag. **69**, 3–12 (2017)

How Spatial Variability of Initial Porosity and Fines Content Affects Internal Erosion in Soils

Jie Yang[1,2], Zhen-Yu Yin[2,3(✉)], Pierre-Yves Hicher[2], and Farid Laouafa[1]

[1] INERIS, 60550 Verneuil en Halatte, France
[2] Research Institute of Civil Engineering and Mechanics (GeM), UMR CNRS 6183, Ecole Centrale de Nantes, 44300 Nantes, France
zhenyu.yin@ec-nantes.fr
[3] Department of Civil and Environmental Engineering, The Hong Kong Polytechnic University, Hung Hom, Kowloon, Hong Kong, China

1 Introduction

Internal erosion occurs when fines are pulled off by seepage forces and transported throughout the matrix of soil particles [1], which results in the degradation of soil strength [2–4]. This phenomenon has been widely studied experimentally and numerically over the last few decades [5–9]. In fact, soil nature is highly variable in space and then heterogenous in magnitude [10]. At the present time, few studies have been conducted to analyze the influence of variations of soil initial state (i.e. the initial porosity and the initial fraction of fines content) on the internal erosion behaviors. In this paper, a thermodynamically consistent continuum model of internal erosion is combined with the random field theory to investigate the erosion behavior of soil with spatially variable initial state of soil in one-dimensional condition.

2 Mathematic Model of Internal Erosion

The saturated porous medium is modelled as a material system composed of 4 constituents [11]: the stable fabric of the solid skeleton, the erodible fines, the fluidized particles and the pure fluid phase. The fines can behave either as a fluid-like (described as fluidized particles) or as a solid-like (described as erodible fines) material. Thus, a liquid-solid phase transition process has been accounted for in the present model by the introduction of a mass production term in the corresponding mass balances for erodible fines and fluidized particles.

A system of the mass balance equations is then given by the following expressions:

$$-\partial_t \phi + div(\mathbf{v}_s) - div(\phi \mathbf{v}_s) = \hat{n} \quad (1)$$

$$\partial_t(f_c) - \partial_t(f_c \phi) + div(f_c \mathbf{v}_s) - div(f_c \phi \mathbf{v}_s) = \hat{n} \quad (2)$$

$$\partial_t(c\phi) + div(c\mathbf{q}_w) + \partial_t(c\phi \mathbf{v}_s) = -\hat{n} \quad (3)$$

$$div(\mathbf{q}_w) + div(\mathbf{v}_s) = 0 \quad (4)$$

In which $\{\phi, f_c, c, \mathbf{q}_w\}$ denote the porosity, the fraction of erodible fines content, the concentration of fluidized particles and the total discharge of liquid; $\mathbf{v}_s$ denotes velocity of the soil skeleton deformation; $\hat{n}$ is the source term describing the exchange between the erodible fines and the fluidized particles.

$$\hat{n} = \hat{n}_e + \hat{n}_f \quad (5)$$

A model for the rate of the eroded mass $\hat{n}_e$ is given by the following relation [12]:

$$\hat{n}_e = -\lambda_e (1 - \phi)(f_c - f_{c\infty})|\mathbf{q}_w| \quad (6)$$

where λ_e is the material parameter, $f_{c\infty}$ is the maximum fines content fraction after long seepage time, assumed to be decreased with the increase in the hydraulic gradient [13]:

$$f_{c\infty} = f_{c0}[(1 - \alpha_1)\exp(-\alpha_2|\mathbf{q}_w|) + \alpha_1] \quad (7)$$

The rate of the filtration is proposed as:

$$\hat{n}_f = \lambda_f \frac{\phi - \phi_{min}}{\phi^\beta} c|\mathbf{q}_w| \quad (8)$$

where λ_f and β are the material parameters, ϕ_{min} is the minimum porosity of soil.
In this study, the flow in the porous medium is governed by Darcy's law:

$$\mathbf{q}_w = -\frac{k(\phi)}{\eta_k \bar{\rho}(c)}(\mathbf{grad}(p_w) - \bar{\rho}(c)\mathbf{g}) \quad (9)$$

where k denotes the intrinsic permeability of the medium (unit: m^2), η_k is the kinematic viscosity of the fluid (unit: m^{2}s^{-1}), p_w is the pore fluid pressure, $\mathbf{g}$ is the vector of the gravity field, and $\bar{\rho}$ is the density of the mixture defined as:

$$\bar{\rho} = c\rho_s + (1 - c)\rho_f \quad (10)$$

The intrinsic permeability k of the porous medium depends on the current porosity and fines content fraction. In our works we take the following expression [14]:

$$k = \frac{k_0}{[1 - f_{c0}(1 - \phi_0)]^{3m}} [1 - f_c(1 - \phi)]^{3m} \tag{11}$$

where m is the permeability parameter.

The mathematic model for a 1D internal erosion has been solved using a finite difference scheme. The computations of the following sections were carried out with $\Delta z = 0.004$ m (100 nodes) and $\Delta t = 1.0$ s to make sure the accuracy and convergence.

3 Probabilistic Analysis

The initial porosity ϕ_0 and the initial fraction of fines content f_{c0} are assumed to be lognormally distributed random variables. They are generated by the local average subdivision method which fully takes into account the spatial correlation, local averaging, and cross correlation. The generated random variables are mapped onto a finite-difference mesh and numerical simulations based on Monte Carlo method are performed.

The following parameters for both ϕ_0 and f_{c0} have been used for the probabilistic analysis in this study: (1) coefficient of variation v: 0.05, 0.075, 0.1, 0.125, 0.15 and 0.175; (2) relative correlation length $\Theta = \theta/L$: 0.25, 0.5, 1.0, 2.0 and 4.0; (3) cross correlation coefficient ρ between ϕ_0 and f_{c0}: -1.0 and 1.0.

To keep accuracy and run-time efficiency, the sensitivity of results to the number of Monte Carlo simulations was examined. The results show that 2000 simulations are enough to give reliable and reproducible estimates.

Figure 1 compares the probability of blockage during erosion with different uncertainties of ϕ_0 and f_{c0}. The probability of blockage during erosion is defined by counting the number of simulations in which the calculated hydraulic conductivity is smaller than its initial value, and then divided by the total number of simulations.

The results indicate that increasing the uncertainty in ϕ_0 and f_{c0} will always increases the probability of blockage during erosion. This can be explained by the fact that greater uncertainty leads to more dramatic changes in soil porosity and permeability at the interface of different layers, which in turn promotes the capture of the fine particles transported in water flow. Moreover, the ragged distributed ϕ_0 and f_{c0} with lower spatial correlation lengths make the layered system more inconsistent, and it therefor accelerates the formation of the filter cakes at the interface.

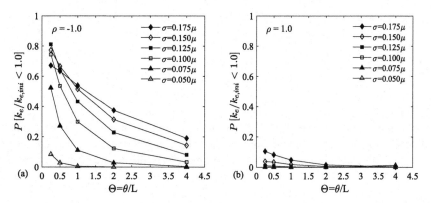

Fig. 1. 1D probabilistic results, comparison of probability of the appearance of blockage during erosion simulations: (a) $\rho = -1.0$; (b) $\rho = 1.0$

4 Conclusions

The influence of the initial state of soil characterized by its porosity and fine content, on the suffusion and filtration behaviors during internal erosion process was investigated in this study. Probabilistic analysis shows that greater uncertainty leads to more dramatic changes of soil porosity and the permeability at the interface of different layers, which in turn promotes the capture of the fine particles transported in water flow. Higher spatial correlation length makes the layered system more uniform. However, the ragged distributed ϕ_0 and f_{c0} with lower spatial correlation lengths make the layered system more inconsistent. It leads to a rougher transition between the high erodible zones and the low erodible zones and therefor accelerates the formation of the filter cakes at the interface and increases the probability of blockage. Moreover, negatively correlated ϕ_0 and f_{c0} are more likely to cause blockage.

References

1. Bonelli, S., Marot, D.: On the modelling of internal soil erosion. In: The 12th International Conference of International Association for Computer Methods and Advances in Geomechanics (IACMAG), 7p (2008)
2. Chang, C.S., Yin, Z.-Y.: Micromechanical modeling for behavior of silty sand with influence of fine content. Int. J. Solids Struct. **48**, 2655–2667 (2011)
3. Yin, Z.-Y., Zhao, J., Hicher, P.-Y.: A micromechanics-based model for sand-silt mixtures. Int. J. Solids Struct. **51**, 1350–1363 (2014)
4. Yin, Z.-Y., Huang, H.-W., Hicher, P.-Y.: Elastoplastic modeling of sand–silt mixtures. Soils Found. **56**, 520–532 (2016)
5. Reddi, L.N., Lee, I.-M., Bonala, M.V.: Comparison of internal and surface erosion using flow pump tests on a sand-kaolinite mixture. Geotech. Test. J. **23**, 116–122 (2000)
6. Sterpi, D.: Effects of the erosion and transport of fine particles due to seepage flow. Int. J. Geomech. **3**, 111–122 (2003)

7. Bendahmane, F., Marot, D., Alexis, A.: Experimental parametric study of suffusion and backward erosion. J. Geotech. Geoenviron. Eng. **134**, 57–67 (2008)
8. Moffat, R., Fannin, R.J., Garner, S.J.: Spatial and temporal progression of internal erosion in cohesionless soil. Can. Geotech. J. **48**, 399–412 (2011)
9. Marot, D., Rochim, A., Nguyen, H.-H., Bendahmane, F., Sibille, L.: Assessing the susceptibility of gap-graded soils to internal erosion: proposition of a new experimental methodology. Nat. Hazards **83**, 365–388 (2016)
10. Dasaka, S., Zhang, L.: Spatial variability of in situ weathered soil. Géotechnique **62**, 375 (2012)
11. Schaufler, A., Becker, C., Steeb, H.: Infiltration processes in cohesionless soils. ZAMM-J. Appl. Math. Mech Zeitschrift für Angewandte Mathematik und Mechanik **93**, 138–146 (2013)
12. Uzuoka, R., Ichiyama, T., Mori, T., Kazama, M.: Hydro-mechanical analysis of internal erosion with mass exchange between solid and water. In: Proceedings of 6th International Conference on Scour and Erosion, Paris, France, pp. 655–662 (2012)
13. Cividini, A., Bonomi, S., Vignati, G.C., Gioda, G.: Seepage-induced erosion in granular soil and consequent settlements. Int. J. Geomech. **9**, 187–194 (2009)
14. Revil, A., Cathles, L.: Permeability of shaly sands. Water Resour. Res. **35**, 651–662 (1999)

CFD Analysis of Free-Fall Ball Penetrometer in Clay

Yuqin Zhang and Jun Liu[✉]

State Key Laboratory of Coastal and Offshore Engineering,
Dalian University of Technology, Dalian 116024, Liaoning, China
junliu@dlut.edu.cn

Abstract. The free fall penetrometer (FFP) is used to measure the soil undrained shear strength by penetrating into the soil with its kinetic energy and self-weight. The free fall ball penetrometer (FFB) takes advantages of both FFP and ball penetrometer. However, the soil-FFB interaction is rather complex, which refers to the soil strain rate effect and drag force. Therefore, it is necessary to analyze the forces acting on the FFB to improve its practicability and the accuracy of soil strength measurement. The dynamic penetration of the ball penetrometer in the uniform clay was simulated in this study using the commercial software ANSYS CFX 17.0, which is based on the computational fluid dynamics (CFD) approach. A thin layer element method was proposed to simulate the ball-soil frictional interaction. In the numerical model, the soil was modeled as non-Newtonian fluid incorporating the strain rate effect. The bearing capacity factor (N_b) and drag coefficient (C_d) were calculated during the ball dynamic penetration within the soil. Different ball velocities, soil strengths and densities, interface frictional coefficients and soil strain rate parameters were considered to investigate their effects on the ball strain rate effect. The fitted formulas of the ball bearing capacity factor, the rate effect coefficient and the drag coefficient for the FFB were established based on the present numerical results.

Keywords: Free fall ball penetrometer · Rate effect · Drag force
Soil undrained strength · CFD

1 Introduction

Strength characterization of the seabed soil is crucial for offshore engineering. Recently, the concept of the free fall penetrometer (FFP) was proposed to improve the testing efficiency compared with traditional static in-situ penetration tests. The FFP penetrates into the seabed by its self-weight and the kinetic energy gained during free fall in water. Morton et al. [1, 2] conducted physical tests to model the dynamic penetration of the free fall ball penetrometer (FFB) within the soil, and the testing results indicated that the measured soil strength agreed well with the static testing results. Compared with the traditional static penetration tests, the dynamic penetration of the FFB is complex as the soil strain rate effect and drag force should be considered under high penetration velocity. To further illustrate dynamic penetration of the FFB in the soil, the computational fluid dynamics (CFD) approach is used in this study to investigate the shear strain rate effect and drag force on the FFB.

2 Numerical Simulation Details

The commercial CFD software ANSYS CFX 17.0 was used in this study. In the simulation, the soil was modelled as non-Newtonian fluid incorporating the soil strain rate effect. A thin layer element method was used to simulate the ball-soil frictional interaction, which has been used by Liu and Zhang [3] in the FFP simulation. The dynamic mesh and subdomain method was applied to improve the computational efficiency and precision as described in Liu and Zhang [3]. Figure 1(a) shows the model dimension and soil shear strength in different domains (α is the frictional coefficient). The ball diameter, D, is 113 mm and the minimum mesh size is $D/80$, which is the same as the thickness of the thin layer element (frictional domain), t. Five layers of prism meshes were set outside the ball to improve the computational precision with a total thickness of 0.25 mm.

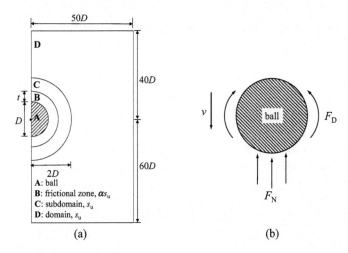

Fig. 1. Problem definition: (a) sketch diagram of domain set up (non-proportional); (b) soil forces acting on the ball.

The model domain is filled with soil and the ball is pre-embedded into the soil initially. The ball moves downwards with constant velocity (see Fig. 1(b)) and the soil undrained shear strength (s_{u0}) is uniform with depth. The soil shear strain rate effect is described in Eq. (1):

$$s_u = \left(1 + \lambda \log \frac{\dot{\gamma}}{\dot{\gamma}_{ref}}\right) s_{u0} = R_f s_{u0} \qquad (1)$$

where λ is the coefficient of the soil strain rate effect, R_f is the rate effect coefficient, $\dot{\gamma}$ is the shear strain rate and $\dot{\gamma}_{\text{ref}}$ is the reference shear strain rate. In this study, $\dot{\gamma}_{\text{ref}}$ is 0.024 s^{-1}. The soil is weightless in this study.

Table 1 shows the values of the studied parameters. Cases with $\lambda = 0$ are set to investigate the ball bearing capacity factor (N_c) and drag coefficient (C_d).

Table 1. Ranges of selected parameters.

Parameter	Value
Penetration velocity, v, m/s	2, 6, 12, 24, 30
Soil undrained shear strength, s_{u0}, kPa	2, 4, 10, 15, 20
Soil density, ρ, g/cm^3	1.4, 1.6, 1.8
Frictional coefficient, α	0.2, 0.6, 1.0
Coefficient of shear strain rate effect, λ	0, 0.1, 0.2, 0.3

3 Results and Discussion

3.1 Ball Bearing Capacity Factor (N_c) and Drag Coefficient (C_d)

When the soil strain rate effect is not considered (i.e. $\lambda = 0$), the forces acting on the ball (Fig. 1(b)) could be expressed as:

$$F = F_N + F_D = N_c s_{u0} A_t + \frac{1}{2} C_d \rho v^2 A_t \qquad (2)$$

where F is the total force acting on the ball, F_N is the bearing force, F_D is the drag force, A_t is the ball projected area. When the ball penetration velocity is rather low (such as 0.02 m/s), the drag force term in Eq. (2) could be ignored. Hence the factor of N_c could be calculated. Figure 2(a) shows the trend of N_c with α and the relationship could be expressed as:

$$N_c = 10.67 + 4.16\alpha \qquad (3)$$

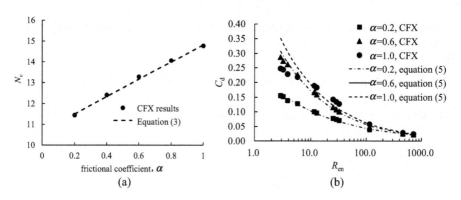

Fig. 2. Numerical results: (a) bearing capacity factor N_c vs. α; (b) drag coefficient C_d vs. R_{en}.

For other cases with high penetration velocity, the value of C_d could be calculated by Eqs. (2) and (3). Figure 2(b) shows the relationship between C_d and the non-Newtonian Reynolds number ($R_{en} = \rho v^2/s_{u0}$), which could be expressed as:

$$C_d = \frac{a}{R_{en}^b}, a = 0.038 + 1.42\alpha - 0.588\alpha^2, b = 0.284 + 0.5\alpha - 0.294\alpha^2, \quad (4)$$

Although the values of C_d for $\alpha = 1.0$ with $R_{en} < 10$ diverge from Eq. (4), the corresponding drag force is low and hence the divergences could be ignored.

3.2 Results of the Rate Effect Coefficient, R_f

When the soil strain rate effect is taken into consideration (i.e. $\lambda > 0$), the forces acting on the ball could be expressed as Eq. (5).

$$F = R_f N_c s_{u0} A_t + \frac{1}{2} C_d \rho v^2 A_t \quad (5)$$

The rate effect coefficient R_f could be calculated by Eq. (3–5). Figure 3 shows the results of the calculated $R_f/R_{f,e}$ ($R_{f,e}$ is the effective rate effect coefficient expressed in Eq. (6)) for $\lambda = 0.2$ and the relationship of R_f and R_{en} could be expressed in Eq. (7).

$$R_{f,e} = 1 + \lambda \log \frac{v/D}{\dot{\gamma}_{ref}} \quad (6)$$

$$R_f = (1 + \lambda \log \frac{v/D}{\dot{\gamma}_{ref}}) \cdot (1.0275 + 0.0125\alpha) \cdot R_{en}^{\lambda(0.0165 + 0.0825\alpha)} \quad (7)$$

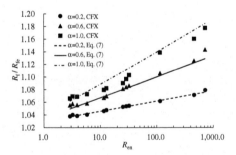

Fig. 3. Results of $R_f/R_{f,e}$ vs. R_{en} for $\lambda = 0.2$.

4 Conclusions

This study simulated the dynamic penetration of the FFB within the uniform soil using the CFD approach. The ball bearing factor, N_c, was calculated and the effects of drag force and soil strain rate effect were also investigated. Both the expressions of C_d and

R_f were established. Further study of continuous dynamic penetration within the soil with varied strength characterizations is in the pipeline.

References

1. Morton, J.P., O'Loughlin, C.D., White, D.J.: Centrifuge modelling of an instrumented free-fall sphere for measurement of undrained strength in fine-grained soils. Can. Geotech. J. **53**(6), 918–929 (2015)
2. Morton, J.P., O'Loughlin, C.D., White, D.J.: Estimation of soil strength in fine grained soils by instrumented free-fall sphere tests. Géotechnique **66**(12), 1–10 (2016)
3. Liu, J., Zhang, Y.: CFD simulation on the penetration of FFP into uniform clay. Chin. J. Theor. Appl. Mech. **50**(1), 1–10 (2018)

Multiscale Modeling of Large Deformation in Geomechanics: A Coupled MPM-DEM Approach

Jidong Zhao[1,2(✉)] and Weijian Liang[1]

[1] The Hong Kong University of Science and Technology, Hong Kong, China
jzhao@ust.hk
[2] Hong Kong University of Science and Technology Shenzhen Research Institute, Shenzhen, China

Abstract. Numerical modeling of geotechnical problems frequently requires the consideration of large deformation soil responses. The Material Point Method (MPM) has gained increasing popularity recently over many other methods such as Finite Element Method (FEM) in continuum modeling of large deformation problems in geomechanics. This study presents a new enrichment of MPM with multi-scale predictive capabilities. We propose a computational multiscale scheme based on coupled MPM and DEM following a similar concept of the FEM-DEM coupling scheme [2, 3]. The MPM is employed to treat a typical boundary value problem in geomechanics that may experience large deformations, and the DEM is used to derive the nonlinear material response required by MPM for each of its material points. The proposed coupling framework helps avoid phenomenological constitutive assumptions in typical MPM, while inherits its advantageous features in tackling large deformation problems over FEM (e.g., no need for re-meshing to avoid highly distorted mesh in FEM). Importantly, it provides direct micro- macro linking for us to understand complicated behavioral changes of granular media over all deformation levels, from initial elastic stage en route to large deformation regime before failure. Several demonstrative examples are shown to highlight the advantages of the new MPM-DEM framework, including the collapse of a soil column, biaxial shear tests and different failure modes observed in a footing problem in geotechnical problems.

Keywords: Hierarchical multi-scale modeling · MPM-DEM coupling Large deformation · Geomechanics

1 Hierarchical Multiscale Coupling of MPM and DEM

Consideration of large deformation behavior of soils is critical for the modeling and analysis of a wide range of geotechnical failures. Recently, the Material Point Method (MPM) [1] has been particularly popular in geomechanics and geotechnical engineering for large deformation analysis. Belonging to a class of mesh-free methods, MPM typically employs a collection of material points to discretize the domain of a boundary value problem while using a regular mesh to interpolate the field variables for each computational step. The deformation mesh is restored to its initial position to resolve the mesh distortion issue commonly encountered by other mesh-based methods such as FEM. MPM has been proven efficient in capturing the complicated behavior of soils undergoing large deformation and has been applied in a wide or geotechnical problems. A major pitfall associated with continuum-based numerical methods, including MPM, is the necessity of assuming proper constitutive relations to describe the material behavior for each material (or integration) point. In order to tackle practical problems with complicated material behaviors, the constitutive models developed for these continuum approaches have become exceedingly complex, always involving phenomenological parameters that lack adequate physical significance and are difficult for calibration.

In this paper, we introduce a novel multiscale modeling approach based on hierarchical coupling of MPM and DEM to resolve the above issues in large deformation simulation of geotechnical problems. Following a similar concept that was based on coupled FEM-DEM [2, 3], we employ MPM to treat a typical boundary value problem in geomechanics that may experience large deformations, and use the DEM to derive the nonlinear material response required by MPM for each of its material points. A schematic of the hierarchical multiscale coupling scheme between MPM and DEM is depicted in Fig. 1. Based on such a hierarchical coupling framework, we may avoid phenomenological constitutive assumptions commonly needed by a continuum-based MPM, while retaining its advantageous features in tackling large deformation problems over FEM (e.g., no need for re-meshing to avoid highly distorted mesh in FEM). The MPM-DEM coupling scheme offers a straightforward way on micro-macro bridging for us to understand complicated behavioral changes of granular media over all deformation levels, from initial elastic stage en route to large deformation regime before failure. We will demonstrate with examples to highlight the advantages of the new MPM-DEM framework, including simulations of the failure of sand samples in biaxial shear tests and the general failure modes in a footing foundation.

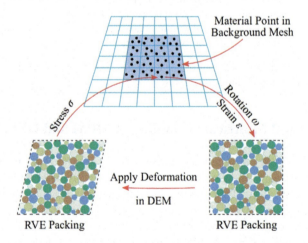

Fig. 1. Hierarchical Multiscale coupling of MPM × DEM for modeling large deformation in geomechanics (c.f., *Guo and Zhao, 2014, 2016a*).

2 MPM-DEM Simulation of Large Deformation Problems

Figure 2 presents the simulation by the coupled MPM-DEM for a dense sand sample under drained biaxial shear, where the progressive development of localization in terms of contours of shear strain, void ratio, cumulative rotation and displacements through different deformation level are depicted. At the final failure state (at a global vertical strain of 15%), the shear strain reaches as high as 250% within the shear bands, while the coupled MPM-DEM scheme remains to provide reasonable predictions. Note that the four quantities in Fig. 2 demonstrate consistent developing patterns at all four strain levels.

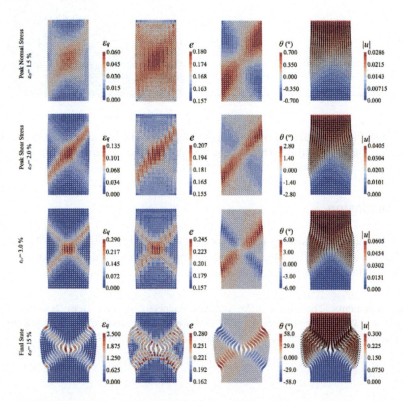

Fig. 2. MPM × DEM multiscale modeling of shear localization in a dense sand sample subjected to drained biaxial compression [4]. From top to bottom showing the the localizing processes at (1) the peak normal stress state (1.5% vertical strain), (2) the peak shear stress state (2.0% vertical strain), (3) 3.0% vertical strain level and (4) the final failure state (15% vertical strain). From left to right are the contours for shear strain, void ratio, total rotation and absolute displacement, respectively.

Figure 3 provides a further coupled MPM-DEM simulation of the problem of rigid footing foundation on dense sand, where we capture the final general failure mode at excessive settlement of the foundation. The maximum shear strain reaches 100% within the shear bands and adjacent to the side-edge of the foundation. The situation is known to be notoriously troublesome for FEM simulations but can be effectively tackled by the MPM-DEM coupling.

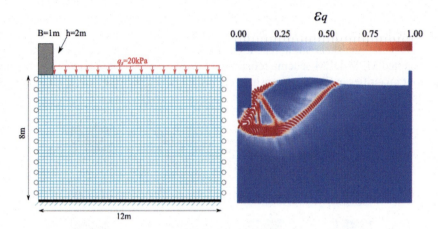

Fig. 3. MPM × DEM multiscale simulation of a footing foundation experiencing general failure at large deformation [4]. (a) MPM background mesh, (b) simulated general failure mode of footing foundation at large deformation in term of shear strain.

3 Conclusion

A novel multiscale modeling scheme based on hierarchical coupling MPM and DEM has been proposed, with demonstrated examples on its predicative capabilities in capturing failure of sand at large deformation.

Acknowledgments. The author acknowledges financial supports from National Science Foundation of China under Project No. 51679207, Research Grants Council of Hong Kong under GRF Project N. 16210017 and a Theme-based Research Project No. T22-603/15N.

References

1. Sulsky, D., Chen, Z., Schreyer, H.L.: Particle method for history-dependent materials. Comput. Meth. Appl. Mech. Eng. **118**(1–2), 179–196 (1994)
2. Guo, N., Zhao, J.D.: A coupled FEM/DEM approach for hierarchical multiscale modelling of granular media. Int. J. Numer. Meth. Eng. **99**, 789–818 (2014)
3. Guo, N., Zhao, J.D.: Parallel hierarchical multiscale modelling of hydro-mechanical problems for saturated granular soils. Comput. Meth. Appl. Mech. Eng. **305**, 37–61 (2016)
4. Liang, W., Zhao, J.D.: Multiscale modeling of large deformation in geomechanics. Int. J. Numer. Anal. Meth. Geomech. (2018, To be submitted)

Effect of Principal Stress Rotation on the Wave-Induced Seabed Response Around a Submerged Breakwater

Hongyi Zhao[1], Jianfeng Zhu[2(✉)], and Dong-Sheng Zheng[1]

[1] College of Harbor, Coastal and Offshore Engineering, Hohai University, Nanjing, China
hyzhao@hhu.edu.cn
[2] Faculty of Architectural Civil Eng. and Environment, Ningbo University, Ningbo, China
zhujianfeng0811@163.com

Abstract. Principal stress rotation (PSR) is one of the main features for the stress condition in the soil element underneath a structure subject to cyclic waves. This paper aims to present an integrated numerical model to investigate the influence of principal stress rotation (PSR) on the wave-induced soil response around a submerged rubble mound breakwater. In the developed model, the VARANS equation is used for governing the flow motion inside and outside the porous media; Biot's dynamic equation is used for linking the soil skeleton and pore fluid interaction; the modified PZIII model considering the impact of PSR is used to reproduce the foundation behavior under cyclic shearing. Numerical results indicate that ignoring the PSR involved in the wave-seabed–structure interactions (WSSI) will significantly underestimate the cumulative shear strains in seafloor and subsequent build-up of pore water pressure, especially in the region underneath the breakwater.

Keywords: Principal stress rotation (PSR) · Sandy seabed
Wave-seabed–structure interactions (WSSI)

1 Introduction

Over the past few decades, a large number of marine structures such as breakwaters, pipelines, caisson structures and oil platform etc. have been constructed in offshore area accompany with the rapid development of worldwide ocean engineering construction. The foundation design of these structures presents a series of difficulties due to the complexity of stress conditions for soil elements subject to cyclic waves. When ocean waves propagate over a porous seabed, the magnitudes of principal stress in the soil element are cyclically altered. This may further lead to a continuous rotation of the principal stress orientation due to the alternating changeover of the horizontal shear stress and the stress difference between the vertical and horizontal normal stresses. Generally, the deformation characteristics of soil deposits are intensively related to the change in the principal stress direction and its magnitude, in particular, may be purely related to the principal stress rotation (PSR) even without a change in principal stress magnitudes under wave loadings [1]. Continuous PSR can also generate excess pore water pressure and cumulative shear strain when the drainage condition is impeded [2, 3], which have been a subject of concern for clarification of liquefaction mechanism around marine structures [4]. Understanding the effects of principal stress rotation on the mechanism of wave-seabed-structure interactions could be beneficial for coastal engineers involved in the stability design of marine structures.

In this study, an integrated numerical model is developed to investigate the nonlinear dynamics of seabed foundation underneath a submerged rubble mound breakwater subjected to ocean waves. In the developed model, the Volume-Averaged Reynold-Averaged Navier-Stokes (VARANS) equations [5] are used to govern wave motion; the dynamic Biot's equations (so-called "u-p" approximation) proposed by [6] is used for linking the solid-pore fluid interaction in a porous seabed foundation; the modified generalized plasticity model PZIII [7] considering the impact of PSR is used to reproduce the nonlinear soil behavior under cyclic shearing, in which the plastic strain generated by the principal stress rotation was considered to be an additional term in the constitutive relationship of soil and the normalized loading direction and plastic flow direction were determined based on the stress tensor invariants. The VARANS equation and Biot's theory are combined together through the continuity of pressure at the interface between the fluid domain and the porous medium [4]. Adopting the developed model, the effects of PSR-induced soil deformations on the seabed response around a submerged breakwater under ocean waves will be examined.

2 Theoretical Model

In this study, only the rotation of ψ in the x-z plane is considered. The partial derivative ψ with respect to stresses can be expressed as

$$\frac{\partial \psi}{\partial \sigma'_x} = \frac{\tau_{xz}}{\left(\sigma'_x - \sigma'_z\right)^2 + \tau_{xz}^2}, \frac{\partial \psi}{\partial \sigma'_z} = -\frac{\tau_{zx}}{\left(\sigma'_x - \sigma'_z\right)^2 + \tau_{xz}^2}$$

$$\frac{\partial \psi}{\partial \sigma'_y} = 0, \frac{\partial \psi}{\partial \tau_{xz}} = \frac{\left(\sigma'_x - \sigma'_z\right)/2}{\left(\sigma'_x - \sigma'_z\right)^2 + \tau_{xz}^2} = \frac{\partial \psi}{\partial \tau_{zx}}$$

(1)

The plastic modulus $H\psi$ for unloading can be assumed to be formulated as

$$H_{\psi u} = \begin{cases} H_{\psi u0} \left[\frac{M_g(\psi)}{(q/p')_u}\right]^{\gamma_u} & \text{for } \left|\frac{M_g(\psi)}{(q/p')_u}\right| > 1 \\ H_{\psi u0} & \text{for } \left|\frac{M_g(\psi)}{(q/p')_u}\right| \leq 1 \end{cases}$$

(2)

where $H_{\psi u0}$ is a material parameter, which can be assumed to equal to H_0 and $(q/p')_u$ represents the effective stress ratio at which unloading takes place. For the unloading process, Eq. (2) is modified as follows

$$\frac{\partial g}{\partial p'} = -\left|\frac{\partial g}{\partial p'}\right| \qquad (3)$$

where $\partial g/\partial q$ and $\partial g/\partial \theta$ are the same as before.

The overall constitutive relations are thus expressed as

$$d\sigma_{ij} = \begin{cases} \left(D^e_{ijkl} - \dfrac{D^e_{ijmn}m_{mn}n_{st}D^e_{stkl}}{H_L + n_{st}D^e_{stkl}m_{kl}}\right)d\varepsilon_{kl}, & \text{for loading} \\ \left(D^e_{ijkl} - \dfrac{D^e_{ijmn}m_{mn}n_{st}D^e_{stkl}}{H_U + n_{st}D^e_{stkl}m_{kl}}\right)d\varepsilon_{kl}, & \text{for unloading} \end{cases} \qquad (4)$$

where D^e_{ijkl} is the elastic stiffness tensor. λ is Lame's constant, K is the body modulus, and Kev0 is the initial value of the body modulus.

3 Results and Discussions

Figure 1 shows the snapshots of pore water pressures in a porous seabed foundation around a submerged breakwater at two typical times t/T = 15 and 25, respectively. The results indicate that the developed model is capable of reproducing the elastoplastic behavior of sandy seabed under cyclic wave loading involving the rotation of principal stress axes. The pore water pressure, not surprisingly, is more likely to build up and can accumulate to a large value at the nodes of standing waves that are formed in front of a breakwater due to wave interferences between incident and reflected waves, compared to those at the antinode. Because PSR becomes increasingly pronounced at the standing wave nodes due to the large shear strains induced by waves, subsequently enhancing the plastic strains and build-up of pore pressure in this area. On the other hand, the original PZIII ignoring the effects of PSR significantly underestimate the build-up of pore pressure, particularly at the standing wave nodes and those underneath the breakwater, which may lead to unsafe design in the construction of offshore structures.

Figure 2 shows the effective stress path, deviatoric stress path and shear stress-strain response at seabed soil underneath a submerged breakwater subject to ocean waves. It is found that, not surprisingly, the effective stresses gradually decreases during the process of cyclic wave; the effective stress paths tend to approach the zero status (liquefaction) after approximately 30 wave repetitions. Unlike the circular stress path of soil in an infinite seabed, the stress path in a seabed of finite thickness considered in this study shows a series of ellipse, indicating that not only the pure rotation of principal stress, the value of principal stress changes as well. The shear stress-strain relationship indicates that the soil become softened (degradation of soil stiffness) and continuously lose its shear resistance during the process of cyclic loading, given rise to the amplitude of shear strain in accordance with the occurrence of soil liquefaction. The accumulated shear strain is significant in the region underneath the breakwater, reaching 0.5% during the later loading stage.

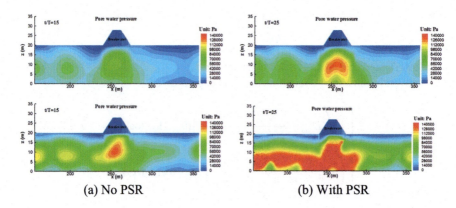

Fig. 1. Snapshots of pore water pressure around a submerged breakwater.

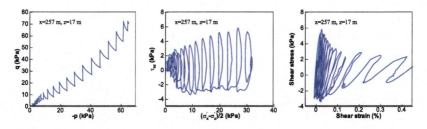

Fig. 2. Distribution of effective stress path, deviatoric stress path and shear stress-strain response in the seabed soil underneath the breakwater.

4 Conclusions

This study presents an integrated numerical model for the wave-seabed-structure interactions around a submerged breakwater. Numerical results indicate that the developed model is capable of capturing the wave-induced dynamic soil response around a submerged breakwater concerning the impact of PSR. Ignoring the PSR-induced soil deformations may significant underestimate the pore pressure development and subsequent liquefaction, especially at the standing wave nodes and those in the region underneath a submerged breakwater.

References

1. Ishihara, K., Towhata, I.: Sand response to cyclic rotation of principal stress directions as induced by wave loads. Soils Found. **23**(4), 11–26 (1983)
2. Towhata, I., Ishihara, K.: Undrained strength of sand undergoing cyclic rotation of principal stress axes. Soils Found. **25**(2), 135–147 (1985)
3. Wang, Z., Yang, Y.M., Yu, H.S.: Effects of principal stress rotation on the wave–seabed interactions. Acta Geotech. **12**(1), 97–106 (2017)

4. Zhao, H.Y., Jeng, D.-S., Liao, C.C., Zhu, J.F.: Three-dimensional modeling of wave-induced residual seabed response around a mono-pile foundation. Coast. Eng. **128**, 1–21 (2017)
5. Jesus, M., Lara, J., Losada, J.: Three-dimensional interaction of waves and porous coastal structures: Part I: numerical model formulation. Coast. Eng. **64**, 57–72 (2012)
6. Zienkiewicz, O.C., Mroz, Z.: Generalized plasticity formulation and applications to geomechanics. In: Mechanics of Engineering Materials. Chichester: John, Cambridge, UK (1984)
7. Pastor, M., Zienkiewicz, O.C., Chan, A.H.C.: Generalized plasticity and the modelling of soil behaviour. Int. J. Numer. Anal. Methods Geomech. **14**, 151–190 (1990)

Numerical Analysis of Spudcan-Footprint Interaction Using Coupled Eulerian-Lagrangian Approach

Jingbin Zheng^(✉) and Dong Wang

Ocean University of China, Qingdao 266100, China
zhengjingbin@ouc.edu.cn

Abstract. The re-installation of jack-up rigs near an existing footprint left by previous installation can be problematic due to the development of horizontal and moment loads on the spudcan and hence on the jack-up leg. This paper reports the results from large-deformation finite-element analyses of spudcan-footprint interaction in non-uniform clay deposits using the Coupled Eulerian-Lagrangian (CEL) approach. A spudcan was first installed and extracted to form the footprint, and was then re-installed at a certain offset distance. The combined effect of strain softening and rate dependency of the clay was incorporated. The performance of the numerical model was verified by simulations of existing centrifuge tests. The reasonable agreement between the numerical and experimental results indicates that the CEL approach is an effective tool in simulating the spudcan-footprint interaction problem.

Keywords: Spudcan · Footprint · LDFE

1 Introduction

With the increasing demand for offshore oil and gas exploitation and windfarm construction, jack-up rigs are often required to return to established sites. However, the re-installation of their spudcan foundations can be problematic due to the footprint left from previous installations. For spudcan penetration near or partially-overlapping the footprint, inclined (horizontal force) and eccentric (moment) loads are induced on the spudcan owing to the slope at the footprint perimeter and the difference in resistance between original soil and the disturbed soil in the footprint area [1].

Nevertheless, only limited numerical studies are available for spudcan-footprint interaction. One of the earliest attempts was made by Jardine et al. [2], but the problem was simplified as plane strain. Not until recently, three-dimensional large-deformation finite-element (LDFE) methods are employed to simulate spudcan re-installation on footprints [3, 4]. However, only one analysis was presented by Hartono et al. [3] reproducing the centrifuge test of reaming procedure, while Zhang et al. [4] only simulated idealized (artificially defined) footprint without considering soil strength heterogeneity caused by spudcan installation and extraction.

In this study, LDFE analyses of centrifuge tests of spudcan-footprint interaction in non-uniform clays were performed using the Coupled Eulerian-Lagrangian (CEL) approach available in the commercial package Abaqus/Explicit. The entire process of spudcan penetration-extraction-repenetration was simulated, with the clay modeled as a strain-softening, rate dependent material. The load-penetration responses from the numerical analyses and centrifuge tests are compared, with the aim to verify the applicability of CEL approach in the spudcan-footprint interaction problem.

2 Numerical Model

The schematic of the problem simulated in the CEL analysis is illustrated in Fig. 1, where the footprint is formed by the initial penetration and extraction of a spudcan with diameter D in the non-uniform clay deposit. The intact undrained shear strength of the clay is $s_u = s_{um} + kz$, where s_{um} is the undrained shear strength at the seabed surface and k is the rate of increase in s_u with the soil depth z. The spudcan is re-installed at an offset distance β. The vertical (V), horizontal (H) and moment (M) loads induced on the load reference point during the re-installation are recorded according to the sign convention suggested by Butterfield et al. [5].

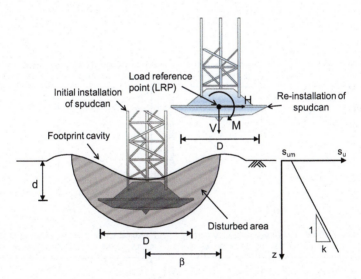

Fig. 1. Schematic and notations for the simulated problem

By considering the three-dimensional loading condition and the symmetry of the problem, a half three-dimensional model was used. Soil was discretized using Eulerian elements. A cuboid of fine mesh was created along the trajectory of the spudcans. The spudcan together with the jack-up leg was modelled as a rigid body. This is equivalent to a jack-up rig with fixed leg-hull connection, and hence the induced horizontal force and moment on the spudcan would be the upper bound in the analyses. Smooth soil-spudcan interface was adopted in the analysis.

The elastoplastic Tresca model was adopted but modified to capture strain rate and strain softening effects of clay deposits, following Einav and Randolph [6]. Details of the model and the adopted constitutive parameters can be found in Zheng et al. [7] who used the same model for spudcan penetration in stratified deposits.

3 Simulation of Centrifuge Tests

Retrospective numerical analyses were performed for centrifuge tests reported by Cassidy et al. [8]. The spudcan diameter and soil parameters for each test are listed in Table 1.

Table 1. Summary of centrifuge tests simulated in this study.

Test	D (m)	s_{um} (kPa)	k (kPa/m)	β/D
OTDR2	18.185	7.5	2	0.5
OTDR3	18.3	7.5	2	0.5
OTDR8	18.185	7.5	2	0.5
OTDR10	18.185	7.5	2	0.5

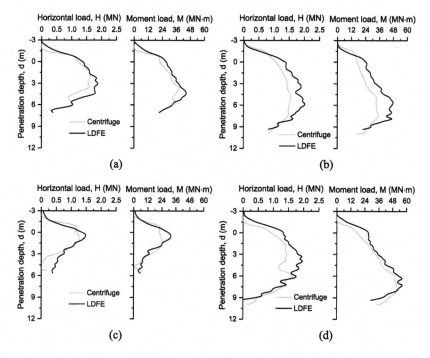

Fig. 2. Measured and computed H and M loads during spudcan-footprint interaction: (a) OTDR2; (b) OTDR3; (c) OTDR8; (d) OTDR10

The numerical and experimental results of horizontal and moment loads induced during spudcan-footprint interactions are compared in Fig. 2. Reasonable agreement is achieved in terms of the form of the H and M profiles, and the maximum H and M values, with the error mostly less than 25%. Generally, the H and M profiles are overestimated by the numerical simulations. This may be caused by the rigid jack-up leg modelled in the numerical analyses, which did not allow any lateral displacement of the spudcan during spudcan-footprint interaction; while the legs in the centrifuge tests have a certain flexural stiffness, which allowed the spudcan to move laterally due to leg deflection under the horizontal force and moment from the soil. It has been suggested that there would be a significant reduction of structural loads on the spudcan as the jack-up leg changes from rigid to flexible (see Fig. 8 of Stewart and Finnie [9]). The above comparisons illustrate that the numerical model proposed can reasonably predict the loads induced on the spudcan during spudcan-footprint interaction.

4 Conclusions

This paper reported the results from LDFE analyses using the CEL approach for spudcan-footprint interaction in non-uniform clay. The aim was to validate the numerical model against existing centrifuge test data. Reasonable agreement was obtained between numerical and experimental results in terms of the form and peak value of the load-penetration profiles. This has confirmed the capability of the CEL approach for undertaking analysis for spudcan-footprint interaction, which provides an effective tool for further and detailed investigation into the soil flow mechanisms and load-penetration responses for spudcan penetration near an existing footprint.

References

1. ISO: Petroleum and Natural Gas Industries - Site-Specific Assessment of Mobile Offshore Units Part 1: Jack-ups. International Organization for Standardization, Geneva, Switzerland (2012)
2. Jardine, R.J., Kovacevic, N., Hoyle, M.J.R., Sidhu, H.K., Letty, A.: Assessing the effects on jack up structures of eccentric installation over infilled craters. In: International Conference on Offshore Site Investigation and Geotechnics Diversity and Sustainability, London (2002)
3. Hartono, H., Tho, K.K, Leung, C.F, Chow, Y.K.: Centrifuge and numerical modelling of reaming as a mitigation measure for spudcan-footprint interaction. In: Offshore Technology Conference-Asia, Kuala Lumpur (2014)
4. Zhang, W., Cassidy, M.J., Tian, Y.: 3D large deformation finite element analyses of jack-up reinstallations near idealised footprints. In: 15th International Conference on the Jack-up Platform Design, Construction and Operation, London (2015)
5. Butterfield, R., Houlsby, G.T., Gottardi, G.: Standardized sign conventions and notation for generally loaded foundations. Géotechnique **47**(5), 1051–1054 (1997)
6. Einav, I., Randolph, M.F.: Combining upper bound and strain path methods for evaluating penetration resistance. Int. J. Numer. Methods Eng. **63**(14), 1991–2016 (2005)
7. Zheng, J., Hossain, M.S., Wang, D.: Prediction of spudcan penetration resistance profile in stiff-over-soft clays. Can. Geotech. J. **53**(12), 1978–1990 (2016)

8. Cassidy, M.J., Quah, C.K., Foo, K.S.: Experimental investigation of the reinstallation of spudcan footings close to existing footprints. J. Geotech. Geoenviron. Eng. **135**(4), 474–486 (2009)
9. Stewart, D.P., Finniem, M.S.: Spudcan-footprint interaction during jack-up workovers. Evol. Ecol. Res. **3**(1), 107–116 (2001)

Upper Bound Solution for Ultimate Bearing Capacity of Suction Caisson Foundation Based on Hill Failure Mode

Wen-bo Zhu[1], Guo-liang Dai[1,2(✉)], Wei-ming Gong[1,2], and Xue-liang Zhao[1,2]

[1] School of Civil Engineering, Southeast University, Nanjing 210096, China
daigl@seu.edu.cn
[2] Key Laboratory of Concrete and Prestressed Concrete Structure of Ministry of Education, Southeast University, Nanjing 210096, China

Abstract. In order to study the upper bound solution of bearing capacity of suction caisson foundations under vertical uplift loads, we introduce the viewpoint of reverse bearing capacity and the Hill failure mode for study. Select the appropriate foundation failure mode and displacement velocity field, it means that the active area of Hill failure mode becomes the passive area under the vertical pullout loads and the logarithmic spiral direction is opposite. At the same time, the factors of soil weight, cohesion, friction coefficient and earth pressure in the upper-bound theorem are put into consideration, and reasonable upper bound theorem is derived for Hill failure mode. The upper bound solutions is calculated by using the Matlab program and compared with the previous experimental data and other upper bound solution, it is proved that the failure mode of the soil and this method are reasonable.

Keywords: Energy dissipation rate · Hill failure mode
Suction caisson foundation · Upper bound theorem · Velocity field

1 Introduction

Compliant offshore structures, such as tension leg platforms (TLP), are usually subjected to considerable uplift forces. Nowadays, suction caisson has become an attractive option for the foundations of offshore wind turbines for the water depth. Detailed study commenced about of the pullout behaviour of suction 4 decades ago, and there have been several published experimental investigations of the pullout behavior of suction caissons in clayey soil and quartz sands. Meanwhile, there have been many research results about ultimate bearing capacity of suction caisson foundation. Deng [1] performed the ultimate lifting capacity of suction caisson foundations under different loading rates and different aspect ratios. Chen [2] used different failure modes to study the upper limit solution of the ultimate bearing capacity of circular foundations. However, very few attempts have been made to study the upper bound solution on the uplift bearing capacity of suction caissons. Wang [3], adopted the concept of "reverse bearing capacity foundation", deduced the upper bound solution for ultimate bearing capacity of suction caisson foundation based on Prandtl failure mode. Due to the Prandtl mechanism is a compression failure mode, the suction caisson foundation is not subjected to compression damage but tensile damage, so the Prandtl destruction mechanism completely used in reference [3] is not ideal.

This paper presents an upper bound method for calculating ultimate bearing capacity of suction caisson foundation with reverse Hill failure mode(the active area becomes the passive area and the logarithmic spiral direction is opposite). In the calculation, parametric studies are presented to explore the effect of soil weight, cohesion, friction coefficient and soil pressure. In the end, paper presents some values of uplift ultimate bearing capacity of suction caisson foundation prove the correctness of the upper bound solution. It provides a reference for the analysis of the uplift limit of suction caisson foundation.

2 Limit Analysis Theorems

The theorems of limit analysis (upper bound solution) provide a powerful tool for solving problems in which limit loads need to be found. The following assumptions are made in this method:

1. The material is rigid-perfectly plastic, with yielding governed by Tresca criterion.
2. The soil pressure on the circular ring of EABD is equivalent to q.
3. The material obeys the associated flow rule.

The configuration of the suction caisson foundation here was described through two parameters- the radius R, the Caisson buried depth L. and the associated velocity field is shown in Fig.1. The wedges OAC, with weights G_1, move with velocities v_0 but making an angle φ, with the linear failure surfaces OC. The Logarithmic spiral AFC, with weights G_2, move with velocities v but making an angle φ, with the curved failure surfaces CF. Soil may slide either along the foundation surface, referred to as interface shear with limiting shear stress ac.

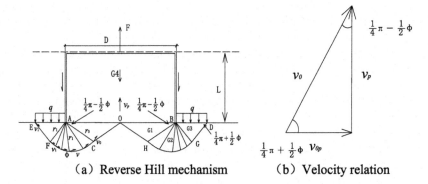

(a) Reverse Hill mechanism (b) Velocity relation

Fig. 1. Analysis of the reverse Hill failure mode.

2.1 Formulation of Upper Bound Solution

Equating the work rates of external loads to the total internal energy dissipation rates, we can obtain the general equation of the ultimate bearing capacity using upper bound method, which is

$$F = -\pi R^2 \tan^2\left(\frac{\pi}{4}-\frac{\varphi}{2}\right)e^{-\pi\tan\varphi}\left[\tan\left(\frac{\pi}{4}-\frac{\varphi}{2}\right)e^{-\frac{\pi}{2}\tan\varphi}+2\right]q + \frac{1}{6}\gamma\pi R^3 \tan\left(\frac{\pi}{4}-\frac{\varphi}{2}\right)$$

$$-\frac{1}{8}\gamma R\pi R^2 \sec^2\left(\frac{\pi}{4}-\frac{\varphi}{2}\right)\left\{\begin{array}{l}-\dfrac{e^{-\frac{3}{2}\pi\tan\varphi}+3\tan\varphi}{1+(3\tan\varphi)^2}+\tan\left(\dfrac{\pi}{4}-\dfrac{\varphi}{2}\right)\dfrac{1-3\tan\varphi e^{-\frac{3}{2}\pi\tan\varphi}}{1+(3\tan\varphi)^2}+\dfrac{1}{3}\sec^2\left(\dfrac{\pi}{4}-\dfrac{\varphi}{2}\right)*\\ \left[\dfrac{1-e^{-2\pi\tan\varphi}}{4\tan\varphi}+\cos 2\left(\dfrac{\pi}{4}-\dfrac{\varphi}{2}\right)\tan\varphi\dfrac{e^{-2\pi\tan\varphi}+1}{1+(2\tan\varphi)^2}-\dfrac{1}{2}\sin 2\left(\dfrac{\pi}{4}-\dfrac{\varphi}{2}\right)\dfrac{1+e^{-2\pi\tan\varphi}}{1+(2\tan\varphi)^2}\right]\end{array}\right\}$$

$$-\frac{1}{6}\gamma R\pi R^2 \tan^2\left(\frac{\pi}{4}-\frac{\varphi}{2}\right)e^{-\frac{\pi}{2}\tan\varphi}\left[\tan\left(\frac{\pi}{4}-\frac{\varphi}{2}\right)e^{-\frac{\pi}{2}\tan\varphi}+2\right]+\gamma\pi R^2 L+\frac{1}{4}c\cos\varphi\pi R^2\sec^2\left(\frac{\pi}{4}-\frac{\varphi}{2}\right) \quad (1)$$

$$+\frac{1}{2}c\pi R^2\sec^2\left(\frac{\pi}{4}-\frac{\varphi}{2}\right)\left(\frac{1-e^{-\pi\tan\varphi}}{2\tan\varphi}-\frac{e^{-\frac{3}{2}\pi\tan\varphi}+3\tan\varphi}{1+(3\tan\varphi)^2}+\tan\left(\frac{\pi}{4}-\frac{\varphi}{2}\right)\frac{1-3\tan\varphi e^{-\frac{3}{2}\pi\tan\varphi}}{1+(3\tan\varphi)^2}\right)$$

$$+\frac{1}{4}c\cos\varphi\pi R^2\sec^2\left(\frac{\pi}{4}-\frac{\varphi}{2}\right)e^{-\pi\tan\varphi}\left[3\tan\left(\frac{\pi}{4}-\frac{\varphi}{2}\right)e^{-\frac{\pi}{2}\tan\varphi}+2\right]$$

$$+\frac{1}{2}c\pi R^2\sec^2\left(\frac{\pi}{4}-\frac{\varphi}{2}\right)\left(\frac{1-e^{-\pi\tan\varphi}}{2\tan\varphi}-\frac{1}{2}\frac{e^{-\frac{3}{2}\pi\tan\varphi}+3\tan\varphi}{1+(3\tan\varphi)^2}+\frac{1}{2}\tan\left(\frac{\pi}{4}-\frac{\varphi}{2}\right)\frac{1-3\tan\varphi e^{-\frac{3}{2}\pi\tan\varphi}}{1+(3\tan\varphi)^2}\right)+2\pi RLac$$

3 Comparison and Analysis of Calculation Results

The upper bound solutions are compared with the test data [4–6] and the upper bound solutions from literature [3], the results are shown in Table 1.

Table 1. The upper bound solution is compared with other methods.

No.	D /mm	L/D	C /kPa	φ/°	test result /kN	Eq. (1) /kN	Diff. 1	literature [3] /kN	Diff. 2
1	110	0.75	2.5	2	0.149	0.114	−0.23	0.5	2.35
2	300	1.0	4	3	2.0	1.89	−0.05	5.9	1.95
3	300	1.33	4	3	3.0	2.39	−0.20	6.3	1.1
4	300	1.67	4	3	4.0	2.88	−0.28	6.7	0.67
5	100	1	3	38	0.165	0.127	−0.23	0.26	0.57
6	100	0.75	3	38	0.13	0.1	−0.23	0.21	0.61
7	100	0.5	3	38	0.075	0.074	−0.01	0.158	1.10

Note: the number 1 is introduced from Deng W, Carter J P.; the number 2-4 are introduced from X. C. SHI; X. N. GONG; the number 5–7 are introduced from, X. B. LU.

The upper bound solutions are compared with the test data [4–6] and the upper bound solutions from literature [3], the results are shown in Table 1. The ultimate bearing capacity of the suction caisson foundation was lifted rapidly in the test data. The test value and the upper bound solution are calculated by Eq. (1) are basically controlled at 20%, the maximum error was 28%, and the minimum error was 1%.

Due to the different reverse failure mode of the foundation, the upper bound solution used in this paper is less than the upper bound solution of completely Prandtl failure mode in [3] and closer to the experimental value.

4 Conclusions

This paper presents an upper bound method for calculating ultimate bearing capacity of suction caisson foundation with reverse Hill failure mode. And the paper presents some values of model test to prove the correctness of the reverse Hill failure mode. The results are shown in Table 1. The ultimate bearing capacity of the suction caisson foundation was lifted rapidly in the test data. The test value and the upper bound solution are calculated by Eq. (1) are basically controlled at 20%, the maximum error was 28%, and the minimum error was 1%. Due to the different reverse failure mode of the foundation, the upper bound solution used in this paper is less than the upper bound solution of completely Prandtl failure mode in [3] and closer to the experimental value. It can provide reference for the ultimate analysis of suction caisson foundation.

Acknowledgements. Project (2017YFC0703408) supported by national key research and development program;Project (51678145) supported by the National Natural Science Foundation; Project (51478109) supported by the National Natural Science Foundation.

References

1. Deng, W. Carter, J.P.: Inclined uplift capacity of suction caissons in sand, In: Proceedings of the 32nd Annual Offshore Technology Conference, Houston, Texas, USA, May 2000
2. Chen, Z.L.: Upper bound limit analysis of circular foundation foundations based on non-linear damage criterion. Central South University, School of Civil Engineering, pp. 36–46 (2008)
3. Wang, Z.Y.: A Study on Uplift Bearing Characteristics of Suction Caisson Foundation in Soft. Dalian: Soil. Dalian University of Technology, School of Civil Engineering, pp. 58–67 (2008)
4. Deng, W., Carter, J.P.: A theoretical study of the vertical uplift capacity of suction caisson. In: Proceeding of the 10th International Offshore and Polar Engineering Conference vol. 2, pp. 342–349 (2000)
5. Shi, X.C., Gong, X.N., Yu, J.L., et al.: Study on buckling resistance of bucket foundation. Build. Struct. **33**(8), 49–51 (2003)
6. Jiao, B.T., Lu, X.B., Zhao, J., et al.: Experimental study on the uplift bearing capacity of suction bucket foundation **21**(3), 27–29 (2006)

Part IV: Laboratory Testing

Soil Liquefaction Tests in the ISMGEO Geotechnical Centrifuge

Sergio Airoldi[1], Vincenzo Fioravante[2], and Daniela Giretti[2(✉)]

[1] ISMGEO Srl, Seriate, BG, Italy
[2] University of Ferrara, Ferrara, Italy
giretti@ismgeo.it

Abstract. Operational since 1988, the ISMGEO geotechnical centrifuge is a 240 g-ton centrifuge with a nominal radius of 2.2 m. The centrifuge can spin up a model of 400 kg up to 600 g. Since 2010 the centrifuge is equipped with a 1D shaking table able to reproduce real earthquake signals under an acceleration field up to 100 g. The special design of this centrifuge allows keeping the shaking table permanently installed on the rotating arm, while the model is driven on the shaking table during the spin-up phase. Physical models are submitted to 1D seismic shear waves, the typical free-field boundary conditions are reproduced by an equivalent shear beam box recently manufactured for dynamic applications.

This paper describes firstly the main features of the facility and of the equivalent shear beam box. Secondly an application example of dynamic tests to simulate soil liquefaction is reported: the model tested reproduced a homogeneous sand column subjected to a real earthquake, properly scaled; specific sensors monitored the seismic excitation within the model, the excess pore pressure development and the soil surface settlements.

The results of test presented here show the capability of the centrifuge to reproduce extreme loading conditions such as liquefaction during an earthquake. Many kinds of soil profiles can be tested, combination of ground motions can be applied to the models, the effectiveness of liquefaction mitigation techniques can also be studied.

Keywords: Liquefaction · Geotechnical centrifuge · Shaking table

1 ISMGEO Geotechnical Centrifuge

The geotechnical **centrifuge** of the Istituto Sperimentale Modelli Geotecnici (ISMGEO, formerly ISMES – Italy), operative since 1988, is a beam centrifuge made up of a symmetrical rotating arm with a diameter of 6 m, a height of 2 m and a width of 1 m, which gives it a nominal radius of 2 m. The centrifuge can reach an acceleration of 600 g (g = gravity acceleration) with a payload of 400 kg. Figure 1 reports a top view photo.

In 2010 the centrifuge was upgraded with a 1D **shaking table** able to reproduce real strong motion at the model scale (i.e. a complete scaled spectrum). The shaker is not integrated into the swinging basket, but it is directly connected to the rigid arm: during the flight the platform which holds the model container is moved into contact with the table and released before dynamic excitation starts. The table works under an acceleration field up to 100 g, it can provide excitations at frequencies up to 1000 Hz and horizontal acceleration up to 50 g (depending on the driving load, whose maximum value is 3.5 kN at 100 g) [1].

A new **Equivalent Shear Beam** (ESB) box [2], has been recently developed (Fig. 2). This type of container reproduces free-field type boundary conditions on the model. The ESB box is composed by twelve rectangular frames with a height of 25 mm each, connected by eleven 3.36 mm thick rubber inter-layers (total height of 337 mm).

Fig. 1. Top view of the centrifuge.

Fig. 2. The ESB box installed in the centrifuge basket

2 Seismic Liquefaction Testing

An application example of dynamic tests to simulate soil liquefaction is reported. Ticino sands is used as testing soil. Ticino sand is a mono granular silica sand with a D_{50} around 0.2 mm. A 28 cm thick homogeneous model was reconstituted by air pluviation at a relative density of 45% and saturated with a pore water fluid with scaled viscosity. The test is performed at a centrifuge acceleration of 50 g, thus simulating a soil column 14 m high. The model is instrumented with a vertical array of pore pressure transducers (ppts) and accelerometers installed at different depths (Fig. 3).

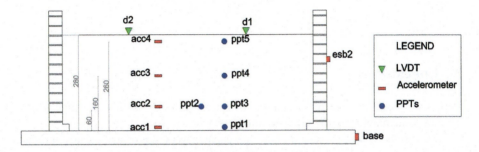

Fig. 3. Model setup (dimensions in mm, model scale)

The ground motion applied at the base of the model has a PGA of 0.21 g (Fig. 4). Figure 4 shows some test results, in terms of pore pressure ratio Ru time history (where Ru is the ratio of excess pore pressure developed during the dynamic excitation to the effective stress level at the depth of the ppts).

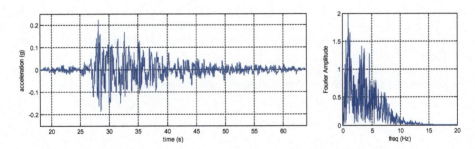

Fig. 4. Ground motion applied to the model (prototype scale)

The shear strains developed during the earthquake are sufficient to bring the soil into non-linear domain and to develop excess pore pressures close to the liquefaction condition (Fig 5).

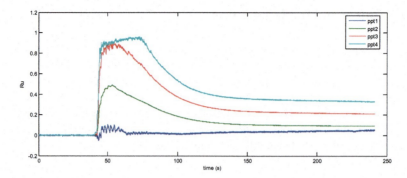

Fig. 5. Excess pore pressure at different depths in Ticino sand (prototype scale)

3 Conclusions

This paper describes the main features of the ISMGEO seismic centrifuge and an example of dynamic tests to simulate soil liquefaction.

The results of test presented show the capability of the centrifuge to reproduce extreme loading conditions such as liquefaction during an earthquake.

These types of tests can conveniently be used to evaluate the effectiveness of techniques for liquefaction risk reduction and can provide reference benchmark for numerical simulations.

References

1. Airoldi, S., Fioravante, V., Giretti, D.: The ISMGEO seismic geotechnical centrifuge. In: Thorel, L., Bretschneider, A., Blanc, M., Escoffier, S. (eds.) Proceedings of the 3rd European Conference on Physical Modelling in Geotechnics (Eurofuge 2016), Nantes, France, 1–3 June, pp 185–190 (2016)
2. Zeng, X., Schofield, A.N.: Design and performance of an Equivalent Shear Beam (ESB) model container for earthquake centrifuge modelling. Geotechnique **46**(1), 83–102 (1996)

ISMGEO Large Triaxial Apparatus

Sergio Airoldi[1], Alberto Bretschnaider[1], Vincenzo Fioravante[2(✉)],
and Daniela Giretti[1(✉)]

[1] ISMGEO Srl, Seriate, BG, Italy
giretti@ismgeo.it
[2] University of Ferrara, Ferrara, Italy

Abstract. The design of earth and rockfill dams is focused on ensuring the stability of the structure under a set of combined mechanical and hydraulic conditions that may occur both during construction and operative life of the dam. The prediction of the hydro-mechanical behavior of construction materials (which range from clayey soils to rockfill) is therefore fundamental. The most important characteristics to be studied are static and dynamic shear strength, shear stiffness at small and intermediate strain and compressibility of the construction materials. Testing of these properties is typically performed using the triaxial apparatus. However, testing large size materials such as rockfill requires the use of large triaxial specimens, whose diameter depends on the particles maximum dimension. Facilities capable of testing large grain size materials are few and sometimes neither the largest apparatuses are big enough to test particles as large as cobbles and boulders. ISMGEO large triaxial apparatus allows the execution of static and cyclic test on 300 mm diameter samples. The maximum particle dimension is 50 mm. Originally designed for testing the gravelly foundation soils of the Messina strait bridge, the apparatus has been recently up-graded in order reach a cell pressure of 2.5 MPa, thus allowing to reproduce the in situ stress conditions within particularly tall earth and rockfill dams.

This paper describes the main features of the facility. Application examples are also shown. In particular, results of motononic compression test and cyclic tests are analyzed, with the main aim of evidencing the apparatus capabilities.

Keywords: Triaxial tests · Large diameter samples · Cyclic tests

1 Apparatus Description

The ISMGEO large triaxial apparatus, shown in Fig. 1, allows testing 300 mm diameter, 600 mm high samples. The rigid reaction frame is designed to support up to 3000 kN. The top and bottom plates, made of high-strength stainless steel, are aligned and fixed through 8 tie bars. The high-strength stainless steel piston is designed to minimize the friction.

A pneumatic apparatus controls the back pressure up to 1 MPa. The confining cell pressure is controlled by a pneumatic apparatus up to 1 MPa and by a hydraulic apparatus from 1 MPa to 2.5 MPa.

In the monotonic tests, the axial load is applied by a servo controlled system which drives a double acting hydraulic piston of 1200 kN capacity, allowing operating both in stress and in strain controlled conditions. The specimens can be isotropically or anisotropically consolidated under stress controlled conditions, while during the shearing stage the deviator stress is applied in a strain-controlled mode, imposing to the piston a constant rate of vertical displacement.

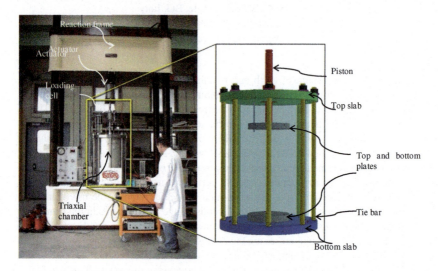

Fig. 1. Picture of the large Tx apparatus.

In the cyclic tests, the same servo controlled system pilots a proportional valve to apply the required axial load amplitude (stress ratio) or a cyclic load at constant rate of strain.

The axial load is measured by a load cell rigidly connected to the actuator. Depending on the expected maximum load to be applied, two load cells are used alternatively, the first with a nominal load = 100 kN and the second of = 1000 kN, both guaranteeing the linearity error $\leq \pm 0.1\%$.

Resistive displacement transducers (nominal displacements 200 mm for monotonic test, 10 mm for cyclic test, with linearity error $\leq \pm 0.2\%$) are fixed to the piston shaft outside the cell to measure the axial displacements. Sample axial deformations can also be measured directly on the specimen (local measurements) using 2 potentiometers placed inside the cell. The volume change of the specimen due to the consolidation and drained loading is measured by an automatic ΔV-apparatus (nominal volume = 10^{-3} m^3, linearity error $\leq \pm 0.1\%$).

The pore water pressure is measured by an absolute pressure transducer (nominal pressure = 2 MPa, linearity error $\leq \pm 0.2\%$).

The transducers are conditioned by a DAQ, connected to a personal computer for the data acquisition.

2 Test Results

2.1 Monotonic Drained Tests

Figures 2a and 2b show the results of 12 drained monotonic tests carried out on gravelly samples (50 mm maximum particle dimension) reconstituted at the same void ratio and consolidated at isotropic pressure ranging from 250 kPa to 2000 kPa. In Fig. 2a the stress ratio q/p' (q = stress deviator, p' = mean effective stress) is plotted as a function of the axial deformation ε_a. In Fig. 2b the void ratio change (current void ratio e minus end of consolidation void ratio e_c) during the drained compression is plotted vs ε_a. The figures show a very good accuracy in the stress and strain measures and also a good repeatability of tests carried out at the same confining pressure.

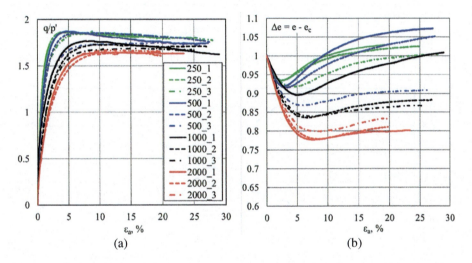

Fig. 2. Example of monotonic test on a gravelly soil: (a) stress ratio and (b) void ratio change as a function of axial deformation.

2.2 Cyclic Tests

The apparatus allows to carry out both tests for the determination of the modulus and damping properties of soils (ASTM D3999, [1]) and liquefaction tests (ASTM D5311, [2]). Figures 3 and 4 show the results of the first kind of test. In Fig. 3 are shown the first 4 loading cycles carried out during a test on a gravelly soil sample consolidated at 500 kPa, in terms of axial deformation plotted against the applied axial load; Fig. 4 reports the test results in terms of degradation of the normalized Young modulus and increase of damping ratio as a function of the axial strain ε_a. As shown in the Figures, axial deformation lower than 0.01% can be measured.

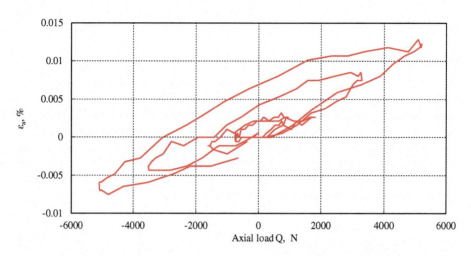

Fig. 3. Gravelly sample isotropically consolidated at 500 kPa: first 4 load cycles of a cyclic test.

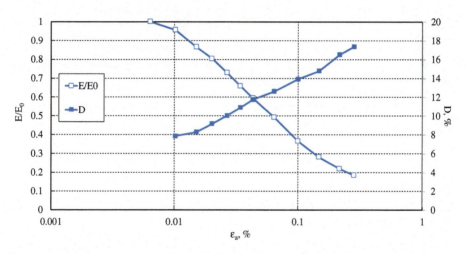

Fig. 4. Gravelly sample isotropically consolidated at 500 kPa: dacay of the stiffness and damping ratio increase as a function of axial strain.

3 Conclusion

This paper describes the main features of the ISMGEO large triaxial apparatus. Two application examples are shown: the results of a series of drained monotonic tests and those of a cyclic test for the determination of the modulus and damping properties of soils. The tested soil is a gravelly material. The results here reported highlight the very good repeatability of the tests and the apparatus capability of measuring axial deformations lower than 0.01% with a good accuracy.

References

1. ASTM D3999: Standard Test Methods for the Determination of the Modulus and Damping Properties of Soils Using the Cyclic Triaxial Apparatus
2. ASTM D5311: Standard Test Method for Load Controlled Cyclic Triaxial Strength of Soil

Experimental Study on Static Liquefaction of Carbon Fiber Reinforced Loose Sand

Xiaohua Bao[✉], Zhiyang Jin, Hongzhi Cui, and Haiyan Ming

Department of Civil Engineering and Underground Polis Academy,
Shenzhen University, Shenzhen 518060, China
bxh@szu.edu.cn

Abstract. Carbon fibers (CF) which have the characteristics of reliable strength, durability and slow bio-degradation rate, was used for static liquefaction resistance of loose sand. The effect of fiber length (3, 6, 10 mm) and fiber content (0.2, 0.5, 0.7, 1.0%) that will influence the liquefaction characteristics of reinforced soils was investigated in detail by a series of undrained triaxial compression tests. The results show that the inclusion of randomly distributed carbon fibers in loose sand is an effective measure for static liquefaction mitigation, and liquefaction resistance increased with the increase of fiber length and fiber content. The stress-strain relationships, development of pore water pressure and liquefaction brittleness index of the tested samples were analyzed. The results show that with the continuous increase of fiber length and content, an effective increase in both the peak and post-peak strength of the reinforced samples was observed. Especially when the content of 10 mm CF up to 1%, the peak deviator stress increased by 65.8% compared with the clean sand in undrained triaxial compression tests at the confining pressure of 100 kPa.

Keywords: Carbon fiber · Loose sand · Liquefaction mitigation

1 Introduction

Static liquefaction typically associated with saturated loose sand and silty sand can cause failures e.g. slopes, earth dams, submarine berms, which are characterized by suddenness, wide range of influence and catastrophe. A large number of studies on liquefaction mitigation of sand mainly focus on dynamic loading [1–3], while there are few literatures on liquefaction mitigation under static loading. Recently, the use of randomly distributed fibers as reinforcement has attracted the attention of researchers and engineers because of its more satisfactory performance. Some researchers [3–5] use polypropylene fiber to alleviate the liquefaction of sand, the experimental results show that the presence of fibers can well maintain the stability of sand structure and the randomly distributed fibers have a significant effect on increasing the static liquefaction resistance of soils. Therefore, in this study a new material, carbon fibers (CF) which have the characteristics of reliable strength, durability and slow bio-degradation rate, was used for static liquefaction resistance of loose sand. The effect of fiber length and fiber content that will influence the liquefaction characteristics of reinforced soils was investigated in detail by a series of undrained triaxial compression tests. Attention was paid to the stress-strain relationship, development of pore water pressure and liquefaction brittleness index of the tested samples.

2 Materials

Fujian standard sand (produced in Fujian, China) is used in the sample preparation. Figure 1 shows the grain size distribution of the sieved Fujian sand. The mean grain size $D_{50}= 0.22$ mm, coefficient of uniformity $C_u= 2.0$, maximum and minimum void ratio $e_{max}= 0.79$, $e_{min}= 0.49$. The used synthetic carbon fiber (CF) with the diameter of 7 μm, made by Toray Group in China, has a high aspect ratio (fiber length over fiber diameter). The tensile strength of fiber is 4900 MPa, whereas its elastic modulus, elongation and density are 230 GPa, 2.1%, and 1.8 g/cm³, respectively. The content of fiber included in a sample is defined as a proportion of dry weight $(W_f/W_s) \times 100\%$, where W_f is the weight of fiber and W_s is the weight of the dry sand.

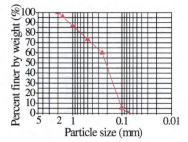

Fig. 1. Grain size distribution.

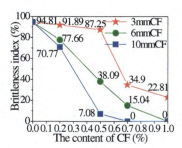

Fig. 2. Liquefaction potentiality.

3 Sample Preparation

Carbon fibers (CF) with different lengths (3, 6, 10 mm) and contents (0, 0.2, 0.5, 0.7, 1%) were added into the loose sand respectively to explore the influence of fiber length and content on the effect of static liquefaction. A series of undrained triaxial compression tests were carried out on 15 saturated samples with the initial relative density $D_r= 10\%$. In order to obtain relatively uniform fiber-sand mixture samples, the required CF was firstly placed in an electronic mixer with dry sand and stirred for 2 min, and then adding 5% deionized water to effectively disperse the CF evenly. The samples were prepared layer by layer (each layer 20 mm) and carefully setup into rubber membrane in the height of 100 mm and diameter of 50 mm using moist tamping specimen preparation technique [4, 5]. After the sample was fully saturated (pore pressure coefficient greater than 0.96), undrained triaxial compression tests were performed under the effective confining pressure of 100 kPa.

4 Tests Results

Figures 2, 3, 4 present the undrained triaxial compression tests results. When the effective stress first reaches the zero-stress state, or the generated excess pore water pressure was equal to the effective confining pressure (100 kPa), the sample was considered to be fully liquefied. The maximum deviator stress (q_{max}) and minimum deviator stress (q_{min}) were used to measure the possibility of liquefaction through undrained triaxial tests. The brittleness index (I_B) defined as ($q_{max} - q_{max}$)/q_{max}, ranging from 0-1, can directly contrasts the mitigating effect of CF reinforced soil. When I_B is equal to 0, it means that the deviator stress of the sample will not decrease throughout shear process, and the sample will appear completely static liquefaction when $I_B = 1$. It can be seen from Fig. 2 that the brittleness index of CF reinforced loose sand with different lengths and contents varies at the confining pressure of 100 kPa. The sample will not liquefy when the content of CF was 1% ($I_B = 0$) except the sample reinforced with 3 mm CF. Moreover, the brittleness index of the samples decreased continuously with the increase of fiber content and fiber length.

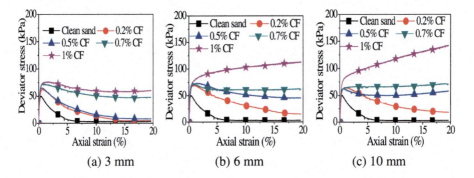

Fig. 3. The stress-strain relationships.

Clearly from Figs. 3 and 4, the presence of fibers, irrespective of the fiber length employed, strongly affected the undrained response. The deviator stress increased monotonically with the increase of fiber contents and an increase in both undrained peak and post-peak shear strengths for fiber reinforced samples compared with the clean sand sample were observed. Moreover, a strain hardening response for higher values of fiber contents (1%) appeared, while the deviator stress still exhibited a strain softening behavior for the fiber content of 0.2%, 0.5%, 0.7% after reaching a peak value (although the peak value is higher than that of clean sand). In another hand, the development of pore water pressure was in consistent with the response of deviator stress. With the increase of the fiber contents, the peak values of the excess pore water pressure became lower. Static liquefaction was fully prevented when the amount of fiber exceeds a certain value comparable to that of the clean sand.

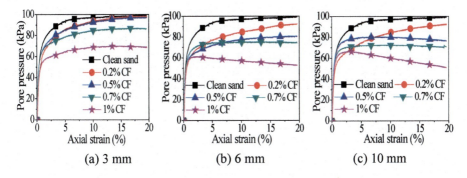

Fig. 4. The development of pore pressure.

Under the same fiber contents with different length, longer fibers in soil lead to a better performance for liquefaction resistance. The results indicated that the longer the used fiber, the less content needed for liquefaction prevention. However, when the fiber length exceeded some value, the increase effect on liquefaction prevention became little compared with the results of fiber length 6 mm and 10 mm samples. Therefore, the suggested optimum CF length for liquefaction prevention is 6 mm considering the fiber dosage.

5 Conclusions

In this study, the liquefaction behavior of loose sand reinforced with randomly distributed carbon fibers were examined by undrained triaxial compression tests. The effect of the fiber length and content on liquefaction resistance was investigated. With the effective confining pressure of 100 kPa and the initial relative density $D_r = 10\%$, the addition of CF can significantly enhance the liquefaction resistance of loose sand. As the CF content and length increased, the peak strength and post-peak strength of samples also increase continuously that effectively restrained the development of pore water pressure within the samples. The static liquefaction of loose sand can be well mitigated by using the randomly distributed carbon fibers.

Acknowledgements. This research is fully supported by the National Natural Science Foundation of China (No. 51678369) and Technical Innovation Foundation of Shenzhen (No. JCYJ20170302143610976).

References

1. Huang, Y., Wang, L.: Laboratory investigation of liquefaction mitigation in silty sand using nanoparticles. Eng. Geol. **204**, 23–32 (2016)
2. Ochoa-Cornejo, F., Bobet, A., Johnston, C.T., Santagata, M., Sinfield, J.V.: Cyclic behavior and pore pressure generation in sands with laponite, a super-plastic nanoparticle. Soil Dyn. Earthq. Eng. **88**, 265–279 (2016)

3. Ye, B., Cheng, Z., Liu, C., Zhang, Y., Lu, P.: Liquefaction resistance of sand rein-forced with randomly distributed polypropylene fibres. Geosynth. Int. **24**(6), 625–636 (2017)
4. Ibraim, E., Diambra, A., Muir Wood, D., Russell, A.R.: Static liquefaction of fibre reinforced sand under monotonic loading. Geotext. Geomembr. **28**(4), 374–385 (2010)
5. Ibraim, E., Diambra, A., Russell, A.R., Wood, D.M.: Assessment of laboratory sample preparation for fibre reinforced sands. Geotext. Geomembr. **34**, 69–79 (2012)

Experimental Study on Effect of Temperature and Humidity on Uniaxial Mechanical Properties of Silty-Mudstone

Jingcheng Chen(✉) and Hongyuan Fu

Changsha University of Science and Technology, Changsha, China
200810150@qq.com

Abstract. In order to study on uniaxial mechanical properties of silty-mudstone under impact of temperature and humidity, a series of uniaxial mechanical tests were carried out. According to these tests, uniaxial mechanical indexes (compressive strength, elastic modulus, splitting strength) of silty-mudstone under different temperature and humidity conditions were obtained, and based on which, its change rule was analyzed. The test results indicate that when considering temperature shift only, uniaxial mechanical indexes of silty mudstone exponential decrease with the increase of cyclic times under same mode of temperature shift, decrease with temperature difference under same cyclic times and direction of temperature shift. After soaking, expansion rate and water content of silty-mudstone increases rapidly at the early stage and tends to be gentle at the later stage, uniaxial mechanical indexes exponential decrease with the increase of water content. After simultaneous variation of temperature and humidity, uniaxial mechanical indexes increase with the increase of temperature difference under negative direction of temperature shift, decrease with the increase of temperature difference under positive direction of temperature shift. Under same conditions, compare influence on mechanical properties: simultaneous variation of temperature and humidity> humidity shift only> temperature shift only.

Keywords: Geotechnical engineering · Silty mudstone
Temperature and humidity · Strength

1 Introduction

In slope excavation, underground mining, underground chamber excavation and other projects, the rock mass of silty-mudstone under the excavation face are easily affected by temperature or humidity, which caused strength degradation and the occurrence of large deformation of the rock mass [1–4], endangering the safety of project. At present, there are some related studies. The micro mechanism of the strength softening of soaked mudstone has been studied [5, 6]; some studies show that the elastic modulus decreases with the increases of temperature [7], the mechanical strength indexes will decrease with the prolongation of the soaking time [8]. However, these studies consider only the impact of singe temperature or singe humidity on rock, and the research objects are usually not silty-mudstone.

Thus, in this paper, a series of experiments have been carried out, the uniaxial mechanical indexes of silty-mudstone under different temperature and humidity conditions have been systematically studied.

2 Methods

Silty-mudstone samples were collected on the field, and a large number of standard cylinder specimens of $\varphi 50 \times 100$ mm and $\varphi 50 \times 25$ mm were produced in the laboratory immediately. Using non-metal ultrasonic detector to screen out the specimens of wave velocity in 2.2–2.3 km/s, then these specimens can be used for uniaxial compression test or splitting test respectively.

3 test schemes were designed to consider the impacts of single temperature, single humidity, both temperature and humidity on the uniaxial mechanical indexes of silty-mudstone:

Test scheme I: In order to analyze the impact of temperature shift, dry the specimens at 25 °C, put the specimens into a thermostat with 0% air humidity. Periodically shift the temperature with different modes in the thermostat. The uniaxial mechanical indexes of silty-mudstone were tested after the end of temperature shift.

Test scheme II: In order to analyze the impact of humidity shift, soak the natural specimens in a water bath, and the water temperature is constant at 25 °C. The water content, uniaxial mechanical indexes of these specimens were tested at different soaked time.

Test scheme III: In order to analyze the impact of both temperature and humidity shift, soak the natural specimens into a water bath, meanwhile periodically shift the water temperature with different modes. After the end of water temperature shift, the water content and uniaxial mechanical indexes of silty-mudstone were tested.

3 Results

3.1 The Impact of Temperature

The test scheme I-1 carries out temperature cycles for different times on the silty-mudstone specimens with the temperature shift mode 7. The test results indicate that uniaxial compressive strength (σ_c), elastic model (E), splitting strength (σ_t) exponential decreases with the increase of temperature cyclic times(n). The fitted curves of the relation between these uniaxial mechanical indexes and temperature cyclic times were obtained:

$$\sigma_c = 7.4527e^{-0.0558n} + 18.7267, R^2 = 0.9901. \tag{1}$$

$$E = 3.9130e^{-0.0827n} + 9.6429, R^2 = 0.9946. \tag{2}$$

$$\sigma_t = 0.6630e^{-0.0457n} + 1.2433, R^2 = 0.9411. \tag{3}$$

The test scheme I-2 carries out 10 temperature cycles on the silty-mudstone specimens with different temperature shift mode. It can be observed that uniaxial compressive strength, elastic model, splitting strength of silty-mudstone are related to temperature difference ($|Vt|$) and direction of temperature shift; see Fig. 1.

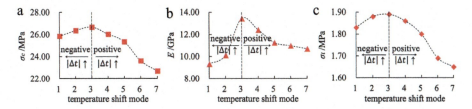

Fig. 1. Uniaxial mechanical indexes of silty-mudstone under different temperature shift mode. Uniaxial compressive (square), elastic modulus (triangle), splitting strength (rhombus).

3.2 The Impact of Humidity

The test results show that water content (w) increases with time, and uniaxial compressive strength, elastic model, splitting strength exponential decreases with the increase of water content. The fitted curves of the relation between these uniaxial mechanical indexes were obtained:

$$\sigma_c = 15.1287 e^{-0.0802w} + 3.6453, R^2 = 0.9882. \qquad (4)$$

$$E = 10.5900 e^{-0.0533w} - 1.1900, R^2 = 0.9902. \qquad (5)$$

$$\sigma_t = 1.1208 e^{-0.0688w} + 0.3913, R^2 = 0.9773. \qquad (6)$$

3.3 The Impact of Both Temperature and Humidity

It can be observed that the specimens of silty-mudstone have different water content in different temperature shift mode; uniaxial compressive strength, elastic model, splitting strength increase with the increase of temperature difference under negative direction of temperature shift, and decrease with the increase of temperature difference under positive direction of temperature shift; the measured value of uniaxial mechanical indexes are different from the calculated value calculated according to the water content, it indicated the water temperature cycle changed these uniaxial mechanical indexes directly; see Fig. 2.

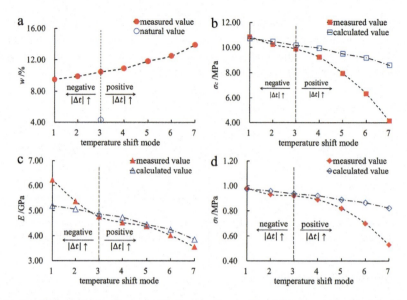

Fig. 2. Uniaxial mechanical indexes of silty-mudstone after different water temperature cycles (soaked 10 days).

4 Conclusions

Our research shows that the strength deterioration of silty-mudstone is related to temperature cyclic times, temperature difference, direction of temperature shift, soaked time after the temperature and humidity influence. The strength deterioration of silty-mudstone caused by simultaneous impact of temperature and humidity, are greater than that of temperature or humidity only. According to this finding, Engineers should pay special attentions to the engineering rock mass of silty-mudstone when it is simultaneous impacted by temperature and humidity.

References

1. Li, K.G., Zheng, D.P., Huang, W.H.: Mechanical behavior of sandstone and its neural network simulation of constitutive model considering cyclic drying-wetting effect. Rock Soil Mech. **34**(2), 168–173 (2013)
2. Yao, H.Y., Zhang, Z.H., Zhu, C.H.: Experiment study of mechanical properties of sandstone under cyclic drying and wetting. Rock Soil Mech. **31**(12), 3704–3708 (2010)
3. Duda, M., Renner, J.: The weakening effect of water on the brittle failure strength of sandstone. Geophys. J. Int. **192**, 1091–1108 (2013)
4. Erguler, Z.A., Ulusay, R.: Water-induced variations in mechanical properties of clay-bearing rocks. Int. J. Rock Mech. Min. Sci. **46**, 355–370 (2009)
5. Liu, C.W., Lu, S.L.: Research on mechanism of mudstone degradation and softening in water. Rock Soil Mech. **21**(1), 28–31 (2000)

6. Huang, H.W., Che, P.: Research on micro mechanism of softening and argillitization of mudstone. J. Tongji Univ.: Nat. Sci. Ed. **35**(7), 867–870 (2007)
7. Yang, Y.M., Ju, Y., Chen, J.L., et al.: Mechanical properties of porus rock media subjected to temperature effect. Chin. J. Geotech. Eng. **35**(5), 856–864 (2013)
8. Zhou, C.Y., Deng, Y.M., Tan, X.S., et al.: Experimental research on the softening of mechanical properties of saturated soft rocks and application. Chin. J. Rock Mech. Eng. **24**(1), 33–38 (2005)

On Shear Strength of Stabilized Dredged Soil from İzmir Bay

İnci Develioglu and Hasan Firat Pulat(✉)

İzmir Katip Celebi University, 35620 İzmir, Cigli, Turkey
hfirat.pulat@ikc.edu.tr

Abstract. Dredging is the process of carrying out work such as protection of inland water and sea lanes, construction/deepening of port basins, opening/deepening of channels and regulation of coasts. As a result of the dredging, the large amount of dredged soil is removed and disposal or reuse of this material is very important for economically and environmentally. This material can be recycled in the civil engineering applications as construction material, road and foundation embankment. However, because of the high organic matter content, dredged soil has very different shear strength characteristic compared to inorganic soil. In this study physico chemical engineering properties and shear strength behavior of natural and stabilized İzmir Bay's dredged soil were examined for four organic matter contents (0%, 4%, 7% and 11%). The index properties of natural samples were defined by sieve analysis, specific gravity, Atterberg limits and pH tests. The direct shear tests were conducted with natural and stabilized dredged soils for saturated loose and dense samples. The shear strength parameters of natural dredged soils were compared with lime and silica fume added (10% and 20%) samples. Test results have shown that specific gravity value decreased, the liquid and plastic limit increased with the organic matter content increasing for natural samples. Silica fume added samples have higher cohesion but lower internal friction angles. The highest internal friction angle belongs to natural dredged soil without organic matter.

Keywords: Dredged soil · Index properties · Shear strength · Stabilized soil

1 Introduction

Recycling and reuse of waste materials is a very important process for sustainable development and ecosystem to minimize environmental damage. Sedimentation of dredged soils on the base of the water bodies affect the productivity and capacity of harbors, water circulation, biological diversity and water quality. For this reason, the sediment deposited on the base of the water bodies is dredged for protection of inland water and sea lanes, construction or deepening of port basins, opening or deepening of channels and regulation of coasts [1]. Dredged soils have been extensively used in various civil engineering applications, such as; building construction, production of ready-mixed concrete, improvement of weak soils, and filling materials for road, embankment and foundation [2]. Numerous studies have been performed to determine the engineering properties of the dredged soil all over the world.

Berilgen et al. examined the index properties and shear strength characteristic of Golden Horn marine clay by laboratory vane test. Unconfined compression tests were conducted with samples obtained from slurry consolidation test. The undrained shear strengths which obtained by unconfined compression test range between 1.9 and 18.4 kPa, by laboratory vane tests range between 3.6 and 134.4 kPa [3].

Schlue et al. investigated the shear strength characteristic of Germany East harbor mud. The large-scale direct shear test was conducted to obtain shear strength parameters. It was found that the cohesion was almost 0 kPa (c' = 0.22 kPa), the peak friction angle was obtained 14° and residual friction angle was found 3°. The residual shear strength and undrained peak shear strength increased with the consolidation stress increment [4].

Malasavage et al. used steel slag fine (SSF) as a soil improvement material. The dredged soil and SSF were mixed 80/20, 60/40, 50/50, 40/60 and 20/80 ratios and laboratory tests were carried out. The liquid and plastic limits were range between 74 and 140, 37 and 49, respectively. The internal friction angle of natural dredged material was 27.3°, the maximum internal friction angle belongs to the mixture of 50/50 ratio ($\phi = 45°$) [5].

In this study, shear strength behavior of İzmir bay's dredged material were investigated for different organic matter content (OM) (0%, 4%, 7% and 11%). In addition to natural dredged soil samples lime and silica fume mixed samples were used in direct shear tests. Samples prepared at loose and dense state, were mixed with 10 and 20% lime (LM) and silica fume (SF) to improve shear strength parameters.

2 Materials and Methods

The dredged material used in this study was obtained from the İzmir Bay (İzmir, Turkey). The OM of dredged soil was determined by ignition at 440 °C in a furnace. After the initial OM was determined as 11%, the dredged soil samples were ignited at 440 °C for various time periods to obtain samples with different organic contents (0%, 4% and 7%). The particle size distribution was obtained using wet sieve analysis. The liquid limit was designated by fall cone method. The dredged soil samples were classified according to Unified soil classification system (USCS). The specific gravity values were determined with pycnometer, vacuum pump and distilled water. The pH values of samples were designated with digital pH meter. The suspension was prepared by mixing 50 g sample and 125 ml distilled water. It was waited 24 h and then pH measurement was conducted two times to check repeatability [6]. All laboratory tests were conducted according to the ASTM standards.

Direct shear tests were started with medium density samples prepared at liquid limit and 15.5 kN/m³. The mixed samples were also prepared in liquid limit but this time two different densities, loose (13 kN/m³) and dense (18 kN/m³), was used. A 0,125 mm/min shear velocity obtained from consolidation tests (t_{90}) was used in this test. Tests were repeated two times to check repeatability.

3 Results and Discussion

Index properties and shear strength parameters of natural dredged soil obtained from various laboratory experiments have been summarized in Table 1.

Table 1. Index properties and shear strength parameters of natural dredged samples

OMC (%)	LL (%)	PL (%)	USCS	G_s	pH	ϕ (°)	c (kN/m^2)
0	32.9	26.4	OL	2.76	8.9	34.0	8.6
4	32.6	27.9	OL	2.64	7.5	40.1	0.0
7	34.7	28.7	OL	2.60	7.2	37.8	9.4
11	39.5	31.0	OL	2.52	7.4	36.2	4.4

It was seen that the LL and PL values tended to increase, specific gravity, pH and internal friction angle tended to decrease as the content of organic matter increased.

Shear strength parameters of lime (LM) and silica fume (SF) mixed samples have been summarized in Table 2. For lime mixed samples, the friction angles of dense samples were smaller than the loose samples; cohesions were bigger than the loose samples. There was no significant change in the friction angle as the amount of additive increases, but cohesions increase. Similar results were also observed in the silica fume mixtured dredged samples.

Table 2. Shear strength parameters of lime (LM) and silica fume (SF) mixed samples

OMC (%)	Additive ratio (%)	Density	LM		SM	
			ϕ (°)	c (kN/m^2)	ϕ (°)	c (kN/m^2)
0	10	Loose	36	0	33	12
0	10	Dense	27	35	32	11
0	20	Loose	35	2	37	4
0	20	Dense	30	23	29	27
4	10	Loose	33	8	32	17
4	10	Dense	31	13	31	16
4	20	Loose	34	6	34	12
4	20	Dense	31	23	29	23
7	10	Loose	33	10	33	8
7	10	Dense	30	28	27	27
7	20	Loose	32	13	31	16
7	20	Dense	29	28	32	13
11	10	Loose	33	6	35	8
11	10	Dense	31	18	32	13
11	20	Loose	37	9	31	14
11	20	Dense	32	22	31	17

It was seen that the Mohr envelopes of the samples with the same additive amount were very similar to each other.

Shear - Normal stress relation of natural samples has been shown in Fig. 1. Mohr envelopes of the natural samples were found to be very similar to each other.

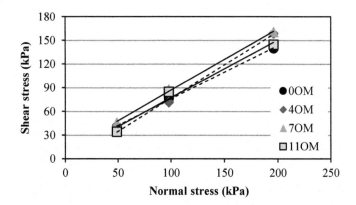

Fig. 1. Mohr - Coulomb failure envelopes of natural dredged samples

4 Conclusion

The laboratory test results have shown that the liquid and plastic limit values tended to increase, specific gravity, pH and internal friction angle tended to decrease as the content of organic matter increased for natural samples. As the amount of additive increases, particle size distribution decrease and the internal friction angle decrease and cohesion increase. It has been found that as the density increased the internal friction angle increased and there was no effect on the cohesion.

References

1. Kang, G., Tsuchida, T., Athapaththu, A.: Engineering behavior of cement-treated marine dredged clay during early and later stages of curing. Eng. Geol. **209**, 163–174 (2016)
2. Rakshith, S., Singh, D.N.: Utilization of Dredged Sediments: Contemporary Issues. J. Waterw. Port Coast. Ocean Eng. **143**(3), 1–13 (2016)
3. Berilgen, S.A., Kilic, H., Ozaydin, K.: Determination of undrained shear strength for dredged golden horn marine clay with laboratory tests. In: Sri Lankan Geotechnical Society's 1st International Conference on Soil & Rock Engineering, pp. 5–11, Colombo (2007)
4. Schlue, B.F., Mörz, T., Kreiter, S.: Undrained shear strength properties of organic harbor mud at low consolidation stress levels. Can. Geotech. J. **48**, 388–398 (2011)
5. Malasavage, N.E., Jagupilla, S., Grubb, D.G., Wazne, M., Coon, W.P.: Geotechnical performance of dredged material—steel slag fines blends: laboratory and field evaluation. Geoenviron. Eng. **8**, 981–991 (2012)
6. Kocasoy, G.: Atıksu arıtma çamuru ve katı atık ve kompost örneklerinin analiz yöntemleri. 2nd edn. Boğaziçi University Publisher, İstanbul (1996)

Instrument for Wetting-Drying Cycles of Expansive Soil Under Loads

Jungui Dong[1(✉)], Guoyuan Xu[1], Hai-bo Lv[2], and Junyan Yang[2]

[1] South China University of Technology, Guangzhou 510641, Guangdong, China
dhunter110@163.com
[2] Guilin University of Technology, Guilin 541004, Guangxi, China

Abstract. In this study, an instrument is developed to study the effect of wetting-drying cycles on the shear strength of expansive soil. The testing error of the developed device is investigated as well. By using this device, the water content of soil sample can be accurately controlled, and the vertical loads at different soil depths can be simulated.

Keywords: Expansive soil · Wetting-drying cycles · Instrument · Loads

1 Introduction

Expansive soil experiences volumetric swelling or shrinkage with adsorption or desorption of water because expansive soil often contains special minerals such as montmorillonite and illite [1]. It is therefore of great importance to study the influence of wetting-drying cycles on shear strength of expansive soil [2].

Numerous studies indicate that the wetting-drying cycles of expansive soil can be divided into two types (i) cycles under no loads and (ii) cycles under simulated loads. Cutting ring sample is used in Lv's [3] study, and different number of cycles are performed under no loads; water content in the test process is calculated by weighing method. Yang et al. [4] stacks weights on sample to simulate the vertical load during wetting-drying cycles. When the water content is calculated using the weighing method, vertical load (weights) must be unloaded prior to weighing. Previous studies have failed to consider the actual load scenarios encountered in practical engineering. In practice, loads always exist at and beyond a certain depth of soil, and there is no unloading at any point in the wetting-drying cycles.

This study focuses on the development of an instrument that simulates the actual load condition encountered in practical engineering. After a detailed description of device design and operation, calculating error of water content and vertical force error are determined and discussed. To date, this instrument has been given patent for invention in China.

2 Instrument Design

The instrument, which has a small density and high strength, consists of a circulating device, lever, and special shear box. The circulating device and its labeled components are diagrammed in Fig. 1. The nuts rotate to move vertically along the columns, enabling the fixed collar to move vertically along the column as well. The upper part of the cover plate, also known as the cross beam, extends approximately 1.0 cm further than the outer edge of the fixed collar. The cross beam employs the dial indicator to measure the vertical deformation of samples. To accelerate the processes of wetting and drying, holes are evenly distributed on both the cover plate and the base plate. The lower edge of the cutting ring is attached by screws to the base plate through the lower of two parts shown in the below aspect. The sample remains whole during wetting-drying cycles due to a hoop that fixes the positions of the two parts of the cutting ring.

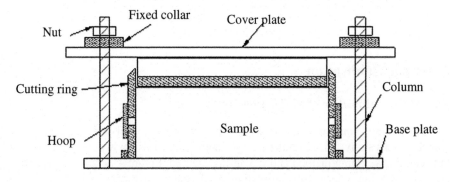

Fig. 1. Device for wetting-drying cycles

Figure 2 illustrates a shear box specially designed to handle the demands of the circulating device. A groove on the bottom of the pedestal maintains the proper position of the circulating device underneath the shear box.

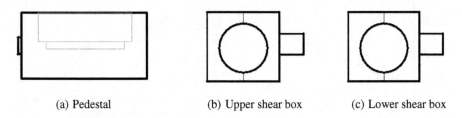

(a) Pedestal　　　　　　(b) Upper shear box　　　　　(c) Lower shear box

Fig. 2. Shear box specially designed for the circulating device

Figure 3 depicts the load simulating mechanisms of the device. A lever, with an initial lever ratio of 1:10, simulates various vertical loads of different depths by changing weights and/or lever ratio. Vertical expansion is measured using a dial indicator installed on the lever.

494   J. Dong et al.

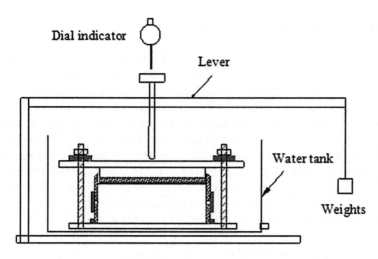

Fig. 3. Device for wetting-drying cycles under loads

3 Error Calibration

3.1 Error Calibration of Water Content

Instead of the drying method, the weighing method was used to calculate moisture content in this study. To substantiate these calculations, this study determines the error between the weighing method and the drying method. Figure 4 demonstrates a strong consistency in the water content measurements derived from the two methods as applied throughout the wetting and drying process. With an error of within 1% between the two methods, weighing method is suitable for calculating, and therein controlling, the range of wetting-drying cycles.

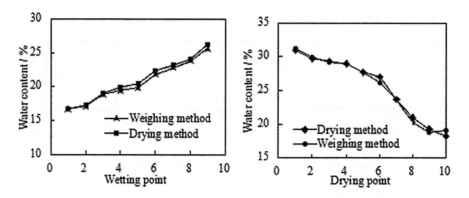

Fig. 4. Comparison of two methods in calculating water content

3.2 Error Calibration of Vertical Force

After the lever is released, to ensure that extrusion pressure from the cover plate and the base plate is equal to the previous vertical force applied by the lever, the relationship between the vertical deformation of the sample and increasing loads, shown in Fig. 5, is studied. In the weighing method, vertical force achieved by screwing the nuts. The screwing ceased when the dial gauge reading changed, usually at deformation of less than 0.01 mm. At a deformation of 0.01 mm, the increase in vertical pressure is less than 0.6 kPa, and, therefore, the error can be ignored.

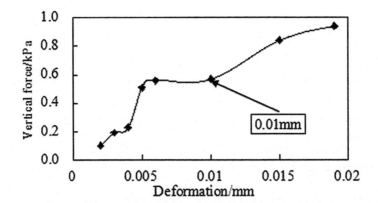

Fig. 5. Relationship between loads and vertical deformation

4 Conclusions

Essential engineering principles are observed by an instrument that can accurately control the water content and simulate vertical loads at specified soil depths. This instrument better approximates practical engineering scenarios as no unloading is required during the simulation or measurement of the wetting-drying cycles and the shear strength test.

References

1. Nayak, N.V., Christensen, J.C.: Swelling characteristics of compacted expansive soil. Clays Clay Miner. **19**(4), 251–261 (1974)
2. Yemadje, P.L., Guibert, H., Chevallier, T., et al.: Effect of biomass management regimes and wetting-drying cycles on soil carbon mineralization in a Sudano-Sahelian region. J. Arid Environ. **127**, 1–6 (2016)
3. Lv, H.B., Zeng, Z.T., Zhao, Y.L.: Preliminary study on the cumulative damage of expansive soil in alternation of wetting-drying environment. J. Nat. Disasters **21**(6), 119–123 (2012)
4. Yang, H.P., Wang, X.Z., Xiao, J.: Influence of wetting-drying cycles on strength characteristics of nanning expansive soil. Chin. J. Geotech. Eng. **36**(5), 949–954 (2014)

Liquefaction of Firoozkuh Sand Under Volumetric-Shear Coupled Strain Paths

S. R. Falsafizadeh and A. Lashkari[✉]

Faculty of Civil and Environmental Engineering, Shiraz University of Technology, Shiraz, Iran
Lashkari@sutech.ac.ir

Abstract. Restriction on volume change during shear may lead to flow liquefaction of loose sands. The behaviors of loose Firoozkuh no. 161 sand specimens subjected to shear with coupling between the volumetric and shear strains are investigated here. The experimental data indicate that the sand behavior is remarkably influenced by the volume change pattern in volumetric-shear coupled strain paths.

Keywords: Sand · Direct simple shear · Flow liquefaction · Strain

1 Introduction

Sand flow liquefaction vulnerability is conventionally investigated through monotonic and cyclic undrained (i.e., constant volume) tests [1]. However, recent field data have confirmed that the undrained assumption is not impeccable because the earthquake-induced excess pore-water pressure may be dissipated partially owing to the high permeability of sands [2]. Thus, it has been suggested that sand behavior in strain paths with coupling of volumetric and shear strains in conjunction with undrained tests can better portray sand flow liquefaction susceptibility [3, 4].

2 Testing Program

Using a NGI type direct simple shear apparatus, influence of coupling between the volumetric and shear strains on the behaviors of Firoozkuh sand was investigated here. For Firoozkuh sand, specific gravity of particles is 2.64 [-], median particle size is 0.276 [mm], and the maximum and minimum void ratios are 1.0 [-], and 0.58 [-], respectively. Figure 1 shows the gradation curve and a SEM image of particles for Firoozkuh no. 161 sand. Specimens with diameter and height of 70 [mm], and 18 [mm], respectively, were prepared using moist tamping method. The critical state friction angle of Firoozkuh no. 161 in the simple shear mode is 27.8° and the critical void ratio is interrelated to the effective normal stress (i.e., σ'_n) through $e_{cs} = 1.1919 - 0.0816 \ln \sigma'_n$ wherein σ'_n must be expressed in kPa.

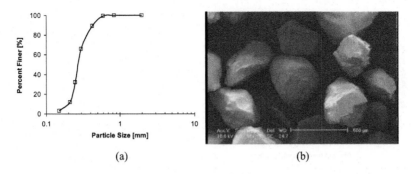

Fig. 1. Firoozkuh sand No. 161: (a) gradation curve; (b) SEM photograph of particles.

3 Tests with Volumetric-Shear Coupling: Results and Discussions

In proportional strain paths, the volumetric strain rate (i.e., $\delta\varepsilon_v$) increases linearly with the shear strain rate (i.e., $\delta\gamma$) through:

$$\delta\varepsilon_v = \zeta\,\delta\gamma \tag{1}$$

where ζ defines proportionality. The behaviors of specimens sheared under proportional strain paths with $\zeta = +0.2, +0.1, 0.0,$ and -0.1 are illustrated in Fig. 2. The undrained ($\zeta = 0$) test showed flow liquefaction following strain-softening beyond $\gamma \approx 2\,[\%]$. In contrast, a non-flow response for the specimen with $\zeta = +0.2$ is attributed to strengthening of microstructure associated with the volumetric-shear coupling. A milder dilation and limited flow liquefaction was attained for the case with $\zeta = +0.1$. Finally, progressive increase in void ratio and loosening of microstructure for $\zeta = -0.1$ case worsens flow liquefaction.

The bi-linear strain paths begin with a proportional path up to a threshold strain (γ_{thr}) beyond which, shearing continues under constant volume condition:

$$\delta\varepsilon_v = \begin{cases} \zeta\,\delta\gamma & \text{if } \gamma \leq \gamma_{thr} \\ 0 & \text{if } \gamma > \gamma_{thr} \end{cases} \tag{2}$$

The behaviors of specimens sheared under bi-linear strain paths ($\gamma_{thr} \approx 20\,[\%]$) are shown in Fig. 3.

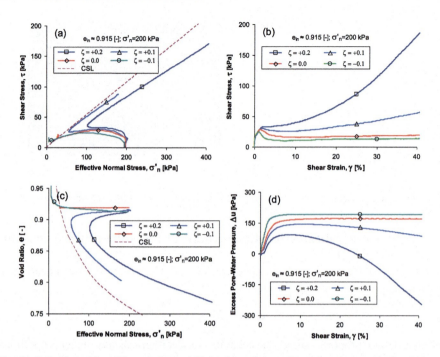

Fig. 2. Behavior of four Firoozkuh no. 161 sand specimens sheared under proportional shear-volumetric strain path: (a) stress paths; (b) shear stress vs. shear strain; (c) void ratio vs. effective normal stress; and (d) excess pore-water pressure vs. shear strain.

The initial contraction turned into dilation around $\gamma \approx 2\,[\%]$ for the specimens sheared under $\zeta = +0.2$ and $+0.1$. However, both specimens demonstrated strain-softening and flow liquefaction in the constant volume regime. A completely different scenario is observed for the specimen sheared under $\zeta = -0.1$ in which a mild dilation is observed in the constant volume phase following a flow liquefaction in the first phase of shearing with $\zeta = -0.1$. To explain the mentioned differences, Fig. 3(c) indicates that the final states of the specimens sheared along $\zeta = +0.2$ and $+0.1$ are located above the Critical State Line (CSL) in the e vs. σ'_n plane at the end of the proportional phase. In an opposite manner, the final state of the specimen sheared along $\zeta = -0.1$ is located slightly below the CSL in the e vs. σ'_n. As a direct consequence, contractive responses are expected for the former specimens and a mild dilation is anticipated for the latter one. Finally, curves for variation of excess pore-water pressure curves with shear strain are plotted in Fig. 3(d) in which the specimens sheared along $\zeta = +0.2$ and $+0.1$ accumulate positive pore-water pressure in the constant volume phase.

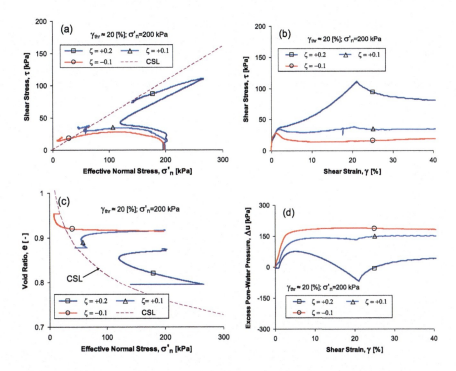

Fig. 3. Behavior of three Firoozkuh no. 161 sand specimens sheared under bi-linear shear-volumetric strain path:(a) stress paths; (b) shear stress vs. shear strain; (c) void ratio vs. effective normal stress; and (d) excess pore-water pressure vs. shear strain.

4 Conclusions

Tests with coupling between volumetric and shear strains can assist describing sands flow liquefaction susceptibility. Experimental data reported here confirmed that the pattern of coupling between the volumetric and shear strains plays strong roles in the mobilization of shear strength, variation of void ratio as an index of the compactness of soil load carrying microstructure, and accumulation of the excess pore-water pressure during shear.

References

1. Lashkari, A., Karimi, A., Fakharian, K., Kaviani-Hamedani, F.: Prediction of undrained behavior of isotropically and anisotropically consolidated Firoozkuh sand: instability and flow liquefaction. ASCE Int. J. Geomech. **17**(10), 04017083 (2017)
2. Malvick, E.J., Kutter, B.L., Boulanger, R.W.: Postshaking shear strain localization in a centrifuge model of saturated sand slope. ASCE J. Geotech. Geoenvironmental Eng. **134**(2), 164–174 (2008)

3. Jrad, M., Sukumaran, B., Daouadji, A.: Experimental analysis of the behavior of saturated granular material during axisymmetric proportional strain paths. Eur. J. Environ. Civ. Eng. **16**(1), 111–120 (2012)
4. Lashkari, A., Yaghtin, M.S.: Sand flow liquefaction instability under shear-volumetric coupled strain paths. Géotechnique (2017, in press). https://doi.org/10.1680/jgeot.17.P.164

Shear Strength of Tailings with Different Void Ratios

Tingting Fan[✉]

School of Civil Engineering and Architecture, Nanchang University,
Nanchang 330031, Jiangxi, China
fan.ttring@qq.com

Abstract. This paper presents a theoretical formulation describing the relationship between shear strength of tailings and their void ratios. The formulation is derived based on the critical state theory of sand. To validate this formulation, a series of laboratory tests, including oedometer and direct shear tests, are carried out on copper and tungsten tailings with various void ratios. The experimental results show that shear strength of sandy soil nonlinearly decreases with the increase of the void ratio, and the strength decreases much slower for soil with larger void ratio. The shear strength of tailings is then simulated using the proposed formulation, which can reasonably describe the shear strength of tailings with different void ratios.

Keywords: Void ratio · Shear strength · Tailings · Critical state
Direct shear test

1 Introduction

The degree of compaction of the soil in the project is one of the most important indicators of its strength. Void ratio is a key factor affecting the mechanical properties of silty sand [1]. The strength index of sandy soil increases with the increase of relative compactness [2, 3]. There are more studies of the influence of void ratio on the shear strength of sand soil. However, there are fewer studies of the theoretical analysis, and the quantitative law of the shear strength of sandy soil with the change of void ratio is still uncertain. Mining and smelting will result in a lot of tailings, which show the strength characters of sandy soil in most cases. The shear strength of tailings is the main factor influencing the scour starting ability of tailings under the water flow overtopping. In addition, the liquefaction and slip of tailings are also related to shear strength. As known to all, the compaction state of sand soil directly affects its shear strength, and it is feasible to obtain the model sand that meets the requirements of shear strength by controlling the void ratio of the model sand.

It is easy to carry out soil shear strength test, but it will be time-consuming and energy-draining if there is no theoretical reference for the test and the initial void ratio of sand soil required by the model test may not necessarily be obtained. The existing studies of the influence of void ratio on sand soils shear strength are limited to the experimental analysis. In order to more effectively find methods of sand kind soil model test, the relationship between the sand soils shear strength and void ratio was analyzed from the perspective of theory.

2 Relationship Between Shear Strength and Void Ratio

The expression of void ratio on shear strength:

$$\tau_f = \frac{M \exp^{\frac{\Gamma_1 - e}{\lambda_1}}}{3 \cos \varphi} \frac{3 - \sin^2 \varphi}{\sqrt{3 + \sin^2 \varphi}} \quad (1)$$

where, Γ_1, λ_1, and M are constants that can be obtained from experiments; e is the void ratio after consolidation under different vertical stresses; τ_f is the shear strength of sand.

The expression was more complicated than coulomb shear strength criterion, and special tests were required to access to parameters such as Γ_1, λ_1 and M in it, but the formula showed the effect of the initial void ratio of sand soils effect on shear strength. The corresponding shear strength could be estimated with the expression according to the void ratio, which was of great practical significance in the similar model test of sand soil.

In the reduced scale similar model test of tailing bam overtopping, the average self-weight stress of the sand was reduced because the height of the tailings dam was reduced according to the geometric similarity ratio in the model test. In fact, under natural condition, self-weight stress made the tailings create a certain state of void ratio. In reduced scale model test, testers could control the state of pores in the process of the dam construction, and make the tailings dam model have similar starting under the action of water flow. How to determine the initial void ratio of model tailings dam was a technical problem. The expression provided a technical reference for reasonably determining the initial state of model sand from a theoretical perspective.

3 Material and Test Program

3.1 Materials

The copper tailings were collected from Wushan Copper Mine and Yongping Copper Mine and tungsten tailings from Xinanzi Tungsten Tin Mine. The number of the sample and the physical and mechanical properties of the rock and soil are shown in Table 1. The reference value of general sandy soil specific gravity is 2.65 to 2.69. Due to the higher proportion of non-ferrous metals, the specific gravity of the copper tailings obtained in the experiment is as high as 2.91. The nonferrous metal content of the tungsten tailings is low, and the measured soil particles are measured. The proportion is similar to that of ordinary sand. All samples are named as **T1**, **T2**, and **W**.

Table 1. Sample number and physical properties of indicators

Sample	Classification and sampling site of tailings	w/%	G_s	C_u	C_c
T1	Copper tailing, Wushan Copper Mine	0.96	2.91	6.18	3.47
T2	Copper tailing, Yongping Copper Mine	0.84	2.94	14.8	5.9
W	Tungsten tailing, Xin'anzi Tungsten-Tin Mine	0.54	2.67	44.44	2.01

Note: w = Water content; G_s=Specific gravity of soil particle; C_u=Uniformity coefficient C_c = Coefficient of curvature

3.2 Test Program

A series of consolidation tests and direct shear tests were carried out with samples. As shown in Table 2, we can get the model parameters from the test results. The model parameters were used to verify the feasibility of predicting the shear strength of sandy soil by theoretical formula, and the quantitative analysis of sandy soil shear strength with different void ratio. The internal friction angle of tailings with different particle sizes, as shown in Table 3.

Table 2. Model parameters obtained by experiment

Sample	0.25 mm–0.5 mm			0.075 mm–0.25 mm			<0.075 mm		
	Γ_1	λ_1	M	Γ_1	λ_1	M	Γ_1	λ_1	M
T1	0.7778	0.0102	1.3741	0.7715	0.0114	1.1820	0.7806	0.0141	1.0929
T2	0.7908	0.0101	1.2663	0.7814	0.010	1.773	0.8032	0.0188	1.0542
W	0.6155	0.0107	1.1051	0.6161	0.0112	0.9238	0.6253	0.0154	0.8458

4 Results

The shear strength criterion of sand soil usually regards the normal stress as independent variable and the stress condition under normal consolidation could correspond to void ratio condition. In the reduced scale geotechnical model test with smaller normal stress, the initial compaction state (void ratio condition) of sand soil is the decisive factor of the shear strength of soil. As shown in Fig. 1, the non-linearity of shear strength decreased with the increase of the void ratio of tailings. When the tailings were in a tighter state (void ratio were relatively small), the shear strength varied considerably with the void ratio. When the void ratio of tailings exceeded a certain number, the shear strength tended to zero. The shear strength of tailings calculated through the formula was in good agreement with the test value, which also proved the rationality of the proposed relation between the void ratio and shear strength. The experimental results verified the theoretical formula was feasible in predicting the shear strength of sand soil.

Table 3. The internal friction angle of tailings with different particle sizes

Sample	0.25 mm–0.5 mm	0.075 mm–0.25 mm	<0.075 mm
T1	39.2°	36.4°	31.7°
T2	34.7°	34.7°	28.7°
W	34.1°	29.1°	27.3°

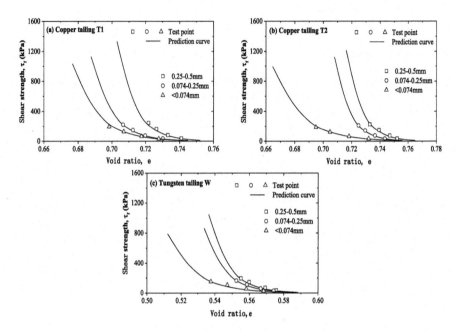

Fig. 1. Test results and prediction curves of shear strength and void ratio (after consolidation)

5 Conclusions

A geotechnical test was conducted for the design of copper tailings and tungsten tailings, and the rationality of the theoretical formula was verified by test results. The following conclusions can be drawn: (1) The nonlinearity of shear strength decreases with the increase of the void ratio of tailings. When the void ratio of tailings exceeded a certain number, the shear strength tended to zero; (2) The formula can reflect the change rule of the shear strength of sand soil with the void ratio, and it is of reference value for the development of similar geotechnical model tests that need to be artificially set up in the future; (3) Particle size significantly affects the engineering mechanics properties of tailings, and fine tailings have obviously weaker engineering properties than coarse particle tailings.

References

1. Thevanayagam, S.: Effect of fines and confining stress on undrained shear strength of silty sands. J. Geotech. Geoenviron. Eng. **124**(6), 479–491 (1998)
2. Zhu, J.G., Shi, J.W., Luo, X.H.: Experimental study on stress-strain-strength behavior of sand with different densities. Chin. J. Geotech. Eng. **38**(2), 336–341 (2016)
3. Cai, Z.Y., Li, X.S.: Development of dilatancy theory and constitutive model of sand. Chin. J. Geotech. Eng. **29**(8), 1122–1128 (2007)

Dynamic Behaviour of Wenchuan Sand with Nonplastic Fines

Tugen Feng[1], Qiuying Qian[2], Fuhai Zhang[1], Mi Zhou[3(✉)], and Kejia Wang[4]

[1] Key Laboratory of Ministry of Education for Geomechanics and Embankment Engineering, Geotechnical Research Institute, Hohai University, Nanjing, China
[2] Department of Grain Engineering, Anhui Vocational College of Grain Engineering, Hefei, China
[3] State Key Laboratory of Subtropical Building Science, South China Institute of Geotechnical Engineering, South China University of Technology, Guangzhou, China
zhoumi@scut.edu.cn
[4] Teaching and Research Center of Civil Air-Defense Engineering, Army Engineering University, Shanghai, China

Abstract. A conflicting result from previous studies for the effect of nonplastic fines on the cyclic resistance of sand is obtained from the literature. Sand deposit in real field typically contains silt content, which is a significant challenge for engineers to quantify the effect of silt content on dynamic behaviour of sand. To provide insight into the effects of nonplastic fines on the dynamic behaviour of Wenchuan sand, a laboratory parametric study of sand utilizing cyclic triaxial tests with a range of confining pressure, silt content percentage and consolidation ratio is performed. This study shows that the silt content can decrease the cyclic resistance of sand. Based on the experimental results, one formula is proposed to quantify the effect of nonplastic fines on the liquefaction behavior of saturated Wenchuan sand.

Keywords: Sand · Silts · Liquefaction · Fines · Cyclic load · Cyclic resistance

1 Introduction

The dynamic characteristics of saturated sand have been studied in the past 50 years in geotechnical engineering, including strength characteristics of saturated sand liquefaction under seismic loads, deformation characteristics, exceed pore water pressure and other features. The effect of silt content on the dynamical behaviour of sand has been studied; however, the conclusions from previous studies appear somewhat contradictory. Some researchers have concluded that the increasing silt content will decrease the sand's resistance to liquefaction [1], while others indicate that the increasing of silt content will increase the sand's resistance to liquefaction [4]. Different to the both views above, Koester [5] has pointed out that there is a critical value of silt content for the behaviour of sand with silt content. Liquefaction resistance of sand decreases with the increasing of silt content when the silt content is lower than the critical value, while the liquefaction resistance of sand increases as the increasing of silt content when the silt content is higher than that critical value.

This study selects Wenchuan sand to conduct a series of experimental study to examine the effect of nonplastic fine on the dynamic behaviour of the saturated sand. Based on the parametric study results of the experimental tests, one formula was proposed to quantify the effect of nonplastic fines on the liquefaction behavior of saturated sand with silt content.

2 Results and Discussions

The sand sample is obtained from Wenchuan area in the Western of China. The grain size distribution is shown in Fig. 1. Parametric study with eight group cases were conducted to examine the effect of silt content on the cyclic resistance, by varying silt content, SC, confining pressure, C_p and consolidation ratio, K_c.

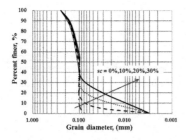

Fig. 1. The grain size distribution   **Fig. 2.** Effect of silt content on void ratio

The maximum and minimum void ratios is shown in Fig. 2. Cubrinovski and Ishihara [2] reported that the void ratio range $e_{max}-e_{min}$ is larger for sand containing more fines than that of pure sand, which is coincided with the curve of void ratio range, $e_{max}-e_{min}$, against silt content.

To explore the effect of silt content on cyclic resistance, a group of experimental tests have been carried out with varying silt content, giving $SC = 0\%$, 10%, 20%, 30%, and the results are plotted in Fig. 3a, b and c. From Fig. 3, it can be seen that the cyclic resistance decreases with the increasing of silt content, and similar findings are also reported by Huang et al. [3] for Mai Liao sand.

The effect of confining pressure on cyclic resistance with varying SC from 0% to 30% and $K_c = 1$ are plotted in Fig. 4a, b, c and d, respectively. It can be found that, with the increasing of silt content, the confining pressure changing has more effect on the increasing of the cyclic resistance. This indicates that the cyclic resistance increases with the increasing of confining pressure.

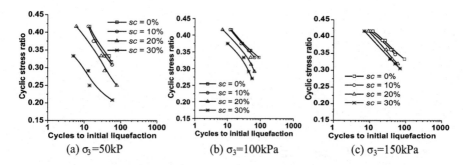

Fig. 3. Effect of silt consent on the Cyclic Resistance with $K_c = 1$

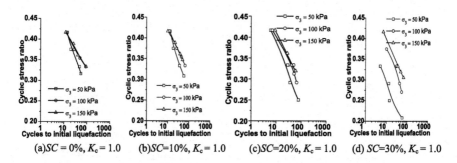

Fig. 4. The effect of confining pressure on Cyclic Resistance with $K_c = 1.0$

Figure 5 shows the effect of consolidation ratio, K_c, on the cyclic resistance. The consolidation ratio has a significant effect on cyclic resistance, and the cyclic resistance increases with the increasing of consolidation ratio. Liu and Chen [6] also reported that the higher value of K_c is, the larger cyclic resistance of Nanjing fine sand.

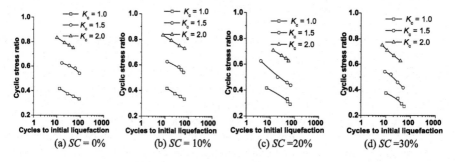

Fig. 5. Effect of K_c on Cyclic Resistance ($\sigma_3 = 100$ kPa)

The above description reveals that silt content, SC, confining pressure, P_c and consolidation ratio, K_c, have an influence on the cyclic resistance of sand containing silt content. Based on the experimental data, a formula is proposed to predict the cycles number of sand to reach the initial liquefaction state. The formula is shown as below.

$$(log 10 N_f)^{-0.3} = CSR/(0.215 \times 0.5^{SC}(\frac{P_c}{P_a})^{0.11} 2^{K_c}) \quad (1)$$

where, P_a is the standard atmospheric pressure of 100 kPa, CSR is the cyclic stress ratio and N_f is the cycles for initial liquefaction.

3 Conclusions

A parametric study of the effect of silt content on the dynamic behaviour of Wenchuan sand was conducted by experimental tests. The following conclusions regarding the effect of silt content on the Wenchuan sand's cyclic resistance were drawn from this study:

(1) The silt content has a significant effect on the cyclic resistance of Wenchuan sand. The test results show that the cyclic resistance decreases with the increasing of silt content.
(2) The silt content SC, confining pressure C_p and consolidation ratio K_c have effect on the Wenchuan sand's dynamic behaviour. A formula was proposed to predict the cycles of Wenchuan sand to reach the state of the initial liquefaction under dynamic loading.

Acknowledgement. The research presented here was undertaken with support from National Nature Science Foundation of China (No. 51009054), Jiangsu Province Nature Science Foundation of China (No. bk2010513) and Major project research funding of the science and technology of the Ministry of Education of China (No. 109077). The research presented here was also undertaken with the support from the China funding Science and Technology Project of POWERCHINA Huadong Engineering Corporation Limited (SD2013-10), the Water Resource Science and Technology Innovation Program of Guangdong Province (2015-17), and the Fundamental Research Funds for the Central Universities of China (D2171820). These supports are gratefully acknowledged.

References

1. Bahadori, H., Ghalandarzadeh, A., Towhata, I.: Effect of non-plastic silt on the anisotropic behavior of sand. Soils Found. **48**(4), 531–545 (2008)
2. Cubrinovski, M., Ishihara, K.: Flow potential of sandy soil with different grain composition. Soils Found. **40**(4), 103–119 (2000)
3. Huang, Y.T., Huang, A.B., Kuo, Y.C., Tsai, M.D.: A laboratory study on the undrained strength of a silty sand from Central Western Taiwan. Soil Dyn. Earthq. Eng. **24**(9), 733–743 (2004)
4. Georgiannou, V.N.: The undrained response of sands with additions of particles of various shapes and sizes. Geotechnique **56**(9), 639–649 (2006)
5. Kuerbis, R., Negussey, D., Vaid, Y.P.: Effect of gradation and fines content on the undrained response of sand. In: Hydraulic Fill Structures, vol. 21, pp. 330–345. ASCE Geotechnical Engineering Division Specialty Publication (1988)
6. Liu, X.Z., Chen, G.X.: Experimental study on influence of clay particle content on liquefaction of Nanjing fine sand. J. Earthq. Eng. Eng. Vib. **23**(6), 150–155 (2003). (in Chinese)

Effects of Induced Anisotropies on Strength and Deformation Characteristics of Remolded Loess

Yongzhen Feng[1(✉)], Wuyu Zhang[1], Lingxiao Liu[1], and Yanxia Ma[1,2]

[1] School of Civil Engineering, Qinghai University,
251 Ningda Road, Xining 810016, Qinghai, China
fengyz8297@163.com

[2] The Key Laboratory of Geomechanics and Embankment Engineering,
Hehai University, Nanjing 210098, Jiangsu, China

Abstract. In order to better explain the influence of induced anisotropy on the strength and deformation characteristics of loess, a complex stress path experiment of induced anisotropy of the remolded loess was carried out by using a hollow cylinder apparatus. Under the same condition of the intermediate principal stress coefficient b, the test carried out directional shear tests in the case of drainage. The effects and regularities of induced anisotropy on the strength and deformation properties of remolded loess were revealed. The results show that the remolded loess has obvious induced anisotropy, which cannot neglect the stress-strain relationship, strength and deformation of soil. The maximum difference in fail strength is 25% under different principal stress direction. Induced anisotropy has great influence on the stress-strain relationship of soil, which is the essence of soil properties. In some engineering practices, the result of stress-strain analysis of soil has great influence. Approximately isotropic remolded loess substantially eliminates inherent anisotropy. Therefore, this experiment will provide a good reference for the further study of soil anisotropy under complicated stress paths such as principal stress axis rotation.

Keywords: Induced anisotropy · Hollow cylinder apparatus · Remolded loess Complex stress path · Strength and deformation characteristics

1 Introduction

Anisotropy is one of the important physical and mechanical characteristics of soils, and it affects the characteristics of soils to a great extent. Generally, it is believed that the anisotropy of soil is divided into inherent anisotropy and induced anisotropy [1]. Among them, the induced anisotropy refers to the soil under different stress conditions, the stress direction or form changes caused by different angles on the soil mechanical properties, permeability and resistance to deformation and other macro-performance differences [2]. In earlier studies of soil anisotropy, most of them were carried out in conventional triaxial or plane strain apparatus. In 1970, Saada [3] first proposed anisotropy research using an instrument that deflects the direction of principal stress. Since the 1980s, the hollow cylinder apparatus (HCA) is currently recognized as a device that can be used to study the anisotropy of soil better. Subsequently, the research of soil anisotropy has been carried out in academia. An empirical study on the anisotropy of loose-sand samples by HCA was conducted by Towhata [4]. Soil anisotropy research has also made a lot of progress by Nakai [5]. Later on, Shen [6] studied the anisotropy of sand.

The present research on anisotropy of loess has already lagged behind. Although some progress has been made in the research on the anisotropy of loess such as Chen [7]. In this paper, Zhejiang University hollow cylinder apparatus, the anisotropic experimental study of remodeled loess, supplemented the theory of loess anisotropy.

2 Materials and Methods

2.1 Test Materials

Table 1 represents the physical and mechanical properties of a preparation of remolded loess sample from a building foundation pit in Xining, Qinghai. The test control sample has a wet density of 1.8 g/cm^3 and a sample size of 200 mm * 10 mm * 60 mm (height * outer diameter * inner diameter).

Table 1. The basic parameters of Qinghai loess samples

Dry density (g/cm^3)	Water content (%)	Relative density	Liquid limit (%)	Plastic limit (%)
1.57	15	2.72	25.0	16.1

2.2 The Instrument Used in the Test

The instrument used in this paper is the hollow cylinder apparatus (HCA) of Zhejiang University. The complex stress path is realized by changing the average principal stress p, principal stress coefficient b, shear stress q and principal stress direction angle α.

2.3 Test Plan

In the experiment, the content of remolded loess sample was $\omega = 15\%$, the average principal stress was 200 kPa and the intermediate principal stress coefficient was $b = 0.5$. The direction angles of major principal stress were 0°, 30°, 45°, 60°, 90° for a total of 5 groups of experiments.

3 Results

3.1 Induced Anisotropy on the Relationship Between Stress and Strain

At low shear stage, the remolded loess has the same initial shear modulus at different angles (Fig. 1). When the shear stress increases to a certain value, the anisotropy of the

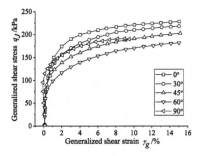
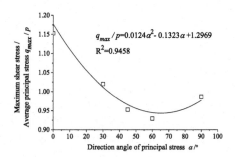

Fig. 1. Stress-strain curves in different α

Fig. 2. The relationship between the normalized strength and α

stress-strain relationship is very obvious. The secant shear modulus decreases with increasing angle, but it rises when the angle at $\alpha = 90°$.

3.2 Effects of Induced Anisotropy on Strength

When the main stress axis is oriented at $\alpha = 60°$ (Fig. 2), the failure strength reaches the minimum, which is 25% different from the maximum at $\alpha = 0°$. The strength of the remolded loess soil shows a quadratic function with the increase of the principal stress direction angle α, and the equation is $q_{max}/p = 0.0124\alpha^2 - 0.1323\alpha + 1.2969$. Furthermore, the shape of the curve is close to the shape of scoop. The maximum difference in fail strength is 25% under different principal stress direction.

3.3 Induced Anisotropy on the Deformation Characteristics

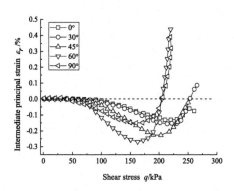

Fig. 3. Variation strains of ε_r with α

Intermediate principal strain increment (Fig. 3) in the early performance are as follows $\Delta\varepsilon_r < 0$. When principal stress direction angle is less than 45°, the intermediate principal strain increment appears at the later stage $\Delta\varepsilon_r < 0$, and $\Delta\varepsilon_r$ showed negative increase rapidly and finally fail.

However, when the direction angle is $\alpha \geq 45°$, the intermediate principal strain increment shows $\Delta\varepsilon_r > 0$. That is, the first negative increase to the late positive increase, and the development rate of $\Delta\varepsilon_r$ is significantly faster than the angle of less than 45° situation. In other words, $\alpha = 45°$ is a threshold value, the direction angle at both sides of $\alpha = 45°$ produces different strain characteristics.

4 Conclusions

(1) The anisotropy of stress-strain relationship is very obvious under the changing of principal stress direction angles. The secant shear modulus decreases with increasing of α, but it rises when the angle is at $\alpha = 90°$.
(2) The relationship between strength of remolded loess and direction angle of principal stress can be fitted into quadratic equation. The relationship curve shape is close to the shape of spoon. The maximum difference in fail strength is 25% under different principal stress direction.
(3) The principal stress direction angle $\alpha = 45°$ is a threshold value and the strain characteristics are different at both sides of $\alpha = 45°$. Remolded loess shows strong anisotropy under the change of the direction of principal stress.

Acknowledgements. This study was supported by the National Natural Science Foundation of China (Grants No. 51768060), the Fundamental Research Program and the Cooperation Program of Qinghai Province (Grants No. 2017-ZJ-792 and No. 2017-HZ-804), the Open Research Fund Program of Hohai University (Grants No. 2016002), the Technological Innovation Service Platform of Qinghai Province (Grants No. 2018-ZJ-T01).

References

1. Li, G.X.: Advanced Soil Mechanics. Tsinghua University Press, Beijing (2004)
2. Liu, Z.Y.: Study on yield characteristics and anisotropic yield surface equation of soft clay. Zhejiang University (2014)
3. Saada, A.S.: Testing of anisotropic clay soils. J. Soil Mech. Found. Div. ASCE **96**(5), (1970)
4. Towhata, I., Ishihara, K.: Undrained strength of sand undergoing cyclic rotation of principal stress axes. Soils Found. **25**(2), 135–147 (1985)
5. Nakai, T., Ftljii, J., Taki, H.: Kinematic of anisotropic hardening model for sand. In: Nakai, T. (ed.) Proceedings of the 3rd International Conference on Constitutive Laws for Engineering Materials, pp. 36–45 (1991)
6. Shen, Y., Zhou, J., Zhang, J.L.: Influence of cyclic rotation of principal stress axis with low shear stress on undisturbed clay. Chin. J. Rock Mech. Eng. **S1**, 123–126 (2008)
7. Chen, W., Zhang, W.Y., Chang, L.J.: Experimental study of anisotropy of compacted loess under directional shear stress path. Chin. J. Rock Mech. Eng. **09**, 4320–4324 (2015)

Experimental Analysis of Over-Consolidation in Subsoil in Bratislava, Slovakia

Zuzana Galliková[(✉)]

Faculty of Civil Engineering, Slovak University of Technology,
Radlinského 11, 810 05 Bratislava, Slovakia
zuzana.gallikova@stuba.sk

Abstract. It is difficult to determine the over-consolidation ratio when conducting practical tasks in the field of geotechnics. Geological layers undergo erosion processes that influence the stresses that exist in subsoil. The paper presents an analysis of erosion thickness via two methods. The first is Baldwin-Butler's method, which is based on compaction curve of many dates. The second is Casagrande's method, which is based on concept of preconsolidation stress. An analysis has been conducted for the central part of Bratislava (Slovakia, Central Europe). The subsoils at this location consist of fine-grained sediments from the Miocene and Pliocene periods. Silt and clay dominate in these Neogene deposits with a minority of fine and clayey sands. The over-consolidation ratio OCR was evaluated based on the results of oedometer tests of undisturbed soil samples. The measured normal stresses reached 7.0 MPa. The resulting thickness of the erosion determined according to Baldwin-Butler's method ranged between 194–272 m, while Casagrande's method produced a range between 81–187 m.

Keywords: Erosion thickness · Solidity · Over-consolidation ratio

1 Introduction

The thickness of the overburden layers that have been gradually eroded or resedimented due to geological processes has influenced the stresses in the subsoil. The determination of the thickness of erosion is complicated by the fact that many processes are complex and thus difficult to define numerically. Besides secondary consolidation, the ongoing ageing process also influences the mechanical behavior of clayey and silty soils. In this analysis, the main objective is to determine the eroded thickness of the overburden using two methods: Baldwin and Butler's [1] and Casagrande's method [2]. The subject of the analysis is the determination of the over-consolidation ratio for the Neogene subsoil, which is also influenced by the processes described above.

2 Geological Conditions in Bratislava

Most of Bratislava lies on Quaternary sediments of various geotechnical types which in most areas are founded on the aforementioned Neogene sediments. The thickness of these Quaternary deposits reaches up to 10.0 to 12.0 m. The predominant part of the geological profile is represented by the Neogene sediments, whose thickness extends for several tens of meters. In particular, the geotechnical report produced by an investigation conducted for the National Bank of Slovakia indicates thickness 150.0 m of Neogene layer. The Neogene sediments are typically fine grained with a pronounced dominance of silt and clay combined with minor proportions of fine and clayey sand. Pellets are blue, blue-gray and gray, with frequent occurrence of lignites and calcium concretions. The Neogene sediments are divided into the horizons of the Volkovce, Beladice and Ivanka formations. The geological age of the fine-grained soils is estimated to be about 10.0 Ma in the Beladice formation. In contrast, the Volkovce formation from which the tested soil samples were taken is considerably younger at around 6.0 Ma.

3 Analysis of Overburden Thickness and Over-Consolidation

3.1 Oedometer Tests on Undisturbed Samples of Soils

The presented laboratory experiments were conducted for samples of silts and clays obtained from three geological boreholes in the locality of central Bratislava. The samples were taken at depths of 19.0 to 32.5 m below the ground. It was necessary to determine preconsolidation pressure on the basis of the compressibility testing of undisturbed specimens. Vertical loads were applied to the samples in oedometers, achieving maximum normal stresses of approximately 7.1 MPa, which is significantly higher than those normally applied in practice (Fig. 1).

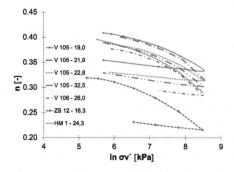

Fig. 1. Oedometer tests on undisturbed specimens – compressibility curves.

3.2 Determination of Erosion Thickness

The determination of the erosion thickness of overbunden requires data on stresses in the subsoil and on sedimentation processes as well as the age classifications of the geological horizon. The process of erosion is analyzed in this study from the perspective of its determination based on the empirical relationship developed by the authors Baldwin and Butler [1], and Casagrande [2].

With the Baldwin and Butler method, the erosion thickness is determined based on the S solidity of the soil, which is expressed as a percentage of the total sediment volume (n = (100 − S)). The porosity of the soil was determined for each test sample that was reconsolidated to the in-situ effective vertical stress σ_v'. The erosion thickness was calculated for the individual depth levels of samples according to:

$$E_1 = 6.02 \cdot S^{6.35} \cdot 1000 - h \quad [m]. \tag{1}$$

In Casagrande's method, the preconsolidation pressure to be used is determined from a one-dimensional compression test carried out in the oedometer. The preconsolidation pressure value obtained using this method is only the quasi-preconsolidation pressure, $\sigma^*_{vmax}{}'$; it is not the real preconsolidation pressure, σ_{vmax}' (Fig. 2). It is evident from the result that the real preconsolidation pressure value is in fact lower than that obtained by Casagrande. During the reduction of porosity Δn, the coefficient of compression of the Neogene clays $C_c = 0.27$ [3] was used. According to the principle in Fig. 2, the erosion thicknesses were calculated for both preconsolidation pressures according to the following relation:

$$E_2 = [\sigma_{vmax}'/(\gamma_{sat} - \gamma_w)] - h \quad [m]. \tag{2}$$

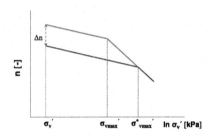

Fig. 2. Ideal compression curve – influence of the change in porosity.

The resulting erosion thicknesses obtained by both methods are summarized in Table 1. In the Baldwin-Butler method, the average erosion thickness is 230.0 m. According to Casagrande's method, it is 171.0 m (E_2^*) when the full value of the preconsolidation pressure $\sigma^*_{vmax}{}'$ is considered in relation (2). In contrast, the reduction in the influence of the change in soil porosity (σ_{vmax}') caused the resulting erosion thickness to be reduced to an average of 98.0 m (in Table 1, column E_2).

Table 1. The thickness of erosion and over-consolidation ratio of Neogene clays.

Depth of samples [m]	E_1 [m]	E_2^* [m]	E_2 [m]	OCR^* [-]	OCR [-]
19.0	194	187	86	5.4	2.8
21.9	242	179	97	5.3	3.1
22.8	246	163	90	4.3	2.6
24.3	221	119	81	4.5	3.3
26.0	272	168	110	4.1	2.8
32.5	203	211	126	3.9	2.6

3.3 Over-Consolidation Ratio OCR

The over-consolidation ratio OCR is determined based on the ratio of the effective vertical stress values σ_{vmax}' acting in the past to the current vertical stress σ_v' (3).

$$OCR = \sigma_{vmax}'/\sigma_v' \quad [-] \tag{3}$$

The limitation of Casagrande's method is that it does not take the process of secondary compression or the ageing of the soil into account. The preconsolidation pressure is thus evaluated using an incorrect approach. The value obtained for the over-consolidation of the erosion thickness was determined in the same way as it was via the use of Casagrande's method. The over-consolidation ratio OCR was expressed for the preconsolidation pressure value, which is defined as the "break point" on the compression curve (OCR^*), and also for the reduced stress value, taking the effect of the secondary compression in the soil (OCR). The resulting values are shown in Table 1.

4 Conclusions

Currently, it is not possible to determine the degree of over-consolidation or quantify the thickness of the eroded overburden numerically with sufficient accuracy. In fine-grained soils, on the one hand there are processes taking place such as primary and secondary compression as well as the effects of ageing, but on the other hand there are also unloading, and reloading effects present throughout the resedimentation process. With regard to the good data available on the geological conditions of the site, it is strongly important to proceed carefully when reducing the excessive preconsolidation pressure value. As Baldwin-Butler's equation is based on a number of data inputs, such as regressional dependence on multiple sites, this can result in the limitation of erosion determination for given subsoil. Both equations are based on the description of mechanical properties and do not allow the effects of secondary compression or the ageing of the soil to be taken into account. The confirmation of the actual thickness of erosion for the Neogene subsoil of Bratislava is currently the subject of further investigation.

References

1. Baldwin, B., Butler, C.O.: Compaction curves. AAPG Bull. **69**(4), 622–626 (1985)
2. Casagrande, A.: The determination of the pre-consolidation load and its practical significance. In: Proceedings of the 1st International Conference on Soil Mechanics and Foundation Engineering, pp. 60–64. Harvard University, Cambridge (1936)
3. Mesri, G., Castro, A.: C_α/C_c concept and K_0 during secondary compression. J. Geotech. Eng. **113**(3), 230–247 (1987)

Study on the Influence of Moisture Content on Mechanical Properties of Intact Loess in Qinghai, China

Anbang Guo[1], Wuyu Zhang[1(✉)], Lingxiao Liu[1], and Yanxia Ma[1,2]

[1] School of Civil Engineering, Qinghai University,
251 Ningda Road, Xining 810016, Qinghai, China
anbang1115@163.com, qdzwy@163.com
[2] The Key Laboratory of Geomechanics and Embankment Engineering,
Hohai University, Nanjing 210098, Jiangsu, China

Abstract. In order to research the effect of different moisture content of intact loess in Qinghai on its strength and deformation characteristics, undrained triaxial compression are performed using the British GDS Standard Stress Path Test System. The results show that under the same moisture content, the failure strength of samples increases with the increase of cell pressure. Under the same cell pressure, the failure strength decreases with the increase of moisture content. The cohesive decreases with the increase of moisture content, but the change of internal friction angle is very small. By fitting the cohesion and internal friction angle with the change of moisture content curve, the formula of Coulomb shear strength cooperating moisture content is obtained. The formula has certain reference value for the design of foundations and ground treatment to practical projects in the Qinghai area.

Keywords: Intact loess · Moisture content · Shear strength parameters

1 Preface

The distribution of loess in the Qinghai area is 24800 km^2, accounting for 3.9% of the total area of the distribution of loess in China [1]. Moreover, seasonal precipitation in the northwest region varies greatly during one year, therefore, it is necessary to study the influence of moisture content on the strength and deformation characteristics of intact loess [2]. In the past ten years, little research has been done on the mechanical properties of undisturbed loess in the Qinghai area [3–5]. This paper utilizes the British GDS Standard Stress Path Test System to research the effect of different moisture content of intact loess in Qinghai on its strength and deformation characteristics through triaxial tests, with a view to provide a reference for the construction of the project.

2 Test Method

The soil samples were taken from the North District of Xining City, Qinghai Province. The depth of soil was 4 m. The soil samples were light yellow in color and uniform in texture. The physical properties are shown/show in Table 1.

Table 1. Physical properties of soil samples.

Depth/m	Natural moisture content/%	Specific gravity	Liquid limit/%	Plastic limit/%
4	5.06	2.72	25.23	14.53

Test equipment using the British GDS Standard Stress Path Test System. Test method using consolidation undrained shear mode. 3 groups of soil samples were taken, with 3 samples in each group. Before the test, the soil sample was taken from the sealed bag, then, the intact soil was made into a cylinder with a diameter of 39.1 mm and a height of 80 mm, loads the sample into the three-segment-die, the sample moisture content is equipped with 8%, 14% and 20% by drip water method, then loads the sample in a humidifier for 3 days to ensure moisture transfer and mix well, consolidation was performed at 100 kPa, 200 kPa and 300 kPa, respectively, a shear rate of 0.08 mm/m was used. In order to study the mechanical properties of intact loess better, take 20% of axial strain as the failure criterion [6].

3 Test Results and Analysis

3.1 Intact Loess Deviator Stress-Strain Characteristics

From Fig. 1(a) we can see, the deviatoric stress-strain curve shows a strain hardening type and the hardening degree increases with the increase of cell pressure. This is because when the cell pressure is increasing, the structure of unsaturated loess will

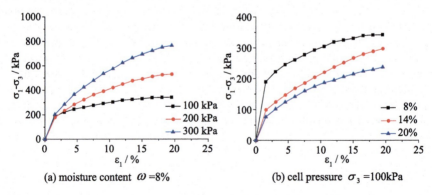

Fig. 1. The deviatoric stress-strain curves of the samples with different cell pressures at different moisture content

decrease. The failure strength increases with the increase of cell pressure, the reason is that the cell pressure increases the density of the sample, thus enhancing the sample strength.

From Fig. 1(b) we can see, with the same cell pressure and strain, the deviatoric stress of sample decreases with the increase of moisture content, because the increase of moisture content increases the water around soil particles, reduce the connection and cohesion of soil particles, the corresponding soil strength also decreases. This shows that the moisture content of the sample plays an important role in the sample strength.

3.2 Sample Damage Morphology

In the experiment, the damaged samples did not show obvious shear bands, but were compacted and had an obvious bloating phenomenon.

3.3 The Effect of Moisture Content on Cohesion and Internal Friction Angle

It can be seen from Fig. 2(a) that the cohesion of the sample decreases with the increase of moisture content, which is more obvious when the moisture content increases from 14% to 20%.

$$\text{Fit equation}: c = 56.4 - 228.58\omega \ (R^2 = 0.94) \tag{1}$$

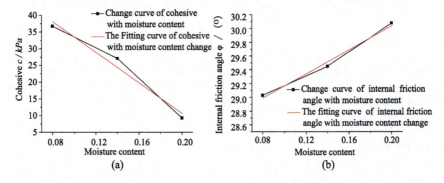

Fig. 2. Change curve of cohesive and internal friction angle with moisture content

It can be seen from Fig. 2 (b) that the internal friction angle of the sample basically does not change much at about 30°, and the moisture content has little effect on the internal friction angle of the original sample.

$$\text{Fit equation}: \varphi = 28.36 + 8.42\omega \ (R^2 = 0.96) \tag{2}$$

Taking above two fitting equations into the soil of the Coulomb shear strength expression ($\tau_f = c + \sigma \tan \varphi$), the formula of Coulomb shear strength cooperating moisture content is obtained:

$$\tau_f = 56.4 - 228.58\omega + \sigma \tan(28.36 + 8.42\omega) \tag{3}$$

τ_f—Shear strength of soil (kPa);
σ—Normal stress acting on the shear plane (kPa).

4 Conclusions

(1). At the same moisture content, with increased of cell pressure, the stress-strain curve changes from weakly hardened to hardened. Under the same cell pressure, the shear strength of undisturbed loess decreases with the increase of moisture content.
(2). There is no obvious shear band in the damaged sample, however, there is a phenomenon of bulging.
(3). Cohesion decreases with the increase of moisture content, the internal friction angle increases with the increase of moisture content, their changing rules can be expressed by fitting Eqs. (1) and (2). Taking fitting Eqs. (1) and (2) into the Coulomb shear strength expression, the formula of Coulomb shear strength cooperating moisture content is obtained: $\tau_f = 56.4 - 228.58\omega + \sigma \tan(28.36 + 8.42\omega)$. This formula has certain reference meaning for some practical engineering foundation designs and foundation treatments in the Qinghai area.

Acknowledgements. This study was supported by the National Natural Science Foundation of China (Grants No. 51768060), the Fundamental Research Program and the Cooperation Program of Qinghai Province (Grants No. 2017-ZJ-792 and No. 2017-HZ-804), the Open Research Fund Program of Hohai University (Grants No. 2016002), the Technological Innovation Service Platform of Qinghai Province.

References

1. Zhang, Z.X.: Loess and loess engineering, 1st edn. Qinghai People's Publishing House, Xining (1998)
2. Wu, M.H., Xu, D.Y., Zhu, X.W.: Deformation characteristics of remolded loess with different moisture content. Subgrade Eng. **01**, 55–59 (2015)
3. Xing, Y.L., Li, T.L., Li, P.: Variation of shear strength with moisture content of loess. Hydrogeol. Eng. Geol. **41**(03), 53–59 (2014)

4. Chang, L.J., Zhang, W.Y., Ma, Y.X.: Experimental study on strength characteristics of unsaturated loess in Qinghai. Railway Eng. **04**, 117–121 (2015)
5. Chen, W., Zhang, W.Y., Ma, Y.X.: Analysis on influencing factors of mechanical properties of remolded loess in Qinghai. Railway Eng. **01**, 82–84 (2014)
6. The Ministry of Water Resource of the People's Republic of China: Standard for Soil Test Method (GB/T 50123—1999). China Planning Press, Beijing (1999)

Undrained Shear Behaviour of Gassy Clay with Varying Initial Pore Water Pressures

Y. Hong, L. Z. Wang$^{(\boxtimes)}$, and B. Yang

Key Laboratory of Offshore Geotechnics and Material of Zhejiang Province,
College of Civil Engineering and Architecture, Zhejiang University,
Hangzhou 310058, China
wanglz@zju.edu.cn

Abstract. The marine sediments often contain relatively large biogas bubbles. The presence of bubbles has been posing additional challenges to the stability of offshore foundation. Although fine-grained gassy soil was found to exhibit different undrained shear strength (c_u) by altering the initial pore pressure u_i (relevant to water depth), systematic studies concerning the effect of u_i on undrained shear behaviour of the soil are still lacking. This study reports a series of undrained triaxial tests aiming to compare and investigate the responses of reconstituted fine-grained gassy soil with the same consolidation pressure (p_c') but at a wide range of varying u_i (0 to 1000 kPa). The shearing-induced excess pore pressure (Δu) in the gassy specimens highly depended on u_i. It can be either smaller than that of the saturated specimen with the same p_c' (due to partial dissipation of Δu into relatively large bubbles at low u_i), or larger than that of saturated specimen (related to collapse of relatively small bubbles at high u_i). Consequently, the presence of bubbles had beneficially increased c_u at relatively low u_i (u_i/p_c' < 0.6), and vice versa. The critical stress ratio of the reconstituted fine-grained gassy soil, however, did not appear to be altered by u_i.

Keywords: Fine-Grained gassy soil · Undrained shear strength
Pore water pressure · Critical state stress ratio · Triaxial test

1 Introduction

Undissolved gas bubbles are widely found in marine sediments throughout the world, due to bacterial activity, thermal alternation or hydrate dissociation [1]. Unlike the conventional unsaturated soils, the gas phase in the gassy sediments is discontinuous, while the water phase remains continuous, with a degree of saturation (s_r) exceeding 90% [2].

The undrained shear behaviour of fine-grained gassy soil, which governs the stability and deformation of various offshore structures founded on gas-charged seabed, was preliminary investigated through triaxial testing [2]. It was founded that the presence of gas may either increase or decrease the undrained shear strength (c_u), depending not only on effective mean normal stress (p') but also on the initial pore water pressure (u_i). The influence of u_i on other undrained shear behaviours (such as stress-strain relationship and pore water pressure), however, was rarely reported. In addition, only two values of initial pore water pressures u_i within a narrow range (0 to 100 kPa) were considered in their experiments. This study systematically investigates the effects of u_i on the mechanical behaviour of fine-grained gassy soil, through a series of undrained triaxial compression tests. The undrained shear behaviour of gassy soils with different u_i are interpreted and discussed with their practical implications.

2 Experimental Programme

To achieve the objectives stated above, a series of undrained triaxial compression tests were carried out on fine-grained gassy soil specimens, which were isotopically consolidated to the same p_c' (200 kPa) but with different values of u_i (0, 120, 250, 600, 1000 kPa). For comparison, one benchmark test on non-gassy saturated specimen (p_c' = 200 kPa, u_i = 120 kPa) was also performed.

3 Soil Sample Preparation and Experimental Procedure

Gassy clay was produced by introducing N_2 to Malaysia kaolin slurry, using the zeolite molecular sieve technique [2]. The gassy slurry was then one-dimensionally consolidated under a maximum vertical stress of 60 kPa. Figures 1(a) and (b) show cross-sectional views of typical gassy and saturated specimens after the consolidation. Each one-dimensionally consolidated specimen was then isotopically consolidated to p' = 200 kPa, but at different back pressures in the triaxial apparatus, followed by an undrained triaxial compression at a strain rate of 1.5% per hour.

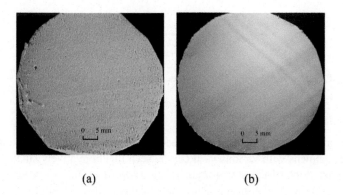

(a)   (b)

Fig. 1. Photographs of (a) gassy clay; (b) saturated clay

4 Interpretation of The Experimental Results

4.1 Effective Stress Path

Figure 2 shows the effective stress paths of the gassy and saturated specimens subjected to undrained shearing. Compared to the saturated specimen, the locus of the undrained effective stress path of the gassy specimen with relatively low u_i (0 kPa) was much closer to the drained effective stress path. This implied the occurrence of partial drainage (through bubble flooding [2]) in the gassy specimen with low u_i, which contained relatively large size of gas bubbles and therefore exhibited a low water entry value. At relatively high u_i (between 250 and 1000 kPa), the undrained effective stress paths of the gassy specimens exhibited a more contractive response than that of the non-gassy saturated specimen, probably due to collapses of the cracks which were initially created around the highly pressurized gas pores [3, 4]. Despite the different effective stress paths at varying u_i, they all appeared to reach a unique critical state stress ratio (M) of 1.

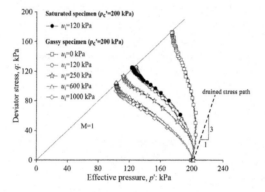

Fig. 2. Effective stress paths of gassy specimens at varying initial pore pressures (u_i)

4.2 Stress-Strain Relationships

Figure 3 compares the relationships between the deviatoric stress and the axial (ε_a) of the gassy specimens with various u_i. It can be seen that at relatively small axial strains ($\varepsilon_a < 3\%$), the slope of the stress-strain relationship of all the gassy specimens were smaller than that of the non-gassy saturated specimen. This is likely to suggest that the presence of the gas bubbles had increased the compressibility of the specimen.

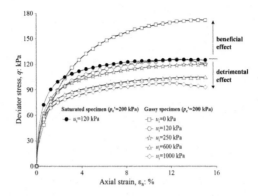

Fig. 3. Stress-strain relationship at varying initial pore pressures (u_i)

Regarding the shear strength, the presence of relatively large gas bubbles (at low u_i) played a beneficial role in increasing c_u, likely due to partial dissipation of pore pressure into the large bubbles. On the other hand, the inclusion of relatively small gas bubbles (at high u_i) became detrimental by reducing c_u, probably related to the collapse of the small gas bubble which had made the material more contractive. The water depth is therefore likely to be a key factor governing the bearing capacities of offshore structures founded on fine-grained gassy marine sediments [5].

5 Conclusions

(1) The shearing-induced excess pore pressure (Δu) of the gassy specimens significantly depended on u_i. It can be either smaller than that of the saturated specimen (due to partial dissipation of Δu into the relatively large bubbles at low u_i), or greater than that of the saturated specimen (due to collapses of the relatively small bubbles at high u_i).
(2) Correspondingly, at a relatively low u_i (0 kPa), the presence of gas had beneficially increased undrained shear strength (c_u) by 33%, than those of the non-gassy saturated specimen. In contrast, the presence of gas at high u_i (1000 kPa) had detrimentally decreased c_u by 22%, as compared to those of the saturated specimen. Water depth, therefore, could be a key factor governing the bearing capacity of offshore structures founded on fine-grained gassy marine sediments.
(3) Despite the distinct undrained shear behavior of the gassy specimens at various u_i, their critical state stress ratio (in the p'-q space) appears to be a constant (i.e., M = 1).

Acknowledgements. The authors gratefully acknowledge the financial support from the National Key Research and Development program of China (No. 2016YFC0800204), the National Natural Science Foundation of China (No. 51779221) and Natural Science Foundation of China of Zhejiang Province (Y17E090016).

References

1. Rebata-Landa, V., Santamarina, J.: Mechanical effects of biogenic nitrogen gas bubbles in soils. J. Geotech. Geoenviron. Eng. **138**(2), 128–137 (2012)
2. Wheeler, S.J.: The undrained shear strength of soils containing large gas bubbles. Géotechnique **38**(3), 399–413 (1988)
3. Sultan, N., De Gennaro, V., Puech, A.: Mechanical behavior of gas-charged marine plastic sediments. Géotechnique **62**(9), 751–766 (2012)
4. Sultan, N., Garziglia, S.: Mechanical behaviour of gas-charged fine sediments: model formulation and calibration. Géotechnique **64**(11), 851–864 (2014)
5. Hong, Y., He, B., Wang, L.Z., Wang, Z., Ng, C.W.W., Masin, D.: Cyclic lateral response and failure mechanisms of a semi-rigid pile in soft clay: centrifuge tests and numerical modelling. Can. Geotech. J. **54**(6), 806–824 (2017)

Creep of Reconstituted Silty Clay with Different Pre-loading

Minyun Hu[1]([✉]), Bin Xiao[2], Shuchong Wu[3],
Peijiao Zhou[1], and Yuke Lu[1]

[1] Zhejiang University of Technology, Hangzhou 310014, China
huminyun@zjut.edu.cn
[2] Zhejiang Tongji Science and Technology Vocational College,
Hangzhou 311231, China
[3] Guangsha College of Applied Construction Technology, Jinhua 321007, China

Abstract. Creep characteristics of reconstituted silty clay were tested under isotropic consolidation and triaxial shearing, respectively, with different pre-loading steps. The test results show that the pre-loading steps have little impact on the creep of sample in isotropic consolidation. However, pre-loading steps have an evident effect on silty clay's creep under triaxial shearing, showing that one-step loading of shearing leads to a higher deviatoric creep strain than the multi-stage loadings along the same loading path. The multi-stage loaded specimens exhibit higher undrained shear strength after creep. The creep behavior was then examined by two different models, the logarithmic model and the hyperbolic model. The hyperbolic model exhibits better agreement with the test data, and the two parameters can be expressed in linear relationship with stress-level.

Keywords: Silty clay · Creep · Loading steps · Isotropic compression
Triaxial shearing

1 Introduction

Soil creeps under constant stress state, resulting in long term deformation/failure of structures built in/on ground especially of predominant clayey soils. In previous studies [1–5], undisturbed and reconstituted soil samples were tested to investigate their creep behaviour under different loading paths, different stress levels, etc. However, the influence of loading steps under a specific loading path, which may relate to the construction schedule and management, were rarely involved.

Silty clay is widely distributed in Hangzhou Bay area, an active area of economy and construction in China. In this paper, series of hydrostatic compression and triaxial shearing tests were carried out on reconstituted silty clay to study its creep character with different loading steps.

2 Material and Method

The soil specimen in this study was from an excavation site in Jiaojiang District, Taizhou, which is located in Hangzhou Bay area, and is classified as silty clay. Its specific gravity was G_s= 2.72, moisture content w = 32.15%, liquid limit w_L= 37.90%, plastic limit w_P= 22.58%, void ratio e = 0.875. The intact sample was in plastic state. The reconstituted cylindrical samples were of $h*D$ = 78*38 mm and were prepared by the moist tamping method (dynamic compaction) [4]. Test program were performed by GDS automatic triaxial testing system. The laboratory temperature was kept constant at 23 ± 1 °C.

The samples were tested under two loading paths: isotropic compression and triaxial shearing. For the isotropic compression tests, samples were loaded at a constant loading rate to an all-round pressure of p = 800 kPa with three different preloading steps: $p = 0 \rightarrow 800$ kPa, $p = 0 \rightarrow 400 \rightarrow 800$ kPa, $p = 0 \rightarrow 200 \rightarrow 400 \rightarrow 600$ 800 kPa. For the triaxial shearing tests, samples were first consolidated under all-around pressure of p = 800 kPa, and then loaded with three different pre-loading steps: $q = s_1 - s_3 = 360$ kPa, $q = 180 \rightarrow 360$ kPa, $q = 90 \rightarrow 180 \rightarrow 270 \rightarrow 360$ kPa. In both load paths, the samples were stressed under drained condition. Figure 1 shows the load paths of the samples and some details are described in Fig. 1 legend.

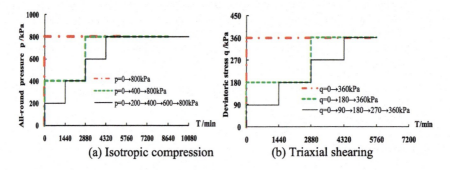

Fig. 1. Load path of test

3 Results and Discussion

3.1 The Isotropic Compression

Figure 2 presents the volumetric strain development during isotropic compressions and the pore pressure responses in all three tests of different pre-loading steps. As it can be seen the pore pressure dissipated to zero during each loading step, the total stress is the final effective stress to the samples. It is clear that although the three samples were compressed in different steps, they exhibit uniform total volumetric strain during the same final pressure, indicating that the pre-loading has little impact on the creep strain under isotropic compression.

3.2 The Triaxial Shearing

Figure 3 presents the deviatoric strain develops with time and the pore pressure responses in three different pre-loading tests. Compared with the isotropically compressed samples, the pre-loading has obvious impact in triaxial shearing system, implying that pre-loading can reduce the time-dependent sheared deformation of silty clay. However, the pre-loading has little effect on the volumetric strain evolution, as shown in Fig. 4. All samples were sheared undrainedly after creep, and the preloading samples exhibit higher undrained shear resistance.

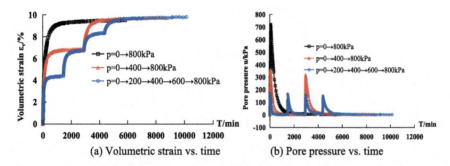

Fig. 2. Volumetric strain/Pore pressure vs. time in isotropic compression

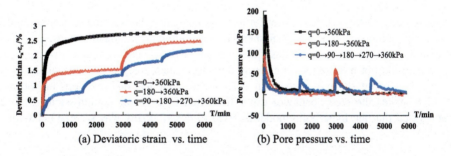

Fig. 3. Deviatoric strain/Pore pressure vs. time in triaxial shearing

3.3 The Creep Model of Reconstituted Silty Clay

The creep behavior was then examined by two nonlinear models, the logarithmic model and the hyperbolic model. The logarithmic model is suitable to describe soil's creep under one-step hydraulic compression, but seemed not fit well with soil's creep under triaxial shearing. Then a hyperbolic model was proposed as

$$\varepsilon_s(t) - \varepsilon_s(t_0) = A_s \frac{t - t_0}{B_s + (t - t_0)} \tag{1}$$

where $\varepsilon_s(t)$ is deviatoric strain at t (min), $\varepsilon_s(t) = \varepsilon_z(t) - \varepsilon_r(t)$; $\varepsilon_s(t_0)$ is the deviatoric strain at t_0 (min), the start time of creep; A_s is the ultimate creep strain $A_s = \varepsilon_s(\infty) - \varepsilon_s(t_0)$; and B_s is a parameter of time, which is the time to have half creep strain evolved.

This model fits well with silty clay's creep under any loading step and the parameters A_s and B_s are in linear relationship with stress level.

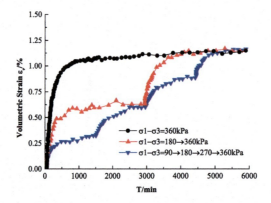

Fig. 4. Volumetric strain vs. time in triaxial shearing

4 Conclusions

In this study, tests were carried out for reconstituted silty clay specimen creeping under isotropic compression and triaxial shearing with different pre-loading steps. The conclusions are: (1) Pre-loading has little impact on silty clay's creep under isotropic compression but has obvious impact on deviatoric stain creep under triaxial shearing. (2) When being sheared, reconstituted silty clay contracts and develops anisotropy, which makes the creep behavior differ from that in isotropic compression. (3) It is suggested that the hyperbolic model be used to describe silty clay's creep behavior.

References

1. Leroueil, S., Kabbaj, M., Tavenas, F., Bouchard, R.: Stress strain-strain rate relation for the compressibility of sensitive natural clays. Geotechnique 35(2), 159–180 (1985)
2. Bishop, A.W., Lovenbury, H.T.: Creep characteristics of two undisturbed clays. In: Proceedings of 7th International Conference of Soil Mechanics and Foundation Engineering, vol. 1, pp. 29–37, Mexico City, Mexico (1969)
3. Bagheri, M., Rezania, M., Nezhad, M.M.: An experimental study of the initial volumetric strain rate effect on the creep behaviour of reconstituted clays. In: International Symposium on Geohazards and Geomechanics (2015)

4. Luo, Q., Chen, X.: Experimental research on creep characteristics of Nansha soft soil. Sci. World J. (2014)
5. Kong, L., Zhang, X.: Creep behavior of Zhejiang strong structured clay by drained traxial test. Chin. J. Rock Mech. Eng. **30**(2), 360–372 (2011). (In Chinese)
6. Ladd, C.C., Preston, W.B.: On the Secondary Compression of Saturated Clays. Department of Civil Engineering, MIT, Cambridge (1965)

Experimental Study on the Influence of Gravel Content on the Tensile Strength of Gravelly Soil

Enyue Ji[1,2(✉)], Shengshui Chen[1,2], Zhongzhi Fu[1,2], and Jungao Zhu[3]

[1] Geotechnical Engineering Department, Nanjing Hydraulic Research Institute, Nanjing 210024, China
eyji@nhri.cn
[2] Key Laboratory of Failure Mechanism and Safety Control Techniques of Earth-Rock Dam of the Ministry of Water Resources, Nanjing Hydraulic Research Institute, Nanjing 210029, China
[3] Key Laboratory of Ministry of Education for Geomechanics and Embankment Engineering, Hohai University, Nanjing 210098, China

Abstract. Tensile strength is one of the most important properties affecting anti cracking performance of earth core rockfill dam. However, the influence of gravel content on the tensile strength of gravelly soil is still unclear. In the paper, based on the self-developed uniaxial tensile test device, a series of tensile tests were performed on gravelly soils with different gravel content. For gravelly soils with different gravel content, the tensile strength decreases linearly with the increase of gravel content at the optimal water content and maximum dry density. In addition, Empirical formula to calculate tensile strength of gravelly soil based on the gravel content is put forward. The relevant conclusions are helpful to improve the anti-cracking design level of actual earth core rockfill dam.

Keywords: Gravelly soil · Tensile strength · Uniaxial tensile test
Gravel content

1 Introduction

High earth-rock dams are widely distributed in China, among which, earth core rockfill dams are frequently used in poor geological conditions due to its excellent ability of coordinating deformation. Usually, in order to increase the modulus of the core, a certain proportion of gravel is incorporated into the clay as the core wall. With the increase of gravel content, the ability of the core to resist shear deformation increases significantly. However, whether the crack resistance of the gravelly soil can still reach the anti-crack design requirements of the dam is worth further study.

At present, many researchers [1, 2] have conducted a large number of uniaxial and triaxial tensile tests on clay and obtained quantitative relations between tensile strength and properties (e.g. dry density, and matrix suction). Besides, some researchers [3, 4] have conducted a series of tensile tests toward some special soils. Zhu [5] studied the variation of tensile strength on gravelly soil with different compaction energy, saturation and water content. In general, there are few studies on the tensile properties of gravelly soil, and the variation of tensile strength with different gravel content on gravelly soil needs to be further investigated.

In this paper, the uniaxial tensile tests of gravelly soil with different gravel content were carried out by using a self-developed tensile device. The variation of tensile strength on gravelly soil with different gravel content was investigated.

2 Device, Soils and Schemes of the Test

2.1 Tensile Test Device

In view of the disadvantages of tensile devices used by the former researchers, a new type of uniaxial tensile device was applied in the test as can be seen in Fig. 1.

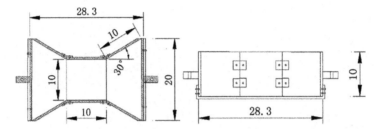

Fig. 1. Schematic diagram of tensile device (unit cm).

2.2 Soils Used in the Test

The clay used in the test was from the core of an earth core rockfill dam. The maximum particle size is 2 mm. The basic parameters of the clay are shown in Table 1.

Table 1. Basic parameters of the core material.

Gs	ω_L/%	ω_P/%	I_p	ω/%	k/cm·sec^{-1}
2.74	30.4	20.5	10	22.2	2.0×10^{-6}

The gravel in the gravelly soil used in the test was from an earth core rockfill dam and the parent rock was granite. The particle size of the gravel is controlled to be 20 mm (1/5 of the longest side of the sample) and the minimum size is controlled to 2 mm.

2.3 Schemes of the Test

First, compaction tests were conducted on gravelly soils with different gravel content to obtain the maximum dry density and the optimum water content. The compaction test results can be seen in Table 2.

Table 2. Optimum water content and maximum dry densities of gravelly soil with gravel content from 0–50%.

Gravel content/%	0%	10%	20%	30%	40%	50%
$\omega_{op}/\%$	17.5	16.4	14.8	13.3	11.5	10.1
$\rho_{max}/\text{g·cm}^{-3}$	1.73	1.82	1.88	1.98	2.01	2.05

The control variable method was used to formulate the test schemes as can be seen in Table 3.

Table 3. Tensile test schemes of gravelly soil.

Classifications	Gravel content/%	$\rho/\text{g·cm}^{-3}$	$\omega/\%$
Clay	0	1.73	17.5,15.5,19.5
		1.73,1.63,1.53	17.5
Gravelly soils	10	1.82	16.4,14.4,18.4
		1.82,1.72,1.62	16.4
	20	1.88	14.8,12.8,16.8
		1.88,1.78,1.68	14.8
	30	1.98	13.3,11.3,15.3
		1.88,1.78,1.68	13.3
	40	2.01	11.5,9.5,13.5
		2.01,1.91,1.81	11.5
	50	2.05	10.1
		2.05,1.95,1.85	10.1,8.1,12.1

3 Test Results

The tensile strength is mainly dominated by the clay content in the gravelly soil. When the gravel content increases, the clay content of the sample will decrease. Theatrically, if the gravel content increases to 100%, the tensile strength of the sample is almost equal to 0 except the partial bite force between the gravels. Figure 2 shows the relation curve of tensile strength versus gravel content.

It can be seen that when the gravel volume increases from 0% to 50%, the tensile strength decreases from 122.6 kPa to 49.8 kPa. The linear relationship can be expressed as:

$$\sigma_t = -1.5741\lambda + 120.96 \tag{3}$$

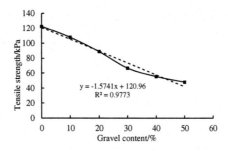

Fig. 2. Relation curve of tensile strength and gravel content.

4 Conclusions

In the paper, based on the self-developed uniaxial tensile test device, a series of tensile tests were performed on gravelly soils with different gravel content. The main conclusions are as follows:

(1) For gravelly soils with different gravel content, tensile strength decreases linearly with the increase of gravel content.
(2) Based on the test results, empirical formula to calculate tensile strength of gravelly soil based on the gravel content was put forward in the paper.

Acknowledgements. The authors acknowledge the financial support from the National Key Research (2017YFC0404806) and Development Program of China (51779152).

References

1. Niu, Z.M., Lu, S.Q.: On some factors influencing the uniaxial tensile strength of cohesive compacted fill. Chin. J. Geotech. Eng. **2**, 35–44 (1983)
2. Tang, G.X., Graham, J.A.: Method for testing tensile strength in unsaturated soils. Geotech. Test. J. **23**(3), 377–382 (2000)
3. Lv, H.B., Zeng, Z.T., Ge, R.D., et al.: Experimental study of tensile strength of swell-shrink soils. Rock Soil Mech. **3**, 615–620 (2013)
4. Tang, C.S., Wang, D.Y., Cui, Y.J., et al.: Tensile strength of fiber-reinforced soil. J. Mater. Civ. Eng. **28**(7), 04016031 (2016)
5. Zhu, J.G., Liang, B., Chen, X.M., et al.: Experimental study on uniaxial tensile strength of compacted soils. J. Hohai Univ. (Nat. Sci.) **2**, 186–190 (2007)

Direct Simple Shear Tests on Swedish Tailings

Qi Jia and Jan Laue[✉]

Division of Mining and Geotechnical Engineering,
Luleå Tekniska Universitet, 97187 Luleå, Sweden
jan.laue@ltu.se

Abstract. Tailings is a waste stream produced by mining industry. It is often stored on the surface in a retaining structure, called tailing impoundment surrounded by tailing dams. Strength property of tailings is important for safety of tailings dam construction. In this study, direct simple shear tests were performed on two Swedish tailings from Malmberget mine and Svappaavare mine in Northern Sweden. The apparatus used is NGI direct simple shear apparatus. The estimated maximal friction angle from shear tests ranged from 15.9 to 24.3°, with cohesions from 0-16.3 kPa. One phenomenon noticed was that for tailings the sample height started decrease after dilatancy. In order to know whether this is a material property, or it is due to the defect of the simple shear apparatus, a series of shear tests were done on Kalix sand. The result showed that this phenomenon happened also to the densely compacted sand, though more rarely. The decreased sample height after reaching the peak values is assumed to be related to both particle breakage and the deficiency of the shear apparatus.

Keywords: Tailings · Direct simple shear · Dilatancy

1 Introduction

Tailings is a waste stream produced by the processes where the desired products are extracted from the mine ore. Tailings consist of ground rock and the size can range from clay to sand. It is often stored on the surface in a retaining structure, called tailings impoundment surrounded by tailings dams. The tailings dams are often constructed of the tailings material itself and raised in stages or continuously over time. The strength properties are important for the stability of tailings dams to avoid dam failure. From environmental point of view, it is prudent to avoid unwanted release of hazardous substances to the environment by ensuring dam safety.

Direct simple shear test is a widely used method to measure shear strength of granular materials thanks to its convenience on sample preparation and less time consumption. Strength parameters can then be estimated from the test result. A series simple shear tests were done on two Swedish tailings, as well as a natural sand from Kalix. This paper presents the test result and compares the result from tailings and natural sand.

2 Material and Method

Direct simpler shear tests were done for two Swedish tailings material from Malmberget mine and Svappaavare mine in Northern Sweden. Undisturbed tailings samples were taken by piston sampler at various depths and locations with dry density varied from 1.22 to 1.7 t/m3. The particle density for tailings varied from 2.80–3.31 t/m3, while the normal value for a natural sand is 2.65–2.7 t/m3. This indicates the high metal content in tailings material and both tailings were from iron ore. Particle size distribution curves (PSD) are shown on Fig. 1. All the tailings samples are classified as clayey silt according to the PSD. A series of tests were also done for a natural sand from Kalix and the PSD curve is shown on Fig. 1.

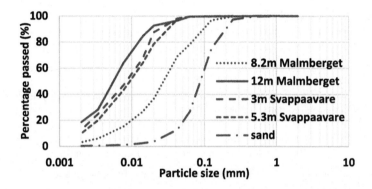

Fig. 1. Particle size distribution curves of the tailings and the sand

Shear tests were performed by the NGI direct simple shear (DSS) apparatus which was developed in the 1960's by Landva and Bjerrum and has been extensively used throughout the world. In Sweden it is common to evaluate strength parameters from DSS tests to be later used in geotechnical modelling. During shear test, a cylindrical soil specimen with a cross-sectional area of 20 cm^2 and a height about 16 mm is enclosed in a reinforced rubber membrane which prevents radial deformation. The specimen is sheared under a certain normal load with selected strain rate, either drained or undrained. All the tests performed in this study were drained test with a low strain rate to prevent pore pressure building up.

3 Result and Discussion

For Malmberget tailings DSS tests were done under normal tresses of 50, 150 and 300 kPa for two samples from different depths. For Svappaavare tailings DSS tests were done under normal stresses of 50, 100 and 200 kPa for two sample from different depths. Stress strain curves as well as change of effective sample height for tailings from Malmberget mine at depth 8.2 m are shown in Fig. 2. Dilatancy appeared at the lowest normal stress of 50 kPa (Fig. 1 right), while higher stress levels impeded particles from overriding each other and therefore resulted in contractency. Strength parameters according to Mohr-Coulombs failure criteria have been evaluated for respective test series (Table 1). Cohesion for tailings sample from Svappaavare at the depth of 3 m resulted in negative value with the help of regression analysis, which was manually chosen to be 0. Friction angle of tailings is generally lower than natural sand which has a value of about 45°.

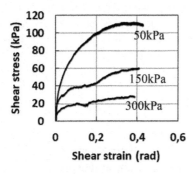
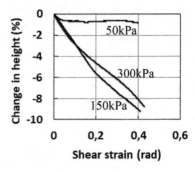

Fig. 2. Stress-strain curves (left) and change in effective sample height (right) from DSS test on Malmberget tailings from depth 8.2 m.

Table 1. Strength parameters for the tailings samples.

Mine	Depth, m	Cohesion c, kPa	Friction angle φ, °
Malmberget	8.2	10.9	18.2
	12	16.3	15.2
Svappaavare	3	0 (−0.6)	22.1
	5.3	3.6	19.7

One phenomenon noticed was that for tailings the sample height started decrease after dilatancy (Fig. 2 right 50 kPa). The phenomenon was also noticed by Bhanbhro (2017) that most of the tests on tailings with dilatancy behaviour showed volume decrease closing to the end of shear test. In order to know whether this is a material property or it is due to the defect of the simple shear apparatus, a serious of shear tests were done on Kalix sand. The sand samples were compacted to a dry density of about 1.6 t/m3. DSS performed under normal stresses of 50, 100,150 and 250 kPa. Figure 3 shows the stress strain curves and the change in effective sample height. The same phenomenon was also found for sand but happened more rarely (Fig. 3, 250 kPa). Bhanbhro (2017) studied particle size degradation after shearing on tailings and found that there was 10–40% particle breakage depending on the original size. Tailings is usually softer than sand and it is less likely that sand particles would break during shearing. Another reason could be that there is a momentum activated when a certain shear strain has reached and therefore the sample starts to be tilted. Thus the phenomenon is attributed to both particle breakage and the apparatus deficiency. However, whether or not momentum is generated has to be evidenced by pressure distribution measurement under the sample.

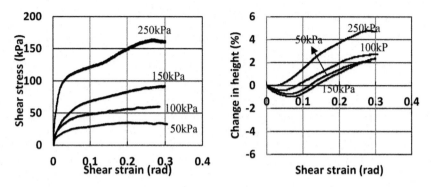

Fig. 3. Stress-strain curves (left) and change in effective sample height (right) from DSS test on Kalix sand.

4 Conclusion

A series of DSS tests were done on two Swedish tailings and a natural sand. The test result showed that tailings have lower friction angles than sand. Some tests with dilatancy behaviour showed sample height decrease closing to the end of the shearing. This was attributed to both particle breakage and apparatus deficiency. However, the apparatus deficiency needs to be evidenced by further tests.

References

Bjerrum, L., Landva, A.: Direct simple shear tests on a Norwegian quick clay. Geotechnique **16**(1), 1–20 (1966). Also NGI Publ. No. 70. Dyvik, R., T. Berre, S

Bhanbhro, R.: Mechanical behaviour of tailings – Laboratory Tests from a Swedish Tailings dam. Ph.D. thesis, Luleå University of Technology (2017)

Concrete-Sand Interface in Direct Shear Tests

Zihao Jin[1(✉)], Qi Yang[1,2], Junzhe Liu[3], and Chen Chen[1]

[1] Central South University, Changsha, Hunan 410075, China
164811089@csu.edu.cn
[2] National Engineering Laboratory for High Speed Railway Construction,
Central South University, Changsha, Hunan 410075, China
[3] Institute of Mountain Hazards and Environment, Chinese Academy
of Sciences, Chengdu 610041, Sichuan, People's Republic of China

Abstract. Large-scale direct shear tests are conducted to study the effects of roughness on strength and dilation for concrete-sand interface. The degree of the strain-softening becomes more considerable with the increase of R. Unlike the tests of interface with random grooves, the shear strength of structure-sand interfaces with regular grooves can be greater than shear strength of sand-sand. Peak interface efficiency increases linearly with R in a semi-logarithmic scale. Meanwhile, increasing normal stress gradually can restrain the behavior of dilation as well as reduce the effect of R on interface dilation.

Keywords: Large-scale direct shear test · Concrete-sand interface
Roughness · Dilation · Interface efficiency

1 Introduction

Roughness is one of significant factors influencing the mechanical properties of sand-structure interface. Accordingly, it is essential to figure out the response of shear strength and dilation for the interface under different levels of roughness. Based on the surface topography, the pattern of surface could be divided into two types, which were "random" and "regular" respectively. For "random" surfaces, Uesugi et al. [1] identified the friction angle of interfaces could be no more than the soil mass friction angle constantly. Furthermore, the bilinear relationship between roughness and interface friction was generally acknowledged by previous research. By contrast, Hryciw and Irsyam [2] pointed out the well geometrical grooves fabricated for "regular" surface could fully mobilize the internal shear strength of sand itself. This is the principal reason why interface friction angle could exceed the internal friction angle of sand, which was reported by Martinez [3]. On the other hand, Zhang and Zhang [4] showed that the particles rolling on the interface gradually became obvious with increasing roughness instead of sliding, which led to the interface dilation observed remarkably.

2 Experimental Apparatus, Material and Program

The experimental shear box consists of four basic components as shown in Fig. 1(a). The dimensions of the shear box are 500 mm × 500 mm × 150 mm. To simplify the roughness of geotechnical structures in situ, concrete slabs were fabricated, including one random surface and four regular surfaces in Fig. 1(b) and (c), respectively. Corresponding geometrical parameters of surface and an improved method for calculating roughness (R = 0, 0.636, 0.652, 0.687, and 0.888 mm) were reported by Jin et al. [5].

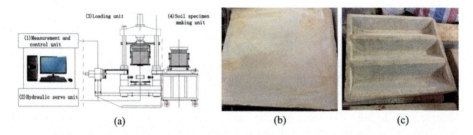

Fig. 1. Test apparatus and concrete slabs (a) TJW–800 large-scale direct shear apparatus (b) random surface (c) regular surface.

In this study, the testing sand is taken from Xiangjiang River in Hunan and the average diameter D_{50} is 0.75 mm. The specimens are compacted to a target initial relative density about 73.3%. In all specimens are subjected to tangential displacement under the fixed rate of 1 mm/min as well as conducted under normal stresses of 50 kPa, 250 kPa and 750 kPa, respectively. Shearing process is terminated when the tangential displacement reaches 50 mm.

3 Experimental Results and Analysis

Figure 2 presents the shear behavior under five roughness levels, but only the results of 50 kPa are enumerated in order to reduce the length of the article. It is clear that in Fig. 2(a) the degree of the strain-softening becomes more remarkable with the increase of interface roughness. Nearly identical phenomenon was observed by Uesugi and Kishida [1] and Hu and Pu [6].

Peak interface efficiency E_p was used to investigate the characteristics of interface shear strength by Dove and Jarrett [7], which can be expressed as:

$$E_p = \tan \delta_p / \tan \varphi_p \qquad (1)$$

where τ_p and σ are peak shear stress and normal stress, respectively. They are obtained from Fig. 2, and φ_p is peak friction angle in terms of soil-soil direct shear tests.

In this study, the improved method proposed by Jin et al. [5] is also utilized to calculate the roughness of steel-sand interface with regular grooves designed by

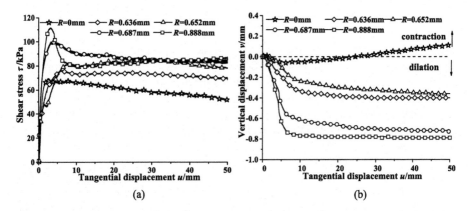

Fig. 2. Curves of shear behavior under normal stress of 50 kPa (a) Shear stress versus tangential displacement (b) vertical displacement versus tangential displacement.

Martinez [3]. In Fig. 3(a), similar to the sand-steel fitting line, the sand-concrete fitting line shows E_p varies linearly with R in a semi-logarithmic scale. Besides, E_p can exceed one, provided that the value of R becomes greater than a threshold (e.g., about 0.70 mm for concrete-sand interface), indicating the development of "passive resistance" contributes to mobilizing the shear strength of sand fully, so the shear strength of sand-sand fails to be the up limit, which is quite well with Martinez [3]. In general, it is reasonable to deduce the failure position could occur within the soil matrix rather than the interface if the value of roughness is above certain threshold.

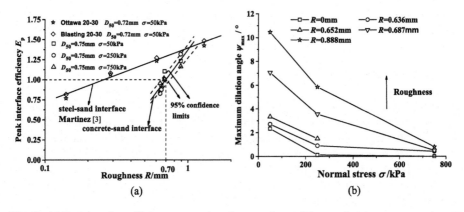

Fig. 3. (a) Peak interface efficiency versus interface roughness (b) normal stress versus maximum dilation angle.

The maximum dilation angles can be calculated from Fig. 2(b) by using the following equation according to Dove and Jarrett [7]; Afzali-Nejad et al. [8]; Lings and Dietz [9]:

$$\psi_{max} = \tan^{-1}\left(-\frac{dv}{du}\right)_{max} \quad (2)$$

where dv and du are the increment in vertical and tangential displacement respectively.

The effect of normal stress versus ψ_{max} are plotted in Fig. 3(b). According to Fig. 3(b), it is evident that ψ_{max} increases with increasing R under the different normal stress, while ψ_{max} decreases with the increase of normal stress under the same roughness levels, which suggests that the high normal stress is likely to diminish the degree of dilation. Meanwhile, the difference of ψ_{max} between $R = 0$ and $R = 0.888$ under the identical normal stress narrows gradually with increasing R, indicating the effect of roughness on dilation is reduced.

4 Conclusions

Based on test results, the main behaviors of the interface are summarized as follows:

(1) The degree of the strain-softening becomes more remarkable with the increase of R. (2) The peak interface efficiency (E_p) can be greater than 1.0 and the shear strength of sand itself was not be the up limit for structure-sand interfaces with regular grooves, thus indicating the failure position could occur within the sand matrix instead of the interface. (3) Increasing normal stress gradually is likely to suppress the dilatancy and reduce the effect of R on interface dilation.

References

1. Uesugi, M., Kishida, H., Uchikawa, Y.: Friction between dry sand and concrete under monotonic and repeated loading. Soils Found. **30**(1), 115–128 (1990)
2. Hryciw, R.D., Irsyam, M.: Friction and passive resistance in soil reinforced by plane ribbed inclusions. Géotechnique **41**(4), 485–498 (1991)
3. Martinez, A.: Multi-scale studies of particulate-continuum interface systems under axial and torsional loading conditions. [Ph.D. Thesis]. Georgia Institute of Technology, Atlanta (2015)
4. Zhang, G., Zhang, J.M.: Monotonic and cyclic tests of interface between structure and gravelly soil. Soils Found. **46**(4), 505–518 (2006)
5. Jin, Z.H., Yang, Q., Chen, C.: Experimental study on effects of roughness on mechanical behaviors of concrete-sand interface. Chin. J. Rock Mechan. Eng. **37**(3), 754–765 (2018)
6. Hu, L., Pu, J.: Testing and modeling of soil-structure interface. J. Geotech. Geoenviron. Eng. **130**(8), 851–860 (2004)
7. Dove, J.E., Jarrett, J.B.: Behavior of dilative sand interfaces in a geotribology framework. J. Geotech. Geoenviron. Eng. **128**(1), 25–37 (2002)
8. Afzali-Nejad, A., Lashkari, A., Shourijeh, P.T.: Influence of particle shape on the shear strength and dilation of sand-woven geotextile interfaces. Geotext. Geomembr. **45**(1), 54–66 (2017)
9. Lings, M.L., Dietz, M.S.: The peak strength of sand-steel interfaces and the role of dilation. Soils Found. **45**(6), 1–14 (2005)

Improvement of Unconfined Compressive Strength of Soft Clay by Grouting Gel and Silica Fume

Mahdi O. Karkush, Haifa A. Ali, and Balqees A. Ahmed

Department of Civil Engineering, University of Baghdad, Baghdad, Iraq
mahdi_karkush@coeng.uobaghdad.edu.iq

Abstract. The effects of cementation gel among particles of soil by permeating grouting on the unconfined compressive strength of soft clay soil have been studied in this work. The soil sample obtained from Al-Jadriah district in Baghdad city which can be classified as soft clayey soil of low plasticity (CL). The soil samples were improved by grouting gel of cement/water with three weight percentages. The percentages of grouting gel are measured in terms of cement to water as; 0.1 C:0.9 W, 0.2 C:0.8 W and 0.3 C:0.7 W. In addition to the grouting get, the soil samples improved with silica fume, where silica fume added in two percentages of 2 and 4%. The improved soil samples were dehydrated after grouting for three periods of time 1, 3, and 7 days. The results of tests indicated increasing the unconfined compression strength and decreasing the axial strain with increasing the percentage of grouting gel. The addition of silica fume to the soft clayey soil improved by cementing agent resulted in significant increase in the unconfined compression strength and significant decrease in axial strain.

Keywords: Soft soil · Grouting · Cement · Shear strength and clay

1 Introduction

Portland cement is used as a grout to fill the voids of the soil media and improve the strength and elastic properties of soil [1–3]. Most grouts with Portland cement required to add small amounts of additives, that is used for specific purposes such as increasing fluidity, retarding sedimentation, and controlling on the set times [4–8]. In the present study, three percentages of grouting gel of cement-water and two percentages of silica fume are used to improve the undrained shear strength of soft soil. Also, the effects of curing period on the growing of undrained shear strength have been studied.

2 Soil and Used Materials

The physical properties of soil sample used in the present work are given in Table 1, measured according to ASTM [9]. The specifications of Portland cement used in the present work are: the specific surface is 2250 cm^2/g and the initial and final hardening times are 1 and 10 h respectively. The compressive strength of cement mortar is 150 kg/cm² after 3 days and 240 kg/cm² after 7 days. In addition, the tensile strength of cement is 21 kg/cm². The used silica fume (Sika Fume-HR) contains extremely fines of (0.1 μm) reactive dioxide of silicon. The soil mixture becomes extremely soft and workable when using silica fume, also, the reactions of silica fume generates a chemical bond with the free calcium components present in the clayey soil. The color of used silica fume is a light gray powder of density 0.65 kg/l.

Table 1. Physical properties of soil sample.

Property	Gs	Coarse (%)	Fines (%)	LL (%)	PL (%)	$\gamma_{dry,max}$ (kN/m³)	ω_{opt} (%)
Value	2.7	31.78	68.22	34	23	15.5	22

3 Stabilization with Grouting Cement Gel and Silica Fume

Grouting gel mixtures are used to improve the unconfined shear strength of soil. The grouting gel is mixture of water/cement, added in three ratios of soil mass (0.9:0.1, 0.8:0.2, and 0.7:0.3)%. The natural soil sample was dried in the oven at 110 °C, then mixed with the optimum moisture content, then placed in layers in the compaction mold. Each layer was compacted with 25 blows distributed uniformly on the surface of a soil sample to achieve the required maximum dry density. From the mold of compaction, a soil sample of dimensions 38.5 mm diameter and 76.8 mm height was extracted. The treatment method can be summarized by injection 1 cm³ of the grouting gel by needle into the soil sample at different locations and reach to the middle height of sample. Then, the mold covered with a polyethylene layer to control the moisture content of soil sample and left for rehydration. The suggested curing periods were 1, 3, and 7 days to measure the effects of curing period on the acquired undrained shear strength. Also, two soil samples are mixed with 2 and 4% of silica fume before injection the grouting gel and curried for 1 day.

4 Results and Discussion

Generally, there was an increase in the undrained shear strength of soil stabilized with grouting gel of cement until the maximum strength reached, and then there was a slight reduction in the strength of the soil as shown in Figs. 1 and 2. The increase in magnitude of unconfined compressive strength ranged from 7 to 82% for the three percentages of grouting mixture and three periods of curing. Increasing the ratio of cement in the grouting mixture causes increasing the undrained strength, but the significant effect is related to the period of curing. Also, there was no change in the axial strain at failure for soil samples treated with grouting gel only. In addition, the unconfined compressive strength increases with increasing the relative compaction and curing time [10]. The delay in mixing and compaction causes decreasing the undrained shear strength of soil. The mixing of soft soil with silica fume before injection the grouting gel causes increasing the undrained shear strength by 32 to 72% after one day of curing. Increasing the cement ratio in the grouting gel causes significant increase in the undrained shear strength. Also, the mixing of soil with silica fume causes a significant decrease in the axial strain ranged 35 to 44%. The low ratios of silica fume alone make the mixture of soil unstable chemically and give negative effects through increasing the moisture content required to produce a homogenous mixture, which causes a decrease in the unconfined compressive strength [11]. The summary of undrained shear strength and axial strains are given in Tables 2 and 3.

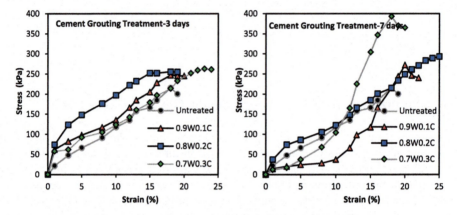

Fig. 1. Stress-strain curves of soil samples treated with grouting gel.

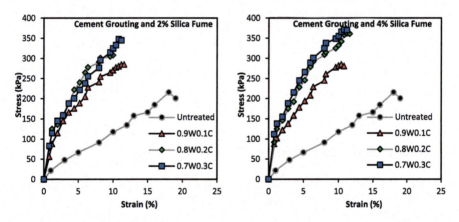

Fig. 2. Stress-strain curve of soil sample treated with cement grouting and silica fume after 1 day of treatment.

Table 2. Results of unconfined compressive strength tests for soil treated with grouting gel.

Soil sample	Untreated Natural	1 day treatment			3 days treatment			7 days treatment		
		0.9 W 0.1 C	0.8 W 0.2 C	0.7 W 0.3 C	0.9 W 0.1 C	0.8 W 0.2 C	0.7 W 0.3 C	0.9 W 0.1 C	0.8 W 0.2 C	0.7 W 0.3 C
c_u, kPa	107.95	116	122	127.5	123.35	127.69	131.39	136	146.5	196.5
ε_a, %	18	9.6	8	9.6	18	18	23	20	25	18

Table 3. Results of unconfined compressive tests-soil treated with grouting gel and silica fume.

Soil sample	Untreated	Cement grouting and 2% silica fume after 1 day treatment			Cement grouting and 4% silica fume after 1 day treatment		
	Natural	0.9 W 0.1 C	0.8 W 0.2 C	0.7 W 0.3 C	0.9 W 0.1 C	0.8 W 0.2 C	0.7 W 0.3 C
c_u, kPa	107.95	142.5	154	174	142.5	181	185.5
ε_a, %	18	11.6	10	10.8	10.4	11.2	10.8

5 Conclusions

The undrained shear strength of soft clayey soil increased with increasing the ratio of cement in the grouting gel and the more significant increase was noticed with increasing the curing period from 1 to 7 days. There was slight change in the axial strain when the soil samples treated with grouting gel of cement and water. The mixing of soil with silica fume then grouted with cement gel causes significant increase in the undrained shear strength and significant decrease in the axial strain. Silica fume accelerate the chemical reactions between soil and grouting gel which gives the hardening to soft soil.

References

1. Karol, R.: Chemical Grouting and Soil Stabilization, 3rd edn. Revised and Expanded. Marcel Dekker Inc., Basel (2003)
2. U.S. Army Corps of Engineers: 1984. Grouting technology. Publication No. EM 1110-2-3506 (2017). http://www.publications.usace.army.mil/Portals/76/Publications/Engineer Manuals/EM_1110-2-3506.pdf
3. Sinroja, J.M., Joshi, N.H., Shroff, A.V.: Ground improvement technique using microfine slag cement grouts. In: Proceeding of IGC, Chennai, pp. 661–664 (2006)
4. Warner, J.: Practical Handbook of Grouting: Soil, Rock, and Structures. Wiley, New York (2004)
5. Dano, C., Hicher, P.Y., Tailliez, S.: Engineering properties of grouted sands. J. Geotech. Geoenviron. Eng. ASCE **130**(3), 328–338 (2004)
6. Akbulut, S., Saglamer, A.: Estimating the groutability of granular soils: a new approach. Tunneling Undergr. Space Technol. **17**, 371–380 (2002)
7. Kazemian, S., Huat, B.B.K.: Assessment and comparison of grouting and injection method in geotechnical engineering. Eur. J. Sci. Res. **27**(2), 234–247 (2009)
8. Santhosh Kumar, T.G., Benny, M.A., Sridharan, A., Babu, T.J.: Bearing capacity improvement of loose sandy foundation soils through grouting. Int. J. Eng. Res. Appl. (IJERA) **1**(3), 1026–1033 (2011)

9. Annual book of ASTM standards, Soil and Rock; Building; Stone; Peats (2003)
10. Karkush, M.O., Ali, H.A: Improvement of unconfined compressive strength of soft clayey soil by using glass wool fibers. In: The Iraqi Journal for Mechanical and Material Engineering, Special Volume, Babylon First International Engineering Conference Issue (B) (2016)
11. Onyelowe, K.C., Onuoha, I.C., Ikpemo, O.C., Okafor, F.O., Maduabuchi, M.N., Kalu-Nchere, J., Aguwa, P.: Nanostructured clay (NC) and the stabilization of lateritic soil for construction purposes. EJGE **22**(10), 4177–4196 (2017)

Shear Modulus and Damping Ratio of Coarse Fused Quartz

Gangqiang Kong[1(✉)], Hui Li[1], Qing Yang[2], Gang Yang[2], and Liang Chen[1]

[1] Hohai University, Nanjing 210098, Jiangsu, China
gqkong1@163.com
[2] Dalian University of Technology, Dalian 116024, Liaoning, China

Abstract. A series of resonant column tests (RCT) and cyclic torsional shear tests (TST) were used to study the dynamic shear modulus, and damping ratios of transparent sand manufactured by coarse fused quartz. The oil-saturated and dry specimens were carried out. The measured results were comparative analysis with previous literatures (Kong et al. 2016, 2018). It was found that the dynamic behaviour of transparent sand manufactured by coarse fused quartz have similar properties with those of fine fused quartz, and fill in the properties of natural sands.

Keywords: Transparent sand · Shear modulus · Damping ratio
Fused quartz

1 Introduction

Transparent sand manufactured by fused quartz and refractive index (RI)-matched oil can be used to mimic the behavior of natural saturated sand [1, 2]. Fused quartz grains have a size, shape, permeability, chemical composition, shear strength and compressibility which are consistent with those of natural angular sand. Fused silica was first identified as a potential transparent sand by Iskander [3]. The static properties of dry and oil-saturated fused quartz were first proposed by Ezzein and Bathurst [1], and the static properties with a sucrose solution-saturated or sodium thiosulfate-treated sodium-iodide (STSI) solution-saturated specimens described by Guzman et al. [4], and Carvalho et al. [5] respectively. The dynamic shear modulus and damping ratios of fine fused quartz (0.5–1.0 mm, 0.1–1.0 mm) through resonant column tests (RCT) and cyclic torsional shear tests (TST) were briefly introduced [6, 7]. The cyclic undrained behavior and liquefaction resistance of fine fused quartz was also studied by Kong et al. [8]. For the further application of coarse fused quartz (0.25–5.0 mm) to substitute sand and to track particle movement inside sand mass during a dynamic event, the aim of this work is to study the dynamic shear modulus, and damping ratios of coarse fused quartz through a series of RCT and TST. The results of the dynamic properties of natural sands are also presented to examine its dynamic properties.

2 Materials and Methods

The fused quartz used in this study was manufactured by the Jiangsu Kaida Silica Co. LTD in China. The particle size distribution of coarse fused quartz used both for the purposes of this paper and in previous research (fine fused quartz) are compared in Fig. 1. The particle size distribution of definition liquefaction-prone sand by Japan Code is also shown in Fig. 1. The grain-size characteristics and classification of transparent granular particle materials and Ottawa sand are plotted in Table 1. The detailed physical and static properties of fused quartz can be seen in Ezzein and Bathurst [1], and Guzman et al. [4]. The mixture oils which is a blend of Norpar® 12 (RI = 1.4235) and Drakeol® 15 (RI = 1.4688) were used in this study. The RCT could amplify the rotational deformation providing a shear strain resolution ranging from 10^{-6} to 10^{-4} and was used in the tests following ASTM D 4015. The TST non-contacting sensor which could provide a shear strain resolution ranging from 10^{-4} to 10^{-1} was used in the tests following ASTM STP 1213. Medium density with 60% (D_r = 60 ± 10%) relative density was controlled for the oil-saturated specimens.

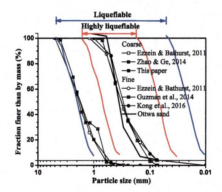

Fig. 1. Particle size distributions of transparent granular materials.

3 Results and Analysis

3.1 Shear Modulus

The normalized shear modulus (G/G_{max}) versus shear strain (γ) of tested specimens under 100 kPa confining pressures is shown in Fig. 2. Many investigators have studied the relationship between G/G_{max} and γ of natural soils. Approximate upper and lower bounds relationship between G/G_{max} and γ reported by Rollins et al. for gravels (with D_r from 40% to 100%), Seed and Idriss for sand (with D_r from 30% to 90%, effective vertical stress σ'_v = 144 kPa, and coefficient of lateral earth pressure at rest k_0 = 0.5), and Vucetic and Dorbry for clay (with OCR from 1 to 15) are shown in Fig. 2 for comparison. Figure 2 also contains the results of fine fused quartz reported by Kong et al. [6]. Most of the G/G_{max} values of the dry and oil-saturated specimens fall within the range for sand and gravel. It is also shown in Fig. 2 that the elastic threshold shear strain level for the fused quartz is then basically in the range of that for the natural soils.

Table 1. Grain-size characteristics and classification of transparent granular particle materials and Ottawa sand.

Type	D_{max} (mm)	D_{50}	C_u	C_c	USCS	Literatures
Coarse	5.00	1.74	3.1	1.02	SP	This paper
Coarse	4.75	1.87	3.2	0.61	SP	Zhao and Ge 2014
Coarse	5.00	1.77	2.8	1.05	SP	Ezzein and Bathurst 2011
Fine	1.20	0.30	3.2	1.15	SP	This paper
Fine	0.85	0.33	2.8	0.97	SP	Guzman et al. 2014
Fine	1.20	0.33	3.9	1.24	SP	Ezzein and Bathurst 2011
Ottawa sand	1.20	0.37	2.2	1.23	SP	ASTM D 2487-00

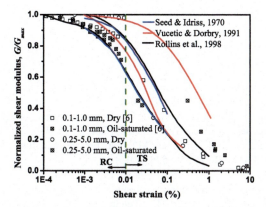

Fig. 2. Curves on $G/G_{max} - \log(\gamma)$ for transparent sand and natural soils.

3.2 Damping Ratio

Damping ratios are also studied for the specimens with two different aggregate graduations. The amplitude decay method was used to determine the damping ratios of specimens according to ASTM D 4015. The results of the tests and reported by Kong et al. [6] are shown in Fig. 2. In general, the damping ratio values increase slowly at first and exhibit exponential growth when the shear strain reaches a value. It is clear that the damping ratio values could be influenced by pore fluid. The damping ratios data for natural soils, reported in the literatures by Rollins et al. for gravel; Vucetic and Dorbry for clay; and Seed et al. for sand, are also shown in Fig. 2. It can be observed that nearly all the plots of the oil-saturated specimens fall into the range of natural sand and gravel. For dry specimens, when the shear strain is smaller than 2×10^{-5}, the majority of the damping ratios fall into the range of natural sand and gravel; when the shear strain is larger than 2×10^{-5}, the majority of the damping ratios under the lower range of the three types of natural soils (Fig. 3).

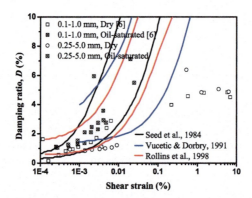

Fig. 3. Curves on D versus $\log(\gamma)$ for transparent sand and natural soils.

Acknowledgements. The research was financially supported by the National Science Foundation of China (51478165).

References

1. Ezzein, F.M., Bathurst, R.J.: A transparent sand for geotechnical laboratory modeling. Geotech. Test. J. ASTM **34**(6), 590–601 (2011)
2. Kong, G.Q., Cao, Z.H., Zhou, H., Sun, X.J.: Analysis of pile under oblique pullout load using transparent soil models. Geotech. Test. J. ASTM **38**(5), 725–738 (2015)
3. Iskander, M., Bathurst, R.J., Omidvar, M.: Past, present, and future of transparent soils. Geotech. Test. J. ASTM **38**(5), 557–573 (2015)
4. Guzman, I.L., Iskander, M., Suescun-Florez, E., Omidvar, M.: A transparent aqueous-saturated sand surrogate for use in physical modeling. Acta Geotech. **9**, 187–206 (2014)
5. Carvalho, T., Suescun, E., Omidvar, M., Iskander, M.: A nonviscous water-based pore fluid for modeling with transparent soils. Geotech. Test. J. ASTM **38**(5), 805–811 (2015)
6. Kong, G.Q., Zhou, L.D., Wang, Z.T., Yang, G., Li, H.: Shear modulus and damping ratios of transparent soil manufactured by fused quartz. Mater. Lett. **182**(1), 257–259 (2016)
7. Kong, G.Q., Li, H., Yang, G., Cao, Z.H.: Investigation on shear modulus and damping ratio of transparent soils with different pore fluids. Granular Matt. **20**(1), 80–88 (2018)
8. Kong, G.Q., Li, H., Yang, Q., Meng, Y.D., Xu, X.L.: Cyclic undrained behavior and liquefaction resistance of transparent sand manufactured by fused quartz. Soil Dyn. Earthq. Eng. **108**(5), 13–17 (2018)

Field and Laboratory Investigation on Non-linear Small Strain Shear Stiffness of Shanghai Clay

Q. Li[1,2(✉)], W. D. Wang[1,2], Z. H. Xu[1,2], and B. Dai[1,2]

[1] Department of Underground Structures and Geotechnical Engineering, Shanghai Underground Space Engineering Design and Research Institute, Arcplus Group PLC, Shanghai, China
qing_li@arcplus.com.cn
[2] Shanghai Engineering Research Center of Safety Control for Facilities Adjacent to Deep Excavations, Shanghai, China

Abstract. Booming infrastructure construction and rapid urban redevelopment in Shanghai demand utilization of more underground spaces, leading to a large number of geotechnical projects. There has been a growing and widespread appreciation that strains of the soil around underground structures such as foundations, excavations and tunnels are, except in some areas that undergo yielding, generally in a small strain range typically of the order of 0.01%–0.1%. In this paper, the small strain stiffness features including the elastic shear modulus G_0 and stiffness degradation curve have been investigated in the field using the suspension wave velocity logging system and in the laboratory using a resonant column apparatus equipped with bender elements. At very small strains, the measured shear moduli obtained from bender element tests are fairly consistent with those from field tests. The elastic shear moduli can be well captured by an empirical power correlation considering stress levels. At small strains, the observed stiffness-strain relationships of typical Shanghai clay layers are highly nonlinear.

Keywords: Nonlinear · Small strain stiffness · Shanghai clay

1 Introduction

Shanghai is located in Yangtze River delta where the ground are stratified. The uppermost 30 m of the soil deposits are generally encountered in geotechnical projects and are typically categorized into six layers as shown in Fig. 1. It has been recognized that occur in the soil around underground structures such as excavations and tunnels are small. Although a large number of underground projects have been constructed over the past few decades, research on small strain stiffness of typical Shanghai clay is rather limited. In this paper, nonlinear stiffness-strain relationship of Shanghai clay at small strains has been investigated in the field using the suspension wave velocity logging system and in the laboratory using a resonant column apparatus equipped with bender elements. Results from field and laboratory tests are compared and discussed.

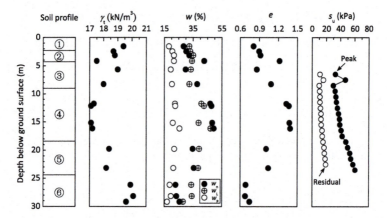

Fig. 1. Typical soil profile and properties in Shanghai (uppermost 30 m blow the surface). ①-Fill; ②-Silty clay; ③-Soft silty clay; ④-Soft clay; ⑤-Silty clay; ⑥-Stiff clay.

2 In-Situ and Laboratory Measurements of Shear Stiffness

The elastic shear modulus was measured in the field using a suspension wave velocity logging system and in laboratory using bender element tests. At the site, three boreholes (BH1–BH3) are selected to measure the velocity of shear waves. Figure 2(a) shows the measured S-wave velocity profile. Data at the top 5 m in all three boreholes were ignored due to the presence of steel casing. To estimate the precision and the reliability of the measurements of the shear wave velocity, statistical analyses were carried out. It can be seen that although there is a general trend of increasing mean velocity with depth, the measurements are scattered, which ma be largely due to the natural variability of the uppermost 30 m clay (Ng et al. 2000).

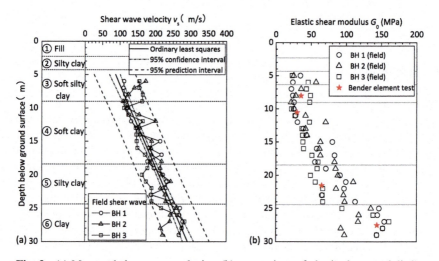

Fig. 2. (a) Measured shear wave velocity; (b) comparison of elastic shear moduli G_0.

After consolidated to the in-situ mean stress, each sample was consolidated to 50 kPa, 100 kPa, 200 kPa and 400 kPa and then bender element tests were carried out. As shown in Fig. 3, the travel time determined using first arrival method is smaller than that of peak-peak method, giving a larger shear wave velocity and elastic shear modulus. Since the first arrival method is simple to use, the G_0 value is determined using this method. As shown in Fig. 2(b), the measured G_0 from bender element tests are comparable to that from the field test, which is considered to be close to real elastic shear modulus.

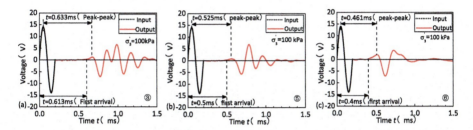

Fig. 3. Typical input and output signals from bender element tests.

3 Elastic Shear Modulus G_0

Figure 4 summaries the elastic shear modulus G_0 deduced from bender element tests. As expected, the elastic shear modulus of each layer increases with an increase of confining stress. An empirical power function is adopted to describe the relationship between G_0 and p'_r. The equation is expressed as $\frac{G_0}{p_r} = S\left(\frac{p'}{p_r}\right)^n$ (Rampellp et al. 1997), where S, n are empirical parameters associated with soil properties and can be derived by regression analysis. The S value of the stiff clay (layer ⑥) is much larger than those of other layers, which may be attributed to the over-consolidation of this layer. On the other hand, the S values for the soft silty clay and soft clay layers are comparable and is the least. The measured n value ranges from 0.45 for the stiff clay to 0.77 for the soft silty clay. Generally, the variations of G_0 with effective mean stress of each layer can be well captured by the above empirical power equation.

4 Variations of Shear Stiffness with Shear Strain

Based on the results of bender element and resonant column tests, a complete stiffness degradation curve is established as shown in Fig. 5. The magnitude of shear stiffness at small strains obtained from resonant column tests agrees well with that from bender element tests, indicating the reliability of the laboratory tests carried out. The shear stiffness is almost constant until $\gamma = 0.001\%$, beyond which the stiffness keeps decreasing. At small strains from 0.001% to 0.1%, the stiffness-strain relationship of each clay layer is highly nonlinear. As shear strain reaches up to 0.1%, the shear modulus G reduces to about 0.4–$0.5G_0$. The degradation curve of each layer is

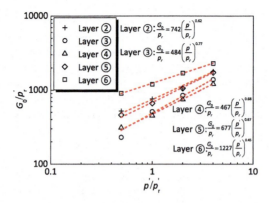

Fig. 4. Normalized elastic shear modulus.

normalized by its corresponding elastic shear stiffness G_0. It is found that the normalized curves of different layers are almost overlapped, giving a $\gamma_{0.7}$ ($\gamma_{0.7}$ is the shear strain where the shear modulus reduces to $0.7G_0$) value of about 2.7×10^{-4}, which falls in the lower range of the database of 21 clays collected by Vardanega and Bolton (2013).

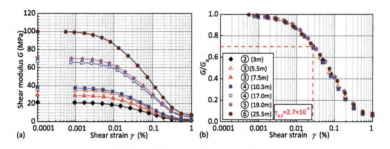

Fig. 5. Variation of shear stiffness $G_0(G_0/p'_r)$ with shear strain γ.

5 Conclusions

Both field and laboratory tests have been carried out to investigate the elastic shear modulus G_0 and non-linear stiffness-strain characteristics of Shanghai clay. The elastic shear moduli increases with mean effective stress and can be captured well by an empirical power function considering stress levels. At small strains, the stiffness-strain relationship are observed to be highly nonlinear, giving a $\gamma_{0.7}$ value of 2.7×10^{-4}.

References

Ng, C.W.W., Pun, W.K., Pang, R.P.L.: Small strain stiffness of natural granitic saprolite. J. Geotech. Geoenviron. Eng. ASCE **126**(9), 819–833 (2000)

Rampellp, S.B., Viggiani, G.M., Amorosi, A.: Small-strain stiffness of reconstituted clay compressed along constant triaxial effective stress ratio paths. Geotechnique **47**(3), 475–489 (1997)

Vardanega, P.J., Bolton, M.D.: Stiffness of clays and silts: normalizing shear modulus and shear strain. J. Geotech. Geoenviron. Eng. **139**(9), 1575–1589 (2013)

Introducing a New Test System for the Rock-Machine Interaction

S. D. Li[1,2,3(✉)], Z. M. Zhou[1,2,3], and W. S. Hou[1,2,3]

[1] Key Laboratory of Shale Gas and Geoengineering, Institute of Geology and Geophysics, Chinese Academy of Sciences, Beijing 100029, China
lsdlyh@mail.iggcas.ac.cn
[2] Institute of Earth Science, Chinese Academy of Sciences, Beijing 100029, China
[3] College of Earth Sciences, University of Chinese Academy of Sciences, Beijing, China

Abstract. Prior to the Cerchar abrasion test, only the loss of the stylus was concerned with CAI value to estimate tool wear in rock excavation applications, without considering the rock scratch. In the actual projects, however, engineers pay more attention to the rate of cutter consumption (i.e. the number of tools consumed by excavating the unit volume of rock) and hope to establish the relationship between tool wear and excavation distance, which are related to the 'rock-machine' interaction. In order to meet these actual needs, in this paper, the Whole Process Cerchar Test method was proposed and the ATA-IGG I servo-control rock abrasion system was developed. This new test can not only test the abrasion of rock against steel stylus, but also the damage degree of the steel stylus to rock. Through the newly proposed method and the new developed system, the regular pattern of interaction between rock and machine can be preliminarily explored.

Keywords: Cerchar abrasion test · 'Rock-machine' interaction · Whole process New development

1 Introduction

The Cerchar abrasion test has been proposed for more than forty years and it has been widely used in coal mining industry (Yarali et al. 2008; Kasling and Thuro 2010) and tunneling industry (West 1989; Gharahbagh et al. 2011; Rostami et al. 2013) and perform very well. However, there are some problems existing in the Cerchar abrasion test such as the stability of steel stylus motion, the influence of the anisotropy of rock samples on the CAI value and so on. To solve these problems, West had improved the Cerchar abrasion test (West 1989), which has been considered as the second-generation machine. The pattern of the stylus moving had been changed in the new device though which the steel stylus could be moved 10 mm on the rock sample surface by turning the handle. In addition, in order to stabilize the movement of the steel stylus and to make the movement speed easily controlled, the speed was reduced to 10 mm/min, and the other parameters were not changed as compared with the first-generation Cerchar abrasion testing machine. The second-generation machine is commercially available from some manufacturers around the world (West 1981), through which lots of multiple regression model were employed to investigate the relationship among the CAI value, quantity of quarz, physical rock properties and so on (Alber 2008; Vassallo et al. 1994; Majeed and Bakar 2016). Moreover, the previous study categorized the effect of some Cerchar test parameters on the CAI values (Hamzaban et al. 2014; Ghasemi 2010). Beste analyzed the scratched shape of the rock under different load (6N/10N/14N) conditions with a total length of 6 cm and a cone angle of 90° steel pin (Beste et al. 2004). The scratches were classified into polished marks, plastic marks, internal cracks, external cracks and so on. Previous studies have rarely studied the scratches of rocks in the Cerchar test.

In the second-generation of Cerchar abrasion test, the horizontal movement of the steel stylus is controlled by turning the handle, which will easily result in uneven and uncontrollable running speed of the steel stylus. In addition, the current Cerchar abrasion test can only characterize the final degree of wear of the rock to the steel stylus and cannot depict the damage process of the steel stylus to the rock. The study presented in this paper is to propose the newly developed whole process Cerchar test with the parameters of controlling and measuring abrasion rate, abrasion distance, abrasion level, abrasion and displacement, scratched microscopic measurement, aimed at meeting the requirement of 'rock-machine' interaction research.

2 A New Method of Whole Process Cerchar Test

In order to meet the "rock-machine" interaction research needs, the Key Laboratory of Shale Gas and Engineering of the Chinese Academy of Sciences has developed the new servo-control rock abrasion tester, named ATA-IGG, as shown in Fig. 1. The new test system mainly introduced advanced experimental techniques: high-frequency dynamic closed-loop servo controller, AC closed loop servo control motor, pull code encoder PRF08, and three-dimensional image processing technology (Surfer). Instead of turning the handle to control the movement of the stylus, the servo control system composed of high-frequency dynamic closed-loop servo controller and AC closed loop servo control motor, reduces the impact of the error from stylus speed and stylus movement distance on the results. Combined with digital electron microscope, the ATA-IGGservo-control rock abrasion tester (Cerchar abrasive instrument) is used to study the wear of steel on steel stylus. And combined with three-dimensional image processing technology (Surfer), two-dimension allaser measurement displacement technology (CCD) is used to study the role of stylus for rock penetration. Other standards (e.g. the hardness of the stylus and rock samples made by saw) refer to the second-generation machine conducted by West (1989).

Fig. 1. The ATA-IGGIservo-control rock abrasion tester

The new test instrument can be implemented in the closed-loop servo control process; the horizontal displacement speed of the pin is 1–100 mm/min. As far as the CCD is concerned, the measurement accuracy is ±0.5 um, with sampling at a rate of 2400/s, which is able to meet the requirements for precise morphology measurement of scratches. In addition, the new tester not only can test the rock Cerchar abrasion value by the digital electron microscope, but also analyze the resultant displacement measured by the PRF08 and morphology measurement of scratches measured by the CCD, since the resultant displacement minus depth of the scratches is the length of a worn stylus, through which the rock Cerchar abrasion value can be calculated with the geometric relations. The data can be collected at a minimum interval of 1 ms, which means that the value of horizontal displacement, the force and the pin wear of the steel pin can be obtained exactly.

The new type of tester can obtain the horizontal displacement value, the force value and the pin length of the steel stylus during the test. It can easily and accurately obtain the Cerchar abrasion index and precisely control the horizontal movement speed and distance of the steel pin. During the test, the horizontal displacement value and the vertical displacement value of the steel pin are measured. The force of the steel pin is measured by the force sensor. The abrasion width of the steel pin is measured by the numerical microscope. The stepped motor closed loop servo controls horizontal movement speed, distance and force. Therefore, the ATA-IGGIservo-control rock abrasion tester can control and measure parameters such as abrasion rate, abrasion distance, abrasion level force, abrasion, displacement, and scratches microscopic measurements based on the existing rock abrasive instrument. In order to verify the reliability of the ATA-IGGIservo-control rock abrasion tester, the same samples were tested in the University of Melbourne, employing the second generation of Cerchar abrasion test. From the Table 1, the ATA-IGGIservo-control rock abrasion tester is convincing.

Table 1. Cerchar abrasion test result of Chinese Academy of Sciences and University of Melbourne, Australia

Specimen	Test values of CAS	Test values of UMA
CDB-1	3.07	3.05
ZCG-1	2.95	2.80
ZCF-1	4.36	4.97
ZCA-1	2.73	2.98

3 Conclusion

In order to meet the needs of the "rock-machine" interaction study, a whole process Cerchar test method was presented and the ATA-IGG I servo-control rock abrasion system was developed. The test method for controlling and measuring the parameters such as abrasion rate, abrasion distance, abrasion level force, abrasion and displacement and scratched microscopic measurement can be controlled and measured on the basis of existing rock abrasive was developed, through which three-dimensional topography can be obtained.

References

Alber, M.: Stress dependency of the Cerchar abrasivity index (CAI) and its effects on wear of selected rock cutting tools. Tunn. Undergr. Space Technol. **23**(4), 351–359 (2008)
Alber, M., Yaralı, O., Dahl, F., Bruland, A., Käsling, H., Michalakopoulos, T.N.: ISRM suggested method for determining the abrasivity of rock by the Cerchar abrasivity test. Rock Mech. Rock Eng. **47**(1), 261–266 (2014)
Bakar, M.Z.A.: Evaluation of saturation effects on drag pick cutting of a brittle sandstone from full scale linear cutting tests. Tunn. Undergr. Space Technol. **34**(1), 124–134 (2013)
Cerchar: The Cerchar Abrasiveness Index. Cerchar-Centred' Etudes et Recherches de Charbonnages de France. 12 S, Verneuil (1986)
Dahl, F., Bruland, A., Jakobsen, P.D., Nilsen, B., Grøv, E.: Classifications of properties influencing the drillability of rocks, based on the NTNU/SINTEF test method. Tunn. Undergr. Space Technol. **28**(1), 150–158 (2012)
Deliormanlı, A.H.: Cerchar abrasivity index (CAI) and its relation to strength and abrasion test methods for marble stones. Constr. Build. Mater. **30**(5), 16–21 (2012)
Er, S., Tuğrul, A.: Estimation of Cerchar abrasivity index of granitic rocks in turkey by geological properties using regression analysis. Bull. Eng. Geol. Environ. **75**(3), 1325–1339 (2016)
Fowell, R.J., Abu Bakar, M.Z.: A review of the Cerchar and LCPC rock abrasivity measurement methods. In: 11th Congress of the International Society for Rock Mechanics, Second Half Century for Rock Mechanics, vol. 1, pp. 155–160 (2007)
Ghasemi, A.: Study of cerchar abrasivity index and potential modifications for more consistent measurement of rock abrasion (2010)
West, G.: A review of rock abrasiveness testing for tunnelling. In: ISRM International Symposium. International Society for Rock Mechanics (1981)

A Parallel Comparison of Small-Strain Shear Modulus in Bender Element and Resonant Column Tests

Xin Liu[1,2(✉)] and Jun Yang[2]

[1] Department of Geology Engineering, Chang'an University, Xi'an, China
liuxinsunny08@126.com
[2] Department of Civil Engineering, The University of Hong Kong, Hong Kong, China

Abstract. This paper presents an experimental study on small strain shear modulus (G_0) of saturated sand specimens using both resonant column and bender element tests, with the aim to explore the impact of testing method. Besides clean sand specimens, non-plastic fines of varying percentage were also added to sand in the experiments in order to investigate the possible interplay of fines with the testing method. It was found that, under otherwise similar conditions, the bender element test may yield significantly greater G_0 values than the resonant column test and this observation is not affected by the fines content. A hypothesis is proposed for the observed effects of testing method, which attributes the discrepancy to the isotropic/anisotropic properties of sand. The study suggests that cautions should be exerted in examining G_0 of sand obtained from different test methods.

Keywords: G_0 · Sand · Bender element test · Resonant column test

1 Introduction

Characterizing the small-strain shear modulus (G_0) of sand is essential in many geotechnical applications, such as in settlement evaluation of machine foundations (Richart et al. 1970), and in ground response analysis in earthquakes (Yang and Yan 2009). In recent years, a fundamental understanding of G_0 in sand has been built using the bender element (BE) test, owning to the advantages of cost-effectiveness and compactness of being equipped with other testing devices (Dyvik and Madshus 1985; Viggiani and Atkinson 1995). Compared with the BE test, estimate of G_0 in the resonant column (RC) test is less challenging. Often in analysis of G_0 in the BE test, a benchmark obtained via RC test was adopted (e.g. Yamashita et al. 2009; Yang and Gu 2013 etc.).

Given a notable difference of principles to determine G_0 in the above tests, a question arises as to whether the impact of test methods emerge in estimate of G_0. However, current understanding on this issue is not adequate. To address this issue, this paper presents an experimental study on G_0 of saturated sand using both the BE and RC tests. The G_0 values yielded from clean sand and sand-fines mixtures were examined, so that the possible interplay of fines with the testing method is discussed. Last but not least, an attempt was also made to explain the discrepancies in G_0 obtained from the BE and RC tests. The findings of this paper are expected to serve as a useful reference in examining G_0 of sand by using different test methods.

2 Materials and Test Methods

In the experiment, uniformed graded ($C_u = 1.4$) quartz sand (Toyoura sand) was used as the base sand, and crushed silica fines (less than 63 μm) of varying percentage (0%–30%) were used as the additive. Both dry tamping and moist tamping method were adopted to prepare the sand specimen. To determine the G_0 values, all experiments were carried out using an apparatus that incorporates both BE and RC functions. In the BE test, by generating a shear wave signal that propagates vertically with horizontal polarization, G_0 of sand is determined in the framework of soil elasticity. To estimate the travel time of shear wave, the start-to-start method (S0 – S2) as suggested by Yang and Gu (2013) on glass beads was adopted in the BE test, which was found to be accurate not only for clean sand (Gu et al. 2015) but also for sand-fines mixtures (Yang and Liu 2016). On the other hand, in the RC test, by applying a series of torsional excitations through magnatic exciters mounted on the top of specimen, an overall response of specimen was recorded. A key step in the RC test is to find the resonant frequency (f_n) of the specimen, afterwards the corresponding shear wave velocity and the associated G_0 are determined on the basis of f_n. Readers may refer to Yang and Gu (2013) for detailed illustrations of the test methods.

3 Test Results and Discussion

In Fig. 1, the G_0 values of clean Toyoura sand obtained from the RC and BE tests are plotted as a function of void ratio. A general trend is found that for a given confining stress G_0 increases with decreasing void ratio, whereas for a given void ratio G_0 decreases with deceasing confining stress. Of more interest, under the same confining stress and the same void ratio, discrepancies are observed between the RC and BE tests for specimens using the moist tamping method (Fig. 1(a)). For instance, at $e = 0.8$ and $\sigma' = 100$ KPa, G_0 in the BE test is about 20% higher than that in the RC test; at a similar void ratio while $\sigma' = 500$ KPa the amount of G_0 in the RC test is found to be 21% lower as compared with the BE test. Note that the significant differences in Fig. 1(a) are unlikely due to errors of signal interpretation, and it can be convinced in Fig. 1(b) for specimens using the dry tamping method: by adopting the same rules in signal interpretations as mentioned above, a unique trendline can be used to characterize the relationship between the G_0 and the void ratio, regardless of the test methods.

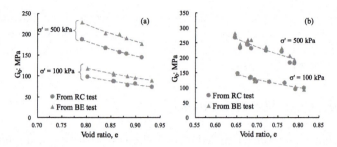

Fig. 1. Variation of G_0 of Toyoura sand with void ratio: (a) moist tamping; (b) dry tamping.

To facilitate the comparison, G_0 obtained in BE and RC test are readily compared. Clearly, for sand specimens using the moist tamping method (Fig. 2(a)), G_0 in the BE test is consistently exceeding about 20% as compared with the RC test. In contrast, by using the dry tamping method (Fig. 2(b)), it shows a good agreement in G_0 between the RC test and the BE test. Recalling that substantial differences in terms of sand fabrics are likely induced for specimens using the moist tamping and dry tamping methods, the above observations lead to a strong implication that the influence of test methods and inherent characteristics of fabrics may collectively affect the magnitude of G_0.

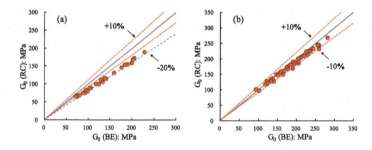

Fig. 2. Comparison of G_0 of Toyoura sand: (a) moist tamping; (b) dry tamping.

Often natural sand is not clean but contains some amount of fines. To examine the possible interplay of fines content with the testing method, G_0 of mixed soils having different fines content were also investigated. Note that only moist tamping method was adopted to prepare the sand-fines mixtures so as to prevent segregation between the coarse and fine grains. As shown in Fig. 3, all data including clean sand extracted from Fig. 2(a) are compared. The differences between the BE and the RC test are about 10% to 20% with higher G_0 in the BE test. Compared with the influence of test method in Fig. 2, this observation indicates that the fines content does not affect the discrepancies of G_0 between the BE and RC tests. Given that the RC test measures the global response of soil, while the BE test measures the local one, the distinct features in both methods implies that the BE test is more sensitive to the fabrics of sand. By using the moist tamping method, it is likely to form an isotropic fabric with no preferred grain orientations (Yang et al. 2008), as a consequence, it induces more grain interlockings. In this connection, it is hypothesized that a greater shear wave velocity in the BE test is likely due to a considerable amount of grain restraints against the particle motion. Of course, further verification of the hypothesis is worthwhile when more experimental evidence becomes available.

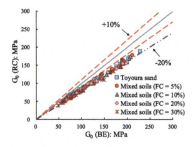

Fig. 3. Comparison of G_0 of sand-fines mixtures using moist tamping method.

4 Conclusions

This paper presents an experimental study on G_0 of sand using the BE and the RC test. The influences of test method and sample reconstitution method were discussed. By using the BE test, a higher G_0 value as much as 20% could be obtained as compared with the RC test, and the observation is consistent in the clean sand and the sand-fines mixtures. The underlying mechanics is explained by inherent characteristics of shear waves that transmit through an isotropic/anisotropic fabric of sand. Most importantly, the findings from this study suggest that cautions should be exerted in examination of G_0 by using different test methods.

Acknowledgements. The work presented in this paper is supported by the Research Grants Council of Hong Kong (No. 17205717 & 17250316) and the Chang'an University through the Fundamental Research Funds for the Central Universities (No. 300102268302) and the "111" Center (B18046). These supports are gratefully acknowledged.

References

Dyvik, R., Madshus, C.: Lab measurements of G_{max} using bender element. In: Khosla, V. (ed.) Advances in the Art of Testing Soils under Cyclic Conditions, New York, pp. 186–196. ASCE (1985)

Gu, X.Q., Yang, J., Huang, M.S., Gao, G.Y.: Bender element tests in dry and saturated sand: signal interpretation and results comparison. Soils Found. **55**(5), 951–962 (2015)

Richart, F.E., Hall, J.R., Woods, R.D.: Vibrations of Soils and Foundations. Prentice-Hall, Englewood Cliffs (1970)

Viggiani, G., Atkinson, J.H.: Interpretation of bender element tests. Géotechnique **45**(1), 149–154 (1995)

Yamashita, S., Kawaguchi, T., Nakata, Y., Mikami, T., Fujiwara, T., Shibuya, S.: Interpretation of international parallel test on the measurement of G_{max} using bender elements. Soils Found. **49**(4), 631–650 (2009)

Yang, J., Gu, X.Q.: Shear stiffness of granular material at small strain: does it depend on grain size? Géotechnique **63**(2), 165–179 (2013)

Yang, J., Liu, X.: Shear wave velocity and stiffness of sand: the role of non-plastic fines. Géotechnique **66**(6), 1–15 (2016)

Yang, J., Yan, X.R.: Site response to multi-directional earthquake loading: a practical procedure. Soil Dyn. Earthq. Eng. **29**(4), 710–721 (2009)

Yang, Z.X., Li, X.S., Yang, J.: Quantifying and modelling fabric anisotropy of granular soils. Géotechnique **58**(4), 237–248 (2008)

Big Data and Large Volume (BDLV)-Based Nanoindentation Characterization of Shales

Shengmin Luo[1], Yucheng Li[1], Yongkang Wu[1], Yuzhen Yu[2], and Guoping Zhang[1(✉)]

[1] Department of Civil and Environmental Engineering, University of Massachusetts Amherst, Amherst, MA 01003, USA
zhangg@umass.edu
[2] Department of Hydraulic Engineering, Tsinghua University, Beijing 100084, China

Abstract. This paper presents a novel nanoindentation technique that obtains massive data on the basis of large volume to extract the mechanical properties of both the individual mineral phases of shales and its bulk rock. Massive data collected on a given surface but extended to a depth of 8–10 μm with the statistical or grid indentation method were processed to extract the mechanical properties of individual mineral phases as well as the dependence of the mechanical behavior upon indentation depth. Large volume-based indentation was then applied at varying depths by the sacrificial removal of the previously indented surface layer. This method is now being implemented as a screening and optimization protocol for various chemical stimulants and additives used in hydraulic fracturing and oil/gas production operations.

Keywords: Nanoindentation · Big data · Large volume · Shale softening

1 Introduction

Oil and gas shales are a type of multi-scale, multi-phase, hybrid inorganic-organic composite materials that possess both frictional and cohesive behavior, and hence it is very challenging to characterize and interpret their engineering behavior. The recovery and extraction of oil and gas from shale reservoirs and other tight formations require hydraulic fracturing, usually assisted or enhanced by other stimulation techniques, to create fracture network or channels for the flow of oil and gas. The aqueous hydro-fracking fluids are usually loaded with various chemical stimulants and additives (e.g., proppants) to enhance and improve oil/gas recovery, maintain the wellbore stability, and minimize deterioration and damage to the formation. Sufficient data obtained from past and existing operations have suggested that some physical and chemical reactions between the constituent mineralogy (e.g., clay, carbonate, pyrite) of shales and the hydrofracking fluids may take place during and after the hydrofracking processes, and such interactions can alter the mechanical properties of individual phases and hence the bulk rock, which may become significant or even severe under the environment of elevated temperatures and pressures. For instance, a widely recognized hypothesis for the rapid drop of gas production rate in shale formations after 6–12 months production is that the clay minerals in fine-grained shales can interact with water via such processes as hydration, absorption, and adsorption, among others, leading to the swelling and softening of the clay matrix as well as the bulk rock, which in turn causes significant proppant embedment and further reduction in fracture conductivity. As a result, the fracture network is negatively impacted. Therefore, understanding how the mechanical properties of individual phases change upon interactions with hydrofracking fluids plays a key role in suppressing shale softening and maximizing the long-term production of hydrocarbons in shale reservoirs. However, traditional macroscopic mechanical testing can yield the mechanical properties of the rock mass as a composite, but no insights can be obtained into the behavior of individual mineralogical phases of shales that may undergo dynamic changes prior to and during the oil/gas extraction process (e.g., hydraulic fracturing, injection of stimulants, rock-fluid interactions). In the past two decades, nanoindentation testing, a non-destructive technique, has been widely used to characterize the mechanical properties of various materials at the micro- to nano-meter scales (Oliver and Pharr 2004). This paper presents a novel big data and large volume (BDLV)-based nanoindentation technique that provides a useful, viable way to accurately characterize the softening behavior of individual phases of shales as well as the bulk rocks at the micron to nanometer scale.

2 Materials and Methods

The studied shale sample was recovered from Longmaxi shale formation of Sichuan basin, China. X-ray powder diffraction (XRPD) results show that clay minerals, quartz, and calcite are the major constituents of the bulk rock. Two adjacent small pieces were first cut from the core sample in the direction parallel to the bedding plane, and then were glued onto a cylindrical sample holder for subsequent sample preparation. Each tested surface with dimensions of 15×10 mm was carefully ground and finely polished with abrasive papers in the order of increasing fineness. To simulate the underground hydrothermal environment during the hydrofracking process and the subsequent oil/gas production, one piece of the shale samples was placed for 24 h in a hydrothermal reactor filled with deionized (DI) water at a temperature of 120 °C and pressure of 2 MPa once a flat and smooth surface was obtained in the previous step, while the other piece was used as a non-treated sample to obtain the baseline data. Nanoindentation experiments were conducted in a G200 nanoindenter (Keysight Technologies, Inc., Santa Rosa, CA) using a Berkovich diamond indenter tip. Indentation loading usually involves pushing the indenter tip into the sample surface, during which the applied load F and the penetration depth h were recorded during the loading and unloading processes, and the mechanical properties (e.g., Young's modulus E and hardness H) can be derived from the load-displacement loading and unloading curves. For highly heterogeneous materials, such as shales and concretes, a grid indentation method was proposed by Constantinides et al. (2006) to probe the mechanical properties of individual constituent phases, from which the statistical methodology is adopted in this study. The newly developed BDLV-nanoindentation technique described in this paper mainly includes two steps, consisting of a big data-based analytics to extract the meaningful results from a massive dataset and a large volume-based experimental approach to obtain more data at different surfaces with varying depths. To employ the big data-based analysis of the nanoindentation results, indentation testing was performed using a continuous stiffness measurement (CSM) method at 500 spots on each sample surface. The mechanical properties in the direction perpendicular to the bedding plane were continuously characterized as a function of indentation depth. Statistical deconvolution was applied to the experimental results (i.e., the Young's modulus) calculated from the E-h curves at different indentation depths. The large volume-based processing was achieved by sequentially removing the softened surface layers of the tested shale sample via polishing and then conducting the big data-based nanoindentation analysis on each of the subsequently prepared, freshly polished surfaces. With such a method, a series of sacrificially removed surfaces can be analyzed to estimate the rate of softening for shales.

3 Results and Discussion

Figure 1 shows the baseline results obtained from the big data-based nanoindentation testing on the untreated sample. Five mechanically distinct phases are determined at the relatively shallow indentation depth (e.g., $h = 0.25$ μm), and they are respectively assigned as organic matter (e.g., kerogen in voids), clay matrix (not individual clay particles), interface between hard-soft minerals (e.g., clay-quartz interface), calcite or other carbonates, and quartz in the order of increasing Young's modulus. With increasing depth, the Young's moduli of all these phases gradually change and eventually converge into one single value at large indentation depth (e.g., $h > 5$ μm). Such depth-dependent elasticity can be affected by several factors. Generally, a 10% rule of thumb, which states that the ratio of maximum indentation depth to characteristic size of the indented phase needs to be smaller than 1/10, can be followed when the mechanical properties of individual phases are of interest in a multi-phase composite. Microscopic observations reveal that coarser and harder silt-size particles are embedded in the layered clay matrix. Therefore, thin film-substrate analogy is applied to interpret the mechanical response of shales that contain solid inclusions and pores within a clay matrix. The data of each mineral phase, except for the interface, were fitted using the model proposed by Wei et al. (2009):

$$E = E_c + \frac{E_p - E_c}{1 + [h/(t\beta_1)]^Y} \quad (1)$$

where t is the characteristic size of the individual phases; Y and β_1 are two constants that can be determined by the curve fitting for each phase; E_p and E_c are the fitted Young's moduli of individual phases and bulk rock, respectively. The results obtained from the big data-based nanoindentation analysis for the pre-treated sample are presented in Fig. 2. Due to the fact that clay minerals tend to become hydrated when they are in contact with water, particularly in a hydrothermal environment with high pressure and temperature, the Young's modulus of clay matrix decreases from 18.8 GPa to 16.3 GPa after a 24-h treatment. In addition, for the Young's modulus of calcite, a reduction from

66.2 GPa to 57.9 GPa is observed, which is caused by partial dissolution of calcite in water under the hydrothermal condition.

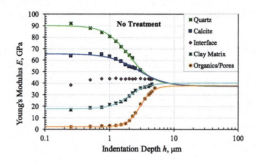

Fig. 1. Baseline data, where solid points represent the Young's moduli of individual phases determined from the big data-nanoindentation analysis.

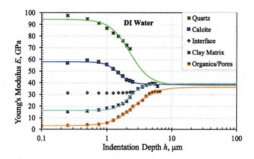

Fig. 2. Big data-nanoindentation results for shale reacted with DI water at HT-HP condition, where a reduction of Young's modulus is observed for both the clay matrix and the calcite.

4 Conclusions

In this paper, the shale-softening behavior is quantitatively characterized using a newly develop big data-based nanoindentation technique. Results obtained from the non-treated sample and DI water-treated sample show that interactions between shale and water significantly weaken the mechanical response of the clay matrix. Since clay matrix usually acts as the major supporting skeleton in shales, the reduction of its Young's modulus increases the potential of proppant embedment. The rate of such a softening process is proposed to be determined by performing the big data and large volume-based nanoindentation analysis after removing the tested surface layers that have been softened, which warrants further investigation in future study.

References

Constantinides, G., Ravi Chandran, K.S., Ulm, F.-J., Van Vliet, K.J.: Grid indentation analysis of composite microstructure and mechanics: principles and validation. Mater. Sci. Eng., A **430**, 189–202 (2006). https://doi.org/10.1016/j.msea.2006.05.125

Oliver, W.C., Pharr, G.M.: Measurement of hardness and elastic modulus by instrumented indentation: advances in understanding and refinements to methodology. Mater. Res. Soc. **19**, 3–20 (2004)

Wei, Z., Zhang, G., Chen, H., Luo, J., Liu, R., Guo, S.: A simple method for evaluating elastic modulus of thin films by nanoindentation. J. Mater. Res. **24**, 816–822 (2009). https://doi.org/10.1557/jmr.2009.0109

Development of a Large Temperature-Controlled Triaxial Device for Rockfill Materials

Hangyu Mao and Sihong Liu[✉]

College of Water Conservancy and Hydropower, Hohai University, No.1, Xikang Road, Nanjing 210098, China
sihongliu@hhu.edu.cn

Abstract. To study mechanical behavior of rockfill materials, lots of apparatuses are developed and applied. However, most of the existing apparatuses cannot control temperature during the experiment. In this study, a large triaxial device for rockfill materials is developed to fulfil temperature control ranging from −20 °C to 70 °C. The device can be used for triaxial test of rockfill materials under any humidity with the external volumetric strain measurement technology. A saturated triaxial test is carried out to determine the accuracy of the external volumetric strain measurement. Finally, the experimental device was tested on metamorphic sandstone at air-dried condition over temperature ranging from −5 °C to 55 °C.

Keywords: Rockfill · Triaxial apparatus · Temperature control

1 Introduction

Rockfill dams have been widely used all over the world, as they are reasonably adaptable to topography, geology and various kinds of rockfill materials [1–3]. Recently, a number of high rockfill dams are under construction in southwest China. These dams are usually under the actions of high water head, high stress and the condition of an alpine climate [3, 4]. The effect of significant temperature variations on the strength and deformation of rockfill materials has aroused great academic attention [5]. Some rockfill test instruments considering temperature effects have been reported [6–8]. However, there have been few reports about temperature-controlled triaxial device for rockfill materials.

In order to study the evolution of the mechanical properties of rockfill materials at different temperatures, a triaxial device with a sample of 30 cm in diameter and 60 cm in height was designed, in which different temperatures ranging from −20 °C to 70 °C can be controlled. A saturated triaxial test was carried out to prove the accuracy of the external volumetric strain measuring system. Finally, the proposed device was tested on air-dried rockfill samples at different temperatures ranging from −5 °C to 55 °C.

2 A Large Temperature-Controlled Triaxial Device

A large temperature-controlled triaxial device is developed in this study. Figure 1 shows a schematic diagram of the newly developed apparatus. It consists of three main parts: loading system, volumetric strain measuring system and temperature control system. The loading system consists of an oil cylinder, a counterforce frame, a pressure hood and three pressure actuators. The pressure oil cylinder can provide the maximum vertical force of 2000 kN and the pressure actuators can provide the maximum pressure of 4 Mpa. The volumetric strain measuring system consists of an internal volumetric strain measuring device and an external volumetric strain measuring device. The temperature control system consists of a spiral copper tube and a temperature-controlled water bath equipment. The spiral copper tube is installed in the pressure hood. The temperature conducting medium in heating bath is heat conduction oil, while in low cooling bath is sodium chloride solution. Different temperatures ranging from −20 °C to 70 °C can be controlled by using the temperature control system. The precision of the temperature control is 0.5%F.S. All controllers and sensors are connected to the signal collection box, which can collect data and control the test process.

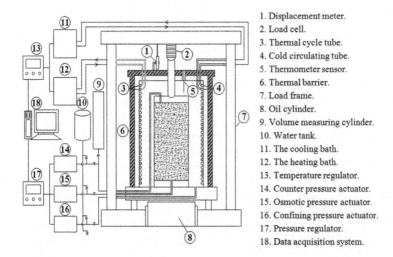

1. Displacement meter.
2. Load cell.
3. Thermal cycle tube.
4. Cold circulating tube.
5. Thermometer sensor.
6. Thermal barrier.
7. Load frame.
8. Oil cylinder.
9. Volume measuring cylinder.
10. Water tank.
11. The cooling bath.
12. The heating bath.
13. Temperature regulator.
14. Counter pressure actuator.
15. Osmotic pressure actuator.
16. Confining pressure actuator.
17. Pressure regulator.
18. Data acquisition system.

Fig. 1. Scheme of the temperature-controlled triaxial device developed

3 Performance of the Triaxial Equipment

3.1 Material Tested

The test material in this research was metamorphic sandstone from Jiangxi Province, China. It is classified as soft rocks and broke easily under loads, with a particle density of 2.64 g/cm³. The density of the sample used in the test was 2.12 g/cm³, and the initial void ratio was 0.245. The largest grain size was always $d_{max} = 60$ mm. The average particle diameter of the sample was $d_{50} = 17.9$ mm, the uniformity coefficient was $C_u = 14.2$, and the curvature factor was $C_c = 1.7$.

3.2 Test of the External Volumetric Strain Measurement System

The measurement of external volumetric strain is a round-about way to obtain volumetric strain [9]. In order to test the feasibility and accuracy of the measurement of external volumetric strain, a triaxial test of saturated sample was carried out at 55 °C. The confining pressure of this test was set to 800 kp. Figure 3 shows the comparison between the measured results of the external volumetric strain and the internal volumetric strain. It can be seen from the comparison that the measurement results of the external volumetric strain are in accordance with the internal volumetric strain. Therefore, the external volumetric strain measurement is reliable.

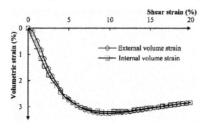

Fig. 3. The relationship between volumetric strain and shear strain in different measurement methods

3.3 Comparison of Test Results at Different Temperatures

Figure 4 shows the comparison of experimental results at different temperature. The temperatures of the three samples were −5 °C, 25 °C and 55 °C respectively. Both of them were air-dried samples and under the confining pressure of 800 kPa.

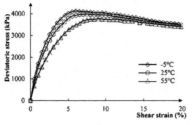
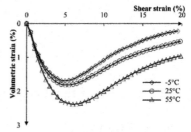

(a) Relationship between shear stress and shear strain

(b) Relationship between volume strain and shear strain

Fig. 4. The relation curves of mechanical properties at different temperatures

By comparison, we find that the strength and deformation properties of rockfill materials are both related to temperature. The test result shows the peak shear strength is incremented as the temperature decrease, while the dilatancy of the rockfill under shear is incremented when the temperature decreases.

4 Summary and Conclusions

A large temperature-controlled triaxial device for rockfill materials was developed. This apparatus can fulfil temperature control in a wide temperature range (-20 °C to 70 °C).

Two different methods were used to record the volumetric strain of a saturated sample. The measurement results of the external volumetric strain are in accordance with the internal volumetric strain. This indicates that the triaxial device can be used for unsaturated rockfill.

Tests were carried out on metamorphic sandstone with a maximum particle size of 60 mm at different temperatures. The results show that the mechanical properties of the rockfill materials are related to the temperature.

Acknowledgements. This work is supported by the Joint Funds of the National Natural Science Foundation of China" (Grant No. U1765205), the Priority Academic Program Development of Jiangsu Higher Education Institutions (PAPD) (Grant No. 3014-SYS1401) and the National Key R&D Program of China (Grant No. 2017YFC0404800). The supports are greatly appreciated.

References

1. Xu, Y., Zhang, L.M.: Breaching parameters for earth and rockfill dams. J. Geotech. Geoenviron. Eng. **135**(12), 1957–1970 (2009)
2. Cooke, J.B.: Progress in rockfill dams. J. Geotech. Eng. **110**(10), 1381–1414 (1984)
3. Ma, H., Chi, F.: Major technologies for safe construction of high earth-rockfill dams. Engineering **2**(4), 498–509 (2016)
4. Huo, J.X., Song, H.Z., Luo, L.: Investigation of groundwater chemistry at a dam site during its construction: a case study of xiangjiaba dam, China. Environ. Earth Sci. **74**(3), 2451–2461 (2015)
5. Shi, B., Cai, Z., Chen, S.: Research progress of rock-fill material degradation in a severe environment. In: Geo-China International Conference, pp. 26–34 (2016)
6. Zhang, B.Y., Zhang, J.H., Sun, G.L.: Deformation and shear strength of rockfill materials composed of soft siltstones subjected to stress, cyclical drying/wetting and temperature variations. Eng. Geol. **190**, 87–97 (2015)
7. Zhang, B.Y., Zhang, J.H., Sun, G.L.: Development of a soft-rock weathering test apparatus. Exp. Tech. **38**(2), 54–65 (2014)
8. Nuszkowski, J., Thomas, A., Hudyma, N., Harris, A.: Development of a test apparatus to determine thermal properties of rock specimens. In: US Rock Mechanics/Geomechanics Symposium (2016)
9. Laloui, L., Peron, H., Geiser, F., Rifa'I, A., Vulliet, L.: Advances in volume measurement in unsaturated soil triaxial tests. Soil Found. **46**(3), 341–349 (2006)

Experimental Study of Small Strain Stiffness of Unsaturated Silty Clay

Tomáš Mohyla(✉) and Jan Boháč

Faculty of Science, Charles University,
Albertov 6, 12843 Prague, Czech Republic
mohylat@natur.cuni.cz

Abstract. The small strain stiffness of saturated soils has been extensively studied in the past decades, and therefore the literature database of small strain stiffness of different soils is vast and contains the necessary information for development and calibration of numerical models. However, there is a lack of data of small strain stiffness in case of unsaturated soils. This paper presents the stiffness of unsaturated silty clay in both small and very small strain regions.

The laboratory tests were carried out in the double-walled cell triaxial apparatus with HAEV (high air entry value) ceramic disk mounted in the base pedestal, and the axis translation technique was used. The reconstituted specimens of 70 mm in diameter were tested and the very small strain stiffness was measured by bender elements, while in the case of the small strain stiffness measurement, the submersible LVDTs were used. Different stress paths were applied, and every stress change was followed by an equalization period. Subsequently, the very small strain shear modulus (G_0) was determined by bender elements, and its decay in the small strain range was determined using short shear probes of up to 30 kPa.

Keywords: Unsaturated soils · Small strain stiffness · Laboratory testing

1 Tested Soil

The tested soil was silty clay (CL) from Central Bohemia with the liquid limit and plasticity index of 29 and 11, respectively. The reconstituted specimens were prepared with vertical load of 100 kPa, and cut to the dimensions of 70 mm in diameter and 140 mm in height.

The saturated specimens were desaturated in the triaxial chamber by applying the suction using the axis translation technique.

2 Testing Program

The objective of this study was to study the influence of suction value, current suction ratio – CSR [1], defined as ratio between maximum historic suction and current suction magnitude, and recent suction history on the soil stiffness.

The testing program consisted of very small strain shear modulus (G0) and shear modulus reduction curve measurements. The maximum shear modulus was measured by bender elements, while the stiffness reduction with increasing strain was measured using LVDT transducers. The testing program was carried out in the means of two independent stress variables – suction and net stress, axis translation technique and the HAEV disk were used. The suction was applied as the difference between the pore-air and pore-water pressures, which were produced by standard pressure controllers.

Two specimens were tested – SM1 and SM2. However, the stiffness reduction curve was determined only for SM2. The Table 1 summarizes the test procedures and the Fig. 1 shows the applied stress paths.

Table 1. SM1, SM2 – stress conditions during stiffness measurement; stress history. P = net stress (kPa), S = suction (kPa), C = CSR, L = last suction change (kPa).

Test ID - Specimen 1	Stress path	Test ID - Specimen 2	Stress path
SM1-1 (P10S50C1L50)	0-1	**SM2-8** (P10S100C1L100)	0-8
SM1-2 (P100S50C1L0)	1-2	**SM2-3A** (P100S100C1L0)	8-3
SM1-3a (P100S100C1L50)	1-2-3	**SM2-6** (P200S100C1L0)	8-3-6
SM1-4 (P100S300C1L200)	1-2-3-4	**SM2-7** (P200S300C1L200)	8-3-6-7
SM1-3b (P100S100C3L200)	1-2-3-4-3	**SM2-9** (P200S200C1.5L100)	8-3-6-7-9
SM1-5 (P100S125C2.4L25)	1-2-3-4-3-5	**SM2-10** (P100S200C1.5L0)	8-3-6-7-9-10
SM1-3c (P100S100C3L25)	1-2-3-4-3-5-3	**SM2-4** (P100S300C1L100)	8-3-6-7-9-10-4
SM1-6a (P200S100C3L0)	1-2-3-4-3-5-3-6	**SM2-3B** (P100S100C3L200)	8-3-6-7-9-10-4-3
SM1-7 (P200S300C1L200)	1-2-3-4-3-5-3-6-7		
SM1-6b (P200S100C3L200)	1-2-3-4-3-5-3-6-7-6		

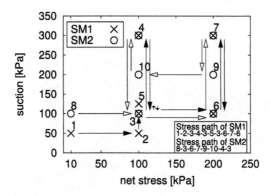

Fig. 1. Stress paths.

3 Results

The results of the laboratory measurements show, that the higher the suction the higher the stiffness – see SM2-6 (SM2_P200S100C1L0) vs. SM2-7 (SM2_P200S300C1L 200), or SM1-2 (SM1_P100S50C1L0) vs. SM1-3a (SM1_P100S100C1L50) and SM1-4 (SM1_P100S300C1L200) (only G0 measurements). Additionally, the soil stiffness is affected by the suction history and CSR – e.g. SM2-3A (SM2_P100S100C1L0) vs. SM2-3B (SM2_P100S100C3L200).

The Figs. 3 and 4 show the stiffness reduction curves at different net stress levels for specimen SM2. The figure show, that degradation of shear modulus starts at much lower strains than is typically reported for high plasticity soils. All the G0 measurements, for both specimens, are summarized in Fig. 2.

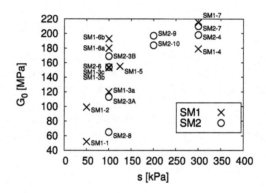

Fig. 2. G_0 measurements at different stress conditions. See Fig. 1 and Table 1.

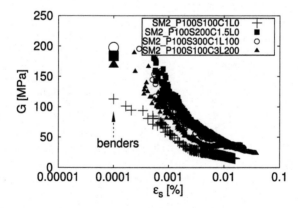

Fig. 3. Stiffness degradation curves of SM2 at net stress of 100 kPa.

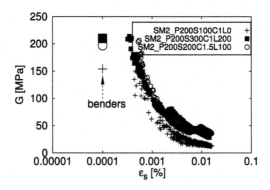

Fig. 4. Stiffness degradation curves of SM2 at net stress of 200 kPa.

4 Conclusion

We have studied the dependence of stiffness of unsaturated silty clay on suction and net mean stress and we observed the influence of recent suction history on small strain stiffness. The laboratory tests produced data about both maximum shear modulus and shear modulus reduction curve. Such data can be used for numerical model calibration.

The tested soil exhibits very rapid stiffness degradation, this behaviour can be found in low-plasticity soils, e.g. [2, 3]. The maximum shear modulus is affected by the current suction value, CSR and recent suction history. If the suction increases, the stiffness will increase too. CSR has similar effect, the higher it is, the higher the stiffness is.

Acknowledgement. The study was supported by the Charles University, project GA UK No. 1160217.

References

1. Ng, C.W.W., Xu, J.: Effects of current suction ratio and recent suction history on small-strain behaviour of an unsaturated soil. Can. Geotech. J. **49**(2), 226–243 (2012)
2. Coop, M.R., Jovičić, V., Atkinson J.H.: Comparisons between soil stiffnesses in laboratory tests using dynamic and continuous loading. In: Proceedings of the Fourteenth International Conference on Soil Mechanics and Foundation Engineering, pp. 267–270. Taylor & Francis, Hamburg (1997)
3. Vucetic, M., Dobry, R.: Effect of soil plasticity on cyclic response. J. Geotech. Eng. **117**(1), 89–107 (1991)

Effects of Soil Fabric on Volume Change Behaviour of Clay Under Cyclic Heating and Cooling

Qingyi Y. Mu[1(✉)], C. W. W. Ng[2], C. Zhou[2], and Hongjian J. Liao[1]

[1] Xi'an Jiaotong University, Xi'an, China
qingyimu@mail.xjtu.edu.cn
[2] Hong Kong University of Science and Technology, Hong Kong, China

Abstract. To investigate the influence of soil fabric on volume changes of soil under cyclic thermal loads, reconstituted, intact and recompacted clay specimens were tested at normally consolidated state using a newly developed temperature-controlled invar oedometer. All specimens showed continuous contraction as the number of heating-cooling cycles increased. The accumulated plastic axial strain of the reconstituted specimen was 38% and 68% larger than those of the intact and recompacted specimens, respectively. On the other hand, the thermal expansion coefficient of the reconstituted specimen was almost three times larger than those of the intact and recompacted specimens. These observed differences are likely attributed to different distributions of clay particles in soil specimens.

Keywords: Soil fabric · Volume changes · Cyclic heating and cooling

1 Introduction

Soil fabric refers to the arrangement of soil particles [1] and significantly influences the volume change behaviour of soil. Most previous studies about the effects of soil fabric on the volume change behaviour of soil were limited to conditions of a constant temperature. The principal objective of this study is to investigate the effects of soil fabric on soil volume changes under cyclic heating and cooling. To meet this objective, a low-plasticity clay was tested at three different states (i.e., reconstituted, intact and recompacted). Through temperature-controlled oedometer tests, thermal axial strains and thermal expansion coefficients were measured and compared.

2 Test Apparatus

A temperature-controlled invar oedometer, which was modified from a conventional oedometer by adding a heating-cooling system, was developed. The heating-cooling system consists of a spiral tube, a K-type thermocouple, a heating-cooling unit, a bath and a pump. During a test, the water temperature in the bath was controlled by the heating-cooling unit. The pump circulated the water in the spiral tube which was installed surrounding the soil specimen. The soil specimen was heated or cooled through heat exchange with the circulating water. Invar material (thermal expansion coefficient: 5×10^{-7} °C^{-1}) was used to minimise the thermal strain of the oedometer ring [2, 3]. More details of the test setup were given by Ng et al. [4]

3 Test Soil and Test Program

A loess taken from Xi'an in China was used in this study. The measured liquid limit and plastic limit were 36% and 19%, respectively. The soil consists of 28% clay and 72% silt particles. According to the Unified Soil Classification System [5], the soil is classified as clay of low plasticity (CL). More physical properties of the test soil were described in detail by Ng et al. [6].

Axial strains and thermal expansion coefficients of the reconstituted, intact and recompacted specimens under cyclic thermal loads were measured and compared. Loess block samples were taken from a trial pit at a depth of 3.5 m, which corresponds to a vertical stress of about 50 kPa. To achieve a similar stress state and soil structure to that of the in situ case, an effective vertical stress of 50 kPa was applied to the reconstituted, intact and recompacted specimens.

4 Interpretation of Experimental Results

4.1 Effects of Soil Fabric on Accumulated Axial Strain Under Cyclic Heating-Cooling

Figure 1 shows the accumulated axial strain with an increasing number of thermal cycles. For the first thermal cycle, the axial strain of the reconstituted specimen is 0.46%, which is 28% and 39% larger than those of the intact (0.36%) and recompacted specimens (0.33%), respectively. For the subsequent thermal cycles, the difference in accumulated axial strain increases with the number of thermal cycles. After thermal stabilization, the accumulated axial strain of the reconstituted specimen is 0.84%, which is 38% and 68% larger than those of the intact (0.61%) and recompacted (0.50%) specimens, respectively. The difference in accumulated axial strain between these three specimens possibly results from the different distributions of clay particles in soil. As revealed in the SEM results [7], clay particles form the soil skeleton of the reconstituted specimen. In the intact and recompacted specimens, most of the clay particles are gathered in a small number of silt-size aggregates. Silt particles play a significant role in the soil skeleton formation of intact and recompacted loess. As reported by Brochard et al. [8], thermal effects on water adsorption and thus thermal strains are more significant for clay-dominated soil skeletons than silt-dominated soil skeletons.

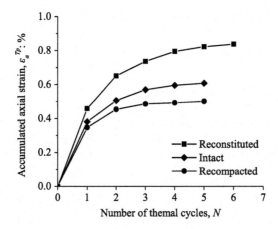

Fig. 1. Accumulated plastic axial strains of the reconstituted, intact and recompacted specimens under cyclic heating and cooling

4.2 Effects of Soil Fabric on the Thermal Expansion Coefficient

Figure 2 shows the thermal expansion coefficients (TEC) of reconstituted, intact and recompacted specimens. The thermal expansion coefficient of each specimen is almost constant under heating and cooling cycles. The average TECs for the reconstituted, intact and recompacted specimens are 1.6×10^{-5} °C^{-1}, 7.0×10^{-6} °C^{-1} and 5.0×10^{-6} °C^{-1}, respectively. The TEC of the reconstituted specimen is almost three times larger than those of the intact and recompacted specimens. It should be noted that some theoretical and experimental studies reported in the literature assume that the TEC of soil depends on its mineral only [9]. Obviously, these theoretical studies should be improved by considering the influence of soil fabric on the TEC.

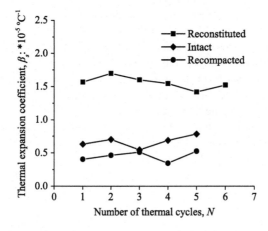

Fig. 2. Thermal expansion coefficients of the intact, recompacted and reconstituted specimens

5 Conclusions

Under cyclic heating and cooling, the axial strains of all specimens accumulated but at a reducing rate. The accumulated axial strain of the reconstituted specimen was about 38% and 68% larger than those of the intact and recompacted specimens, respectively. The TEC of the reconstituted specimen, which was determined from the measurements in the cooling phases, was almost three times larger than that of the intact and recompacted specimens. As revealed by the SEM results, these observed differences are likely attributed to differences in the soil fabric among the three types of specimens. For the reconstituted specimen, clay particles filled the spaces between silt particles and the soil fabric was homogeneous in general. On the contrary, most of the clay particles in the intact and recompacted specimens were gathered in a small number of silt-size aggregates.

Acknowledgements. The authors would like to acknowledge the financial support provided by the Research Grants Council of the Hong Kong Special Administrative Region (HKSAR) through the research grants 616812, 16209415 and 16216116 and the National Science Foundation of China through the research grant 51509041 and 41630639.

References

1. Mitchell, J.K., Soga, K.: Fundamentals of Soil Behavior, 3rd edn. Wiley, New York (2005)
2. Di Donna, A., Laloui, L.: Response of soil subjected to thermal cyclic loading: experimental and constitutive study. Eng. Geol. **190**, 65–76 (2015)
3. Vega, A., McCartney, J.S.: Cyclic heating effects on thermal volume change of silt. Environ. Geotech. **2**(5), 257–268 (2015)
4. Ng, C.W.W., Mu, Q.Y., Zhou, C.: Effects of boundary conditions on cyclic thermal strains of clay and sand. Géotech. Lett. **7**(1), 73–78 (2016)
5. ASTM. Standard practice for classification of soils for engineering purposes (unified soil classification system). American Society for Testing and Materials, West Conshohocken (2006)
6. Ng, C.W.W., Mu, Q.Y., Zhou, C.: Effects of soil structure on the shear behaviour of an unsaturated loess at different suctions and temperatures. Can. Geotech. J. **54**(2), 270–279 (2016)
7. Ng, C.W.W., Mu, Q.Y., Zhou, C.: Effects of specimen preparation method on the volume change of clay under cyclic thermal loads. Géotechnique (2018). https://doi.org/10.1680/jgeot.16.p.293
8. Brochard, L., Honório, T., Vandamme, M., Bornert, M., Peigney, M.: Nanoscale origin of the thermo-mechanical behavior of clays. Acta Geotech. **12**, 1261–1279 (2017)
9. Campanella, R.G., Mitchell, J.K.: Influence of temperature variations on soil behavior. J. Soil Mech. Found. Eng. Div. **94**(SM3), 709–734 (1968)

Comparing Water Contents of Organic Soil Determined on the Basis of Different Oven-Drying Temperatures

Brendan C. O'Kelly[1(✉)] and Weichao Li[2]

[1] Department of Civil, Structural and Environmental Engineering,
Trinity College Dublin, Dublin, Ireland
bokelly@tcd.ie
[2] Department of Geotechnical Engineering, Tongji University, Shanghai, China

Abstract. In order to prevent possible charring, oxidation and (or) vaporization of substances other than the pore water, some researchers adopt oven-drying temperatures in the range of 60–90 °C for water content determinations on peat and other highly organic soils. The growing consensus, however, is that the standardized oven-temperature ranges of 105–110 °C and 110 ± 5 °C used for inorganic soil are also appropriate for routine water content determinations for these soils. As described in this paper, correlations between water content and other material properties are affected when lower oven-drying temperatures are adopted. In this regard, the paper describes approaches for computing and comparing water content values determined for different drying temperatures, with additional experimental oven-drying data presented herein to validate them.

Keywords: Correlation · Drying temperature · Organic content
Water content

1 Introduction

As summarized in the papers by O'Kelly (2014) and O'Kelly and Sivakumar (2014), there has been much debate regarding an appropriate drying temperature for water content determinations on peat and other highly organic soils using the oven-drying method. Some charring, oxidation and (or) vaporization of susceptible organic matter may occur for the standard oven-drying temperature ranges of 105–110 °C (BS 1377-2 1990) and 110 ± 5 °C (ASTM D2216 2010) employed for inorganic soil. The associated reductions in the specimen equilibrium dry mass, which are incorrectly interpreted as evaporation of pore water from the viewpoint of performing water content calculations, result in apparent higher values than the correct water content value (O'Kelly 2004). On this basis, numerous researchers have recommended over the years using lower drying temperatures in the range of 60–90 °C for water content determinations on these soils, with oven-drying temperatures in this range still routinely used in some commercial and research laboratories for this purpose (O'Kelly 2014, 2016; O'Kelly and Sivakumar 2014).

However, as demonstrated experimentally for very different organic soils in the papers by O'Kelly (2004, 2005a, 2005b, 2014), O'Kelly and Sivakumar (2014) and Skempton and Petley (1970), all of the relevant water is not fully evaporated from the test specimen at the lower oven-drying temperature values. Nevertheless, these researchers concluded that for practical purposes, the standard oven-drying temperature ranges of 105–110 °C and 110 ± 5 °C employed for inorganic soil are also appropriate (and preferable than using lower over-drying temperature values) for routine water content determinations on these soils.

The impact of adopting different oven-drying temperatures on correlations developed between water content and other material properties was investigated in the papers by O'Kelly (2014) and O'Kelly and Sivakumar (2014) for peat and other organic soils, including biosolid, sewage sludge and water-treatment residue materials. For instance, as demonstrated in Fig. 1, the use of an oven-drying temperature value of 80 °C (rather than the standard oven temperatures of either 105 °C or 110 °C) has the knock-on effect in translating the undrained strength – water content correlation.

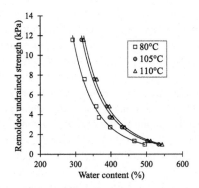

Fig. 1. Undrained strength correlations for water-treatment residue material ($N = 57\%$) demonstrating effect of oven temperature employed for water content determinations (O'Kelly 2014).

Hence, the use of a standardized oven-drying temperature is highly desirable when correlating water content with other material properties for peat and other highly organic soils (O'Kelly 2014, 2016; O'Kelly and Sivakumar 2014). For consistency, it is recommended that the values of water content corresponding to the pertinent standardized oven-drying temperature range are deduced in cases where measured or reported water content values have been determined on the basis of lower oven-drying temperatures (O'Kelly 2014; O'Kelly and Sivakumar 2014). For this purpose, the methods presented in the papers by Skempton and Petley (1970) and O'Kelly (2004, 2005a) for relating water content values determined for the same material but on the basis of different oven-drying temperatures can be used.

2 Computing and Comparing Water Content Values Determined on the Basis of Different Drying Temperatures

According to the method after O'Kelly (2004, 2005a), the water content value corresponding to the oven-drying temperature of 105 °C (i.e., $w_{105\,°C}$) can be computed from the water content value deduced for the lower oven-drying temperature of t°C (i.e., $w_{t°C}$) according to the following expression:

$$w_{105\,°C} = (w_{t°C} - \alpha_{105\,°C} + 1)/\alpha_{105\,°C} \tag{1}$$

where $w_{105\,°C}$ and $w_{t°C}$ are dimensionless values of water content (i.e. not their % values) and 105 °C is the water content parameter [$105\,°C \leq 1$ for $t < 105$ °C], defined as (O'Kelly 2004, 2005b):

$$\alpha_{105\,°C} = m_{105\,°C}/m_{t°C} \tag{2}$$

where $m_{105\,°C}$ and $m_{t°C}$ are the specimen equilibrium dry masses corresponding to the standard oven temperature of 105 °C and the lower temperature of t°C, respectively.

The $\alpha_{105\,°C}$ parameter is closely connected to the apparent moisture content ratio parameter utilized in the method presented in Skempton and Petley (1970). Similar expressions to those given by Eqs. 1 and 2 can be written for computing water content values corresponding to the oven-drying temperature value of 110 °C (i.e., $w_{110\,°C}$).

O'Kelly (2005a) reported that the amount of pore water that remains within the voids for $t < 100$ °C, and the level of susceptibility of the organic matter to charring for $t > 85$ °C, are approximately related to the organic content, which can be quantified in terms of the loss in dry mass on ignition (N) value. Adding to the database presented in O'Kelly (2005a), O'Kelly and Sivakumar (2014) investigated the relationship between the gradient values of the $\alpha_{105\,°C}$ against oven-drying temperature trend-lines (i.e., $\beta_{105\,°C}$) for 18 very different organic soils and their N values (see Fig. 2). The deduced relationship is given by Eq. 3, with the calculated $\beta_{105\,°C}$ values less than 0.0015 °C^{-1} for all but two of the 18 soils considered in their investigation.

$$\beta_{105\,°C} = 0.00026 \times \ln(N) - 0.00023 (n = 16) \tag{3}$$

The validity of generally applying the empirical Eq. 3 for other organic soils is supported by additional experimental oven-drying data presented in a forthcoming article entitled 'Water Content Determinations for Peaty Soil Using the Oven-Drying Method' by Li, W. et al. for four highly decomposed peaty soils ($N = 32.5$–73.4%), having von Post classification numbers ranging H_8–H_{10}. These soils, which were mainly comprised of decomposed *phragmites* plants, were sampled from depths of between 6.8 m and 9.2 m at a site located on the northeast shore of the Dianchi Lake in Kunming city, Yunnan, China. As evident from the Fig. 2, there is excellent agreement between the best-fit trend lines based on the original and extended databases.

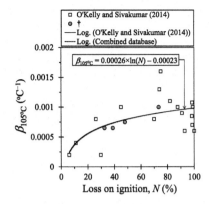

Fig. 2. Parameter $\beta_{105\,°C}$ against loss on ignition. † From forthcoming article entitled 'Water Content Determinations for Peaty Soil Using the Oven-Drying Method' by Li, W. et al.

3 Discussion and Conclusions

Comparisons and standardization of water content values determined on the basis of different oven-drying temperatures for peats and other highly organic soils can be made using Eq. 1 provided that the soil-dependent value of the $\alpha_{105\,°C}$ parameter can be evaluated for the material under investigation. As described in O'Kelly (2005a), the pertinent $\alpha_{105\,°C}$ value can be determined for the soil from its experimental plot of the reduction in specimen equilibrium dry mass for step increases in the oven-drying temperature. When such data are not available, an alternative approach is to compute the $\alpha_{105\,°C}$ value from the $\beta_{105\,°C}$ value deduced using the soil's measured or reported N value and the empirical correlation given by Eq. 3. Using either of these approaches, valid comparisons of experimental water content values determined on the basis of non-standard oven-drying temperatures can be made. Further, all such water content determinations can also be standardized using these approaches for an oven-drying temperature of 105 °C or 110 °C, as recommended in the papers by O'Kelly (2014, 2016) and O'Kelly and Sivakumar (2014).

References

ASTM D2216: Standard test methods for laboratory determination of water (moisture) content of soil and rock by mass. ASTM International, West Conshohocken (2010)

BS 1377-2: Methods of test for soils for civil engineering purposes (classification tests). BSI, London (1990)

O'Kelly, B.C.: Accurate determination of moisture content of organic soils using the oven drying method. Drying Technol. **22**(7), 1767–1776 (2004)

O'Kelly, B.C.: Method to compare water content values determined on the basis of different oven drying temperatures. Géotechnique **55**(4), 329–332 (2005a)

O'Kelly, B.C.: New method to determine the true water content of organic soils. Geotech. Testing J. **28**(4), 365–369 (2005b)

O'Kelly, B.C.: Drying temperature and water content-strength correlations. Environ. Geotech. **1**(2), 81–95 (2014)

O'Kelly, B.C., Sivakumar, V.: Water content determinations for peat and other organic soils using the oven-drying method. Drying Technol. **32**(6), 631–643 (2014)

O'Kelly, B.C.: Geotechnics of municipal sludges and residues for landfilling. Geotech. Res. **3**(4), 148–179 (2016)

Skempton, A.W., Petley, D.J.: Ignition loss and other properties of peats and clays from Avonmouth. King's Lynn Cranberry Moss. Géotech. **20**(4), 343–356 (1970)

Planar Granular Column Collapse: A Novel Releasing Mechanism

Gustavo Pinzon and Miguel Angel Cabrera[✉]

Department of Civil and Environmental Engineering, Universidad de los Andes,
Carrera 1 Este No. 19A - 40, Bogota, Colombia
{ga.pinzon1496,ma.cabrera140}@uniandes.edu.co

Abstract. The granular column collapse is a simplified system of the complex dynamics observed in gravity-driven natural mass flows and granular media manufacture systems. The granular column collapse has become a benchmark case for the study of granular materials under transitional flow regimes. In the current paper, we study the granular collapse of ceramic beads in a fully two-dimensional experimental set-up with the inclusion of a novel gate mechanism, avoiding any disturbance on the granular media at release. The system is backlighted by a high-intensity LED panel, allowing a clear distinction between beads and their surroundings. The collapse runout and kinematics are studied by means of digital image analysis, combining an edge-detection technique with the particle image velocimetry method. The column mobility is studied as a function of the initial aspect ratio $a = H_0/R_0$, with H_0 and R_0 as the initial column height and width, respectively. Results are in agreement with previous experimental studies, recognizing the influence of a on the final column deposit and validating the experimental set-up for further studies.

Keywords: Granular column · Granular flows · Hele Shaw
Experimental modelling

1 Introduction

Granular flows are a common phenomenon encountered in nature as rock avalanches, landslides, debris flows, and pyroclastic flows. There is an absence of a generalized constitutive set of equations able to describe the transitional motion and behaviour of granular flows, from its initiation, flow, and final deposition. The latter is due to the remarkable behaviour of granular media as a gas, a liquid, or a solid, depending on the scenario (i.e., initial state, boundary conditions, and stress-strain rate solicitations) [1]. The granular column collapse is a benchmark model able to reproduce transitional granular flows on a small scale, allowing the phenomenon-dynamics simplification. In this system, a granular column with initial aspect ratio $a = H_0/R_0$ is quickly released over a horizontal surface and collapses by self-weight, with H_0 and R_0 as the initial column height and width, respectively. Previous work on granular column collapses in axisymmetric and quasi-two-dimensional configurations found a clear dependency on a as a control parameter on the transition from short to tall columns [3,4].

In this paper, we revisit the experiments of Lacaze et al. [2], propose an optimized gate mechanism that releases all particles instantaneously, and study the column collapse dynamics, from release to deposition, by means of digital image analysis. Finally, we study the corresponding scaling principles related to the runout distance and aspect ratio.

2 Experimental Setup

Lacaze et al. presented a planar 2D model, similar to a Hele-Shaw cell, were the internal deformation and kinematic field are easily measured, within a 1.0 to 1.8 particle diameters model thick [2]. This configuration had a top-swinging gate as the opening mechanism. When opened slowly, the bottom particles are released sooner than the top particles, and when opened quickly, the gate interacts with the surrounding media, generating drag and secondary fluxes that may alter the granular collapse dynamics.

The current paper presents a similar setup but with a gate that slides off the motion plane, allowing all particles to be set free simultaneously. The experimental set-up is conformed by two Plexiglass (PMMA) square windows of 450 mm side and 10 mm thick. The latter are separated by a 1.2 mm gap, where a 1 mm thick PMMA hollow square stands on. The inner length of the hollow square is 390 mm (see Fig. 1).

For building the granular column, a sliding gate passes through a 0.2 mm thick slot, at 55 mm from the lateral wall (R_0). The movement of the sliding gate is performed by a linear pneumatic actuator, operating under a pressure of 4 bar. The speed and configuration of the opening mechanism guarantee that all particles are liberated simultaneously. The system is backlighted by a LED panel with 4000 lm in intensity. The granular collapse is recorded by a Mikrotron MotionBLITZ Cube 4 camera, using a frame rate of 800 fps at a resolution of 720 × 530 px (see Figs. 2a–c).

Experiments are performed with 1 mm diameter ceramic beads, of $\rho_b =$ 3600 kg/m^3 of bulk density, and manufactured by Sigmund-Lindner GmbH. The particles present a repose angle of $\varphi = 28.15° \pm 0.75°$, measured under the constrained experimental conditions, by releasing the particles in free-fall from a height of 200 mm and measuring the final repose angle after 5 repetitions. Each test starts by introducing the gate into the slot and building the column up to the desired initial column height H_0. Then, the granular column is shuddered assuring a flat top. The model is fixed to a flat base and the sliding gate is connected to the pneumatic actuator. The illumination system is turned on, along with the pressure system, and the high-speed camera. At this point, the opening mechanism is activated, releasing the particles and recording the movement of the granular mass. For each aspect ratio, seven repetitions were performed in order to assure statistical significance.

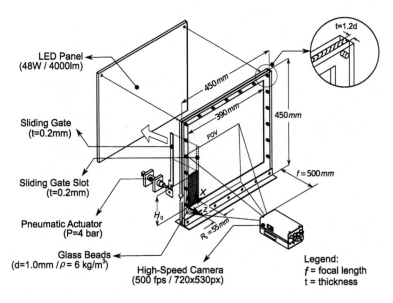

Fig. 1. Experimental Setup. Note that the granular column release is performed by the gate displacement along the z-axis.

3 Digital Image Analysis

The high-speed recordings are employed as direct measurements of the column collapse. The experimental spatial domain is calibrated by single marks placed on the left-hand-side wall. The marks are employed in a depth correction algorithm, generating an orthonormal image plane. Given the lighting conditions, it is possible to abstract the column shape variations during collapse. For this purpose, each frame is binarized with an intensity threshold between 120 and 140 in a 0–255 grayscale. Subsequently, an edge detection algorithm is implemented, merging all neighbour pixels with similar intensity. At this point, single particles areas ejected from the flowing mass are discarded from the column body. The resulting column-body coordinates are used to create a mask that delineates the column shape at each frame. This shape identification allows for a direct measurement of the runout distance R_i and column height H_i during collapse and at deposition (H_f, R_f).

The relative displacement between frames can be used as a descriptor of the velocity field transitions during collapse. The velocity field of the 2-D column collapse is measured with the Particle Image Velocimetry (PIV) technique (PIVlab [6]). The column shape, obtained by the edge detection algorithm, is employed as the computation domain on PIV. Figures 2d–f presents the velocity field transition for a column with aspect ratio of $a = 3.2$. Immediately after the release, the overall granular mass displaces, presenting higher velocities at the lateral free-surface (Fig. 2d). After that, the column transitions to a heap flow, in which the surface-layer thickness decreases in time until reaching deposition (Figs. 2e and f).

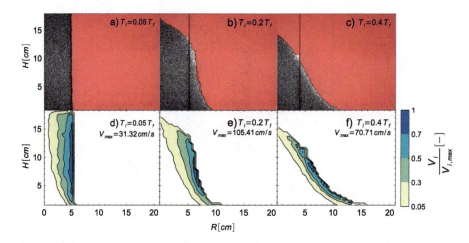

Fig. 2. (a–c): Raw images of the column during collapse with the mask (red area), and (d–f): Velocity profiles from the PIV analysis. Note that the darker areas in (d–f) mark the zones with higher velocities relative to the velocity field at each frame.

4 Scaling Principles

Previous works highlighted the governing role of a on the column collapse dynamics [3,4]. A clear distinction is found between short and tall columns proportional to a, where a portion of the column collapses as a landslide for the first case, and in the second case the overall mass free-falls prior to spreading. The final deposition runout R_f and height H_f are the most common indicators used in the description of the collapsing column dynamics.

These effects are commonly presented as a function of the final normalized column mobility $R^* = (R_f - R_0)/R_0$ and the normalized final deposit height $H^* = H_f/R_0$ (see Fig. 3). The distinction between short and tall columns has been regarded to be the inflexion point at which the best agreement of a power-law-fitting, in the (R^*, a) plane, is obtained (see Fig. 3a). The dashed trends in Figs. 3a and b are the reported values found by [2–4].

Overall, our experimental results match the experimental trends obtained by Lube et al., Lajeunesse et al. and Lacaze et al. [2–4], but differ on the identification of the transition point between short and tall columns at an aspect ratio of $a^* = 1.45$ in our experiments against values of $a^* = 1.7$, $a^* = 0.74$, and $a^* = [2.5 - 4.1]$, respectively. This disagreement has been acknowledged to be the result of inherent measurement errors between experimental setups or by the number of particles that conform the column [7], or be caused by the material response when using different particles (i.e., shape, roughness, density) [5]. Nevertheless, the possibility of simulating the complex dynamics of 3D systems, as in the axysymmetric columns, and of directly measuring the column shape transition allows the validation of the current system and poses it as an alternative for the study of the granular column collapse in further scenarios (i.e., submerged, under centrifugal acceleration).

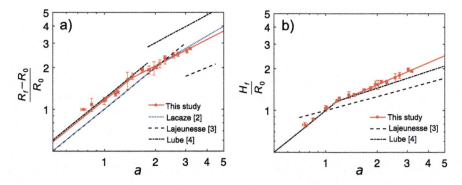

Fig. 3. Normalized runout distance and normalized deposit height as a function of the aspect ratio a. In (a) the change in slope between trends is referred as the transition point a^* from short to tall columns.

5 Conclusions

This paper presented a 2D planar experimental set-up for the simulation of the granular column collapse, with a novel gate mechanism that releases the column simultaneously. The model allows the direct and accurate measurement of the column collapse mobility (e.g., shape transitioning and kinematics) and final deposit height and runout, H_f and R_f, respectively. The results on the collapse deposit of dry ceramic beads are in agreement with previous works in the identification of the collapsing trends for short and tall columns as a function of a, validating the effectiveness of the system in capturing complex 3D granular dynamics. Undergoing work is focused on the study of the granular column collapse in submerged conditions and its relation to volumetric variations during collapse.

Acknowledgements. M.A.C. received funding from the Universidad de los Andes, Early-stage Researcher Fund (FAPA) under Grant No. PR.3.2016.3667.

References

1. Andreotti, B., Forterre, Y., Pouliquen, O.: Granular Media: Between Fluid and Solid. Cambridge University Press, Cambridge (2013)
2. Lacaze, L., Phillips, J.C., Kerswell, R.R.: Planar collapse of a granular column: experiments and discrete element simulations. Phys. Fluids **20**(6), 063302 (2008)
3. Lajeunesse, E., Mangeney-Castelnau, A., Vilotte, J.P.: Spreading of a granular mass on a horizontal plane. Phys. Fluids **16**(7), 2371–2381 (2004)
4. Lube, G., Huppert, H.E., Sparks, R.S.J., Hallworth, M.A.: Axisymmetric collapses of granular columns. J. Fluid Mech. **508**, 175–199 (2004)
5. Staron, L., Hinch, E.J.: The spreading of a granular mass: role of grain properties and initial conditions. Granular Matter **9**(3), 205 (2006)
6. Thielicke, W., Stamhuis, E.: PIVLab - towards user-friendly, affordable and accurate digital particle image velocimetry in MATLAB. J. Open Res. Softw. **2**(1), e30 (2014)
7. Warnett, J.M., Denissenko, P., Thomas, P.J., Kiraci, E., Williams, M.A.: Scalings of axisymmetric granular column collapse. Granular Matter **16**(1), 115–124 (2014)

Effect of Non-plastic Fines on Cyclic Shear Strength of Sand Under an Initial Static Shear Stress

Daniela Dominica Porcino$^{(\boxtimes)}$, Valentina Diano, and Giuseppe Tomasello

University Mediterranea, Reggio Calabria, Italy
daniela.porcino@unirc.it

Abstract. In the present study, a series of undrained cyclic simple shear (*CSS*) tests were performed on clean Ticino sand (*TS*) and *TS* mixed with 10% of non-plastic fines content (*FC*) under two values of static horizontal shear stress ratio ($\alpha = 0$ and 0.1) to simulate level ground and sloping ground conditions before earthquake loading, respectively. The results in terms of cyclic resistance were analyzed assuming the global void ratio (*e*) as control parameter. One of the most significant findings is that the liquefaction resistance of silty sand is lower than that of clean sand due to the presence of fines and this detrimental effect is even more evident in presence of static shear stress. The concept of equivalent granular void ratio (*e**), proposed earlier to take into account the simultaneous effect of void ratio and fine content based on symmetrical cyclic tests on silty sand, is valid for non-symmetrical tests as well.

Keywords: Initial static shear stress · Non plastic fine · Cyclic liquefaction

1 Introduction

The presence of initial static shear stress can have a significant effect on the undrained cyclic shear strength of sandy soils such as in many practical applications involving earth dams and slopes. The sustained shear stress level is characterized by α, as:

$$\alpha = \tau_{stat}/\sigma'_{vo} \tag{1}$$

where σ'_{vo} is the vertical effective consolidation stress and τ_{stat} is the initial static shear stress applied on the horizontal plane of the sample before shearing. Addition of non-plastic fines into clean sands, depending on the density parameters used for comparison, was found to cause either an increase or a decrease in liquefaction resistance of sands [1], and any results appear somewhat contradictory. Furthermore, it is worthwhile noticing that there have been only a few experimental studies of the cyclic behaviour of silty sands in sloping ground conditions (i.e. [2]). The objective of this paper is to have an insight into the liquefaction resistance in sand-silt mixtures (*FC* = 0, 10%) under two different initial static shear stress levels. Results will be analyzed based on the equivalent granular void ratio- based concept.

2 Materials and Procedures

Specimens of Ticino clean sand (*TS*) (D_{50} = 0.56 mm; G_s = 2.66) and a mixture of *TS* with 10% of non-plastic fines (*TS10*) (D_{50} = 0.025; G_{sf} = 2.72) were prepared at different values of post-consolidation void ratios (0.59 – 0.78) by using moist tamping method. Testing was performed using a modified automated simple shear (*SS*) apparatus (*NGI* type) where cyclic shear stresses were applied around zero (τ_{stat} = 0) or non-zero ($\tau_{stat} \neq 0$) value of an initial static shear stress, providing symmetrical ($\alpha = 0$) and non-symmetrical ($\alpha = 0.1$) tests, respectively. For the purpose of describing the triggering of "liquefaction", a criterion of 3.75% shear strain (in single amplitude) was adopted in all tests.

3 Test Results

3.1 Effect of Non-plastic Fines and Initial Static Shear Stress

The effect of initial static shear stress on the failure patterns is shown in Fig. 1 for *CSS* tests carried out on *TS10*. In addition to the types of cyclic softening and dilatant re-stiffening shown in Fig. 1a, the tests also exhibit cyclic "ratcheting" or progressive accumulation of shear strains in the direction of driving shear force (Fig. 1b).

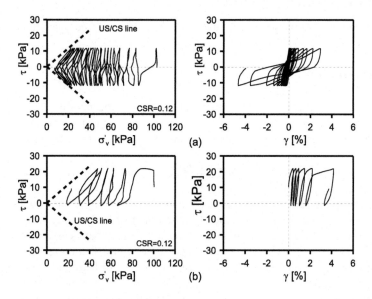

Fig. 1. Cyclic response of *TS10* under (a) $\alpha = 0$ and (b) $\alpha = 0.1$ ($e = 0.68$, $\sigma'_{vo} = 100$ kPa).

To quantify the influence of *FC* on the cyclic resistance ratio of silty sands, a correction factor, K_{FC} [3] can be introduced:

$$K_{FC} = \mathrm{CRR}_{FC \neq 0} / \mathrm{CRR}_{FC=0} \quad (2)$$

where $CRR_{FC\neq0}$ and $CRR_{FC=0}$ are liquefaction resistance ratios when $FC \neq 0$ and $=0$, respectively. As shown in Fig. 2 for specimens prepared at the same (global) void ratio (e), K_{FC} values were found to be 0.73 at $\alpha = 0$ (level ground) and 0.61 at $\alpha = 0.1$ (gently sloping ground). It may be seen that, both in presence and in absence of an initial pre-shearing stress, undrained cyclic strength of sand silt mixture decreases with respect to clean sand.

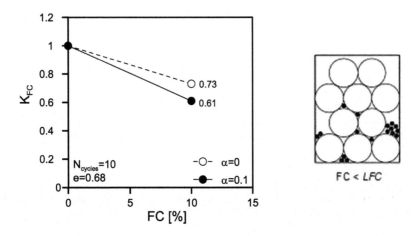

Fig. 2. Factor of cyclic resistance for the effects of fines content with different initial shear stress conditions (α values) ($e = 0.68$; $\sigma'_{vo} = 100$ kPa).

3.2 Application of the Equivalent Granular Void Ratio-Based Approach

The concept of equivalent granular void ratio (e^*) (Eq. 3) [4] was shown to be effective in predicting the undrained cyclic resistance of sands with non-plastic fines starting from CRR values of clean sands. It is defined as:

$$e^* = \frac{e + (1-b)FC}{1 - (1-b)FC} \qquad (3)$$

where b represents the fraction of fines that are active in force structure, e the global initial void ratio and FC the considered fines content.

This approach was valid for amount of fines less than the so called limiting fines content (LFC) (Fig. 2), which was estimated around 30% for the tested sand-silt mixture [5]. Figure 3a shows the cyclic shear strength (CRR) against initial void ratio (e) for TS and $TS10$ in symmetrical and non-symmetrical CSS tests. Four distinct trends for clean TS and $TS10$ can be clearly identified in both symmetrical and non-symmetrical tests. However, if the same experimental data are re-plotted in Fig. 3b in terms of e^*, only two linear relations between CRR and e^* were achieved, irrespective of fines content and void ratio. It means that the concept of e^* can be extended to be applied to silty sands even in presence of a sustained shear stress. The main advantage of this approach is that makes it possible to predict the undrained behaviour of sands in a unique way for all fines contents and densities, without the need for performing many tests at different fines content but using only clean sand data.

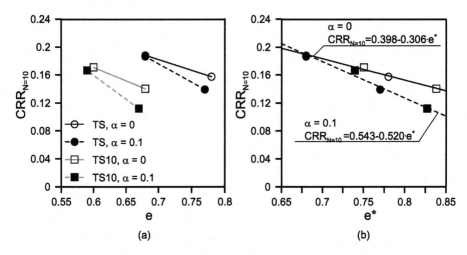

Fig. 3. Cyclic resistance ratio in correspondence of 10 cycles of loading versus (a) initial global void ratio and (b) equivalent granular void ratio.

4 Conclusions

The paper presents selected results from an experimental investigation concerning the combined effect of initial static shear and fines content on liquefaction behaviour of sand. The main findings are summarized as follows:

(a) when tested at the same initial (global) void ratio, the silty sand exhibits a lower undrained cyclic resistance ($CRR_{N=10}$) with respect to clean Ticino sand which ranges from about 39% to 27% in $\alpha \neq 0$ and $\alpha = 0$ tests, respectively.
(b) the concept of equivalent granular void ratio (e^*), in lieu of e, was found to be more appropriate for predicting the behaviour of sand-silt mixtures in both symmetrical and non-symmetrical cyclic simple shear tests.

References

1. Polito, C.P., Martin II, J.R.: Effects of non-plastic fines on the liquefaction resistance of sands. J. Goetech. Geoenviron. Eng. **127**(5), 408–415 (2001)
2. Wei, X., Yang, J.: The effect of initial static shear stress on liquefaction resistance of silty sand. In: 6th International Conference on Earthquake Geotechnical Engineering, Christchurch, New Zeland (2015)
3. Bouckolavas, G.D., Andrianopuolos, K.I., Papadimitriou, A.G.: A critical state interpretation for the cyclic liquefaction resistance of silty sands. Soil Dynam. Earthq. Eng. **23**(2), 115–125 (2003)

4. Thevenayagam, S., Shenthan, T., Mohan, S., Liang, J.: Undrained fragility of clean sands, silty sands and sandy silts. J. Geotech. Geoenviron. Eng. **128**(10), 849–859 (2002)
5. Porcino, D., Diano, V.: The influence of non-plastic fines on pore water pressure generation and undrained shear strength of sandy-silt mixtures. Soil Dynam. Earthq. Eng. **101**, 311–321 (2017)

Shear Band Between Steel and Carbonate Sand Under Monotonic and Cyclic Loading

Shengjie Rui[1(✉)], Zhen Guo[1], Lizhong Wang[1], Wenjie Zhou[1], and Kanmin Shen[2]

[1] Key Laboratory of Offshore Geotechnics and Material of Zhejiang Province, College of Civil Engineering and Architecture, Zhejiang University, Hangzhou 310058, China
11712026@zju.edu.cn
[2] Powerchina Huadong Engineering Corporation Limited, Hangzhou 310058, China

Abstract. Compared to other silica sands, carbonate sands show special engineering properties, i.e., irregular particle shape and high crushability. For pile, anchor or caisson foundations, it is important to precisely assess the interface friction between the structure and the carbonate sand for the design of foundation capacity. Thus, a series of monotonic and cyclic interface shear experiments between steel and carbonate sand have been carried out based on a large ring shear apparatus. In the experiments, the effects of normal stress level, particle size, monotonic and cyclic loadings on shear band thickness were studied. As comparisons, similar monotonic interface shear tests for the standard quartz sand were performed as well. The result shows that the carbonate sand forms thicker shear band than quartz sand because its lower hardness. The shear zone gradually develops with the increase of shear displacement, and the thickness of shear zone develops faster in the initial 2 m shear displacement. The larger the median particle size is, the larger the thickness of the shear band is, but the t/d50 (the ratio of shear band thickness to median size) is gradually reduced. Monotonic or cyclic shear also has certain effect on the thickness of shear zone.

Keywords: Interface shear · Carbonate sand · Shear band thickness Monotonic and cyclic

1 Introduction

As to foundations, the interface interaction between sands and the structure have a great impact on the installation and bearing performance of the foundation. At present, there are some experimental studies on the interface shear zone between quartz sand and steel or concrete interface, such as Jardine et al. [1], Frost et al. [2], Dove et al. [3] and others. Ho [4] pointed out that the shear band thickness was related to the median particle size. Yang [5] investigates the relationship between the thickness of the shear band, normal stress and shear displacement through ring shear test.

In this paper, through interface ring shear test, the relationship between the shear zone thickness of the carbonate sand and the steel interface with the normal stress, grain size, shear displacement scale and monotonic or cyclic shear was studied, which could provide some reference for the understanding of shear zone.

2 Description of Tests

In this experiment, the ring shear apparatus made by Zhejiang University is used. The diameter of the outer ring and inner ring is 300 mm and 200 mm, and the ring width is 50 mm. The lower interface arrangement is adopted. The carbonate sands used in the experiment were taken from an island reef in South China Sea. The quartz sand used for comparison was China's standard sand. The sands were sieved and divided into two groups of 0.1–0.25 mm (fine sand) and 0.5–1 mm (coarse sand). The relative density of sands were controlled about 72%, and the sample height is 20 mm. The steel interface used in the experiment is ordinary carbon steel. The annular steel interface was air-abraded before each test. The value of the interface roughness before the experiment was controlled between 3.25 ± 0.05 μm.

The normal stress exerted in experiments is 50 kPa, 100 kPa, 200 kPa, 300 kPa, and the shear rate is 2.5 degrees/min (about 5.45 mm/min). The detail of tests performed is shown in Table 1.

Table 1. The detail of tests performed.

Test code	Category of sand	Normal stress (kpa)	Shear displacement (m)	Number of cycles
M1–M4	Carbonate coarse sand	50;100;200;300	2	
M5–M8	Carbonate fine sand	50;100;200;300	2	
M9–M12	Quartz coarse sand	50;100;200;300	2	
M13–M16	Quartz fine sand	50;100;200;300	2	
M17–M20	Carbonate coarse sand	100	0.5;1;4;8	
M21–M24	Carbonate fine sand	100	0.5;1;4;8	
M25–M27	Quartz coarse sand	100	1;4;8	
M28–M30	Quartz fine sand	100	1;4;8	
C31–C34	Carbonate coarse sand	50;100;200;300	2	200
C35–C38	Carbonate fine sand	50;100;200;300	2	200

The thickness of the shear band is then measured with caliper. 6 such points are measured evenly on the ring interface, and the average is calculated in the end.

3 Experimental Result

3.1 Category of Sand, Size of the Grain and Normal Stress

Under the shear displacement of 2 m, the experimental results of four different sands under normal stress are shown in Fig. 1. The result shows that carbonate coarse sand is more likely to form thicker shear zones than quartz coarse sand as well as the fine ones. The thickness of the shear zone formed by carbonate sand is more sensitive to normal stress. The greater the grain size, the thicker the thickness of the shear zone is.

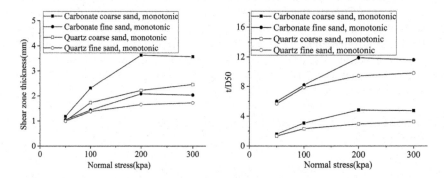

Fig. 1. Variation of shear zone thickness with normal stress

3.2 Category of Sand, Size of the Grain and the Shear Displacement

The relationship between the thickness of shear band and shear displacement under the normal stress of 100 kPa is shown in Fig. 2. According to the experimental results, the thickness of shear band increases with the shear displacement. In the initial 2 m shear displacement, shear zone thickness of fine sand with the displacement increased faster, and calcareous sand is basically stable after 4 m, while quartz sand is still slightly increased; for coarse sand, with the increase of shear displacement, thickness of shear band increases even under 8 m shear displacement. The shear band thickness of coarse

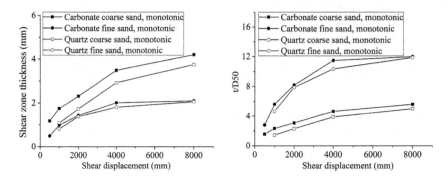

Fig. 2. Variation of shear zone thickness with shear displacement

sand should be 4–5 times larger than that of medium grain size, and the shear band thickness of fine sand should be 10–12 times the medium grain size.

3.3 Size of the Grain, Normal Stress and Monotonic or Cyclic Shear

Under the shear displacement of 2 m, the relationship between the shear zone thickness and the normal stress under monotonic or cyclic shear is shown in Fig. 3. The results show that the thickness of shear zone is influenced by monotonic or cyclic shear. For both carbonate coarse and fine sand, monotonic shear is more likely to form thicker shear zones.

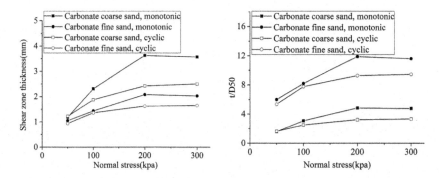

Fig. 3. Comparison of shear zone thickness under monotonic or cyclic shear against normal stress

4 Conclusion

The shear zone thickness formed by sand and steel is related to the type of sand, normal stress, shear displacement, particle size and monotonic or cyclic shear.

(1) Under the limited displacement conditions, the thickness of the shear band is positively correlated with the normal stress, and the carbonate sand is more sensitive to normal stress than that of the quartz sand.
(2) The shear zone gradually develops with the increase of shear displacement, and the thickness of shear zone develops faster in the initial 2 m shear displacement.
(3) The thickness of the shear band is related to the median particle size. The larger the median particle size is, the larger the thickness of the shear band is, but the $t/d50$ is gradually reduced.
(4) Monotonic or cyclic shear also has certain effect on the thickness of shear zone. For both coarse and fine sand, the shear band formed by monotonic shear is thicker than cyclic shear.

References

1. Jardine, R.J., Lehane, B.M., Everton, S.J.: Friction coefficients for piles in sands and silts. In: Proceedings of 4th International Conference on Offshore Site Investigation and Foundation Behaviour, pp. 661–677. Publisher, London (1992)
2. Frost, J.D., Hebeler, G.L., Evans, T.M., DeJong, J.T.: Interface behaviour of granular soils. In: Proceedings of Earth and Space Conference, pp. 65–72. Publisher, Houston (2004)
3. Dove, J.E., Frost, J.D.: Peak friction behaviour of smooth geomembrane-particle interfaces. J. Geotech. Geoenviron. Eng. **125**(7), 544–555 (1999)
4. Ho, T.Y.K., Jardine, R.J., Anh-Minh, N.: Large-displacement interface shear between steel and granular media. Géotechnique **61**(3), 221–234 (2011)
5. Yang, Z.X., Jardine, R.J., Zhu, B.T., Foray, P.: Sand grain crushing and interface shearing during displacement pile installation in sand. Géotechnique **60**(6), 469–482 (2010)

Comparative Study on the Compressibility and Shear Parameters of a Clayey Soil

Erik Schwiteilo[(✉)] and Ivo Herle

Institut für Geotechnik, Technische Universität Dresden, Dresden, Germany
{Erik.Schwiteilo,Ivo.Herle}@tu-dresden.de

Abstract. Oedometer and direct shear tests on one and the same soil were performed in several independent laboratories. The observed differences are analysed. In order to obtain homogeneous and reproducible samples, a clayey soil was re-molded and pre-consolidated. The handling of the soil specimens and detailed testing procedures were described and communicated to each participating laboratory in advance. The evaluated soil properties scatter in a relatively wide range. The analysis of the results shows that a large part of the scatter originates from the experimental conditions like specimen size and frictional forces on the boundaries. It can be concluded, that a better standardization and further unication of experimental methods would be advantageous.

1 Introduction

One open question in the work of geotechnical laboratories is the scatter of the measured results and the derived soil parameters respectively. To which extent the same soil material under the same standardized testing procedures yields identical results? To check this aspect of the laboratory testing, a series of oedometer and direct shear tests was conducted in different laboratories. The same reconstituted soil material was sent to 7 participating geotechnical laboratories all over Germany. The preparation of the specimens was done according to a predefined checklist by the technicians in the laboratories (the specimens had to be cut for the particular test devices). To guarantee a comparable procedure, an instruction guideline for the test procedure was communicated in advance. More details can be found in [1].

2 Soil Material and Sample Preparation

A reconstituted fine grained soil was used for the testing. The soil was mixed with deionized water to a water content $> 2.0 \cdot LL$. This suspension was consolidated in a tube (D = 10 cm) stepwise to a load of 250 kPa. This resulted in a sufficiently stiff sample which could be posted to the participating laboratories.

The drainage during the consolidation took place mainly in the radial direction through a permeable tube. Because of a relatively short drainage path in the radial direction the consolidation time for one sample could be shortened to three weeks.

To check, that the posted soil material is comparable, the grain size distribution and the Attaberg limits were analyzed for each posted batch. The grain size corresponded in all cases to a silty clay with a clay fraction of 65%. The plasticity index I_P had an average value of 39.6% (with the limit values 43.7% and 36.5%), the average w_L was 63% (with the limits 65.2% and 62.3%).

3 Oedometer Tests

The oedometer tests were conducted according to the German standard DIN 18135:2012 [3]. After the delivery of the samples, the specimens should be installed in the test devices in a short period of time to avoid any drying of the soil. The geometry (diameter, height) and the wet mass of the specimens should be documented. From the laboratory no. 7 two test were conducted, thus two results (a and b) from this laboratory 7 are presented. The loading sequence corresponded to 9 and unloading to 7 steps. In each load step the primary settlement had to finish or the next step followed after 24 h, respectively. To avoid a drying of the soil, the specimen was flooded at the load of 100 kPa. As shown in Fig. 1 all specimens compressed non-linearly during the loading. The measured vertical strains at an effective vertical stress of $\sigma_v = 2000$ kPa are between 18 and 29%. The strains were determined conventionally as $\Delta h/h_0$. In the laboratory no. 1 the specimen could only be loaded to a vertical effective stress of $\sigma_v = 1200$ kPa, in the laboratory no. 3 to 1000 kPa and in the laboratory no. 7-a to 1800 kPa. After reaching the maximum load the specimens were unloaded stepwise. As seen in Fig. 1, a similar non-linear stress-strain trend can be observed for all tests.

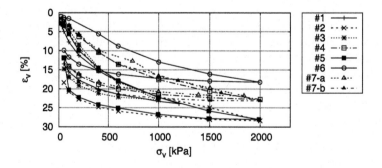

Fig. 1. Stress-strain curves of the oedometer tests.

Based on the compression curves in Fig. 1 two groups of curves can be identified. The reason for the difference is probably the geometry of the specimens.

In Fig. 2 the void ratio e is plotted over σ_v in logarithmic scale. These results show that specimens with a diameter of 70 mm (dashed lines) reach lower void ratios during the loading than specimens with a diameter of 50 mm (solid lines). One exception is the curve from the laboratory no. 6 where the top plate was blocked at the beginning of the test. The height of the specimens corresponded to 20 mm.

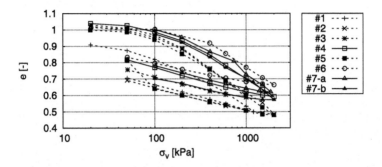

Fig. 2. Influence of the oedometer ring diameter; d = 50 mm - solid, d = 70 mm - dashed.

4 Direct Shear Tests

The direct shear tests were conducted according the German standard DIN 18137:2002 [4]. Each participant had to conduct a test series with three single tests at different vertical loads. All samples were consolidated to a vertical load of 1000 kPa, subsequently the load on two specimen was reduced to 400 and 600 kPa, respectively. In the laboratory no. 1 the vertical loads corresponded to 500, 300 and 200 kPa due to the limitations of the used testing device. In the laboratory no. 6, two test series were performed (named 6-a and 6-b). The shear tests were conducted with a velocity of 0.001 mm/min and an initial shear gap of 0.6 mm.

One goal of this comparative study was to evaluate the scatter in the derived parameters. In Fig. 3 the measured shear stresses for the overconsolidated state and for the normally consolidated state, respectively (vertical load of 1000 kPa) are shown. These results enable to define φ' and c' as parameters of the Mohr-Coulomb criterion for the overconsolidated state and φ^* for the normally consolidated state.

As can be seen in Fig. 3-a for the overconsolidated state, a mean friction angle of 12.7° and a mean cohesion of 31 kPa can be determined. However, there is a scatter of 16% which means ±2° for the friction angle and a scatter of 55% which means ±17 kPa for the cohesion. In the normally consolidated state (see Fig. 3-b) the mean friction angle is 14° with a standard deviation of 9.3%. This corresponds to the scatter of ±1.3°.

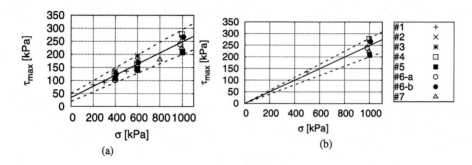

Fig. 3. Shear strength in overconsolidated state (a), Shear strength in normally consolidated state (b).

The large scatter in the shear strength parameters φ' and c' results probably from the different constructions of the used testing devices. Especially the unknown wall friction in the vertical direction can introduce a substantial error in the assumed vertical load at the shear surface [2]. A further observation is, that the maximum values of the shear stresses are located more or less on a straight line. However, the inclination of the lines from the different laboratories are different.

5 Conclusion

The results of a comparative study on the compressibility and shear parameters of a clayey soil was presented. Both, oedometer and direct shear test results show greater fluctuations than expected. Different sources for this scatter can be assumed, although a major factor seems to be different constructions of the testing devices.

References

1. Schwiteilo, E., Herle, I.: Vergleichsstudie zur Kompressibilitt und zu den Scherparametern von Ton aus Oedometer-und Rahmenscherversuchen. Geotechnik **3**, 204–217 (2017). Wiley Online Library
2. Kostkanov, V., Herle, I.: Measurement of wall friction in direct shear tests on soft soil. Acta Geotechnica **7**, 333–342 (2012). Springer
3. DIN 18135: Baugrund – Untersuchung von Bodenproben – Eindimensionaler Kompressionsversuch DIN Deutsches Institut fr Normung e. V (2012)
4. DIN 18137: Soil, investigation and testing, Determination of shear strength, Direct shear test 3 (2002)

Digital Image Measurement System for Soil Specimens in Triaxial Tests

Longtan Shao, Xiaoxia Guo[✉], and Boya Zhao

State Key Laboratory of Structural Analysis for Industrial Equipment,
Department of Engineering Mechanics, Dalian University of Technology, Dalian 116024, China
hanyuer@dlut.edu.cn

Abstract. In order to measure the deformation distribution over the entire surface of soil specimens in triaxial tests, a digital image technique is developed by use of a complementary metal-oxide-semiconductor (CMOS) camera and two mirrors to obtain the deformation field and the strain field of the whole surface of the specimen. Two sets of circular light-emitting diode (LED) lights are placed on the top and the bottom of the pressure chamber. A shading box seals the camera and the plate glass in front of the chamber to ensure an unchanged lighting environment. The process of data processing involves the error check, the expanding and the splicing of the images, and the calculation of the strain in each element. The measurement precision of the strain is approximately $10^{-4} \sim 10^{-5}$. And then, the application of this digital image processing technique is introduced in the study of shear band problem, membrane embedding problem, the heterogeneous soil, the end contact and end constraint, and so on. At last, based on the digital image measurement system, a series of geotechnical test instruments have been developed, including: (1) Advanced triaxial compression test apparatus; (2) Bidirectional dynamic triaxial apparatus; (3) High pressure triaxial compression apparatus; (4) Plane strain compression apparatus; (5) Unsaturated soil triaxial compression apparatus; (6) Frozen soil triaxial apparatus; (7) Asphalt concrete stress-strain apparatus; (8) Hydrate triaxial apparatus; (9) Soft material testing machine.

Keywords: Digital image measurement system · Deformation field
Measurement precision · Geotechnical test instruments

1 Measurement of the Deformation Distribution over the Entire Surface of the Specimen

The digital image technique is applied to deformation measurements on the surface of the specimen. In this approach, a complementary metal-oxide-semiconductor (CMOS) camera and two mirrors are used to capture images of the entire surface of the specimen. The specimen is wrapped in a black rubber membrane, on which white squares are printed, i.e., the surface of the specimen is meshed, and the nodes of the black and white grid on the images can be recognized using a sub-pixel corner detector algorithm. The hardware of the measurement system is composed of a redesigned pressure cell, the CMOS camera, a camera bracket and a shading box (shown in Fig. 1). The coordinates of the element nodes on the entire surface of the specimen are recorded during the test. The recorded data are processed in five steps [1, 2].

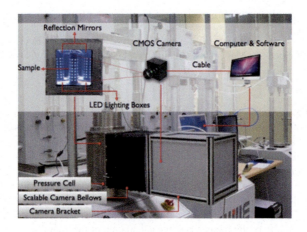

Fig. 1. Hardware of the measurement system.

To confirm the reliability of the deformation measurement results, the measurement precision was estimated by comparing the strain measurement results from the digital image system with that from the strain gauges and comparing the volume change results with that of the drainpipe. The measurement precision of the strain is approximately $10^{-4} \sim 10^{-5}$.

2 Application of This Digital Image Processing Technique

The application of this digital image processing technique is introduced in the study of shear band problem, membrane embedding problem, the heterogeneous soil, the end contact and end constraint, the constitutive model, and so on.

2.1 Shear Band Problem

The deformation distribution over the entire surface of the specimen is obtained by digital image measurements during triaxial testing. The strain distribution is derived from the measured deformation, and the stress distribution is then calculated based on the strain, Young's modulus and Poisson's ratio. Furthermore, the stress level S of any point on the surface of the specimen is evaluated. Figures 2, 3 and 4 show that the shear band began at a point, then expanded gradually, and become the penetrated shear band.

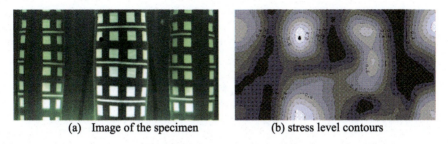

(a) Image of the specimen (b) stress level contours

Fig. 2. Image of the specimen and stress level contours in the pre-failure state.

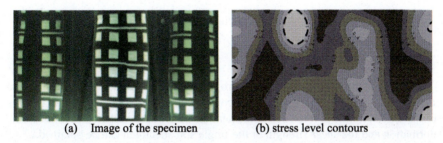

(a) Image of the specimen (b) stress level contours

Fig. 3. Image of the specimen and stress level contours in the in-failure state.

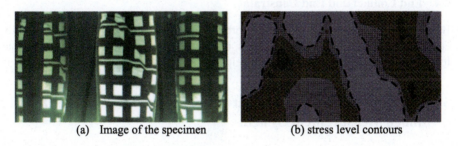

(a) Image of the specimen (b) stress level contours

Fig. 4. Image of the specimen and stress level contours in the post-failure state.

2.2 Membrane Embedding Problem

The problem of membrane embedding can be studied quantitatively based on digital image system. The relation between the membrane embedding and its mainly influence factor, such as σ3, d50, as shown in Fig. 5. The formula of membrane embedding volume is put forward.

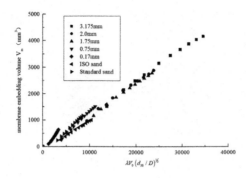

Fig. 5. Membrane embedding volume and $\lambda V_0 (d_{50}/D)^{1/2}$.

2.3 Heterogeneous Soil

A systematic study on the mechanical properties of the layered soil specimen under different conditions can be conducted. By comparing different test results from the single soil specimen and the layered soil specimen, the effect of the layered specimen on the deformation and the shear strength of the single soil specimen can be obtained.

2.4 End Contact and End Constraint

Based on qualitative analysis of local deformation characteristics, strain non-uniformity coefficient is defined and calculated, this, quantitative comparisons are then given out for local deformations along specimen height.

3 A Series of Geotechnical Test Instruments Based on the Digital Image Measurement System

Based on the digital image measurement system, a series of geotechnical test instruments have been developed, including: (1) Advanced triaxial compression test apparatus; (2) Bidirectional dynamic triaxial apparatus; (3) High pressure triaxial compression apparatus; (4) Plane strain compression apparatus; (5) Unsaturated soil triaxial compression apparatus; (6) Frozen soil tri-axial apparatus; (7) Asphalt concrete stress-strain apparatus; (8) Hydrate tri-axial apparatus; (9) Soft material testing machine.

References

1. Liu, X.: Method of entire surface deformation measurement for soil specimen in triaxial tests and its application (Doctoral dissertation). Dalian University of Technology, Dalian (2012)
2. Shao, L.T., Liu, G., Zeng, F.T., Guo, X.X.: Recognition of the stress-strain curve based on the local deformation measurement of soil specimens in the triaxial test. Geotech. Test. J. **39**(4), 20140273 (2016)

Open Science Interface Shear Device

Hans Henning Stutz[1(✉)], Ralf Doose[2], and Frank Wuttke[2]

[1] Department of Engineering, Aarhus University,
Inge Lehmanns Gade 10, 8000 Aarhus C, Denmark
hhs@eng.au.dk
[2] Kiel University, Ludewig-Meyn-Str. 10, 24229 Kiel, Germany
{Ralf.Doose,Frank.Wuttke}@ifg.uni-kiel.de

Abstract. Different scientific communities face the challenge of the so called reproducibility crisis. In engineering and geotechnical engineering in particular this is not discussed in detail. However, research in soil mechanics is often done with highly specialized and expensive equipment. This research can not repeated or the testing methodology is seldom applied with a different soil type due to the availability of the testing protocol and equipment. Nowadays, the increasing accessibility of 3-D printing facilities, single-board computer and micro-controllers lead to new opportunities and ways to overcome this challenges.

The aim of this contribution is to shed some light on this facts and discuss opportunities to improve the scientific situation with respect of open science in soil mechanics.

To the authors best knowledge this is a novel application of Open Science in geotechnical engineering. The developed device is called Open Science interface shear device (OSInterface), which will act as a pilot project to demonstrate the opportunities using methods that are applied for example for open source software development in open hardware development with a focus to geotechnical engineering. Basically, this interface shear device differs not much from existing shear devices but the development and the construction is accessible to everyone.

The different aspects of such a development based on the OSInterface are discussed and also the challenges are addressed. The idea is to motivate many different researchers to rebuild the OSinterface device and share their own developments.

1 Introduction

The interface shear behaviour is an important physical phenomena that must be described and studied in detail to achieve a good design and or prediction of geotechnical structures [10]. Often the interface shear tests is conducted as specialised direct shear test. Many different test configurations for interface shear testing exists (e.g. [8,11]). In these tests, different aspects as monotonic and cyclic loading paths [3,6], temperature influences [5], surface roughness conditions [9] were studied. However, many of these tests have a special set up and testing methodology, which is often not available or clearly documented. To this aim, such an interface shear device that can be manufactured from the sources given in the Appendix is presented in this publication. The idea is that the device is planned, manufactured and distributed as Open Science Hardware development. Therefore the abbreviation of this device will be OSInterface (OpenScience Interface shear device) in the future.

Open Science is well known in geotechnical engineering for Open Source Software or Scripts (e.g. Soilmodels.com [7]). But this is only one part of Open Science. However, Open Hardware development is an important part of scientific Open Science developments in the future. Especially in soil mechanics it seems to be important to the authors because soil mechanical experiments needs hardware as a vital part for experimental studies. Therefore, the hardware devices are essential to repeat the results from these experiments, but these hardware is often not accessible. The same account for the testing protocol or methodology that is used.

Beside the scientific reasons it must be kept in mind that the most experimental research in geotechnical engineering is covered by public funding. As part of a new initiative the OSInterface is manufactured and the designed is shared via Open Science principles. In addition to opening the design and the control software the device will be available for interesting researchers free of charge. The idea is that the Open hardware development is one aspect but Open Lab space is important to consider for free access to scientific knowlegde and developments. The only exception that is made with use of the OSinterface device; the experimental data that are measured will be available after an embargo time as open data.

2 Open Science Interface Shear Device

The OSInterface device is developed considering two different interface shear box test set ups at University of California Davis [4] and University of Western Australia [1]. The general layout of these test set ups is used with some modifications to make it cheaper. The main development for the OSinterface consider the basic use for interface shear testing under monotonic and cyclic loads under the constant normal load and stiffness conditions. The OSInterface device cost approximately 5500 excluding the workshop costs.

The design of the device is simple and robust with possible standard materials and components. The mechanical parts of the device are manufactured out of stainless steel. The device is equipped with two load cells, one for the normal load and one for the shear force measurement, two linear potentiometer to measure the travel way of the shear box and optional the vertical displacement. Figure 1 shows parts of the construction and details of the construction. To establish a detailed investigation of the shear behaviour directly at the interface on both sides of the interface shear box two glassy observation windows are manufactured at the shear box. This was important to have the opportunity to do some Particle Image Velocities measurements. The loading and the displacement is applied via

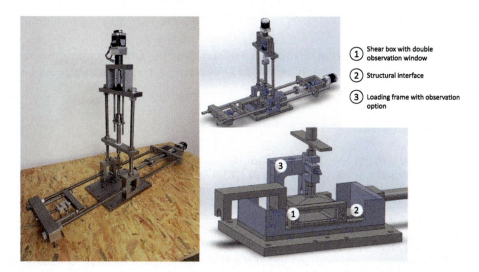

Fig. 1. Current manufacturing process (left) and detail drawings of the OSinterface device (right)

two stepper motors that can apply maximum load of 150 Ncm which will lead to admissible stresses of approximately 800 kPa. The spindle and the encoder of the motor are chosen to allow shear velocities of minimal 0.0005 mm/min to maximal 25 mm/min.

The acquisition and control of the device is realized with the graphical programming language LabView. This software will be available as .vi-file that can be modified or as an executable file for standard application of the OSInterface device.

Another important issue with Open Hardware is the license for it. The authors choose the CERN OHL v1.2 [2] license for the OSInterface device.

3 Remarks and Conclusions

Some remarks and comments are added with respect to Open Science approaches considering the development of lab hardware. From the experience made with the OSinterface device the authors can say that building Open Hardware is a higher effort compared the "standard approach". Considering the scientific benefit, the authors hope that everyone will put a small effort to open specific aspects research to allow everyone will spent more time with research than with hardware development in geotechnical engineering.

Acknowlegements. The first author acknowledge the support by Wikimedia Deutschland e. V., Stifterverband and the Volkswagen Foundation within the Open Science Fellows Program. In addition the author would like to acknowledge the help and excellent manufacturing team of the mechanical workshop at Kiel University, Institute of Geo-Science.

A Appendix

Additional information about the current status of the OSinterface device and the OpenGeoLab initiative under: https://github.com/HansStutz/OpenGeoLab and http://www.OpenGeoLab.com.

References

1. Boukpeti, N., White, D.J.: Interface shear box tests for assessing axial pipesoil resistance. Gotechnique **67**(1), 18–30 (2017)
2. CERN Open Hardware Licence v1.2 can be found under. https://www.ohwr.org/documents/294
3. DeJong, J.T., Randolph, M.F., White, D.J.: Interface load transfer degradation during cyclic loading: a microscale investigation. Soils Found. **43**(4), 81–93 (2004)
4. DeJong, J.T., Westgate, Z.J.: Role of initial state, material properties, and confinement condition on local and global soil-structure interface behavior. J. Geotech. Geoenviron. Eng. **135**(11), 1646–1660 (2009)
5. Di Donna, A., Ferrari, A., Laloui, L.: Experimental investigations of the soil-concrete interface: physical mechanisms, cyclic mobilisation and behaviour at different temperatures. Can. Geotech. J. **53**(4), 659–672 (2015)
6. Evgin, E., Fakharian, K.: Effect of stress paths on the behaviour of sandsteel interfaces. Can. Geotech. J. **33**, 853–865 (1996)
7. Gudehus, G., et al.: The soilmodels.info project. Int. J. Numer. Anal. Methods Geomech. **32**(12), 1571–1572 (2008)
8. Ho, T.Y.K., Jardine, R.J., Anh-Minh, N.: Large-displacement interface shear between steel and granular media. Gotechnique **61**(3), 221–234 (2011)
9. Martinez, A., Frost, J.D., Hebeler, G.L.: Experimental study of shear zones formed at sand/steel interfaces in axial and torsional axisymmetric tests. Geotech. Test. J. **38**(4), 409–426 (2015)
10. Stutz, H., Mašín, D., Sattari, A.S., Wuttke, F.: A general approach to model interfaces using existing soil constitutive models application to hypoplasticity. Comput. Geotech. **87**, 115–127 (2017)
11. Uesugi, M., Kishida, H.: Frictional resistance at yield between dry sand and mild steel. Soils Found. **26**(4), 139–149 (1986)

Effect of Soil Characteristics on Shear Strength of Sands

Danai Tyri[1(✉)], Polyxeni Kallioglou[2], Kyriaki Koulaouzidou[2], and Stefania Apostolaki[3]

[1] ANDRA, INSA Lyon, GEOMAS, Villeurbanne, France
danai-panagiota.tyri@insa-lyon.fr
[2] Department of Civil Engineering, UTH, Volos, Greece
[3] Department of Civil Engineering, AUTH, Thessaloniki, Greece

Abstract. The paper presents the results of a laboratory investigation into the influence of (a) mineral composition and angularity of sand grains and (b) fines type and content on the strength parameters of sandy soils tested at various densities and normal stresses in a direct shear apparatus under drained conditions.

Keywords: Shear strength · Critical state · Dilatancy · Sands · Fines Relative density

1 Introduction

The shear strength of sand is usually characterized by the peak friction angle, φ'_{max}, and the critical state friction angle, φ'_{crit}. The latter is considered when the sand is shearing at constant volume conditions and is generally lower than the former, especially in the case of high relative density, D_r. Bolton (1986) investigated the effect of D_r on the dilatancy related component of strength, $\varphi'_{max} - \varphi'_{crit}$, of clean sands and suggested simple correlations for the estimation of the friction angles. The purpose of the work presented in this paper is to clarify the effect of sand grains characteristics and fines type and content as well as relative density on the shear strength parameters of sands by means of direct shear tests, which were carried out in the Laboratory of Geotechnical Engineering of University of Thessaly.

2 Tested Soils and Testing Procedure

The tested soils fall into three groups: clean sands of similar grading (group 1), sands containing non-plastic fines (group 2) and sands containing plastic fines (group 3). Sands A and B are comprised totally of quartz whereas sand C is comprised of 77% quartz and 23% feldspar. Only sand B consists of angular grains. Both group 2 and group 3 were prepared by mixing the same clean sand A with a non-plastic sandy silt with $D_{50} = 0.04$ mm and $Cu = 6$ (group 2), or kaolin with $PI = 18\%$ (group 3). Table 1 summarizes the physical properties of the tested soils.

Table 1. Physical properties of tested soils

Group	Tested soil	FC (%)	CF (%)	Cu	D_{50} (mm)	D_{max} (mm)	e_{min}	e_{max}	G_s	USCS
(1)	A	–	–	1.8	0.35	0.85	0.645	0.907	2.659	SP
	B	–	–	1.8	0.38	0.50	0.600	0.994	2.659	SP
	C	–	–	1.3	0.21	0.35	0.617	0.988	2.650	SP
(2)	A-S-1	8	2	2.5	0.32	0.85	0.606	0.879	2.657	SP-SM
	A-S-2	17	2	9.1	0.31	0.85	0.414	0.687	2.653	SM
(3)	A-K-1	11	4	7.3	0.31	0.85	0.568	0.820	2.657	SP-SM
	A-K-2	18	8	11.4	0.30	0.85	0.412	0.664	2.652	SC-SM

FC: Fines content < 75 μm CF: Clay fraction < 2 μm Cu: Coefficient of uniformity.

The specimens had dimensions of D = 62.5 mm and H = 25 mm, and were prepared at optimum moisture content directly into the shear box using the undercompaction method (Ladd 1978). Specimens of uniform structure at various densities were prepared from each soil and tested in the direct shear apparatus at a range of normal stresses σ'_v = 50–400 kPa under drained conditions according to ASTM D3080 (2011). The strength parameters at both peak and critical state were determined in order to evaluate the dilatancy of the tested sands.

3 Results

Due to the space restriction, the results only at σ'_v = 100 kPa are presented in Figs. 1, 2 and 3 for soil groups 1, 2 and 3 respectively. The results show the variation of peak angle, φ'_{max}, and critical angle, φ'_{crit}, of soil resistance as well as the dilatancy related component of strength, $\varphi'_{max} - \varphi'_{crit}$, with relative density, D_r. The values of φ' were determined based on the results for the whole range of the applied σ'_v (50–400 kPa) using the Mohr – Coulomb criterion. The linear correlations between the above parameters are also given. For soil group 3 (sands with plastic fines) the variation of cohesion, c′, and shear stress, τ, with relative density, D_r, is also presented.

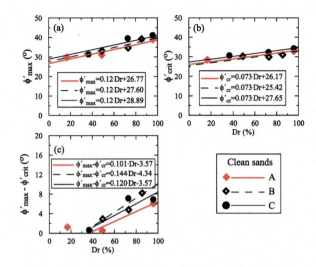

Fig. 1. Variation of (a) peak friction angle, φ'_{max}, (b) critical friction angle, φ'_{crit}, and (c) $\varphi'_{max} - \varphi'_{crit}$ with relative density, D_r, for soil group (1) at $\sigma'_v = 100$ kPa (Apostolaki 2016).

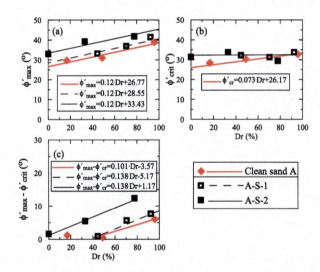

Fig. 2. Effect of non-plastic fines on the variation of (a) peak friction angle, φ'_{max}, (b) critical friction angle, φ'_{crit}, and (c) $\varphi'_{max} - \varphi'_{crit}$ with relative density, D_r, for soil group (2) at $\sigma'_v = 100$ kPa (Tyri 2016).

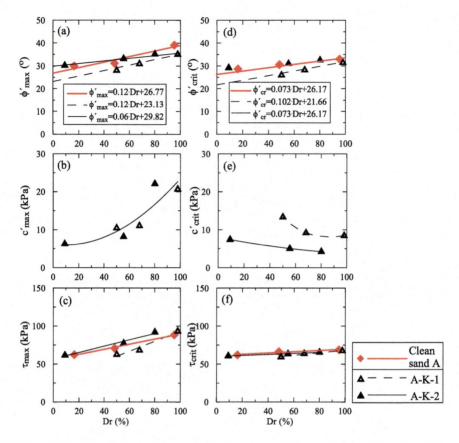

Fig. 3. Effect of plastic fines on the variation of (a) peak friction angle, φ'_{max}, (b) peak cohesion, c'_{max}, (c) peak shear strength, τ_{max}, (d) critical friction angle, φ'_{crit}, (e) critical cohesion, c'_{crit}, and (f) critical shear strength, τ_{crit}, with relative density, D_r, for soil group (3) at $\sigma'_v = 100$ kPa (Koulaouzidou 2016).

4 Conclusions

The following conclusions can be drawn from the results presented herein:

(a) Simple correlations between D_r and φ'_{max}, φ'_{crit} and $\varphi'_{max} - \varphi'_{crit}$ are supported for the tested soils, showing the increase of φ' values with the increasing D_r, especially in the case of $\varphi'_{max} - \varphi'_{crit}$.
(b) The differences among the φ' values of the tested clean sands at the same level of D_r are small due to their similar gradings and may be attributed to the effect of grain angularity and soil mineralogy, e.g. the higher values are pertaining to sand containing portions of feldspar or angular grains.
(c) For the mixtures of sand with non-plastic fines, an increase of φ' values is observed with the increase of FC and D_r, except for φ'_{crit} value which remains constant.

(d) For the mixtures of sand with plastic fines, a development of cohesion, c′, which depends on D_r, is observed in both peak and critical states but in an opposite way, whereas φ′ values increase with increasing D_r. The effect of FC on strength components, c′ and φ′, is not clear.

References

Apostolaki, S.: Shear strength of clean sands in direct shear tests. Diploma Thesis. Department of Civil Engineering, University of Thessaly, Greece (2016). (in Greek)
ASTM D3080: Standard Test Method for Direct Shear of Soils Under Consolidated Drained Conditions. ASTM, vol. 04.08 (2011)
Bolton, M.: The strength and dilatancy of sands. Géotechnique **36**(1), 65–78 (1986)
Koulaouzidou, K.: The effect of fines on the shear strength of sands. Diploma Thesis. Department of Civil Engineering, University of Thessaly, Greece (2016). (in Greek)
Ladd, R.: Preparing test specimens using undercompaction. Geotech. Test. J. **1**(1), 16–23 (1978)
Tyri, D.: Shear strength of sands in direct shear tests. Diploma Thesis. Department of Civil Engineering, University of Thessaly, Greece (2016). (in Greek)

Mechanical Behaviors of GRPS Track-Bed with Changing Water Levels and Loading Cycles

Han-Lin Wang[1,2,3] and Ren-Peng Chen[1,2(✉)]

[1] MOE Key Laboratory of Building Safety and Energy Efficiency, Hunan University, Changsha, China
chenrp@hnu.edu.cn
[2] College of Civil Engineering, Hunan University, Changsha, China
[3] Department of Civil and Environmental Engineering, Faculty of Science and Technology, University of Macau, Macau, China

Abstract. This paper presents a full-scale model study of the mechanical behaviors of geosynthetic-reinforced pile-supported (GRPS) railway track-bed under coupled effects of changing water levels and large number of loading cycles. Four testing procedures were performed: water level increasing, loading at high water level, water level lowering and loading at low water level. The results indicate that with the water level increasing and loading at high water level, the differential settlement between the subsoil (simulated by water bag) and pile cap increased, leading to more significant soil arching effect. When enough loading cycles were applied at high water level, a stable soil arching was developed. At the stable state of soil arch, the distributions of dynamic soil stress were slightly influenced by the lowered water level and loading at low water level. In these two procedures, the overall settlement of the model varied slightly and the differential settlement stayed nearly unchanged.

Keywords: Geosynthetic-reinforced pile-supported track-bed · Water level · Cyclic loading · Soil arching · Settlement

1 Introduction

In the last decades, the geosynthetic-reinforced pile-supported (GRPS) track-bed is widely used in the constructions of railways [1–3]. Assessing the mechanical behaviors of the track-bed becomes necessary to the long-term maintenance of the railways [4–6]. To date, the mechanical behaviors of the GRPS embankment has been widely studied and some analytical models about this issue have been reported [7–13]. In these models, the coupled effects of changing water content in the track-bed materials and cyclic loading were not considered. Furthermore, extreme climate conditions (such as heavy precipitation) were frequently identified throughout the world, which might deteriorate the performance of the track-bed [14–18]. To the authors' knowledge, the mechanical behaviors of the GRPS track-bed with changing water levels and loading cycles have not addressed yet.

In this study, a full-scale model of the GRPS railway track-bed was established, with the water bags simulating the subsoil and concrete slabs simulating the pile caps. Four testing procedures were conducted: water level rising, loading at high water level, water level lowering and loading at low water level. The results allow the evolutions of settlement and soil arching effect to be analyzed.

2 Materials and Methods

The full-scale GRPS railway track-bed model was built, consisting of superstructure and substructure. The superstructure is composed of rails and fasteners, track slab, cement asphalt mortar (CAM) layer and concrete base. The substructure includes subgrade surface layer (GW), subgrade bottom layer (GC) and subsoil (GP). At the bottom of subsoil, 15 concrete slabs are arranged to simulate the pile caps. Surrounding the pile caps are the polyvinyl chloride water bags. The differential settlement between the subsoil and piles can be simulated by the drainage of the water bags. In this model, several soil pressure sensors and displacement sensors are installed.

Before this study, the soil arching effect was developed and stabled by the complete drainage of water bag and 1,200 k cycles of train moving loads. In this study, four testing procedures were performed: water level rising, loading at high water level, water level lowering and loading at low water level. During the testing processes, the accumulative settlement and soil stress were measured.

3 Accumulative Settlement

3.1 Testing Results

For the overall accumulative settlement of the track-bed, it increases rapidly in the beginning of loading at high water level. As the loading cycles increase, the increasing rate of the accumulative settlement slows down, suggesting a relatively stable state of the track-bed. When the water level decreases or loading at low water level, the overall accumulative settlement slightly increases.

In terms of the accumulative settlement inside the subgrade, the sensors located above the water bag present higher values than those above the pile caps at a given elevation in the beginning of loading at high water level. This indicates that the differential settlement continues to develop in this procedure, leading to a strengthened soil arch. By contrast, with the water level lowered and loading at low water level, the accumulative settlement of each sensor increases slightly, while the differential settlement stays unchanged.

3.2 Calculation Using Constitutive Model

Using the constitutive model of permanent deformation of coarse-grained soil [19], the accumulative settlement of the part above the water level is estimated for the case of loading at high water level. The results show that only minority of the settlement develops above the water level.

4 Evolution of Soil Arching

The dynamic soil stress is defined as the stress amplitude under cyclic loading for each sensor. With the water level rising and loading at high water level, the soil arching effect is strengthened, with part of the dynamic stress transferred from the soil above the water bag to that above the pile cap. As the loading cycle increases, a stable soil arch is developed, which is consistent with the settlement measurement. On the other hand, the stable state of soil arching is developed earlier near the crown of the soil arch, while the soil arching effect is still more significant with deeper depth. At this stable soil arch, the water level lowering and loading at low water level influence the distributions of the dynamic stress in the track-bed slightly. This observation is also strongly supported by the negligible variation of the differential settlement measured inside the track-bed.

5 Conclusions

In this study, the mechanical behaviors of the GRPS track-bed model were investigated, following four testing procedures: water level rising, loading at high water level, water level lowering and loading at low water level. With the water level increasing and cyclic loading at high water level, the overall accumulative settlement increases rapidly and the differential settlement develops in the beginning. As a result, the soil arching is strengthened till the differential settlement becomes stable. At the stable state of soil arching, the water level lowering and loading at low water level affect the mechanical behaviors of the track-bed slightly.

References

1. Chen, R.P., Chen, J.M., Wang, H.L.: Recent research on the track-subgrade of high-speed railways. J. Zhejiang Univ.-Sci. A **15**(12), 1034–1038 (2014a)
2. Chen, R.P., Jiang, P., Ye, X.W., Bian, X.C.: Probabilistic analytical model for settlement risk assessment of high-speed railway subgrade. J. Perform. Constructed Facil. **30**(3), 04015047 (2016a)
3. Chen, R.P., Wang, Y.W., Ye, X.W., Bian, X.C., Dong, X.P.: Tensile force of geogrids embedded in pile-supported reinforced embankment: a full-scale experimental study. Geotext. Geomembr. **44**, 157–169 (2016b)
4. Wang, H.L., Cui, Y.J., Lamas-Lopez, F., Dupla, J.C., Canou, J., Calon, N., Saussine, G., Aimedieu, P., Chen, R.P.: Effects of inclusion contents on the resilient modulus and damping ratio of unsaturated track-bed materials. Can. Geotech. J. **54**, 1672–1681 (2017)
5. Wang, H.L., Cui, Y.J., Lamas-Lopez, F., Dupla, J.C., Canou, J., Calon, N., Saussine, G., Aimedieu, P., Chen, R.P.: Investigation on the mechanical behavior of track-bed materials at various contents of coarse grains. Construction Build. Mater. **164**, 228–237 (2018b)

6. Wang, H.L., Cui, Y.J., Lamas-Lopez, F., Dupla, J.C., Canou, J., Calon, N., Saussine, G., Aimedieu, P., Chen, R.P.: Permanent deformation of track-bed materials at various inclusion contents under large number of loading cycles. J. Geotech. Geoenviron. Eng. **144**(8), 04018044 (2018c)
7. Chen, R.P., Chen, Y.M., Han, J., Xu, Z.Z.: A theoretical solution for pile-supported embankments on soft soils under one-dimensional compression. Can. Geotech. J. **45**, 611–623 (2008)
8. Hewlett, W.J., Randolph, M.F.: Analysis of piled embankments. Ground Eng. **21**, 12–18 (1988)
9. Low, B.K., Tang, S.K., Choa, V.: Arching in piled embankments. J. Geotech. Eng. **120**(11), 1917–1938 (1994)
10. Terzaghi, K.: Theoretical Soil Mechanics. Wiley, New York (1943)
11. van Eekelen, S.J.M., Bezuijen, A., van Tol, A.F.: An analytical model for arching in piled embankments. Geotext. Geomembr. **39**, 78–102 (2013)
12. Zhou, W.H., Chen, R.P., Zhao, L.S., Xu, Z.Z., Chen, Y.M.: A semi-analytical method for the analysis of pile-supported embankments. J. Zhejiang Univ. Sci. A **13**(11), 888–894 (2012)
13. Zhuang, Y., Wang, K.Y., Liu, H.L.: A simplified model to analyze the reinforced piled embankments. Geotext. Geomembr. **42**, 154–165 (2014)
14. Chen, R.P., Wang, H.L., Hong, P.Y., Cui, Y.J., Qi, S., Cheng, W.: Effects of degree of compaction and fines content of the subgrade bottom layer on moisture migration in the substructure of high-speed railways. Proc. Inst. Mech. Eng. Part F J. Rail Rapid Transit **234**(4), 1197–1210 (2018)
15. Chen, Y.M., Wang, H.L., Chen, R.P., Chen, Y.: A newly designed TDR probe for soils with high electrical conductivities. Geotech. Test. J. **37**(1), 35–45 (2014b)
16. Wang, H.L., Chen, R.P., Luo, L., Wu, J.: Numerical modeling of moisture migration in high-speed railway subgrade. In: Zhao, S., Liu, J., Zhang, X. (eds.) International Symposium on Systematic Approaches to Environmental Sustainability in Transportation, Fairbanks, Alaska, USA, pp. 349–363. ASCE, Reston, VA (2015)
17. Wang, H.L., Chen, R.P., Qi, S., Cheng, W., Cui, Y.J.: Long-term performance of pile-supported ballastless track-bed at various water levels. J. Geotech. Geoenviron. Eng. **144**(6), 04018035 (2018a)
18. Wang, H.L., Chen, R.P., Cheng, W., Qi, S., Cui, Y.J.: Full-scale model study on variations of soil stress in the geosynthetic-reinforced pile-supported track-bed with water level change and loading cycles. Can. Geotech. J. (2018d, in press)
19. Gidel, G., Hornych, P., Chauvin, J.J., Breysse, D., Denis, A.: A new approach for investigating the permanent deformation behaviour of unbound granular material using the repeated load triaxial apparatus. Bulletin de Liaison des Laboratoires des Ponts et Chaussées **233**, 5–21 (2001)

Analysing the Permanent Deformation of Cohesive Subsoil Subject to Long Term Cyclic Train Loading

Natalie M. Wride(✉) and Xueyu Geng

School of Engineering, University of Warwick, Coventry CV4 7AL, UK
n.m.wride@warwick.ac.uk

Abstract. Subgrade soils of railway infrastructure are subjected to a significant number of load applications over their design life. The use of slab track on existing and future proposed rail links requires a reduced maintenance and repair regime for the embankment subgrade, due to restricted access to the subgrade soils for remediation caused by cyclic deformation. It is therefore important to study the deformation behaviour of soft cohesive subsoils induced as a result of long term cyclic loading. In this study, a series of oedometer tests and cyclic triaxial tests have been undertaken to investigate the undrained deformation behavior of soft kaolin. Combined with the examination of excess pore pressures and strains obtained from the cyclic triaxial tests, the results have been compared with an existing analytical solution for long term settlement considering repeated low amplitude loading. Modifications to the analytical solution have been made based on the laboratory analysis that show good agreement with further test data.

Keywords: Cyclic loading · Traffic loading · Creep behaviour · Soft soil

1 Introduction

Transport links are fundamental to the management of population growth. Expansion of urban areas has led to construction over ground previously avoided by contractors due to its' low bearing capacity and high compressibility. The variable compressive nature of cohesive soils on which infrastructure routes can be founded is an important consideration for transportation earthworks where linearity is often prescribed. As a result, the topic of transportation infrastructure construction over compressible deposits has progressively become a significant area of interest for engineers due to the importance of sustainability over the whole design life of proposed earthworks schemes.

Cohesive soils exhibit time-dependent deformations due to the dissipation of pore water pressure from within the soil skeleton and due to the viscous deformation of the soil particles [1]. This process can be described as an elastic visco-plastic soil behavior [2]. The time-dependent stress strain behavior of a soil is important for the determination of anticipated post-construction settlements that occur over the whole lifetime of an earthworks scheme. To address the time-dependent behaviour of a soil for the design of earthworks, the conventional solution is to employ static calculation methods; where the dynamic effects are taken into account by multiplying the static load by a dynamic amplification factor determined based on the train speed, quality of vehicle and quality of the track. However, a number of case studies on highways and rail networks, [3–5], have demonstrated that a significant amount 'traffic-induced' settlement occurs, underestimated by static analysis. This is particularly important for considering rail routes on concrete slab track where strict tolerances in post-construction settlement are inherent. Due to the susceptibility of these linear routes to differential settlement problems, traffic loading is therefore a fundamental consideration to address for the design of future transport schemes.

In this study, a series of undrained and partially drained cyclic triaxial tests on remoulded saturated clay are presented. The undrained tests demonstrate an increase in axial strain with a decrease in the frequency of loading. The cyclic threshold stress of the soil is between a cyclic stress ratio of 0.6 and 0.8. In comparison with partially drained triaxial tests, the axial strain reduces significantly as a result of the dissipation of generated pore pressure. The ratio between undrained and partially drained axial strains increases with a decrease in cyclic stress ratio. These cyclic triaxial tests have compared with an existing cyclic elastic visco-plastic solution, subsequently modified to include the initial stress distribution beneath slab track embankments.

2 Laboratory Testing

2.1 Test Samples

The samples used in this study were remoulded kaolinite soil. The physical properties of this clay are: a specific gravity of $G_s = 2.68$; a plastic limit of $w_p = 25\%$; a liquid limit of $w_L = 55\%$ and a plasticity index of $I_p = 30\%$.

A standard method for soil sample preparation was selected to ensure kaolin samples were consistent for testing. The kaolin clay was oven dried and ground to a powder. The kaolin powder was then mixed to a slurry with a water content of 1.5 times the liquid limit w_L and then preconsolidated. Following preconsolidation, the kaolin blocks were trimmed to cylindrical samples of 38 mm diameter and 76 mm height.

2.2 Testing Apparatus and Procedure

Samples were tested using the GDS Enterprise Level Dynamic Triaxial Testing (ELDYN) System. Once placed in the triaxial cell, samples were saturated at a back pressure of 300 kPa with an effective mean pressure of 10 kPa followed by a "B-check" to ensure sample saturation was greater than 97% (a B value of 0.97). Samples were then anisotropically consolidated to an initial vertical effective principal stress of 120 kPa with a radial effective stress of 72 kPa, representing the confining conditions of a soil element at approximately 6.0 m below ground level with a K_0 consolidation ratio of 0.6. Cyclic triaxial tests were completed for frequencies ranging from 0.1 Hz to 5 Hz for a cyclic stress ratio (CSR) of 0.4, 0.6 and 0.8 for a duration of 6,000 cycles (0.1 Hz) and 30,000 cycles or failure otherwise. The ratio is defined by q_{cyc}/q_s where q_{cyc} is the cyclic deviatoric stress and q_s is the static deviatoric stress at failure.

2.3 Results

The development of axial strain with number of load cycles is presented in Fig. 1 for cyclic stress ratios of 0.6 and 0.8. It should be noted that for a CSR of 0.6, the samples remained stable at the end of the test, whilst for samples with a cyclic stress ratio of 0.8, there was a rapid development in axial strain and the samples failed in a reduced number of load cycles. It can therefore be demonstrated that the threshold cyclic stress ratio is between 0.6 and 0.8, previously acknowledged by [6–8].

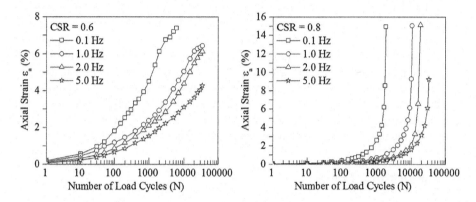

Fig. 1. Development of axial strain with number of load cycles for undrained cyclic triaxial tests: CSR = 0.6 and CSR = 0.8.

Partially drained (PD) cyclic triaxial tests were performed for the same loading conditions stated above for undrained (UD) tests. The results for the development of axial strain, considering stable samples (CSR = 0.4 and 0.6) at a frequency of 1 Hz and 5 Hz (Fig. 2) show the cyclically induced axial strain reduces during partially drained tests. In the case of CSR of 0.4, the axial strain reduces as a result of partial drainage by a factor of approximately 4. However, with an increase in CSR and therefore closer proximity to the threshold

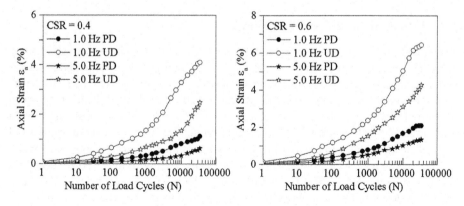

Fig. 2. Development of axial strain for partially drained cyclic triaxial tests at 1 Hz and 5 Hz: CSR = 0.4 and CSR = 0.6.

cyclic stress ratio, this factor decreases to approximately 3. This consideration is important for considering traffic loading, where the loading period is not constant and, as a result, pore pressures are simultaneously generated and dissipated.

3 Numerical Analysis

Based on the undrained cyclic triaxial tests completed, the elastic visco-plastic model developed by [2, 9] and employed by [10] to include the response under low amplitude semi-sinusoidal loading was used. The model was programmed in MATLAB and adapted to include an initial stress distribution generated in ABAQUS. This initially included the distribution of a triaxial sample and has subsequently been extended to a rail slab track embankment system. Overall, the initial model predictions are in good agreement with further cyclic triaxial testing undertaken. The authors intend to apply the model to case study for further examination.

References

1. Ladd, C., Foot, R., Ishihara, K., Schlosser, F., Poulos, H.: Stress-deformation and strength characteristics. In: Proceedings of the 9th International Conference on Soil Mechanics and Foundation Engineering, pp. 421–494 (1997)
2. Yin, J.H., Graham, J.: Viscous–elastic–plastic modelling of one-dimensional time-dependent behaviour of clays. Can. Geotech. J. **26**(2), 199–209 (1989)
3. Chai, J., Miura, N.: Traffic-load-induced permanent deformation of road on soft subsoil. J. Geotech. Geoenviron. Eng. **128**(11), 907–916 (2002)
4. Sakai, A., Samang, L., Miura, N.: Partially-drained cyclic behavior and its application to the settlement of a low embankment road on silty-clay. Soils Found. **43**(1), 33–46 (2003)
5. Cui, X., Zhang, N., Li, S., Zhang, J., Wang, L.: Effects of embankment height and vehicle loads on traffic-load-induced cumulative settlement of soft clay subsoil. Arab. J. Geosci. **8**(5), 2487–2496 (2015)
6. Sangrey, D.A., Henkel, D.J., Esrig, M.I.: The effective stress response of a saturated clay soil to repeated loading. Can. Geotech. J. **6**, 241–252 (1969)
7. Zhou, J., Gong, X.: Strain degradation of saturated clay under cyclic loading. Can. Geotech. J. **38**(1), 208–212 (2001)
8. Ni, J., Indraratna, B., Geng, X.Y., Carter, J.P., Chen, Y.L.: Model of soft soils under cyclic loading. Int. J. Geomech. **15**(4), 401–406 (2015)
9. Yin, J.H., Graham, J.: Viscous–elastic–plastic modelling of one-dimensional time-dependent behaviour of clays: reply. Can. Geotech. J. **27**(2), 262–265 (1990)
10. Hu, Y.Y.: Long-term settlement of soft subsoil clay under rectangular or semi-sinusoidal repeated loading of low amplitude. Can. Geotech. J. **47**(11), 1259–1270 (2010)

Dynamic Tensile Failure of Rocks Under Triaxial Stress State

Bangbiao Wu[1] and Kaiwen Xia[2]

[1] School of Civil Engineering, Tianjin University, Tianjin 300350, China
[2] Department of Civil Engineering, University of Toronto, Toronto M5S1A4, ON, Canada
kaiwen.xia@utoronto.ca

Abstract. Tensile failure of rocks is a main problem in underground engineering projects, in which rocks are subjected to dynamic disturbances while under in situ stresses. When disturbed by dynamic loads from blasting, seismicity or rock bursts, the underground structures under static pre-stress would be vulnerable to tensile failure. A modified split Hopkinson pressure bar (SHPB) system is adopted to load Brazilian disc (BD) specimens statically simulating the in-situ stress state, and then exert dynamic load generated by impact. Several groups of specimens are tested under triaxial stress state. The dependence of dynamic tensile strength of the rock material on the static pre-stress and loading rate is established. The result shows that the dynamic tensile strength of the rock increases with the loading rate, revealing the rate dependency. It also shows that the dynamic tensile strength decreases with the increasing pre-tension but increases with the hydrostatic confinement.

Keywords: Tensile failure · Rock dynamics · SHPB · Triaxial stress state

1 Introduction

Underground rocks are vulnerable to tensile failure when disturbed from blasting, seismicity and rock bursts. The rocks near the underground opening are prone to pre-tension, the far zone rocks are subjected to hydrostatic pre-stress, and rocks in the intermediate zone are under triaxial stress states. As shown in Fig. 1, the discontinuities in the intermediate zone rocks lead to tensile stress with hydrostatic confining pressure locally. When the seismicity or rock bursts occur, the rock is vulnerable to tensile failure. Therefore, it is the purpose of this paper to investigate the dynamic tensile failure of rock materials under triaxial stress states.

Over the past few years, significant progress has been made in the characterization of rock failure under high loading rates [1–4]. In these tests, rock specimens are stress-free before being subjected to dynamic loading. However, in rock engineering practice, such as underground rock blasting, rocks are under various tectonic and lithostatic stress. Under the superposed dynamic loads with the static tectonic stress, the rock behaves differently as compared to the material subjected solely to either static stress or impact loading. Gong and Malvern [5] conducted passively confined tests of axial dynamic

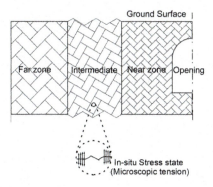

Fig. 1. Potential sources of tensile failure under hydrostatic stress states

compressive strength of concrete, Chen and Ravichandran [6] introduced an experimental technique for imposing dynamic multiaxial compression with mechanical confinement, Song et al. [7] investigated the confinement effects on the dynamic compressive properties of an epoxy syntactic foam, and Frew et al. [8] developed a dynamic triaxial SHPB system to study the dynamic failure of brittle materials under hydrostatic confining pressure. In addition to the studies of the compressive dynamic properties under pre-stress mentioned above, the tensile failure of rock under static pre-stress has also been investigated recently [9, 10].

2 Experimental Setup and Results

2.1 Experimental Design

The BD specimen is adopted for the measurement of dynamic tensile strength of rock loaded by the modified SHPB system introduced in Fig. 2. Through the modified SHPB system, the hydrostatic confinement and the pretension are exerted on the rock specimen and maintained before the dynamic load is applied. The preload is applied through a hydraulic press, and the dynamic loading is exerted using a low speed gas gun, similarly to a conventional SHPB system.

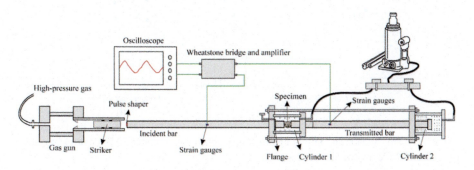

Fig. 2. Schematics of the modified SHPB system for Tri-axial test

A hydrostatic stress is exerted on the BD specimen through the confinement cylinders and the press by opening the two valves at the same time. When the hydrostatic stress is achieved, the valve controlling the lateral confinement is closed, and the cylinder at the transmitted bar end is used to provide more axial load to the specimen for pre-tension so that the specimen is under a general triaxial stress condition. Two groups of hydrostatic stresses are investigated, with four pretension conditions for each hydrostatic stress. The hydrostatic stresses are set at 5 MPa and 10 MPa, and the pretensions are 20%, 40%, 60% and 80% of the tensile strength under the corresponding hydrostatic stress state, as shown in Fig. 3.

Hydrostatic Stress: 5 MPa, 10 MPa   Pretension: 20%, 40%, 60%, 80% of the strength

Fig. 3. Schematics of the modified SHPB system for Tri-axial test

2.2 Experimental Results

Existing studies [11, 12] have demonstrated that BD tests are valid when the stress equilibrium condition is satisfied. Figure 4 illustrates the correlation between tensile strength and the loading rate when the rock specimens are under 5 MPa and 10 MPa hydrostatic stress respectively. It can be observed that the tensile strength increases with the loading rate linearly, revealing the phenomenon of rate dependency that is common for brittle materials, within the loading rate up to 1400 GPa/s.

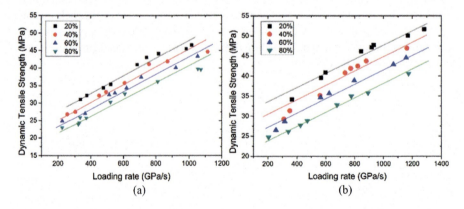

Fig. 4. Dynamic tensile strength of confined rock specimen (a) under 5 MPa hydrostatic stress, (b) under 10 MPa hydrostatic stress.

3 Conclusions

In this paper, the modified SHPB system with hydraulic cylinders is adopted to measure the dynamic tensile strength under triaixal stress states. The confining pressures are set at 5 MPa and 10 MPa, and the pre-tension is 20%, 40%, 60% or 80% of the static tensile strength under the set hydrostatic confining pressure. The experimental results reveal the familiar phenomenon that there is a significant enhancement in tensile strength with the increase of the loading rate, and that the dynamic tensile strength increases with the hydrostatic confinement and decreases with increasing pre-tension.

References

1. Huang, S., Xia, K.: Effect of heat-treatment on the dynamic compressive strength of Longyou sandstone. Eng. Geol. **191**, 1–7 (2015)
2. Gao, G., Yao, W., Xia, K., Li, Z.: Investigation of the rate dependence of fracture propagation in rocks using digital image correlation (DIC) method. Eng. Fract. Mech. **138**(1), 146–155 (2015)
3. Zhang, Q.B., Zhao, J.: A review of dynamic experimental techniques and mechanical behaviour of rock materials. Rock Mech. Rock Eng. **47**(4), 1411–1478 (2014)
4. Xia, K.: Review of laboratory measurements of dynamic strength and fracture properties of rock. In: 1st International Conference on Rock Dynamics and Applications. ISRM, Lausanne, Switzerland (2013)
5. Gong, J.C., Malvern, L.E.: Passively confined tests of axial dynamic compressive strength of concrete. Exp. Mech. **30**(1), 55–59 (1990)
6. Chen, W., Ravichandran, G.: An experimental technique for imposing dynamic multiaxial-compression with mechanical confinement. Exp. Mech. **36**(2), 155–158 (1996)
7. Song, B., Chen, W., Yanagita, T., Frew, D.J.: Confinement effects on the dynamic compressive properties of an epoxy syntactic foam. Compos. Struct. **67**(3), 279–287 (2005)
8. Frew, D.J., Akers, S.A., Chen, W., Green, M.L.: Development of a dynamic triaxial Kolsky bar. Meas. Sci. Technol. **21**(10), 105704 (2010)
9. Wu, B., Chen, R., Xia, K.: Dynamic tensile failure of rocks under static pre-tension. Int. J. Rock Mech. Min. Sci. **80**, 12–18 (2015)
10. Wu, B., Yao, W., Xia, K.: An experimental study of dynamic tensile failure of rocks subjected to hydrostatic confinement. Rock Mech. Rock Eng. **49**, 1–10 (2016)
11. Dai, F., Huang, S., Xia, K., Tan, Z.: Some fundamental issues in dynamic compression and tension tests of rocks using split Hopkinson pressure bar. Rock Mech. Rock Eng. **43**(6), 657–666 (2010)
12. Zhou, Y.X., Xia, K., Li, X.B., Li, H.B., Ma, G.W., Zhao, J.: Suggested methods for determining the dynamic strength parameters and mode-I fracture toughness of rock materials. Int. J. Rock Mech. Min. Sci. **49**(1), 105–112 (2012)

Experimental Study of Dynamic Shear Module and Damping Ratio of Intact Loess from Xining, China

Wenju Wu[1], Wuyu Zhang[1(✉)], Lingxiao Liu[1], and Yanxia Ma[1,2]

[1] School of Civil Engineering, Qinghai University, 251 Ningda Road, Xining, Qinghai, China
wuwenju92@163.com
[2] The Key Laboratory of Geomechanics and Embankment Engineering, Hohai University, Nanjing 210098, Jiangsu, China

Abstract. There is no common understanding on the influences of loess anisotropy on dynamic modulus and damping ratio. A series of laboratory tests were carried out on Xining intact loess using the GDS bidirectional dynamic triaxial test system. This study aimed to investigate the dynamic shear modulus and damping ratio of the Xining intact loess at different angles associated with the index plane. In addition, the difference between loess in Xining and other regions was also discussed. The following conclusions could be drawn based on the test results. Firstly, the anisotropy of the loess had a significant influence on the development of dynamic shear deformation and the dynamic shear modulus, indicating that it is necessary to keep the deposition direction of intact soil unchanged during sampling. Secondly, the increase of consolidation pressure reduced the effect of the anisotropy of loess. Higher confining pressure and a closer sampling angle to the natural deposition direction of soil resulted in higher dynamic shear modulus and slower rate of the development of soil dynamic shear deformation. In other words, the ability of soil to resist dynamic shear deformation became stronger. Finally, the damping ratio of Xining loess increased with the increase of dynamic shear strain. The increase rate of damping ratio of Xining loess was much quickly than that in other western regions, which could provide reference for the research on seismic zoning.

Keywords: Intact loess · Dynamic triaxial test · Anisotropy
Dynamic shear module · Damping ratio

1 Introduction

Loess in the northwest of China due to geological environment is very fragile and prone to geological disasters in the earthquake vibration, at the same time as the gradual expansion of the size of the area of the construction project, the designers pay more attention on the dynamic loess foundation and slope engineering instability destruction problem. The dynamic shear modulus and damping ratio of soils play an important role in seismic analysis and site safety assessment. They are the basis for studying the design and calculation of geotechnical structures under the action of earthquake and other dynamic loads [1]. At present, there are many studies on the dynamic shear modulus and damping ratio of soil [2–4].

The present research on dynamic loess is only limited in a specific region, the study sites scattered results are not perfect. This paper studies the dynamic shear modulus and damping ratio, and provides a theoretical reference for the loess dynamic analysis and construction of disaster prevention and mitigation.

2 Test Method

The soil sample is taken from a loess building site in north of Xining, and the sampling depth is 4 m. The preparation methods are referenced from relevant regulation [5]. The physical properties of the loess are shown in Table 1. To study the anisotropy of the soil, samples were prepared by cutting in an angle θ associated with the index plane ($\theta = 0° / 30° / 45° / 60° / 90°$).

Table 1. Physical Properties of Loess

Specific gravity of soil particle	Natural moisture content (%)	Natural dry density (g/cm³)	Liquid limit (%)	Plastic index (%)	Compression coefficient (a_{1-2}/MPa^{-1})
2.72	10.41	1.50	26.59	11.79	0.28

The GDS dynamic triaxial test system was used, setting confining pressure of 100 kPa, 200 kPa and 300 kPa with isotropic consolidation. After that, we applied the dynamic load with a sine wave ($f = 1$ Hz, $N = 10$). Increasing the load step by step until the sample reaches the failure (axial strain $\geq 3\%$) or is obviously deformed.

3 Dynamic Deformation

Figure 1 shows that the cutting angle of loess has significant influence on the development of dynamic shear deformation, the development is relatively slow when $\theta = 0°$.

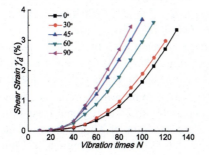 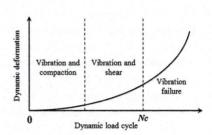

Fig. 1. Shear strain of loess in different cutting angles (100 kPa) (left)

Fig. 2. Development curve of soil dynamic deformation (right)

Also, it could be found the increase of consolidation pressure causes the anisotropy of the loess to weaken gradually from all the data. Under the action of cyclic loading, the development of dynamic deformation of soil mainly goes through 3 stages (Fig. 2) [6].

Xining loess almost follows this law. The limit cycle times of vibration compaction stage and vibration shear stage is 40 or 50 times, the increase of confining pressure and the 0 angle sample obviously prolong the vibration and shear stage of the loess which means the increase of "Nc" made the occurrence of vibration failure of soil.

4 Dynamic Shear Modulus

The treatment method of equivalent dynamic shear modulus [6] is adopted in this paper.

Figure 3 shows there are obvious differences in shear modulus - dynamic strain curves of loess samples at different angles, the dynamic shear modulus at 0° is always greater than that of other angles.

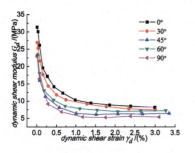
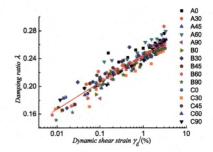

Fig. 3. Dynamic shear modulus of loess in different cutting angles (100 kPa). (left)

Fig. 4. Damping ratio and the change of the dynamic shear strain. (right)

5 Damping Ratio

The relationship between the damping ratio and the change of the dynamic shear strain is shown in Fig. 4. (Which: A, B and C represents σ_3 = 100 kPa / 200 kPa / 300 kPa. 0, 30, 45, 60 and 90 represents θ = 0° / 30° / 45° / 60° / 90° respectively.)

The damping ratio of loess has a good linear correlation in the semi logarithmic coordinate with the change of the dynamic shear strain [4, 7]. The formula is used to fit the relationship between the damping ratio - dynamic shear strain of Xining loess, as shown as follow: $\lambda = 0.03851 \lg(\gamma_d) + 0.24397$, which could provide reference for the dynamic analysis of loess in Xining area. Compared with other regional loess in the northwest region [4], the maximum dynamic shear modulus of Xining loess is small, the damping ratio of Xining loess is higher than that in the western regions of China, but less than the central region represented by Chengde [7].

6 Conclusions

The anisotropy of loess on dynamic shear deformation - modulus has significant effect, it is recommended relevant researches involved in the field of intact soil sampling and handling process, to ensure the sample does not turn over.

The increase of consolidation pressure reduced the effect of loess anisotropy. Higher confining pressure and a closer sampling angle to natural deposition direction of soil resulted in higher dynamic shear modulus and slower development rate on deformation means ability soil to resist dynamic shear deformation became stronger.

The damping ratio of Xining loess increased with the increase of shear strain. The increase rate of damping ration of Xining loess was much quickly than that in western regions, which could provide reference for the research on seismic zoning of China.

Acknowledgement. This study was supported by the National Natural Science Foundation of China (No. 51768060), the Fundamental and Cooperation Research Program of Qinghai (No. 2017-ZJ-792, No. 2017-HZ-804), the Open Research Fund Program of Hohai University (No. 2016002), the Technological Innovation Service Platform of Qinghai (No. 2018-ZJ-T01).

References

1. Hardin, B.O., Drnevich, V.P.: Shear modulus and damping in soils: measurement and parameter effects. J. Soil Mech. Found. Div. **98**(6), 603–624 (1972)
2. Li, Z., Li, H.E., He, Y.J.: Experimental study on dynamic shear module and damping ratio of Shanxi area loess. J. Disaster Prev. Mitig. Eng. **34**(4), 523–527 (2014)
3. Li, R.S., Chen, L.W., Yuan, X.M., Li, C.C.: Experimental study on influences of different loading frequencies on dynamic modulus and damping ratio. Chin. J. Geotech. Eng. **39**(1), 71–79 (2017)
4. Wang, Z.J., Luo, Y.S., Wang, R.R., Yang, L.G., Tan, D.Y.: Experimental study on dynamic shear modulus and damping ratio of undisturbed loess in different regions. Chin. J. Geotech. Eng. **32**(9), 1464–1469 (2010)
5. JGJ/T87-2012: Technical specification for engineering geological prospecting and sampling in constructions. 2nd edn. China construction industry press, Beijing (2012)
6. Xie, D.Y.: Soil Dynamics, 1st edn. Higher Education Press, Beijing (2011)
7. Wei, L.Y., Li, H., Dong, L.Y., Feng, L.: Study on experiment of dynamic shear module and damping ration of loess in Chengde of Hebei. J. Seismol. Res. **39**(3), 513–517 (2016)

True Triaxial Tests on Natural Shanghai Clay

Guan-lin Ye[✉], Shuo Zhang, Yun-qi Song, and Jian-hua Wang

Department of Civil Engineering, Shanghai Jiao Tong University, Shanghai 200240, China
ygl@sjtu.edu.cn

Abstract. Drained true triaxial tests on undisturbed and remolded Shanghai silty clay were carried out to investigate the influence of intermediate principle stress on the strength and deformation of natural clay. Test results show that not only the shear strength, but also the volume change of Shanghai clay decreases with the increase of Lode angle θ. The shear strength and volume change of the undisturbed samples are larger than those of the remolded ones. In the π plane, the strength of the Shanghai clay basically obeys the SMP (spatial mobilized plane) criterion.

Keywords: True triaxial test · Natural clay · Shear strength · Volume change

1 Introduction

Most of natural soils involved in geotechnical engineering are subjected to three independent principal stresses. True triaxial tests are useful for investigating the mechanical behaviour of soils in this general stress conditions.

Most available tests for clays were carried out under undrained conditions. However, the mean effective stress cannot remain in the same π plane in such conditions. Several drained true triaxial tests on remolded clay can be found [1–4]. Drained tests on natural clays with focus on the influence of structure is very limited.

In current study, drained true triaxial tests on undisturbed and remolded Shanghai clay with different Lode angle were conducted to investigate the influence of structure on mechanical behavior of clay under general stress conditions was investigated by comparing test results of undisturbed and remolded Shanghai clay. The test results will be useful for constitutive modelling of natural clays.

2 Test Method and Material

The undisturbed samples were collected by the block sampling method in a braced excavation site near Long-hua Tower in Shanghai. The samples come from the Layer 4 Shanghai clays in the depth of 12 m. The Layer 4 clay is a typical marine clay. The remolded samples are prepared using the same batch of block samples. The size of the rectangle specimen is 85 mm * 85 mm * 53 mm, which is 3 mm larger than the standard size, due to the fact that the specimen will contract about 3 mm during isotropic consolidation.

A mixed rigid-flexible boundary true triaxial apparatus [3] is used in this study. In the true triaxial cell, the rubber membrane covers both the vertical and the horizontal platens. This design allows the stresses from the rigid platens to be smaller than the cell water pressure. σ_1 is driven by a motor in a strain-controlled mode, at a rate of 0.0039 mm/min, so as to ensure a fully dissipation of excess porewater pressure when the specimen is close to or during failure. σ_2 is applied by the cell pressure. And σ_3 is applied through an air cylinder inside of the triaxial cell. Consolidation and shearing are automatically controlled by an in-house developed program written in visual basic.

A series of drained tests on undisturbed and remolded specimens are carried out with different Lode angles (0, 15, 30, 45, and 60°) in the same π-plane were conducted. In each test, sample is initially isotropically consolidated under an effective confining stress of 100 kPa. A back pressure of 100 kPa is applied to obtain a Skempton's B value larger than 0.95 for ensuring a full saturation.

3 Testing Results and Discussion

3.1 Stress-Strain Relationship

Figures 1 and 2 shows the measured curves of (a) deviatoric stress ratio q/p versus shear strain ε_d, (b) volumetric strain ε_v versus shear strain ε_d, from the drained true triaxial tests on the undisturbed and remolded clays.

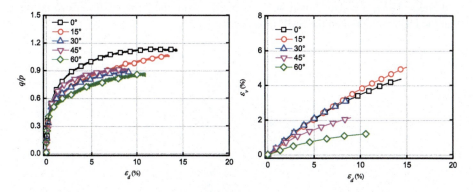

Fig. 1. Test results of undisturbed clay in terms of stress-strain relation and volumetric strain

It can be seen that both undisturbed and remolded specimens have a consistent trend of decreasing stress ratio q/p with the increasing θ values. The decreasing gradients of shear strength undisturbed and remolded clays are almost the same. This demonstrates that the intermediate principal stress has a significant effect on the strength of both undisturbed and remolded clays. In addition, the contractive volumetric strain also shows a decreasing trend with the increasing θ values.

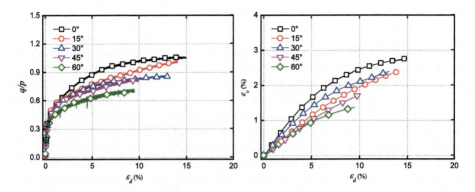

Fig. 2. Test results of remolded clays in terms of stress-strain relation and volumetric strain

The effects of structure on the stress-strain behavior of the clay is analyzed by comparing the tests results of undisturbed and remolded specimens with same θ–value. It is found that the strength of undisturbed clay is slight higher than remolded clay in all Lode angles. The volumetric contraction of the undisturbed clay is larger than remolded clay.

3.2 Failure Criterion

Figure 3 show the normalized effective principal stresses at failure points which are plotted on the π plane. In addition, the curves form the SMP failure criteria (Matsuoka and Nakai 1974) are also shown on the π-plane. From the comparison, it can be seen that the test results match the SMP failure criterion.

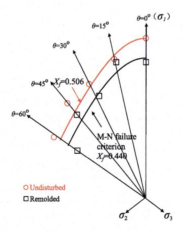

Fig. 3. Failure points in π plane of Shanghai soft clay

4 Conclusions

A series of drained true triaxial tests have been performed to determine the stress–strain and volumetric deformation behavior of the natural Shanghai clay under three-dimensional stress conditions. It is found that the intermediate principal stress has a significant influence on both strength and volume change of undisturbed and remolded clays. And the strength and the volumetric contraction of undisturbed clay is slight higher than remolded clay in all Lode angles. Both undisturbed and remolded clays match the SMP failure criterion.

References

1. Kumruzzaman, M., Yin, J.-H.: Influence of the intermediate principal stress on the stress–strain–strength behaviour of a completely decomposed granite soil. Géotecnique **62**(3), 275–280 (2012)
2. Nakai, T., Matsuoka, H.: A generalized elastoplastic constitutive model for clay in a three-dimensional stress. Soils Found. **26**(3), 81–89 (1986)
3. Ye, G.L., Sheng, J.R., Ye, B., Wang, J.H.: Automated true triaxial apparatus and its application to over-consolidated clay. Geotech. Test. J. **35**(4), 517–528 (2012)
4. Ye, G.L., Ye, B., Zhang, F.: Strength and dilatancy of overconsolidated clays in drained true triaxial tests. J. Geotech. Geoenviron. Eng. **140**(4), 80–90 (2013)

Study on Strength of Expanded Polystyrene Concrete Based on Orthogonal Test

Quan You, Linchang Miao[✉], Chao Li, Shengkun Hu, and Huanglei Fang

Institute of Geotechnical Engineering, Southeast University, Nanjing, China
Lc.miao@seu.edu.cn

Abstract. Expanded polystyrene concrete has the potential to be used in subway ballast and other underground constructions; however, the strength of expanded polystyrene concrete cannot be well controlled since expanded polystyrene is hydrophobic and tends to float during the preparation process. The orthogonal table L18 (3^7) was adopted to analyze effects of several factors on the compressive strength of expanded polystyrene concrete. Factors including water cement ratio, beads incorporation volume ratio, superplasticizer usage, dosage of additive I and II, sand ratio and fiber content were considered. Test results illustrate that the water cement ratio and beads incorporation volume ratio are two key factors that affect the compressive strength of expanded polystyrene concrete. Recommended mixture proportion is also proposed to realize the optimum strength of expanded polystyrene concrete. The orthogonal test can help systematically investigate effects of mixture materials on the strength of expanded polystyrene concretes and optimize the design of mixture proportion.

Keywords: Expanded polystyrene concrete · Optimum mixture proportion Compressive strength · Orthogonal test

1 Introduction

As a kind of lightweight aggregate concrete, expanded polystyrene concrete is a new type of building material that has the potential to be used in underground construction. It is made by mixing polystyrene into concrete, and has advantages like lightweight, heat preservation and good absorbed energy behavior. Comprehensive experimental researches on expanded polystyrene concrete began in 1972 [1], and mainly focused on mixing process and strength.

However, there are still no universally accepted design methods for the mixture proportion of expanded polystyrene concrete except for few guidelines. Experimental results revealed that the strength of concrete was not a simple linear relationship with proportion of material [2]. Under the influence of multiple factors, it was difficult to find the law how performance of concrete varied with the amount of investigated admixtures. The standardized table used in orthogonal test was proved to be a scientific method in the theory and application, which could help design tests and consider various factors at the same time.

In this paper, the orthogonal table L_{18} (3^7) was used to design mixture proportion tests considering effect of factors including water cement ratio, beads incorporation volume ratio, superplasticizer usage, dosage of additive I and II, sand ratio and fiber content. Dominant factors of compressive strength were obtained based on the orthogonal analysis, and the optimization mixture proportion of the expanded polystyrene concrete in compressive strength was also proposed.

2 Test Material and Mix Design

2.1 Test Material

Material used in the preparation of lightweight expanded polystyrene concrete included polystyrene beads, cement, river sand, stone, silica fume, superplasticizer, fiber, additive I and additive II.

Diameter of the polystyrene bead was 2 to 3 mm. Cement was P.O 425 Portland cement. Fiber was polyethylene fiber with diameter of 18–45 μm. Additive I and II could improve the hydrophilicity of polystyrene concrete.

2.2 Mix Design and Testing Method

Seven factors were selected to study their effects on the strength of polystyrene concrete. Each factor was set at three levels (see Table 1) and standard L18 (3^7) orthogonal table was adopted to design the tests in total of 18 mixture proportions [3].

Table 1. Form of factors and levels

Levels	A	B	C	D	E	F	G
	Water cement ratio	Beads ratio	Superplasticizer usage	Additive I	Additive II	Sand ratio	Fiber content (kg/m^3)
1	0.30	4:1	1.5%	1.0%	0.06%	29%	0.9
2	0.32	3:1	2.0%	1.5%	0.08%	32%	1.4
3	0.34	7:3	2.5%	2.0%	0.10%	35%	2.7

Lightweight polystyrene concrete was prepared by referring to sand and stone binding method. After curing for 7 and 28 days, unconfined compressive strengths of the test specimens were measured.

3 Results and Discussions

3.1 Results of Orthogonal Test

According to the procedure described above, the average density and compressive strength of expanded polystyrene concrete in each mixture proportion were measured. Calculated results are tabulated in Table 2.

Table 2. Average compressive strength of design specimens

Number	1	2	3	4	5	6	7	8	9
7d compressive strength (MPa)	31.7	26.9	20.1	29.5	20.0	19.5	27.3	25.7	22.6
28d compressive strength (MPa)	37.0	34.0	27.0	32.3	26.1	25.7	37.8	33.7	25.3
Number	10	11	12	13	14	15	16	17	18
7d compressive strength (MPa)	36.3	31.5	26.0	35.0	27.8	25.4	32.5	29.1	22.7
28d compressive strength (MPa)	39.5	33.0	27.6	32.2	31.1	27.7	32.3	32.4	30.0

3.2 Intuitive Range Analysis on Compressive Strength

Range analysis is carried out on the compressive strength of expanded polystyrene concrete. Table 3 shows the influence extent of investigated factors on 7-day compressive strength: beads ratio (B) > water cement ratio (A) > superplasticizer usage (C) > additive II (E) > fiber content (G) > additive I (D) > sand ratio (F). The influence extent changes after 21 days, and the rank of factors that affect 28-day compressive strength of expanded polystyrene concrete is B > A > F > D > G > E > C.

Table 3. Range analysis of 7-day and 28-day compressive strength

Factors	A	B	C	D	E	F	G
Range of 7d compressive strength	2.553	9.336	1.989	0.906	1.698	0.649	1.352
Range of 28d compressive strength	3.871	7.958	0.267	2.045	1.395	2.931	1.902

From the analysis of the range value, either for 7-day or 28-day compressive strength, the range value of factor B and A is the largest, indicating that polystyrene beads ratio and water cement ratio are determinant factors for the strength of expanded polystyrene concrete. Superplasticizer usage, additive II, and fiber content show relatively larger effects on 7-day strength than 28-day strength, while the sand ratio and additive I have significant influence on 28-day compressive strength.

3.3 Trend Diagram Analysis on Compressive Strength

Figure 1 depicts the tends of compressive strengths for each factor. Both 7-day and 28-day compressive strength reach the trough at the middle value of water cement ratio. When the value of beads ratio increases, the corresponding compressive strength decreases. It should be noted that the superplasticizer usage has nearly no effect on 28-day strength while 7-day strength reaches the trough at superplasticizer usage of 2.0%.

The optimum mixture proportions (A_1, B_1, C_1, D_1, E_2, F_2 and G_1) for compressive strength can be obtained considering comprehensively changes of both 7-day and 28-day comprehensive strengths under various factors. Specific mixture proportion is shown in Table 4.

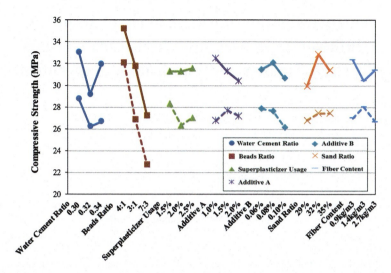

Fig. 1. Trend of compressive strength for each factor. The dotted lines are marked as 7-day compressive strength and the solid ones are 28-day strength.

Table 4. Optimum mixture proportion for compressive strength based on orthogonal tests

Factors	A	B	C	D	E	F	G
Optimum level	0.30	4:1	1.5%	1.0%	0.08%	32%	0.9 kg/m^3

4 Conclusions

By performing orthogonal tests on expanded polystyrene concrete, some conclusions can be drawn: The key factors affecting the compressive strength of expanded polystyrene concrete are beads incorporation ratio and water cement ratio. With decreased water cement ratio and beads incorporation ratio, the compressive strength of polystyrene concrete increases. Mixture proportion A_1, B_1, C_1, D_1, E_2, F_2 and G_1 with compressive strength higher than 35 MPa could realize optimum state of the expanded polystyrene concrete that fulfills the requirement of subway ballast.

References

1. Cook, D.J.: Expanded polystyrene beads as lightweight aggregate for concrete. School of Civil Engineering, University of New South Wales, Australia (1972)
2. Li, B.P.: Orthogonal test of mixture ratio of three-blended concrete. Concrete **4**, 48–50 (2004). (in Chinese)
3. Hsu, C.C., Chien, C.S., Wei, T.Y., Li, Y.T., Li, H.F.: Taguchi DoE for ceramic substrate SMT defects improvement. In: 11th International Microsystems, Packaging, Assembly and Circuits Technology Conference (IMPACT), CA, USA, pp. 381–384. IEEE (2016)

An Investigation on the Effects of Rainwater Infiltration in Granular Unsaturated Soils Through Small-Scale Laboratory Experiments

R. Darban[1(✉)], E. Damiano[1], A. Minardo[1], L. Olivares[1], Lei Zhang[2], L. Zeni[1], and L. Picarelli[1]

[1] Department of Engineering,
University of Campania Luigi Vanvitelli, Aversa, Italy
reza.darban@unicampania.it
[2] Department of Geoengineering and Geoinformatics, Nanjing University,
Nanjing, People's Republic of China

Foreword. This paper reports some results of experiments carried out at the Università della Campania in order to investigate on the triggering mechanisms of flowslides in air-fall pyroclastic soils. The research includes flume tests on small-scale slopes aimed at capturing the main effects of rainfall on the soil response. Besides traditional devices, the soil is instrumented with optical fiber sensors (OFS) which allow measuring strain profiles all along the slope.

1 Introduction

The shallow unsaturated pyroclastic covers of hills and mountains which surround the Campania Plain are recurrently subjected to rainfall-induced landslides. The increase of the saturation degree induced by precipitations may lead to slope failure, which generally occurs abruptly, preceded by small deformations and often followed by rapid, destructive flow-like movements caused by static soil liquefaction [1, 2]. The goal of the research is to focus on the deformation process leading to slope failure. To this aim, distributed optical fiber technology allowing continuous monitoring of strain is being adopted. The investigation consists in reproducing the mechanism of slope failure through small-scale physical slope model subjected to artificial rainfall and in the monitoring of following movements by both traditional instruments (laser sensors and video-cameras) and optical fiber sensors (OFS) [3].

The soil adopted in this study is taken from the Cervinara slope which in 1999 experienced a catastrophic flowslide. The deposit is constituted by alternating layers of non-plastic volcanic ashes and of pumices which mantle a 35–45° sloping fractured limestone located at a depth of a few meters from the ground surface [1]. The volcanic ash is a silty sand with a porosity around 70%. Its saturated hydraulic conductivity is about $1 \cdot 10^{-6}$ m/s, but in unsaturated conditions it drops to values of more than two order of magnitude less. The friction angle is 38° and the effective cohesion nil. Suction plays a beneficial effect on the shear strength leading to an apparent cohesion of more than 10 kPa for suction values of 80 kPa [1].

Besides conventional sensors (Fig. 1), the flume is instrumented with optical fibres (OF) which are buried in the soil. The set-up is schematically sketched in Fig. 1c. The scheme implements a conventional Brillouin Optical Frequency Domain technique (BOFDA) [3]. The light from a laser is split into distinct branches for pump/probe generation. In the upper (probe) branch, the laser beam is firstly modulated by means of an intensity electro-optic modulator (IM1) driven by an RF synthesizer, then filtered by an FBG, and finally launched into one end of the OF. In the lower (pump) branch, the laser beam is modulated by means of another electro-optic modulator (IM2) driven by the RF output of a vector network analyzer (VNA), then it is launched into the opposite end of the same OF. The backscattered light caused by Brillouin scattering is fed into a high-bandwidth (12 GHz) photodetector. The extraction of deformation profiles has been done by applying to the acquired data the high-pass filtering method [4, 5].

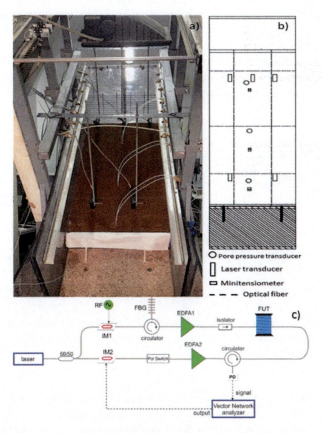

Fig. 1. (a) The flume; (b) a sketch of instrumentation; (c) optical fiber setup: IM, intensity modulator; FBG, fiber-Bragg grating; EDFA, erbium-doped fiber amplifier; Pol. Switch, polarization switch; PD, photodetector; FUT, fiber under test.

2 Experimental Results

Experimental results show that the soil saturation induced by artificial rainfall usually leads to some volumetric collapse due to the drop in suction which occurs at the arrival of the wetting front [6]. In the experiments, the collapse is evidenced by laser sensor measurements generally indicating an average volumetric strain of 3–5% based on records of the soil surface settlement. This phenomenon is also revealed by OFS as reported in Fig. 2a which illustrates the results provided by both sensors during the first infiltration stage of an experiment. As a result of influx, suction drops to values of a few kPa. Consequently, the soil deformation leads to formation of tension cracks. Figure 2b reports strain measurements along the two longitudinal strands of the fiber up to slope failure. The same experiment shows that a sudden increase of tensile strains along the fiber occurs about 10 min before slope failure as a result of a crack, normal to the slope direction, in the uppermost part of the slope. Measured strain corresponds to a fiber elongation of 5 mm, which is more or less the size of the crack opening a few minutes prior to general slope failure. Such results suggest that the OFS can detect both the general deformation process induced by infiltration and local effects due to cracking. The experiments show that loose ash at failure may experience static liquefaction, which causes an abrupt drop in the shear strength and consequent acceleration of the soil mass while the movement style turns into flow [2].

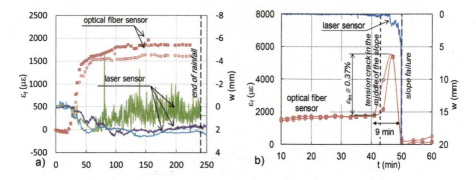

Fig. 2. Temporal evolution of strain (ε_f) measured along the two longitudinal sections of the buried fiber compared with settlements (w) at ground surface (a) during the first infiltrateon stage; (b) towards slope failure.

A further experiment has been done in order to focus on the strain field prior to failure, which may be conditioned by natural morphological features. To this aim the inclination of the uppermost part of the slope (40 cm in length) has been increased of about 4° leading to higher local shear stresses. Such an experiment lasted about 160 min, but a general slope failure cannot occur. However, thanks to the updated version of the OF interrogator described above, some interesting data have been obtained. The OF interrogator allows obtaining a high spatial and time resolution (respectively less than 5 cm and 17 s). The strain distribution along the two longitudinal sections of the OF are shown in Fig. 3. In detail, Fig. 3a shows a 2D map of the strain field at a depth of 5 cm in the slope 80 min after the beginning of rain; in turn, Fig. 3b reports the strain evolution along the longitudinal fiber strands. As a result of the adopted geometry, the slope experienced a highly variable deformation profile with a maximum extension strain occurring closely around the cross section which marks the changing slope angle. Downslope to this section the strain gradually decreases attaining negative (compression) strains in the toe zone. This suggests that the chosen geometrical slope profile favors a mechanism of progressive failure characterized by a highly not homogenous strain field characterized by compression of the slope toe. The strain field in Fig. 3a also indicates that the uppermost part of the slope is subjected to some stress release, which leads to formation of a tension crack at about 20 cm from the slope top. This process can be revealed by OF readings well before the appearance of any externals sign.

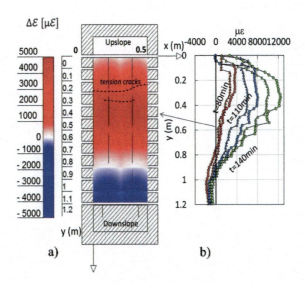

Fig. 3. (a) Strain field induced by raining; (b) longitudinal strain profile along the fiber strands.

3 Conclusions

The mechanism of slope deformation induced by rainfall in unsaturated sloping granular soils up to failure is being investigated through experiments on well instrumented small-scale model slopes. To this aim, optical fibers are being adopted. The effectiveness of these sensors in revealing pre-failure strains makes them a potentially powerful tool for continuous time and space strain readings in laboratory applications and, in perspective, for implementation in early warning systems.

References

1. Olivares, L., Picarelli, L.: Shallow flowslides triggered by intense rainfalls on natural slopes covered by loose unsaturated pyroclastic soils. Géotechnique **53**(2), 283–288 (2003)
2. Olivares, L., Damiano, E.: Post-failure mechanics of landslides: a laboratory investigation of flowslides in pyroclastic soils. J. Geotech. Geoenviron. Eng. **133**(1), 51–62 (2007)
3. Garus, D., Krebber, K., Schliep, F., Gogolla, T.: Distributed sensing technique based on Brillouin optical-fiber frequency-domain analysis. Opt. Lett. **21**, 1402–1404 (1996)
4. Zeni, L., Catalano, E., Coscetta, A., Minardo, A.: High-pass filtering for accurate reconstruction of the Brillouin frequency shift profile from Brillouin optical frequency domain analysis data. IEEE Sens. J. **18**(1), 185–192 (2018)
5. Farahani, M.A., Castillo-Guerra, E., Colpitts, B.G.: Accurate estimation of Brillouin frequency shift in Brillouin optical time domain analysis sensors using cross correlation. Opt. Lett. **36**, 4275–4277 (2011)
6. Damiano, E., Avolio, B., Minardo, A., Olivares, L., Picarelli, L., Zeni, L.: A laboratory study on the use of optical fibers for early detection of pre-failure slope movements in shallow granular soil deposits. Geotech. Test. J. **40**(4), 529–541 (2017)

Adaptive Strength Criterion of Sandy Gravel Material Based on Large-Scale True Triaxial Tests

Yuefeng Zhou^(✉), Jiajun Pan, Zhanlin Cheng, and Yongzhen Zuo

Key Laboratory of Geotechnical Mechanics and Engineering of Ministry of Water Resources, Changjiang River Scientific Research Institute, Wuhan 430010, China
zhou.yuefeng@163.com

Abstract. A series of large-scale true triaxial tests was performed on sandy gravel material from a hydropower station in southwest China. Adopting a special friction-reduction technique, the strength and deformation characteristics of the material were investigated at different stress levels and different intermediate principal stress ratios. The stress, deformation and strength were then analyzed under the framework of generalized deviatoric stress, mean effective stress and Lode angle. The tested strength parameters with the suggested friction-reduction technique were compared with several classic strength criterions, showing a good fitness with the Lade-Duncan criterion and an underestimation by the Mohr-Coulomb criterion. With the increase of coefficient of intermediate principal stress, the material shows an increase of the internal friction angle, but a reduction on the q-p stress ratio. The strength envelope in the π plane shrinks with the increase of stress level.

Keywords: Strength criterion · True triaxial test · Sandy gravel material · π plane

1 Introduction

There is a vast reservoir of clean non-polluting hydropower resources in China's vast, including the Yalongjiang River, the Daduhe River, the Lancangjiang River and the Jinshajiang River etc. As a widely used construction material for earth-rock dam projects, coarse-grained soil has two typical features below. Firstly, the mechanical behavior is complicated, including barotropy, dilatancy and rhethology etc. Secondly, the stress condition is complex, i.e. the unequal pressures in three principal stresses directions and different stress paths following construction and impoundment periods.

Currently, most laboratory investigations on the coarse-grained soil are based on conventional triaxial tests or plane strain tests, both of which are special simplification of complicated field stress conditions. In true triaxial tests, a specimen could be at a stress state of different principal stresses in three directions, which is valuable to investigate soil behavior and constitutive theory. With consideration of the influence of size effect of coarse-grained soil, it is still necessary to understand the mechanical properties under complicated stress conditions.

2 True Triaxial Tests

In this study, a series of large-scale true triaxial tests were performed on a sandy gravel material. The adopted large-scale true triaxial apparatus (Fig. 1) has a specimen size of 30 cm × 30 cm × 60 cm, which is the largest one within reviewed articles. A major advantage of the true triaxial apparatus is the adopted friction-reduction technique. The contact surface between lateral plate and specimen is composed of distributed slide blocks, which are underlined by sets of grooves with embedded balls. When the specimen deforms laterally, the friction force transfers from sliding form into rolling form, and physically reduces greatly.

Fig. 1. (a) The loading system, (b) the controlling system and the specimen of the low-friction large-scale true triaxial apparatus

The tested material was retrieved from a hydropower station in southwest China, which belongs to sandy gravelly material. Considering specimen size, the adopted maximum particle size for sample preparation is 6 cm. The strength and deformation characteristics of the material were investigated at different stress levels, i.e. 0.2 Mpa, 0.4 Mpa, 0.6 Mpa and 0.8 Mpa; and at different intermediate principal stress ratios, i.e. 0.0, 0.25, 0.50 and 0.75.

3 Test Results and Analyses

3.1 Adopted Stress Space and Strength Criteria

There are three independent stress parameters needed to describe the stress state at a certain point in stress space. For example, the major principal stresses components $(\sigma_1', \sigma_2', \sigma_3')$, the stress invariables (I_1, I_2, I_3), the mean stress and deviator stress $(q, p', b$ or $\theta)$ etc. Several commonly adopted strength criteria are Lade-Duncan criterion, the SMP criterion (i.e. the Matsuoka-Nakai criterion) and the Mohr-Coulomb criteria etc.

Lade-Duncan criterion:

$$I_1^3/I_3 = k_{f1} \qquad (1a)$$

where, k_{f1} is a material constant, which could be determined from conventional triaxial tests.

After stress space transformation, the equation can be deduced as below.

$$27p^3 - k_{f1}\left[p + \frac{(2-b)q}{3\sqrt{b^2 - b + 1}}\right]\left[p + \frac{(2b-1)q}{3\sqrt{b^2 - b + 1}}\right]\left[p + \frac{(1+b)q}{3\sqrt{b^2 - b + 1}}\right] = 0 \quad (1b)$$

Matsuoka-Nakai criterion:

$$I_1 I_2 / I_3 = k_{f2} \quad (2a)$$

where, k_{f2} is a material constant, which could be determined from conventional triaxial tests. The equation can be deduced as below.

$$9p^3 - pq^2 - k_{f2}\left[p + \frac{(2-b)q}{3\sqrt{b^2 - b + 1}}\right]\left[p + \frac{(2b-1)q}{3\sqrt{b^2 - b + 1}}\right]\left[p + \frac{(1+b)q}{3\sqrt{b^2 - b + 1}}\right] = 0 \quad (2b)$$

3.2 Test Results

Adopting Eqs. 1a, 1b, 2a and 2b the sagittal length and the Lode angle at different coefficient of intermediate principle stress can be calculated. Taking tests at initial consolidation pressure of 0.2Mpa as examples, the strength envelopes could be plotted on the π plane (Fig. 2). The failure stress states following different stress paths are shown together. In can be seen that the Lade-Duncan criterion is most fitted for the utilized coarse-grained soil. At different b values, the SMP criterion and the Mohr-Coulomb criterion underestimated the soil strength at different degrees.

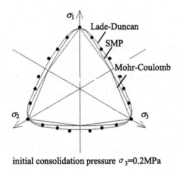
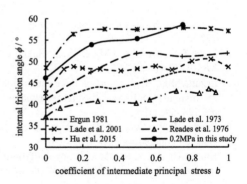

Fig. 2. π plane from the tested results (left)

Fig. 3. Comparison of the tested and others' results (right)

The tested internal friction angle φ is calculated using a nonlinear equation for coarse-grained soil proposed by Duncan.

$$\varphi = \varphi_0 - \Delta\varphi \log(\sigma_3 / p_a) \quad (3)$$

where, φ_0, $\Delta\varphi$, σ_3 and p_a are the initial internal friction angle, the incremental angle, the consolidation pressure and the atmospheric pressure.

Lade-Duncan, SMP and Mohr-Coulomb strength envelopes for higher stress level can be plotted as well. Similar to the reduction of internal friction angle with stress level in Eq. 3. The strength envelope in the π plane shrinks with the increase of stress level.

Utilizing Eq. 3, the tested internal frictions are calculated and further compared with other reported strength parameters for cohesionless soils in Fig. 3. Similar phenomenon can be found that the internal friction angle rises rapidly initially with b value, followed by an obvious slower increase when b > 0.25.

4 Conclusions

Three major conclusions can be drawn in this study. Firstly, the self-designed low-friction large-scale true triaxial apparatus with the specimen size of 30 cm × 30 cm × 60 cm functions well for coarse-grained soil. Secondly, for the tested soil with maximum particle size of 6 cm, Lade-Duncan criterion matches well with the tested internal friction angle. The SMP criterion and the Mohr-Coulomb criterion underestimated the soil strength at different degrees. Thirdly, the strength envelope in the π plane shrinks with the increase of stress level.

Acknowledgements. The authors acknowledge the financial support of the National Key Research and Development Program of China (project No. U1765203) and National Science Foundation of China (project No. 51509018).

References

1. Lade, P.V., Wang, Q.: Analysis of shear banding in true triaxial tests on sand. J. Eng. Mech. **127**(8), 762–768 (2001)
2. Lade, P.V., Duncan, J.M.: Cubical triaxial tests on cohesionless soil. J. Soil Mech. Found. **99**(10), 793–812 (1973)
3. Ergun, M.U.: Evaluation of three-dimensional shear testing. In: Proceedings of the 10th International Conference on Soil Mechanics and Foundation Engineering, Stockholm, Sweden, pp. 593–596 (1981)
4. Reades, D.W., Green, G.E.: Independent stress control and triaxial extension tests on sand. Géotechnique **26**(4), 551–576 (1976)
5. Hu, P., Huang, M.S., Ma, S.K., et al.: True triaxial tests and strength characteristics of silty sand. Rock Soil Mech. **32**(2), 465–470 (2011)

Part V: Geotechnical Monitoring, Instrumentation and Field Test

Validation of In-Situ Probes by Calibration Chamber Tests

Sergio Airoldi[1], Alberto Bretschneider[1], Vincenzo Fioravante[2], and Daniela Giretti[1(✉)]

[1] ISMGEO Srl, Seriate, BG, Italy
giretti@ismgeo.it
[2] University of Ferrara, Ferrara, Italy

Abstract. This paper describes the ISMGEO Large Calibration Chamber and its application for the validation of in-situ testing probes such as Cone Penetration, Nuclear Cone Penetration, geophones for seismic wave propagation. The calibration chamber is active since the 1980s, during this period thousands of tests have been carried out demonstrating the effectiveness of such experimental facility.

Keywords: Calibration chamber · CPT · Nuclear cone

1 Introduction

The development of new technologies for in-situ geotechnical testing requires a strong calibration based on the comparison with the existing methods. The measure of in-situ physical and mechanical parameters by classical geotechnical field methods can found strong benefit from laboratory tests. Large scale laboratory tests are useful to reproduce real-like stress conditions for soils. In this frame the use of calibration chambers is an outstanding method to reproduce site conditions.

The geotechnical calibration chamber is a test system capable of providing the environmental necessary to simulate full scale in situ tests in the laboratory. The dimensions of the soil specimen and the boundary conditions are such that experimental data can be readily interpreted and applied to in situ conditions. The purposes for which the calibration chamber apparatus was developed include the following:

(a) establishment of a wide range of correlations between test result and geotechnical parameters of the soil, to better understand the typically empirical in situ geotechnical tests; further, given the controlled conditions in the chamber, it is possible to check the correlations currently suggested for the interpretation of in situ test results;
(b) verification of the mechanical operation of in situ geotechnical instrumentation;
(c) development of new in situ geotechnical instrumentation;
(d) testing of model piles, tie rods, etc.

2 ISMGEO Geotechnical Calibration Chamber

The ISMGEO calibration chamber (Fig. 1) has been in operation since the 1980s [1] and since then it has been extensively used to test and calibrate geotechnical field probes. It houses a 1.42 m high by 1.2 m diameter specimen. The equipment consists of a flexible wall chamber, a loading frame, the apparatus for sand deposition and the saturation system.

Fig. 1. The ISMGEO geotechnical calibration chamber

Vertical and horizontal stresses can be independently applied in a controlled manner to the boundaries of the sample. Vertical stresses are applied to the specimen through a piston (positioned at the bottom of the chamber), the horizontal stresses are applied by the pressure of water surrounding the specimen. The cylindrical specimen is enclosed at the sides and base by a membrane; the top of the membrane is sealed around an aluminium plate which confines the specimen at the top and transfers the thrust of the chamber piston from the specimen to a top lid and loading frame to counteract the piston force and to hold the hydro-mechanical press which pushes the probe into the chamber.

To control both the stresses and the strains during the saturation and consolidation phase, the following parameters are measured by the data acquisition system: vertical stress, pore pressure within the sample, water pressure in the inner cell (i.e. horizontal stress on the sample), vertical displacement of the chamber piston, horizontal displacements of the specimen (measured by 7 LVDTs placed at seven different heights of the samples) and specimen volume change.

Soil samples are prepared by air pluviation and can be reconstituted at the desired relative density values. Internal and external sensors can be installed on soil specimen to measure soil and water pressure, waves propagation, specimen deformation.

3 Testing Examples

A classic approach is the use of calibration chamber testing for Cone Penetration Test data interpretation to evaluate the mechanical properties of sand deposits (relative density, shear strength). An example of Relative density of NC and OC siliceous sands is reported in Fig. 2, from [2].

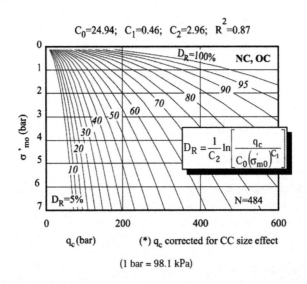

Fig. 2. Relative density of NC and OC siliceous sands from [2]

Another application of calibration chamber was the validation of nuclear CPT probes to measure soil density by gamma-ray. This method can be used to obtain a continuous measure of density with depth in penetrometer tests. Even if the method is known since some decades [3], it is not yet largely applied in common practice.

CPTs on carbonate sands have also been extensively studied. Carbonate sands have crushable grains and can be significantly more compressible than silica sands, so CPT-based correlations for silica sands are not applicable and a soil-specific calibration are needed for density, shear strength and compressibility parameters evaluation [4].

The propagation of seismic waves has been also studied on silica and carbonate sands. As an example seismic wave propagation tests have been performed on dry carbonate oolitic sand from Kenya [5]. Sand specimen were reconstructed at relative densities of 30% and 85%. Geophones were embedded during the specimen preparation and were used as source and receiver of seismic waves (Fig. 3).

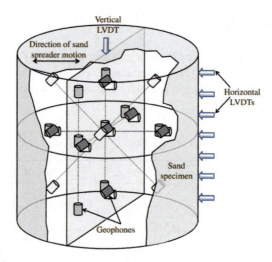

Fig. 3. Propagation of seismic waves, experimental setup.

Propagation of compression and shear waves were measured along the vertical, horizontal and inclined planes. The tests allowed to quantify and describe the effects of the fabric, of the stress induced anisotropy as well as the stress history on the velocity of propagated seismic waves and to calibrate semi-empirical correlations useful to analyze body wave velocity profiles measured in situ and to evaluate the state parameters and anisotropy properties of carbonate sand deposits.

References

1. Bellotti, R., Bizzi, G., Ghionna, V.N.: Design, construction and use of a calibration chamber. In: Proceeding of the Second International Seminar of Penetration Testing, Amsterdam, vol. 2, pp. 439–446 (1982)
2. Jamiolkowski, M., Lo Presti, D.C.F., Manassero, M.: Evaluation of relative density and shear strength of sands from CPT and DMT. In: Conference Paper in Geotechnical Special Publication, January 2003 (2003). https://doi.org/10.1061/40659(2003)7
3. Shibata, T., Mimura, M., Shrivastava, A.K.: Use of RI-CONE penetrometer in foundation engineering. In: XIII ICSMFE, New Delhi, India, pp. 147–150 (1994)
4. Giretti, D., Been, K., Fioravante, V., Dickenson, S.: CPT calibration and analysis for carbonate sand. Géotechnique (2017). http://dx.doi.org/10.1680/jgeot.16.P.312
5. Fioravante, V., Giretti, D., Jamiolkoswki, M.: Small strain stiffness of carbonate Kenya Sand. Eng. Geol. **161**, 65–80 (2013)

Cyclic Tests with a New Pressuremeter Apparatus

Soufyane Aissaoui[1(✉)], Abdeldjalil Zadjaoui[1], and Philippe Reiffsteck[2]

[1] Department of Civil Engineering, RisAM Laboratory, Tlemcen University, Tlemcen, Algeria
aissaouisoufyane@yahoo.fr
[2] IFSTTAR Paris, 14-20 Boulevard Newton, 77447 Marne la Vallée Cedex 2, France

Abstract. As part of the validation work for a new pressuremeter apparatus, cyclic expansion tests were carried out at the laboratory level on a sandy material of known behavior and characteristics. This communication presents the first results obtained and highlighted the relevance of the developed material.

Keywords: Cyclic · Lateral deformation · Pressuremeter

1 Introduction

In order to improve the estimation of soil deformability, it seemed advisable to deepen the path of in situ tests. Among these, the pressuremeter and in particular the self-boring Pressuremeter are devices capable of meeting the requirements of a weak or controlled reshuffle, simple use and good repeatability. Moreover, in comparison with all other in situ tests, the Pressuremeter test can be interpreted theoretically in an easy way. The load applied, the geometry of the apparatus and the conditions at the limits of the problem make relatively simple the theoretical problem to be solved [1].

The routine interpretation of the Pressuremeter Ménard test by following the standard operating procedure NF EN ISO 22476-4 [3], is to pull a module pressuremeter EM. This module is determined in the pseudo-elastic range of the curve produced by the test, according to [4]. This module cannot be applied without reflection to any problem, in particular, it is clear that when interested in the movements associated with small deformations of the soil, this module cannot be considered representative of the behavior of the ground.

The problem mentioned above to revive the conduct of pressuremeter tests with unloading loop-reloading according to the standard [5] allows to determine a cyclic modulus of deformation. The values obtained are intermediate between the small deformation modules and the traditional Ménard modules. They are often used to calculate the deformation of the structures by adopting a linear elastic behavior to model the soil [6]. However, a single cycle is insufficient to determine the evolution of soil characteristics under cyclic loading [7].

This contribution consists of cyclic tests carried out with a newly developed pressuremeter apparatus (a probe), has the peculiarity of being equipped with a feeler to ensure the deficiencies mentioned above.

2 Experimental Device

The principle of the test is to measure the deformation of the soil when applying pressure cycles. The idea being to be able to measure the radial deformation of the soil according to the cycles using a probe equipped with a Hall effect sensor instead of inferring the movements from the volume changes of a fluid (Fig. 1).

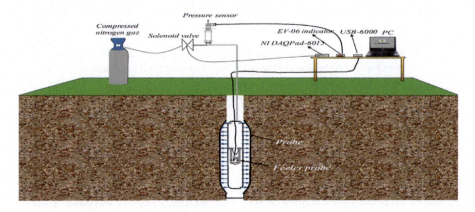

Fig. 1. Diagram of principle of a cyclic test with the probe developed.

3 Test Method

The principle of these tests is to cyclically request the Pressuremeter probe by a pressure that varies according to a sinusoidal signal given by the following formula [8]:

$$P_{cav} = P_0 \cdot (1 + R_c \cdot \sin(\omega t)); \; \omega = \frac{2\pi}{T} \quad R_c = \frac{\Delta P_{cyc}}{P_{cav,i}} \tag{1}$$

P_{cav} the pressure in the cavity; $P_0 = P_{cav,i}$ the initial horizontal pressure given by [2]; R_c the cyclical stress report; ω the heartbeat; T the period; t the time and ΔP_{cyc} the half-amplitude pressure. The variable that changes during the tests is the half-amplitude pressure ΔP_{cyc}. It is chosen according to the cyclic stress ratio. Cyclic loading can be alternated, or the stress diverter cycle between $\pm \Delta p_{cyc}$ or non-alternating, if the enslaved quantity does not change a sign during the solicitation as shown in Fig. 2.

4 Validation Tests and Results

Validation tests have been carried out in a laboratory in a sand tank, the tank used is filled with sandy soil of the Seine available at the IFSTTAR laboratory in France. All tests are carried out with the Jean Lutz controller, this device allows a numerical control with enslavement of all the parameters of the test. The piloting is carried out using a specific software (Prevo), developed specifically for the conduct of cyclic tests, the different loading cases tested are shown in Fig. 3.

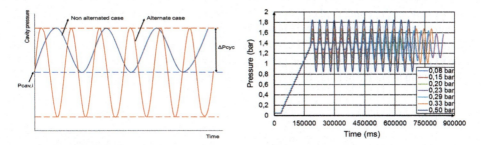

Fig. 2. Parameters of the cyclic stress. **Fig. 3.** Loading for multiple values of ΔP_{cyc}.

It can be observed that when it comes to the small loading cycles, for example in the case of a half amplitude equal to 0.08 bar, it is difficult to identify the loops in the evolution of the pressure as a function of the radial deformation. The latter represents almost no cycles as shown in Fig. 4. However, when it comes to relatively high cycles, as in the case of a half-amplitude equal to 0.50 bar, the loops are clearly recorded in the two graphical presentations (Fig. 5).

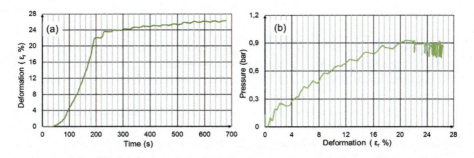

Fig. 4. (a) The deformation variation over time, (b) Pressure-strain curve for $\Delta P_{cyc} = 0{,}08$ bar.

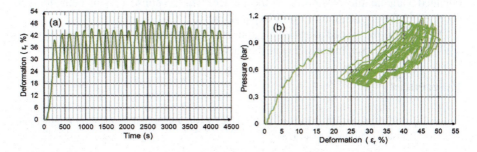

Fig. 5. (a) The deformation variation over time. (b) Pressure-strain curve for $\Delta P_{cyc} = 0{,}50$ bar.

The results of the rough and corrected pressures on the basis of deformation are shown in Fig. 6, we can tell from the sight of the tests carried out that the results found showed the proper functioning of the probe developed and allow to carry out multi-cyclic tests under controlled conditions. However, some difficulties still remain in addition to the sound of the recorded signals, this difficulty is directly related to the acquisition system. For the whole of the tests, a monotonous part which extends from zero to the theoretically uncorrected creep pressure, followed by a cyclic phase which manifests itself as much as the number of cycles around the chosen half-amplitude.

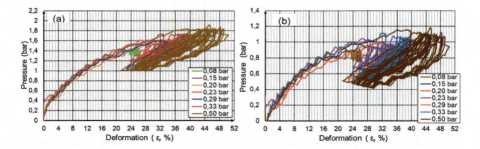

Fig. 6. Multi-cycle tests results. (a) Rough curves; (b) Curve corrected.

5 Conclusion

Multi-cycle tests have been carried out in good conditions in line with the engineer's practices in this field. This shows the relevance and performance of the probe developed. This confirmation is fully justified by the cyclic expansion tests of allure, trend and measurement interval.

This invention on the probe facilitates access to radial deformation instead of making calculations after measuring the volume variation. It follows a certain precision given the sensor used and by consequence of shear modules to small deformations.

In this communication we were able to validate, check the operation and the performance of the probe developed to carry out multi-cycle tests. The experimental curves obtained validate the feasibility of the device. The response of soil behavior was observed in the application of cyclic expansions.

This new probe is used to establish parameters that quantify the behavior of a cyclic-loading soil and economical sizing of the structures.

References

1. Baguelin, F., Jézéquel, J.F., Shields, D.H.: The Pressuremeter and Foundation Engineering, 618p. Transtech Publications (1978)
2. Briaud, J.L.: The Pressuremeter. A. A. Balkema, Rotterdam (1992)
3. AFNOR: Essai pressiométrique Ménard – Partie 1 – Essai sans cycle. Norme NF P94–110-1, Reconnaissance et essais, 43 (2005)

4. Combarieu, O., Canépa, Y.: Léssai cyclique au pressiomètre, no. 233, pp. 37–65. BLPC, France (2001)
5. AFNOR: Essai pressiométrique Ménard – partie 2 Essai avec cycle. NF P94-110-2, Reconnaissance et essais, 43 (1999)
6. Borel, S., Reiffsteck, Ph.: Caractérisation de la déformabilité des sols au moyen déssais en place. ERLPC, Série Géotechnique, 132p. (2006)
7. Dupla, J.C., Canou, J.: Cyclic pressuremeter loading and liquefaction properties of sands. Soils Found. **43**(2), 17–31 (2003)
8. Dupla, J.C.: Application de la sollicitation déxpansion de cavité cylindrique à l'évaluation des caractéristiques de liquéfaction d'un sable. Ph.D. thesis ENPC, France (1995)

A New in Situ Measurement Technique for Monitoring the Efficiency of Expansive Polyurethane Resin Injection Under Shallow Foundations on Clayey Soils

Hossein Assadollahi[1,3(✉)], L. K. Sharma[2], Anh Quan Dinh[3], and Blandine Tharaud[4]

[1] Navier Laboratory, IFSTTAR, CNRS, École des Ponts ParisTech, Champs-sur-Marne, 77454 Marne-la-Vallée, France
hossein.assadollahi@insa-strasbourg.fr
[2] Department of Earth Sciences, Indian Institute of Technology, Bombay, India
[3] DETERMINANT R&D SARL - Department of Engineering and Consulting, 75008 Paris, France
[4] DETERMINANT SARL – Solinjection, Department of Geotechnical Engineering and Investigation, 13590 Meyreuil, France

Abstract. The present study investigates the efficiency of the polyurethane resin injection in clayey soils formation. The polyurethane resin is an expansive type of material that densifies the soil and increases its bearing capacity. This technique is mainly adapted in repairing damaged buildings in presence of sensitive soils with shallow foundations. The uncertainty of the distribution of the resin into the soil does not make it easy to know when to stop the injection process to avoid any additional structural damage. A field measurement technique was adapted for a case of a cracked and damaged residential building due to the presence of shrink-swell clays under the foundation. A Geotechnical field investigation survey was carried out before and after the treatment process. Four optical fiber sensors were installed on different angles of the building in order to measure the local strain while the injection process was taking place. The in-situ soil dynamic resistance was also investigated before and after the treating process by the PANDA test. This monitoring technique allows to determine a limit for the resin injection and to prevent any additional damage to the structure.

Keywords: Polyurethane resin · Monitoring · Optical fiber sensors PANDA test

1 Introduction

The shrink-swell phenomenon has been studied by some researchers in the past years by taking into account the coupled and uncoupled behavior of clays interacting with the atmosphere or in contact with building constructions [1, 2]. Constructions with shallow foundations are much more sensitive to these conditions and in most cases, cannot properly transfer the upper loads to the ground resulting in structural damages. Cracks are common problems affecting the stability of buildings if they are not treated quickly. However, geotechnical repairing techniques should be adapted before treating structural elements. The method recently adapted and developed in order to repair damaged and cracked constructions, is the injection of expansive polyurethane resin into the supporting soil. Although this method is cost effective and is not time consuming but its efficiency stays a big question. Several experimental and numerical studies have been carried out on this material [3–7]. In situ investigation of its expansion into the soil has been verified by using 3D electrical tomography technique [8]. But none of these studies have verified the efficiency of the polyurethane resin in the soil and the upper structure characteristics simultaneously. Furthermore, there is no reasoning on the time in which the injection should be stopped. To avoid structural damage, in situ monitoring is proposed to be executed on the structure. The general scheme of the procedure is illustrated on Fig. 1(a). Therefore, this paper presents the results and developments achieved from in situ measurements and monitoring of a residential building, damaged and cracked by the shrinkage and swelling of clayey soils.

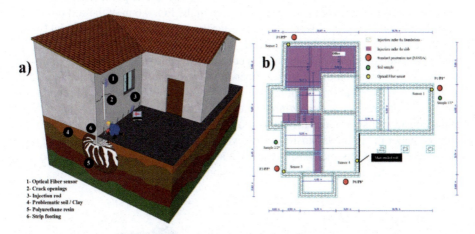

Fig. 1. (a) The schematic representation of the monitoring process and the foundation soil treatment of damaged buildings, (b) The general map of the studied building with the test points.

2 Methodology

In situ PANDA tests were carried out in order to determine the dynamic strength of the soil in depth at different points around the building allowing to have a general idea of the mechanical characteristics of the geological formation (Fig. 1(b)). The repairing solution of the building was to inject expansive polyurethane resin under the foundations and some parts of the floor slabs mainly on the north side of the building. In order to monitor the effects on the injection process in real time, optical fiber sensors were used. These sensors were used in dynamic mode with measurement time steps at each 20 ms with a resolution of 20 μm and they were installed on four different angles of the building (Fig. 1(b)). Each strain measurement was made when the injection was taking place at that specific point. One of the sensors (sensor 4) was installed on the largest crack on the building and the others were installed on the walls. Secondary PANDA tests were carried out after the injection phase in order to verify the improvements in the dynamic resistance in the soil profile near the monitored points.

3 Results and Discussion

Results of the in-situ monitoring by fiber optics along with the dynamic strength of the soil are presented in Fig. 2. It can be seen from Fig. 2(d) that the injection under the monitored wall with the sensor number 1, has caused a slight tension of the optical sensor after the injection of 25 kg of resin. It was the first test and it showed a stability of the structure during the injection. The wall was slightly compressed (0.02 mm) during the injection and when the injection was completed, it went back to its initial state (generating a tension of 0.01 mm which seems linked to the evolution of the temperature during the monitoring process). The results of the PANDA test at this point (P1/P1*) show a significant improvement of the dynamic strength in the first 0.6 m of the soil profile but a decrease from 0.6 to 2 m depth. The test after the injection has stopped quickly than the one carried out before the injection. It is generally concluded that the wall at this point was rising up stably and the soil has been improved in the first meters. Figure 2(c) shows a general tension of the optical fiber from the beginning of the injection. 35 kg of resin was injected at this point but the results of the optical fiber measurements did not show a compression at the beginning meaning that the resin was not being distributed at the considered spot under the foundations. It was then observed that the resin was diffused under the floor/slab instead. However, a compression phase and a rising up of the wall is captured by the sensor number 2 at the end of the monitoring period. The injection was stopped right after the compression phase was observed.

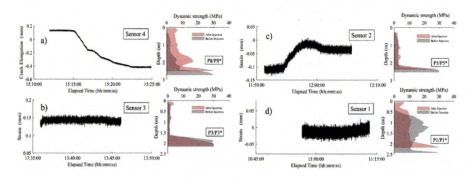

Fig. 2. Results obtained from the optical fiber sensors at 4 different position of the building along with the results of in-situ PANDA tests before and after the injection process.

The results of the PANDA test at P5/P5* did not show any changes in the soil profile as expected because the resin was not distributed in the considered spot. Figure 2(b) shows that the injection under the wall at this point has led to an overall rise in the structure resulting in a stable response of the optical fiber sensor. 25 kg were injected under this wall without causing any damage. Results of the PANDA test in the vicinity of this point (P3/P3*) show a slight improvement in the top layer of the profile and almost no change in the dynamic strength after the injection phase. It can be derived that the control PANDA test after the injection (P3*) was not able to capture the effects of the injection but the stability of the structure in rising up the wall was confirmed with the optical fiber sensor. Figure 2(a) shows the results of the crack elongation observed by the 4th optical fiber sensor installed on the largest external crack and particularly the most damaged part of the building. This unique curve shows that the resin injection under the foundation soil resulted in a closure of the crack in a range of 0.5 mm. It should be noted that the closing of the largest crack was observable with the naked eye during the injection phase. 27 kg of resin were injected at this point and the results from the PANDA test carried out before and after the injection phase (P8/P8*) show significant improvements in the whole soil profile by at least two times higher. The first stage of the strain response corresponds to the resin filling phase and a stable state. The second stage corresponds to the primarily sealing of the cracks in which the resin is taking effect. The third stage shows the secondary sealing of the cracks along with a stable rising up of the structure. The last stage shows the stabilized phase in which the crack is completely sealed and the structure does not move anymore, therefore the injection phase could be stopped. It can be concluded that the injection can continue until sensors are showing stable strain values and until a tension has not taken place (i.e. a downward movement). If sensors are installed on cracks, the injection should be stopped after a complete compression and a stabilization phase is shown on the sensor results. Complementary tests results (not reported here) would give more credit to this study.

References

1. Hemmati, S., Gatmiri, B., Cui, Y.J., Vincent, M.: Thermo-hydro-mechanical modelling of soil settlements induced by soil-vegetation-atmosphere interactions. Eng. Geol. **139–140**, 1–16 (2012)
2. Fernandes, M., Denis, A., Fabre, R., Lataste, J.F., Chrétien, M.: In situ study of the shrinkage of a clay soil cover over several cycles of drought-rewetting. Eng. Geol. **192**, 63–75 (2015)
3. Buzzi, O., Fityus, S., Sasaki, Y., Sloan, S.: Structure and properties of expanding polyurethane foam in the context of foundation remediation in expansive soil. Mech. Mater. **40**, 1012–1021 (2008)
4. Buzzi, O., Fityus, S., Sloan, S.: Use of expanding polyurethane resin to remediate expansive soil foundations. Can. Geotech. J. **47**(6), 623–634 (2010)
5. Valentino, R., Romeo, E., Stevanoni, D.: An experimental study on the mechanical behaviour of two polyurethane resins used for geotechnical applications. Mech. Mater. **71**, 101–113 (2014)
6. Dei Svaldi, A., Favaretti, M., Pasquetto, A., Vinco, G.: Analytical modelling of the soil improvement by injections of high expansion pressure resin. Bulletin für Angewandte Geologie **10**(2), 71–81 (2005)
7. Nowamooz, H.: Resin injection in clays with high plasticity. C. R. Mecanique **344**, 797–806 (2016)
8. Santarato, G., Ranieri, G., Occhi, M., Morelli, G., Fischanger, F., Gualerzi, D.: Three-dimensional electrical resistivity tomography to control the injection of expanding resins for the treatment and stabilization of foundation soils. Eng. Geol. **119**(1–2), 18–30 (2011)

Monitoring and Early Warning System in Earth Management - Based on Self-organized Fusion Sensor Networks

Rafig Azzam[1], Hui Hu[2], and Herbert Klapperich[3(✉)]

[1] RWTH Aachen & CiF e. V., Aachen, Germany
[2] Ruhrtec, Hangzhou, China
[3] TU Bergakademie Freiberg & CiF e. V., Freiberg, Germany
klapperich@cif-ev.de

Abstract. In this paper, an advanced self-organized monitoring sensor network in earth management is presented and an appropriate evaluation concept for the safety state of slopes based on data fusion and data mining methods to minimize false alarms and improve the criteria under which an early warning can be issued.

Keywords: Monitoring · Early warning system
Self-organized fusion sensor networks · Geohazards

1 Introduction

Geohazards, such as landslides induced by rainfalls floods, earthquakes and human activities are dramatically increased worldwide. Apart from socio-economic factors, like increasing population and concentrations of settlements on endangered areas, extreme weather conditions are the main reasons for this ascent. But these occurrences are not only concentrated on the high mountain ranges with steep slopes and strong relief. In February 2003, a landslide in the middle of Germany near the village of Wolfstein-Roßbach damaged some houses (one of them totally). Another example is the Manshiet Nasser failure in Cairo in September 2008, where a large rock tilt buried many houses. This few examples show the devastating effect of geohazards in settlement areas and the need for precise monitoring and early warning systems to protect human life and property.

Existing monitoring systems for early warning in areas prone to geohazards are often monolithic systems that are very cost-intensive, considering installation as well as operational expenses.

In this presentation, a new type of early warning and monitoring system in earth management using low-cost sensors from industry in a wireless Ad-hoc, Multi-Hop sensor network will be introduced. The self-organizing and self-administrating structure of such a system allows the setup of a very flexible sensor network, with independently working nodes due to their own energy supply. Micro sensors (MEMS) are used for tilt, acceleration, height and elongation measurements to observe surface deformation and joint's opening. Due to the fast data transfer, a special data infrastructure for processing and visualization process, monitoring and also warning in real time is possible. Sensor and data fusion are realized to minimize the false alarms and improve the criteria under which an early warning can be issued. Threshold values, as they have been widely used as criterion to issue an alarm, are not reliable in landslide monitoring, since deformation values do not indicate the safety state of a slope directly. Hence, a new advanced concept has been used in combination with this monitoring system.

2 Self-organized Sensor Network

In the frame of the special program "GEOTECHNOLOGIEN" of the German Federal Ministry of Education and Research (BMBF), the joint project "Sensor-based Landslide Early Warning System" (SLEWS) aims at the development of a prototypic Alarm and Early Warning system (EWS) for different types of landslides using wireless sensor networks (WSN) for real-time monitoring (Arnhardt et al. 2007). The WSN consists of a number of so called sensor nodes and a data collecting point, the gateway (Fig. 1).

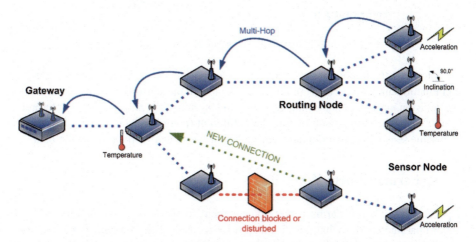

Fig. 1. Structure of a self-organizing (Ad-Hoc) wireless sensor network.

The solar powered gateway is connected either directly or via GSM/GRPS to the internet and subsequently to a data infrastructure to process the sensor data. Each node has a sensor board were the measuring sensors and the communication and processing unit are integrated. Special features of the Network are the real-time ability, self-organization and self-healing capacity, energy efficiency, bidirectional communication skills and data interfaces regarding OGC (Open Geospatial Consortium) specifications. The bidirectional structure of the system enables data transfer not only from each node to the spatial data infrastructure (SDI), but also to transmit commands or software-updates to individual or to a group of nodes (Bill et al. 2008). Special sensor nodes for the monitoring of surface deformations due to landslides, measuring acceleration, tilting or extension, were developed and tested (Fernandez-Steeger et al. 2008). Figure 2 shows the type of sensors, so called MEMS, that are used in the system and their properties to fulfill the requirement of the system.

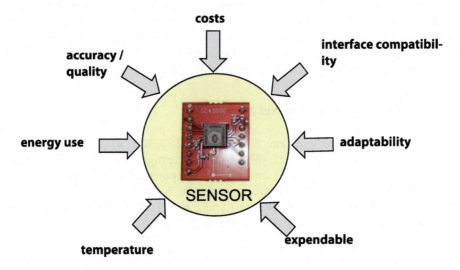

Fig. 2. Used in the wireless network and their requirements.

Another very important aspect of the project is the sensor fusion. The fusion (i.e. combination and comparison) of all sensor data from the sensor network contributes to the decision making of alarm and early warning systems and allows a better interpretation of data. The comparison of data from different (complementary sensor fusion) but also same sensor-types (redundant sensor fusion) permits a verification of the data.

3 Criteria of Early Warning

Up today criteria for alarms in geo-management focus on threshold evaluation of a certain parameter. But what does it mean if we measure a certain deformation, does it mean that the slope will fail, and when it would fail? How do we know that the measured deformation is dangerous? If we give always false alarms, would the people trust us after a while? Do we have other options to deal with this problem?

Yes, if we use sensor and data fusion, the detection of measurement errors and malfunctions will be more reliable. A warning system that shows many false alarms lose its acceptance by users, operators and public. Therefore, it is very important to detect errors before the whole warning chain is activated. Sensor fusion of many sensors of the same type, but also different sensors that can detect the same parameter, can be used to find errors and uncertainties and thus avoids the trigger of a false alarm. The extracted data can be processed in a special data infrastructure that allows a comprehensive stability analysis and further simulation to obtain an objective evaluation of the safety state. Depending on the expected failure mechanism the stability analysis can be modified and adapted to the real situation. At the end, the time dependency of the safety factor

Monitoring and Early Warning System in Earth Management 675

is subject to a time series analysis that allows the prediction of a time slot for possible failure. As this analysis is continuously repeated after every measurement that shows value changes, the time slot is constantly actualized. An alarm can be issued if the prediction shows that a failure could occur in the next 24 h (Figs. 3 and 4). To allow such prediction many geotechnical information including numerical modelling are needed and have to be integrated into the system prior to monitoring.

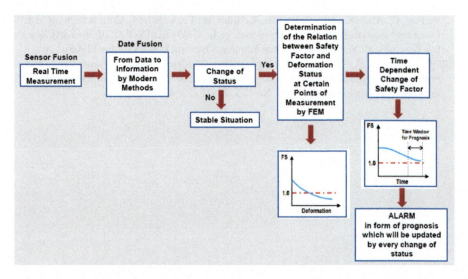

Fig. 3. Flow chart for the evaluation of the safety state of a slope that is under destabilization.

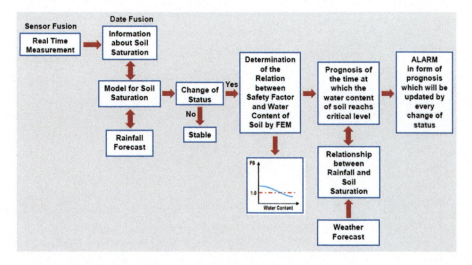

Fig. 4. Flow chart for the evaluation of the safety state of a slope that is destabilized by rain fall.

References

Arnhardt, C., Fernandez-Steeger, T.M., Walter, K., Kallash, A., Niemeyer, F., Aazzam, R., Bill, R.: Usage of wireless sensor networks in a service based spatial infrastructure for landslide monitoring and early warning. In: American Geophysical Union Fall Meeting 2007, 10–14 December 2007, San Francisco (USA), Abstract IN 11A – 0101 (2007)

Fernandez-Steeger, T.M., Arnhardt, C., Niemeyer, F., Haß, S., Walter K., Homfeld, S.D., Nakaten, B., Post, C., Asch, K., Azzam, R., Bill, R., Ritter, H., Toloczyki, M.: Current status of SLEWS - a sensor based landslide early warning system. In: GEOTECHNOLOGIEN – Early Warning Systems in Earth Management, Status-Seminar, Osnabrück, 8–9 October (2008)

Bill, R., Niemeyer, F., Walter, K.: Konzeption einer Geodaten- und Geodiensteinfrastruktur als Frühwarnsystem für Hangrutschungen unter Einbeziehung von Echtzeit-Sensorik. In: GIS - Zeitschrift für Geoinformatik 2008, Nr. 1, S. 26–35 (2008)

Monitoring the Foundation Soil of an Existing Levee Using Distributed Temperature Fiber Optic Sensors

Giulia Bossi[1(✉)], Luca Schenato[1], Alessandro Pasuto[1], Silvia Bersan[2], Fabio De Polo[3], Simonetta Cola[4], and Paolo Simonini[4]

[1] CNR-IRPI – National Research Council of Italy,
Research Institute for Geo-Hydrological Protection, Padua, Italy
giulia.bossi@irpi.cnr.it
[2] CRUX Engineering BV, Delft, Netherlands
[3] Hydraulic Department, Autonomous Province of Bolzano, Bolzano, Italy
[4] Department of Civil, Environmental and Architectural Engineering,
Università degli Studi di Padova, Padua, Italy

Abstract. Increasing reliability of levees is one of the main strategies to follow for flood risk reduction. Through monitoring it is possible to assess the response of levees to floods and to identify areas that may be subject to sand boiling. In this work a new monitoring system based on distributed temperature fiber optic sensors is presented. A hybrid cable that embeds optical fibers and electric wires has been installed at the toe of a 350 m long stretch of an operational levee in Northern Italy along with traditional sensors such as piezometers and spot temperature sensors. Through the interrogation of the fiber it is possible to acquire temperature measures every 1 m with an accuracy of 0.5 °C. Besides, by heating the cable through the electrical wires, the transient behavior in response to the active heating and consequent passive cooling along the fiber is assessed. Since heat is mostly transferred by advection, this procedure enables to estimate the presence of groundwater flow in different zones along the levee stretch. First results show a good correspondence between areas with larger presence of sandy levels and the fiber data. Further developments will arise when a significant flood event will provide an intense seepage driving force within the levee foundation to be detected by the sensors.

Keywords: Distributed fiber optic sensors · Levees monitoring
Internal erosion

1 Introduction

Artificial levees are long stretching structures that confine rivers and reduce the risk of floods in the surrounding areas. According to European Commission [1], risk is a function of the probability of occurrence and the impact on natural and anthropic elements. In the framework of climate change, the probability of occurrence of floods in the temperate regions is increasing, so efforts must be focused on reducing impacts, thus vulnerability. Considering that levees are already present in most of Europe, vulnerability can be reduced by increasing resilience and reliability of existing structures [2, 3]. Therefore, long-term levees health monitoring is crucial to define the functional status of the structure and its resilience capacity, and the development of innovative monitoring methods is important to ameliorate the efficiency of the risk mitigation strategy.

In this work, we present a distributed monitoring system in optical fiber that measures soil saturation and water flow in order to identify areas prone to internal erosion and eventually sand boiling.

2 The Test Area

The test area is located in North East Italy in the Autonomous Province of Bolzano (Italy) (46°14′11.0″N 11°10′52.8″E) along a 350 m-long stretch of an operational levee along the river Adige. The levee is 4 to 4.5 high with a 1.5 m high berm on the landside. The levee has been subject to sand boiling after the flood events of 1997. The sand boils have been georeferenced and are indicated in Fig. 1. To mitigate the internal erosion process after the flood the levee has been improved with a 10 m deep jet-grouting seepage cut-off diaphragm wall.

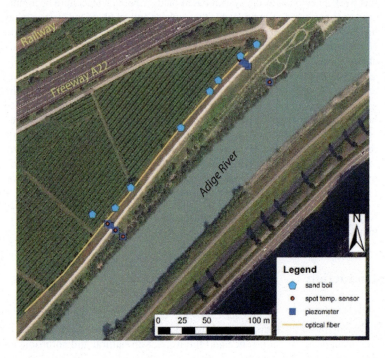

Fig. 1. Study area with location of the sand boils developed during the 1997 flood and the monitoring system.

The geotechnical investigation have indicated that the material forming the 4.5 m levee body is tout-venant placed above a 1.5 m level of silty sand (SM), followed by 2 to 3 m of silty gravel (GM) on top of clayey sand (SC).

3 The Monitoring System

Piezometers and temperature probes have been deployed in two cross-sections of the levee along with one kilometer of optical fiber cable installed in three levels in a 1.8 m deep trench excavated at the toe of the dike. During the excavation of the trench some small scattered levels of coarser material were found in the SM layer.

The distributed fiber optic monitoring system measures soil saturation and water transport using temperature as a proxy. To our knowledge, this is the first time in Italy that a distributed optical fiber sensor is installed at the toe of a river embankment (Fig. 2).

Fig. 2. Installation of the fiber cable in the trench at the embankment toe.

The 1 100 m-long cable was set up at three different depths in the trench and acts as a distributed temperature sensor employing Raman Backscattering Reflectometry with a spatial resolution of 2 m (distance among sampling points of 1 m) and an accuracy of 0.5 °C. In this installation, a hybrid cable that embeds optical fibers and four copper wires is used: the electric wires heat the cable enabling the measurement of transient behavior in response to the active heating and consequent passive cooling along the fiber (active thermometric or heat-up method). By measuring the heating and cooling time in different zones along the fiber, it is possible to detect the seepage water flow around the cable since areas with higher water transport will rise to lesser temperatures due to the heat-advection of groundwater flows [4].

Although passive measurements are preferred for structures as long as river embankments, the active method offers higher chances of success in case of insufficient natural temperature gradients along the seepage path. The installation of the system has been completed in June 2016. Initial data show that temperatures measured by the single-point probes are consistent with data obtained from the distributed fiber optic sensors (DFOS).

4 First Results and Future Developments

The monitoring system has been acquiring data in real time from the piezometers and spot temperature sensors since June 2016. Moreover, six measures have been gathered from the fiber optic monitoring system in different times of the year. Figure 3 shows two examples of the temperature field measured during the summer and winter season. As one can note, the temperature "fingerprint" is evident, and it can be observed that the upper layer is most sensitive to seasonal temperature fluctuation, with several degrees of variation from winter to summer. However, since the installation, no high discharge event has occurred in the Adige River: therefore, no measures are available with an intense seepage driving force.

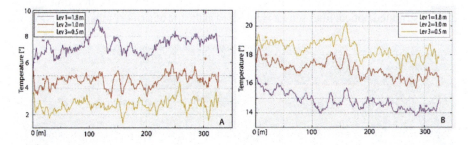

Fig. 3. Temperature measures along the three levels of the optical fiber: (A) winter time, (B) summer time.

Data collected so far show that, even in peace time, the temperatures measured by the fibers are influenced by the heterogeneity of the soil (more-coarse lenses within a finer matrix), that have been visually identified during the excavation of the trench. Thus, from our first evidence, they may be used as a proxy for identification of areas prone to backward erosion. Accordingly, the measurements performed in the Salorno area will be valuable to calibrate and assess the effectiveness of distributed temperature FOSs for the early detection of piping.

References

1. European Commission: Risk Assessment and Mapping Guidelines for Disaster Management, Bruxelles (2010)
2. Park, J., Seager, T.P., Rao, P.S.C., Convertino, M., Linkov, I.: Integrating risk and resilience approaches to catastrophe management in engineering systems. Risk Anal. **33**, 356–367 (2013)
3. Hui, R., Jachens, E., Lund, J.: Risk-based planning analysis for a single levee. Water Resour. Res. **52**, 2513–2528 (2016)
4. Schenato, L.: A review of distributed fibre optic sensors for geo-hydrological applications. Appl. Sci. **7**, 896 (2017)

Development of IoT Sensing Modulus for Surficial Slope Failures

Wen-Jong Chang[1(✉)], An-Bin Huang[2], Shih-Hsun Chou[1], and Jyh-Fang Chen[3]

[1] National Cheng Kung University, Tainan 70101, Taiwan
wjchang@mail.ncku.edu.tw
[2] National Chiao Tung University, Hsinchu 30010, Taiwan
[3] Harbor and Marine Technology Center, Taichung 43542, Taiwan

Abstract. To improve the limitations of rainfall-based slope warning system, a new system that integrates the hydro-mechanical slope analysis and wireless sensing module for surficial ground response monitoring is under development. The proposed system aims to establish a customized, time-dependent warning system for shallow slope failures triggered by rainfalls. A coupled hydro-mechanical analysis considering both the hydraulic infiltrations and mechanical responses of unsaturated soils in slope stability analysis is adopted. A real-time, wireless sensing module adopting the internet of things (IoT) technology is developed. The sensing module integrates the micro-electro-mechanical system (MEMS) sensing components with wireless communication modules. The module measures in-situ surface inclination and water content profile and uploads the data to cloud storage. The wireless sensing modules have been deployed in a potential landslide site for more than one year and current progress shows that feasibility of a customized, time-dependent warning system is promising.

Keywords: Internet of Things · Slope failure
Coupled hydro-mechanical analysis

1 Introduction

Slope failure triggered by rainfall is the most frequent geotechnical disasters in natural slopes. Based on previous case histories, empirical correlations among rainfall characteristics, geological conditions, and failure occurrence have been proposed to develop rainfall-based warning system. These techniques are widely adopted because of simplicity and easy cooperation with rainfall forecasting. Limited by available data and significant global climate changes, the reliability of rainfall-based warning techniques is marginally accepted and the threshold values require continuous update. To improve the reliability of rainfall-based warning, a time-dependent warning system that integrating rigorous analysis and real-time field monitoring is under development.

It is well recognized that field hydrological and geomechanical properties and conditions are the key elements controlling the stability of a slope under the influence of rainfall (e.g., [1, 2]). Rainfall induced landslides often occurred as relatively shallow failure surfaces orientated parallel to the slope surface in areas where a residual or colluvial soil profile has formed over a bedrock interface [3]. These characteristics allow the use of infinite slope assumptions to analyze the slope stability. For surficial slope failure, the reduction of shear resistance in unsaturated soil due to rainfall infiltration is the major triggering mechanism [4]. Coupled hydro-mechanical analysis has been proposed to predict the critical depth and failure time [1]. The rigorous analysis is integrated with sensing techniques based on internet of things (IoT) to develop an innovative monitoring and warning system for surficial slope failures.

2 Coupled Hydro-Mechanical Analysis of Surficial Slope Failure

Collins and Znidarcic [1] showed a relationship among the critical depth (d_{cr}) where the FS = 1, slope angle, soil strength parameters and pore-water pressure head in a shallow landslide as an infinite slope failure. Because the positive pore pressure head in fine grain soil is small before the advancing of the wetting front to the depth, the relationship can be approximated as,

$$d_{cr} = \frac{c' + \gamma_w \cdot h_c \cdot tan\emptyset^b}{\gamma \cdot cos^2\beta \cdot (tan\beta - tan\emptyset')} \tag{1}$$

where β = slope angle; c' = drained cohesion of saturated soil; γ = unit weight of soil; γ_w = unit weight of water; $\emptyset'$ = drained friction angle of saturated soil; h_c = suction (negative pore-water pressure) head; $\emptyset^b$ = friction angle with respect to matric suction (= $\gamma_w h_c$). The soil parameters can be obtained from laboratory tests (i.e., c', $\emptyset'$, $\emptyset^b$ and γ) and field measurement/monitoring (i.e., β, h_c). Equation (1) defines the stability envelop in unsaturated, infinite slopes subjected to rainfall infiltration.

The suction head profile can be predicted from Richard's equation, which describes the water flow in unsaturated soil [5]. For the vertical infiltration condition in an unsaturated, homogeneous layer, the Richard's equation can be simplified as,

$$M(h_c)\frac{\partial h_c}{\partial t} = \frac{\partial}{\partial t}\left[k(h_c)\frac{\partial h_c}{\partial z}\right] \tag{2}$$

where M = retention capacity of soil; z = depth; k = hydraulic conductivity of soil. Both M and k are functions of the suction head in unsaturated soil. Combing the stability envelope and suction head profiles, the time-dependent failure mechanism can be predicted as shown in Fig. 1, in which the initial groundwater table was assumed at 4 m below surface, γ = 20 kN/m³, c' = 3kPa, $\emptyset'$ = 30° and $\emptyset^b$ = 26°. Implementing the results of Fig. 1a and soil water characteristic curve (SWCC) that related the water content and matric suction, a time-dependent warning based water content variations at a specific depth is developed as shown in Fig. 1b.

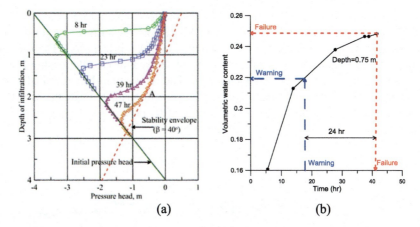

Fig. 1. Time-dependent warning scheme: (a) Infiltration results for fine grain soil with superimposed stability envelope [1]; (b) warning based on moisture content

3 IoT Sensing Module and Field Test

The new scheme involves the use of a low cost MEMS (Micro-Electro-Mechanical Systems) sensor stick capable of monitoring the field water content at multiple depths and tilt angle on the ground surface. The technologies for IoT were adopted in the development of the sensing module. An open source microcontroller unit (MCU) placed in the container box on top of the sensor stick was used to control the sensors, data logging and transmission by firmware. Multiple soil moisture sensors were mounted on the sensor stick, which is made of a 28 mm diameter PVC pipe. The tilt sensor, with a full range of $\pm 15°$, provided a redundancy and direct measurements of the slope angular movement as a final check of the slope stability. Multiple sensor sticks were connected by local wireless communication system and the data were uploaded to cloud server via the 3G system used for cell phone data transmission. The transmitted data were simultaneously recorded in the data logger as a backup.

The moisture sensor was based on capacitance measurements and the fact that soil dielectric constant (and thus capacitance) had a positive but non-linear relationship with the soil volumetric water content. The moisture sensors used in the sensor stick had an operating frequency of 16 MHz, which is less sensitive to soil conductivity, and was calibrated in the soil taken from the test site to minimize the effects of soil grain characteristics. All the IoT modules are powered by solar panel and rechargable Li-ion batteries to reduce the installation efforts.

A test site was setup in a natural slope at Jiaxian district in Kaohsiung city, Taiwan. The site had a slope angle of 33 to 42° covered by silty soil with a thickness of about 2.0 m and the current surface was a result of earlier landslides. Four sensor sticks were installed in the site since June 2016. Moisture sensors were embedded at depths of 25 and 75 cm from surface with sampling interval of 15 min.

Figure 2 shows the continuous moisture records of Node 1 for the first period of rain season in this area. The rainfall data were acquired from the weather station about 1 km away. The lower moisture sensor at depth of 75 cm showed high water content because of the field ground water table is only about 1.5 m from the surface and the recorded volumetric water content agreed with the void ratio of 0.3 of local soil. The variations of the upper moisture sensor at a depth of 25 cm showed strong relationship between the water content and rainfall characteristics. The inclination data revealed that no significant deformation had occurred and the field condition agreed with the data. Based on the soil parameters and field conditions, a customized, time-dependent warning system of the Jiaxian site can be developed.

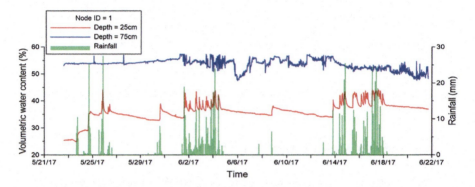

Fig. 2. Field moisture records

4   Concluding Remarks

An innovative warning scheme based on coupled hydro-mechanical analysis and real-time monitoring for prediction surficial slope failures triggered by rainfall infiltration is proposed. Sensor sticks adopting IoT technologies have been developed to cooperate with the warning scheme. The sensor stick allows volumetric water content at multiple depths and surface angle be monitored. The water content readings are used as the primary input to assess the stability of an unsaturated soil slope. Field test results revealed that the field monitoring system meets the design goals and the improvement of the warning framework is promising.

References

1. Collins, B., Znidarcic, D.: Stability analyses of rainfall induced landslides. J. Geotech. Geoenviron. Eng. **130**(4), 362–372 (2004)
2. Springman, S., Thielen, A., Kienzler, P., Friedel, S.: A long term study for the investigation of rainfall-induced landslides. Geotechnique **63**(14), 1177–1193 (2013)
3. Rahardjo, H., Lim, T., Chang, M., Fredlund, D.: Shear-strength characteristics of a residual soil. Can. Geotech. J. **32**(1), 60–77 (1995)
4. Cho, S., Lee, S.: Evaluation of surficial stability for homogeneous slopes considering rainfall characteristics. J. Geotech. Geoenviron. Eng. **128**(9), 756–763 (2002)
5. Freeze, R., Cherry, J.: Groundwater. Prentice-Hall, Englewood Cliffs (1979)

Evaluation of Free Swelling of Expansive Soil Using Four-Electrode Resistivity Cone

Ya Chu[1,2(✉)], Songyu Liu[1], Guojun Cai[1], and Hanliang Bian[3]

[1] Institute of Geotechnical Engineering, Southeast University, Si-pai-lou 2, Nanjing, China
136235507@qq.com, 101004004@seu.edu.cn
[2] Department of Civil, Architectural and Environmental Engineering,
Missouri University of Science and Technology, Rolla, USA
[3] Institute of Geotechnical and Rail Transport Engineering, Henan University, Kaifeng, China

Abstract. Current swell characterization models are limited by lack of standardized tests. Electrical resistivity measurements provide potentially powerful tool for detection of expansive soil, which is on the rise in the whole world. A four-electrode resistivity cone penetrometer (RCPT) test, which are non-destructive, continuous and demonstrates strong correlation with subsurface information of soil properties, was developed to estimate the free swelling of expansive soil layers. The mixtures of kaolinite and bentonite, which were loaded in a calibration chamber, were used to investigate the relationship between electrical resistivity and free swelling rate with varying physical property in this study. The four-electrode resistivity cone penetrometer (RCPT) test performed in calibration chamber shows that there is a decreasing tendency of resistivity with the increase of bentonite fraction by weight (BF). Comparing with the electrical resistivity and the free swelling rate of expansive soils, it is found that, under controlled moisture content, the resistivity of expansive soils decreased with the increase of free swelling rate. Furthermore, increasing moisture content of the mixtures clay led to a reduction in the electrical resistivity of clay, which indicates that altering the moisture content of clay particle changed the current movement within the clay's interlayer. It shows that the RCPT can be used effectively for estimating the free swelling potential of expansive soil.

Keywords: Resistivity cone penetrometer test · Resistivity · Free swelling Calibration chamber

1 Introduction

A four-electrode resistivity cone penetrometer (RCPT) test was developed to estimate the free swelling of soil layers in this study. The most influential property of expansive soil on volume change is its mineralogical properties (*i.e.* montmorillonite content, clay content and specific surface) (Zheng et al. 2008; Li et al. 2014), which electrical response is also sensitive to (Santamarina et al. 2001; Su and Momayez 2017). Laboratory tests are carried out on mixtures of bentonite and kaolinite. On the basis of results, the electrical resistivity was used to estimate the free swelling potential of expansive soil.

2 Materials and Experimental Methods

2.1 Calibration Chamber

The calibration chamber system used in this research was self-development by writers, which includes penetration system, measurement system and sample system, as shown in Fig. 1.

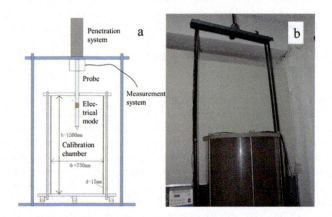

Fig. 1. Calibration chamber test system: (a) schematic of system; (b) picture of system

2.2 Materials and Methods

Expansive soils used in the test were prepared by mixing the commercial kaolin and bentonite. Index properties of KA and BE are listed in Table 1. Distilled water was used as the control samples.

Table 1. Index properties of kaolinite and bentonite

Parameter	Bentonite	Kaolinite
w_{opt} (%)	7.10	12.95
w_L (%)	281.2	65.8
w_p (%)	39.4	34.5
Clay content (%)	40.26	22.19
G_s	2.67	2.70

In order to research the swelling characteristics, the mixed Bentonite-Kaolinite soils were prepared with varying bentonite fractions, the bentonite was added to the kaolinite at a bentonite fraction by weight (BF = weight of bentonite/total weight) of 0%, 10%, 20%, 30%, 40%, 50%, 60% and 70%. When loading mixed soil samples, the design thickness of each layer is 330 mm and has a controlled density (1.5 t/m^3). In order to analyze the influence of moisture content and swelling characteristics on soil parameters, experimental design with different parameters was carried out.

3 Results and Discussion

3.1 Results at Controlled Moisture Content

Figure 2 shows that the resistivity data of mixed expansive soils with varying BF at controlled moisture content. It can be illustrated that increasing BF, which means the increasing of free swelling rate, cause a decrease in resistivity. Due to two important parameters (i.e. moisture content w and density), which were the main factors of resistivity, were controlled. The change of resistivity of mixed expansive soils was caused by the mineralogical properties (such as BF and specific surface) of soil sample. The colorful lines in Fig. 2 was the interface area of layers, therefore, the resistivity in these area shows an irregular fluctuation.

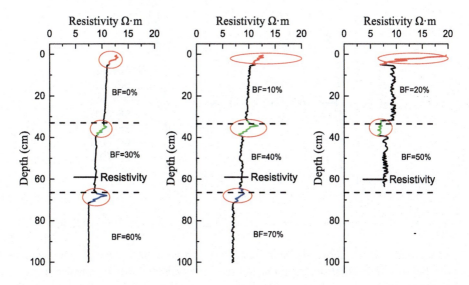

Fig. 2. Resistivity curve of mixed expansive soil at varying BF with depth ($w = 30\%$)

It was shown in Fig. 2 that the resistivity of mixed soil at BF = 0% range from 10–12 $\Omega \cdot$ m, whereas at BF = 70%, the resistivity decease to 6–7 $\Omega \cdot$ m. This phenomenon was caused by the hydrophilic of soil particle. With the addition of BF, the swelling potential and montmorillonite content increased, therefore, the soil particle will adsorb more water, which caused the increase of current ability (i.e. decrease of resistivity).

3.2 Comparison of Resistivity at Controlled BF or Moisture Content

Figure 3 shows comparison of resistivity at controlled BF or moisture content. It can be illustrated in Fig. 3(a) that resistivity decreased with the increase of moisture content at a given BF (BF = 40%). With the addition of moisture content, the current pathway increased, therefore, the resistivity decreased. It should be noted in Fig. 3(b) that expansive soil with different BF (varying swelling potential) shows different current potential. There are two important factors (BF and moisture content w) affecting the resistivity of expansive soil. For moisture content, increasing w cause the increase of current path, which means the water (conductive medium) in porosity increased, therefore, the resistivity decrease. For BF (swelling potential), increasing BF cause the increase of special surface of soil, which means more water molecular were absorbed around soil particle, therefore, the current potential of soil particle increased.

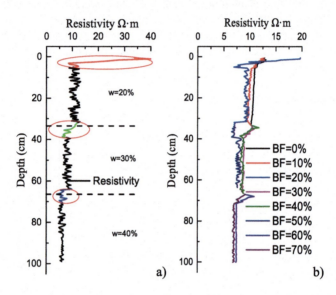

Fig. 3. Resistivity curve of mixed expansive soil at controlled station: (a) BF = 40%, (b) w = 30%

According to the above research, resistivity can be used to estimate the swelling potential of soil, in addition, the impact of moisture content must be considered.

References

Zheng, J.L., Zhang, R., Yang, H.P.: Validation of a swelling potential index for expansive soils. In: Unsaturated Soils. Advances in Geo-Engineering: Proceedings of the 1st European Conference, p. 397. CRC Press, Durham (2008)

Li, J., Cameron, D.A., Ren, G.: Case study and back analysis of a residential building damaged by expansive soils. Comput. Geotech. **56**, 89–99 (2014)

Santamarina, J.C., Klein, A., Fam, M.A.: Soils and waves: particulate materials behavior, characterization and process monitoring. J. Soils Sediments **1**(2), 130 (2001)

Su, O., Momayez, M.: Indirect estimation of electrical resistivity by abrasion and physico-mechanical properties of rocks. J. Appl. Geophys. **143**, 23–30 (2017)

Evaluation of the Variation of Deformation Parameters Before and After Pile Driving Using SCPTU Data

Wei Duan, Guojun Cai[(✉)], Jun Yuan, and Songyu Liu

Institute of Geotechnical Engineering, Southeast University, Nanjing, China
focuscai@163.com

Abstract. The applicability of seismic piezocone test technology (SCPTU) as a standard in situ test method for geotechnical investigation in offering valuable information on soil types and soil behavior for the purpose of site characterizations, foundation designs, and hazard analyses. Compared with some traditional investigation methods such as drilling, sampling and field inspecting method or laboratory test procedures, SCPTU can save time and cost. In a view to evaluate soil properties, SCPTU tests, undisturbed sampling from an adjacent borehole, and laboratory tests were performed at Yizheng sites in Jiangsu province of China. The determination of soil profiles and the variations of main geotechnical properties before and after pile driving are facilitated based on SCPTU test that can offer four independent sounding readings. The results indicate that the feasibility of SCPTU tests to interpret and evaluate the engineering properties of pile foundation.

Keywords: Seismic piezocone (SCPTU) · Pile driving
Undrained shear strength · Deformation modulus

1 Introduction

The structure built on soft cohesive soils are often faced with many geotechnical issues. The high compressibility of the clays causes significant settlements. With the requirement of reducing construction time, deformation is a primary task. Therefore, different soil improvement methods are used to accelerate and decrease the settlement. The drilling sampling and laboratory tests will cause soil disturbance and the testing results have difficult in charactering the true situation of the field. However, in-situ testing is a convenient, accurate and effective method to measure the performance of geotechnical engineering and obtain true geotechnical parameters. cone penetration test (CPT) is the main one of the in-situ test techniques [1, 2]. Of growing interest is the utilization of piezocone penetration test (CPTU) data, since continuous stratigraphic profiling and multiple readings are obtained in a single sounding [3]. The seismic piezocone (SCPTU) is a new in situ technology, which provides single point continuous, or multipoint simultaneous measurement of tip resistance (q_t), sleeve friction (f_s), pore pressure (u_2), and shear wave velocity (V_s) appears to be an extremely promising tool to interpolate or extrapolate soil property parameters for preliminary design [4, 5]. In addition, SCPTU has the similar mechanism to jacked-in pile [6].

This study focuses on characterizing the soil profile and estimate the main geotechnical properties before and after pile driving based on the SCPTU data. A series of SCPTU tests have been conducted and analyzed The final goal of this study is to provide an in-situ test calibration for an extensive characterization of the ground at the Yizheng city.

2 Field Investigation

The test site situates in the city of Yizheng in Jiangsu province, China. As shown in the test layout of Fig. 1, a set of CPTU tests were conducted before and after pile installation to assess the performance of pile installation. It is necessary that the CPTU soundings should be conducted near one another from before and after pile installation. It is therefore that the comparisons are made sense. According to the local design experience, water depth at the test locations varied in the range 1.0–2.5 m. The soil stratum consists of surface fill, silty clay, silty clay, and gravel sand.

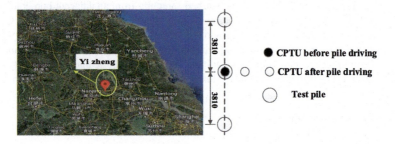

Fig. 1. Test layout of in-situ and pile load tests

Figure 2 depicts typical CPTU profiles obtained at neighboring area for the two CPTU data sets. It was evident from the Fig. 2 that q_t values gradually increase and the f_s and u_2 values shows similar variation to the q_t profile. The soil state changed denser after pile driving. The result indicates that the pile driving has a significant effect.

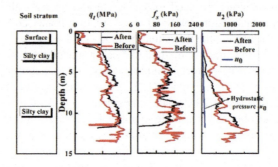

Fig. 2. Typical CPTU profiles before and after pile driving at the Yizheng site

3 Interpretation of SCPTU Results

3.1 Undrained Shear Strength

Figure 3 presents the variation of s_u values before and after pile driving. It can be seen from the Fig. 3 that the s_u values have an increase in comparison with before pile driving, indicating the increase of strength of soils.

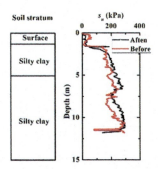

Fig. 3. The variation of s_u values before and after pile driving

3.2 Deformation Modulus

The parameter of deformation moduli is basic parameter to calculate settlements and deformation characteristics of fine-grained soils. The one-dimensional constrained modulus, M, and small strain shear modulus, G_{max} are the widely used in practice for fine-grained soils. Previous works have illustrated that the magnitude of soil moduli is affected by the stress history, stress and strain level, drainage condition, and so on [2]. Although there are many complex influence factors, the deformation characteristics of clay with respect to constrained modulus, small-strain shear modulus can be interpreted from the SCPTU tests.

Figure 4 gives the variation of M values before and after pile driving. It was evident from the Fig. 4 that the M values tend to increase by the pile treatment. The results demonstrate that the bearing capacity of foundation soil has increased.

Figure 5 shows the variation of G_0 values before and after pile driving. It can be observed that the G_{max} values after pile driving are larger than before, and there are small differences at some depth, such as 5 m, 7 m, 8 m. In all, the amplitude of variation of G_0 is less than M. This means that the variation of small strain shear modulus is less sensitive than the change of compression modulus.

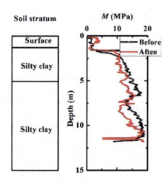

Fig. 4. The variation of M values before and after pile driving

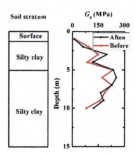

Fig. 5. The variation of G_0 values before and after pile driving

4 Conclusions

The soil profiles and the variations of main geotechnical properties before and after pile have been discussed in the paper. The results illustrate the great potential of seismic piezocone as an in-situ testing technique to characterize the Yizheng site soil properties and give soil stratigraphic profiling. The soils are main silty clays, and the undrained shear strength and deformation modulus have increased. The variation of small strain shear modulus is less sensitive than the compression modulus in deformation modulus. The bearing capacity of foundation soil has increased with the pile treatment. Additionally, it also illustrate the feasibility of using the SCPTU tests to characterize the performance of pile foundation.

References

1. Lunne, T., Robertson, P.K., Powell, J.J.M.: Cone Penetration Testing in Geotechnical Practice. Blackie Academic and Professional, London (1997)
2. Robertson, P.K.: Interpretation of cone penetration tests - a unified approach. Can. Geotech. J. **46**(46), 1337–1355 (2009)

3. Mantaras, F.M., Mantaras, F.M., Odebrecht, E., Odebrecht, E., Schnaid, F., Schnaid, F.: Using piezocone dissipation test to estimate the undrained shear strength in cohesive soil. Can. Geotech. J. **52**(3), 1–8 (2015)
4. Cai, G., Liu, S., Tong, L.: Field evaluation of deformation characteristics of a lacustrine clay deposit using seismic piezocone tests. Eng. Geol. **116**(3), 251–260 (2010)
5. Cai, G., Liu, S., Puppala, A.J.: Liquefaction assessments using seismic piezocone penetration (SCPTU) test investigations in Tangshan region in China. Soil Dyn. Earthq. Eng. **41**, 141–150 (2012)
6. Niazi, F.S., Mayne, P.W.: CPTU-based enhanced unicone method for pile capacity. Eng. Geol. **212**, 21–34 (2016)

Development of an FBG-Sensed Miniature Pressure Transducer and Its Applications to Geotechnical Centrifuge Modelling

An-Bin Huang[1(✉)], Kuen-Wei Wu[2], Mohammed Z. E. B. Elshafie[3], Wen-Yi Hung[4], and Yen-Te Ho[5]

[1] National Chiao Tung University, Hsinchu 30010, Taiwan
huanganbin283@gmail.com
[2] Oxford University, Oxford, UK
[3] Cambridge University, Cambridge, UK
[4] National Central University, Taoyuan City, Taiwan
[5] Citpo Technologies, Taipei 11167, Taiwan

Abstract. Because of the elevated gravity and dimensional constraints, miniaturized sensors are required to measure pore pressure within soil specimens in geotechnical centrifuge testing. Electrical sensors have been used for the purpose. The small dimensions and their mechanical configurations made the available electrical pore pressure sensors fragile. They can be easily damaged in specimen preparation and testing. The electrical sensors tend to have a nonlinear and non-repeatable behavior in negative pressure range. Taking advantage of the optical fiber Bragg grating (FBG), the authors developed an FBG based miniature pore pressure sensor. The FBG sensors are immune to moisture and electromagnetic interference. The readings are linear in both positive and negative pressure range. A specially made FBG readout unit capable of taking readings at 10 kHz was used to monitor the pore pressure within the soil specimen in dynamic centrifuge tests. The paper describes and design of the miniature FBG pore pressure transducer system and its applications in static and dynamic physical model tests at the Cambridge University and National Central University geotechnical centrifugal testing facilities.

Keywords: Geotechnical centrifuge test · Pore pressure sensing · FBG

1 Introduction

Pore pressure measurement is an important part of centrifuge modeling. Due to scaling effects, the pore pressure transducer (PPT) has to be small. Druck PDCR81, a MEMS based electrical PPT has been widely used for centrifuge modeling [1]. The sensor is prone to electromagnetic interference often induced by the centrifuge actuator [2]. Muraleetharan and Granger [3] indicated that the Druck's PDCR81 PPT after a centrifuge test that involves negative pressure range, the calibration factor may change and loose its linearity. When a clay specimen is used in centrifuge modeling, negative pore pressure can be induced by the removal of the preconsolidation pressure.

The optical fiber Bragg grating (FBG) is immune to electromagnetic interference, lighting and moisture. Multiple sensors can be connected to a given fiber. Due to these unique capabilities, the use of FBG sensors for monitoring civil infrastructures and laboratory experiments is gaining popularity. The FBGs have been used as strain sensors in centrifuge modelling [2, 4, 5]. FBG based pressure sensors have been successfully used for measuring pore pressure in the field for ground monitoring and laboratory soil element tests [6, 7]. The currently available FBG pressure sensors are however, too big to be deployed in centrifuge modeling.

An FBG sensed miniature PPT was developed. The new PPT was experimented in a clay specimen for static pore pressure monitoring at the centrifuge facility in Cambridge University, UK. Pore pressure readings taken at 10 kHz were made in a saturated sand specimen under cyclic loading conditions in centrifuge testing at National Central University, Taiwan. In both cases, concurrent readings were also taken using a Druck's PDCR81 PPT. This paper describes the basic principles of FBG sensing, design of the FBG based PPT and the available test results from the centrifuge tests at Cambridge University and National Central University.

2 The FBG Sensed PPT

Optical fibers are made of silica, with a diameter about the size of a human hair, and can transmit light over long distances with very little loss of fidelity. The outside diameter of a typical optical fiber with the acrylic coating is 250 μm. An FBG is made by a periodic variation of fiber core refractive index. The typical length of an FBG is 1 to 20 mm long. When the FBG is illuminated by a wideband light source, a fraction of the light is reflected back upon interference by the FBG. The wavelength of the reflected light, is linearly related to the longitudinal strains of the FBG, thus making FBG an ideal strain gage. The returned signal from every FBG carries a unique domain of wavelength, making it possible to have multiple FBG elements on the same fiber. The multiplexing among various sensors on a single fiber can be accomplished by wavelength division addressing. Figure 1 shows a schematic view and photograph of the miniature FBG PPT. The FBG was used to sense the deflection of a metallic diaphragm inside of the PPT due to pressure variations. A separate FBG was placed behind the transducer to monitor temperature fluctuations. Readers are referred to [7] for the method to nullify temperature effects using the FBG temperature sensor. A typical FBG readout unit is capable to detect FBG wavelength shifting by 1 pm (10^{-12} m). An FBG breaks when stretched by a strain equivalent to approximately 8000 to 10000 pm in wavelength variation. Depending on the required safety margin, the maximum allowable pressure was designed to correspond to 1000 to 6000 pm of FBG wavelength variation.

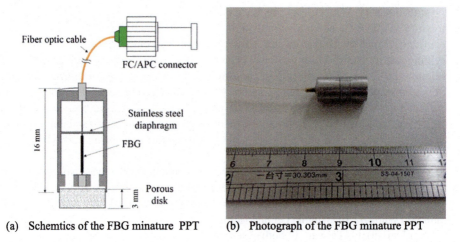

(a) Schemtics of the FBG minature PPT   (b) Photograph of the FBG minature PPT

Fig. 1. Schematic view and photograph of the miniature FBG PPT.

3   Comparisons with the Druck PDCR81 PPT

Three pairs of FBG and Druck PPTs were placed in a centrifuge package for comparison of pore pressure measurements in a saturated Speswhite Kaoline clay specimen. Readings were taken as the g-level increased from 1 g to 10, 20, 40, 60, 80 and 100 g, as the clay specimen underwent consolidation. The FBG PPT readings were taken at 1 Hz. Figure 2 shows the pore pressure readings from FBG and Druck PPT, both located at depth of 235 mm (SCALE).

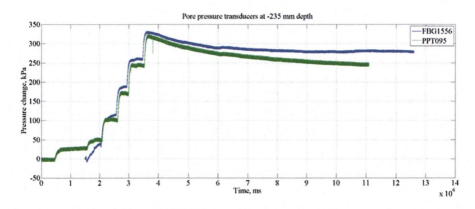

Fig. 2. Comparison of FBG and Druck PPT readings in a clay specimen.

Comparisons were also made in sand under seismic loading conditions in a centrifuge model. Three pairs of FBG and Druck PPT sensors along with accelerometers were placed in a saturated, uniformly graded sand ($D_{50} = 0.193$) specimen. The completed model was placed on a shaker and the centrifuge was spun to 50 g in 10 g increments.

A seismic motion with an amplitude of 0.22 g was then applied to the base of the model for 16 s. Figure 3 shows the acceleration and pore pressure history recorded during and following the seismic event at a depth of 125 mm (model scale). Figure 3b depicts the fluctuation and increase of excess pore pressure during the period of excitation. Figure 3c demonstrates the dissipation of excess pore pressure after the seismic motion stopped. The FBG PPT readings were taken at a frequency of 10 kHz.

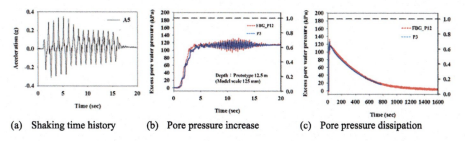

Fig. 3. Acceleration and pore pressure readings.

4 Concluding Remarks

It is believed that this was the first time the FBG PPTs were used in centrifuge testing. Early results showed that FBG PPT can be a promising tool to measure pore pressure in high g conditions while having the required small dimensions. Consistent and comparable pore pressure readings were obtained between the FBG and Druck PPTs in both the clay and sand specimens. The FBG sensors are likely to be more stable and durable. To fulfill its full function in dynamic centrifuge modeling however, the FBG PPT readings taken at a frequency of over 100 kHz would be more desirable. This remains to be a challenge.

References

1. Take, W.A., Bolton, M.D.: A new device for the measurement of negative pore water pressures in centrifuge models. Presented at International Conference on Physical Modelling in Geotechnics, ICPGM 2002, St. John's, Newfoundland, Canada, pp. 89–94 (2002)
2. Weng, X., Zhao, Y., Lou, Y., Zhan, J.: Application of fiber bragg grating strain sensors to a centrifuge model of a jacked pile in collapsible loess. ASTM Geotech. Test. J. **39**(3), 362–370 (2016)
3. Kapogianni, E., Sakellariou, M., Laue, J.: Experimental investigation of reinforced soil slopes in a geotechnical centrifuge, with the use of optical fibre sensors. Geotech. Geol. Eng. **35**(2), 585–605 (2016)
4. Zhang, D., Xu, Q., Bezuijen, A., Zheng, G., Wang, H.-X.: Internal deformation monitoring for centrifuge slope model with embedded FBG arrays. Landslides (2016). https://doi.org/10.1007/s10706-016-0127-2
5. Muraleetharan, K.K., Granger, K.K.: The use of miniature pore pressure transducers in measuring matric suction in unsaturated soils. ASTM Geotech. Test. J. **22**(3), 226–234 (1999)

6. Huang, A.-B., Lee, J.-T., Ho, Y.-T., Chiu, Y.-F., Cheng, S.-Y.: Stability monitoring of rainfall-induced deep landslides through pore pressure profile measurements. Soils Found. **52**(4), 737–747 (2012)
7. Lee, J.T., Tien, K.C., Ho, Y.T., Huang, A.B.: A fiber optic sensored triaxial testing device. ASTM Geotech. Test. J. **34**(2), 103–111 (2011)

Evaluation of Engineering Characteristics of South China Sea Sedimentary Soil in Sanya New Airport Based on CPTU Data

Kuikui Li[1], Wencheng Liang[2], Guojun Cai[1(✉)],
Songyu Liu[1], Yu Du[2], and Liuwen Zhu[2]

[1] Southeast University, Nanjing 210096, China
focuscai@163.com
[2] Institute of CCCC, FHDI Engineering Co., Ltd., Guangzhou 510275, China

Abstract. The penetrating of the piezocone penetration testing (CPTU) is the process of soil strength destruction. The measured data of CPTU can directly reflect the strength characteristics of soil. This paper describes the marine CPTU and its test methods and equipment. Aiming at the artificial island revetment project of Sanya new Airport, sleeve friction, cone tip resistance and pore water pressure are obtained through the CPTU test at sea, and the soil layer in the area is divided based on these parameters. The data of CPTU are used to interpret the natural density, undrained shear strength and sensitivity of clayey soils, as well as the compaction degree and internal friction angle of sandy soils. An analysis of the consistency is given through comparing the data of marine CPTU test with the results of drilling, laboratory tests, standard penetration test (SPT). It provides reference for engineering practice in the coastal areas.

Keywords: Marine soft soil · Engineering characteristics · Marine CPTU

1 Introduction

The penetrating of the piezocone penetration testing (CPTU) is a modern in situ testing method. The CPTU test has the characteristics of reducing disturbance, accurate and reliable, as well as high efficiency [1]. The penetrating of CPTU is the process of soil strength destruction. The probe is pressed into the soil by mechanical uniform pressure at a certain level. Then, cone tip resistance q_c, side friction f_s and pore water pressure u_2 are obtained. Due to the measurement of pore water pressure, the cone tip resistance is corrected, which More truly reflect the characteristics of soil strength [2].

2 Sanya New Airport Project Overview

Sanya New Airport project is in Hongtangwan, which is the west of Sanya City. The seabed elevation of the site is about −15−−25 m, mainly silt, cohesive, sandy and mixed soils.

In the CPTU test, a plurality of test holes is set in the land formation area and the revetment area of the reclamation project respectively, and each hole corresponds to at least one drilling hole as comparisons.

3 Test Results and Analysis

3.1 Soil Layering Stratigraphy

Robertson (2009) points out that mechanics classification is more attractive in geotechnical practice than indoors based on plasticity classification, because the mechanical properties of soils in engineering are more important than their physical properties. The method of soil layering adopts the method of "In-situ Test Regulations for Railway Engineering Geology" (TB10018-2003).

Combining with the drilling data, the stratigraphic classification is conducted based on the above static penetrating stratification method, and the specific description of each strata is as follows (Table 1):

Table 1. Soil classification and description

Layer	q_c (MPa)	f_s (KPa)	R_f (%)	U_2 (KPa)	q_t (MPa)	Thickness (m)
②1 Silt	0.20	9.25	4.65	45.73	0.22	0.30–2.34
②2 Muddy soil	0.68	27.60	4.12	103.72	0.74	0.62–0.82
②3 Fine sand	1.07	27.45	3.01	84.38	1.12	0.44–1.16
③1 Clay ~ silty clay	1.49	54.39	3.86	185.47	1.56	0.48–5.06
③2 Clay ~ silty clay	3.05	103.58	3.54	395.21	3.19	0.50–6.50
③3 Clay ~ silty clay	4.27	147.25	3.63	423.02	4.40	0.30–4.98
③4 Fine sand	2.20	36.02	1.76	80.06	2.22	0.36–3.00
③5 Fine sand	8.66	94.72	1.19	−16.07	8.65	0.20–4.44
③6 Fine sand	14.90	142.16	1.01	−110.53	14.86	0.08–4.62
③8 Medium sand, Grit, Gravel sand	10.23	68.44	0.72	−19.31	10.22	0.42–3.90
③9 Medium sand, Grit, Gravel sand	15.68	82.89	0.56	48.70	15.70	0.02–4.06
④2 Clay ~ silty clay	3.46	112.97	3.33	1064.38	3.87	0.50–5.34
④3 Clay ~ silty clay	4.02	136.83	3.54	928.47	4.34	0.48–11.6
④4 Fine sand	6.37	126.26	2.02		6.37	0.40–0.56
④5 Fine sand	10.10	141.70	1.54	−7.04	10.10	0.44–4.78
④6 Fine sand	16.52	167.13	1.10	−145.08	16.45	0.02–8.62
④8 Medium sand, Grit, Gravel sand	14.79	120.79	0.92	−10.11	14.78	0.80–2.48
④9 Medium sand, Grit, Gravel sand	19.31	116.24	0.64	−15.57	19.30	0.02–4.54

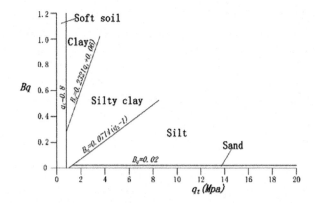

Fig. 1. CPTU soil classification map (TB10018-2003)

Through the interpretation of the CPTU test data, the result of the stratigraphic division in this area is consistent with the drilling test. At the same time, CPTU test explored what the drilling did not reveal (Fig. 1).

3.2 Comparative Analysis of CPTU and SPT Index of Sandy Soil

Relative Density and Soil Compaction. According to "In-situ Test Regulations for Railway Engineering Geology" (TB10018-2003). The degree of compaction of the sand layer in the project is classified. And it is in good agreement with the drilling demarcation of the sand layers. Only in layer ④8, the result of sand quarrying is slightly larger than that of drilling results and shows a compact state.

Internal Friction Angle. According to the CPTU data, the internal friction angle of sands is calculated using the formula in Appendix D.2 of Eurocode 7 -Geotechnical Design-Part 2: Ground investigation and testing. At the same time, according to "4.3.7 of Guangdong Province Building Foundation Design Code" (DBJ15-31-2003), the internal friction angle of sand is estimated by SPT test. The analysis results show that the internal friction angles obtained by the two in-situ test methods are consistent.

3.3 Undrained Shear Strength of Cohesive Soil

The undrained shear strength of cohesive soil CPTU is calculated by CPTU. At the same time, taking the average UU test and the unconfined compressive strength test index of revetment drilling for bank protection as a comparison.

The result shows that the undrained shear strength obtained from the CPTU test data is obviously larger than that obtained from the UU test and unconfined compressive strength test, which is about 1.3 to 1.8 Times. Considering that CPTU is the test data of soil under in-situ stress state, there may be perturbation when sample is taken from laboratory soil and the stress is released after sampling.

3.4 Natural Density and Sensitivity

Calculate the natural density and sensitivities of each clay layer using the formula for calculating soil bulk density in the Guide to Cone Penetration Testing for Geotechnical Engineering.

From this, the natural density and sensitivity of each clay soil layer can be calculated and compared with those obtained by the laboratory test. The consistency is well.

4 Conclusions

The above shows the accuracy of using CPTU technology to divide soil layers in geotechnical engineering, especially in offshore geotechnical engineering, for sampling is difficult. CPTU technology not only saves time and labor, but also can avoid to disturb the samples. And results have good consistency with drilling test and indoor soil test.

Acknowledgements. Majority of the work presented in this paper was funded by the National Key R&D Program of China (2016YFC0800200) and the National Natural Science Foundation of China (Grant No. 41672294).

References

1. Shen, X.K., Cai, Z.Y., Cai, G.J.: Applications of in-situ tests in site characterization and evaluation. China Civ. Eng. J. **49**(2), 98–120 (2016)
2. Liu, S., Cai, G.J., Tong, L.Y.: The Theory and Engineering Application of Modern Multifunction CPTU. Science Press, Beijing (2013)
3. Robertson, P.K.: Interpretation of cone penetration tests - a unified approach. Can. Geotech. J. **46**(11), 1337–1355 (2009)
4. Robertson, P.K.: Soil classification using the cone penetration test. Can. Geotech. J. **27**(1), 151–158 (1990)
5. Robertson, P.K., Cabal, K.L.: Guide to Cone Penetration Testing for Geotechnical Engineering, 4th edn. Gregg Drilling & Testing, Signal Hill (2010)
6. Institute of CCCC-FHDI Engineering Co., Ltd.: CPTU Analysis Report of Reclamation Project of Sanya New Airport and Lingkong Industrial Park (2017)

Simplified Vibration Control Technique of Monitoring Devices in Operating Tunnels

Bangan Liu[(✉)] and Yongbin Wei

China State Construction Engineering Corporation, Beijing, China
liubangan@cscec.com

Abstract. Automated TS (total stations) have been widely utilized in the displacement monitoring of underground structures during neighboring constructions. However, in operating tunnels the vibration of both walls and grounds due to running trains may cause severe disturbance to the installed TS and make it impossible to obtain reliable data. It is therefore necessary to design a vibration controlling technique to reduce the vibration of the monitoring devices. This paper proposed utilizing dampers installed on the total stations to reduce the vibration, which could also be applied to other kinds of monitoring devices which needed low vibration environments. The parameters of the dampers could be designed by a simplified single degree analytical method and finite element dynamic analysis. The method and the dampers were applied in the monitoring of an operating neighboring tunnel during the construction of the Changsha subway line #4 project, Hunan province, China. Workability of the total stations were guaranteed; reliable measuring data were obtained during the construction period and this technique proved applicable.

Keywords: Tunnel · Monitoring devices

1 Introduction

Recently, construction of urban railway systems is getting rapid expansion in cities of China, which may have strong influence on the neighboring operating railway systems. It is thus necessary to monitoring the deformation of the neighboring operating tunnels during the construction period. Automation total stations are efficient, flexible and with high precision, it is applicable in the automatic monitoring the deformation of the operation tunnels since it could run without human interference for months. However, in some tunnels, the side walls and the tunnel ground on which the total stations should be installed may be vibrated severely by the running trains. If the total stations are also vibrated by the walls or ground and could not be stopped timely, which is to say, before the next train coming, the total station was unable to work. Thus, it is necessary to reduce the vibration of the total stations so that it could clam down rapidly and perform effective measurement. The problem of the vibration of the total station was met during the construction of the Changsha #4 subway and one neighboring tunnel (#2 subway) vibrated severely during the daily time. This paper tried to use dampers to reduce the vibration of the total stations. To calculate the parameter of the dampers, a simplified and approximate method together with a finite element modal analysis was performed.

Then the damper was utilized in the monitoring of the railway line and monitoring was guaranteed during the construction time of the Changsha #4 subway line.

Vibration controlling techniques have been utilized widely in structural engineering, however in construction measuring of tunnel engineering, usually the total station was installed on the side walls and no vibration controlling techniques were needed and thus the experience was in lack. In this project the side wall of an operating tunnel vibrated because of running trains. Thus, a quick decision should be made during the construction period so that the construction would not be stopped because of lacking of monitoring data of neighboring tunnels. This report presented a simple and practical controlling technique of the vibrating of installed total stations.

2 Calculation of the Damper Parameter

2.1 Single Degree of Freedom Assumption

The total station and the base installed on the wall could be simplified to a single degree of freedom system [1]. If a coulomb damper was installed with an angle θ below the total station. Assuming the vibration could be reduced by 20% in each vibration cycle, then by approximate calculation, the kinetic energy should be

$$W_D = \frac{1}{2}m\dot{y}^2 = \frac{1}{2}m(A\omega \cos \omega t)^2 \qquad (1)$$

The loss of energy during one cycle could be simplified as

$$W_F = 4AF' \qquad (2)$$

After each cycle, assuming the vibration amplitude before and after each cycle is A_0 and A_1

$$\left(\frac{A_1}{A_0}\right)^2 = 1 - \frac{4AF'}{\frac{1}{2}mA^2\omega^2} = 1 - \frac{8AF'}{mA^2\omega^2} \approx 0.8^2 \qquad (3)$$

Thus the damper parameter could be calculated approximately.

2.2 FE Modeling

The parameter of the damper is related to the vibration frequency of the system, thus finite element analysis (model analysis) of the total station and the steel base was performed. The total station is made of aluminum, plastic and glasses and it is difficult to model the inside of the total station [2]. Thus the total station was modeled as a homogeneous solid with the same weight of the real total station. Then the material for the total station was assumed to have the same elastic modulus and Poisson's ratio of aluminum. The base was assumed to be steel and side of the steel was attached to the side wall of the tunnel. All the bolts were assumed to be bonded. The equivalent density of the total station was calculated 1183 kg/m^3 (Tables 1 and 2).

Table 1. Parameters of the FE model.

Parts	Parameters	Value
Total station (Equivalent)	Density (kg/m^3)	1183
	Elastic modulus (Gpa)	73.1
	Poisson's ratio	0.35
Base (Steel)	Density (kg/m^3)	7800
	Elastic modulus (Gpa)	200
	Poisson's ratio	0.3

Table 2. Calculated modal frequency.

Result	Frequency
Calculated modal frequency (Hz)	118.00
	146.67
	339.46
	535.03
	877.66
	1067.1
	1283.4
	1523.3
	……

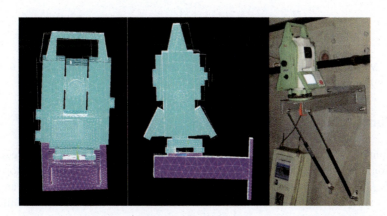

Fig. 1. Modal shape of the first and second modal frequency (f1 = 118.0 Hz, f2 = 146.67 Hz) and the installation of the dampers in field.

2.3 Calculation of the Damper Parameter

The weight of the total station is 7.6 kg. Assuming the installation angle 60°, the frequency be 146.67 Hz, and the vibration amplitude be 1 mm. Then the damper parameter could be calculated. If two dampers were utilized.

$$F = \frac{0.36mA\omega^2}{2 \times 8\sin\theta} = 125.77N \qquad (4)$$

Thus dampers with force around 12.83 kg could be utilized in this measurement. In this project, two dampers of 10 kg were utilized.

2.4 Application of the Damper

The dampers were utilized as shown in Fig. 1, the measuring data before and after utilization of the dampers were presented in Fig. 2. It could be observed that the oscillating displacement before installing of the dampers reached almost 1.5 mm, which will affect the safety judgement during the construction period. The measuring data proved applicable after installation of the dampers. Though some negative displacement occurred at the beginning of the testing measuring period, the overall displacement response was stable.

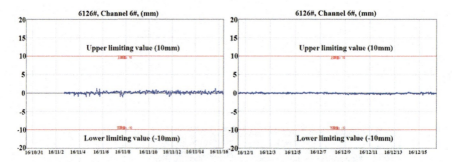

Fig. 2. Measuring data before installation of the dampers (Displacement data, No. 6 module, unit: mm).

3 Conclusion

A simplified calculation method of the damper parameters was proposed to reduce the vibration of the automatic total station and base system during monitoring in operating tunnels. The method was based on a single-degree-freedom simplified model and finite element analysis. The damper parameters calculated and two dampers were utilized in the monitoring projects and proved applicable.

References

1. Chopra, A.: Dynamics of Structures. Prentice Hall, New Jersey (1995)
2. Leica Geosystems: Leica MS50/TS50/TM50 User Manual (2013)

Optimization of Location of Robotic Total Station in Tunnel Deformation Monitoring

Bangan Liu[✉] and Yongbin Wei

China State Construction Engineering Corporation, Beijing, China
liubangan@cscec.com

Abstract. This paper described an optimization method of deciding the location of RTS (robotic total station) in tunnel deformation monitoring of a series of points. Normally the method of resection and polar coordinate was utilized in deformation monitoring by RTS, in which two datum points should be installed in immovable location and the robotic total stations could be installed in movable location. Usually the locations were decided from human experience. This paper presented methods of evaluating the systematic accuracy, which were functions of the locations of the RTS and two datum points. Then the systematic accuracy could be optimized by some optimization methods. This method was applicable in installation of the RTS system in tunnel deformation monitoring.

Keywords: Robotic total station · Optimization · Deformation monitoring

1 Introduction

RTS (Robotic total stations) are widely utilized in the engineering measuring such as tunneling engineering, structural engineering, etc. The method of resection and polar coordinate is a usual method in the installation of the robotic total stations, in which one robotic total station and two datum points are installed. In this method two datum points should be installed in immovable location and the robotic total stations could be installed in movable positions. Usually the two datum points and the robotic total stations were installed based on human experience. This paper presents a method of evaluating the systematic accuracy of the installation of the RTS and two datum points. Based on the evaluating method, a method of minimizing the systematic errors was presented based on the optimization method.

Usually only the position of the RTS and two datum points need to be determined before engineering measuring. For each point there are x, y and z positions, thus there are 9 unknowns in the measuring systems. Generally, for a RTS, the measuring error of the distance is A mm ± 1 ppm form and the error of the angle is B" form (as for one commercial robotic total station, the measuring error of the distance is 0.6 mm ± 1 ppm and the measuring error of the angle is 0.5"). Thus, there a best position, which has minimum system error during measuring for one project.

There are two possible evaluating methods for the total systematic error of measuring projects. First is that the total adding together error is minimum of all measuring points. Second is that the maximum error of one single measuring point is minimum of the system. This paper gives the possible methods of calculating the two different optimized installing points for RTS and datum points.

2 Method

Assume the positions of all the measuring points are already known. The position of the RTS and two datum points needs to be determined. Also assume the total systematic error of the measuring project is only related to the measuring distances and angles. Other errors such as temperature, sunlight, wind etc. are not considered in this paper. The measuring process of one single point has two steps: first, is the decision of the position of the RTS and the initial direction of the horizontal dial by resection method from two datum points; second, is the decision of the measuring point by polar coordinate method from the RTS.

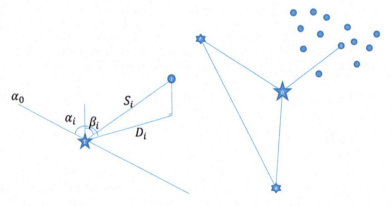

Fig. 1. Illustration of the polar coordinate method and resection method (The position of the Total station is O; two datum points are A and B; the measuring points are i)

$$\begin{cases} x_o = x_A + S_{OA} \sin \beta_{OA} \sin(\angle BAO + \alpha_{AB}) \\ y_o = y_A + S_{OA} \sin \beta_{OA} \cos(\angle BAO + \alpha_{AB}) \\ z_o = z_A + S_{OA} \cos \beta_{OA} \\ \alpha_0 = \angle AOy + C \end{cases} \quad (1)$$

$$\begin{cases} dx_o = dx_A + \sin \beta_{OA} \sin(\angle BAO + \alpha_{AB})dS_{OA} \\ \qquad + S_{OA} \cos \beta_{OA} \sin(\angle BAO + \alpha_{AB})d\beta_{OA} \\ \qquad + S_{OA} \sin \beta_{OA} \cos(\angle BAO + \alpha_{AB})d\angle BAO \\ dy_o = dy_A + \sin \beta_{OA} \cos(\angle BAO + \alpha_{AB})dS_{OA} \\ \qquad + S_{OA} \cos \beta_{OA} \cos(\angle BAO + \alpha_{AB})d\beta_{OA} \\ \qquad - S_{OA} \sin \beta_{OA} \sin(\angle BAO + \alpha_{AB})d\angle BAO \\ dz_o = dz_A + dS_{OA} \cos \beta_{OA} - S_{OA} \sin \beta_{OA} d\beta_{OA} \\ d\alpha_0 = d\angle BAO \end{cases} \quad (2)$$

While S means slope distance and D means horizontal distance. α and β mean horizontal and vertical angles. α_o means an initial angle and C is a constant. O represented the position of RTS, A and B represented two datum points, and i represented each of the measuring points. The position of all the points was shown in Fig. 1. Coordinates of unknowns could be expressed from the known parameters. The total differential could be calculated. The mean square error could be determined by the total differential.

$$\begin{bmatrix} m_{x_o}^2 \\ m_{y_o}^2 \\ m_{z_o}^2 \\ m_{\alpha_o}^2 \end{bmatrix} = \begin{bmatrix} (y_{OA}/S_{OA})^2 & ((z_{OA}y_{OA}/\rho D_{OA}))^2 & (x_{OA}/\rho)^2 & 0 \\ (x_{OA}/S_{OA})^2 & ((z_{OA}x_{OA}/\rho D_{OA}))^2 & (-y_{OA}/\rho)^2 & 0 \\ (z_{OA}/S_{OA})^2 & (D_{OA}/\rho)^2 & 0 & 0 \\ 0 & 0 & 0 & 1 \end{bmatrix} \begin{bmatrix} m_{S_{OA}}^2 \\ m_{\beta_{OA}}^2 \\ m_{\angle BAO}^2 \\ m_{\angle BAO}^2 \end{bmatrix} \quad (3)$$

Then, the error of position of measuring points could be determined,

$$\begin{cases} dx_i = dx_o + \sin\beta_i \cos(\alpha_i + \alpha_o)dS_i + S_i \cos\beta_i \cos(\alpha_i + \alpha_o)d\beta_i \\ \qquad - S_i \sin\beta_i \sin(\alpha_i + \alpha_o)d\alpha_i - S_i \sin\beta_i \sin(\alpha_i + \alpha_o)d\alpha_o \\ dy_i = dy_o + \sin\beta_i \sin(\alpha_i + \alpha_o)dS_i + S_i \cos\beta_i \sin(\alpha_i + \alpha_o)d\beta_i \\ \qquad + S_i \sin\beta_i \cos(\alpha_i + \alpha_o)d\alpha_i - S_i \sin\beta_i \cos(\alpha_i + \alpha_o)d\alpha_o \\ dz_i = dz_o + \cos\beta_i dS_i - S_i \sin\beta_i d\beta_i \end{cases} \quad (4)$$

Also, the mean square errors are,

$$\begin{bmatrix} m_{x_i}^2 \\ m_{y_i}^2 \\ m_{z_i}^2 \end{bmatrix} = \begin{bmatrix} 1 & 0 & 0 & (-y_i)^2 & (x_i/S_i)^2 & (-y_i/\rho)^2 & (z_ix_i/\rho D_i)^2 \\ 0 & 1 & 0 & (x_i)^2 & (y_i/S_i)^2 & (-x_i/\rho)^2 & (z_iy_i/\rho D_i)^2 \\ 0 & 0 & 1 & 0 & (z_i/S_i)^2 & 0 & (-D_i/\rho)^2 \end{bmatrix} \begin{bmatrix} m_{x_o}^2 \\ m_{y_o}^2 \\ m_{z_o}^2 \\ m_{\alpha_o}^2 \\ m_{S_i}^2 \\ m_{\alpha_i}^2 \\ m_{\beta_i}^2 \end{bmatrix} \quad (5)$$

Thus, the mean square of each direction of position of every single measuring point could be determined. There are two possible methods of evaluation of the overall error of all the measuring points. Assume the mean error of single measuring point is

$$m_{p_i} = \pm\sqrt{m_{x_i}^2 + m_{y_i}^2 + m_{z_i}^2} \quad (6)$$

The overall systematic error of the measuring system could be evaluated by

$$\min\left(\sum m_{p_i}^2\right) \quad (7)$$

Or by

$$\min\left(\max\left(m_{p_i}^2\right)\right) \tag{8}$$

Basically, this optimization problem is a constrained optimization problem of 9 degrees of freedom. The best position of total station and two datum points could be calculated by optimization method [1, 2].

3 Application

This paper chose the least square method to be applied to the decision of the position of total station and two datum points in tunnel engineering. Since total station and the two datum points are always installed on the side wall of the tunnel and the problem could be simplified into a one-dimension problem, i.e. only the X coordinates (Assuming the X direction is the tunnel direction) of the three points needed to be decided. The testing case is as follows: Width of the tunnel is assumed to be 6 m; The deformation extent is from 0–150 m; Inside measuring points should be installed every 10 m at both sides of the tunnel (totally 30 points); Two datum points should be installed outside the deformation zone; Total station could be install in deformation zone; All measuring points, datum points and total station are assumed to be the same height. The calculated positions of the total station and two datum points were illustrated in Fig. 2. The calculated mean square errors are 4.28 mm².

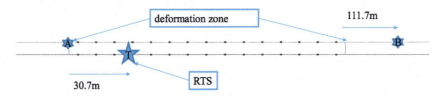

Fig. 2. Illustration of the best position of the robotic total station and two datum points.

4 Conclusion

This paper presented methods of evaluating the systematic error of deformation measuring by RTS, which were functions of the locations of the RTS and two datum points. The error could be optimized by some optimization methods. This method was preliminary tested in installation of the RTS system in tunnel deformation monitoring and valid result could be obtained.

References

1. Nelder, J., Mead, R.: A simple method for function minimization. Comput. J. **7**(4), 308–313 (1965)
2. Yan, H., Yang, W.: Surveying Principles and Methods. Science Press, Beijing (2009)

Life-Cycle Analysis of Foundation Structures Using Sensor-Based Observation Methods

Arne Kindler and Karolina Nycz[✉]

Stump Spezialtiefbau GmbH, 10243 Berlin, Germany
karolina.nycz@stump.de

Abstract. Modern distributed fiber optic sensing technology allows measurements with very high precision of about 1 μm/m and a special resolution up to 10 mm or higher. Although it is an established measurement method in many fields of engineering, such as aircraft construction, offshore technology or pipeline oil industry it still is considered with some doubt in geotechnical engineering. For the purpose of establishing this modern measuring technology in geotechnical engineering a series of investigations on different geo-technological structures such as bored and driven piles, ground anchors or diaphragm wall was carried out. The measuring technology was integrated as a part of the respective structure and the measurements were mostly conducted during the test loading of the structures. For a comparison as well as verification of the obtained results the structures were equipped with classical instrumentation for strain measurements. In the paper the basics of the measuring technology will be discussed. Selected case studies will be presented and discussed from the instrumentation of the structural elements to the evaluation of the measurement results.

Keywords: Optical fiber · Distributed strain measurements
Distributed temperature measurements

1 Measurement Technology

1.1 Fundamentals

A chain of single-point-measurement-sensors is inefficient for measurements along larger distances and vaster areas or volumes. Moreover, in order to register a phenomenon, the exact location of the occurrence has to be known in advance. In order to measure and determine physical quantities with high spatial resolution in workable and profitable manner and over longer periods of time a distributed sensing technology is required. Fiber optic measurement technology gives the possibility of a distributed strain and temperature sensing (DSS & DTS) and opens new ways for recording and monitoring space-time distribution of these physical quantities.

Distributed sensing uses strain and temperature dependence of certain optical properties of optical fibers to measure strain and temperature along an optical fiber. A measuring system consists of a fiber-optic measuring cable, a laser light source and an optical receiver with an evaluation unit, whereby the latest three are combined in one device controlled by a PC. During a measurement, laser light impulse is coupled into the optical fiber and gets scattered onto the molecules of the fiber optic cable. As a result, a smaller part of light is backscattered. In the process the intensity of the backscattered light and the propagation time are measured. As a result, an integral value of strain or temperature of the length section from which the backscattered light originates within a time window is obtained. By linking the measurement result of the physical quantities with the measurement of the propagation time of the light, the distribution of the measured variable over the entire length of optic fiber is obtained. Depending on the type of backscattering process, three measurement methods can be distinguished:

1. Rayleigh Scattering – for the high-precision measurement of the strain along the fiber-optic cable with spatial resolution up to 1 mm for measuring distances smaller than 100 m.
2. Raman Scattering – to measure the temperature along the fiber-optic cable with special resolution up to 12.5 cm for measuring distances smaller 12 km.
3. Brillouin Scattering – to measure the strain along the fiber-optic cable with spatial resolution up to 20 cm for measuring distances smaller than 30 km.

2 Case Studies

2.1 Anchor Monitoring on the Example of a Grouted Anchor Using DSS

The stranded anchors for the test field were manufactured in company-owned factory which has made the instrumentation process with the measuring cables much simpler. In the free steel length range the measuring cable was mounted onto steel strand. In the area of the grout body the fiber optic cable was guided in the middle of steel strands (middle of the spacer) and hence in the middle of the grouted body. The measuring cable was fixed under minimal tension so that no movement of the cable during installation in the ground would occur. Altogether five anchors with five steel stands each and a total length of 35 m (grout body length 4 m) each anchor were produced. After insertion of the anchors in to the ground they were subjected to a standard suitability test according to DIN EN 1537. The measurements were performed with the ODiSI (optical distributed sensor integrator). The data acquisition and evaluation took place on site so that the deformation of the grout body could be analyzed simultaneously with individual load stages of the suitability test. For each load step the data has been recorded at the beginning, in the middle and towards the end of the load stage. Due to lack of conventional measurement technology able to record data in the grout body of the anchor comparative measurement data could not be obtained. After assessment of the recorded data, the conclusion was made that a characteristic strain curve over the entire grout body can be depicted. With increasing load, the characteristic of the curve changes. The curve seems to portray a chaotic crack pattern that is imprinted into the grout body. As long as the effective load stays below the test load, the curve remains unchanged in its characteristic (Fig. 1).

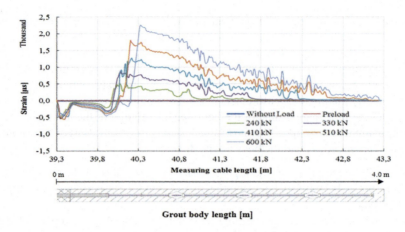

Fig. 1. Characteristic strain curves of test anchor 5 for all load levels.

2.2 Static Pile Load Test Using DSS

The applicability of the fiber optic sensing technique for static pile load tests has been investigated and verified on several pile systems such as large bored pile, micro pile, Atlas pile or Franki pile. The obtained measuring results are more accurate and extensive than the data recorded by the means of classical strain monitoring devices. Furthermore, the instrumentation of the pile with fiber optic measurement technology is simpler than the instrumentation with conventional measuring sensors. The measuring cable is attached to the reinforcement cage of the pile and led up and down in loops over the entire length of the pile. In the course of the results analysis, measured strain is converted into the force acting in the pile. The calculated force curves of the individual load stages are than displayed graphically as a function of depth. As an example the results of the measurement on a Franki pile are shown below. The pile was approx. 12.5 m long with a diameter of 0.51 m and has been subjected to maximal test load of 4000 kN. The pile was also equipped with a cladding pipe from the head to a depth of approx. 7.2 m. The influence of the cladding pipe on the force distribution in the pile can be clearly deduced from the measurement. From the load stage of 4300 kN upwards, the composite stress in the pipe is overcome. This is characterized by the change in the force curve from approx. 3.0 m. The force progression in range of approx. 7.2 m to 12.4 m shows a decrease of the acting force in the pile due to skin friction influence.

3 Outlook of Future Studies

Further projects using optical measurement technology are currently in progress. Following a small outlook of ongoing projects is presented: deformation measurements of a diaphragm wall, convergence measurements of a tunnel construction, temperature and strain measurements of a thermo-active-pile, dynamic pile load test using fiber optic sensor chains, pile integrity tests using Distributed Temperature Sensing (DTS).

After completion of the experiments gained knowledge will be summarized and published in relevant literature (Fig. 2).

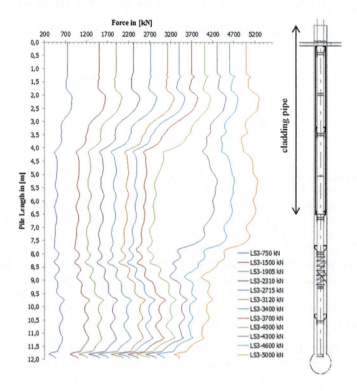

Fig. 2. Normal force curves of static pile test load for the Franki pile.

References

1. Kindler, A., Schaller, M.-B., Glötzl, J.: Nachweis der Ankertragfähigkeit auf Grundlage der faseroptischen Messtechnik. Bautechnik **94**(2), 144–151 (2017)
2. Kindler, A., Schaller, M.-B., Glötzl, J.: Großwig, S.: Static pile load test using fibre-optic strain measurements with a very high local resolution. In: Pile Foundation Symposium 2017, Braunschweig (2017)
3. Kechavarzi, C., Kenichi, S., de Battista, N., Pelecanos, L., Elshafie, M., Mair, R.: Distributed Fibre Optic Strain Sensing for Monitoring Civil Infrastructures. ICE Publishing, London (2016). ISBN: 978-0-72776-055-5
4. Kindler, A., Nycz, K., Grosswig, S., Schaller, M.-B.: Distributed fibre optic sensing technology in the extreme areas of civil engineering. In: 8th International Conference on Structural Health Monitoring of Intelligent Infrastructure Proceedings on Structural Health Monitoring for Infrastructure Sustainability, Brisbane (2017)
5. Kindler, A., Großwig, S.: Distributed strain sensing in geotechnical engineering – recommendations for distributed fiber optic strain measurements in geotechnical engineering. Bautechnik **95**(5) (2018, to be published)

From Mobile Measurements to Permanent Integrity Control: New Geotechnical Monitoring Instruments for Leakage Detection and Localization

Mike Priegnitz[1](✉) and Ernst Geutebrück[1,2]

[1] Texplor Exploration & Environmental Technology GmbH, Potsdam, Germany
{mike.priegnitz,ernst.geutebrueck}@texplor.com
[2] Texplor Austria GmbH, Vienna, Austria

Abstract. Underground constructions and deep excavations below the groundwater table often bear technical and financial risks. Failures in the sealing system can cause severe damages. Texplor has developed and patented accurate and reliable measuring and monitoring devices for the quality control of various sealing systems. These devices use electrical conductivity of liquids to identify weakness zones of sealing constructions. Quality Control of sealing systems can be carried out either as single investigations at a given time or as a continuous monitoring with automated leak alarm functions. The non-corrodible sensors for the MSS® Permanent Leak Monitoring System are specially developed for PE liners and have a certified lifetime of 100 years. The risks of environmental damages and financial losses are minimized by advanced and/or continuous quality control.

Keywords: Leak detection · Sealing systems · Quality control and monitoring

1 Introduction

An increasing number of underground structures in urban areas often requires deep excavations below the groundwater table and therefore entails high technical and financial risks. Many problems arise during the construction of ancillary structures, like stations and service shafts for public transport, underground parking areas or similar construction parts.

The requirements for best quality and extended lifetime are very high for these types of constructions. A failure in the sealing system can cause severe damages on the construction itself and on neighboring buildings. Furthermore, substantial financial losses might occur due to an impairment of the building process and expensive repair works. Leakages in sealing walls, joints and bottom seals, e.g. natural geological sealing horizons, injection bases etc., are factors, which can substantially influence the technical and financial success of a construction project.

Two categories of quality control of the sealing functionality can be distinguished:

- mobile investigation (one-time measurement) and
- permanent integrity control.

© Springer Nature Switzerland AG 2018
W. Wu and H.-S. Yu (Eds.): *Proceedings of China-Europe Conference on Geotechnical Engineering*, SSGG, pp. 716–719, 2018.
https://doi.org/10.1007/978-3-319-97112-4_160

Mobile measurements are performed if a construction needs a quality control to prove the structural integrity at a given time or if an actual problem occurs in the sealing system. Permanent integrity control devices are integrated into the construction and continuously monitor the structural tightness. Those monitoring systems are used for potentially risky environments such as storage basins for hazardous liquids, waste treatment sites, or in metro stations, tunnels, below floor constructions, etc.

Texplor has developed and patented accurate and reliable measuring and monitoring devices, which are adaptable to the specific requirements of each client. These measuring devices use electrical conductivity of liquids to identify weakness zones of sealing constructions.

The basic principle of Texplor's technologies is to create a specific electrical signal on one side of the sealing construction. On the other side of the sealing construction, the spatial distribution of the incoming signal is observed. This incoming signal is measured by sensors, placed in a regular grid within the construction area. High amplitudes are caused by high incoming signals and thereby correlate with flow paths for liquids and, thus, weaknesses in the sealing construction. The recorded data is processed and displayed in a 2D top-view map of the construction site in a corresponding coordinate system localizing the weak areas in the sealing system.

2 Technologies for Mobile Measurements

For mobile measurements, the Flexible Groundwater Monitoring Technology (FGM®), which is a multi telemetric system with non-polarizable sensors, is used.

Generally, inside the construction, e.g. building pit, basin, etc., the sensors are placed individually according to the measurement strategy to suit any given field geometry. Outside the construction, an electrical tracer is placed into one or more wells to create an artificial electrical signal (see Fig. 1).

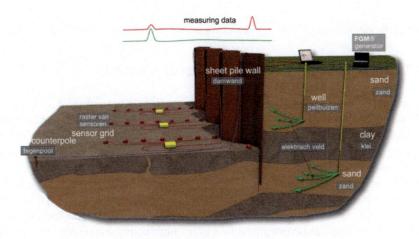

Fig. 1. FGM® measuring principle of a sealing construction.

Usually, three different energy levels are used: a passive (self-potential) and two active levels measure the voltage distribution in the ground caused by natural and/or artificial groundwater flows in order to localize leakages. The multi telemetric system allows simultaneous measurements of all sensors at their respective position.

An example of an FGM® measurement of a metro station in Bangkok, Thailand, is shown in Fig. 2. At this construction pit of approx. 2000 m^2, severe water inflow occurred through both the horizontal and vertical sealing constructions. To identify the problematic areas, a total number of 287 FGM® sensors were placed inside the pit, covering the entire area of the natural horizontal sealing layer as well as the artificial (diaphragm walls) vertical sealing layers. As shown in the results in Fig. 2, the survey identified a number of weakness zones, which required rehabilitation procedures such as grouting.

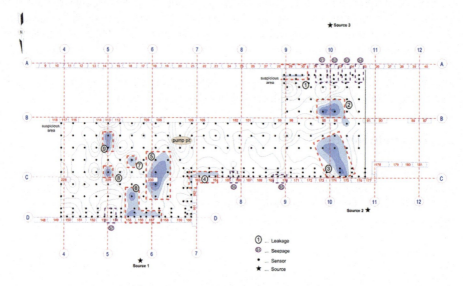

Fig. 2. Result of an FGM® survey in a metro station construction pit in Bangkok, Thailand.

3 Technologies for Permanent Integrity Control

In cases, where a mobile measurement at a given moment is not sufficient for the client, a permanent quality control of the construction integrity can be installed and integrated into the sealing construction. This need for a permanent control can arise from economic interests, ecological requirements or legislative regulations. Texplor has developed a Multi Sensor System (MSS®) which can monitor any kind of sealing, e.g. concrete, asphalt or impermeable geomembrane. These liners are usually made of polyethylene (PE) and are installed in storage basins, construction pits, waste deposits, landfill covers and basements, below floor constructions, etc. Texplor has designed special MSS® CombiModules that can be integrated into every liner construction; furthermore, post installations in existing constructions are possible.

The MSS® CombiModules consist of a PE body with a sensor on one side (e.g. top) and a source on the other side (e.g. bottom). Sensor and source are made of graphite/gold, are non-corrodible and have a certified lifetime of ~100 years. They can also resist environments of pH-values 1–14. The CombiModules are integrated into the liner by extrusion welding. After the welding process, vacuum testing of all extrusion seams is compulsory to ensure the tightness of the installation. The MSS® technology can be used in single, double or even triple liner constructions (see Fig. 3).

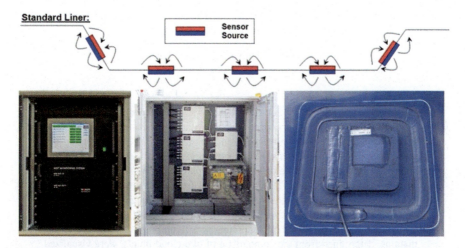

Fig. 3. Schematics of the MSS® Permanent Leak Monitoring System by Texplor, MSS® Central Unit, MSS® Outdoor Measuring Cabinet, MSS® CombiModule.

The MSS® Monitoring technology uses a Central Unit for all operations and for an automated leak localization. The data of all MSS® CombiModules are collected in a database and evaluated permanently. The MSS® operational software defines measurement intervals, energy levels and alarm types. A display shows overview maps, the status of tightness, alarm situations and more.

The measurements are carried out automatically according to the set measurement intervals and energy levels. A tracer signal is emitted at the source side of the MSS® CombiModules. In a tight situation, the incoming signal at the sensor side of the CombiModules is low. If a leakage appears, the amplitude of the incoming signal is rising dramatically – the higher the magnitude, the closer the MSS® CombiModule is situated to the leak location.

By placing a sufficient amount of MSS® CombiModules in a regular grid within the construction, the MSS® technology not only provides reliable information on the structural integrity of the construction, but also allows an automated leak localization based on the permanently recorded data.

In a long term, even small leakages might create a big problem in the construction: physical damages, building freeze, delay of construction works, and substantial financial losses are potential consequences. Texplor's MSS® Permanent Leak Monitoring of sealing constructions as strategic tool, minimizes or even avoids these risks.

Visual and Quantitative Investigation of the Settlement Behavior of an Embankment Using Aerial Images Under Large-Size Field Loading

Ali Alper Saylan, Okan Önal(✉), Ali Hakan Ören, Gürkan Özden, and Yeliz Yükselen Aksoy

Dokuz Eylül University, İzmir, Turkey
okan.onal@deu.edu.tr

Abstract. In this study, the settlement behavior of the embankment due to the surcharge loading have been visually and quantitatively investigated. The spatial variation of settlements of an oversized embankment have been monitored using aerial images taken at intervals. In order to achieve the deformations due to the surcharge loading in the field, two topographical three-dimensional models (before and after surcharge loading), derived from aerial images and ground control points, have been generated and compared. The aerial images were acquired by an unmanned aerial vehicle with autonomous flight capabilities. The topographies were compared by generating two cloud point models. The deformations, determined from the comparison of two cloud points, were visualized by assigning an intensity value, which was calculated based on the three-dimensional variations of each data point. The surficial settlement profiles nearby the surcharge load have been computed. It has been concluded that centimeter level accuracy on surficial settlement has been achieved by low altitude aerial images, although the illumination condition of the area has an influence on the measurements.

Keywords: Structure From Motion · Geotechnical field test · Settlement monitoring

1 Introduction

The field monitoring in geotechnical projects usually rely on mechanical gauges, electronical sensors and tachometric surveying. Although being accurate, these measurement techniques are inadequate for high resolution spatial monitoring. In recent decades, laser scanners and robotic total stations have been in use in some exclusive projects for comprehensive three-dimensional (3D) monitoring. However, these techniques are still costly and are affected by the obstacles throughout their scan directions. On the other hand, recent advances in the digital image processing algorithms enable detailed digital representation of objects or scenes using a process called Structure From Motion (SFM).

SFM technique is a relatively low cost photogrammetric method for high resolution topographic reconstruction [1]. In this technique, a 3D structure can be resolved from a series of overlapping images where the geometry of the scene, camera positions and orientation are solved by an iterative procedure without the need to specify a known network of target coordinates [2]. SFM technique requires to extract similar features in the overlapping images. In order to achieve this task, an object recognition algorithm called Scale Invariant Feature Transform (SIFT) was developed and implemented by Lowe [3] in 1999 in SFMToolkit3. SIFT algorithm recognizes features in each image that are invariant to the image scaling and rotation and partially invariant to changes in illumination condition and 3D camera viewpoint [4]. The matching features in each image i.e. tie points are used to solve the geometry of the scene, camera location and orientation simultaneously.

In this study, SFM technique was implemented to a full scale geotechnical field test as an instrumentation and monitoring tool. The monitoring was performed on an oversized embankment, on which some units of an industrial plant will be constructed. The total height of the embankment was projected as 30 m, where the total surface area of the completed embankment would be 42500 m^2 (Fig. 1a). During the progress of the embankment fill a surcharge test was made at mid-height in order to assess deformation characteristics of the fill by analyzing spatial variations of the induced settlements obtained via image processing of the aerial digital photographs using an unmanned aerial vehicle (UAV). The total height of the surcharge fill was 2.7 m with a surface area of 1750 m^2. Estimated surcharge load as a result of the test fill is 5.7 t/m^2.

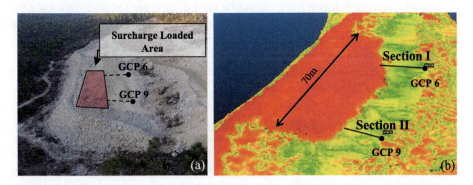

Fig. 1. General view of the embankment (a), the spatial variances in the inspected area by color visualization (b).

2 Materials and Methods

2.1 Unmanned Aerial Vehicle and Flight Planning

The UAV, utilized in this study is an autonomous quadcopter, which was fully assembled by the researchers of this study. Effective brushless motors and 15-in. carbon fiber props have been used with a capacity of 16000 mAh to sustain approximately 45 min of flight time with a total weight of 3179 g. A modified version of GoPro Hero 4 Silver Edition action camera was used with a two-axis gimbal stabilization on the copter. The camera has a capacity to snapshot two 12 Megapixel frame per second with 5.4 mm non-distortion narrow angle lens.

In this study, Mission Planner software was used for flight planning and real-time monitoring during the flight was made via 433 MHz telemetry connection. The flight plan ended up a 6:55 min of flight time with 5 m/s aerial speed covering 14000 m^2 surface area.

2.2 Modelling of the Embankment Topography Using SFM Technique

Once the images are geotagged, they have been imported in a stand-alone software product (Agisoft Photoscan) that performs photogrammetric processing of digital images and generates 3D spatial data. This software accepts full camera calibration matrix, including non-linear distortion coefficients for more accurate processing.

The processing of the aerial images begins by aligning all images in 3D space by examining the identical features in the image matrixes. This process detects the tie points between the overlapping images and results in a sparse cloud point. At this level, the topography may be seen in 3D and the erroneous tie points can be deleted manually.

2.3 Calibration of the 3D Topography

The generated topography consists of 160 million of points in a 3D environment with an inherent scale and orientation. The exact locations of the calibration targets, called as Ground Controls Points (GCP), should be known, in order to locate and scale the topographical model. In this study, 12 GCP were located around the surcharge loading in order to monitor the settlements of the embankment. The calibrations of topographical models (i.e. prior and after the test) were made based on GCP coordinates. Post calibration average errors between RTK field measurements and image analysis coordinates were computed as 1.5 cm and 2.1 cm for the GCP of two topographical models.

2.4 Comparison of Two Topographical Models

The comparison of these cloud points was made by using CloudCompare an open source 3D point cloud and mesh processing software. Since the focus is on the settlements of the embankment, only the differences in z direction were analyzed. In Fig. 1b, the disparities were visualized by assigning color values, where red zones are the differentiated regions in z direction. It is obvious that, the settlements below the surcharge load cannot be computed, since the presence of the surcharge load intervened the visual appearance of the settled surface below the fill. However, the settlement profiles near the surcharge load were easily determined.

3 Results and Discussions

The surficial settlements due to surcharge loading were interpreted in two cross-sections (Fig. 1b). These cross-sections were deliberately selected to include the GCP in order to evaluate the disparities at these points. Although the error margins in the calibrated models are on the order of couple centimeters (1.5 –2.1), it is observed that the computed elevational differences between the pre-surcharge and post-surcharge z-coordinates of at GCP6 and GCP9 in these sections are 2 mm and 1 mm for Section I and Section II, respectively (Fig. 1b), indicating these control points can be regarded as the reference for determination of the settlement profile. The cross-sectional profiles are shown in Fig. 2 with the end-points of the analyzed sections are marked. It can e concluded that, in both cross-sections, the settlements vary in a range of 0–6 cm.

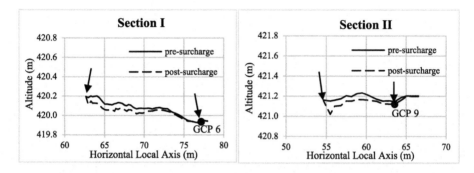

Fig. 2. The settlement profiles of Section I and Section II.

4 Conclusion

The surficial settlements of an embankment due to surcharge loading was visually and quantitatively investigated by processing aerial photographs. SFM technique was used for model creation. The initial and loaded surfaces of the embankment were modeled in order to compare the surficial settlements quantitatively. The surficial settlements were determined in the range of 0–6 cm. The fact that induced settlements can be determined at sub-centimeter levels is an indication for the applicability of the proposed methodology as an instrumentation and monitoring tool to full scale geotechnical field loading tests on embankments.

Acknowledgements. This study would like to acknowledge the TÜBİTAK project 116M506, under the supervision of Assoc. Prof. Dr. Okan Önal, for their equipment and financial support.

References

1. Westoby, M.J., Brasington, J., Glasser, N.F., Hambrey, M.J., Reynolds, J.M.: Structure-from-motion photogrammetry: a low-cost, effective tool for geoscience applications. Geomorphology **179**, 300–314 (2012)
2. Snavely, N., Seitz, S.M., Szeliski, R.: Modeling the world from internet photo collections. Int. J. Comput. Vis. **80**(2), 189–210 (2008)

3. Lowe, D.G.: Object recognition from local scale-invariant features. In: The Proceedings of the Seventh IEEE International Conference, vol. 2, pp. 1150–1157 (1999)
4. Lowe, D.G.: Distinctive image features from scale-invariant keypoints. Int. J. Comput. Vision **60**(2), 91–110 (2004)

Real-Time Monitoring of Large-Diameter Caissons

Brian Sheil[1(✉)], Ronan Royston[1,2], and Byron Byrne[1]

[1] Department of Engineering Science, University of Oxford, Oxford, UK
brian.sheil@eng.ox.ac.uk
[2] Ward and Burke Construction Ltd, Bourne End, UK

Abstract. Large-diameter open caissons are a widely-adopted solution for deep foundations, underground storage and attenuation tanks, pumping stations, and launch and reception shafts for tunnel boring machines. The sinking process presents a number of challenges including maintaining verticality of the caisson, controlling the rate of sinking, and minimizing soil-structure frictional stresses through the use of lubricating fluids. A bespoke monitoring system has been developed at University of Oxford to provide early warning of adverse responses during the sinking phase (e.g. excessive soil-structure interface friction). The monitoring system was trialled on a recent pilot project in the UK involving the construction of a 32 m internal diameter, 20 m deep reinforced concrete caisson. This paper describes the monitoring system that was developed and its impact on the construction process of the pilot project. Early indications are that real-time feedback of live construction data has a major impact on the efficiency and safety of the construction process.

Keywords: Construction · Caisson · Monitoring · Soil-structure interaction

1 Introduction

Monolithic reinforced concrete caissons have a large number of geotechnical applications. The construction process involves a complex system of concurrent casting of the concrete walls and excavation of the soil inside and beneath the caisson thereby allowing the caisson to sink into the ground under self–weight. Surprisingly, there is a paucity of literature on the design and performance of large-diameter caissons. Tomlinson (1986) and Nonveiller (1987) represent the earliest investigations into the sinking process. Case histories have been documented by Puller (2003), Safiullah (2005) and Abdrabbo and Gaaver (2012). More recently, the mechanisms associated with the sinking process have been explored using numerical simulations (Wang et al. 2014) and laboratory testing (Royston et al. 2016). This paper describes the challenges associated with the construction of large-diameter open caissons. A bespoke monitoring system was developed at University of Oxford to address these challenges and the impact of the monitoring system on a UK pilot project is described.

2 Design and Construction Challenges

The sinking stage of the construction process for large-diameter caissons involves subtle balancing between the downward weight of the caisson walls, w, and upward resistance generated from soil-structure interaction. The latter comprises bearing stresses, v, acting on the angled 'cutting face' at the base of the caisson, and frictional stresses, τ, acting on the exterior surface of the caisson (see Fig. 1). A steel 'leading edge' is used to create an annulus surrounding the structure, which is filled with lubricant to limit friction build-up during the construction process. The development of frictional contact stresses is therefore dependent on (a) the (short- and long-term) stability of the lubricant-filled annulus, and (b) the effectiveness of the lubricant in minimizing friction in the event that the annulus collapses. An accurate estimation of these contact stresses is essential to optimize design (e.g. wall thickness) and reduce risk (e.g. failure to sink caisson). However, a rational and experimentally-verified procedure to measure, calculate or predict soil-structure contact stresses, and their development, along the soil-structure interface does not exist for this problem.

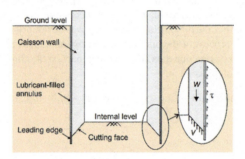

Fig. 1. Problem definition

The caisson sinking process is mainly controlled through excavation of the 'internal level'. Uniform excavation of the internal level is desirable to maintain verticality of the caisson. This is achieved by excavating different 'sectors' iteratively within the caisson. Alternatively, excavation can be focused in specific areas to re-correct tilts that have been induced by sudden drops in caisson elevation. A prerequisite for this process is timely feed-back of the caisson elevation/verticality. Traditionally, however, feedback takes the form of manual surveys of the caisson resulting in infrequent feed-back which is also susceptible to human error.

3 Pilot Monitoring System: Anchorsholme Park, UK

A caisson monitoring system was developed at University of Oxford to measure and monitor (a) the settlement and tilt of the caisson, (b) frictional and normal contact stresses that develop on the exterior face of the concrete walls, and (c) bearing pressures and pore water pressures that develop on the cutting face during the sinking process. The monitoring system was deployed on a live construction site at Anchorsholme Park, UK (see Fig. 2). This project involved the construction of a 32 m internal diameter, 20 m deep reinforced concrete caisson in dense sand.

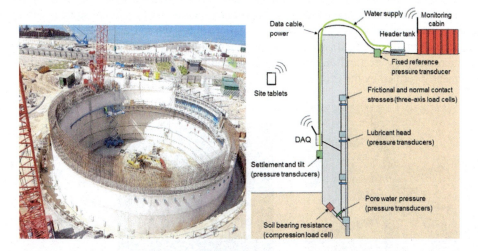

Fig. 2. Layout of pilot monitoring system at Anchorsholme Park, UK

All instrumentation were cast directly into the concrete walls for protection against mechanical damage while the data acquisition system (DAQ) was located within an IP68 enclosure mounted on the caisson wall. The caisson level detection system comprised four pressure transducers (PTs) located at quarter points around the caisson circumference. These PTs were connected, via wire-reinforced hosing, to a header tank and a fixed reference PT at ground level (see Fig. 2). The header tank and hosing was filled with a mixture of de-aired water, anti-freeze and dye (the lattermost aided identification of air in the system). Pressure measurements relative to the fixed reference point outside the caisson enabled tilt and average elevation to be determined.

Normal and frictional contact stresses on the external wall face were monitored using three-axis load cells fitted within bespoke waterproofed aluminium housings with a 'sensing' area of 100 mm × 100 mm. Lubricant pressures were monitored using PTs connected to open 'ports' in the wall. Measurements of the stress acting on the tapered cutting face beneath the caisson wall included the total bearing (normal) stress and the pore water pressure. An 'S–type' compression load cell was used to measure the total bearing stress. The load cell was fitted within a waterproofed aluminium housing with a 150 mm × 150 mm bearing area. The pore water pressure was measured using a PT fitted with a porous stone filter cap which had a high resistance to air entry.

A local wireless network was set up to provide real-time (1 Hz) feed-back of the live construction data to site engineers and operatives via tablets. Caisson tilts and elevation (accurate to < 1 mm) were provided directly to the excavator operators thereby eliminating the need for an otherwise dedicated site engineer for this purpose (see Fig. 3). Intermittent manual surveys were still conducted to verify accuracy. At formation level (i.e. end of sinking), the maximum tilt of the caisson was ~0.14°. Feedback of the frictional stresses and lubricant pressures on the external surface of the caisson were used to monitor conditions in the annulus surrounding the caisson. Measurements of the lubricant pressures showed immediate decay after initial pumping into the annulus. This was due to filtration of the lubricant into the surrounding soil indicating an incorrect lubricant viscosity. Annulus closure can lead to a build-up of frictional stresses during the sinking process; the consequences of a failure of this type are enormous (caisson can become entirely wedged). Monitoring feedback was used to refine the viscosity to ensure lubricant pressure (and therefore annulus stability) was maintained. This was subsequently verified by the frictional contact stress measurements.

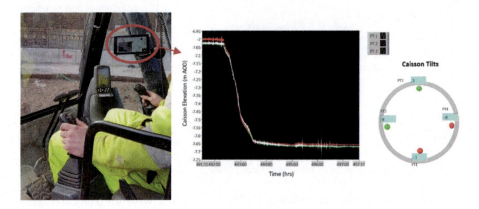

Fig. 3. Feed-back of caisson elevations via tablet in excavator

4 Conclusions

A bespoke monitoring system has been developed at University of Oxford to inform the construction of large-diameter caissons. For the pilot project considered here, the monitoring system had a significant impact on the economy and safety of the construction process. Additional developments underway include integration of wireless sensor networks, event-driven decision-making, and anomaly detection.

References

Abdrabbo, F., Gaaver, K.: Challenges and uncertainties relating to open caissons. DFI J. J. Deep Found. Inst. **6**(1), 21–32 (2012)

Nonveiller, E.: Open caissons for deep foundations. J. Geotech. Eng. **113**(5), 424–439 (1987)

Puller, M.: Deep Excavations: A Practical Manual. Thomas Telford (2003)

Royston, R., Phillips, B.M., Sheil, B.B., Byrne, B.: Bearing capacity beneath tapered blades of open dug caissons in sand. In: Proceedings of CERI 2016 conference, Ireland (2016)

Safiullah, A.M.M: Geotechnical problems of bridge construction in Bangladesh. In: Japan-Bangladesh Joint Seminar on Advances in Bridge Engineering, vol. 10, pp. 135–146 (2005)

Tomlinson, M.J.: Foundation Design and Construction. Longman, Harlow (1986)

Wang, J., Chai, L.S., Wu, H.: Numerical simulation on the sinking process of open caisson with particle flow code (PFC). Adv. Mater. Res., 831–834 (2014)

Influences on CPT-Results in a Small Volume Calibration Chamber

F. T. Stähler[✉], S. Kreiter, M. Goodarzi, D. Al-Sammarraie, T. Stanski, and T. Mörz

MARUM - Center for Marine Environmental Sciences, University of Bremen,
Leobener Straße 8, 28359 Bremen, Germany
fstaehler@marum.de

Abstract. The cone penetration test (CPT) is a standard method for determining geotechnical subsoil properties. Different CPTs were performed in a small volume calibration chamber to quantify the influence of soil type, mean effective stress, relative density, boundary condition, dry or wet samples and vibratory CPT on the cone resistance. The results contribute to the interpretation of field and laboratory CPT-results.

Keywords: Cone penetration test · Calibration chamber · Cone resistance

1 Introduction

The foundation of wind-power plants in North Sea grounds requires an accurate determination of geotechnical subsoil properties. The cone penetration test (CPT) is a standard method and leads, on the basis of empirical correlations on the cone resistance, to a determination of the desired parameters [4]. Most empirical correlations were established in calibration chambers and the best studied material is Ticino Sand. These correlations on Ticino Sand may not be suitable for the interpretation of CPT-results in North Sea Sand. The cone resistance may be influenced by the soil type and the grain angularity, mica content, void ratio or percentage of fines [2, 8, 10]. Some studies suggest that dry and fully saturated wet sand lead to similar CPT-results [10, 11]. It is generally agreed that CPT-results are influenced by the mean effective stress [5, 9].

Displacement controlled vibratory CPTs are intended as a prediction tool for vibratory pile driving and have lower cone resistances in sandy soil compared to quasi-static standard CPTs [3, 14]. Finite sample size and boundary artifacts are an issue for laboratory CPTs in a calibration chamber Fig. 1 [9, 13]. One part of this study therefore compares the results of a constant stress boundary (BC1) and a simulated field boundary (BC5). A comparison with literature is used to determine which boundary condition generates reliable results. The other part of this study compares CPT-results between Ticino and North Sea Sand, dry and wet samples, vibratory and static penetration and different mean effective stresses in order to contribute to the interpretation of different sands.

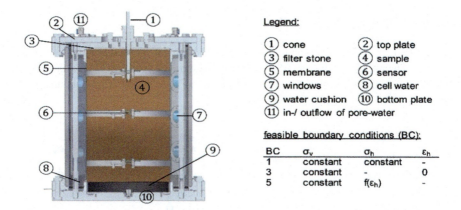

Fig. 1. MARUM calibration chamber (MARCC), where BC is the boundary condition, $\sigma_{v,h}$ is the stress in vertical or horizontal direction and ε_h is the strain in horizontal direction.

2 Methodology

Two different sands are used in this study. Ticino Sand from Northern Italy is a uniform-coarse to -medium sand with angular to sub-rounded grains and mainly consists of feldspar, quartz and mica [5, 10]. Cuxhaven Sand from Northern Germany is a medium to fine sand with angular to sub-angular grains and has a quartz content of at least 95% [6]. Cuxhaven Sand is used as an analogue to North Sea Sand.

All CPTs were performed in the MARUM calibration chamber (MARCC) with a sample height of 54.5 cm, a sample diameter of 30 cm and a sample to cone diameter ratio R_d of 25 [6]. The sample is surrounded by a flexible latex membrane and changes in diameter are measured by three displacement sensors. The vertical stress is applied by a water cushion at the bottom of the sample and the horizontal stress is induced by water pressure in the surrounding cell.

Eight Ticino Sand samples were air pluviated to relative densities between 0.5 and 1. Four samples were tested dry and were consolidated to a vertical effective stress σ'_v of 200 kPa, a horizontal effective stress σ'_h of 90 kPa and to a mean effective stress p' of 126.67 kPa. Further four samples were tested wet and were consolidated to σ'_v of 100 kPa and σ'_h of 45 kPa with p' of 63.33 kPa. The consolidation stresses were kept for 90 min and all samples were then penetrated by a miniature cone of 12 mm diameter. The penetration was performed under BC1 and the penetration rate was adjusted to 20 ± 0.3 mm/s. The steady state cone resistance was usually reached between 100 and 300 mm penetration depths.

Nineteen static and vibratory CPTs with BC1 and BC5 are published elsewhere and are compared with the CPT-results of this study [7, 13, 14]. The vibration of the MARUM vibratory CPT is oriented vertically and was operated with a single-amplitude of 0.5 mm and a frequency of 20 Hz.

3 Results and Discussion

An increase in the cone resistance is observed with an increase in relative density (Fig. 2). Changes from BC1 to BC5, vibratory to static penetration, wet to dry samples and from Ticino to Cuxhaven Sand led to higher cone resistances. Higher mean effective stress also leads to higher cone resistance.

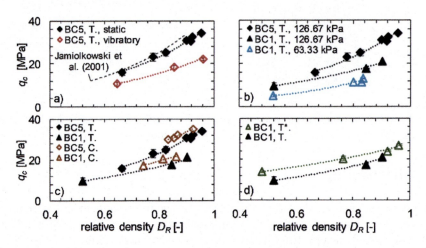

Fig. 2. (a) Influence of quasi-static and vibratory CPTs, (b) mean effective stress, (c) soil type and boundary condition, and (d) dry or wet sand on the cone resistance q_c as a function of the relative density D_R; where T is wet Ticino Sand, T* is dry Ticino Sand and C is wet Cuxhaven Sand.

The results with BC5 are in good agreement with the previous study of Jamiolkowski et al. [10], who performed CPTs in a large calibration chamber and corrected the results to field conditions (Fig. 2a). Differences in the cone resistance between BC1 and BC5 indicate boundary artifacts in CPT-results in the MARCC with BC1 Fig. 2b, c, [7, 13]. This might be related to a sample to cone diameter ratio R_d of 25, which is less than the recommended ratio of $R_d \geq 33$ or $R_d \geq 100$ for loose or very dense sand, respectively [1]. Hence reliable results either need very high R_d with BC1 or simulated field conditions BC5.

Vibratory CPTs reduce the cone resistance which might be caused by an increase in pore pressure around the cone or a change in the grain fabric Fig. 2a [14]. As expected, the mean effective stress is positively correlated with the cone resistance Fig. 2b, [9]. Ticino Sand has a lower cone resistance than Cuxhaven Sand, possibly because of the higher mica content (Fig. 2c). This observation is in agreement with Robertson and Campanella (1983) [12], who described a lower cone resistance for Ticino Sand than for Ottawa quartz Sand. A decrease in the cone resistance of wet samples in Fig. 2d is in contrast to previous studies of Jamiolkowski et al. [10] and Pournaghiazar et al. [11]. This might be explained by water acting as a lubricant at grain contacts.

Acknowledgements. We acknowledge the MARUM – Center for Marine Environmental Sciences, University of Bremen and the Federal Ministry for Economic Affairs and Energy (BMWi) for financially supporting the project "Vibro Drucksondierungen" (FKZ: 0325906A). We are grateful for technical assistance and help of Marc Huhndorf, Wolfgang Schunn, Joann Schmid and Amr Elshafei.

References

1. Ahmadi, M.M., Robertson, P.K.: A numerical study of chamber size and boundary effects on CPT tip resistance in NC sand. Sci. Iranic. **15**(5), 541–553 (2008)
2. Ahmed, S.M., Agaiby, S.W., Mayne, P.W.: Sand compressibility aspects related to CPT. In: 3rd International Symposium on Cone Penetration Testing (CPT 2014), Las Vegas (2014)
3. Bonita, J., Mitchell, J.K., Brandon, T.L.: The effect of vibration on the penetration resistance and pore water pressure. In: Mayne, V.D.F. (ed.) Geotechnical and Geophysical Site Characterization, Porto, 19–22 September 2004. Mill Press. Rotterdam (2004)
4. DIN EN ISO 22476-1: Geotechnical investigation and testing – Field testing – Part 1: Electrical cone and piezocone penetration test. German Standards (2013)
5. Fioravante, V., Giretti, V.: Unidirectional cyclic resistance of Ticino and Toyoura sands from centrifuge cone penetration tests. Acta Geotech. **11**(4), 953–968 (2015)
6. Fleischer, M., Kreiter, S., Mörz, T., Huhndorf, M.: A small volume calibration chamber for cone penetration testing (CPT) on submarine soils. In: Lamarche, G., et al. (eds.) Submarine Mass Movements and Their Consequences. Springer, Cham (2016)
7. Goodarzi, M., Stähler, F.T., Kreiter, S., Rouainia, M., Kluger, M., Mörz, T.: Numerical simulation of cone penetration test in a small-volume calibration chamber: the effect of boundary conditions. In: 4th International Symposium on Cone Penetration Testing (CPT 2018), Delft (2018, in press)
8. Hara, T., Kokusho, T., Tanaka, M.: Liquefaction strength of sand materials containing fines compared with cone resistance in triaxial specimen. In: World Conference on Earthquake Engineering, Beijing (2008)
9. Hsu, H.-H., Huang, A.-B.: Calibration of cone penetration test in sand. Proc. Nat. Sci. Counc. ROC(A) **23**(5), 579–590 (1999)
10. Jamiolkowski, M., Lo Presti, D.C.F., Manassero, M.: Evaluation of relative density and shear strength of sands from CPT and DMT. In: Soil Behavior and Soft Ground Construction (2001)
11. Pournaghiazar, M., Russel, A.R., Khalili, N.: Development of a new calibration chamber for conducting cone penetration tests in unsaturated soils. Can. Geotech. J. **48**, 314–321 (2011)
12. Robertson, P.K., Campanella, R.G.: Interpretation of cone penetration tests. Part I: Sand. Can. Geotech. J. **20**, 718–733 (1983)
13. Stähler, F.T., Goodarzi, M., Kreiter, S., Al-Sammaraie, D., Fleischer, M., Stanski, T., Ossig, B., Mörz, T.: A small volume calibration chamber for cone penetration tests under simulated field conditions, Messen in der Geotechnik 2018, Braunschweig (2018)
14. Stähler, F.T., Kreiter, S., Goodarzi, M., Al-Sammaraie, D., Mörz, T.: Liquefaction resistance by static and vibratory cone penetration tests. In: 4th International Symposium on Cone Penetration Testing (CPT 2018), Delft (2018, accepted)

Estimation of Relative Density in Calcareous Sands Using the Karlsruhe Interpretation Method

Franz Tschuchnigg[1(✉)], Johannes Reinisch[1], and Robert Thurner[2]

[1] Institute of Soil Mechanics, Foundation Engineering and Computational Geotechnics, Graz University of Technology, Graz, Austria
franz.tschuchnigg@tugraz.at
[2] Keller Grundbau, Vienna, Austria

Abstract. The paper describes the challenges that arise during execution and quality control of deep vibro-compaction works on land reclamation projects realized in the Middle East. For quality control, the achieved relative density throughout the hydraulic fill can be measured indirectly by performing Cone Penetration Tests (CPT). Commonly used empirical correlations between the cone resistance q_c and the relative density D_r are not suitable to reproduce the characteristics of carbonate sands, because these sands show significantly lower q_c-values than silica sands under similar conditions. An alternative to traditional correlation methods to derive the q_c-requirement is the use of the semi-empirical Karlsruhe Interpretation Method (KIM) which is based on numerical analyses of spherical cavity expansion (SCE) problems and results of calibration chamber tests. This paper explains the general procedure when applying the KIM with a focus on the specific characteristics of carbonate sands.

Keywords: KIM · CPT · Land reclamation · Deep vibro-compaction · Calcareous sands

1 Introduction

The relative density D_r achieved by vibro-compaction works can be measured indirectly with a correlation between the cone resistance q_c obtained from a Cone Penetration Test. Schmertmann [1] and Baldi [2] developed such correlations based on calibration chamber tests performed on a wide range of different silica sands. However, as most of the material used for land reclamation in the UAE consists, at least partially, of calcareous sand, it is questionable if these correlations are able to reflect the characteristics of these sands.

Calcareous sands consist of varying amounts of skeletal remains of marine organisms such as fragmented sea-shells and corals and show a high tendency towards grain crushing. Almeida et al. [3] have shown that, compared to silica sands, calcareous sands produce significantly lower q_c-values even in a similar soil state.

As an alternative to the existing correlations, the Karlsruhe Interpretation Method (KIM) [4] has the advantage that the required parameters are determined on sand samples taken from the actual site which allows a site-specific representation of the material and can take the crushability of calcareous sand into account.

2 Influence of the Carbonate Content on Soil Properties

On a database of 11 sand samples it was found that basic geotechnical parameters like the limit void ratios e_{min} and e_{max}, the critical friction angle φ'_c and the grain density ρ_s increase with an increasing carbonate content of the sand. It is assumed that this behavior is caused by the angular shape of sea-shells and their fragments which form strongly interlocking grain skeletons with a high number of voids. Figure 1 shows the increase of the void ratios (e_{min} and e_{max}) with the carbonate content for the sand samples in the database. The grain density is influenced by the mineralogy of the sand. Sea-shells and other skeletal remains of marine organisms consist of calcite or aragonite with a density of 2.7 g/cm³ and 2.9 g/cm³ respectively and therefore cause the grain density to increase as well.

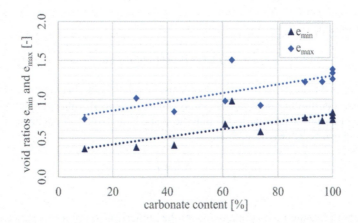

Fig. 1. Increase of the void ratios e_{min} and e_{max} with the carbonate content

Especially the void ratios and the critical friction angle have a significant influence on the soil's unit weight γ and therefore also on the resulting mean effective stress p'_0, which is used in the equation of the KIM. With an increasing carbonate content, γ and p'_0 decrease and affect the resulting q_c-requirement calculated with the KIM.

3 CPT Interpretation Methods by Schmertmann and Baldi

The method developed by Schmertmann [1] is an empirical relationship between the cone resistance q_c and the relative density D_r which was developed based on the results of 80 calibration chamber tests on silica sands. Equation (1) shows the equation for interpreting CPT measurements:

$$q_c = C_0 * \sigma_v'^{(C_1)} * \exp^{(D_r * C_2)} \tag{1}$$

The C_i-values by Schmertmann [1] and Baldi [2] listed in Table 1 do not account for grain crushing or the lower compressibility of calcareous sands which are the main reasons for the differences between the cone resistance in silica and calcareous sands.

Table 1. C_i constants for CPT interpretation methods by Schmertmann [1] and Baldi [2]

Correlation	C_0	C_1	C_2
Schmertmann [1]	12.31	0.71	2.91
Baldi [2]	157	0.55	2.41

4 The Karlsruhe Interpretation Method

The Karlsruhe Interpretation Method (KIM) was developed by Cudmani [4] in 2001 and is a semi-empirical method based on numerical simulations of a spherical cavity expansion (SCE) problem. Equation (2) shows the basic form of Cudmani's equation for the cone resistance q_c:

$$q_c = k_q(D_r) * p_{LS}(D_r, p_0') \tag{2}$$

Term k_q is the shape factor and establishes the relationship between the cone resistance q_c and the limit pressure p_{LS}. By analyzing the results of calibration chamber tests on 9 different sands, Cudmani found that the shape factor solely depends on the relative density and can be approximated with:

$$k_q = 1.5 + \frac{5.8 * D_r^2}{D_r^2 + 0.11} \tag{3}$$

The limit pressure p_{LS} is obtained from the numerical solution of a spherical cavity expansion (SCE) problem. In the same way as the cone resistance, the limit pressure depends on the state of the soil and increases with an increasing relative density D_r and mean effective stress p_0'.

$$p_{LS} = a * p_0'^b \tag{4}$$

$$a = a_1 + \frac{a_2}{D_r + a_3} \tag{5}$$

$$b = b_1 + \frac{b_2}{D_r + b_3} \tag{6}$$

The a_i and b_i parameters are determined by curve-fitting the results of a series of solutions of the SCE for different initial conditions with Eqs. (5) and (6). For these simulations, it is necessary to calibrate the hypoplastic constitutive equation (to model the mechanical behavior of the sand present on the site). The hypoplastic model used in the KIM was proposed by von Wolffersdorff [5] and requires eight material parameters. It has to be mentioned that both, the determination of the input parameters for the hypoplastic constitutive model and the KIM parameters entail a certain sensitivity which extends into the resulting q_c-requirement.

This is one reason why the input parameters for the hypoplastic model and the KIM parameters (and therefore also the resulting q_c-value) might show some differences even though the sands basic properties (i.e. e_{min}, e_{max}, φ'_c, ρ_s) are very similar.

5 Results and Conclusion

Figure 2 depicts the relationship between the carbonate content and the cone resistance q_c and clearly shows that the results decrease with an increasing carbonate content. This proves that the characteristics of calcareous sands can be incorporated in the KIM.

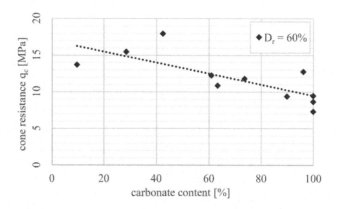

Fig. 2. Relationship between the carbonate content and the cone resistance q_c

However, for samples with a carbonate content below 50% the resulting values of the cone resistance increase. Based on the studies and the available data it is assumed that the carbonate sand fraction only starts to influence the sand's mechanical behavior once it surpasses a certain threshold of around 50%. Further research is ongoing to investigate this behavior in more detail and additional samples in the database are needed to underpin the findings so far. In an attempt to clarify the connection between the soil properties, the input parameters for the hypoplastic model and the KIM parameters, an extensive sensitivity analysis will be carried out as a next step.

Concluding this paper, it is stated that even though there are certain sensitivities connected to the determination of the parameters, the KIM is considered to be a valid method to calculate a q_c-requirement in calcareous sands.

References

1. Schmertmann, J.: An updated correlation between relative density Dr and Fugro-Type electric cone bearing, qc, Contract Report DACW 39-76 M 6646 WES, Wicksburg, Mississippi, Technical report (1976)
2. Baldi, G., Bellotti, R., Ghionna, N., Jamiolkowski, M., Pasqualini, E.: Interpretation of CPTs and CPTUs, 2nd Part: drained penetration of sands. In: 4th International Geotechnical Seminar, Field Instrumentation and In-Situ Measurements, pp. 143–155, Singapore (1986)
3. Almeida, M., Jamiolkowski, M., Peterson, R.: Preliminary result of CPT tests in calcareous Quiou sand. In: Huang, A.-B. (ed.) Proceedings of the First International Symposium on Calibration Chamber Testing, pp. 41–53. Elsevier Ltd., Potsdam, New York (1991)
4. Cudmani, R.: Statische, alternierende und dynamische Penetration in nichtbindigen Böden. Ph.D. thesis, Veröffentlichungen des Instituts für Bodenmechanik und Felsmechanik, Universität Karlsruhe, Heft 152 (2001)
5. von Wolffersdorff, P.A.: A hypoplastic relation for granular materials with a predefined limit state surface. Mech. Cohesive-Frictional Mater. **1**(3), 251–271 (1996)

Investigation on Precursor Information of Granite Failure Through Acoustic Emission

C. S. Wang[1], H. W. Zhou[1(✉)], W. G. Ren[1], S. S. He[1], H. Pei[1], and J. F. Liu[2]

[1] School of Mechanics and Civil Engineering,
China University of Mining and Technology, Xuzhou, China
zhw@cumtb.edu.cn
[2] School of Water Resources and Hydropower,
Sichuan University, Chengdu, Sichuan, China

Abstract. Dynamic disaster often occur in the underground engineering, so it is important to research the precursor information of rock failure, triaxial test was performed to get acoustic emission (AE) characteristics, correlation dimension was used to research sequential characteristics, this study shows that (1) AE events have good fractal characteristics and correlation dimension can describe the process of granite failure; (2) Comprehensive correlation can be divided into falling stages and rising stages; (3) When correlation dimension is increasing, surrounding rock is stable, when correlation dimension is decreasing, surrounding rock will be destroyed.

Keywords: Beishan granite · AE · Fractal dimension · Precursor information

1 Introduction

AE is a phenomenon that energy was rapid released and elastic wave instantly generated in the process of loading, it usually produces by crack generation, expansion and failure and is paid much attention to in engineering rock. It is very important to provide theoretical basis for on-site monitoring by studying the characteristics of AE during rock failure through laboratory experiments for the complexity of site engineering conditions. The relationship between AE and stress-strain in rock failure process under different stress states have been studied by many scholars, such as Vil-helm et al. [1] found the parameters of the autocorrelation function of the AE event series change significantly near failure; Chmel et al. [2] studied micro cracking in impact-damaged Westerly granite and Karelian quartzite based on evidence derived from fractoluminescence and AE; Přikryl et al. [3] researched AE characteristics and failure of uniaxially stressed granitic rocks. Meng et al. [4] analyzed effects of AE and energy evolution of rock specimens under the uniaxial cyclic loading and unloading compression. Bunger et al. [5] show that simple breakage of a crystalline rock in tension begets further breakage of rock in the area around the first crack is self-sustaining and spontaneous and detected via sustained AE; Zhang et al. [6] Compared AE characteristics of different rocks during the damage evolution process; Lavrov and Shkuratnik [7] analyzed the main directions of the development of models for the AE in geological materials. Xie [8] expound and proved fractures of rocks from micro-crack to broken all have fractal characteristics, and fractal dimension gradually reduce in the process of failure. Correlation dimension [9] of AE have been widely used in the study of fractal characteristics on rock failure.

2 Experiment Method

The test samples were taken from the Beishan surface granite in the preselected area of China's high-level radioactive waste geological repository. The density was 2.6 g/cm3. Triaxial compression experiments were carried out on the MTS815 Flex Test GT rock mechanics test system manufactured in the United States. All the testing procedures were programmed by computer. Based on "standard test method for engineering rock mass", the samples were processed into standard size with Φ50 × H100 mm, the allowable deviation of the diameter is less than 0.2 mm, the allowable deviation of the unevenness of the two end faces is less than 0.05 mm, the vertical deviation of the end face and the axis does not exceed ± 0.25°.

Circumferential deformation was used to control loading, axial extensometer and circumferential extensometer were used to measure axial deformation and circumferential deformation respectively. PCI-II Acoustic Apparatus manufactured by American Acoustics Company was used to records AE events in the process of loading, sampling frequency was 200 kHz, threshold of AE was set to 28 dB. At the beginning of the experiment, the AE system and the loading device were turned on at the same time, so that the two devices start recording data at a uniform time. During the test, the confining pressure is set to 5, 10, 15 and 30 MPa.

3 Experiment Results and Analysis

The method calculating AE fractal dimension through G-P relationship is shown in Eq. (1).

$$D = \lim_{r \to 0} \frac{\lg C(r)}{\lg r} \tag{1}$$

Where, r is distance of reconstruction space, $C(r)$ is numbers ratio of AE events pairs whose distance is less than r to the total pairs. Obviously, $C(r)$ is related to r value chosen, and r value is too large or too small to reflect the internal properties of the system. When D increases, AE events are more concentrated in the reconstruction space, indicating that AE energy rates tend to be consistent; When D value decrease, AE events were scattered in the reconstruction space, indicating abnormal AE events increase.

In this paper, the time series of energy release rate of AE is taken as the research object. Reconstruction dimension is 15, delay time is 1 s, correlation dimension is calculated every 500 s, $\lg r$ and $\lg C(r)$ was fitted by linear (Fig. 1).

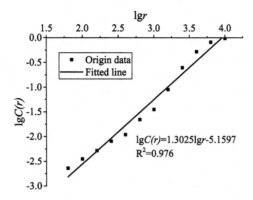

Fig. 1. Fitting curve of lg$N$ and lg$Q$

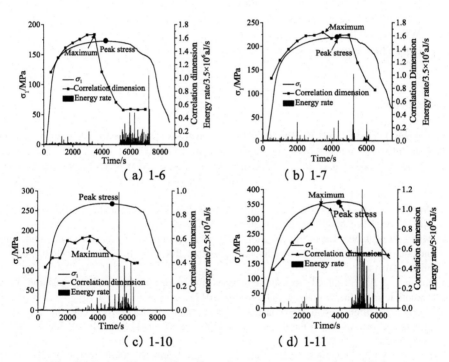

Fig. 2. Curves of correlation dimension, energy rate and stress with time

The relationship between D value and stress, time was shown in Fig. 2, D can be divided into rising and falling stages, rising stages is from starting loading to later of crack unstable expansion, proportion of short distances AE events pairs increases during this process, that indicate AE energy rate is relatively stable and few abnormal events occur, it is consistent with the rule of energy rate in Fig. 2. Therefor few abnormal cracks occur during rising stage. When the correlation dimension reaches the maximum, it begins to decline continuously. The falling stage begins at later of crack unstable expansion period and then continues to decline throughout the experiment, proportion of long distances events pairs increases during falling stage, that shows AE energy rate appear abnormality in size and time sequence, therefore, many abnormal cracks occur in this stage. So D can be chosen as precursor information, When correlation dimension is increasing, surrounding rock is stable, when correlation dimension is decreasing, surrounding rock will be destroyed.

4 Conclusion

Through laboratory experiment, time sequence characteristics was researched based on fractal theory, conclusions as following:

(1) AE events have good fractal characteristics and correlation dimension can describe the process of granite failure;
(2) Comprehensive correlation can be divided into falling stages and rising stages;
(3) When correlation dimension is increasing, surrounding rock is stable, when correlation dimension is decreasing, surrounding rock will be destroyed.

References

1. Vilhelm, J., et al.: Application of autocorrelation analysis for interpreting acoustic emission in rock. Int. J. Rock Mech. Min. Sci. **45**(7), 1068–1081 (2008)
2. Chmel, A., Shcherbakov, I.: Microcracking in impact-damaged Westerly granite and Karelian quartzite based on evidence derived from fractoluminescence and AE. Int. J. Rock Mech. Min. Sci. **80**, 181–184 (2015)
3. Přikryl, R.: AE characteristics and failure of uniaxially stressed granitic rocks: the effect of rock fabric. Rock Mech. Rock Eng. **36**(4), 255–270 (2003)
4. Meng, Q.: Effects of acoustic emission and energy evolution of rock specimens under the uniaxial cyclic loading and unloading compression. Rock Mech. Rock Eng. **49**(10), 3873–3886 (2016)
5. Bunger, A.P.: Sustained acoustic emissions following tensile crack propagation in a crystalline rock. Int. J. Fract. **193**(1), 87–98 (2015)
6. Zhang, Z.: Differences in the acoustic emission characteristics of rock salt compared with granite and marble during the damage evolution process. Environ. Earth Sci. **73**(11), 6987–6999 (2015)
7. Lavrov, A.V., Shkuratnik, V.L.: Deformation-and fracture-induced acoustic emission in rocks. Acoust. Phys. **51**(1), S2–S11 (2005)
8. Xie, H.: Introduction to fractal rock mechanics. Science Press, Beijing (1996)
9. Costin, L.S.: A microcrack model for the deformation and failure of brittle rock. J. Geophys. Res. Solid Earth **88**(B11), 9485–9492 (1983)

An Internet Based Intelligent System for Early Concrete Curing in Underground Structures

Yongbin Wei, Jindi Lin, Wei Zhao, Chao Hou, Zhenhuan Yu,
Maowei Qiao, and Bangan Liu(✉)

China State Construction Engineering Corporation, Beijing, China
liubangan@cscec.com

Abstract. This paper described a newly developed internet based intelligent early concrete curing system for underground structure construction. The system was composed of field measuring and controlling subsystem (including temperature, humidity and wind speed sensors, solenoid valve, water pump and spray et al.), remote displaying and controlling subsystem (displaying and controlling through PC software or smart phone app) and remote server (data storage and controlling strategy). This system could monitor the wind speed, temperature and humidity condition in field and transfer the data to remote server, which, based on some input parameters, could automatically generate and send the controlling instruction to the field controlling subsystem to control the water pump and spray time. If heat preservation is needed, messages might be sent to the field so that workers would perform the covers maintenance. This system was utilized in a side wall concrete curing project in a station of the Changsha #4 subway, Hunan, China. The inside and surface temperature difference of the concrete is around 3.12 °C. The surface and environmental temperature difference of the concrete is around 9.36 °C. Humidity is maintained at no less than 85%. The site photos shows pitted surfaces and drying shrinkage cracks would appear more easily for concrete under manual work than which under this system. The system could avoid the non-uniformity caused by manual work and also reduce the labor cost.

Keywords: Concrete intelligent maintenance · Internet of things
Atomizing spray stream

1 Introduction

With the development of economy and the accelerating process of urbanization, all major cities were energetically building rail transit. During the construction of rail transit, concrete cracks might appear in the project because of the temperature differences of concrete and shrinkage stress. Good maintenance could make concrete good performance, and it also had an important influence on the durability of concrete. At present, the traditional way of concrete curing was implemented by constructor according to the personal experience of the field. In this way, it could cause many problems, such as irregular sprinkler, inaccurate control of temperature and humidity of concrete, insufficient water conservation and low maintenance efficiency [1–4].

In order to ensure the quality of concrete, this paper designed an intelligent concrete curing device, which was utilized in a side wall concrete curing project at the Changsha #4 subway.

2 Curing System Design and Application

The concrete curing system used a distributed architecture and was composed of field measuring and controlling subsystem, remote displaying and controlling subsystem and remote server. The overall architecture of the system is shown in Fig. 1.

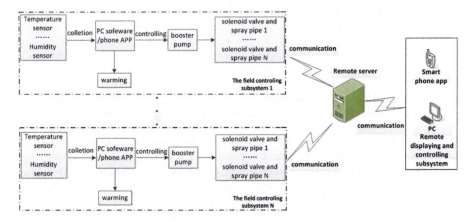

Fig. 1. System structure.

The intelligent early concrete curing system can be used for concrete maintenance in multiple field areas at the same time [5]. Each field acquisition control system is composed of displaying and controlling through PC software or smart phone app, wind speed sensor, temperature and humidity sensor, booster pump, solenoid valve and spray pipe.

The field control system could monitor the wind speed of the environment, the surface and internal temperature and humidity. According to the controlling strategy (according to the specifications of the concrete curing index and the situation of the site, the parameters of the control terminal were configured and the corresponding control strategy is formed), the field control system could control the water pump and spray time. And fine curing of concrete could be realized. If the temperature between the surface and interior was large, the maintenance needed to be carried out in the field. And the messages were sent to the field so that workers would perform the covers maintenance.

The field control system uploaded the monitoring data and the corresponding control results to the remote server in real time. All the monitoring data were save in the database on the server. Related managers could viewed data through remote control terminals, and checked the concrete maintenance in real time. Administrators could configure and manage each field control system through remote control terminals, and they could operate remotely.

The field controlling system was encapsulated in a stainless steel box, which could be waterproof. The internal system was composed of pressure pump, signal conditioning board, data acquisition and output control module, solenoid valve, solid-state relay, leakage protection switch, remote data transmission module and human-computer interaction module, the inlet and outlet pipeline interface. The system structure is shown in Fig. 2. The controlling and viewing could be performed on a PC software or a mobile application. The software supported multiple projects at the same, and it could implement hierarchical authority management.

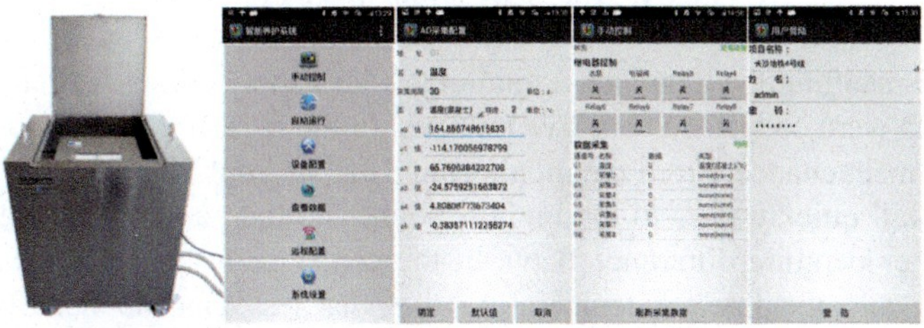

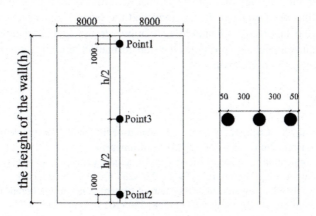

Fig. 2. Field controlling system and the mobile application.

The system was applied to the side wall maintenance project of Changsha Metro Line 4 in June 2016. Three temperature sensors were arranged on the plane center line of the side wall (as shown in Fig. 3). In addition, there was one temperature sensor and one humidity sensor on the outer surface of the test point, which were used to measure the air temperature and humidity outside the side wall. Two sets of automatic sprinkler lines on the side of the wall was fixed on the scaffold (as shown in Fig. 4). The spacing of the nozzle was 1 m, and the distance from the side wall was 0.5 m–1.5 m.

Fig. 3. Sensors placement position (facade and profile) (mm).

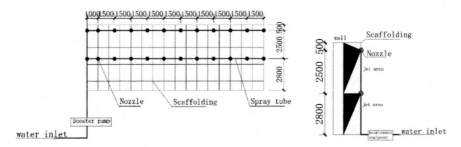

Fig. 4. The spray layout of wall (mm).

After dismantling the template, a total of 28 days spray curing was carried out. All kinds of data were collected every five minutes. In the face of the wall, no cracks were found after maintenance. From the monitoring data, we could find that the temperature inside the wall quickly rises to the highest point, and then a slow decline to a steady trend. The temperature difference between the center point and the surface was less than 10 °C, and the internal temperature rise was less than 45 °C. Moreover, the surface temperature of the wall was less than 40 °C, and the surface humidity was kept above 87%RH.

3 Conclusions

This paper introduced the design and application of concrete intelligent maintenance system. Through on-site application, the real-time monitoring of the internal temperature and humidity of the concrete was realized, and the concrete could be sprayed automatically so that the moisture content of the concrete surface was guaranteed. The system could avoid the non-uniformity caused by manual work and the formation of cracks, and improved the quality of concrete maintenance. This system is also a typical application of the Internet of things in construction. The real-time maintenance by management has been realized. It has improved the level of construction information and reduced the labor cost. Meanwhile, it improves the level and efficiency of the project management and also promotes the transformation of civil engineering from the extensive type to the refinement.

References

1. Yang, J., Peng, X., Chen, K., Wang, J.: Research of the intelligent maintenance system of concrete. Concrete **36**(4), 155–158 (2015). (in Chinese)
2. Zhao, Y.: Application research of the system of cement concrete intelligent maintenance in the bridge engineering. Highw. Eng. **39**(2), 284–287 (2014). (in Chinese)
3. Chen, K.: Application research of the system of cement concrete intelligent maintenance in the bridge engineering. Highway **38**(11), 218–220 (2013). (in Chinese)
4. Min, T., Luo, L., Xaia, W., Chen, K.: Research on key technology and design of the system of cement concrete intelligent maintenance. Highw. Eng. **39**(2), 158–162 (2014). (in Chinese)
5. Hou, C., Wei, Y.: Design and application of remote intelligent temperature monitoring system for mass concrete. Constr. Technol. **43**(21), 94–96 (2014). (in Chinese)

Application of Multivariate Data-Based Model in Early Warning of Landslides

Hongyu Wu, Mei Dong(✉), and Xiaonan Gong

Research Center of Coastal and Urban Geotechnical Engineering,
Zhejiang University, Hangzhou, China
mdong@zju.edu.cn

Abstract. The evolution of landslides is influenced by internal geological conditions and external trigger factors such as the precipitation and the change of groundwater level. It's difficult to establish a precise physical model to predict the happening of landslides on account of the complex and unobtainable internal mechanisms. Meanwhile, development of advanced real-time monitoring system has challenged engineers to detect and analyze valuable features from high amount of data of the status of slopes. Multivariate data-based models have been applicable to the early warning of landslides. In the current paper, cumulative displacement was decomposed into trend components and periodic components by Ensemble Empirical Mode Decomposition (EEMD) method. Based on abundant monitoring data, the trend displacement was predicted by polynomial function, while the periodic displacement was predicted by Particle Swarm Optimization and Support Vector Machine (PSO-SVM) model. A high and steep slope along the express way in Zhejiang province, China was taken as a case study. The model proposed in the current paper has been demonstrated to be highly accurate in predicting the displacement of rainfall-induced landslides.

Keywords: Displacement prediction · Early warning · PSO-SVM
Rainfall-Induced landslides

1 Introduction

The evolution of landslides is influenced by internal geological conditions and external trigger factors such as the precipitation and the change of groundwater level. It's difficult to establish a precise physical model to predict the happening of landslides on account of the complex and unobtainable internal mechanisms. Therefore, data-based models have been widely used in practical engineering.

Predicting the displacement of landslides is a significant work in the early warning system. Numerous data-based models have been proposed to solve this problem, such as Artificial Neural Network (ANN) [1], Support Vector Machines (SVM) [2] and Extreme Learning Machine (ELM) [3]. Among these models, SVM is widely used and has achieved good performance in predicting the displacement [4].

In this paper, a short-term prediction was conducted by SVM models using daily data rather than monthly data. A case study was taken to estimate the performance of models with different input data.

2 Methodology

In the current paper, a high and steep slope along the express way in Zhejiang province, China was taken as a case study. The cumulative displacement of this landslide was decomposed into trend components and periodic components by Ensemble Empirical Mode Decomposition (EEMD) method. Based on abundant monitoring data, the trend displacement was predicted by polynomial function, while the periodic displacement was predicted by three SVM models (Grid-SVM, GA-SVM, PSO-SVM). Each model contains 4 types of combination of input data as shown in Table 1. Root mean square error (RMSE) and correlation coefficient (R^2) were taken as the indexes to estimate the performance of these models.

Table 1. Input data of different models

Models	Input data
Grid-SVM-1, GA-SVM-1, PSO-SVM-1	Displacement in the prior 3 days only
Grid-SVM-2, GA-SVM-2, PSO-SVM-2	Displacement in the prior 3 days and rainfall in the prior 3 days
Grid-SVM-3, GA-SVM-3, PSO-SVM-3	Displacement in the prior 3 days and rainfall in the prior 4 days
Grid-SVM-4, GA-SVM-4, PSO-SVM-4	Displacement in the prior 3 days and rainfall in the prior 5 days

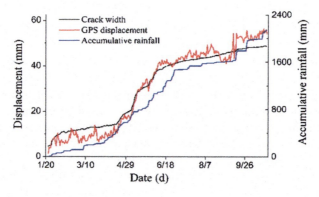

Fig. 1. Monitoring data of a landslide by the express way in Zhejiang province, China

3 Results and Data Analysis

The landslide we researched in this study is a typical rainfall-induced landslide as we can see in Fig. 1. The displacement developed slowly in the dry season, while grew rapidly in the wet season, especially in May and June. Rainfall is the crucial trigger factor of the evolution of this landslide. 100 sets of daily data (from 2016/3/6 to 2016/6/13) were selected to test the models. The first 80 sets were training data and the rest were test data.

Figure 2 shows the predicted periodic displacement of 12 models. With input data of only displacement in the prior 3 days, Grid-SVM-1 and GA-SVM-1 model cannot fit the original curve very well. Grid-SVM-4 and GA-SVM-4 have the best performance of prediction as shown in Table 2. The accuracy increased with more input data of rainfall.

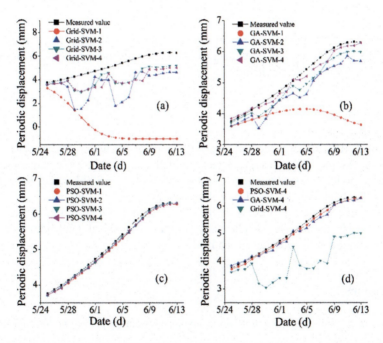

Fig. 2. Comparison of predicted periodic displacement of different models: (a) Periodic displacements predicted by Grid-SVM models. (b) Periodic displacements predicted by GA-SVM models. (c) Periodic displacements predicted by PSO-SVM models. (d) Periodic displacements predicted by Grid-SVM-4, GA-SVM-4, PS-SVM-4.

Compared to the Grid-SVM model and GA-SVM model, PSO-SVM model has better forecast capacity. Four PSO-SVM models all have accurate prediction of periodic displacement. Similarly, the accuracy of PSO-SVM models also increased slightly with more input data of rainfall as shown in Table 2.

Table 2. Comparison of the parameters and performance of different models

Models	c	g	RMSE	R^2
Grid-SVM-1	0.57	84.45	0.4595	0.6124
Grid-SVM-2	9.19	1.00	0.1596	0.0524
Grid-SVM-3	27.86	0.33	0.0951	0.3916
Grid-SVM-4	9.19	0.33	0.1076	0.3366
GA-SVM-1	23.17	2.26	0.1241	0.0010
GA-SVM-2	95.73	0.08	0.0421	0.9229
GA-SVM-3	99.58	0.07	0.0260	0.9610
GA-SVM-4	99.50	0.03	0.0073	0.9985
PSO-SVM-1	100	0.1	0.0068	0.9981
PSO-SVM-2	100	0.1	0.0058	0.9967
PSO-SVM-3	100	0.1	0.0053	0.9967
PSO-SVM-4	100	0.1	0.0060	0.9974

In summary, the PSO-SVM model has better forecast capacity than the Grid-SVM model and GA-SVM model. Input data with rainfall in the prior 5 days can increase the accuracy of SVM models evidently.

4 Conclusion

Multivariate data-based model is a powerful tool in predicting the displacement of landslides. As the crucial trigger factor, rainfall data can increase the accuracy of prediction models evidently. The forecast capacity of PSO-SVM model is better than Grid-SVM model and GA-SVM model especially in the wet season.

The whole process proposed in this study with PSO-SVM model has a high accuracy in predicting the displacement of rainfall-induced landslides which can contribute to the early warning of landslides.

References

1. Mayoraz, F., Vulliet, L.: Neural networks for slope movement prediction. Int. J. Geomech. **2** (2), 153–173 (2002)
2. Zhou, C., Yin, K., Cao, Y., Ahmed, B.: Application of time series analysis and PSO–SVM model in predicting the Bazimen landslide in the Three Gorges Reservoir. China. Eng. Geol. **204**, 108–120 (2016)
3. Huang, F., Huang, J., Jiang, S., Zhou, C.: Landslide displacement prediction based on multivariate chaotic model and extreme learning machine. Eng. Geol. **218**, 173–186 (2017)
4. Zhu, X., Xu, Q., Tang, M., Nie, W., Ma, S., Xu, Z.: Comparison of two optimized machine learning models for predicting displacement of rainfall-induced landslide: a case study in Sichuan Province, China. Eng. Geol. **218**, 213–222 (2017)

Establishment of a Health Monitoring Assessment System for Hong-Gu Tunnel in China

Xiangchun Xu[1(✉)], Songyu Liu[1], Liyuan Tong[1], Jun Guo[2], and Xing Long[3]

[1] Institute of Geotechnical Engineering,
Southeast University, Nanjing 210096, China
xxc_geo@foxmail.com, 101004004@seu.edu.cn
[2] Nanchang Municipal Public Group, Nanchang 330000, China
[3] Jiang Xi FREESUN Technology Co., Ltd, Nanchang 330000, China

Abstract. Based on Nanchang Hong-gu tunnel, which is the largest inland immersed tunnel in China, some research works on its health monitoring and assessment system were conducted. System platform was established with a mature structure including sensor subsystem, data acquisition subsystem, data transmission subsystem, database subsystem, data processing and control subsystem, health assessment and pre-warning subsystem. Monitoring index selection, sensor selection and sensor layout was fully considered in health monitoring design and technology scheme. AHP-Fuzzy evaluating method was employed in establishing the assessment model, and index weight determining method and grading control criterion was investigated and given. Then, functions of software system were achieved with advanced cloud platform technology. In summary, establishment of Nanchang Hong-gu tunnel health monitoring, assessment and smart management system was described. This system realized and made it more convenient to monitor the tunnel in real-time, health assessment, pre-warning and other functions were implemented as well. Managers can operate the system in PCs and cell phones as well. In addition, analysis works in this system would potentially provide a theoretical basis for other immersed tunnels.

Keywords: Immersed tunnel · Health monitoring · System platform

1 Engineering Survey

Total length of Nanchang Hong-gu tunnel was about 2650 m, including 1305 m immersed tube section. 9 in 12 immersed tube segments were 115 m separately, other 3 were 90 m separately. Construction clearance was 11.5 m × 4.5 m, cross section size was 30 m × 8.3 m. Nanchang Hong-gu tunnel was the largest inland river immersed tunnel in China with complex geological and fracture condition. Runoff of Ganjiang river concentrated in April ~ June, accounted for 49.6% of the year, caused a highly fluctuating water level from rainy season to dry season.

2 Analysis of Hong-Gu Tunnel Health Monitoring and Assessment System

2.1 Overview of Hong-Gu Tunnel Health Monitoring and Assessment System Platform

Hong-gu tunnel health monitoring and assessment system consists of hardware system and software system. The hardware system includes sensor subsystem, data acquisition subsystem and data transmission subsystem. The software system includes database subsystem, data processing and control subsystem, health assessment and pre-warning subsystem. Functions of the whole system were realized by coordination of each subsystem.

2.2 Hong-Gu Tunnel Health Monitoring Scheme

Monitoring indexes were selected by fully considering of tunnel design scheme and geological condition [1, 2]. Figure 1 shows detail of monitoring sensors layout.

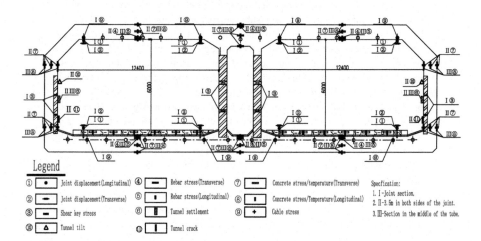

Fig. 1. Sensors layout

2.3 Analysis on Hong-Gu Tunnel Health Assessment Model

(1) Establishment of assessment model. Comprehensive assessment of immersed tunnel transformed the tunnel structure into a number of key indicators, then, evaluation criteria of indicators was established through theoretical research, engineering experience and existing specification. Mathematical model of comprehensive evaluation methods was introduced to calculate the reference value F, which reflected the tunnel health status.

Analytic hierarchy process method was adopted to establish the Hong-gu tunnel health assessment model. Figure 2 shows detail of the model.

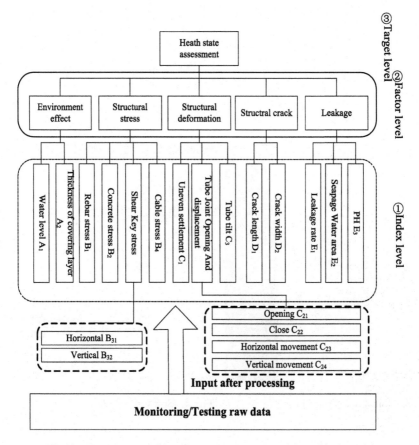

Fig. 2. Hong-gu tunnel health assessment system hierarchical model

(2) Weight determining. The subjective weight of evaluation index was determined by improved analytic hierarchy process (AHP)-multiplication scale method, and the objective weight of evaluation index was determined by entropy weight method. Then, the fusion weight ω_i was calculated with following function

$$\omega_i = 0.575\omega_{1i} + 0.425\omega_{2i} \tag{1}$$

in which, ω_{1i} was subjective weight, ω_{2i} was objective weight.

(3) Assessment criteria of tunnel health. The overall criteria of tunnel health assessment were described in Table 1 [3].

2.4 Hong-Gu Tunnel Health Monitoring and Assessment Software

The software system of this project adopted hierarchical B/S structure, .NET development platform, and ASP.NET as well as ADO.NET as the core technology. Figure 3 shows software interface in APPs, Fig. 4 shows software interface in PCs.

Table 1. Quantitative grade classification of tunnel disease

Grade	Disease status	Comprehensive assessment value F
A	Not damaged or slightly damaged	$1.5 \geq F > 1.0$
B	Damaged	$2.5 \geq F > 1.5$
C	Relatively serious damage	$3.7 \geq F > 2.5$
D	Seriously damaged	$4.0 \geq F > 3.7$

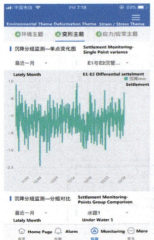

Fig. 3. Software interface in APPs

Fig. 4. Software interface in PCs

References

1. Iwaki, H., Shiba, K., Takeda, N.: Structural health monitoring system using FBG-based sensors for a damage-tolerant building. In: Smart Structures and Materials 2003: Smart Systems and Nondestructive Evaluation for Civil Infrastructures, pp. 392–399. International Society for Optics and Photonics, San Diego (2003)
2. Shi, B., Xu, X., Wang, D.: Study on BOTDR-based distributed optical fiber strain measurement for tunnel health diagnosis. Chin. J. Rock Mech. Eng. **24**(15), 2622–2628 (2005)
3. Liu, Z., Huang, H., Zhao, Y.: Immersed tube tunnel real-time health monitoring system. Chin. J. Undergr. Space Eng. **4**(6), 1110–1115 (2008)

Monitoring a Flexible Barrier Under the Impact of Large Boulder and Granular Flow Using Conventional and Optical Fibre Sensors

Jian-Hua Yin[✉], Jie-Qiong Qin, Dao-Yuan Tan, and Zhuo-Hui Zhu

Department of Civil and Environmental Engineering,
The Hong Kong Polytechnic University, Kowloon, Hong Kong, China
cejhyin@polyu.edu.hk

Abstract. Flexible barriers, as one effective way to mitigate rock-fall hazards, have obvious advantages compared to the conventional concrete check-dams. It has been proved that a flexible barrier can perform well in trapping debris flow. In this study, an improved large-scale physical model for studying the impact of debris flow on a flexible barrier system is designed and constructed in Hong Kong. Some mechanical systems of the model are carefully designed to minimize the friction in door opening process. A comprehensive monitoring system has been specially designed and installed in this physical model. The system consists of conventional sensors (accelerometers, large tension-link force transducers, strain gauges on posts), special optical fibre sensors (small FBG tension-link force transducers, FBG basal friction transducers, FBG earth pressure cells), and high-speed cameras, high-frequency data-loggers, and computers. A single boulder impact test and a granular flow impact test were conducted using this large-scale physical model in order to evaluate the performance of the flexible barriers. In this paper, we will present the design and installation of this comprehensive monitoring system and selected test results.

Keywords: Flexible barriers · Monitoring system · Impact
Optical fibre sensors

1 Introduction

Flexible barriers are widely considered as one effective way to mitigate rock-fall hazards by trapping falling rock avalanches [1]. Recently, the use of flexible barriers has been extended to landslide and debris flow mitigation. It has been proved that the flexible barriers have good performance in arresting debris avalanches [2, 3]. Compared to the conventional concrete check-dams, flexible barriers have the advantages of cost-effective design, slight site disturbance and easy construction on natural terrain. Moreover, the large deformation of flexible barriers prolongs the duration of impact process and reduces the impact forces [2]. However, the behavior of flexible barrier system upon debris impact has not been investigated systematically. Therefore, in this study, a large-scale physical model for debris flow with flexible barrier system is designed and constructed in Hong Kong (see Fig. 1) for better understanding of the performance of flexible barrier system.

A flexible barrier system is typically composed of ring-net barrier, steel wire ropes, energy dissipaters, steel posts, anchorages and foundation. Upon debris impact, the impact energy is primarily absorbed by the barrier with large deflection behavior. Force transmission among the components of the system occurs subsequently, which results in that the forces in components are complicated and hence may not be estimated straightforward. Thus, a comprehensive monitoring system has been specially designed and installed in the large-scale physical model in Hong Kong to investigate the flow-barrier interaction and the force transmission within the components of flexible barrier system. The monitoring system comprises diverse types of sensors, high-speed cameras, high-frequency data-loggers and computers.

Fig. 1. A large-scale physical model for the impact of debris flow on flexible barriers in Hong Kong

2 Design and Installation of the Monitoring System

2.1 Model Setup

The large-scale physical model in this study has two main components: upper storage container and slope. The storage container is mounted above the slope with a volume of 5 m^3. Perspex is used as transparent sidewalls to form a channelized slope. The slope has a channel width of 1.5 m and a length of 6.8 m, and is inclined at 33°.

A pair of flexible barrier steel posts with a length of 2.74 m are installed at the most downstream end of the slope and perpendicular to the slope. Each post is hinged to the foundation, using a special designed clevis and connected with two anchor cables at the upper end (see Fig. 1). The steel wire rope is suspended by the posts to form a frame with a width of 2.48 m and a height of 1.48 m. The barrier, composed of 300 mm diameter steel rings, is mounted on the ropes. In order to evaluate accurate impact loads, energy dissipaters are not equipped in this flexible barrier system.

2.2 Sensors and High-Speed Cameras

Conventional sensors include accelerometers, large tension-link force transducers and strain gauges on posts. Accelerometers are installed on the upper ends of posts and the barrier to measure the acceleration during the impact and then obtain the velocities and displacements of posts and barrier by integration. Each post is equipped with two accelerometers recording the acceleration along the flow direction and perpendicular to the Perspex sidewall respectively. Moreover, three accelerometers are installed on the rings along the horizontal centerline of the barrier for measuring the upper intermediate, lower intermediate and bottom acceleration (see Fig. 2).

To measure tension force in the framed rope, a large tension-link force transducer with a measuring capacity of 50 kN is installed at the ends of the rope (see Fig. 2). Meanwhile, four large tension-link force transducers are connected between the anchor cables and foundation for estimating the tension forces in cables. The clevises for supporting the posts are specially designed as transducers. Strain gauges are adhered to the clevises. The strains measured during impact process can be used to calculate the forces along axes and the moments about x-axis. Hence, the impact forces transmitted to the posts can be estimated.

All developed optical fibre sensors are based on fibre Bragg grating (FBG) sensing technology, including small FBG tension-link force transducers, FBG basal friction transducers and FBG earth pressure cells.

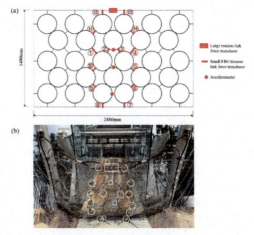

Fig. 2. (a) Schematic diagram and (b) photo of the flexible barrier with sensors

Small FBG tension-link force transducers are developed to investigate the interaction between rings. The transducers are installed between rings as connectors, presented in Fig. 2. The measured forces can be applied to estimate the impact load through back-calculation. A FBG basal friction transducer consists of a thin plate as sensing element with a pair of FBG sensors attached on the inner surface. The transducers are fixed on the slope at the upstream end and free at the downstream end. Hence, the shear stresses between approaching debris and slope base are measured. Several FBG earth pressure cells with a diameter of 200 mm are placed on the upstream, midstream and downstream of the slope. The pressure cells on downstream record the normal stresses induced by deposited debris, thus providing the bulk density. Furthermore, the measured normal stresses on the upper part of slope can be employed to obtain the basal friction angle by analyzing together with basal shear stresses detected by corresponding basal friction transducers.

Two high-speed cameras, with frame rates ranging from 523 to 200,000 frames per second, were used to take videos of the impact of rock falls and debris flow on a flexible barrier. They were placed in front and flank of the channeled slope respectively, directed at the location of flexible barrier. The velocities of approaching debris, heights of deposited debris, impact duration, deflection of the barrier, *etc.* can be obtained from the records by high-speed cameras.

3 Single Boulder Impact Test and Granular Flow Impact Test

A single boulder impact test was carried out to evaluate the dynamic behavior of the flexible barrier and verify the performance of the comprehensive monitoring system. The boulder with a diameter of 400 mm is made of granite with surface roughening. It was released from the storage container at a height of 3.7 m. According to the records by high-speed cameras, the velocity of the approaching boulder is 7.8 m/s at the location of barrier before the impact and the maximum deflection of the barrier subjected to the impact load is 1.08 m

A volume of 4 m^3 gravel was used to simulate granular flow in this test. The gravel was composed of sub-angular particles with a size range from 15 mm to 25 mm. The internal friction angle and the friction angle between gravel and channel slope base are 26° and 31° respectively. Different from the impact by a boulder, granular flow impact is more gradual and prolonged. The accumulation of gravel behind the flexible barrier was observed in the granular flow and the normal stress induced by the deposited gravel were measured by FBG earth pressure cells.

4 Conclusions

In this paper, a comprehensive monitoring system is presented in detail and verified by a single boulder test and a granular flow impact test. The velocity of approaching flow, deflection of the barrier, load distribution on ring nets, force transmission within the components, etc. are investigated by using our monitoring system, providing useful insights on the performance of a flexible barrier system.

Acknowledgements. The work in this paper is supported a CRF project (Grant No.: PolyU12/CRF/13E) from Re-search Grants Council (RGC) of Hong Kong Special Administrative Region Government of China.

References

1. Escallón, J.P., Wendeler, C.: Numerical simulations of quasi-static and rockfall impact tests of ultra-high strength steel wire-ring nets using Abaqus/Explicit. In: 2013 SIMULIA Community Conference (2013)
2. Wendeler, C., Volkwein, A., Roth, A., Denk, M., Wartmann, S.: Field measurements and numerical modelling of flexible debris flow barriers. In: Debris-Flow Hazards Mitigation: Mechanics, Prediction, and Assessment, pp. 681–687. Millpress, Rotterdam (2007)
3. Kwan, J.S.H., Chan, S.L., Cheuk, J.C.Y., Koo, R.C.H.: A case study on an open hillside landslide impacting on a flexible rockfall barrier at Jordan Valley, Hong Kong. Landslides **11**(6), 1037–1050 (2014)

Quantifying Fiber-Optic Cable–Soil Interfacial Behavior Toward Distributed Monitoring of Land Subsidence

Cheng-Cheng Zhang[1,2], Bin Shi[1(✉)], Su-Ping Liu[1], Hong-Tao Jiang[3], and Guang-Qing Wei[4]

[1] School of Earth Sciences and Engineering, Nanjing University, Nanjing, China
shibin@nju.edu.cn
[2] Department of Civil and Environmental Engineering, University of California, Berkeley, Berkeley, CA 94720, USA
[3] School of Geographic and Oceanographic Sciences, Nanjing University, Nanjing, China
[4] Suzhou NanZee Sensing Technology Co., Ltd., Suzhou 215123, Jiangsu, China

Abstract. Land subsidence is often associated with strata compaction, which can be measured by InSAR, GPS or extensometers. Despite considerable efforts, recording detailed subsurface deformation using these methods has proven to be difficult in some cases. The distributed fiber-optic sensing (DFOS) technique may overcome this dilemma by distributed strain measurement at kilometer scales and beyond. However, one crucial issue remains to be solved is the mechanical coupling between borehole backfill and embedded fiber-optic (FO) strain-sensing cable, which affects the quality of strain measurements. Here we perform laboratory pullout tests using a self-devised apparatus to investigate the interaction mechanism between FO cable and soil under confining pressures (CPs) up to 1.6 MPa. Our results indicate a critical CP that ensures a good cable–soil coupling, which has a direct application to the analysis of groundwater extraction-induced land subsidence in Shengze (Suzhou, China). A simple estimation of the CP exerted on the borehole-embedded cable suggests a strong coupling of the cable to the backfill for the main compaction zone. Taken together, these findings provide a basis for monitoring land subsidence and the associated strata compaction using the DFOS technique.

Keywords: Land subsidence · Distributed fiber optic sensing (DFOS) Confining pressure · Coupling

1 Introduction

Groundwater extraction-induced land subsidence is a global environmental issue [1, 2]. The resulted compaction of susceptible aquifer systems can be measured using InSAR, GPS or extensometers [2]. The former two techniques are utilized to measure elevation changes of the land surface. Although the latter are capable of capturing displacements at different depths, the measuring points are often limited. This dilemma may be overcome by using the distributed fiber-optic sensing (DOFS) technique, which realizes distributed strain measurement along a low-cost fiber-optic (FO) cable at kilometer scales and beyond. Hence, it offers great potential for monitoring geohazards [3]. An early trial was reported by Kunisue and Kokubo concerning the use of DFOS to monitor formation compaction accompanying the exploitation of natural gases in Japan [4]. More recently, our group successfully applied the DFOS technique to monitor strata deformation induced by groundwater withdrawal in Shengze (Suzhou, China) [5]. A schematic diagram of DFOS-based land subsidence monitoring is shown in Fig. 1. The basic idea is that FO cables are directly installed in a borehole and then, the borehole is backfilled with, for instance, sand–gravel–clay mixtures. After the backfill is sufficiently consolidated, strain measurements can be performed routinely. Compared with borehole extensometers, a complete strain profile along the depth can be obtained using the DFOS technique.

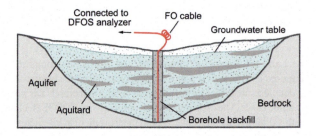

Fig. 1. Schematic of distributed sensing of strata compaction using the distributed fiber-optic sensing (DFOS) technique (modified from [6]).

One crucial issue lies in whether the FO cable and the borehole backfill are mechanically coupled. Early contributions focused on examining the progressive failure behavior of the cable–soil interface using laboratory pullout tests, typically under low confining pressures (CPs) [7–9]. For monitoring strata deformation up to a depth of several hundreds of meters, it will be of great interest to know whether there is a critical CP under which a good cable–soil coupling can be assured. To this aim, we devised a laboratory pullout apparatus capable of investigating the cable–soil interaction under CPs up to a few megapascals. We performed extensive tests to determine the critical CP and discussed its implications for land subsidence monitoring. The results presented here are based on one of our earlier contributions [10].

2 Laboratory Pullout Tests

A displacement-controlled pullout apparatus was devised to investigate the cable–soil interaction (Fig. 2). The apparatus consists of three parts: a pressure chamber, a DFOS analyzer and a pullout tester. A 2 mm-diameter tight-buffed FO cable was embedded in a 60 mm-diameter heat-shrinkable tube filled with a compacted sand–clay mixture (8% in water content and 1.85 g/cm^3 in density). Then the specimen was installed in the pressure chamber and the cable was pulled out. The CPs investigated here varied between 0 and 1.6 MPa. For each displacement step of 0.973 mm, the strain distribution along the cable was captured using a Neubrex NBX-6050A BOTDA analyzer with a spatial resolution of 0.1 m and a sampling interval of 0.05 m.

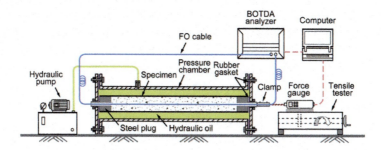

Fig. 2. The devised pullout apparatus for investigating fiber-optic (FO) cable–soil interaction.

Typical strain profiles are shown in Fig. 3. Under a low CP, the strain increased and propagated toward cable toe with increasing displacements. In contrast, the strain propagation was restrained around the cable head under a high CP, indicating a strong mechanical coupling between the cable and the soil.

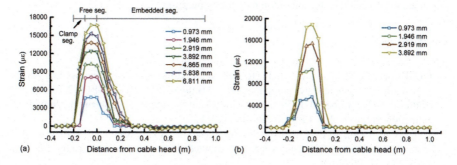

Fig. 3. Typical strain profiles along the FO cable with increasing pullout displacements: (a) confining pressure (CP) = 0 MPa and (b) CP = 1.6 MPa.

3 Implications for Land Subsidence Monitoring

A coefficient ζ_{c-s} is proposed to quantitatively describe the mechanical coupling between cable and soil during a pullout test, i.e.,

$$\zeta_{c-s} = \frac{\int \varepsilon(x)\mathrm{d}x}{u_0} \qquad (1)$$

where $\varepsilon(x)$ is the strain along the cable and u_0 is the pullout displacement applied at cable head. Note that the coefficient is defined for a cable with a unit length (i.e., 1 m). For long-term monitoring purposes, the maximum working strain is often set as 10,000 με. The effect of CP on the coefficient ζ_{c-s} under this strain value is shown in Fig. 4. If a ζ_{c-s}

of 0.9 is regarded as an indicator of a good cable–soil mechanical coupling, the critical CP is calculated to be 0.17 MPa. Taking the Shengze land subsidence (China) as an example, the main pumping aquifer is 74.4–87.7 m in depth [5], corresponding to a CP of ca. 0.8 MPa. Hence, our results suggest that the field monitoring data of the main compaction zone is reliable. Combined, these findings provide a basis for distributed sensing of land subsidence using the DFOS technique.

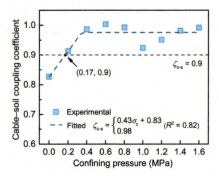

Fig. 4. Influence of CP on the cable–soil coupling coefficient ζ_{c-s}. The maximum working strain is assumed to be 10,000 με for long-term monitoring purposes.

References

1. Bawden, G.W., Thatcher, W., Stein, R.S., et al.: Tectonic contraction across Los Angeles after removal of groundwater pumping effects. Nature **412**(6849), 812–815 (2001)
2. Galloway, D.L., Burbey, T.J.: Review: regional land subsidence accompanying groundwater extraction. Hydrogeol. J. **19**(8), 1459–1486 (2011)
3. Zhang, C.-C., Zhu, H.-H., Liu, S.-P., Shi, B., Zhang, D.: A kinematic method for calculating shear displacements of landslides using distributed fiber optic strain measurements. Eng. Geol. **234**, 83–96 (2018)
4. Kunisue, S., Kokubo, T.: In situ formation compaction monitoring in deep reservoirs using optical fibres. In: Carreón-Freyre, D., Cerca, M., Galloway, D.L., Silva-Corona, J.J. (eds.) Land Subsidence, Associated Hazards and the Role of Natural Resources Development (EISOLS 2010), vol. 339, pp. 368–370. IAHS Publications, Wallingford (2010)
5. Wu, J., Jiang, H., Su, J., et al.: Application of distributed fiber optic sensing technique in land subsidence monitoring. J. Civ. Struct. Heal. Monit. **5**(5), 587–597 (2015)
6. Galloway, D., Jones, D.R., Ingebritsen, S.E.: Land subsidence in the United States. U.S. Geological Survey Circular 1182, Denver, CO, USA (1999)
7. Iten, M., Puzrin, A.M., Hauswirth, D., et al.: Study of a progressive failure in soil using BEDS. In: Jones, J.D.C. (ed.) Proceedings of the SPIE 7503, 20th International Conference on Optical Fibre Sensors, p. 75037S (2009)
8. Zhang, C.-C., Zhu, H.-H., Shi, B., et al.: Interfacial characterization of soil-embedded optical fiber for ground deformation measurement. Smart Mater. Struct. **23**(9), 95022 (2014)
9. Zhang, C.-C., Zhu, H.-H., Shi, B.: Role of the interface between distributed fibre optic strain sensor and soil in ground deformation measurement. Sci. Rep. **6**, 36469 (2016)
10. Zhang, C.-C., Shi, B., Liu, S.-P., et al.: A study of the mechanical coupling between borehole backfill and fiber-optic strain-sensing cable. Chin. J. Geotech. Eng. (2018, in press)

Monitoring the Horizontal Displacement by Soil-Cement Columns Installation

Wei Zhao[1(✉)], Longcai Yang[1], Binglong Wang[1], Congcong Xiong[1], Ye Zhang[2], and Xilei Zhang[3]

[1] Key Laboratory of Road and Traffic Engineering of the Ministry of Education, Tongji University, Shanghai 201804, China
zhaowtj@163.com
[2] Zhejiang Yueqing Bay Railway Co., Ltd., Zhejiang 325600, China
[3] China Railway 16th Bureau Group Co., Ltd., Zhejiang 313000, China

Abstract. It is generally accepted that the installation of soil-cement columns causes almost no soil compaction in surrounding strata. Combined with the field test, the disturbance of the soil-cement columns installation to the surrounding strata is introduced in this paper. The displacement of surrounding strata is monitored by inclinometer on site and the monitoring results show that: The installation of soil-cement columns have a certain effect to surrounding strata; The direction of horizontal displacement of lateral ground caused by soil-cement columns installation is along the side of soil-cement columns; The horizontal displacement of lateral ground decreases with the increase of horizon distance and depth; The influence range along the depth is approximately the same as the length of column, maximum displacement occurs at the surface of soil, the horizontal displacement is close to zero below the bottom of the columns; The installation effect of soil-cement columns can be reduced by reasonable control of the construction sequence.

Keywords: Soil-cement columns · Installation · Field test
Horizontal displacement field

1 Introduction

Soil-cement columns is an effective form for reinforcement of soft soil foundation. More and more research results in bearing capacity and settlement character of soil-cement columns composite foundation have been achieved by scholars from various countries after years of research [1, 2]. However, there are relatively few studies on the disturbance of surrounding strata caused by installation of soil-cement columns [3, 4]. Based on field test, the horizontal displacement field of surrounding strata under the installation disturbance is obtained, and the construction influence range of soil-cement columns is also analyzed in this paper.

2 Engineering Survey

2.1 Field Geological Conditions

Yueqing is located in the southeast of Zhejiang Province, China. The soil layer in this area is characterized by high water content, strong compressibility, low strength and poor permeability, which is very unfavorable for engineering construction. Table 1 shows the statistics of the physical and mechanical properties of the soil at the construction site.

Table 1. Geotechnical parameters of construction site.

Soil parameters	Natural water content ω (%)	Natural density ρ (g/cm³)	Natural porosity ratio e	Internal friction angle Φ (°)	Cohesive force c (kPa)	Compression modulus Es (MPa)	Layer thickness (m)
Plain filling soil	–	–	–	–	–	–	2.5
Silt soil	66.63	1.58	1.92	4.03	5.58	1.47	2.9
Fine gravel soil	29.50	1.92	0.84	15.50	10.00	4.90	2.3
Gravelly clay	38.70	1.82	1.09	12.60	28.67	1.82	3.5
Coarse gravel soil	34.40	1.86	0.99	20.70	27.00	4.76	8.8

2.2 Construction Parameters of Soil-Cement Columns

There are 28 rows of columns have been installed on site. The installation sequence of columns and the plane position are shown in Fig. 1. The installation order of each row consistent with the direction of the arrow in the figure. The length of the soil-cement column is 6.5 m, the diameter of the pile is 0.5 m, and the pile spacing is 1.5 m. The cement input ratio is about 15% and the injection pressure of cement slurry is about 0.6 MPa.

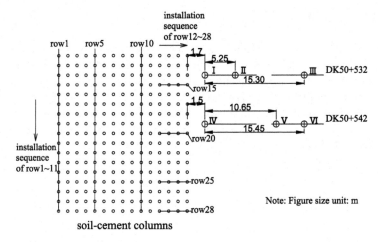

Fig. 1. The installation sequence of columns and the plane position

2.3 Field Monitoring

The horizontal displacement of the surrounding strata within existing high speed railway subgrade DK50+525 ~ DK50+735 is mainly monitored during the installation of soil-cement columns. The DK50+532 and DK50+542 sections were selected as the horizontal displacement monitoring sections of the surrounding strata. The monitoring points of the inclinometer I, II and III are arranged in the DK50+532 section. Similarly, IV, V and VI were arranged in the DK50+542 section. The location of each monitoring point is shown in Fig. 2.

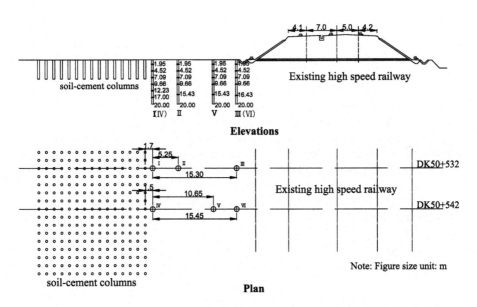

Fig. 2. The location of each monitoring point

3 Analysis and Discussion of Monitoring Results

The variation of the horizontal displacement at the surface of soil during the soil-cement columns installation is shown in Fig. 3(a). The horizontal displacement of monitoring points I, II and III at the surface of soil caused by the installation is 10.38 mm, 6.34 mm and 1.29 mm. The direction of horizontal displacement of lateral ground is along the side of soil-cement columns, the horizontal displacement of surrounding strata increased gradually during the construction period, and it is decreased with the increase of horizon distance. At the same time, it is more sensitive to the installation of soil-cement columns for a smaller horizontal distance.

Figure 3(b) shows the horizontal displacement of the monitoring points I, II and III along the depth of the soil. It can be seen that the horizontal displacement of surrounding strata decreases with the increase of depth. The influence range along the depth is approximately the same as the length of column, maximum displacement occurs at the

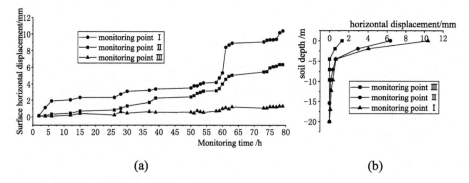

Fig. 3. Horizontal displacement field of surrounding strata caused by soil-cement columns installation: (a) Variation of the horizontal displacement at the surface of soil; (b) The horizontal displacement along the depth of the soil.

surface of soil, the horizontal displacement is close to zero below the bottom of the columns, and it is almost unaffected by the installation of soil-cement columns.

The distance between the monitoring point I and the nearest soil-cement column is 1.7 m, and the monitoring point IV is 1.5 m. They are most affected by the installation of soil-cement columns. Figure 4 shows that the trends of surface horizontal displacement changes of the monitoring point I and IV, it can be divided into three segments. There is a section of accelerating, which is due to the close installation of soil-cement columns.

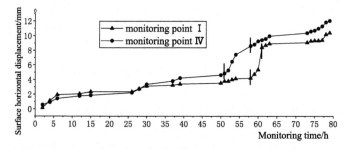

Fig. 4. The trends of surface horizontal displacement changes of the monitoring point I and IV

References

1. Wang, Z., Shen, S.: Investigation of field-installation effects of horizontal twin-jet gro. Can. Geotech. J. **50**(3), 288–297 (2016)
2. Yan, C.: Experimental study on horizontal bearing capacity of reinforced cement-soil piles. Chin. J. Geotech. Eng. **35**(S2), 730–734 (2013)
3. Ma, H.: Settlement characteristic of composite foundation for cement mixing piles in soft soil. J. Central South Univ. **40**(03), 803–807 (2013)
4. Chai, J., Miura, N., Koga, H.: Lateral displacement of ground caused by soil–cement column installation. J. Geotech. Environ. Eng. **319**(131), 623–632 (2005)

Long-Term Field Monitoring of Additional Strain of Vertical Shaft Linings in East China

Guoqing Zhou, Lianfei Kuang(✉) , Guangsi Zhao,
Hengchang Liang, and Xiaodong Zhao

State Key Lab for Geomechanics and Deep Underground Engineering,
China University of Mining and Technology, Xuzhou 221116, Jiangsu, China
klfcumt@126.com

Abstract. The vertical additional force (VAF) is the fundamental cause of the rupture of hundreds of vertical shaft linings (VSL) in East China, long-term monitoring of additional strain which caused by VAF is conducive to well control the performance of VSL during operating, curing and after treatment. Based on long year filed monitoring results, the long-term strain of VSL can be separated into temperature strain and additional strain, and the former varies sinusoidal with the seasonal temperature, while the latter shows a linear cumulative. The cumulative rate of vertical strain accelerates when VSL are on the verge of rupture, and followed by suddenly decrease in the process of crack. The strain monitoring system can also be used to control the process of stratum-grouting as an electronic eye, and the mechanism of curing the rupture of VSL is that the VAF could be not only released but also restrained by cement grouting. Furthermore, the time of rupture can be predicted based on the long-term strain monitoring data, for example, by R/S method. The suggested methods here can be used for preventing and curing rupture of VSL in other similar mines.

Keywords: Vertical additional force · Shaft lining · Rupture · Stratum-grouting Monitoring

1 Introduction

East China is a very important coal production base, in this region, there are lots of large mining arears with more than 300 VSL and 100 million tons' coal output per year. A common feature of these VSL is that the alluviums which they cross are generally thicker, the thickness of these deep thick alluvium is generally 160–753 m.

Since the mid-1980s, VSL ruptures of over 100 working shafts have occurred in this region, and the position of the ruptures is mostly in the area of the interface between bedrock and the bottom of the deep thick alluvium. The characteristics of shaft rupture present an annular zone, in which the concrete is spalled, reinforced steel is convex-exposed, and the diameter of the VSL is obviously shrunk. A large number of theoretical analysis, numerical simulations and laboratory physical model test results show that once the bottom aquifer dewaters, the compression deformation of the deep thick alluvium leads to an additional force apply on the outside surface of the VSL [1]. This additional force has not been taken into account during the early design of the shaft, once the total force increases to a certain value and exceeds the strength of the concrete, ruptures will occur. Therefore, long-term monitoring of addition strain of VSL is helpful for understanding the real-time stress state of VSL, grasping the development of VSL force, predicting the time of VSL rupture and guiding the curing rupture of VSL.

2 Field Monitoring Methodology

The field monitoring system is composed of sensors, microcontroller unit, data transfer unit, cloud computing platform and so on. System framework is sketched in Fig. 1.

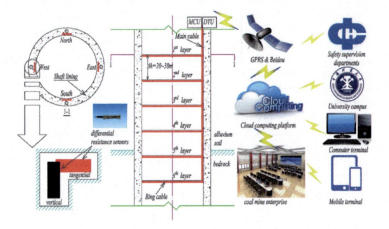

Fig. 1. Framework of shaft monitoring system

The differential resistance sensors are embedded in the shaft concrete in two perpendicular directions, vertical and tangential, through the way of slotting. These two sensors are located in the same slot, and in general, four positions of east, south, west and north will be set along the circumference of VSL at the same layer. And 5–8 layer sensors are usually deployed along the longitudinal direction of VSL, where 1–2 layer is usually placed below the bedrock interface. The key part of monitoring, the region of the interface will also lay 1–2 layer sensors, the remaining 4–5 layer sensors are usually set up above the interface at an interval of 20–30 m. Eight strain gauges of four positions in each layer are all collected through a ring cable to the main cable, Main cable transmissions the signal of each layer sensors to the ground module box.

With the help of the automatic data acquisition unit and the remote GPRS module, the monitoring data can be further transmitted to the safety supervision departments, the coal mine enterprise and our university campus in real time. The collected monitoring data will first have a deep processing and analysis by the cloud computing platform, once a dangerous signal is found, early warning and prediction will carry out immediately, and the responsible person will receive an alarm message or even a phone call.

3 Results and Discussion

3.1 Long-term Evolution of VSL Strain

The typical additional strain curve can be decomposed into a curve of drop line superposition with sine curve, as shown in Fig. 2(a), and can be expressed as $y = a + bx + A\sin(\pi(x - x_0)/w)$, where y is strain, x is time, others are all undetermined constant, as shown in Fig. 2(a), The linear part $y_1 = a + bx$ is the contribution of the VAF, and the last item $y_2 = A\sin(\pi(x - x_0)/w)$ is the contribution of seasonal temperature [2]. Based on the field results, the rate of addition strain accumulation is between 0.015 – 0.068 uɛ/d, and the linear VAF is the fundamental cause of VSL rupture.

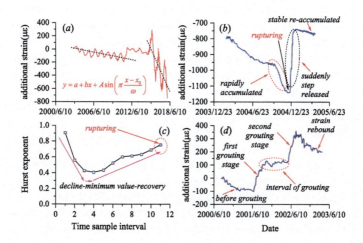

Fig. 2. Monitoring results of VSL

3.2 Strain Features on VSL Rupturing

The strain features on VSL rupturing of a shaft at the end of 2004 is displayed in Fig. 2(b). In the 1–2 months before the ruptures, the strain compression rate is faster, and the cumulative rate is between 35–70 uɛ/m. When the VSL ruptures, the strain is released substantially, and the quantitative value of the release is between 85–370 uɛ. The whole process will last about 10–15 d. After rupture, the strain restarts to accumulate at a new level [3]. On the whole, before and after the ruptures, the vertical strain is shown as accumulated, rapidly accumulated, suddenly step released at short notice, and stable re-accumulated.

3.3 Fracture Prediction of VSL

Our research show that R/S method can be applied to dynamic analyze the addition strain and warn fracture in VSL [4]. Further study on the Hurst exponent (H) of addition strain before fracture of the above mentioned ruptured VSL is made, and Fig. 2(c) shows the tread of H with adding the sample data every half month until the fracture moment. The trend of H with time shows an abnormal variation before fracture (2–5 months), and can be described as "decline-minimum value-recovery". Most important is that rupture occurs in the process of H rising, therefore, H has a certain mid- or short-term precursory warning significance for the fracture prediction of VSL.

3.4 Stratum-Grouting Control of VSL

The vertical average addition strain curve of another ruptured VSL during the stratum grouting is shown in Fig. 2(d) [5]. Before grouting, the addition strain is almost linearly increase with the drainage of the aquifer. With the injection of slurry, the stratum moves upward, the VAF is sustained release and the strain decreases about 200 uε. In the interval of grouting, the strain is basically maintaining stability. In the next second grouting stage, the strain is also released about 200 uε, but after this stage, the strain rebound is obvious. Therefore, the mechanism of curing the rupture of VSL is that the VAF could be not only released but also restrained by stratum grouting, and the monitoring system is the precondition of all control.

4 Conclusions

In this paper, based on a large number of VSL (≥ 14) field monitoring results over a long period (≥ 18 y), we comprehensive discussed the achievements and progresses of research on the health evaluation of shaft structure. It included the long-term and rupture moment features of VSL addition strain, the R/S prediction method of fracture, and the key role of monitoring system in grouting. As a universal method, the relevant techniques introduced here can also be applied to the health monitoring and evaluation of other structures, such as dams, bridges, tunnels and so on.

Acknowledgements. The study presented here are financially supported by the National Natural Science Foundation of China (51504245, 41502271, 41772338, 51408595), and Jiangsu Provence Postdoctoral Science Foundation (1402009B).

References

1. Cui, G.X., et al.: The Frozen Wall and Shaft Lining in Deep Thick Alluvium. China University of Mining and Technology Press, Xuzhou (1998)
2. Liang, H.C., et al.: An in-situ long period observation of additional strain in shaft linings. J. China Univ. Min. Technol. **38**(6), 794–799 (2009)
3. Liang, H.C., et al.: Strain characters of shaft lining crack. J. China Coal Soc. **35**(2), 198–202 (2009)
4. Zhao, G.S., et al.: Application of R/S method for dynamic analysis of additional strain and fracture warning in shaft lining. J. Sens. (2015). Article Number: 376498
5. Zhou, G.Q., et al.: Application and effect of grouting in surrounding soil on releasing and restraining additional stress of shaft lining. Chin. J. Geotech. Eng. **27**(7), 742–745 (2005)

Part VI: New Geomaterials and Ground Improvement

Sacrificial Anode Protection for Electrodes in Electrokinetic Treatment of Soils

Abiola Ayopo Abiodun[✉] and Zalihe Nalbantoglu

Department of Civil Engineering, Eastern Mediterranean University, Via Mersin 10, Gazimağusa, North Cyprus, Turkey
abiola.abiodun@cc.emu.edu.tr

Abstract. The electrokinetic (EK) treatment is proven to be an effective method in laboratory and field scale studies. The EK technique has lots of drawbacks such as electrode corrosion, cell setup, uncertain energy usage and cost, loss of energy with time, choice of materials, and no standard procedures. The metal electrodes corroded rapidly over a short period of time, cause voltage loss, reduce energy efficiency, and contaminate soil, thus, result in complexity and variability within the test system. Thus, EK method is seldom used as an alternative soil treatment with its enormous potentials. This paper discusses the challenges, potential remedies, and recommendations for EK technique, such as using a sacrificial anode to prevent, control or minimize the corrosion of the electrodes. The choice of electrodes is significant in improving the index and engineering properties of the soil and prevent the electrode corrosion or contamination of the soil, control variability and maintaining the electrical efficiency of the system. This implies that cheap and available metals can be used effectively in EK technique, both in the laboratory and in-situ bench-scale studies.

Keywords: Corrosion · Electrokinetic · Electrode · Sacrificial anode Soft soils

1 Introduction

The electrokinetic (EK) treatment is the use of electric current, I on the soil via electrodes placed into or its adjacent ends to migrate charged ions from the electrolyte by pore fluids through the very low permeable soil and invariably altered the soil mineralogy to desirable properties. The EK treatment of the soil is solely based on the basic EK processes. EK finds application in geotechnical engineering. Despite the enormous potentials of EK technique, there are a lot of drawbacks. This paper discusses the challenges, remedies, and recommendations for the efficiency of the EK processes. The EK-treatment method can be very effective provided its challenges are accurately defined and suitable EK setup is used for appropriate deficient soils [1].

2 Principles of Electrokinetic Phenomena

The EK technique involves electrochemical processes which control the physico-chemical changes in the soil. It includes electrolysis, electroosmosis, electrophoresis, electromigration, sedimentation potential, streaming potential [2–4].

2.1 Controlling Factors of EK Applications in Soils

The efficacy of EK treatment depends on the soil and material dependent factors. The soil factors are the pH, initial concentration, electrical conductivity, salinity, temperature, water content, total dissolved salt, zeta potential, and the soil index properties such as density, porosity, void ratio, etc. The material factors are the electrode type and electrolyte solution, energy supply device, and the entire EK cell setup [5].

3 EK Test Testing Model

The EK test set up has a big unit for the soil to be tested and the two smaller units at the adjacent ends, which houses the electrolytes. The electrodes are placed into or at the adjacent ends of the soil, connected to a D.C power source. The EK test tank (Fig. 1). It is made up of transparent plexiglass. The walls in between the big and small units are perforated to allow the chemical ion to migrate via the soil mass. The electrolytes are Na_2SiO_3 and $CaCl_2$, and electrodes, Al, and stainless steel. The energy source for voltage supply ranges from 0.5–1.0 $V.cm^{-1}$ in the soil mass [6–9].

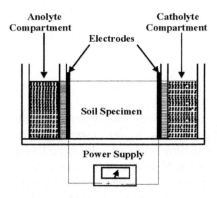

Fig. 1. EKT test tank model configuration

3.1 Applications and Challenges of Electrokinetic Treatment Method

The EK technique has more useful applications over the conventional ground improvement methods. It is efficient to improve not easily drained, low permeable soils. It is applicable both in ex-situ and in-situ applications, cost-effective, easy to operate, rapid, technically simple and short duration. It has a flexible operation and adaptable to various ground conditions. It is a non-noise operation, no site disturbance, and relatively short treatment duration. It is applicable to remediate deficient soils under existing structures such as light building, pipeline, roads, and railways. EK has potential to mitigate large ground, both in space and in depth. It is effective in mitigating liquefaction and slope failures. EK is used for consolidation, decontamination, dewatering, soil improvement, strengthening, and remediation [10, 11]. EK does not have standardized procedures due to the complexity of its controlling factors and lack of adequate understanding of its electro-, hydro-, and physicochemical effects on soil-liquid interactions [1]. There are limited theoretical formulations to define the EK processes. The flow path, seepage, and leaching issues are among the drawbacks why EK method is considered as a last resort for large-scale in situ applications. The on-site limitations are irregular topography, slopes, buried metals, complexity in soil deposits. Also, drawbacks such as trial and error methods, energy and cost uncertainty, loss of energy efficiency and electrode corrosion are common. The slurry or compacted soil sample is used to replicate the in-situ soil. The reconstituted soil sample does not provide the accurate the in-situ conditions in terms of its engineering, index, and physicochemical properties [12]. After the complete EK test setup, leakage could occur through the jointed ends, the adjacent soil side walls, or the soil bottom path. The leakage might cause irregular flow path without migrating through the soil mass. Also, unwanted excess heat is generated at the electrode sites. The heat causes cracks and fissures in the soil. Also, undesirable gaseous compounds are often formed at the electrode sites during the electrolytic reactions [13]. The inconsistent pH may affect the solubility, viscosity, concentration of the soil pore fluid. This causes the variability in the soil during the EK method, thus, affect the accuracy of the results. The inappropriate selection of electrolytes and electrodes may affect EK process. Electrode corrosion and electrolyte decay are common causes of various changes in pH and other inputs. The metal electrode corrosion initiates standard electrode potential essential for electron transfer from the anode to the cathode. The corroded electrode contaminates the treated soil. The in-situ anomalies such as buried metals, ground openings, groundwater, and slopes can reduce the efficacy of the EK process [14, 15].

3.2 Remedies for Challenges in Electrokinetic Technique

To avoid challenges during EK treatment, the comprehensive preliminary studies: theoretically and practically before the main EK set up are required. Factors such as the choice of materials, which are cheap, durable, inexpensive and suitable should be considered. Such precaution is to avoid the conventional trial and error procedures which are time, energy and resources wasteful. Thus, the careful selection of appropriate test tank materials, electrolyte, electrode, the power device is essential for good results. To eliminate the time-consuming preparation of slurry or compacted soil sample, an extraction of in situ soil samples may be considered. The in-situ soil samples will represent the exact soil properties. The proper design of EK setup will prevent short-circuiting, electric shock, leakage, and irregular flow path. The introduction of a more active metal, a sacrificial anode system, such as a zinc metal, will prevent corrosion at the active anode and less active cathode. The more active metal will sacrifice itself and get corroded while the main two electrodes will remain protected as indicated in Fig. 2. With this concept, very active metals on electrochemical series, which are quite cheap and readily available, can be selected for EK treatment. The aluminum anode and stainless-steel cathode can be protected by the more active sacrificial zinc metal, which loses its electrons to other metals, oxidized and corroded, while other metals are protected. The suitable chemical stabilizers can be used to prevent generation of hydrogen ions within a short duration. A controlled temperature can prevent the unnecessary build-up of heat, to avoid desiccation, fissure or cracking. Also, adequate testing devices are vital to measuring all input and output parameters. The optimization of all variables such as voltage, pore fluid concentration, energy used and operating costs, set up materials are essential to achieving optimum EK processes and overall cost-effectiveness of the EK application [16, 17].

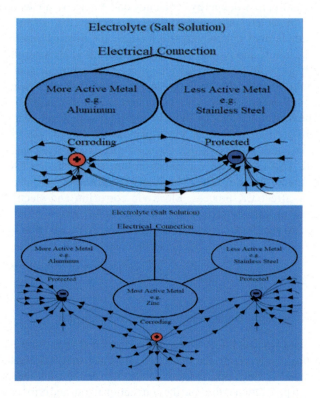

Fig. 2. Diagrammatic illustration of an electrode configuration in a sacrificial anode system.

References

1. Malekzadeh, M., Julie, L., Nagaratnam, S.: An overview of electrokinetic consolidation of soils. Geotech. Geol. Eng. **34**(3), 759–776 (2016)
2. Casagrande, L.: Electro-osmotic stabilization of soils. Trans. Boston Soc. Civ. Eng. **39**, 51–83 (1951)
3. Jayasekera, S.: Stabilizing volume change characteristics of expansive soils using electrokinetic: a laboratory-based investigation. In: Sri Lankan Geotechnical Society's First International Conference (2007)
4. Mosavat, N., Erwin, Oh, Gary, C.: A laboratory investigation of physicochemical in Kaolinite during electrokinetic treatment subjected to enhancement solutions. Electron. J. Geotech. Eng. **19**, 1215–1233 (2014). Bund. E.
5. Iyer, I.: Electrokinetic remediation. Part. Sci. Technol. **19**, 219–228 (2001)
6. Liaki, C., Rogers, C.D.F., Boardman, D.: Physico-chemical effects on clay due to electromigration using stainless steel electrodes. J. Appl. Electrochem. **40**(6), 1225–1237 (2010)
7. Delgado, A.V., González-Caballero, F., Hunter, R.J., Koopal, L.K., Lyklema, J.: Measurement and interpretation of electrokinetic phenomena. J. Colloid Interface Sci. **309**(2), 194–224 (2007). 1225–1237
8. Delgado, A.V., González-Caballero, F., Hunter, R.J., Koopal, L.K., Lyklema, J.: Measurement and interpretation of electrokinetic phenomena (IUPAC technical report). Pure Appl. Chem. **77**(10), 1753–1805 (2005)
9. Wall, S.: The history of electrokinetic phenomena. Curr. Opin. Colloid Interface Sci. **15**(3), 119–124 (2010)
10. Fourie, A.B., Johns, D.G., Jones, C.J.F.P.: Dewatering of mine tailings using electrokinetic geosynthetic. Can. Geotech. J. **44**(2), 160–172 (2007)
11. Mosavat, N., Erwin, Oh, Gary, C.: A review of electrokinetic treatment technique for improving the engineering characteristics of low permeable problematic soils. Int. J. Geomate **2**(2), 266–272 (2012)
12. Malekzadeh, M., Lovisa, J., Sivakugan, N.: An overview of electrokinetic consolidation of soils. Geotech. Geol. Eng. **34**(3), 759–776 (2016)
13. Rittirong, A., Shang, J.Q., Mohamedelhassan, E., Ismail, M.A., Randolph, M.F.: Effects of electrode configuration on electrokinetic stabilization for caisson anchors in the calcareous sand. J. Geotech. Geoenviron. Eng. **134**(3), 352–365 (2008)
14. Ivliev, E.A.: Electro-osmotic drainage and stabilization of soils. Soil Mech. Found. Eng. **45**(6), 211–218 (2008)
15. Nordin, N.S., Saiful, A.A.T., Aeslina, A.K.: Stabilization of soft soil using electrokinetic stabilization method. Int. J. Zero Waste Gener. **1**(2), 5–12 (2013)
16. Cho, J.M., Kim, D.H., Yang, J.S., Baek, K.: Electrode configuration for the electrokinetic restoration of greenhouse saline soil. Sep. Sci. Technol. **47**(11), 1677–1681 (2012)
17. Mohamedelhassan, E.: Electrokinetic strengthening of soft clay. Proc. Inst. Civ. Eng. Ground Improv. **162**(4), 157–166 (2009)

The Effects of Colemanite and Ulexite Additives on the Geotechnical Index Properties of Bentonite and Sand-Bentonite Mixtures

Şükran Gizem Alpaydın and Yeliz Yukselen-Aksoy[✉]

Dokuz Eylül University, Buca Izmir, Turkey
yeliz.yukselen@deu.edu.tr

Abstract. Bentonite-sand mixtures are generally used in landfill liners. The properties of bentonite-sand mixtures should not change by time. For that reason, it is necessary to improve the properties of bentonites and bentonite-sand mixtures. In this study, the boron minerals namely colemanite and ulexite were added to the bentonite and bentonite-sand mixtures in order to improve their consistency limits and compaction characteristics. The boron additives were added 5, 10, 15, 20% to the bentonite and the liquid limits and plastic limits were determined. The boron additives were added to the 10% bentonite + 90% sand and 20% bentonite + 80% sand mixtures and compaction characteristics were determined in the presence of colemanite and ulexite.

As the ulexite ratio increased, the liquid limit value decreased, but as the colemanite ratio increased, the liquid limit value increased. The 20% ulexite and colemanite addition, decreased the liquid limit values of bentonite 8.4% and 13.8%, respectively. As the concentration of boron additives increase, the optimum water content of the mixtures increases. This effect is more significant at 10% bentonite + 90% sand. As ulexite and colemanite content increases the maximum dry unit weight of the bentonite-sand mixtures increase. The effect is more significant at the colemanite additive and at 10% bentonite + 90% sand mixture.

Keywords: Boron · Soil improvement · Consistency limits · Compaction

1 Introduction

The volume change and strength loss may cause structural deformations at structures. There is a need for durable soil layers under the all structures, especially at the nuclear waste landfills. Bentonite-sand mixtures are generally used in landfill liners. The properties of bentonite-sand mixtures should not change by time. Bentonites are highly swelling clay and shrinks when becomes dry. For that reason, engineering properties of soil needs to be improved when bentonites are present. Soil improvement is the alteration of any property of a soil to improve its engineering performance. One of the soil improvement technique is the mixing the soil by another material. Boron is ranked 51st among the elements commonly found in the earth crust. Boron never appears as a free element in nature. Almost 230 types of boron minerals are known to be present in nature. Natural borates are used to define concentrated boron ores such as; tincal, colemanite, and ulexite. Previous studies on the reaction of boron with clay structure have shown that the boron is held strongly by the aluminum or silicon tetrahedron portion in the clay structure [1]. In this study, the effect of colemanite and ulexite on the consistency limits of bentonite and compaction characteristics of 10% bentonite + sand, 20% bentonite + sand samples were investigated.

2 Material Characterization and Methods

The bentonite samples were gathered from Eczacıbaşı Esan Company and it is Na-bentonite. All soil samples were oven dried (105 °C), crushed and sieved through 0.425 mm. Grain size distribution, specific gravity of the samples were determined according to [2, 3], respectively. Liquid and plastic were determined according to [4]. According to the results the specific gravity and clay fraction values 2.73 and 79%, respectively. The compaction characteristics of the samples were determined according to [5].

3 The Effects of Ulexite and Colemanite on the Consistency Limits and Compaction Characteristics

3.1 Consistency Limits

The liquid limit and plastic limit values of the bentonite were determined in the presence 5%, 10%, 15% and 20% ulexite and colemanite, respectively. The results have shown that ulexite addition to the bentonite decreases the liquid limit of bentonite (Table 1). The 20% ulexite addition decreases the liquid limit value by 8.5%. As ulexite content increases, the plastic limit value of bentonite decreased.

Table 1. Consistency limits of the bentonite samples

Boron amount	Boron type	w_L (%)	w_P (%)
0%		427.0	60.6
5%	Ulexite	419.2	58.0
	Colemanite	448.0	55.0
10%	Ulexite	413.7	57.0
	Colemanite	472.0	55.0
15%	Ulexite	393.9	53.0
	Colemanite	438.0	52.0
20%	Ulexite	390.9	49.0
	Colemanite	405.0	51.3

When compared with ulexite, the colemanite addition has reverse effect on the liquid limit values. The liquid limit values of bentonite increased in the presence of colemanite. This effect is more significant at in the presence of 10% colemanite. As colemanite content increased, the plastic limit values of bentonite decreased. The 20% colemanite decreased the plastic limit of bentonite approximately 15%.

3.2 Compaction Characteristics

The compaction characteristics of the 10% bentonite and 20% bentonite mixtures were determined with colemanite and ulexite additives. The Fig. 1 shows the compaction characteristics of the 10% bentonite in the presence of colemanite. The effect of colemanite in the presence of clearly can be seen form the compaction curves. The colemanite addition increase the maximum dry unit weight (γ_{drumax}) and decreases the optimum water content (w_{opt}). As colemanite content increases, the effect increases, too. However, it should be noted that when the colemanite content increased from 10% to 15% colemanite, the γ_{drumax} value did not change.

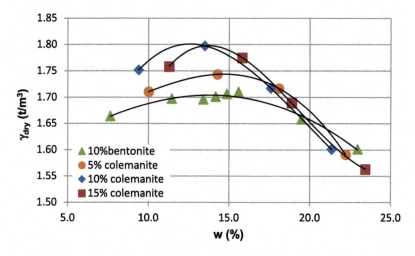

Fig. 1. Compaction characteristics of the 10% bentonite samples in the presence of colemanite

The results of the all compaction test results are given in Table 2. The ulexite addition increases the (γ_{drymax}) and decreases the optimum water content (w_{opt}) similar to colemanite for 10% bentonite + sand mixture. However, the effect of colemanite is more significant than ulexite. As bentonite content increased, the optimum water content starts to increase in the presence of ulexite and colemanite.

Table 2. Compaction characteristics of the samples

Boron amount	Boron type	10% bentonite		20% bentonite	
		γ_{dmax}	w_{opt}	γ_{dmax}	w_{opt}
0%		1.71	14.0	1.72	14.5
5%	Ulexite	1.78	13.8	1.73	15.2
	Colemanite	1.80	13.1	1.73	15.2
10%	Ulexite	1.79	13.2	1.71	15.5
	Colemanite	1.80	12.6	1.73	15.0
15%	Ulexite	1.79	13.2	1.70	16.0
	Colemanite	1.80	13.8	1.73	15.5

4 Conclusions

The effect of colemanite and ulexite on the consistency limits of bentonite and compaction characteristics 10% bentonite + sand and 20% bentonite + sand mixtures were investigated. The colemanite and ulexite have reverse effects on the liquid limit values of bentonite. The ulexite decreased the liquid limit value, while colemanite increased. The plastic limit of bentonite decreased in the presence of both ulexite and colemanite. The γ_{drymax} of the 10% and 20% bentonite + sand mixtures increased and the optimum water content values of 10% bentonite + sand mixtures decreased. However, for the 20% bentonite + sand mixture the optimum water content increased in the presence of ulexite and colemanite. The 5% ulexite or colemanite is the optimum content for compaction.

References

1. Gillott, J.E.: Clay in Engineering Geology. Elsevier Science Publishers B.V., Amsterdam (1987)
2. ASTM: D 422-63: Standard Test Method for Particle-Size Analysis of Soils. ASTM International, West Conshohocken (1999)
3. ASTM: D 854-92: Standard Test Method for Specific Gravity of Soils. ASTM International, West Conshohocken (1999)
4. ASTM: D 4318-98: Standard Test Method for Liquid Limit, Plastic Limit, and Plasticity Index of Soils. ASTM International, West Conshohocken (1999)
5. ASTM: D698-12: Standard Test Methods for Laboratory Compaction Characteristics of Soil Using Standard Effort. ASTM International, West Conshohocken (1999)

Strength Development of Lateritic Soil Stabilized by Local Nanostructured Ashes

Duc Bui Van[1(✉)], Kennedy Chibuzor Onyelowe[2], Phi Van Dang[1], Dinh Phuc Hoang[1], Nu Nguyen Thi[1], and Wei Wu[3]

[1] Hanoi University of Mining and Geology,
North Tu Liem District, Hanoi 100000, Vietnam
buivanduc@humg.edu.vn
[2] Michael Okpara University of Agriculture,
Umuahia 440109, Abia State, Nigeria
[3] Institute of Geotechnical Engineering,
University of Natural Resources and Life Sciences, Vienna, Austria

Abstract. The research performs a series of geotechnical tests to estimate the strength behavior of lateritic soil stabilized by nanostructured ash materials. The synthesized ash materials are mixed with the soil in varying proportions of 3%, 6%, 9%, 12% and 15% by weight of dry soil. The obtained results show that unconfined compressive strength and CBR of stabilized soil are remarkably improved, implying that the ash materials are good admixtures in the stabilization of lateritic soils. The main aims of this work are: (i) to convert these solid wastes to usable engineering materials and (ii) to evaluate the effect of some selected solid waste materials ashes on the behavior of stabilized soil for use in pavement and environment geotechnics works.

Keywords: Local nanostructured ash · Solid wastes · Soil stabilization

1 Materials and Methods

Selected solid waste materials, e.g. palm bunch, waste paper, palm kernel shell, vehicle tyre, coconut shell, snell shell, etc. were collected, sun dried, burnt and completely pulverized. The ashes were further nanosized by passing the collected powder through a nano filter or sieve of size 200 nm [1]. The above nanosized ashes were subjected to characterization by the UV-Vis spectrophotometric test to determine the following parameters: peak wavelength of machine at test (λ), degree of particle absorbance, full width of half maximum (β) and photometric diffraction angle (θ) [2]. The Debye-Scherrer's formula was thereafter applied to determine the average particle size (D) of the ashes under study, which many researchers have applied previously though on the synthesis of the well-known conventional nanomaterials [3, 4].

$$D = \frac{0.9\lambda}{\beta \cos \theta} \quad (1)$$

β is obtained from the corrected FWHM by convoluting Gaussian profile which models the specimen broadening β_r as follows [4, 5]:

$$\beta_r^2 = \beta_0^2 - \beta_i^2 \qquad (2)$$

Where, β_0 is observed broadening and β_i is instrumental broadening.

The chemical compound composition test was also conducted on the ash materials to determine their degree of pozzolanicity; aluminates, silicates and ferrites contents and expected reactions between the compounds and soil elements. The strength characteristic tests were conducted according to [6, 7]. Particularly, both natural and stabilized soil samples were undergone CBR and Unconfined compressive strength tests with 3%, 6%, 9%, 12% and 15% proportions of nano palm bunch ash in accordance to [6–8]. The UV/VIS Spectrophotometer test at 25 °C was conducted in accordance with [7] on both the soil and the admixture to determine their particle absorbance and average particle sizes, respectively.

2 Results and Discussions

The grain size distribution curve and basic geotechnical properties of studied disturbed lateritic soil are displayed in Fig. 1a. Figure 1b shows the variation of ash particle absorbance against wavelength for the Nanostructured ash particles using UV/VIS Spectrophotometer at 25 °C.

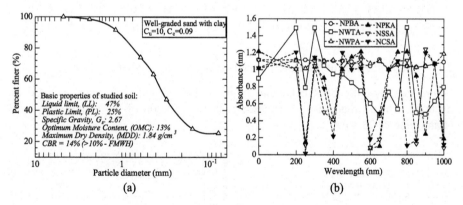

Fig. 1. Particle size distribution (a), (b) variation of Absorbance against wavelength for the Nanostructured Ash Particles using UV/VIS Spectrophotometer at 25 °C.

From the exercise, it can be deduced that the maximum absorbance varied at different wavelengths at a full width of half maximum of 150 nm and exposure angle of 65°. The average particle sizes were estimated by Debye Scherrer expression (Eq. 1) shown in Table 1. Table 2 shows the compound composition and bonding potentials of the nanosized waste material ashes. The result of the analyzed ashes shown in Table 2 shows that the percentage of (SiO_2 + FeO_3 + Al_2O_3) is greater than 70%, which makes the admixture samples of highly pozzolanic materials except NPKA and NCSA. This property was of great advantage because it brought about a high degree of interaction and bonding between the studied soil samples and the admixtures.

Table 1. Average particle size from UV/VIS Spectrophotometer at 25 °C by Debye Scherrer.

Ashes	Absorbance, nm	Wavelength (λ), nm	Average particle size, nm	Remarks
NPBA	1.120	650	9.223	nanosized palm bunch ash
NWTA	1.498	800	11.358	nanosized waste tyre ash
NWPA	1.191	1000	14.198	nanosized waste paper ash
NPKA	1.217	800	11.358	nanosized palm kernel shell ash
NSSA	1.245	900	12.778	nanosized snell shell ash
NCSA	1.207	200	2.840	nanosized coconut shell ash

Table 2. Chemical composition test on the selected waste ashes, (%wt).

Ash	CaO	MnO	MgO	ZnO	CuO	Fe_2O_3	Al_2O_3	SiO_2	Na_2O	P_2O_5	K_2O
NPBA	12.7	0.13	0.01	0.78	Trace	0.95	20.12	64.45	0.71	0.64	0.14
NWTA	12.57	0.13	0.01	3.21		2.55	31.12	44.45	4.11	1.81	0.04
NWPA	9.21	3.67	Trace	Trace		1.11	17.23	52.11	7.41	7.99	1.27
NPKA	11.25	4.44	7.86	4.67	12.76	3.89	17.77	23.88	9.01	2.91	1.64
NSSA	17.83	1.16	Trace	Trace	Trace	2.44	6.11	65.01	2.31	5.12	0.02
NCSA	12.31	2.87	7.82	4.66	11.99	3.98	19.89	25.19	9.09	1.11	1.09

2.1 The Influences of Nanostructured Ashes on the CBR Behavior

Figure 2a shows the CBR behaviour of the stabilized soil with nanostructured ash materials which progressively increased with the addition of the additives at moist curing time of 28 days.

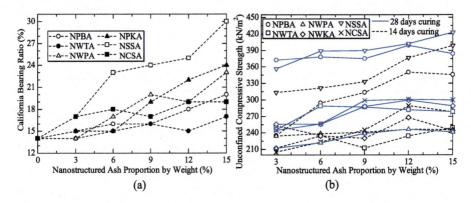

Fig. 2. Strength behaviours of stabilized soils

This behavior may be attributed to the admixtures' increased reactive surface area, their highly pozzolanic behavior and lower density as a result of nanosization. The soil + 12% ash and soil + 15% ash mixtures passed the minimum CBR value of 20–30% specified by [6] for materials suitable for use as base course materials when determined at MDD and OMC. This is close to the findings of [9], which stated that the minimum CBR value of 20–30% is required for sub-bases when compacted at OMC.

2.2 The Influences of Nanostructured Ashes on the UCS Behavior

Figure 2b shows the UCS behaviour of the stabilized soil under the addition of nanostructured ash materials. There was consistent improvement of the UCS with the addition of variable proportions of ash materials. This behaviour is also attributed to the pozzolanic behaviour of the additive and its increased reactive surface that encouraged bonding. A further observation was that the presence of the admixtures in the soil increased the frictional angle of the stabilized mixture attributed to the physicochemical and highly pozzolanic properties of the admixture and to their ability to reduce adsorbed water thereby making soils with higher clay content to behave like granular soil. These values satisfy the "very stiff" material condition for use as sub-base materials for pavement sub-grade and sub-base construction.

3 Conclusions

- The CBR gave good results that met the requirements for the stabilized soil to be used as a sub-base material in pavement construction.
- The UCS test results show great improvements in the stabilized soil with the nanostructured ash materials. The CBR and UCS improved results may be due to the increased reactive surface.
- The compaction and strength of the stabilized lateritic soil improved remarkably at different percentages of the ash materials by weight of stabilized mixture.

References

1. Anitha, P., Pandya, H.M.: Comprehensive Review of preparation methodologies of nano hydroxyapatite. J. Environ. Nanotechnol. **3**(1), 101–121 (2014)
2. Bykkam, S., et al.: Extensive studies on X-ray diffraction of green synthesized silver nanoparticles. Adv. Nanoparticles **4**(1), 1–10 (2015)
3. Ahmad, K.D., et al.: Synthesis of nano-alumina powder from impure kaolin and its application for arsenite removal from aqueous solution. J. Environ. Health Sci. Eng. **11**(1), 19 (2013)
4. Bao, Y., et al.: Synthesis of carbon nanofiber/carbon foam composite for catalyst support in gas phase catalytic reactions. New Carbon Mater. **26**(5), 341–346 (2011)
5. Ali, N., et al.: The effects of CuO nanoparticles on properties of self compacting concrete with GGBFS as binder. Mater. Res. **14**(3), 307–316 (2011)
6. BS 1924: Methods of Tests for Stabilized Soil. British Standard Institute, London (1990)

7. BS 1377-2: Methods of Testing Soils for Civil Engineering Purposes. British Standard Institute, London (1990)
8. BS 5930: The Code of Practice for Site Investigation. British Standard Institute, London (2015)
9. Gidigasu, M.D., Dogbey, J.L.: Geotechnical characterization of laterized decomposed rocks for pavement construction in dry sub-humid environment. In: Proceedings of the 6th South-East Asian Conference on Soil Engineering, Taipei, vol. 1, pp. 493–506 (1980)

Laboratory Carbonation Model of Single Soil-MgO Mixing Column in Soft Ground

Guanghua Cai[1], Songyu Liu[2(✉)], and Guangyin Du[2]

[1] Nanjing Forestry University, Nanjing 210037, China
[2] Southeast University, Nanjing 210096, China
liusy@seu.edu.cn

Abstract. The carbonation of reactive MgO-treated soil are considered as an innovative treatment technology in the realm of soft ground improvement. With this in view, the indoor mixing column model tests were carried out under different initial water content and CO_2 pressure through the artificial dig-hole column method. The study shows that the temperature of mixing column firstly increases and then decreases, and the temperature can reach the highest in less than 2 h. The temperature peak is the highest at the initial water content of 20%, the second at 15% and the lowest at 30%. In addition, the temperature peak increases with CO_2 pressure. The strength decreases with increasing initial water content or decreasing CO_2 pressure, and it decreases exponentially with the increase of water content. Nesquehonite peaks are slightly higher at water content of 30% than that at 15%. The peaks of nesquehonite and dypingite or hydromagnesite are relatively higher at 200 kPa than those at 25 kPa. At low water content of 15%, there are amounts of prismatic nesquehonite and few flake dypingite/hydromagnesite. At the high water content of 30%, there are amounts of carbonation products with loose connection. At low CO_2 pressure of 25 kPa, the pores are larger, and there are lots of flocculent brucite and few nesquehonite. Based on XRD and SEM analyses, the existences of nesquehonite hydromagesite/dypingite are the main reason for strength development.

Keywords: Reactive MgO · Carbonation · CO_2 · Soil-MgO mixing model

1 Introduction

Curing materials such as Portland cement (PC) and lime have been extensively utilized for treating the soft ground or problematic soil to meet the bearing needs. The stabilized soil exhibits higher strength, lower permeability or compressibility as compared to its untreated counterparts, which is beneficial for reducing settlement and increasing stability. The strength development of PC-treated soils is primarily attributed to the formation of cementitious materials. However, the formation of hydration products and consequent strength gain are relatively slow. And it is reported that the significant negative environmental impacts, associated with the production of PC, have drawn great attention in terms of huge energy consumption and nonrenewable resources as well as

CO_2 emissions, being one of the major sources of the CO_2 concentration increase in atmosphere [1]. With this in view, PC is partially or completely replaced by alternative binders such as geopolymers, calcium carbide residues, reactive MgO and ground granulated blast-furnace slag in soil treatment processes [1, 2]. In recent years, reactive MgO starts to draw people's attention due to the following reasons: (a) reactive MgO is generally from the calcination of magnesite at a low temperature as compared to the PC or is synthetically produced from seawater and lake brines, reducing fuel consumption [3]; (b) when reactive MgO is mixed alone as the binder in soils or concretes, it can absorb amounts of CO_2 to produce hydrated magnesium carbonates in short term, generating higher strength, finer microstructure and lower permeability [3–6].

Previous studies have revealed that the reactive MgO-admixed soil is able to fulfill the rapid and significant enhancement of strength under the condition of CO_2 [3–6]. The hydration or carbonation products generated could effectively cement soil particles and reduce the porosity of soils compared with un-treated soils [6]. Therefore, the reactive MgO and the carbonation of reactive MgO-treated soil are respectively considered as an innovative additive and a prospective treatment technology in the realm of soft ground improvement. With this in view, the indoor mixing column model tests were carried out under different initial water content and different CO_2 ventilation pressure through the artificial dig-hole column method.

There are four holes in every model ground for parallel measurement (Fig. 1(a)). After the mixtures were filled in the model holes and compacted, the temperature sensor of thermal resistance was inserted into the columns (Fig. 1(b)). After inserting the fine plastic tube into the model ground and covering the model ground by the sealing film, the weights are piled on the seal foundation to prevent the plump of the sealing film, and the loading weight was appropriately increasing with the CO_2 ventilation pressure increasing (Fig. 1(c and d)). The ventilation pressure was 25 kPa, 50 kPa, 100 kPa and 200 kPa, and the carbonation time is 3.0 h. After carbonation, the columns were dug out (see Fig. 2).

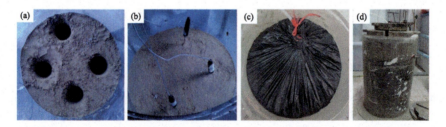

Fig. 1. Model charts (a) digging holes, (b) making columns, (c) intubating and sealing, (d) preloading and aerating.

Fig. 2. Schematic diagram of carbonation (a) the model foundation after excavation, (b) two MgO contents, (c) influence of initial water content, (d) influence of CO_2 pressure.

2 Conclusions

The temperature of mixing column shows the change tendency of first increase and then decrease during the carbonation process, and the temperature can reach the highest in less than two hours' carbonation. The temperature peak is the highest at the initial water content of 20%, the second at 15% and the lowest at 30%. In addition, the temperature peak increases with the increase of CO_2 pressure. The of carbonated MgO-mixing column decreases with water content while increases with increase of CO_2 pressure, and the strength of 20% MgO is higher than that of 15% MgO (Fig. 3). The strength decreases with the increase of water content, and the relationship between strength and water content can be fitted by the exponential function (Fig. 4).

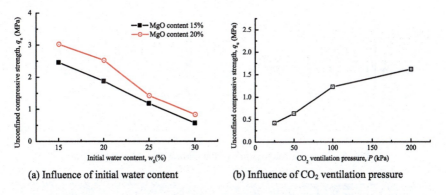

(a) Influence of initial water content

(b) Influence of CO_2 ventilation pressure

Fig. 3. Unconfined compressive strength under different conditions.

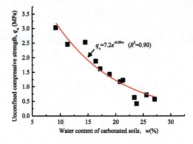

Fig. 4. Relationship between unconfined compressive strength and water content.

When initial water content increases from 15% to 30%, MgO peaks become weak while brucite peaks turn strong, and nesquehonite peaks are slightly higher at water content of 30% than that at 15% (Fig. 5(a)). The MgO peaks are basically same at the same angle for different CO_2 pressure, and the peaks of carbonation products are relatively higher at 200 kPa than those at 25 kPa (Fig. 5(b)). At low water content of 15%, there are amounts of prismatic nesquehonite and few MgO particles as well as dypingite/hydromagnesite; the carbonation products connect closely and small pores are fewer though there is larger pore. At the high initial water content of 30%, there are small pores, amounts of carbonation products while carbonation products connect loosely. At low CO_2 pressure of 25 kPa, the pores are larger, and there are loose carbonation products, lots of flocculent brucite and few nesquehonite (Fig. 6). Based on XRD and SEM analyses, the existences of prismatic nesquehonite and lamellar hydromagesite/dypingite are the main reason for the strength development. Carbonation model test will provide theoretical guidance for the MgO carbonating technology in the engineering application of soft foundation reinforcement. However, the present study of carbonated mixing column is limited to silt, CO_2 gas is diffused to the whole soil layer and the CO_2 effective utilization is lower. The quality of the column bottom is poorer and the carbonation depth is very limited. Therefore, there is a need to extend the proposed methodology to other soils (viz., clay) in further study, and the shallow overall carbonation technology will be used in the foundation treatment in order to increase the CO_2 utilization while improving the strength of the soil.

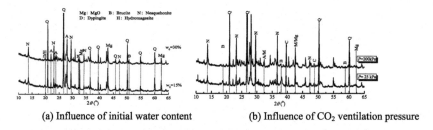

(a) Influence of initial water content   (b) Influence of CO_2 ventilation pressure

Fig. 5. X-ray diffractograms of the carbonated mixing column.

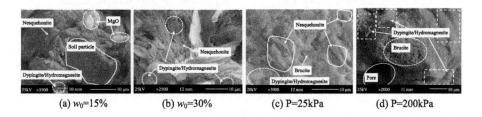

(a) $w_0=15\%$ (b) $w_0=30\%$ (c) P=25kPa   (d) P=200kPa

Fig. 6. SEM images of the carbonated mixing column.

References

1. Andrzej, C., Karin, H.C.: Effects of reactive magnesia on microstructure and frost durability of portland cement-based binders. J. Mater. Civil Eng. **25**, 1941–1950 (2013)
2. Liska, M., Al-Tabbaa, A.: Performance of magnesia cements in porous blocks in acid and magnesium environments. Adv. Cem. Res. **24**(4), 221–232 (2012)
3. Cai, G.H., Liu, S.Y., Du, Y.J., Zhang, D.W., Zheng, X.: Strength and deformation characteristics of carbonated reactive magnesia treated silt soil. J. Cent. South Univ. **22**(5), 1859–1868 (2015)
4. Cai, G.H., Du, Y.J., Liu, S.Y., Singh, D.N.: Physical properties, electrical resistivity and strength characteristics of carbonated silty soil admixed with reactive magnesia. Can. Geotech. J. **52**(11), 1699–1713 (2015)
5. Cai, G.H., Liu, S.Y., Du, Y.J., Cao, J.J.: Influences of magnesia activity index on mechanical and microstructural characteristics of carbonated reactive magnesia-admixed silty soil. J. Mater. Civil Eng. 2017 **29**(5), 04016285(1–12) (2016)
6. Yi, Y., Liska, M., Akinyugha, A., Unluer, C., Al-Tabbaa, A.: Preliminary laboratory-scale model auger installation and testing of carbonated soil-MgO columns. Geotech. Test. J. **36**(3), 1–10 (2013)

Enhancing the Strength Characteristics of Expansive Soil Using Bagasse Fibre

Liet Chi Dang[✉] and Hadi Khabbaz

School of Civil and Environmental Engineering, University of Technology Sydney,
Ultimo, NSW 2007, Australia
liet.dang@uts.edu.au

Abstract. This paper presents an experimental investigation on the compressive and shear strength characteristics of expansive soil reinforced with randomly distributed bagasse fibre. Bagasse fibre, an agricultural waste by-product left after crushing of sugar-cane for juice extraction, was employed in this investigation as reinforcing components for expansive soil reinforcement. To comprehend the bagasse fibre reinforcement effects on the strength of reinforced soils, a series of experimental investigations was carried out. They include unconfined compressive strength (UCS) tests, using a conventional compression machine, and shear strength tests, using advanced triaxial compression apparatus, conducted on non-reinforced and fibre reinforced soil samples with different percentages of randomly distributed bagasse fibre from 0% to 2%. Following the compression tests, scanning electron microscopy (SEM) analysis was conducted on selected soil samples to evaluate the micromechanical reinforcement between soil particles and fibre surface. The obtained test results indicated that bagasse fibre reinforcement not only significantly improved the compressive strength, the initial deformation modulus, and the shear strength of reinforced soils, but it also considerably transformed the reinforced soil behaviour from strain softening to strain hardening by curtailing the post-peak shear strength loss.

1 Introduction

Expansive soils are fine-grained soil or decomposed rocks, showing significant volume change when exposed to the moisture content fluctuations. Swelling and shrinkage phenomena are most likely to take place near the ground surface where it is directly affected by environmental and seasonal changes. The expansive soils are almost certainly to be unsaturated state and have dominantly montmorillonite clay minerals. The severity of the damage to residential buildings and other civil engineering structures built on top of expansive soils depends on the amount of monovalent cations absorbed into the clay minerals. Chemical stabilisation of expansive soil using lime or cement is the most commonly ground modification method for controlling the shrink-swell behaviour of expansive soil due to seasonal variations. Lime reacts with expansive clay in the presence of water and changes the physicochemical properties of expansive soil, which in turn modifies the engineering properties of treated soil. However, chemical stabilisation of expansive soil is not always acceptable in Australia since it can cause adverse effects on the environment by contaminating the quality of ground-water and affecting the normal growth of vegetation. To overcome these impacts, an eco-friendly stabiliser is required to improve the shrink-swell behaviour, the strength, and the durability of expansive soil without causing any adverse effects on the environment. In recent years, a number of researchers [1, 2] have conducted several experimental investigations on expansive soil modifications using green geomaterials, low-cost geofibres such as bagasse fibre, hay fibre, polypropylene fibre, and recycled carpet fibre, to examine the beneficial effects of fibre reinforcement on the engineering properties of expansive soil. In this investigation, the strength characteristics of reinforced soils were investigated through conducting an array of extensive laboratory tests such as UCS tests and shear strength tests. Those geotechnical tests were performed on expansive soil samples reinforced with different contents of bagasse fibre ranging from 0% to 2% after 7 days of curing. The findings of this investigation are analysed and discussed to comprehend the influence of bagasse fibre reinforcement on the strength properties of expansive soil. SEM tests were also conducted on selected soil samples to examine the fibre reinforcement of the soil strength.

2 Materials and Experimental Program

The soil samples and bagasse fibre used in this experimental study were collected from Queensland, Australia, which were thoroughly described in [1, 2]. The UCS tests were conducted on nonreinforced and reinforced soil samples after 7 days of curing using a conventional compression machine at a loading rate of 1 mm/min in accordance with the standard procedures specified in AS 5101.4 (2008). Meanwhile, an advanced triaxial compression apparatus was adopted for isotropically consolidated-undrained (CU) tests at a relatively small shearing rate of 0.017 mm/min to ensure the equilibrium of excess pore water pressure throughout the soil samples during the shearing process in accordance with the standard procedures specified in AS 1289.6.4.2 (1998). SEM analysis was carried out on a fraction of UCS samples after testing to examine the microstructural strength development of bagasse fibre reinforced soils.

3 Results and Discussion

3.1 Effects of Bagasse Fibre on Stress-Strain Behaviour

Figure 1 displays the effects of bagasse fibre reinforcement on the axial stress-strain behaviour of expansive soil, deprived from the unconfined compressive strength tests. As illustrated in Fig. 1, the peak stress of reinforced soils significantly increased with an increase in bagasse fibre content from 0% to 2%. For example, the peak axial stress increased significantly from 137 kPa for 0% bagasse fibre reinforced soil (non-reinforced soil) to around 194 kPa for soil reinforced with 2% bagasse fibre at the same curing period of 7 days, resulting in the corresponding improvement in the peak stress of approximately 41%. As expected, the post-peak stress also increased as bagasse fibre content increased up to 2% as clearly depicted in Fig. 1. The highest post-peak stress (residual strength) was visually observed for the soil reinforced with 2% bagasse fibre. The improvement in the residual strength corroborates that bagasse fibre reinforcement could transform the behaviour of reinforced soils from strain softening to strain hardening by reducing the loss of the post-peak stress. This confirms the beneficial effect of bagasse fibre reinforcement on the axial stress and strain of expansive soil. Since the effect on the peak compressive strength (peak strength) of soils reinforced with different bagasse fibre contents is not clearly captured in Fig. 1. Hence, the peak strength and the initial deformation modulus (secant modulus) of reinforced soils are plotted in Fig. 2 against bagasse fibre content for further evaluation and comparison. It can be noted that the secant modulus is defined as the slope of initial stress-strain curves between the origin and the point of 50% failure stress. As observed in Fig. 2, the peak strength increased significantly with a small amount of 0.5% bagasse fibre firstly introduced into the soil matrix, followed a slight increase in the peak compressive strength as fibre reinforcement increased up to 2%. Meanwhile, the secant modulus of the reinforced soil increased significantly from about 7.2 MPa to 14 MPa (i.e., the corresponding increment was about 94%) with an increase in fibre content from 0% to 1%. However, the secant modulus of soils reinforced with bagasse fibre decreased slightly as further increase in the fibre content exceeded 1%. This phenomenon indicates that an excessive use of bagasse fibre reinforced soil reduced the stiffness of the fibre-soil matrix.

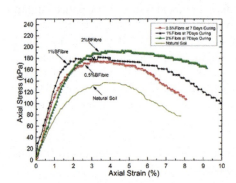

Fig. 1. Axial stress-strain relationship of expansive soil reinforced with different bagasse fibre contents

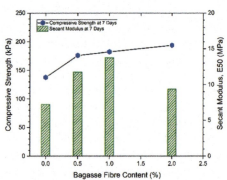

Fig. 2. Variations of unconfined compressive strength and secant modulus of expansive soil reinforced with different fibre contents

3.2 Effects of Bagasse Fibre on the Shear Strength Parameters

Figure 3 exhibits variations of the peak shear strength parameters including the internal friction angle and cohesion along with bagasse fibre contents. As observed in Fig. 3, the peak internal friction angle increased considerably when bagasse fibre content increased from 0% to 2%. For instance, the peak friction angle increased from approximately 19.5° for 0% bagasse fibre reinforced soil (natural soil) to 28° for 2% bagasse fibre reinforced soils, which results in the corresponding increment of about 44%. However, as evident in Fig. 3, the reinforcement of expansive soil with bagasse fibre appeared to have less notable effect on the cohesion of reinforced soils in comparison with the corresponding improvement of the peak internal fiction angle. For example, the peak cohesion of reinforced soils remained almost constant (around 10 kPa) as bagasse fibre content increased from 0% to 1%, but then it began to increase up to 14 kPa with a further increase in bagasse fibre content up to 2%. It is assumed that the high tensile strength of bagasse fibre contributes to the improvement in the cohesion of reinforced soils due to enhancing the tension resistance of the fibre-soil matrix during post-peak shearing in progress. Meanwhile, the frictional interaction and interlocking mechanism between the fibre surface and the soil matrix formed by compaction energy, as visually depicted in Fig. 4, are responsible for the improvement in the peak internal friction angle of soils reinforced with bagasse fibre.

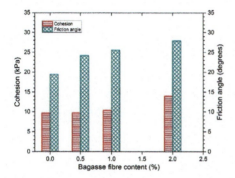

Fig. 3. Shear strength parameters of expansive soil reinforced with different fibre contents

Fig. 4. SEM image of interaction and interlocking mechanism between fibre-soil

4 Conclusions

The reinforcement of expansive soil with bagasse fibre significantly improved the initial stiffness, the compressive strength, and the shear strength properties of reinforced soils. Meanwhile, it transformed the behaviour of reinforced soils from strain softening to strain hardening by diminishing the loss of the post-peak shear strength. The findings of this investigation provide a deeper insight into making use of agricultural by-products (e.g. bagasse fibre) as a green solution for sustainable construction development.

Acknowledgments. The experiment results presented in this paper are part of an ongoing research at University of Technology Sydney (UTS) supported by the Australian Technology Network (ATN), Arup Pty Ltd., Queensland Department of Transport and Main Roads (TMR), ARRB Group Ltd and Australian Sugar Milling Council (ASMC). The authors gratefully acknowledge their support.

References

1. Dang, L.C., Khabbaz, H., Fatahi, B.: An experimental study on engineering behaviour of lime and bagasse fibre reinforced expansive soils. In: 19th ICSMGE, pp. 2497–2500, South Korea (2017)
2. Dang, L.C., Fatahi, B., Khabbaz, H.: Behaviour of expansive soils stabilized with hydrated lime and bagasse fibres. Procedia Eng. **143**, 658–665 (2016)

Engineering Properties and Microstructure Feature of Dispersive Clay Modified by Fly Ash

HengHui Fan[✉], YingJia Yan, XiuJuan Yang, Lu Zhang, and HaiJun Hu

College of Water Resources and Architectural Engineering, Northwest A&F University,
Yangling 712100, Shaanxi, China
13572555506@163.com

Abstract. Dispersive clay has the very low erosion resistance capacity and it causes serious threatening to the engineering safety such as the hydraulic engineering. It is one of the problematic soils which have been concerned in the geotechnical engineering field. The pinhole, crumb, compression, SEM test had been investigated in different fly ash contents and curing time, so as to study the effect of fly ash on the dispersivity, deformation and microstructure characteristics. The test results showed that the fly ash had a significant influence on the dispersivity of dispersive clay. The dispersivity was reduced with the increasing of fly ash contents and curing time. The SEM tests showed that the quantity of hydration products changed significantly with the increasing of fly ash contents and curing time, leading to an increase in contact area between framework grains, a reduction in intergranular pore and a transformation from point-contact to surface-contact.

Keywords: Dispersive clay · Fly ash · Improvement

1 Introduction

Dispersive soil is typical clay in nature, which has led to failure incidents throughout the world affecting the dams, embankments and roads [1]. In dispersive clays, when saturated with low salty water or pure water, the cohesion between particles disappeared quickly thus causing weaknesses in the bonds. The lime and cement, as additives, are common used to improve the nature of the dispersion of soil [2]. The last 20 years, scholars have carried out a serious research on fly ash as a modifier to improve the soil's stability [3–6]. The main aim of this study was to evaluate the change in the microstructure of the fly ash treated dispersive soil.

2 Materials and Methods

2.1 Dispersive Clay and Fly Ash

Based on the pinhole test, crumb test, double hydrometer test, pore water cation test and exchangeable sodium ion percentage test, the soil was considered to be D (dispersive).

The fly ash used in this test was taken from the ash storage filed of Xianyang Weihe Power Plant.

2.2 Methodology

The geotechnical tests or analyses carried out on the mixed specimens were the pinhole test, the crumb test, the SEM test.

3 Results and Discussion

3.1 Effect of Fly Ash on Dispersivity of Soil Samples

The dispersibility of treated dispersive clay specimens are shown in Table 1. As indicated in Table 1, under the 50 mm water head, the pinhole was quick erosion and the pinhole expanded. When treated with 6% fly ash and cured for 0 day, under the 50 mm water head, the pinhole is constant. However, under the 180 mm water head, the pinhole eroded. When treated with 10% fly ash and cured for 0 days, under the 1020 mm water head, the pinhole is constant.

Table 1. Results of the pinhole and crumb test modified with fly ash

Fly ash content	Curing times (day)					
	0		5		7	
	Pinhole test	Crumb test	Pinhole test	Crumb test	Pinhole test	Crumb test
0%	D	D	D	D	D	D
4%	D	D	I	I	ND	ND
6%	I	I	ND	ND	ND	ND
8%	ND	ND	ND	ND	ND	ND
10%	ND	ND	ND	ND	ND	ND

Note: D: dispersive, ND: non-dispersive, I: intermediate

As indicated in Table 1, the dispersive clay in the crumb tests collapsed quickly and at the bottom of the beaker it is full of the rubber particles, and the water is cloudy. When treated with 6% fly ash, the treated samples' dispersibility is intermediate. However, when treated with 8% fly ash, the treated samples are non-dispersive. The use of the 4% fly ash with 7 days of curing time has changed the designation of samples from D to ND, and with 8% fly ash, the treated samples are also ND. Generally speaking, the dispersibility of the soil gradually decreases with the increase of the curing time and additive contents.

3.2 Effect of Flay Ash on Micro-features of Soil Samples

There are typical pictures of the specimens in Fig. 1(a–f). The microstructure of the dispersive soil has the following characteristics. The contact form between the skeleton units of the soil particles is mainly point-contact and surface-contact and the skeleton units has a clear boundary with each other. The pores in the dispersive soils are mainly overhead pores with irregular shape and no filling in them.

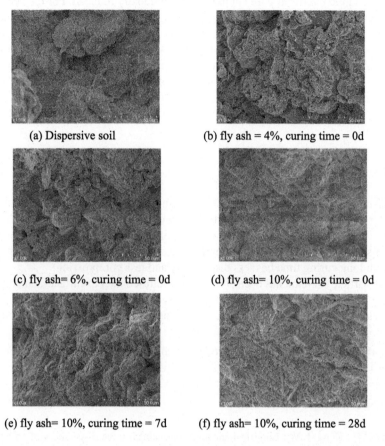

Fig. 1. Microstructure photos of dispersive clay modified with fly ash (Mag = 1000)

It can be observed that when the fly ash additive is 4%, the contact form between the skeleton unit bodies of the modified soil is still point-contact and surface-contact, and also with obvious boundary between the skeleton unit body. While on the surface of the soil particles there are some cementing materials which generated from the reaction of the fly ash and the soil. In that case, the cohesive force between the particles has been enhanced. However, the overhead pores are still in the soils with a little filling in them. When the additive of fly ash is 6%, the contact between the skeleton unit bodies of the modified soil is mainly surface-contact, and the boundary between the skeleton unit bodies become blurred. The overhead pores inside the soil are reduced and the cohesive force between the modules is enhanced. When the increment of fly ash is 10%, the contact form between the skeleton units of the modified soil is mainly surface-contact, and no obvious boundary between the skeleton units. The overhead pores in the soil reduced, while the contact area between the soil particles increased, and also the water stability enhanced. Therefore, with the increase of the fly ash additive, the contact form between the skeleton unit bodies of the modified soil is changed from point-contact to surface-contact, and the overhead pores in the soil are gradually reduced. For the cementation effect between the soil particles gradually increases, soil's water stability enhanced step by step.

When the curing time is 0d, there are still overhead pores in the modified soil, but the particles in the soil are mainly in surface-contact, and the cement particles are filled with the intergranular pores. With the increase of curing time, the hydration reaction of fly ash and dispersive soil can produce hydration products such as CSH and C_3AH_6. In that case, the cementing affection increased. As a result, the soil became dense and stabilized.

4 Conclusions

The dispersivity of the treated soil using fly ash have been obviously changed. Due to the interaction between the fly ash and the dispersive soil, the dispersivity of the treated soil has been reduced along with the increase of the additive percent. The curing time has considered being the significant parameter of the stabilization process. The treated soil with long curing time has less dispersivity and less compression.

The particle skeleton has been altered from point-contact to surface-contact with the increase of the additive. In the touching process, the cementation effect between the soil particles enhances gradually. In that case, the dispersive soils come to be stabilized.

Acknowledgements. This study was partly funded by the National Natural Science Foundation of China No. 51579215 and No. 51379177.

References

1. Fan, H.H., Kong, L.W.: Empirical equation for evaluating the dispersivity of cohesive soil. Can. Geotech. J. **50**(9), 989–994 (2013)
2. Ma, X.Y., Xu, Y.J.: Behavior of dispersive clay and its lime stabilized test in Guanlu reservoir in Qingdao. Chin. J. Geotech. Eng. **22**(4), 441–444 (2000)
3. Mir, B.A.: Some studies on the effect of fly ash and lime on physical and mechanical properties of expansive clay. Int. J. Civil Eng. **13**(3), 203–212 (2015)
4. Wang, M., Bai, X.H., Liang, R.W., et al.: Micro study on soft foundations reinforcement with lime-fly ash piles. Rock Soil Mech. **22**(1), 67–70 (2001)
5. Engr, J.G., Galupino, J.D.: Permeability characteristics of soil-fly ash mix. ARPN J. Eng. Appl. Sci. **10**(15), 6440–6447 (2015)
6. Horpibulsuk, S., Phetchuay, C., Chinkulkijniwat, A.: Soil stabilization by calcium carbide residue and fly ash. J. Mater. Civil Eng. **24**(2), 184–193 (2012)

Effect of Strain Rate on Interface Friction Angle Between Sand and Alkali Activated Binder Treated Jute

Shashank Gupta, Anasua GuhaRay[(✉)], Arkamitra Kar, and V. P. Komaravolu

Department of Civil Engineering, BITS Pilani Hyderabad Campus, Hyderabad 500078, India
guharay@hyderabad.bits-pilani.ac.in

Abstract. The interface friction angle between reinforcement and soil is a significant property that defines the suitability of geotextile for several applications such as reinforced retaining wall and slope stability. However, it is not an intrinsic property and varies with several experimental factors such as relative density of sand and shearing strain at which shear tests are conducted. Recent literature shows the wide application of jute geotextile in geotechnical constructions such as slope stability, river bank protection, and subgrade stabilization. However, its application is limited due to low durability under the soil. Therefore, to improve the resistance of jute geotextile against the biological degradation, it has been treated with the fly ash-based treatment solution. This study makes an attempt to investigate the effect of strain rate on the interface friction angle between sand and alkali activated binder treated jute geotextile. The tests are conducted in large shear box apparatus specifically assembled to determine interface shear properties. The jute geotextile is treated with alkali activated binder of four different water to solid ratios, each of them is cured at the temperature of 40 °C. The results obtained are then collated with those obtained from untreated jute geotextile. This study further delineates the effect of the degree of compaction on the interface friction between the reinforcement and sand. Hence, the interface shear properties apropos to the relative density of sand, shear strain, and treatment composition are compared and the obtained trends along with the optimum values are presented.

Keywords: Jute geotextile · Alkali activated binder · Large shear box · Ground improvement

1 Introduction

Geotextiles have become an essential part of geotechnical engineering applications due to their ability to reinforce, filter, separate, and drain soil. Due to their outstanding mechanical properties, synthetics geotextiles are extensively used worldwide [1]. However, synthetic geotextiles production consumes high energy which leads to significant CO_2 footprint [2]. Therefore, researchers are exploring natural alternatives to synthetic geotextile like jute, coir, and flax. In India, the vast laboratory studies and more than 260 field trials have been conducted with jute geotextile (JGT) [3]. However, the low durability limits the use of JGT in various geotechnical applications. Therefore, several researchers have developed the treatment procedures to improve the mechanical properties and durability of jute. Scouring, bleaching, acetylation, transesterification, and bitumen treatment are some of these treatment procedures [4–6]. However, the existing chemical treatment methods require either expensive chemicals or sophisticated procedure that is impractical to implement in the field construction. This study introduces the treatment of JGT with alkali activated binders (AAB). Alkali activated binder is prepared by mixing, fly ash, sodium silicate, sodium hydroxide, and water [7]. This study focuses on the shear interaction of AAB treated JGT with sand at different shear strain rates.

2 Experimental Investigations

2.1 Materials

The geotextile used in this study is "Tossa" jute (*Corchorus Olitorius*) which is commercially available in India. The properties of jute used in this study are provided in Table 1. For the preparation of treatment mix, Class F fly ash, sodium silicate, sodium hydroxide, and water are required. In this study, food grade sodium hydroxide pellets of 99% purity are used. The composition of sodium silicate solution is 55.9% water, 29.4% SiO_2, and 14.7% Na_2O. The river sand used for the experiment is obtained from the banks of Godavari river, India. According to Unified Soil Classification System, the sand is classified as poorly graded sand (SP). The properties of sand are mentioned in Table 2. The relative density of sand is kept 40% for all the experiments.

Table 1. Properties of jute geotextile

Property	Value
Thickness (mm)	0.69 mm
Mass density (kg/m^2)	260.17 gm/m^2
Tensile strength (kN/m)	10.6 kN/m
Aperture (mm)	1.22 mm × 1.63 mm

Table 2. Properties of sand

Property	Value
Coefficient of Uniformity (C_u)	2.5
Coefficient of Curvature (C_c)	0.9
Classification	SP
Maximum Void Ratio	0.637
Minimum Void Ratio	0.411

2.2 Preparation of Treated Geotextile

The treatment mix is prepared by mixing aluminosilicate precursor and alkali activator solution. The precursor used in this mix is Class-F fly ash and the alkali activator solution is composed of sodium silicate, sodium hydroxide, and water. In all the treatment mixes, the ratio of fly ash to sodium silicate to sodium hydroxide is kept as 400:129.43:10.57 [7]. However, the amount of water is varied to prepare treatment solutions of four different water to solid ratios viz. 0.35, 0.40, 0.45, and 0.50. The activator solution is first prepared by mixing sodium silicate solution, sodium hydroxide pellets, and water. Then the solution is mixed thoroughly with fly ash. This treatment mix is then applied on the JGT by paintbrush such that the treatment coating covers the complete surface area of JGT. Following the AAB treatment, the samples are kept in a humidity controlled chamber at a temperature of 40 °C for 24 h for initial strength gain.

2.3 Experimental Methodology

A HEICO automatic large-size direct shear apparatus is used to study the interface friction angle between reinforcement and sand. The specifications given by ASTM D5321/D5321M-17 (2017) are followed during the execution of experiments. The size of each, upper and lower, shear boxes is 300 mm × 300 mm × 150 mm. The geotextile is fixed at the top of the lower box and the upper shear box is filled with sand at the relative density of 40%. The sand in the shear box is compacted with a hammer by hand tamping and then the normal load is applied. The motor powered gears control the motion of the lower shear box. The experiment is terminated when the shear displacement reaches the approximate value of 25 mm. The tests are conducted for three different normal loads of 1.25 kg/cm^2, 1.75 kg/cm^2, and 2.25 kg/cm^2. The shear strains are varied from 1 mm/min to 2.54 mm/min to find the correlation between the strain rate and interface friction angle.

3 Results and Discussions

The interface friction angle between sand and treated geotextiles is provided in Fig. 1. At lower stain rates, it is observed that the interface friction angle between 0.50 w/s AAB treated JGT and sand is higher than rest of the samples. However, at higher strain rates, untreated JGT and JGT treated with AAB of w/s ratio of 0.50 and 0.45 have approximately equal interface angles. It is also observed that the interface friction reduces with increasing strain rates for all geotextiles. The possible explanation for the better performance of JGT treated with high w/s ratio, is the chemical interaction in alkali activated binder. The extent of reaction between fly ash and activator solution increases with water content. Hence the amount of unreacted fly ash reduces with increasing w/s ratio. Fly ash particles are spherical in nature [7] and therefore offer less friction. Hence, the high amount of unreacted fly ash in JGT treated with 0.35 w/s and 0.40 w/s AAB results in lower friction angle compared to untreated JGT.

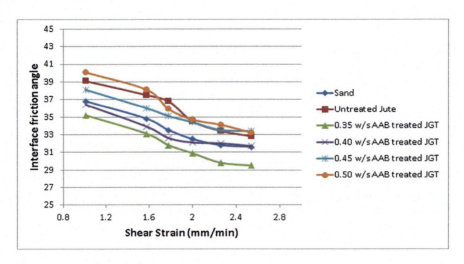

Fig. 1. Interface friction angle vs Shear Strain for different JGT samples

4 Conclusions

The present study devises a new technique of treating jute geotextile with AAB solution of different w/s ratio (0.35–0.50) in order to improve the strength characteristics of soil. It is observed that the geotextile treated with AAB having high w/s ratio have high shear strength compared to untreated jute. Also, with increase in strain rate, the interface friction angle decreases for all w/s ratio.

References

1. Shtykov, V.I., Blazhko, L.S., Ponomarev, A.B.: The performance of geotextile materials used for filtration and separation in different structures as an important part of geotextiles requirements. Procedia Eng. **189**, 247–251 (2017)
2. Wambua, P., Ivens, J., Verpoest, I.: Natural fibres: can they replace glass in fibre reinforced plastics? Compos. Sci. Technol. **63**(9), 1259–1264 (2003)
3. Sanyal, T.: Jute Geotextiles and Their Applications in Civil Engineering. Springer, Heidelberg (2017)
4. Wang, W.M., Cai, Z.S., Yu, J.Y., Xia, Z.P.: Changes in composition, structure, and properties of jute fibers after chemical treatments. Fibers Polym. **10**(6), 776–780 (2009)
5. Saha, P., Roy, D., Manna, S., Adhikari, B., Sen, R., Roy, S.: Durability of transesterified jute geotextiles. Geotext. Geomembr. **35**, 69–75 (2012)
6. Andersson, M., Tillman, A.M.: Acetylation of jute: Effects on strength, rot resistance, and hydrophobicity. J. Appl. Polym. Sci. **37**(12), 3437–3447 (1989)
7. Kar, A.: Characterizations of Concretes with Alkali-activated Binder and Correlating Their Properties from Micro-to Specimen Level. West Virginia University (2013)

Enhancement of Mechanical Properties of Expansive Clayey Soil Using Steel Slag

Anurag S. Hirapure and R. S. Dalvi[✉]

College of Engineering, Shivajinagar, Pune, Maharashtra, India
rsd.civil@coep.ac.in

Abstract. Dumping wastes from industries and respective by-products have become a serious issue nowadays due to less dumping sites and stringent environmental regulations. Which caught the attention of the many researchers working in the field of Environmental Geotechnical engineering. Steel slag a by-product of steel making industry which is long been used in various applications in the construction industry. It is used as aggregates in road construction, railway ballast, and hydraulic protection structures but the use of steel slag in the soil stabilization is a modern approach. The focus of this research is to study the effect of steel slag on strength characteristics of the clayey soil. The steel slag is blended with clayey soil at 10%, 15%, 20%, 25% and 30% of dry weight. The optimum content of steel slag was obtained based on the maximum unconfined compressive strength. Series of laboratory tests were performed on the optimum mix to evaluate its suitability as a stabilizer material. The tests were also conducted with conventional stabilizing materials with clayey soil and the optimum mix to compare the results. Soil-slag treated with 5% lime mix and 10% fly ash yielded much improvement as compared to a soil-slag mix.

Keywords: Steel slag · Clayey soil · Soil stabilization · Strength parameter

1 Introduction

In India, nearly 20% of the soil has been covered with the expansive soil namely black cotton soil. These soils have the tendency of volume change when water is added or removed from it. These are essentially clay deposits having low shear strength, high plasticity, compressibility and possess very high volumetric change when saturated due to this reason they do not meet existing standard requirements for use in civil engineering projects.

According to Federal Highway Administration (FHWA), steel slag is a by-product of steel making industry and produced during the separation of molten steel from impurities in steel making furnace. Earlier steel slag has been used for various civil engineering application but its use in the soil engineering is limited due to the presence of mineral phase that causes volumetric instability. Akinwumi et al. (2012) investigated the effect of adding pulverized steel slag on some geotechnical properties of lateritic soil and reported that steel can be profitably used as a stabilizer for lateritic soil. Soil mixed with

Steel Slag, Lime, and Rice Husk Ash can effectively improve the soft subgrade soil from poor to excellent (Ashango and Patra 2016). Athulya et al. (2016) concluded that steel slag-soil mixes acquired good strength, drainage, and plasticity characteristics to act as a potential material for highway embankment construction. Laboratory investigation of two different types of clay and three types of BOF slag fines along with two activators were carried out by Poh et al. (2006) and reported the improvement in terms of strength, volume stability and durability.

In the present study steel slag mix with clayey soil and the optimum content of steel slag has been obtained based on unconfined compressive strength. The effect of different stabilizers has also been studied on the optimized soil-slag mix.

2 Materials

The clayey soil used in the present study was collected from the agriculture land near Pune, Maharashtra, India. The geotechnical properties of the soil are given in the Table 1. According to unified classification system the soil has been classified as silt with high compressibility having the free swell index of 73%.

Table 1. Geotechnical properties of soil

Tests	Characteristics	Values
Particle size distribution	Gravel (%)	1
	Sand (%)	6
	Silt and Clay (%)	93
	Classification of Soil	MH
Physical Properties	Specific Gravity	2.65
	Liquid Limit (%)	72
	Plastic Limit (%)	41
	Plasticity Index (%)	31
	Shrinkage Limit (%)	16
Engineering Properties	Optimum Moisture Content (%)	34.2
	Maximum Dry Density (kN/m^3)	13.01
	Unconfined Compressive Strength (kN/m^3)	373
	Free Swell (%)	73

In this study Steel slag is obtained from Rambo Industries Coimbatore, Tamil Nadu, India. The colour of the steel slag is greyish with specific gravity of steel slag is 3.19. C_u and C_C values are 2.09 and 1.64 respectively.

3 Result and Discussion

The clayey soil has been mixed with different percentages of steel slag 10%, 15%, 20%, 25% and 30% and based on maximum unconfined compression strength (UCS) value optimum mix has been found out as 25% steel slag. The increase in the value may be attributed due to the frictional resistance offered by the steel slag due to a presence of wide range of particles in the steel slag. As the 75% soil +25% steel slag yielded the maximum UCS value it has been taken as optimum mix ratio. It is seen that UCS value of optimum mix increases by 38% than that of clay. Furthermore, it is also seen that liquid limit, plastic limit and plasticity index of optimum mix decrease to 18%, 22% and 13% than that of clay soil (Hirapure and Dalvi 2017).

3.1 Stabilization of Soil-Slag Mix

Addition of the steel slag reduce the expansion but did not fully suppressed it and also in case of the soaked CBR the values obtained were on the lower side (Hirapure and Dalvi 2017); for this reason, optimum mix (75%S + 25%SS) sample has been treated with the 5% lime and 10% fly ash. Figure 1 shows maximum value of dry density for optimum mix treated with 5% lime and 10% fly ash than that of soil treated with lime and fly ash. Fly ash being finer material, when added to the mix acts as the filler material which fills up the voids between soil and steel slag thus increasing the maximum dry density.

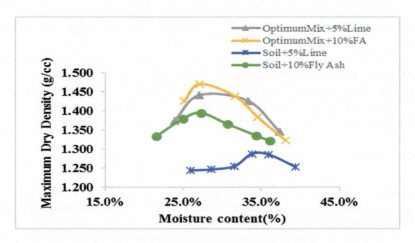

Fig. 1. Compaction curve with different stabilizers.

3.2 California Bearing Ratio Test

Figure 2 shows the variation in soaked and unsoaked CBR values for different combination of mixes. CBR result shows the improvement in the soaked CBR value with the addition of the lime to the optimum mix. Hydration phenomenon during soaking resulted in the formation of C-S-H gel that improved the strength of the mix. The expansion ratio is also reduced to 0.46% with the addition of 5% lime to the optimum mix. The addition of lime to soil-slag mix decrease the attraction of water because of formation of double layered calcium ions around clay particles and the pozzolanic compounds present in the steel slag binds the mix due to which swelling is reduced. Addition of lime also initiates the floc formation for clay particles.

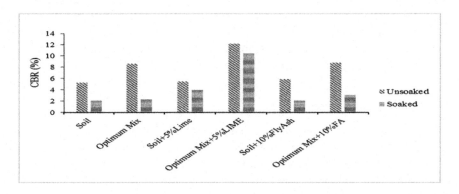

Fig. 2. Variation of CBR value for different mix.

4 Conclusion

The clayey soil has been mixed with different percentages of steel slag. The UCS value of mix increases with increase in steel slag content and its maximum for an optimum content of 25% steel slag. Soil-slag treated with 5% lime mix and 10% fly ash yielded much improvement as compared to a soil-slag mix. The improvement observed was also more than that of natural clayey Soil. Mix also shows the improvement in the soaked CBR value and expansion ratio which indicates its potential for swelling reduction. Addition of lime has the better effect on optimum soil-slag mix than fly ash.

References

Akinwumi, I., Adeyeri, J., Ejohwomu, O.: Effect of steel slag addition on the plasticity, strength and permeability of lateritic soil. In: International Conference on Sustainable Design Engineering, and Construction, pp. 457–464 (2012)

Ashango, A., Patra, N.: Behavior of expansive soil treated with steel slag, rice husk ash, and lime. J. Mater. Civ. Eng. **28**(7), 1–4 (2016)

Athulya, G., Dutta, S., Mandal, J.: Performance evaluation of stabilized soil-slag mixes as highway construction material. Int. J. Geotech. Eng. **10**, 51–61 (2016)

Hirapure, A., Dalvi, R.: Influence of steel slag addition on strength characteristics of clayey soil. In: Sixth Indian Young Geotechnical Conference 6IYGEC, pp. 137–142 (2017)

Poh, H., Ghataora, G., Ghazireh, G.: Soil stabilization using basic oxygen steel slag fines. J. Mater. Civ. Eng. **18**(2), 229–240 (2006)

Breakage Effect of Soft Rock Blocks in Soil-Rock Mixture with Different Block Proportions

Xinli Hu[1(✉)], Han Zhang[1], Chuncan He[1], and Wenbo Zheng[2]

[1] Faculty of Engineering, China University of Geosciences, Wuhan, Hubei 430074, People's Republic of China
huxinli@cug.edu.cn
[2] School of Engineering, University of British Columbia, Kelowna, BC, Canada

Abstract. Nowadays soil-rock mixture (S-RM) with highly weathered soft rock blocks is widely used as a backfill material for slopes. S-RM samples with different block proportions by weight (WBP) are simulated by DEM direct shear tests considering the breakage of rock blocks. The breakage characteristics of the soft rock blocks, and the breakage effect on the shear mechanical properties of S-RM with different WBPs are investigated. The breakage meso-ratio and breakage energy of rock blocks increase with the increment of horizontal displacement and WBP. The breakage of rock blocks induces the rapid increment of breakage energy. The breakage of rock blocks has a negative impact on the increment of internal friction angle, while it may contribute to the cohesion of S-RM with the increment of WBP. The width of strain localization zone decreases with increment of higher WBP (46%–60%) because of larger breakage degree of rock blocks.

Keywords: Soil-rock mixture · Soft rock blocks · Discrete element method Breakage characteristics · Shear mechanical properties

1 Introduction

Soil-rock mixture (S-RM) is a highly heterogeneous geomaterial mixed with high strength rock blocks and fine-grained soil [1]. Most studies focused on the shear mechanical behaviors of S-RM by laboratory and in-situ tests [2–4]. However, the studies on the shear mechanical behaviors and damage mechanisms of S-RM mainly focus on the S-RM with high strength rock blocks, which is not easy to break. For the S-RM with soft rock blocks with unconfined compression strength (UCS) less than 15 MPa, the blocks are much easier to break during the shearing process thus affects the mechanical properties of S-RM [5]. Discrete element modelling (DEM) has been widely used in the study of meso-mechanical behaviors for S-RM [6–8]. However, typically rock blocks in S-RM were considered unbreakable in these studies. The mesoscale failure characteristic in S-RM is still unknown when the breakage of rock blocks is considered. For filling slopes contain the S-RM with soft rock blocks, it is significant to study the breakage characteristics and shear mechanical properties of the S-RM with soft rock blocks.

2 DEM Direct Shear Test of S-RM with Soft Rock Blocks

The studied slope filled with S-RM (Fig. 1(a)) for a ±500 kV converter station is located in Funing County, Yunnan Province, China. The material matrix is mainly sandy soil mixed with highly-weathered argillaceous shales (soft rock blocks). The UCS of the soft rock blocks is about 1.8–6.8 MPa. In this study, the sizes of the soft rock blocks in the S-RM range from 10 to 100 mm.

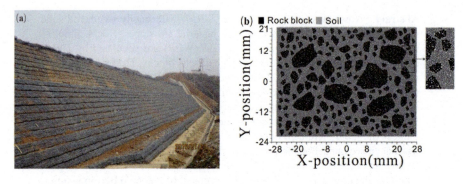

Fig. 1. (a) Filling slope and (b) Numerical model of S-RM with 46% WBP

DEM tests are performed using PFC^{2D}. Six S-RM samples with different WBPs are used for DEM shear tests: 20%, 30%, 40%, 46%, 50% and 60% respectively. The soil particles radius is simplified as 1 mm–2.5 mm, while the rock block particles radius is range 1 mm–2 mm (Fig. 1(b)). The parallel bond is chosen to simulate the breakable rock blocks, while the contact bond is chosen to simulate the soil. Base on the lab test of soil and S-RM with 46% WBP, the meso-parameters of soil and rock block are back-calculated (Table 1). The gravitational acceleration is set to 10 m/s^2 and the damping coefficient of particles is set to 0.1 in the DEM shear test. The loading rate for shearing is set to 0.01 m/s. The breakage of soft rock blocks is obviously observed during the shear process.

Table 1. Meso-parameters of soil and rock block in DEM

	Particle modulus/Pa	Density/ kg/m^3	Stiffness ratio	Normal bond strength/Pa	Shear bond strength/Pa	Friction coefficient
Soil	1.0e7	1900	1.0	1.0e5	1.0e5	0.5
Block	5.0e7	2200	1.0	1.5e6	1.5e6	1.5

3 Breakage Meso-Characteristics of Soft Rock Blocks

The breakage meso-ratio of rock blocks B_{mr} is used to evaluate the breakage degree of rock blocks in the DEM.

$$B_{mr} = \frac{N_c}{N} \times 100\% \tag{1}$$

where the N_c is the total number of rock meso-cracks generated in the shear process; N is the total number of parallel bonds on block particles at the beginning of shearing. The elastic strain energy stored in the parallel bond will lose as the breakage energy dissipation E_{br} and convert into surface energy when the bond is broken. The B_{mr} increased with the increment of horizontal displacement (Hd) and WBP (Fig. 2). When the horizontal displacement reached point A, the B_{mr} and E_{br} had a rapid increment, indicating the breakage of rock blocks induced the rapid increment of E_{br} (Fig. 2(a)). With the increment of WBP, the E_{br} increased as the B_{mr} (Fig. 2(b)).

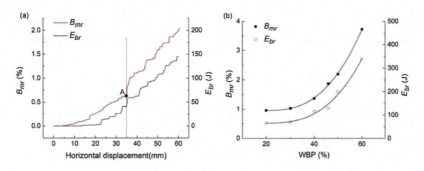

Fig. 2. (a) Evolution of B_{mr} and E_{br} of 46% WBP S-RM and (b) The B_{mr} and E_{br} vs WBP

4 Shear Mechanical Properties of S-RM with Different WBPs

4.1 Shear Strength of S-RM

The rock blocks are mainly hard rock blocks with high UCS in the researches [1, 4]. In the research [5], the rock blocks contain not only hard rock blocks but also soft rock blocks with only 5.7 MPa average UCS. The internal frictional angle of S-RM increased linearly with the increment of WBP in this paper. But the increment with each WBP is smaller than the S-RM contains hard rock blocks. Therefore, the breakage of rock blocks has a negative effect on the increment of internal friction angle (Fig. 3(a)). The cohesion of S-RM increased linearly with the increment of WBP, opposite to the researches [1, 4], indicating the breakage of rock blocks may contribute to the cohesion of S-RM (Fig. 3(b)).

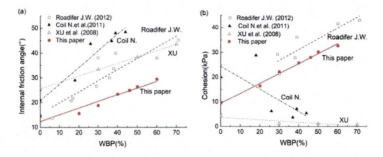

Fig. 3. (a) Internal friction angle and (b) Cohesion vs WBP

4.2 Strain Localization of S-RM

The rock blocks mainly broke in the strain localization zone. In the strain localization zone, the average horizontal displacement in each y-position interval of particles increased rapidly with the increment of y-position (Fig. 4(a)). Because of the larger breakage degree of rock blocks, the width of strain localization zone decreased with the increment of WBP when the WBP of samples is higher (46%–60%) (Fig. 4(b)).

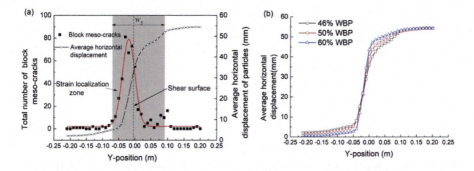

Fig. 4. Strain localization of S-RM with (a) 46% WBP and (b) higher WBP

References

1. Xu, W.J.: Study on meso-structural mechanics (M-SM) characteristics and stability of slope of soil-rock mixtures (S-RM). Ph.D. dissertation, Institute of Geology and Geophysics, Chinese Academy of Science, Beijing (2008)
2. Lindquist, E.S.: The strength and deformation properties of mélange. Ph.D. dissertation, University of California, Berkeley (1994)
3. Xu, W.J., Xu, Q., Hu, R.L.: Study on the shear strength of soil-rock mixture by large scale direct shear test. Int. J. Rock Mech. Min. Sci. **48**, 1235–1247 (2011)
4. Coli, N., Berry, P., Boldini, D.: In situ non-conventional shear tests for the mechanical characterisation of a bimrock (BimTest). Int. J. Rock Mech. Min. Sci. **48**, 95–102 (2011)

5. Roadifer, J.W., Forrest, M.P.: Characterization and treatment of mélange and sand stone foundation at Calaveras Dam. In: Proceedings of GeoCongress 2012 State of the Art and Practice in Geotechnical Engineering, Oakland, California, pp. 3362–3371 (2012)
6. Alessandro, G., Claudio, R., Tatiana, R.: Characterization and DEM modeling of shear zones at a large dam foundation. Int. J. Geomech. **12**(6), 648–664 (2012)
7. Xu, W.J., Wang, S., Zhang, H.Y., Zhang, Z.L.: Discrete element modelling of a soil-rock mixture used in an embankment dam. Int. J. Rock Mech. Min. Sci. **4**(86), 141–156 (2016)
8. Yan, Y., Zhao, J.F., Ji, S.Y.: Discrete element analysis of the influence of rock content and rock spatial distribution on shear strength of rock-soil mixtures. Eng. Mech. **34**(6), 146–156 (2017). (in Chinese)

Comparative Experimental Study on Solar Electro-Osmosis and Conventional Electro-Osmosis

Jianjun Huang[1,2] and Huiming Tan[1,2(✉)]

[1] Key Laboratory of Coastal Disaster and Defence of Ministry of Education, Hohai University, Nanjing, China
thming2008@163.com
[2] College of Harbour Coastal and Offshore Engineering, Hohai University, Nanjing, China

Abstract. As an effective method for accelerating the consolidation of soft soils, electro-osmosis has not been widely used in practice because of its high electric energy consumption. With the development of renewable solar energy, the solar energy is applied to electro-osmosis method and this paper has proposed a new solar electro-osmosis technology. By monitoring and comparing the electric current, voltage, water discharge, water content, shear strength, energy consumption and carbon emission, this paper presents the performance of solar electro-osmosis method itself to improve the consolidation of soft clay based on the laboratory experiment as well as the different electro-osmosis behaviors between conventional and solar electro-osmosis. The test results show that both electric current and voltage in solar electro-osmosis are not stable and vary periodically in day time and the solar electro-osmosis can accelerate consolidation of soft clay effectively with a less energy consumption of unit water drainage compared to conventional electro-osmosis. Besides, the solar electro-osmotic shear strength appears more uniform than that of conventional electro-osmosis. Moreover, the corresponding carbon emission of solar electro-osmosis is much less as well. Thus, with the comparison to conventional electro-osmosis, the solar electro-osmosis technology shows better drainage efficiency and environmental benefits.

Keywords: Electro-osmosis · Solar energy · Drainage · Energy consumption Carbon emission

1 Introduction

The disposal of dredged soft clay (i.e. mud) generated from harbour, waterway and coastal engineering is a challenge in practice. Most dredged soft clay not only contains a high water content but also has a low permeability. Several methods are proposed to accelerate silt, such as the preloading method and the vacuum method. Compared to the improvement methods above, the electro-osmosis method has a better drainage behavior as well as a less time-consuming. The principle of this method is to drive the pore water from the anode to the cathode when an electrical field is applied.

Electro-osmosis technology has been used since 1930s for removing water from clay and silt soils [1]. Since then, several successful applications have been reported by researchers [2, 3]. At the same time, a series of studies on electro-osmosis have been conducted by scholars, of which the energized form is one of the main factors of electro-osmosis consolidation [4, 5]. Moreover, other energized forms are further put forward, including current intermittence [6, 7] and electrode conversion [8].

However, there are limited studies on the energized form effects on treatment of soils by electro-osmosis and most of them are only focused on the consolidation result of electro-osmosis [9, 10] but its influence on environment around. In this work, electro-osmosis with solar energy which is a green energy has been put forward to research on the characteristics itself as well as its environmental benefits. This paper presents the unsteady voltage as well as non-constant electric current of solar energy and the performance of solar electro-osmosis to improve the consolidation of soft clay by laboratory experiment. Based on the measured data, the energy consumption and carbon emission have also been discussed further.

2 Materials and Methods

2.1 Experimental Apparatus and Soil

The experimental apparatus is consisted of two identical main cells, one solar cell panel, one D.C. power supply and 6 pairs of electrodes. One of the three main cells was connected to a solar cell panel while one cell was connected to the D.C. power supply. The soft clay used for electro-osmosis experiments is from Nanjing City of China. The soft clay is saturated with the density of 1.54 g/cm^3 and the initial water content is 77.03%.

The main section of the apparatus is the cell that consists of a rectangular tank that holds the silt sample with 250 mm height. It was made of plexiglass plates with 1 cm thickness and had an inner dimension of 300 × 300 × 350 mm (length × width × height). Three pairs of anode and cathode were added on the two sides of the main cell, of which anode is made of graphite in order to decrease erosion and cathode is made of steel pipe with small holes 5 mm in diameter. Besides, the distance of homopolar is 100 mm while the distance of heteropolar is 200 mm. In order to prevent the small holes of cathode from jamming and measure the voltage in certain time intervals during the process of the test, cathode was wrapped up with wet geotextile as reversed filter and a number of voltage probes made of copper were installed at the surface of the cell at distances of 1, 7, 13 and 19 cm from the anode. The solar cell panel and D.C. power supply were used as power supply of each electro-osmosis experiment. The cell dimensions are 1580 × 808 mm. The stable voltage of D.C. energy supply is 34 V and the open circuit voltage of solar cell panel is between 0 to 36 V respectively.

2.2 Experimental Procedures

Two types of test were performed in laboratory. The procedure is described in 6 steps as follows.

(1) Before the test, three pairs of anode and cathode were added on the two sides of each rectangular tank in solar and conventional electro-osmosis condition.
(2) Two rectangular tanks are filled with soft clay until 250 mm high.
(3) In order to monitor the water discharge, the graduated cylinder is put under the drainage pipe of each tank.
(4) The anode and the cathode of solar electro-osmosis condition are connected to solar cell panel while that of conventional electro-osmosis condition are connected to D.C. power supply and then each electro-osmosis starts working.
(5) The electric current, voltage and water discharge are measured and recorded in each hour of each test.
(6) After the schedule working time, two experiments are stopped and the soft clay soils at different places are sampled from each testing tank for the succeeding geotechnical property tests, including water content and shear strength.

3 Results

(1) As shown in Fig. 1, the electric current of solar electro-osmosis appears an intermittent variation with rising stage, steady stage and descent stage during a day. Besides, a sudden increase of current is appeared at the beginning of steady stage in the next 24 h compared to the previous steady stage ending in the last 24 h. In total, solar electro-osmosis method presents an unsteady voltage as well as non-constant electric current.
(2) As shown in Fig. 2, with the comparison between solar and conventional electro-osmosis, it can be easy to find that there are 3 stages in conventional electro-osmosis including rapid increase stage, low increase stage and stable stage. Relatively, drainage of solar electro-osmosis is distributed in the shape of steps, of which the drainage is increased with time during the day while it is very little or none at night. Moreover, the total drainage of solar and conventional electro-osmosis is almost equal while the actual electro-osmosis working time of them is not.
(3) The total shear strength after conventional electro-osmotic consolidation is higher than that of solar electro-osmotic consolidation. However, because of the water redistribution of solar electro-osmosis when power off at night, the solar electro-osmotic shear strength appears more uniform than that of conventional electro-osmosis.
(4) The energy consumption coefficient of solar electro-osmosis was 0.35 w • h/ml while that of conventional electroosmosis was 0.41 w • h/ml. Solar electro-osmosis shows a better and more cost-effective drainage performance.
(5) According to the assessment of life cycle, the carbon emission unit drainage of solar electro-osmosis is much smaller which is 0.03×10^{-3} kg/ml than that of conventional electro-osmosis which is 1.18×10^{-3} kg/ml and it is to say that solar electro-osmosis has a better environment benefit compared with conventional electro-osmosis.

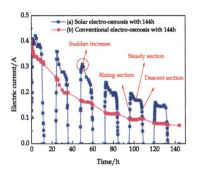
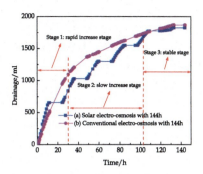

Fig. 1. Evolution of current with time.   **Fig. 2.** Variation of drainage.

References

1. Bruell, C.J., Segall, B.A., Walsh, M.T.: Electroosmotic removal of gasoline hydrocarbons and TEC from clay. J. Environ. Eng. **118**(1), 68–83 (1992)
2. Casagrande, L.: Electro-osmosis stabilization of soils. J. Boston Soc. Civ. Eng. **39**(1), 51–83 (1952)
3. Bjerrum, L., Moum, J., Eide, O.: Application of electro-osmosis to a foundation problem in a Norwegian quick clay. Géotechnique **17**, 214–235 (1967)
4. Hamir, R.: Some aspects and applications of electrically conductive geosynthetic materials. Ph.D. thesis, University of Newcastle upon Tyne, UK (1997)
5. Yoshida, H., Shinkawa, T., Yukawa, H.: Comparison between electroosmotic dewatering efficiencies under conditions of constant electric current and constant voltage. J. Chem. Eng. Jpn. **13**(5), 414–417 (1980)
6. Sprute, R.H., Kelsh, D.J.: Limited field tests in electrokinetic densification of mill tailings. Report of Investigations 8034, USBM (1975)
7. Micic, S., Shang, J.Q., Lo, K.Y., Lee, Y.N., Lee, S.W.: Electrokinetic strengthening of a marine sediment using intermittent current. Can. Geotech. J. **38**(2), 287–302 (2011)
8. Lo, K.Y., Inculet, I.I., Ho, K.S.: Electroosmotic strengthening of soft sensitive clays. Can. Geotech. J. **28**(1), 74–83 (1991)
9. Wu, H., Hu, L.M., Wen, Q.B.: Electro-osmotic enhancement of bentonite with reactive and inert electrodes. Appl. Clay Sci. **111**, 76–82 (2015)
10. Xue, Z.J., Tang, X.W., Yang, Q.: Influence of voltage and temperature on electro-osmosis experiments applied on marine clay. Appl. Clay Sci. **141**, 13–22 (2017)

Stress-Dilatancy Relationship for Fiber-Reinforced Soil

Yuxia Kong(✉)

Institute of Geotechnical Engineering, Nanjing Tech University, Nanjing, China
kongyuxia@njtech.edu.cn

Abstract. Polypropylene fiber inclusion clearly increases soil shear strength and reduces the loss in strength post-peak. Stress-dilatancy relationship plays an essential role in constitutive modeling for soils. The increase in the effective confining stress provided by the fibers can be deduced from comparing the stress-strain and strength behavior of reinforced soil and pure soil. When a polypropylene fiber-soil assembly dilates in response to applied shear deformations, the work done by the driving stress will be dissipated by not only particle sliding but also the fiber deformation. The stress-dilatancy relationships in conventional triaxial compression, extension conditions were discussed in this study. The stress-dilatancy relationship is validated against a series of triaxial tests on Nanjing sand mixed with discrete polypropylene fibers.

Keywords: Fiber-reinforcement · Sand · Stress-dilatancy
Constitutive modeling

1 Experimental Observations

Randomly distributed polypropylene fibers act to interlock particles of the granular materials thus improve the share strength of the reinforced soils [1, 2]. Using random discrete flexible fibers mimics the behavior of plant roots and gives the possibility of improving the strength and the stability of near surface soil layers [6].

The material tested in this study was made of Nanjing sand and discrete polypropylene fibers. Nanjing sand was sampled from the beach of Yangtze River in the region of Nanjing, China. Figure 1 shows the shear strength envelopes of pure Nanjing sand and fiber-reinforced sand in the $p - q$ plane. The critical state points for pure sand define a straight line, while bilinear failure envelope for other sand-fibre combinations. Before the kink point, failure is governed by fiber stretching, slippage and mobilization in the matrix. After the kink point, failure is governed by tensile yielding, stretching of the fiber and mobilization. Strains will develop in the fibres because of the strains that occur in the soil around the fibres. In their interaction with the soil, as the strain increase, the fibres may pullout of the soil or may reach their tensile strength.

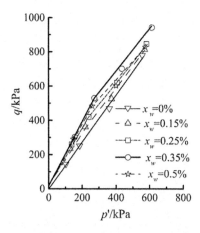

Fig. 1. Shear strength envelopes of Nanjing sand and fiber-reinforced sand

2 Stress-Dilatancy Relationship for Fiber-Reinforced Sand

The physical manifestation of dilatancy was first identified by Reynolds [3], and long afterward, Rowe introduced a stress-dilatancy theory for assembly of particles in contact [4]. The "bridge" effect at the interface between fiber surface and soil matrix were analyzed by Tang et al. [5]. The "bridge" effect of fiber reinforcement will increase the effective confining stress and the increment should be considered in the stress-dilatancy relationship. Rowe's stress-dilatancy relationship states that for the soil sample that is being sheared, the ratio of the work done by the driving stress to the work done by the driven stress in any strain increment should be a constant K:

$$\frac{\text{work put in by driving stress}}{\text{work taken out by driving stress}} = -K \quad (1)$$

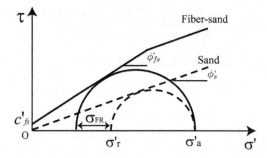

Fig. 2. The reinforcing effect of fiber on the effective confining stress in Mohr-circle space

The tensile strains in the sand try to stretch the fibers and the stretched fibers try to resist extension. Thus, the fibers tend to increase the normal stress on the soil and contribute directly to the shear stress. σ_{FR} (Fig. 2) plays a role as driven stress and contributes to the work done on the fiber-sand. Using σ_{FR}, the "bridge" effect of fiber reinforcement could be considered in the stress-dilatancy relationship. In conventional triaxial compression (CTC), the axial stress σ'_a is the driving stress (with an associated compressive strain increment $\delta\varepsilon_a$), the radial stresses σ'_r and σ_{FR} are the driven stresses (with an associated tensile strain increment $-\delta\varepsilon_r$). Considering σ_{FR}, the stress-dilatancy relationship for fiber-sand for triaxial compression then sates that,

$$\frac{\sigma_{a'}\delta\varepsilon_a}{-2(\sigma_{r'}+\sigma_{FR})\delta\varepsilon_r} = K_{fs} \qquad (2)$$

where K_{fs} is the counterpart of K for fiber sand.

The stress-dilatancy for triaxial compression can be rewritten as following:

$$\frac{\delta\varepsilon_p}{\delta\varepsilon_q} = \frac{3\eta(K_{fs}+2-2AK_{fs}) - 9(K_{fs}-1-AK_{fs})}{2\eta(K_{fs}-1-2AK_{fs}) - 3(2K_{fs}+1+2AK_{fs})} \qquad (3)$$

For fiber sand, it is convenient to fix the value of ϕ'_f at the ultimate critical state value ϕ'_{fs}. Then, the constant K_{fs} becomes

$$K_{fs} = \frac{3+2M_{fs}}{3-M_{fs}} \qquad (4)$$

where $M_{fs} = 6\sin\phi'_{fs}/(3-\sin\phi'_{fs})$.

For conditions of triaxial extension, the radial stress and the increase in effective confining stress σ_{FR} are now the driving stresses and the axial stress is the driven stress, and the stress-dilatancy becomes

$$\frac{2\sigma_{r'}\delta\varepsilon_r}{-(\sigma_{a'}+\sigma_{FR})\delta\varepsilon_a} = K_{fs} \qquad (5)$$

The equivalent of Eq. 5 for triaxial extension is then

$$\frac{\delta\varepsilon_p}{\delta\varepsilon_q} = \frac{3\eta(2K_{fs}+1+2AK_{fs}) + 9(K_{fs}-1+AK_{fs})}{2\eta(1-K_{fs}-AK_{fs}) - 3(K_{fs}+2+AK_{fs})} \qquad (6)$$

and the constant K_{fs} for triaxial extension is becomes

$$K_{fs} = \frac{3+M_{fs}}{3-2M_{fs}} \qquad (7)$$

3 Simulation

For comparison, the presented stress-dilatancy relationship and Rowe's original stress-dilatancy relationship were also plotted in Fig. 3. The addition of fibers caused the soil behavior to diverge from the dilatancy trend from the pure sand. Although these values

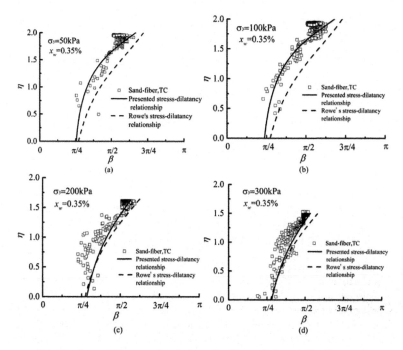

Fig. 3. Stress-dilatancy simulations for fiber-reinforced Nanjing sand

of b have been calculated from total strain increments, it is accepted that the stress-dilatancy relationship for fiber sand provides a reasonable description of the plastic flow of fiber-reinforced sand. The proposed stress-dilatancy relationship has met with the data better than Rowe's original stress dilatancy relationship.

Acknowledgments. The authors would like to acknowledge the supports provided by the National Natural Science Foundation of China (Grant No. 11402109).

References

1. Consoli, N.C., Casagrande, M.D.T., Coop, M.R.: Performance of a fibre-reinforced sand at large shear strains. Géotechnique **57**(9), 751–756 (2007)
2. Gray, D.H., Ohashi, H.: Mechanics of fiber reinforcement in sand. J. Geotech. Eng. **109**(3), 335–353 (1983)
3. Reynolds, O.L.: On the dilatancy of media composed of rigid particles in contact with experimental illustrations. Philos. Mag. Ser. 5 **20**(127), 469–481 (1885)
4. Rowe, P.W.: The stress-dilatancy relation for static equilibrium of an assembly of particles in contact. Proc. R. Soc. Lond. A Math. Phys. Eng. Sci. **269**(1339), 500–527 (1962)
5. Tang, C., Shi, B., Gao, W., Chen, F., Cai, Y.: Strength and mechanical behavior of short polypropylene fiber reinforced and cement stabilized clayey soil. Geotext. Geomembr. **25**(3), 194–202 (2007)
6. Wood, D.M., Diambra, A., Ibraim, E.: Fibres and soils: a route towards modelling of root-soil systems. Soils Found. **56**(5), 765–778 (2016)

Improvement of Soft Soils Using Bio-Cemented Sand Columns

Aamir Mahawish[1(✉)], Abdelmalek Bouazza[1], and Will P. Gates[2]

[1] Monash University, Melbourne, VIC 3800, Australia
aamir.mahawish@monash.edu
[2] Deakin University, Burwood, VIC 3125, Australia

Abstract. This paper discusses the results of small-scale model tests that were undertaken to investigate the behaviour of un-cemented and bio-cemented coarse sand columns installed within a soft clay bed. The tests also focused on studying the effect of biocementation to reduce excessive radial expansion and settlement that may occur during installation or loading of sand columns. A multi-phase biochemical treatment strategy was used to obtain a partial bio-cementation, in particular at the upper section of the sand column where bulging usually occurs. The results show that the partially bio-cemented sand column substantially reduced vertical strains which were in the order of 71% and 97% compared with un-cemented sand column and kaolin clay, respectively. The bulging of the bio-cemented sand column was associated with the bottom section of the column where there were no or less bio-cemented materials.

Keywords: Biocementation · Sand column · Soft soil

1 Introduction

A sand or stone column tends to bulge laterally when its length exceeds 4 times its diameter [1, 2]. Biocementing of sand columns (i.e., as used in vibro-replacement) has been considered in the current study especially for the specific case of a sand column having a length greater than its critical column length [3] and with the view of mitigating the lateral expansion, which occurs in the upper section of the columns. Small-scale laboratory tests of model sand columns were undertaken to investigate the behaviour of a soft clay treated with un-cemented and bio-cemented coarse sand column. A multi-phase percolation treatment technique was used to induce calcium carbonate precipitation in the upper section of the sand column where multiple thinner layers of bacterial suspension and cementation solution were sequentially percolated to achieve targeted cementation.

2 Materials and Methods

2.1 Preparation of Bacterial Suspension and Cementation Solution

Sporosarcina pasteurii (American Type Culture Collection, ATCC® 11859) was used to invoke the hydrolysis of urea. The bacterial strains were cultivated using an ammonium-yeast extract medium containing, 20 g yeast extract, 10 g $(NH_4)_2SO_4$ and 130 mM tris buffer (pH = 9.0) per litre of distilled water at 30 °C. The cultivation procedures were performed according to [4]. The cementation solution used in this study consisted of 1M of urea and 1M of calcium chloride.

2.2 Sand

The sand used to represent typical materials used in stone columns was a coarse sand with an average grain size of 1.6 mm and classified as SP [5]. The sand had the following characteristics: specific gravity (G_s = 2.64), coefficient of curvature (C_c = 0.97), coefficient of uniformity (Cu = 1.35), maximum and minimum void ratios (e_{max} = 0.84 and e_{min} = 0.55).

2.3 Sand Column Treatment Procedures and $CaCO_3$ Precipitation Quantification

A percolation method, described in [6], was used to bio-cement coarse sand columns with an internal diameter of 51 mm and a height of about 300 mm. The protocol that has been adopted throughout this study was a 12-phase percolation treatment strategy with the sand column subjected to up to 12-biochemical treatment cycles. After completion of the small-scale test, $CaCO_3$ content was quantified using a gravimetric acid washing (2M HCl) technique as described in [6].

2.4 Kaolin Clay

Powdered kaolin (Grade HR1F) was mixed with water to induce a slurry with an initial moisture content of around 115% to avoid the formation of air bubbles within the slurry and to increase its workability. The slurry was prepared for each enlarged (large-diameter) consolidation cell through thoroughly mixing the predetermined amount of kaolin powder (6700 g) with 7705 ml of water in a large mixing bowl using an electric mixer for about 3 h. A vertical pressure of 120 kPa, in the enlarged consolidation cells, was applied to consolidate the slurry from about 480 mm high to about 300 mm. This preparation process led to an undrained shear strength (C_u) of about 16 kPa. The properties of the consolidated soft clay bed are presented in Table 1 Small-scale laboratory testing instrumentation and operation.

Small-scale laboratory testing was performed based on the description (including methodology, i.e. loading and unloading, operation, cells dimensions, string transducer and pore pressure transducers and scale factors) presented in [5]. The clay bed was consolidated to the consistency of soft clay (C_u = 16 kPa). Then, the clay samples were unloaded, and a thin wall aluminium tube, having a 51 mm outside diameter, was pushed slowly through the sample to the base of the cell using a fabricated guide attached to the top of the cylinder, creating a cylindrical cavity of 51 mm diameter at the centre of the clay bed. The cavity was filled with a 51 mm diameter frozen sand column, which was left to thaw for 3 h. In the case of the bio-cemented sand column, after completing the biochemical treatment cycles, the column was rinsed with deionised water. The bio-cemented column was then placed in the cavity of the second cell in the same manner. Both un-cemented and bio-cemented sand columns were loaded with the pistons, creating a unit-cell condition. The columns were initially loaded with the same consolidation pressure (120 kPa) to ensure intimate contact between columns and surrounding clay. The cells were then loaded in stages at an increment of around 50 kPa until reaching a maximum pressure of 375 kPa. The samples were allowed to undergo full consolidation as determined from pore pressure and settlement measurements at each load stage. After the tests completion, the clay samples were unloaded and allowed to swell and then they were extruded from the column cells using the piston.

Table 1. Properties of consolidated kaolin clay.

Parameter	Unit	Value
Specific gravity	–	2.64
Plastic limit	%	29
Liquid limit	%	62
Compression index, Cc	–	0.8
Recompression index, Cr	–	0.09
Saturated unit weight	kN/m^3	16.2
Moisture content	%	58
Average undrained shear strength, Cu	kPa	16*

* Tested using a vane shear apparatus.

3 Results and Discussions

3.1 Vertical Stress-Strain Behaviour

The vertical stress-strain variation of un-cemented and bio-cemented sand columns are presented in Fig. 1. It can be inferred from Fig. 1 that the inclusion of un-cemented sand column within the soft clay bed is associated with a reduction in vertical strains. The reduction of vertical strain for un-cemented column compared with kaolin specimen varied with the range of applied stresses, ranging between around 9% with low pressure and up to around 15% with high stresses (around 300 kPa). The partially bio-cemented columnar inclusion within a soft clay bed showed a substantial reduction in the vertical strain as shown in Fig. 1. The averaged vertical strain reduction across the range of applied stresses was about 71% and 97% compared with un-cemented sand column and kaolin clay, respectively. This is a large improvement in strength for a low replacement ratio of 11% as the bio-cemented sand column behaved as a stiff column. Radial expansion in the un-cemented sand column occurred at the section equating to about 4× the diameter of the sand column, ranging between 20% near to the top of the column to about 3% close to the bottom with an average expansion of about 8.7%. By contrast, there was a significant reduction in the radial expansion where there is a negligible expansion (less than 0.5%) at the upper section. This could be associated with the distribution and precipitation of calcium carbonate within the sand matrix. However, the radial expansion was observed in sections where there were no or less bio-cemented materials (data not shown).

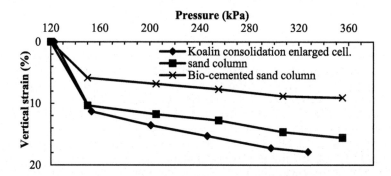

Fig. 1. Vertical stress-strain relationship for column group tests.

4 Conclusion

This study was aimed at investigating the effect of biocementation on sand column behaviour using small-scale laboratory models. A significant increase in sand column stiffness and a substantial reduction in column strain was observed for the bio-cemented sand column, with strains reductions in the order of 71% and 97% compared with un-cemented sand column and kaolin clay, respectively. Bulging seems to occur in the sections where no or less bio-cement materials were present.

References

1. Barksdale, R.D., Bachus, R.C.: Design and Construction of Stone Columns: Final Report SCEGIT, pp. 83–104. Federal Highway Administration, Washington D.C (1983)
2. Black, J.A., Sivakumar, V., Madhav, M.R., Hamill, G.: Reinforced stone columns in weak deposits – a laboratory model study. J. Geotech. Geoenviron. Eng. **133**(9), 1154–1161 (2007)
3. Hughes, J., Withers, N., Greenwood, D.: A field trial of the reinforcing effect of a stone column in soil. Geotechnique **25**(1), 31–44 (1975)
4. Mahawish, A., Bouazza, A., Gates, W.P.: Effect of particle size distribution on the bio-cementation of coarse aggregates. Acta Geotech., 1–7 (2017)
5. Gniel, J., Bouazza, A.: Improvement of soft soils using geogrid encased stone columns. Geotext. Geomembr. **27**(3), 167–175 (2009)
6. Mahawish, A., Bouazza, A., Gates, W.P.: Biogrouting coarse materials using soil-lift treatment strategy. Can. Geotech. J. **53**(12), 2080–2085 (2016)

Stabilization of Expansive Black Cotton Soils with Alkali Activated Binders

S. Mazhar, A. GuhaRay(✉), A. Kar, G. S. S. Avinash, and R. Sirupa

Department of Civil Engineering,
BITS Pilani Hyderabad Campus, Hyderabad 500078, India
guharay@hyderabad.bits-pilani.ac.in

Abstract. The black cotton soil is mainly composed of clay minerals of Smectite group and is highly expansive when exposed to moisture. The present paper proposes a method of geo-polymerizing the black cotton soil with alkali activated binders (AAB). AAB is produced by the reaction between an aluminosilicate precursor (primarily Class F fly ash and/or slag) and an alkali activator solution of sodium hydroxide and sodium silicate. The water to solids ratio is maintained at 0.3. Mineralogical and microstructural characterizations through X-ray diffraction (XRD) and Fourier-transform infrared spectroscopy (FTIR) are carried out for both untreated and treated soil to identify the changes in chemical composition and surface morphology. The index properties, compaction characteristics, unconfined compressive strength (UCS) and swelling characteristics of both untreated and treated soil are carried out. It is observed that the AAB reduces the plasticity index and free swell of black cotton soil by 15–25%, while the UCS values are increased by 10–12%. Recommendations on practical implementation of this technique for stabilization of expansive soils are proposed based on the findings of this study.

Keywords: Expansive soil · Stabilization · Alkali activated binders Ground improvement

1 Introduction

Black cotton soil (BCS) is a high plastic soil containing clay minerals of smectite group. This gives rise to large swelling and shrinkage, leading to serious problems in civil engineering. Industrial by-products including fly ash, lime, solid wastes and their combination are used to stabilize expansive soils [1]. Natural pozzolans are used to stabilize the low-swelling clayey soil, which can be conceptualized as a type of geopolymerization [2]. Miao et al. [3] proposed a geopolymerization method for stabilising expansive soils with alkalis of calcium hydroxide or potassium hydroxide. However, the reaction of potassium hydroxide is exothermic; hence it is difficult to control the rate of reaction at site, which may lead to internal micro-cracking. The present study proposes a method of geo-polymerizing the black cotton soil with alkali activated binders (AAB), produced by the reaction between an aluminosilicate precursor and an alkali activator solution of sodium hydroxide and sodium silicate. Sodium hydroxide is available economically and the rate of reaction can be easily controlled at the site.

2 Experimental Investigations

2.1 Materials

In the present study, the black cotton soil (Table 1) is collected from Nalgonda district of Telangana state in the southern part of India. The BCS is sun-dried and ground prior to all the tests.

Table 1. Properties of black cotton soil.

Property	Value
Liquid Limit (LL)	52%
Plastic Limit (PL)	20%
Plasticity Index (PI)	32%
Soil Classification (according to USCS)	CH
Free Swell Index (FSI)	68.5%
Maximum Dry Density (MDD)	16.5 kN/m^3
Optimum Moisture Content (OMC)	24%
Unconfined Compressive Strength	180

The alkali activated binder (AAB) is produced by the reaction of fly ash using an activating solution containing sodium silicate, sodium hydroxide, and water. The sodium silicate solution is composed of 55.9% water, 29.4% SiO_2, and 14.7% Na_2O. The water to solids ratio is maintained at 0.3. The addition of sodium hydroxide to sodium silicate is exothermic. Hence, the activator solution is prepared at least 24 h before using it to react with the fly ash in order to dissipate any residual heat that can hasten the alkali-activation of fly ash and lead to a flash The BCS is mixed uniformly with 5%, 10% and 15% of AAB. The curing period was maintained at 1, 3, 7, 21 and 28 days. It is after this aging period that the geopolymer is finally produced.

3 Results and Discussions

3.1 Chemical Characterization

XRD analyses are performed using a RIGAKU Ultime IV diffractometer to identify the minerals present. The powdered samples are examined through CuKα rays generated at 40 mA and 40 kV. The operating 2θ range is from 0° to 100° with a step of 0.02° @ 2 secs/step. The raw BCS consisted of clay minerals such as Montmorillonite (M), Quartz, (Q) Illite (I) and Muscovite (Ms). The diffractogram for treated BCS (Fig. 1) shows additional peaks corresponding to Quartz (Q), Mullite (Mu), Portlandite (P), and Hydroxy sodalite (S) which are characteristic of hardened AAB. Additional peaks corresponding to Augite (Au) are visible which formed due to reaction between BCS and AAB. FTIR spectroscopy of untreated and treated BCS are collected using a JASCO FTIR 4200 setup using KBr pellets, over a spectral range of 4000–400 cm^{-1}.

FTIR spectroscopy show the presence of IR transmittance by O-H stretching vibrations between 3435 and 3616 cm^{-1}. Untreated BCS shows peaks at 2924 cm^{-1} for C-H (asymmetric stretching), at 2350 for C = O (symmetric), 1633 for C = O (stretching), 1439 for C-H (bending), 1033 for Si-O-Si (antisymmetric stretch), for Al-O (stretching) at 785, and at 527 for Si-O-Al (bending) vibrations (Fig. 2).

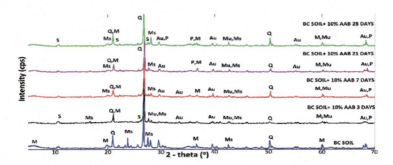

Fig. 1. X-Ray diffractogram of BCS treated with 10% AAB for various durations.

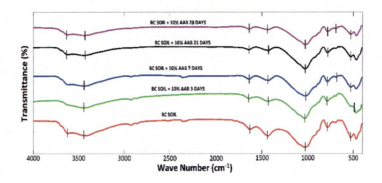

Fig. 2. FTIR spectroscopy of BCS treated with 10% AAB for various durations.

3.2 Geotechnical Characterization

Figure 3a shows the variation of Atterberg limits of pulverized BCS mixed with 10% AAB at different time periods. With addition of 10% AAB, the LL of BCS decreases from 52% to 38%, while PI decreases from 32% to 14% after 21 days. It may be inferred that fly ash, being a non-plastic ash and a pozzolan, diminishes the LL and PI of BCS. With addition of AAB, the free swell index (FSI) decreases rapidly from 70% to 25% (Fig. 3b). The optimum moisture content of BCS, obtained by Modified Proctor's test, is found to decrease by 20% and the maximum dry density increased to 11% when mixed with 10–12% of AAB after 3 days. The increase in MDD may be attributed to the higher density of fly ash. The unconfined strength of BCS reaches a value of 600 kN/m^2 from 180 kN/m^2 on addition of 10% AAB after 28 days of geopolymerisation.

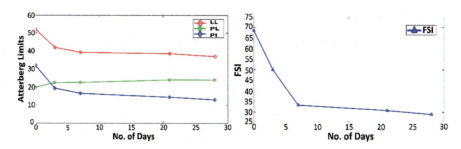

Fig. 3. (a) Atterberg Limits and (b) FSI of BCS treated with 10% AAB.

4 Conclusions

It is observed that BCS treated with approximately 10% AAB gives the maximum strength and least swelling and plasticity index.

References

1. Maneli, A., Kupolati, W.K., Abiola, O.S., Ndambuki, J.M.: Influence of fly ash, ground-granulated blast furnace slag and lime on unconfined compressive strength of black cotton soil. Road Mater. Pavement **17**(1), 252–260 (2016)
2. Hossain, K.M.A., Mol, L.: Some engineering properties of stabilized clayey soils incorporating natural pozzolans and industrial wastes. Constr. Build. Mater. **25**(8), 3495–3501 (2011)
3. Miao, S., Shen, Z., Wang, X., Luo, F., Huang, X., Wei, C.: Stabilization of highly expansive black cotton soils by means of geopolymerization. J. Mater. Civ. Eng. ASCE **29**(10), 04017170 (2017)

Formation of Biomineralized Calcium Carbonate Precipitation and Its Potential to Strengthen Loose Sandy Soils

Sangeeta Shougrakpam and Ashutosh Trivedi

Department of Civil Engineering, Delhi Technological University, Delhi, India
sangeetashougrakpam@dtu.ac.in, atrivedi@dce.ac.in

Abstract. Microbially induced calcium carbonate precipitation (MICP) is considered as a sustainable and environmentally friendly technique for the strengthening of loose sandy soils. MICP is triggered as a result of a biomineralized reaction product of bacteria and a cementation reagent in soils. In the present study laboratory experiments were performed on sand columns to investigate the MICP potential to consolidate and strengthen the loose and collapsible sand specimens using the ureolytic bacteria *Sporosarcina pasteurii*. The sand columns were treated initially using bacterial cell solution and followed by percolating a cementation reagent solution that consists of urea and calcium chloride. In the MICP process, urease enzymes released by the bacteria plays a major role in urea hydrolysis. The results showed that MICP could effectively strengthen the sand columns through calcium carbonate precipitation as calcite to bind the sand grains. The improvements, however, varied with treatment conditions, soil types and other environmental factors such as pH and temperature of the local environment. The results from Scanning Electron Microscope (SEM) analysis confirmed the precipitation of calcite by bridging the sand grains. This study suggests that calcite precipitation through biomineralization as a metabolic product of ureolytic bacteria are highly effective and may provide as a green construction material as a sealing agent for filling the gaps, cracks and fissures for civil engineering structures. Further, another effective technique of MICP is for sequestration of CO_2 present both in the atmosphere and in the soil.

Keywords: Biomineralization · Calcite · CO_2 sequestration · MICP · Urease Urea hydrolysis

1 Introduction

The microbially induced calcium carbonate precipitation (MICP) is a natural process that harnesses the bacterial metabolism (i,e, ureolytic activity) to precipitate calcium carbonate or calcite as an in situ cementation agent within sand grains as an induced biomineralized by-product. Calcite as biocementing material binds or encapsulates the sand grains, thereby, increases the strength and stiffness of loose saturated sand matrix either by bio-cementation or bio-clogging approached (Ivanov and Chu 2008; Mitchell and Santamarina 2005). During MICP, a continuous reaction between the urease-producing bacterial (UPB) cells and the outside environment to hydrolyze urea to increase the alkalinity of the pore fluid where variation in pH, dissolved inorganic carbons (DIC) and Ca^{2+} can occur. The cell surfaces act as effective nucleation sites for carbonate precipitation (Achal et al. 2009). Cell walls with negatively charged functional groups, such as carboxyl are able to adsorb Ca^{2+} present in the soil. Further, extracellular polymeric substances (EPS) also play a vital role to enhance the microbial cementation process.

The laboratory results presented herein are based on soil column experiments augmented, herein, utilizes urea hydrolysis catalyze by urease enzymes released by ureolytic bacteria *Sporosarcina pasteurii* (MTCC 1761), to precipitate calcium carbonate (as calcite) as a biocementing agent. The formation of biomineralized calcium carbonate precipitation is enhanced by using a mixture solution of calcium chloride ($CaCl_2$), urea ($CO(NH_2)_2$), and urease-producing bacteria (UPB) which hydrolyzed urea with the production of carbonate and an increase of the pH level. CO_2 release during the biochemical reaction will react with hydroxyl ions (OH^-) and forms carbonate that required for the precipitation of $CaCO_3$ (Gurbuz et al. 2011). From ureolysis and $CaCO_3$ formation stoichiometry, in order to hydrolyze 1 mol of urea sequestrates 1 mol of CO_2. Consequently, the amount of CO_2 sequestrated is directly proportional to the amount of $CaCO_3$ precipitated by MICP as reported by (Okwadha and Li 2010).

2 MICP Based Cementation in Sand Column Experiments

The MICP-based experiments were conducted in six-sand columns (55-mm diameter and 100-mm height) by applying on top, a bacterial solution (BS) and a cementation reagent solution (CRS). The average particle diameter of the sand (D_{50}) is 0.26, C_u = 1.75 and C_c = 1.08, wherein, bacterial sizes of 0.3 to 3 μm may move freely. The bacterial cells used in BS were at a cell concentration of 10^7 cells/ml (OD_{600} of 0.7–1.0). The BS consists of UPB cells, 20 g/l of $CO(NH_2)_2$, 11.25 g/l of $CaCl_2$, 10 g/l of NH_4Cl, 2.12 g/l of $NaHCO_3$ and 3 g/l of nutrient broth. The urease activity of the bacterial solution was 10–15 mM/min. Initially, treatment process was started by applying 50 ml of BS on top of each sand column and was retained for 8 h before applying of CRS to ensure the UPB cells were evenly attached and distributed within the columns. Two urea-calcium (1:1 molar ratio) based CRS of 0.5 M and 0.75 M were tested for influence on ureolytic-driven calcium carbonate precipitation. Then, three columns each were treated in equal doses of each 0.5 M and 0.75 M for 24 h. Later, a lower CRS concentration of 0.25 M mixed with fresh doses of BS (50 ml/l) were applied in successive treatments to maintain a saturated condition to enhance the active biogeochemical reaction till 14, 21 and 28 days. The efficiency of calcite formation in lower concentration (0.05–0.25 M) is more efficient than in higher concentration (0.5–1.0 M) of urea-$CaCl_2$ (Okwadha and Li 2010; De Muynck et al. 2010). All the treatments were performed at room temperature (25 ± 5 °C). Treated sand columns after oven-dried as shown in Fig. 1 were used for unconfined compressive strength (UCS) test under strain-controlled conditions at a uniform loading rate of 1.0 mm/min as shown in Fig. 2.

Fig. 1. (a) MICP treatment; (b) Oven-dried specimen; (c) MICP-treated sand column

Fig. 2. UCS test (a) Mounted sand column; (b) Applied axial load (c) Failure of sand column

To determine precipitated $CaCO_3$ in the soil specimens, acid washed method was followed as were explained in Rebata-Landa (2007).

3 Results and Discussions

Fresh doses of BS and CRS were applied in sand columns to maintain pH within 7–8. During treatment, pH of the effluents was varied from 6.5 to 8. The treated sand grains have grown in sizes indicating it as a function of MICP. In the present study, the 14 days treated sand have gained UCS values of 1.44 MPa and 2.17 MPa that were treated at 0.5 M and 0.75 M CRS respectively. Both UCS values and measured % CaCO3 content were found to decrease after 14 days in the saturated state as in Table 1.

Table 1. UCS based on $CaCO_3$ content of MICP- treated saturated sands

Saturation for days	Molarity of CRS	Axial strain (%)	UCS (MPa)	% $CaCO_3$
14	0.5 M	2.5	1.44	6.0
21		2.0	1.10	4.0
28		2.3	1.10	4.5
14	0.75 M	1.5	2.18	8.0
21		2.0	2.17	7.5
28		2.5	1.10	4.3

The UCS values were found to vary from 1.1 to 2.18 MPa based on 4.0 to 8.0% calcite precipitation by weight. The present study shows longer treatment duration days has no additive strength or calcite content to the sand columns as the most effective treatment duration was within 48 h. Further, Feng et al. (2016) also found that the UCS of 3–5 days was more than 5–7 days of treatment duration.

The SEM images of 28 days MICP-treated sands in Fig. 3 shows the calcite encapsulation on the sand particle surface, infilling and bridging the particles.

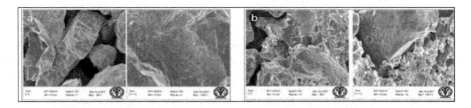

Fig. 3. (a) MICP treated sand in 0.75 M CRS (b) MICP treated sand in 0.5 M CRS

4 Conclusion

After the pores are filled up with precipitated calcite crystals, coating or bridging of the individual soil grains have increased the UCS and stiffness within the soil matrix. Hence, calcite may be a green construction material in civil engineering structures.

References

Achal, V., Mukherjee, A., Basu, P.C., Sudhakara, R.M.: Strain improvement of *Sporosarcina pasteurii* for enhanced urease and calcite production. J. Ind. Microbiol. Biotechnol. **36**(7), 981–988 (2009)

De Muynck, W., De Belie, N., Verstraete, W.: Microbial carbonate precipitation in construction materials: a review. Ecol. Eng. **36**, 118–136 (2010)

Feng K., Montoya B.M.: Influence of confinement and cementation level on the behavior of microbial-induced calcite precipitated sands under monotonic drained loading. J. Geotech. Geoenviron. Eng. https://doi.org/10.1061/(ASCE)GT.1943-5606.0001379

Okwadha, G.D., Li, J.: Optimum conditions for microbial carbonate precipitation. Chemosphere **81**, 1143–1148 (2010)

Gurbuz, A., Sari, Y.D., Yuksekdag, Z.N., Cinar, B.: Cementation in a matrix of loose sandy soil using biological treatment method. Afr. J. Biotechnol. **10**(38), 7432–7440 (2011)

Ivanov, V., Chu, J.: Applications of microorganisms to geotechnical engineering for bioclogging and biocementation of soil in situ. Rev. Environ. Sci. Biotechnol. **7**(2), 139–153 (2008)

Mitchell, J.K., Santamarina, J.C.: Biological considerations in geotechnical engineering. J. Geotech. Geoenviron. Eng. **131**, 1222–1233 (2005)

Rebata-Landa, V.: Microbial activity in sediments: Effects on soil behavior. Ph.D. thesis, Georgia Institution of Technology, Atlanta (2007)

Experimental Study on Calcium Carbonate Precipitates Induced by Bacillus Megaterium

Xiaohao Sun, Linchang Miao[(✉)], and Chengcheng Wang

Institute of Geotechnical Engineering, Southeast University, Nanjing, China
Lc.miao@seu.edu.cn

Abstract. Compared with Sporosarcina pasteurii, the Bacillus Megaterium shows higher enzyme activity in low temperature. Experiments were conducted with Bacillus Megaterium in 30 °C of temperature. In the process of culturing bacteria, urea was added to the medium, and urea concentration varied from 0 to 20 g/L. The concentration of urea and calcium acetate remained 0.5 M in gelling solution in mineralization reaction. The effect of urea concentration on Bacillus Megaterium inducing calcium precipitation was studied. Then the ion concentration in gelling solution varied from 0.25 to 1.25 M with the urea concentration of 10 g/L, and the effect of ion concentration in gelling solution on Bacillus Megaterium inducing calcium precipitation was researched as well. Results show that adding urea to the medium can significantly accelerate the reaction and increase the amount and productive rate of precipitation. The amount of calcium precipitation grows with the higher urea concentration, while there is relatively small difference of productive rate between 10 g/L of urea concentration and 20 g/L. With less than 1 M of ion concentration, the higher the ion concentration in gelling solution, the more the precipitation gets. When the ion concentration is greater than 1 M, due to high ion concentration inhibiting the reaction, there is no obvious increase in calcium precipitation. Therefore, the optimum ion concentration for MICP with Bacillus Megaterium is 0.5–0.75 M. The conclusions lay the ground for subsequent engineering application with Bacillus Megaterium.

Keywords: Bacillus Megaterium · Factors · Calcium carbonate

1 Introduction

Microbially Induced Calcium Carbonate (MICP) [1] has received worldwide attention and has been widely used in engineering fields [2–5]. The ordinary bacteria for MICP is Sporosarcina pasteurii (S. pasteurii), but S. pasteurii has low activity and low precipitation rate (the ratio of the amount of practical precipitation and the theoretical value) in low temperature conditions. Therefore, Bacillus megaterium (B. megaterium) is considered to use in low temperature due to higher activity. However, relative research about B. megaterium in MICP field is rarely reported in detail at home and abroad. The paper analyzed the effect of the urea concentration in medium and ion concentration in the gelling solution on B. Megaterium inducing calcium precipitation.

B. megaterium (ATCC14581) was used in this paper, and the formula of medium was as follows: 5 g/L of Yeast extract (YE), 10 g/L of sodium chloride (NaCl), 10 g/L of Peptone. The pH was 7.0 and the medium was put in constant temperature incubator with 30 °C and 120 r/min for 36 h after aseptic inoculation 1% (v/v).

To study the effect of the concentration of urea in medium on precipitation rates of calcium carbonate, the concentration of urea in different groups was 0, 10 and 20 g/L respectively. As for the concentration of urea and calcium acetate in gelling solution, both of them were 0.5 M and the temperature was remained at 30 °C. The absorbance of bacterial solution [1] was measured every four hours and the productive rates of calcium carbonate at 2, 4, 6, 12, 24, 48, 72, 96 and 120 h were obtained.

The effect of Ca^{2+} concentration on precipitation rates was studied through varying the concentration of Ca^{2+} in gelling solution from 0.25 to 1.25 M and maintaining the urea concentration unchanged, at 0.5 M. The amount of urea in medium was 10 g/L and the temperature was 30 °C. Eventually, productive rates at 24, 48, 120 and 168 h were obtained.

2 Results

2.1 The Influence of Urea Concentration in Medium on MICP

From growth curves of B. megaterium in Fig. 1, it can be easily found that adding urea in medium had significant effect on the growth characteristic. With the concentration of urea of 10 g/L or 20 g/L, microorganisms entered stable period at about 24 h, while stable period was reached at around 36 h for those groups without adding urea. However, due to more NH_4^+ in medium, the higher the concentration of urea in medium, the lower the microbial concentration got. Therefore, adding urea in medium might inhibit bacterial reproduction.

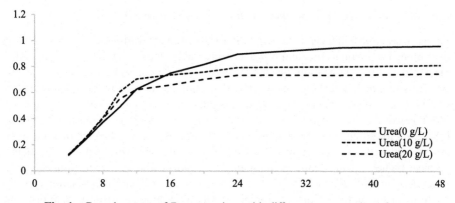

Fig. 1. Growth curves of B. megaterium with different concentration of urea.

Figure 2 shows that between 0 and 4 h, the productive rates of groups with urea concentration of 10 g/L and 20 g/L were much higher than those without urea and thus adding urea could create a favorable initial sedimentation condition for MICP, which was beneficial to improve the early productive rate. In 2th day, the productive rate of calcium carbonate with 10 g/L and 20 g/L was 52.9% and 60.0% respectively, which was 1.8 and 2.1 times of the non-urea counterpart (29%). After that, the productive rate with urea concentration of 10 g/L or 20 g/L decreased, which was mainly because calcium carbonate prevented microbes from absorbing nutrients, making urea decomposition rate in solution decline and effective nucleation sites reduce [6]. The productive rates of samples without urea rose, but the total amount of calcium carbonate was still the lowest. Adding suitable amount of urea, therefore, was helpful to promote initial mineralization reaction, and improve early productive rates and the final ones, which had a positive effect on quick and efficient repair of concrete cracks and reducing engineering cost.

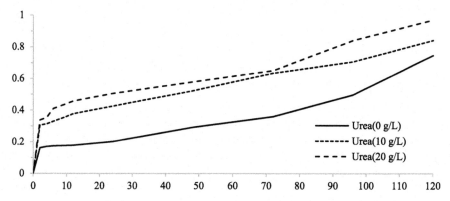

Fig. 2. Productive rates of B. megaterium with different urea concentration.

2.2 The Influence of the Concentration of Ca^{2+} on MICP

Figure 3 describes changes in precipitation amounts with various Ca^{2+} concentration over time. The amount of calcium production in all groups increased with time. After the 3rd day, the precipitation speed of groups with ion concentration of 0.25 M significantly dropped, indicating that the MICP reaction began to slow down. After the 5th day, Ca^{2+} in solution totally transferred into calcium precipitation and at this time, the MICP reaction in the solution had stopped, the final amount of calcium carbonate being 0.57 g. As for groups with concentration of 1.00 M and 1.25 M, the precipitation speed almost remained unchanged from 1st to 7th day, at about 0.12 g/d. It was mainly because higher calcium concentration made calcium carbonate more unlikely to be generated around the cells, reducing the effect of calcium precipitation on the microorganism and allowing microbial activity to last longer [6]. In addition, the amount of calcium precipitation about 1.00 M and 1.25 M was similar in each period, and the final precipitation amount was 1.25 g and 1.29 g respectively. Therefore, growing ion concentration in gelling solution could increase the final amount of calcium precipitation, but too high ion concentration might not have better effect.

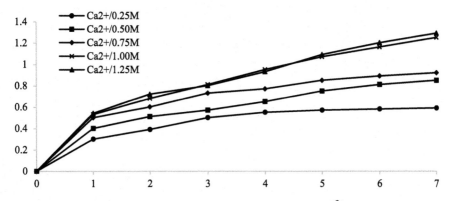

Fig. 3. The amounts of calcium precipitation with various Ca^{2+} concentration.

3 Conclusion

MICP can be affected by several factors, and in this paper, B. Megaterium was used as raw bacteria for MICP reaction. Following conclusions can be come to through studying the effects of urea concentration in medium and Ca^{2+} concentration in gelling solution with 30 °C. (1) Adding urea in medium can promote initial MICP reaction of B. Megaterium, increasing the early and eventual productive rates of calcium precipitation, but higher urea concentration might inhibit bacterial reproduction. Therefore, the optimum amount of urea is 10 g/L. (2) Growing ion concentration in gelling solution enables the final amount of calcium precipitation to rise, but with higher than 1 M of ion concentration, such method cannot continue to achieve better effect.

References

1. Whiffin, V.S.: Microbial CaCO3 precipitation for the production of biocement. Murdoch University, Perth (2004)
2. Rivadeneyra, M.A., Delgado, R., Quesada, E., et al.: Precipitation of calcium carbonate by Deleya halophila in media containing NaCl as sole salt. Curr. Microbiol. **22**(3), 185–190 (1991)
3. Ruiz, C., Monteoliva-Sanchez, M., Huertas, F., et al.: Calcium carbonate precipitation by several species of Myxococcus. Chemosphere **17**(4), 835–838 (1988)
4. Ehrlich, H.L.: Geomicrobiology: its significance for geology. Earth-Sci. Rev. **45**(1–2), 45–60 (1998)
5. Hammes, F., Verstraete, W.: Key roles of pH and calcium metabolism in microbial carbonate precipitation. Rev. Environ. Sci. Biotechnol. **1**(1), 3–7 (2002)
6. De Muynck, W., Verbeken, K., De Belie, N., et al.: Influence of urea and calcium dosage on the effectiveness of bacterially induced carbonate precipitation on limestone. Ecol. Eng. **36**(2), 99–111 (2010)

Effect of Fly Ash on Swell Behaviour of a Very Highly Plastic Clay

İ. Süt-Ünver[1(✉)], M. A. Lav[1], and E. Çokça[2]

[1] Department of Civil Engineering,
İstanbul Technical University, İstanbul, Turkey
incisut79@hotmail.com
[2] Department of Civil Engineering,
Middle East Technical University, Ankara, Turkey

Abstract. Swelling of expansive soils causes excessive heave. Resulting expansion can lead to considerable distress to pavements and/or lightweight structures which are settled on that kind of subsoils. Several methods can be applied to control the swelling problem. One of the most commonly used method is the addition of stabilizing agents, such as lime or fly ash to the expansive soil. Soil samples that are very highly plastic have been taken from clay deposits of Ankara in Turkey in this study. Both disturbed and undisturbed specimens have been obtained in order to perform laboratory tests. Fly ash has been used as the chemical stabilizer. Firstly, reference tests have been carried out on natural soil samples. Then the clay is mixed with fly ash at different percentages of dry weight of the soil and the engineering properties of the stabilized soil samples have been determined. Index properties and swelling characteristics of the soil samples have been found. Results are evaluated and the changes in these properties especially with regard to swell potential, have been interpreted in this paper. Optimum values have been determined for swell potential. The correlations between the index properties, the swell percentage and the swell pressure are presented.

Keywords: Expansive clay · Fly ash treatment · Swell potential

1 Introduction

Light buildings and pavements are damaged frequently due to expansive soils around the world. Swelling clay minerals make soils expansive. As they get wet, the clay minerals absorb water molecules and expand, and soils can undergo loss of shear strength. Conversely, as they dry they shrink, leaving large voids in the soil. Soils with smectite clay minerals, such as montmorillonite, exhibit the most profound swelling properties. As clays swell with moisture increases the resulting uplift pressures can damage surficial and/or lightweight structures. There are several stabilization methods to eliminate the damages due to expansive soils. Fly ash stabilization has been used successfully on some of the projects to minimize swelling potential of clay soils.

Both disturbed and undisturbed soil samples have been taken from clay deposits of the Bilkent Atatürk Hospital district of Ankara, near Ankara-Eskişehir State Highway in Turkey. In order to identify the soil, reference tests were performed and as a result of these tests soil specimens are classified as very highly plastic (plasticity index, PI = 76%). The clay fraction of the soil has been determined as 45%. Therefore, the activity is 1.7 (A = 1.7). As a result of several correlations published in the literature the soil can be considered as highly expansive. In order to treat the expansive nature of the soil, fly ash has been used as the chemical stabilizer. Within the scope of laboratory studies, reference tests on untreated soils have been conducted first. Then the clay soil is mixed with the fly ash agent at different percentages of dry weight of the soil (5%, 10%, 15%, 20% and 25%) and engineering properties of the treated soil samples have been determined. Results are evaluated and the changes in several soil properties, especially with regard to swell potential, have been interpreted in this paper. Optimum values have also been evaluated for swell potential. The correlations between the index properties, the swell percentage and the swell pressure are presented.

2 Clay Properties

Soil properties of the Bilkent Clay are given in Table 1 below. Laboratory experiments have been conducted both on disturbed and undisturbed clay samples. As seen in Table 1, the clay is very highly plastic (PI = 76%). Clay-size fraction is 45%. Also, it is noticed that the clay is quite active (i.e., activity, A = 76/45 = 1.7). Therefore, swell potential of Bilkent Clay is very high based on some empirical charts considering the Atterberg Limits, clay-size fraction and activity. This is also confirmed by the swell pressure and percentage values found by odometer tests (Table 1). It should be noted here that natural water content is 41%. Since this water content is more than the optimum water content ($w_{opt.}$ = 29%), the swell percentage and swell pressure of the undisturbed sample are relatively low (i.e., swell percentage = 4%, swell pressure = 50 kPa). However, swell percentage and swell pressure are quite high for the remoulded samples prepared at optimum water content and maximum dry density as given in Table 1.

3 Laboratory Studies and Conclusions

Atterberg Limits and swelling properties of the treated soil samples have been examined by increasing the fly ash content steadily. The changes are given in the following figures (Figs. 1, 2 and 3). Tests have been performed according to the ASTM Standards (ASTM D 427, ASTM D 4318, ASTM D 4546).

When Fig. 1 is evaluated, it is realized that the plasticity index values are decreased continuously in a stable manner by addition of fly ash. In each fly ash increment of 5%, the decrease in plasticity index is between 3% and 13%.

According to several empirical criteria in the literature, the sample natural clay soil in this study has been classified as having "very high" swell potential based on index properties. This conclusion is consistent with the swell percentage test results (Fig. 2).

The allowable swell percentage for surficial structures is generally accepted as 0.5% as given in Highway Standards (KGM 2005) in Turkey. However, the maximum fly ash content (25%) in this study is not adequate to obtain allowable swell percentages for surficial structures, as shown in Fig. 2. There are reductions in swell percentages as the fly ash content increases. But, the swell percentage is still 4.5%, which is inadequate, at the highest amount of fly ash (25%). It can be stated here that fly ash stabilization of a very highly plastic clay is not possible with reasonable fly ash contents as far as the swell percentage criterion is concerned.

Table 1. Properties of Bilkent clay

Sieve analysis			Atterberg limits				Proctor test		Expansion test	
+No.200 (%)	−No.200 (%)	Clay Size (%)	LL (%)	PL (%)	PI (%)	SL (%)	γ_{drymax} (g/cm^3)	w_{opt} (%)	SP (kPa)	SP % (%)
0.5	79.2	45	109	33	76	9	1.369	29	374.6	26.9

Note: LL: Liquid Limit, PL: Plastic Limit, PI: Plasticity Index, SL: Shrinkage Limit. SP: Swell Pressure. SP%: Swell Percentage; Experiments have been performed on remoulded samples prepared at optimum water content (w_{opt}) and maximum dry density (γ_{drymax}).

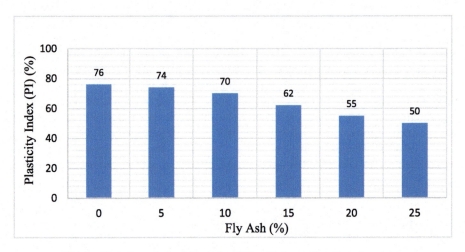

Fig. 1. Variation of plasticity index (PI) with increasing fly ash content

Swell pressure remains almost same for 5% fly ash amount (Fig. 3). Then, swell pressure is decreased quite significantly from 5% to 10% fly ash content (about 72% decrease). These decreases are moderate in later fly ash amounts.

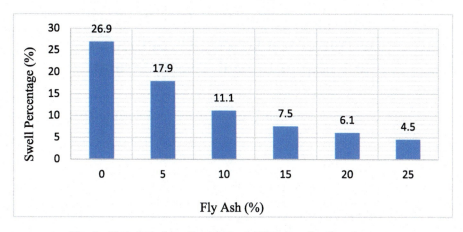

Fig. 2. Variation of swell percentage with increasing fly ash content

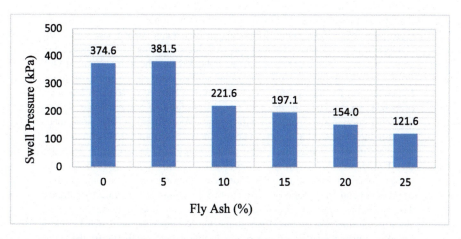

Fig. 3. Variation of swell pressure with increasing fly ash content

References

ASTM D 427: Standard Test Methods for Shrinkage Factors of Soils by the Mercury Method, ASTM International, United States (1998)

ASTM D 4318: Test Methods for Liquid Limit, Plastic Limit, and Plasticity Index of Soils, ASTM International, United States (2010)

ASTM D 4546: Standard Test Methods for One-Dimensional Swell or Collapse of Cohesive Soils, ASTM International, United States (2008)

KGM: Technical Specification for Lime Improvement (in Turkish), General Directorate of Highways, Ankara, Turkey (2005)

Analysis of Pore Size Distributions of Nonwoven Geotextiles Subjected to Unequal Biaxial Tensile Strains

Lin Tang[1(✉)], Songtai Sun[2], Xiaowu Tang[3], and Ruixuan Zhang[1]

[1] Harbin Institute of Technology at Weihai, Weihai 264209, China
tanya_tang3@163.com
[2] State Key Laboratory for Disaster Reduction in Civil Engineering,
Tongji University, Shanghai 200092, China
[3] Research Center of Costal and Urban Geotechnical Engineering,
College of Civil Engineering and Architecture, Zhejiang University,
Hangzhou 310058, China

Abstract. Nonwoven geotextiles are highly porous structures consisting complex shapes of pores. The filtration applications of nonwoven geotextiles are typically subjected to unequal biaxial tensile strains, which can cause the changes in pore sizes and result in soil clogging or leaking in the protection of banks or slopes. Some previous studies indicated that the pores decrease with biaxial tensile strains, whereas the other studies showed the opposite trend. Depending on the theoretical model of Rawal, a theoretical solution has been proposed that accounts for the pore size distributions of nonwoven geotextiles subjected to unequal biaxial tensile strains. A comparison has been made between the theoretical and experimental pore size distributions at defined levels of equal biaxial tensile strains. Both the theoretical and experimental pore size distributions increase with the equal biaxial strains. When it comes to the unequal biaxial tensile strains, the theoretical solution indicates that the changes of pores depend on the ratio of strains in two directions and the values of strains. The larger the ratios of strains in two directions are, the more fibers reorient towards the direction of the larger strain, the narrower the pore will be. Nevertheless, under the same ratio of strains in two directions, the larger the stains are, the larger the pore will be stretched. The solution may explain the inconsistencies among the previous results.

Keywords: Pore size distributions · Nonwoven geotextiles
Unequal biaxial tensile strains

1 Introduction

Nonwoven geotextiles are applied extensively as filtration and drainage material in various field. However, there are few studies about changes of nonwoven geotextiles pore size under biaxial stretch. As far as experiments, Fourie and Addis (1997) drew a conclusion utilizing water-powered sieve method that equivalent pore size of non-strained nonwoven geotextiles was 157 μm and when biaxial stress was 0.19 × 0.37 kN/m, pore size decreases to 137 μm, when stress of 0.54 × 0.37kN/m, pore size increasing to 159 μm, that is, pore size would alter at various levels of stress. Wu and Hong (2016) measured the mean discharge velocity of needled-punched nonwoven geotextiles under equal biaxial tensile strains. The mean discharge velocity decreased with an increase in strain subjected to strain up to 5%. This trend reversed subjected to a strain that was great than 5%. Discrepancies still exist in the studies, and thus, further studies are still necessary.

2 Theory

The unequal biaxial tensile stress of nonwoven geotextiles is taken as a plane stress problem. Depending on the theoretical model of Rawal (2010), a theoretical solution of the pore size distributions (PSD) of nonwoven geotextiles subjected to unequal biaxial tensile strains is shown below.

$$F_f(d)(\varepsilon_{bi}) = 1 - \left[\left(1 + \omega(\varepsilon_{bi})d + \frac{\omega^2(\varepsilon_{bi})d^2}{2}\right)e^{-\omega(\varepsilon_{bi})d}\right]^N \quad (1)$$

$$\omega(\varepsilon_{bi}) = \frac{4V_f(\varepsilon_{bi})K_\alpha(\beta_f)}{\pi d_f} \quad (2)$$

$$V_f(\varepsilon_{bi}) = \frac{\mu}{\lambda \rho_f t_{GT}} \quad (3)$$

$$K_\alpha(\beta_f) = \int_{-\frac{\pi}{2}-\alpha}^{\frac{\pi}{2}-\alpha} |\cos\beta_f| \chi(\beta_f) d\beta_f \quad (4)$$

$$N = \frac{t_{GT}}{d_f} \quad (5)$$

$$\lambda = (1+\varepsilon_x)(1+\varepsilon_y)\left[1 - \frac{\nu(\varepsilon_x+\varepsilon_y)}{1-\nu}\right] \quad (6)$$

$F_f(d)(\varepsilon_{bi})$ is the cumulative probability of a particle with diameter d, passing through the layers of nonwoven. N is the number of layers. εx and εy are the unequal biaxial tensile strains, d_f is the fibre diameter, μ is the initial unit area mass, λ is the volume change rate, t_{GT} is the nonwoven thickness, ρ_f is the fiber density, ν is the Poisson's ratio of the nonwoven. $K_a(\beta_f)$ is the direction parameter, β_f is the orientation angle of fibers. The details of the other parameters are explained by Rawal (2010).

3 Mechanism and Trial Calculation

Through the microscope experiments, the fibers were found to reorientate towards the larger loading direction, which results that the pores become narrower, as shown in Fig. 1(a). The larger the ratios of strains in two directions are, the smaller the pore size will be. Nevertheless, the fibers are set apart in the smaller loading direction. The larger the stains are, the larger the pore will be stretched, as shown in Fig. 1(b). Whether the pore will increase or decrease depends on the ratio of strains in two directions and the values of strains, as shown in Fig. 1(c). The direction parameter K_α in the PSD theoretical solution can simulate this mechanism. The value of K_α ranges from 0 to1. As the fiber orientation into the larger loading direction, K_α increases and the pore size distribution curve moves towards the direction of the small pore size.

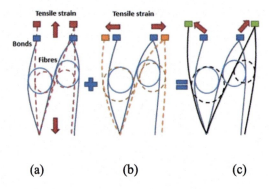

Fig. 1. Change of pore size under reorientation of fibers

Table 1. Properties of needle-punched nonwoven geotextiles

Physical mechanics parameters	
Mass per unit area (g/m^2)	107
Thickness (mm)	0.86
Breaking strength (kN/m)	2.5
Fiber density (g/cm^3)	1.32
Fiber size (μm)	23
Poisson's ratio	2.6

The parameters of a nonwoven geotextile are provided in Table 1, which are used in trial calculation to show the mechanism. The calculated pore size distributions under different biaxial tensile strain are presented in Fig. 2. Assuming that under no strain state K_α is equal to 0.63 (Rawal 2010), the biaxial strain increases by 1:2, and when the strain is 1%:2%, if K_α increases from 0.63 to 0.8, from the trial, the pore size

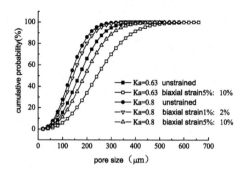

Fig. 2. Calculated pore size distributions under biaxial tensile strain

distribution curve moves towards the pore size reduce direction. Whereas if the strain is 5%:10%, K_α is equal to 0.8, the pore size distribution curve moves towards the pore size increment direction. Taking the equivalent pore size O_{95} reads from the theoretical PSD curve as an example, if K_α increases from 0.63 to 0.7 (which increase by 11.1%), the theoretical O_{95} reduces from 310 μm to 273 μm (which decrease by 11.9%). Therefore, the theoretical solution is sensitive to the K_α value, and a reasonable determination of K_α is important to the prediction of PSD.

Thus, as for the prediction of the PSD theoretical solution, the pore size will undergo the effect of the dominant change depending on whose effect is stronger when concerning the fiber direction and the biaxial strain. This theoretical solution is expected to explain the phenomenon that pore size or water permeability alters when nonwoven geotextiles under biaxial tension and different stress levels. Its accuracy needs further tests to verify and the change of K_α value remains to be precisely determinated in micro-test.

4 Conclusions

The trails show that the PSD theoretical solution depends on the fiber direction parameter K_α and the biaxial tensile strain; the stronger the fiber orientation is, the larger the value K_α and the smaller the pore size are; the larger the biaxial strain is, the larger the pore size is. The formula still need to be further validated in unequal biaxial tensile tests. And unequal biaxial tensile would exert significant changes in pore size, so it needs further study through the filter tests to validate whether these influences would make a difference.

Acknowledgments. This work was supported by the National Natural Science Foundation of China (51708160).

References

Fourie, A.B., Addis, P.: The effect of in-plane tensile loads on the retention characteristics of geotextiles. ASTM Geotech. Test. J. **20**(2), 211–217 (1997)

Wu, C.S., Hong, Y.S.: The influence of tensile strain on the pore size and flow capability of needle-punched nonwoven geotextiles. Geosynth. Int. **23**(6), 422–434 (2016)

Rawal, A.: Structural analysis of pore size distribution of nonwovens. J. Text. Inst. **101**(4), 350–359 (2010)

Effect of Fine Particles on Cement Treated Sand

Ganapathiraman Vinoth[1], Sung-Woo Moon[2], Jong Kim[2], and Taeseo Ku[1(✉)]

[1] National University of Singapore, 1 Engineering Drive 2, 117576 Singapore, Singapore
ceekt@nus.edu.sg
[2] Nazarbayev University, Qabanbay Batyr Avenue 53, 010000 Astana, Kazakhstan

Abstract. Cement treated sand improves mechanical properties through the cementitious bonding between cohesionless particles, thus allowing several geotechnical applications such as soil stabilization against slope failure and liquefaction. Since pure sand without any fine particles is seldom available in nature, this study aims to investigate the effect of fine particles (kaolin) in a very small proportion (<5%) on cement treated sand. Two types of cements are used: (i) Ordinary Portland cement (OPC) and (ii) Calcium sulfoaluminate cement (CSA). OPC is the widely used cementitious binder whereas CSA is a rapid hardening cement that is becoming popular due to its low carbon foot print. Three different cement contents (3%, 5% and 7%) and four different fine contents (0%, 1%, 3% and 5%) for each cement content are chosen. The stiffness and strength of the cement treated sands are measured through shear wave velocity and unconfined compressive strength respectively, after 1-day and 7-day curing periods. The results show that the influence of fine particles is visible even with fine content as low as 1%. However, the effect is different between the two types of cements used and between low and high cement contents. As the fine content increases, the increase in strength and stiffness is more for OPC than CSA and more significant at low cement content (3%) than high cement content (5% and 7%).

Keywords: Calcium sulfoaluminate · Cemented sand · Fine particles
Shear wave velocity · Unconfined compressive strength

1 Introduction

Improving the engineering properties of sand through cementation is a commonly adopted technique [1]. Various studies on (1) cemented soil with poorly graded or well graded pure sand and (2) uncemented soil with fine particles have been conducted [2, 3]. Commonly, pure sand is seldom available in nature. It is accompanied with either silt or clay at varying proportions. The fine particles (clay or silt) are generally ignored in both soil classification and geotechnical design, when they constitute less than 5% of the total mass of uncemented sand. The fine particles have little effect on the behavior of uncemented sand at very low proportions [4]. Some studies have covered the applicability of silty and clayey sand for cement treatment [5]. However, the effects of fines in a small proportion on cemented sand are relatively unexplored. It is also expected that different kinds of binder shall produce different effects. This study aims to explore the effects of fine particles, in a small proportion, on strength and stiffness of cemented sand treated with two different types of cement.

2 Materials and Methodology

A uniformly graded pure sand (D_{50} = 0.71 mm; C_u = 1.78) is replaced with different proportions (0% (i.e. pure sand), 1%, 3% & 5%) of commercially available kaolin ($Al_2Si_2O_5(OH)_4$). The grain size distribution curves after replacing sand with fines are shown in Fig. 1a. Two types of cements are used: (i) Ordinary Portland Cement (Type-I) (OPC), (ii) Calcium Sulfoaluminate Cement (CSA). Three different cement contents (3%, 5% & 7%) are used for each cement type and a constant water content (by total mass of solids) of 10% is used for all samples.

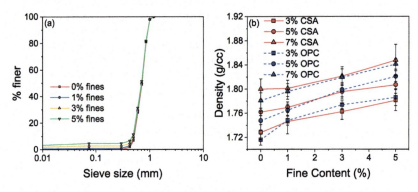

Fig. 1. (a) Grain size distribution curves with different amount of fines, (b) initial density of 1-day and 7-day samples.

CSA cement is a rapid hardening cement that produces ettringite as the hydration product as shown in Eq. 1 which contributes to the high early strength [6]. In addition to high early strength, the carbon emission during its manufacture and hydration is also considerably lesser which makes CSA cement a sustainable alternative to OPC.

$$4CaO \cdot 3Al_2O_3 \cdot SO_3 + 2CaSO_4 \cdot 2H_2O + 32H_2O$$
$$\rightarrow 6CaO \cdot Al_2O_3 \cdot 3SO_3 \cdot 32H_2O + 4Al(OH)_3 \qquad (1)$$

Samples are prepared in the following manner. After mixing sand, kaolin, cement and water for 15 min in Hobart mixer, cylindrical samples (50 mm diameter; 100 mm height) are prepared in PVC molds by tamping in three layers and their initial mass is measured. The top and bottom of the molds are wrapped in polyethylene sheets and kept in water and cured for 1-day and 7-days. After curing, samples are extracted and shear wave velocity (V_s) is measured through bender elements after which their unconfined compressive strength (UCS) is measured.

3 Results and Discussion

The bulk density of samples, measured right after preparation, increases with cement content. At a particular cement content, it increases with fine content (see Fig. 1b) which is expected as the finer particles lubricate the sand and occupy the voids between sand grains resulting in denser mixtures.

3.1 V_s and UCS

It can be observed in Fig. 2 that increase in fine content results in increase in V_s. Similar trends can be observed with UCS as well in Fig. 3 for different cement contents. However, the rate of strength gain is not parallel between two cement types. As a rapid hardening cement, CSA treated samples gain significantly larger strength in 1-day compared to OPC treated samples.

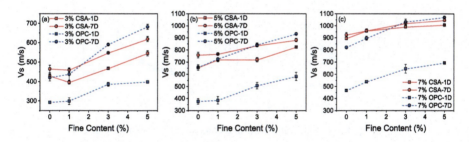

Fig. 2. Shear wave velocity with fine content (a) 3% (b) 5% and (c) 7%.

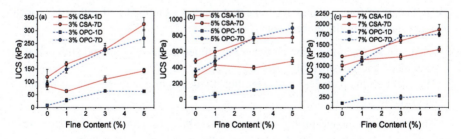

Fig. 3. Unconfined compressive strength with fine content (a) 3% (b) 5% and (c) 7%.

At 7 days, CSA samples of 5% and 7% cement content with no fines have higher strength than corresponding OPC treated samples. However, as the fine content increases, OPC samples gain more strength, comparable to that of CSA samples, at 7 days. Variation of V_s also shows the same trend in Fig. 2.

3.2 UCS Ratio and Correlation Between V_s and UCS

In addition to the type of binder, the effect of fine particles varies with respect to cement content as well. Figure 4a shows the UCS of all cement content at 7-day curing period normalized by the corresponding UCS of samples with no fines. From this figure, it is clear that the increase in strength and stiffness with increasing fine content is more pronounced with OPC rather than CSA and with lower cement content than higher cement content. Due to the presence of fine particles, the 7-day strength is increased by 1.5 to 3.2 times and 1.1 to 2.2 times in OPC treated samples and CSA treated samples, respectively. Figure 4b shows the relationship between V_s (km/s) and UCS (MPa). Two different relationships are fitted with the data: (i) Power Function ($UCS = 1.44V_s^{3.66}$) (ii) Exponential Function ($UCS = 0.019\exp(4.32V_s)$. The curves fit very well with the scatter with R^2 values of 0.951 and 0.954, respectively.

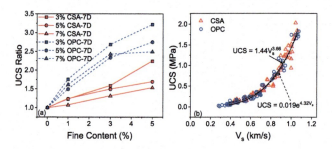

Fig. 4. (a) UCS ratio with fine content (b) V_s versus UCS.

4 Conclusions

The influence of fine particles on cemented sand is examined and following conclusions are made from the current study:

1. Even a small fraction of fine particles can influence the strength and stiffness of the cemented sand considerably.
2. The effect of fine particles varies significantly depending upon both the cement type and cement content.

The results encourage further study on this subject by considering different water contents and curing conditions to understand the effects more thoroughly.

References

1. Clough, G.W., Sitar, N., Bachus, R.C., Rad, N.S.: Cemented sands under static loading. J. Geotech. Geoenviron. Eng. **107**, 799–817 (1981)
2. Thevanayagam, S., Mohan, S.: Intergranular state variables and stress-strain behavior of silty sands. Geotechnique **50**(1), 1–23 (2000)
3. Fernandez, A., Santamarina, J.: Effect of cementation on the small-strain parameters of sands. Can. Geotech. J. **38**, 191–199 (2001)
4. Kuerbis, R., Negussey, D., Vaid, Y.P.: Effect of gradation and fines content on the undrained response of sand. In: Hydraulic Fill Structures. ASCE pp. 330–345 (1988)
5. Stavridakis, E.I.: A solution to the problem of predicting the suitability of silty–clayey materials for cement-stabilization. Geotech. Geol. Eng. **24**(2), 379 (2006)
6. Ukrainczyk, N., Frankoviæ Mihelj, N., Šipušić, J.: Calcium sulfoaluminate eco-cement from industrial waste. Chem. Biochem. Eng. Q. **27**, 83–93 (2013)

A Geochemical Model for Analyzing the Mechanism of Stabilized Soil Incorporating Natural Pozzolan, Cement and Lime

Ba Thao Vu[1], Van Quan Tran[2(✉)], Quoc Dung Nguyen[1,2], Anh Quan Ngo[1,2], Huu Nam Nguyen[1,2], Huy Vuong Nguyen[1,2], and Hehua Zhu[1,2]

[1] Department of Geotechnical Engineering, Hydraulic Construction Institute, Vietnam Academy for Water Resources, Hanoi, Vietnam
vubathao@gmail.com
[2] Hydraulic Construction Institute, Vietnam Academy for Water Resources, Hanoi, Vietnam
vanquan.tran.ec@gmail.com

Abstract. In this study, a geochemical model is proposed to analyze of strength increase mechanism of stabilized soils incorporating natural pozzolan, cement and lime. Modelling considered kinetics and thermodynamics of mineralogical transformations. The pozzolanic reaction between lime, cement, natural pozzolan and soil is successfully simulated by the geochemical model. The model shows that the quantity of C-S-H and C-A-S-H precipitation are suitable for the strength development of stabilized soils. The compressive and tensile strength are depended on the stabilized soils mixing ratios for soil, natural pozzolan, cement and lime. The good agreement between the numerical and experimental results for optimal mixing ratio shows that the geochemical model could be considered as a tool to assess the optimal mixing ratio, optimal content of natural pozzolan is 15%. Moreover, the obtained results demonstrate that the soil of Dak Nong province in Central Highlands of Vietnam is inert in the process of soil stabilization, the natural pozzolan and lime are necessary to increase the amount of C-S-H as well as the strength.

1 Introduction

The process of stabilization is known, basing on a set of consecutive reaction: dissolution/precipitation, cation exchange, surface complexation, pozzolanic reactions. However, to the authors' knowledge, only one geochemical model is proposed to simulate the reactions between lime and bentonite by De Windt et al. [1]. In order to investigate the mechanism of stabilization and find out the optimal mixing ratio of lime, natural pozzolan, cement and soil, the geochemical model is proposed to simulate the reactions of the process. In this paper, the first section depicts the laboratory test which is carried out on a sample of a test soil chosen from Dak Nong Province, many mixing ratios are employed. The second section introduces the governing equations used in the proposed geochemical model, such as the principle of thermodynamic equilibrium, the mineral dissolution/precipitation reactions and their equilibrium constants. The experimental and numerical results are analyzed in the third section.

2 Materials and Laboratory Tests

Soils. The soil used in the present study were picked up from a rural road base in Gia Nghia city, Dak Nong province, Central Highlands Area of Vietnam (cf. Fig. 1). The disturbed soil was excavated, placed in plastic bags, and transported to the laboratory for preparation and testing. Laboratory tests were carried out to classify type of soil, analyse the chemical properties, primary minerals, and engineering properties.

Natural Pozzolan. The natural pozzolan used in this investigation was collected from Quang Phu district, Dak Nong province. The natural pozzolan was transported to the laboratory to test the specific surface area and chemical composition.

Lime and Cement. The lime and cement used were commercially available lime and cement typically used for construction purposes.

A series of laboratory tests consisting of compaction, unconfined compressive strength and splitting tensile strength were conducted on the unstabilized soil and stabilized soils. Three mixing ratio in percentage of stabilized soil: soil/natural pozzolan/cement/lime 83/10/4/3, 78/15/4/3 and 75/20/4/3 were carried out. Proctor standard compaction test (ASTM D-698 2000) was applied to determine the maximum dry density (MDD) and the optimum moisture content (OMC) of the unstabilized soil and stabilized soils. Unconfined compressive strength and splitting tensile strength tests (cf. Fig. 1). Each specimen used in the unconfined compressive strength (ASTMD1633) test and splitting tensile strength (ASTM C496) were compacted in a cylindrical mould at optimum

Fig. 1. Amount of CSH, CSH+CASH, Compressive strength and splitting tensile as a function of natural pozzolan quantity in the stabilized soil

moisture content and maximum dry density. Specimens were cured in a plastic bag to prevent moisture change. Tests were performed at curing ages of 14 days. For each type of mixtures, the strength value was obtained as the average of three tests.

3 Modelling Approach

The interaction between the ionic species and the mineral species leads to precipitation/dissolution of minerals. The mineral saturation ratio Ω_m can be expressed as:

$$\Omega_m = K_{s,m}^{-1} \prod_{j=1}^{N_c} (\gamma_j C_j)^{\nu_{mj}} \quad m = 1, ..., N_p \quad (1)$$

where m is the indice of the mineral species, $K_{s,m}$ is the equilibrium constant. C_j is the molal concentration of primary species in the solution $(mol.kg^{-1})$. ν_{mj} is the stoichiometric coefficient of the primary species, γ_j is the activity coefficient of the ion j, N_c, N_p are the number of corresponding primary species and mineral species. The state of equilibrium (or disequilibrium) of mineral species in the solution is controlled by the mineral saturation index IS_m, as follows:

$$IS_m = \log \Omega_m \quad (2)$$

For a given mineral species, the solution is in equilibrium with the mineral species if $IS_m = 0$. The solution is under-saturated and the mineral species can still dissolve if $IS_m < 0$. Finally, the solution is super-saturated and the mineral species

Table 1. Initial composition and equilibrium state composition of materials

Material composition (g/100 g)	Phase	Formula	Composition (g/100 g)	
			Initial	Equilibrium state
Soil	Quartz	SiO_2	4	4
	Kaolinite	$Al_2Si_2O_5(OH)_4$	23	23
	Gibbsite	$Al(OH)_3$	61	59
Lime	CaO	CaO	100	0
Natural pozzolan	Diopside	$CaMg(SiO_3)_2$	29	42
	Forsterite	Mg_2SiO_4	23	0
	Cristobalite	SiO_2	1	0
	Albite	$NaAlSi_3O_8$	27	0
	Quartz	SiO_2	4	4
Unhydrated cement	Alite	Ca_3SiO_5	16.75	0
	Belite	Ca_2SiO_4	54.45	0
	Aluminate	$Ca_3Al_2O_6$	14.08	0
	Ferrites	$Ca_4Al_2Fe_2O_{10}$	8.52	0

may be precipitated if $IS_m > 0$. The details of modelling approach are described in Tran's work [2]. Mineralogical composition of the initial materials: Soil, lime, natural pozzolan was identified by powder X-ray diffraction (XRD). The Bogue calculation to determine the unhydrated clinker phases in the ordinary Portland cement (OPC) is more detailed in the Tran's work [2]. The initial mineralogical composition of the mixture is given in Table 1.

4 Results and Discussion

The reactions of soil-natural pozzolan-hydrated of Portland cement and lime are carried out by the geochemical code Phreeqc [3].

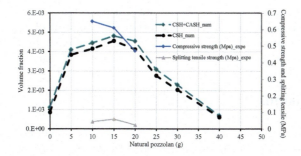

Fig. 2. Amount of CSH, CSH+CASH, Compressive strength and splitting tensile as a function of natural pozzolan quantity in the stabilized soil

The numerical calcium silicate hydrate profiles as a function of natural pozzolan can be described in two zones. In the first of these, the amount of calcium silicate hydrates increases proportionally with increased natural pozzolan quantity. The amount of calcium silicate is maximal when 15% natural pozzolan is mixed with the 85% soil, 4% lime and 3% cement. After the maximum value, the amount of calcium silicate hydrates decreases progressively even if the quantity of natural pozzolan is continued to increase in the second zone. The experiment result (cf. Fig. 2) shows the behavior of splitting tensile strength development as the function of natural pozzolan quantity is the same as amount of change of calcium silicate which is less identical than the behavior of compressive strength. However, the compressive strength and the amount of calcium silicate hydrates are decrease if the used natural pozzolan quantity is greater than 15% of stabilized soil. The calcium silicate hydrates are primarily responsible for the strength (compressive strength...) of cement based materials [4]. Therefore, the mechanism of strength development of stabilized soils depends importantly on the amount of calcium silicate hydrates. The numerical result shows that the geochemical model seems to predict relatively the amount of calcium silicate as the function of natural pozzolan quantity. The geochemical model seems to be a tool to assess the optimal mixing ratio.

In detail of the optimal mixing ratio of soil/natural pozzolan/cement/lime: 78/15/3/4, the composition in the equilibrium state is shown in the Table 1. The minerals of soil are relatively stable. The mineral of lime and natural pozzolan, unhydrated cement are more active than the minerals of soil. This result shows that the soil of Dak Nong Province is inert in the process of soil stabilization. Therefore, the process depends importantly on natural pozzolan, cement and lime. A part of calcium silicate hydrates is arisen from the hydration of cement. The second part of calcium silicate hydrates is resulted of pozzolanic reactions. In detail, lime releases Ca^{2+} ions by CaO, portlandite and calcite dissolution, natural pozzolan releases $H_4SiO_{4(aqueous)}$ (cf. Table 1) by forsterite, cristobalite and albite dissolution. The lime and natural pozzolan dissolution are indispensable to precipitate the second part of C-S-H.

Amount of Al^{3+} ions is primarily released by the dissolution of albite. Amount of albite accounts for 27% of natural pozzolan. However, thank to the forsterite, cristobalite and albite dissolution, the amount of $H_4SiO_{4(aqueous)}$ is higher than the amount of Al^{3+} ions. The amount of calcium aluminate silicate hydrate is lower than the amount of calcium silicate hydrate.

5 Conclusions

The numerical and experimental results show that the soil of Dak Nong Province is inert in the process of soil stabilization, the natural pozzolan and lime are absolutely necessary to increase the amount of C-S-H. That leads to increase the mechanical properties of stabilized soil. When the quantity of cement and lime present in stabilized soils are constant: 4% and 3%, respectively, the optimal mixing ratio of stabilized soil is found with 15% natural pozzolan. That is found by the model and the experiments. Therefore, the geochemical model could be considered as a tool to find the optimal mixing ratio of stabilized soil. The model and experiments analyzed only one case of cement and lime quantity. More cases need to be investigated in the future: different quantity of lime and cement.

Acknowledgements. This work was funded by the independent national project "Study of used natural pozzolan in construction and maintenance works of rural transport, irrigation in the province of Dak Nong, project code: DTDL.CN-55/16. "The Ministry of Science and Technology of Vietnam" assigned the project for the Hydraulic Construction Institute to host and implement.

References

1. De Windt, L., Deneele, D., Maubec, N.: Kinetics of lime/bentonite pozzolanic reactions at 20 and 50C: batch tests and modeling. Cem. Concr. Res. **59**, 34–42 (2014)
2. Tran, V.Q.: The contribution toward understanding of mechanisms of depassivation of steel in concrete exposed to sea water: theory and thermochemical modeling Ecole Centrale de Nantes (2016)

3. Parkhurst, D.L., Appelo, C.A.J.: Description of input and examples for PHREEQC version 3–a computer program for speciation, batch-reaction, one-dimensional transport, and inverse geochemical calculations A43 U.S. Geological Survey book 6, 497 (2013)
4. Amer, A.A., El-Sokkary, T.M., Abdullah, N.I.: Thermal durability of OPC pastes admixed with nano iron oxide. HBRC J. **11**, 299–305 (2015)

Compaction Characteristics and Shrinkage Properties of Fibre Reinforced London Clay

Jianye Wang[✉], Andrew Sadler, Paul Hughes, and Charles Augarde

Department of Engineering, Durham University, Durham, UK
jianye.wang@durham.ac.uk

Abstract. This paper presents an experimental study of changes in the engineering properties of London Clay when reinforced with polypropylene fibres. Standard Proctor tests and linear shrinkage tests were carried out to evaluate changes to optimum moisture content, bearing capacity and shrinkage properties due to the inclusion of fibres. The experimental results indicate that as the fibre inclusion ratio increases, both optimum moisture content and maximum dry density of the soil are reduced. Fibre length was also shown to influence maximum dry density but not optimum moisture content. The introduction of fibres also reduced linear shrinkage of the soil by as much as 40% when compared to unreinforced soil.

Keywords: Fibre reinforcement · Compaction · Linear shrinkage

1 Introduction

Utilising fibres as reinforcement in fine-grained soils, is a potentially promising method for improving the engineering behaviour of soils. Compared to traditional stabilisation methods such as lime treatment or conventional geosynthetic reinforcement (strips or fabrics), there are a number of advantages of fibre-soil composites: (1) the inclusion of fibre only changes the physical properties of soil and has no impact on the surrounding environment or structure. (2) Fibres not only improve the strength of the soil, but also offer greater toughness and ductility and reduce the loss of peak strength [1]. (3) The mixing and construction of the fibre-soil composite can be done with conventional equipment. Currently, polypropylene (PP) fibre is the most widely used fibre in laboratory assessments of soil reinforcement. Soganci [2] studied the effects of polypropylene fibre on compaction and swelling characteristics of an expansive soil. Test results indicated that inclusion of fibre reduced maximum dry density and swell potential. Tang et al. [3] investigated the desiccation cracking behaviour of polypropylene fibre reinforced clay and found that desiccation cracking was significantly reduced, and crack resistance improved with fibre inclusion. Here we report on a compaction and linear shrinkage tests carried out on London Clay reinforced with fibres studying the influence of fibre inclusion ratio and fibre length.

2 Materials and Methods

The soil used in this study is London Clay, which was excavated from Clapham, London, UK, and which has been widely characterized in research activities in the UK. The classification properties of the soil were determined in accordance with BS-1377, part 2 [4], with the results shown in Table 1. PP fibres of two different lengths (6 mm and 12 mm) used in the tests were produced by ADFIL [5]. The physical properties of the fibres are given in Table 2.

Table 1. Classification properties of the London Clay.

Property	Specific gravity	Liquid limit (%)	Plastic limit (%)	USCS classification	Grain size		
					<20 μm (%)	<6 μm (%)	<2 μm (%)
Value	2.72	58.2	20.9	CH	86.2	69.8	55.9

Table 2. Physical properties of PP fibre used (ADFIL, 2017).

Property	Fibre type	Length (mm)	Diameter (μm)	Acid resistance	Specific gravity
Value	Monofilament	6 & 12	22	High	0.91

The soil was air dried and then passed through a 2 mm sieve (for compaction specimens) and a 425 μm test sieve (for linear shrinkage test specimens). Fibres were weighed and mixed with the soil in small increments (10% of the total weight) by hand. After fibre/soil mixing, distilled water was added to the mixture until the target moisture content was achieved. Initial mixing of the wet material was performed with a pallet knife, while further mixing was implemented using a mechanical mixer to ensure uniform distributions of water and PP fibres. The fibre inclusion ratio (ρ) is defined herein as $\rho = W_f/W$, where W_f is the mass of fibre, W is the mass of dry soil. The inclusion ratios of fibre (ρ) in the compaction and linear shrinkage tests were selected as 0.3%, 0.6% and 0.9% for both lengths of fibre.

Standard Proctor tests were performed to determine the optimum moisture content (OMC) and maximum dry density (MDD) of the unreinforced and reinforced London Clay in accordance with BS-1377, part 4 [6]. Soil was mixed with fibres on a glass plate and sufficient water added to achieve the liquid limit. This smooth homogeneous paste was then placed in the mould as per BS-1377, part 2 (BSI, 1990). Filled moulds were then placed in a position free from draughts to let the sample shrink away from the walls of the mould slowly. The drying was completed in a laboratory oven first at 65 °C then at 105 °C. After cooling the length of each soil bar was measured.

3 Results and Discussion

The variation of MDD and OMC with respect to the fibre content is plotted in Fig. 1. It can be seen that with the increase of fibre percentage from 0% (i.e. unreinforced soil) to 0.9%, for a given fibre length, the addition of fibre decreases MDD which can be attributed partly to the decrease of the average unit weight of the solids in the soil fibre mixture, and also the fibres preventing efficient particle packing. The OMC also decreases on fibre addition, a finding in agreement with Senol [7] but inconsistent with Ayyappan et al. [8]. Further investigation is needed here. It can also be concluded that for the same fibre content, soil reinforced with longer fibres tends to have a higher MDD, though this phenomenon is becomes less clear when the fibre inclusion ratio increases to 0.9%. This may be due to fibre interweaving. Fibre length is not however seen to have a significant impact on OMC. Figure 2 shows the effects of PP fibre addition on linear shrinkage. It can be seen that linear shrinkage reduces significantly with the increase in PP fibres though the rate of decline decreases as the fibre inclusion ratio increases. When the fibre inclusion is 0.9%, the linear shrinkage percentage is reduced to less than 40% of the value in the unreinforced soil. It is notable that the 6 mm fibre is more effective than the 12 mm fibre when the inclusion ratio is relatively low, and there is no obvious difference for the two lengths of fibre when the inclusion ratio increases to above 0.6%. The improvement of the linear shrinkage properties in the reinforced materials is likely due to the development of interaction between fibre surfaces and soil particles, with fibres acting as a frictional and tension resistant element in the mixture to prevent the shrinkage of the sample.

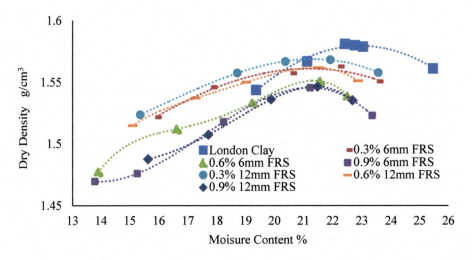

Fig. 1. Compaction curves of unreinforced and fibre reinforced soil (FRS)

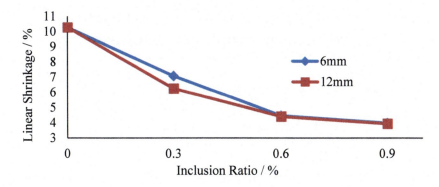

Fig. 2. Linear shrinkage of London clay with and without fibres.

4 Conclusions

Based on the experimental results presented here, it can be concluded that fibre reinforcement reduces both the MDD and OMC of London Clay. For a given fibre content, longer fibres produce higher MDD but fibre length does not have a clear influence on OMC. In linear shrinkage of reinforced soil is reduced significantly through the addition of fibres and this trend becomes slower as the fibre contents increase. Shorter fibres show greater improvement in shrinkage behaviour at inclusion ratios of 0.3% but a similar effect to the longer fibres on shrinkage behaviour above this value.

References

1. Park, S.: Unconfined compressive strength and ductility of fiber-reinforced cemented sand. Constr. Build. Mater. **25**(2), 1134–1138 (2011)
2. Soğancı, A.S.: The effect of polypropylene fiber in the stabilization of expansive soils. Int. J. Environ. Chem. Ecol. Geol. Geophys. Eng. **9**(8), 994–997 (2015)
3. Tang, C.-S.: Desiccation cracking behavior of polypropylene fiber–reinforced clayey soil. Can. Geotech. J. **49**(9), 1088–1101 (2012)
4. British Standards Institution. BS 1377- Part 2 Classification tests, Methods of test for soils for civil engineering purposes. BSI, London (2004)
5. ADFIL, Product Data Sheet. http://www.adfil.com/products/micro-syntheticfibres/monofilament-fibres/. Accessed 28 Feb 2018
6. British Standards Institution. BS 1377- Part 4 Compaction related tests, Method of test for soils for civil engineering purposes, BSI, London (2004)
7. Şenol, A.: Effect of fly ash and polypropylene fibres content on the soft soils. Bull. Eng. Geol. Environ. **71**(2), 379–387 (2012)
8. Ayyappan, S.: Investigation of engineering behaviour of soil, polypropylene fibers and fly ash-mixtures for road construction. Int. J. Env. Sci. Dev. **1**(2), 171–175 (2010)

Flow Behavior of Clays Amended by Superhydrophobic Additives

Yongkang Wu[1], Dongfang Wang[1], Yuzhen Yu[2], and Guoping Zhang[1(✉)]

[1] Department of Civil and Environmental Engineering,
University of Massachusetts Amherst, Amherst, MA 01003, USA
zhangg@umass.edu
[2] Department of Hydraulic Engineering, Tsinghua University,
Beijing 100084, China

Abstract. A novel superhydrophobic, chemical-resistant, hybrid inorganic-organic, polymer additive was recently synthesized as soil amendments to finely tune the flow behavior of various soils for functional applications. This paper presents an experimental investigation of the flow and conduction behavior of clays amended by the superhydrophobic additive with the consideration of water as the permeant. Results show that the superhydrophobic additive can transform a wetting, permeable, but dry clay powder into a non-wetting, impermeable barrier at low hydraulic gradients. However, when mixed with wet clay slurry, it can significantly accelerate the consolidation process of clays. The improvement to the water-based flow and conduction behavior enables the amended soils to function as an ideal material for many viable yet challenging applications.

Keywords: Superhydrophobicity · Clay · Flow behavior

1 Introduction

Permeability is one of the three pillars (the other two are compressibility and shear strength) supporting soil's engineering applications in practice. In many cases, the permeability of a soil has to be finely tuned in order to meet specific engineering requirements. For instance, transforming a highly permeable sand, the only locally available material, into a less permeable or even non-permeable barrier that can function as a traditional compacted clay liner, can bring much economical advantage as well as engineering benefits (e.g., increased internal angle of friction); On the other hand, increasing the permeability and consolidation rate of clays or other fine-grained slurries (e.g., red mud, dredged sediments) can accelerate the dewatering process and hence also bring constructional and economical advantages. To fulfill these practical needs, a novel superhydrophobic, chemical-resistant, hybrid inorganic-organic polymer additive was recently synthesized as soil amendments to finely tune the flow behavior of various soils for functional applications.

Superhydrophobicity is typically used to describe material surfaces with a static water contact angle (WCA) of >150° [1, 2]. Due to this special nature, the hybrid organic-inorganic composite materials play a key role in the development of advanced functional materials [3]. However, the newly developed superhydrophobic material is not an inorganic-organic composite, but a hybrid material in itself that contains an inorganic polymeric backbone with attached organic functional groups.

This paper presents an experimental investigation of the flow and conduction behavior of some typical clays amended by the superhydrophobic additive with the consideration of water as the permeant. The change in the water-based permeability will make the amended soils an ideal material for many viable yet challenging applications pertinent to the flow behavior of soils.

2 Materials and Methods

2.1 Materials

The Boston Blue Clay (BBC) was used as the model material to study the flow behavior. Both pure BBC and the same clay amended by the superhydrophobic inorganic-organic polymer (IOP) additive were studied. In addition, a dry, highly pure kaolinite clay without and with IOP was also studied to investigate whether a small percentage of IOP added to the pure kaolinite can significantly alter its hydrophobicity.

The mineralogical composition of the soil samples as well as the IOP additive was analyzed by using X-ray diffraction (XRD) method. The XD patterns show that the BBC mainly comprises quartz, illite, biotite, albite, kaolinite, with small amounts of chlorite, orthoclase, and hornblende, while the pure kaolinite clay contains primarily kaolinite with traces of quartz. The IOP additive is almost entirely an amorphous material with < 5 wt% quartz.

2.2 Contact Angle Measurement

The effect of different percentages of IOP additive on the hydrophobicity of the dry pure kaolinite clay was examined by the change in the water contact angle (WCA). The WCAs were measured with the sessile drop method in an OCA 15EC contact angle goniometer (Dataphysics, Filderstadt, Germany). A liquid droplet was deposited via a 1.0 mL syringe pointed vertically down onto the sample surface, and the contact angle between the droplet and sample surface was captured by a high-resolution camera and subsequently analyzed by the analysis software.

2.3 Oedometer Test

Oedometer tests were performed to determine the consolidation and permeability properties of the pure BBC as well as the IOP-amended BBC. To prepare the resedimented samples for oedometer tests, pure BBC slurry was firstly prepared at a water content twice of the liquid limit (which is 45.6% as measured by the ASTM standard method). For the amended soil, IOP additive was evenly mixed at a fraction 20 wt% of the dry clay into the clay slurry to prepare the IOP-BBC mixture for subsequent testing.

The Sigma-1 ICON automated consolidation system from GEOTAC (Trautwein Soil Testing Equipment, Inc., Houston, Texas) was used to perform the ASTM-specified incremental loading consolidation tests. The samples were consolidated under a series of effective axial stresses (i.e., 5 kPa, 10 kPa, 20 kPa, 50 kPa, 100 kPa, 200 kPa, 500 kPa, 1000 kPa, and 1600 kPa) with a load increment ratio of ~ 1.0 and a constant loading duration of 24 h. For each load increment, the deformation of the sample was recorded with time. Both the Casagrande's logarithm time method and Taylor's square root of time method was used to estimate the coefficient of consolidation [4].

3 Results and Discussion

3.1 Contact Angle

The static WCAs of the IOP additive and pure kaolinite with and without IOP additive are shown in Fig. 1. It can be seen that the IOP additive is superhydrophobic with a static WCA of 163.9°. The kaolinite clay is hydrophilic, as indicated by its WCA of $\sim 30°$. However, adding 5.0 wt% IOP additive to the pure kaolinite significantly increases the WCA to 161.99°. Therefore, the IOP additive can turn a wetting, permeable kaolinite clay into a non-wetting, virtually impermeable barrier at low hydraulic gradients.

Fig. 1. Water contact angles of IOP, pure kaolinite, and the kaolinite-IOP mixture with 5 wt% IOP added to the pure clay.

3.2 Rate of Consolidation

Figure 2 presents the coefficient of consolidation, c_v, of the pure BBC and the IOP-amended BBC with 20 wt% additive. The coefficient of consolidation was estimated using both the Casagrande's and Taylor's methods. The coefficients of hydraulic conductivity of the two samples under different axial consolidation stresses can also be estimated.

It is clear to see that the consolidation coefficient of the IOP-BBC mixture is significantly higher than that of pure BBC at relatively higher consolidation stress. Correspondingly, the coefficient of hydraulic conductivity of the mixture is also higher than that of pure BBC. The reason may be that, for clays with low permeability, due to the superhydrophobicity of the additive, the IOP particles in the mixture tend to form aggregates, and these inter-linked or in-contact aggregates in turn become superhydrophobic flow passages, which in turn increases the flow velocity due to the much-reduced surface attraction to water. However, when the consolidation stress is low, the difference between the two samples is not significant. Both samples start from a slurry consistency (i.e., at a water content of 2.0 times the liquid limit) where particle contacts may not be so significant. At low stresses, the additive aggregates may not be able to form interlinked, continuous flow passages. Another possible reason is relatively low excess pore pressure and hence the hydraulic gradients that may not be enough to break through the possible capillary barriers caused by the superhydrophobic additive.

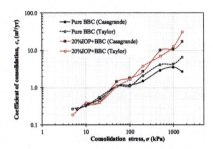

Fig. 2. Comparison of test results between Pure BBC and IOP-BBC mixtures

4 Conclusions

The flow and conduction behavior of two typical clays amended by the superhydrophobic IOP additives were investigated in this study. Results show that a small fraction of IOP additive can significantly increase the contact angle of dry clay, which indicates that the superhydrophobic additive can potentially transform a wetting, permeable, dry clay into a non-wetting, impermeable barrier at low hydraulic gradient. In addition, the IOP additive can significantly increase the conduction behavior of wet clays via increasing the coefficient of consolidation by one order of magnitude. This newly developed additive can expectedly bring transformative impacts to geotechnical engineering practice and may find many viable yet challenging applications such as turning permeable sand into an impermeable barrier and accelerating the dewatering process of many fine-grained sediments and industrial tailings.

References

1. Wenzel, R.N.: Resistance of solid surfaces to wetting by water. Ind. Eng. Chem. **28**(8), 988–994 (1936)
2. Cassie, A.B.D., Baxter, T.S.: Wettability of porous surfaces. Trans. Faraday Soc. **40**, 546–551 (1944)
3. Kickelbick, G.: Introduction to hybrid materials. In: Hybrid Materials: Synthesis, Characterization, and Applications. Wiley-VCH Verlag GmbH & Co. KGaA, Weinheim (2007)
4. ASTM D2435/D2435 M-11 Standard test methods for one-dimensional consolidation properties of soils using incremental loading, ASTM International, West Conshohocken, PA (2011)

Mechanical and Thermal Behaviour of Cemented Soil with the Addition of Ionic Soil Stabilizer

Fei Xu[1,2(✉)], Hua Wei[1], Wenxun Qian[1], and Yuebo Cai[1,3]

[1] Materials and Structural Engineering Department,
Nanjing Hydraulic Research Institute, Nanjing, China
babyshaq@126.com
[2] College of Water Conservancy and Hydropower,
Hohai University, Nanjing, China
[3] Dam Safety Management Center of the Ministry of Water Resources,
Nanjing, China

Abstract. The cement soil is widely used in the aspects of chemical reinforcement of soft soil. However, the poor early strength of the cemented soil retards its application. To improve the mechanical property of cemented soil, Ionic soil stabilizer (ISS) is often used as an additive. Nevertheless, mechanism of ISS addition remain controversial. In the paper, based on direct compressive strength tests, optimum dosage of ISS was obtained, and thermal gravity analysis (TGA) was conducted for hydration products evaluation. Through TGA, the influence mechanism of ISS on cement hydration was concluded. In addition, the determination procedures of hydration products appear to be helpful for additive selection of cemented soil.

Keywords: Cemented soil · ISS · Thermal analysis · Hydration process

1 Introduction

Using cementitious materials to treat soft soil, regarded as soil stabilization, has been a widely adopted method in Geotechnical engineering for decades. In practice, such method has the major advantages of broad resource of raw materials, rapid hardening process and convenience for mechanical operation. However, a serious of issues [1], like the slow development of early strength and the massive cement consumption of certain soils, were exposed through engineering practice, which retard its application in practice.

To combat such issues, extensive researches have adopted types of organic and inorganic additives for cemented soil. With respect to organic components, it has been proven that ISS owns the widest applicability of organic additives [2]. As concluded by many researches [3, 4], the major mechanisms of ISS addition are concluded as thinning double ions layers covered on the soil particles and transferring hydrophilic soil particles to hydrophobic by functional groups of stabilizer. However, similar double ions layers are proved to be exist in cement particles as well, but limited researches are related to the effects of ISS on cement hydration. Accordingly, the mechanisms of ISS on cemented soil are further investigated through directed compression tests and thermos-gravity analysis (TGA) in this paper.

Table 1. Basic properties of silty soil

Basic soil properties		Value
Liquid limit (%)		61.24
Plastic limit (%)		32.75
Optimum water content (%)		34.8
Maximum dry density (g/cm^3)		1.65
Particle size	D10 (μm)	29.33
	D50 (μm)	101.1
	D90 (μm)	229.1

Table 2. Chemical compositions (%) of soil, OPC

Sample	SiO$_2$	Al$_2$O$_3$	Fe$_2$O$_3$	CaO	MgO	SO$_3$	TiO$_2$	K$_2$O	Na$_2$O	P$_2$O$_5$	LOI
Soil	61.44	14.21	2.38	2.34	1.72	0.04	0.36	1.34	1.43	0.089	13.63
OPC	21.97	7.48	3.45	58.43	1.3	1.71	0.33	0.65	0.099	0.11	4.54

2 Materials and Methodology

2.1 Specimen Preparation

The cement used in this study was commercial Ordinary Portland Cement (OPC) 42.5[#], and the soil was silty clay, collected from Nanjing suburb. Basic properties of tested soil are given in Table 1, and the compositions of the OPC and the soil were examined by XRF, listing in Table 2.

The soil and cement were mixed with the ratio of 9:1, and the water to solid ratio was 0.35 for conducting convenience. The ISS was diluted in distill water in advance with the ratios 1/100, 1/150, 1/200, 1/250 and 1/300. The moulds for specimens were of 40 × 40 × 160 mm, and the compactness was kept as 97%. All specimens were covered with plastic wrap, and kept in the curing chamber under the conditions of 20 °C and RH 95%.

2.2 Test Methodology

Direct compressive tests were conducted using digital compression test machine, under a loading rate of 2 mm/min. The curing periods were 1, 3, 7, 14 days. After compressive test, specimens were soaked in anhydrous ethanol for 48 h to stop hydration process, then dried in oven under 40 °C for 48 h to evaporate free water. The thermo-gravity analysis was conducted on Linseis STA PT-1600 with a temperature range from 50 to 800 °C, and a heating rate of 10 °C/min.

3 Test Results

Results of compressive strength tests of all specimens are given in Fig. 1. US and CS are corresponded to untreated soil and cemented soil. As revealed, cement addition greatly enhanced soil strength. Compared to CS, the strength of ISS treated cemented soil (named as ICS in context) was greatly promoted, especially in early ages. In ICS samples, the dilution of 1/150 exhibited the best strength in every period, which was regarded as the optimum dosage.

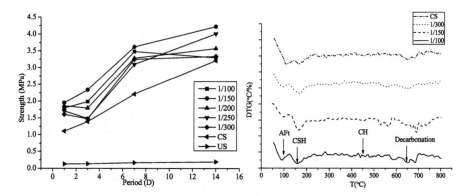

Fig. 1. Direct compressive test results of all specimens against curing periods. (left)

Fig. 2. DTG curves for selected samples of 7d (right)

Selecting CS, 1/100, 1/150 and 1/300 of 7d as the representative samples. The DTG results of selected samples were given in Fig. 2. As presented, double endothermic peaks under 200 °C were observed for all curves, which were attributed to the dehydration of CSH($\sim$160 °C) and Aft($\sim$80 °C). With respect to the temperature range 550$\sim$750 °C, it was probably due to calcite and soil mineral decarbonation. Notably, minor peaks around 450 °C, relating to portlandite (CH) that produced along with CSH, were presented ICS samples. The semi-quantification calculation of hydration products was following Eqs. (1) and (2).

$$W_i = W_{Li} \frac{M_i}{m_i} \tag{1}$$

$$W_{AFt} = (W_{50\sim 200\,°C} - W_{CSH}) \frac{M_{AFt}}{m_i} \tag{2}$$

In Eqs. (1) and (2), W_i and W_{AFt} were the mass of product i and AFt, $W_{50\sim 200°C}$ and W_{Li} was weight loss during the specific temperature range. M_{AFt}, M_i and m_i were the molecular masses of AFt, product i and water, respectively. Since the low crystallinity of CSH, its weight fraction was obtained through CH calculation results, following the stoichiometry relationships obtained in previous research [5, 6]. The calculation results were given in Fig. 3.

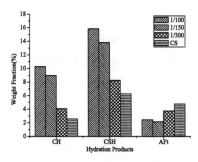

Fig. 3. Estimation results of hydration products

4 Conclusions

In this paper, based on direct compressive tests and TGA analysis, the mechanical property effect of ISS was analyzed and hydration products of cemented soil was obtained. The main conclusions are as follows:

(1) With the addition of ISS, the hardening process of cemented soil was greatly accelerated. Before reaching the optimum dosage, the strength increases with the increasing dilution, then decreased with the excessive addition.
(2) Compared to CS, amounts of CSH of hydration products increased significantly, with the AFt decreased by about 50%, implying that ISS mainly accelerated the hydration process of C_3S and C_2S in cement.

Acknowledgments. The authors acknowledge the financial support from the National Key Research (2016YFC0401610) and Central Public-interest Scientific Institution Basal Research Fund (Y417015).

References

1. Fan, H.H., Gao, J.N., Wu, P.T.: Prospect of researches on soil stabilizer. J. Northwest Sci-Tech Univ. Agric. For. (Nat. Sci. Ed.) **34**(2), 141–174 (2006). (in Chinese)
2. Pei, L.I., Yang, W., Deng, Y.: Status quo and trend of soil stabilizer development. Subgrade Eng. **3**, 1–8 (2014). (in Chinese)
3. Katz, L., Rauch, A., Liljestrand, H.: Mechanisms of soil stabilization with liquid ionic stabilizer. Transp. Res. Rec. **1757**(1), 50–57 (2001)
4. Liu, Q.B., Wei, X., Zhang, W.F.: Experimental study of ionic soil stabilizer-improves expansive soil. Rock Soil Mech. **30**(8), 2286–2290 (2009). (in Chinese)
5. Karen, S., Ruben, S., Barbara, L.: A Practical Guide to Microstructural Analysis of Cementitious Materials. CRC Press, Boca Raton (2016)
6. Zeng, Q., Li, K., Fen-Chong, T.: Determination of cement hydration and pozzolanic reaction extents for fly-ash cement pastes. Constr. Build. Mater. **27**(1), 560–569 (2012)

Experimental Study on the Relation Between EDTA Consumption and Cement Content of Cement Mixing Pile

Canhong Zhang[1(✉)], Baotian Wang[1], and Enyue Ji[2]

[1] Key Laboratory of Ministry of Education for Geomechanics and Embankment Engineering, Hohai University, Nanjing 210098, China
2008zhangcanhong@163.com
[2] Geotechnical Engineering Department, Nanjing Hydraulic Research Institute, Nanjing 210024, China

Abstract. To investigate the applicability that using EDTA (Ethylene Diamine Tetraacetic Acid) titration detects the cement content of cement mixing pile, a series of laboratory EDTA consumption tests under the condition of different curing ages and different cement content distributions are performed. The results indicate that the relationship between the cement content distribution and EDTA consumption approximate linear when cement content is low (less than 10%). The curve between EDTA consumption and cement content presents nonlinear characteristic when the increasing rate of EDTA consumption reduces gradually with the increasing of cement content. Hyperbolic fitting effect is perfect according to the correlation analysis between the cement content of cement mixing pile and EDTA consumption. The standard curve can be obtained from the linear relationship between the reciprocal of the cement content and the reciprocal of EDTA consumption, which is applied to actual test on site.

Keywords: Cement mixing pile · Cement content · EDTA titration
EDTA consumption · Correlation analysis

1 Instruction

EDTA (disodium oxalate tetra acetate) titration method is a standard method for rapid determination of cement or lime content on site [1]. With the method, not only the cement content in the cement stabilized soil can be accurately determined, but also the uniformity of cement and soil mixing can be checked. The EDTA titration method in the specification is more suitable for the detection of cement content in cement stabilized soil filling in highway engineering with a cement content less than 6.0%. Tests in highway engineering show that the cement content has a linear relationship with EDTA consumption. According to the consumption of EDTA, the cement content can be obtained quickly. The cement content of the cement mixing piles ranges from 12% to 20% [2], which exceeds the scope of highway regulations. The results calculated with this linear relationship is smaller than the true data. Therefore, it is necessary to correct the results of the traditional EDTA titration test to obtain a general curve of cement content and EDTA consumption.

In this paper, the relationship between cement content and EDTA consumption is experimentally studied. An improved method is proposed and a more accurate fitting curve between the improved EDTA consumption and the cement content is given.

2 Soil and Schemes of the Test

2.1 Physical and Mechanical Properties of Soil Samples

The tested silt clay soil is taken from Yanglin ship lock site in Suzhou. Its physical properties are shown Table 1. Ordinary portland cement P.O 42.5 is used.

Table 1. Indexes of physical properties for mucky soil

Depth (m)	Natural moisture content (%)	Specific gravity	Void ratio	Plasticity index (%)	Liquidity index (%)
3~5	47.0	2.67	1.34	20.6	1.14

2.2 Schemes of the Test

Standard samples are prepared with different cement content at 2%, 4%, 6%, 10%, 15%, 20%, 25%, 30%, 35%, 40%, 50%, 70% and 100% (2% refers to there is 2 g cement added in 100 g soil). These samples are cured under standard maintain conditions. Then each 100 g sample's consumption of EDTA are tested at the different curing age of 3h, 6h, 12h, 24h, 1d, 3d, 5d, 7d, 14d, 28d, 60d and 90d.

3 Soil and Schemes of the Test

The relationship between the cement content and EDTA consumption under different curing ages are plotted on the Figs. 1 and 2.

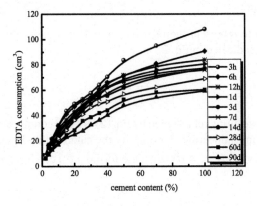

Fig. 1. Relationship between EDTA consumption and cement content under different ages.

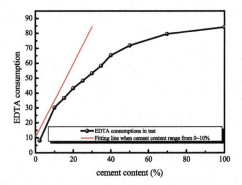

Fig. 2. The curve of EDTA consumption and cement content after curing 12 h.

Based on the chemical principle of EDTA titration, $Ca(OH)_2$ generated with Hydration reaction in the cement can be dissolved by 10% NH_4Cl [3]. The equation listed as follow:

$$Ca(OH)_2 + NH_4Cl = CaCl_2 + 2NH_3H_2O \uparrow \qquad (1)$$

Then Ethylene Diamine Tetraacetic Acid fixes the free Ca^{2+}, the consumption of EDTA solution reflects the amount of cement content. When the cement content is high, the NH_4Cl cannot dissolve the Ca^{2+} thoroughly. Therefore, the increase rate of EDTA consumption will decline. This solution is reflected in the Figs. 1 and 2: the EDTA consumption curve gradually deviates from the curve of linear fitting, and the higher the cement content, the greater the deviation is. Figure 2 shows that the critical point of dissolution of the cement content is about 10% in these experiments. The design cement content is 12% to 20% of cement mixing pile, and there are uneven mixing problems in it. The EDTA titration in Technical Specification for Construction of Highway Roadbase is no longer applicable.

3.1 Discussion on the Relationship Between Cement Content and EDTA Consumption

Hyperbola is used to represent the relationship between cement content x and EDTA consumption y, as indicated in Eq. 2.

$$\frac{1}{y} = a + b\frac{1}{x} \qquad (2)$$

Hyperbolic curve fitting represented by Eq. (2) is more complicated than the straight-line fit, especially in engineering. To simplify the calculation, the curve is converted to a straight line. Let $y' = 1/y$, $x' = 1/x$, then the formula (2) can be rewritten as Eq. (3):

$$y' = a + bx' \qquad (3)$$

In the above formula, x' and y' represents the reciprocal of cement content and EDTA consumption, respectively. Because of the Eq. (3) is a linear equation, it is convenient for application. To test the applicability of the equation, a scatter plot of the reciprocal of the cement content and the reciprocal of the EDTA consumption is given in Fig. 3.

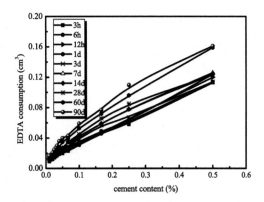

Fig. 3. The linear fitting curve of the reciprocal of the cement content and the reciprocal of EDTA consumption under different ages.

The correlation coefficients are all near to 1.0. The fitting effect of Eq. (3) is perfect, in other words, the hyperbolic Eq. (2) is appropriate. Compared with the conventional linear curve, the hyperbolic standard curve has a higher accuracy. When the consumption of the EDTA is tested. Then the reciprocal of the cement content can be achieved according to the Fig. 3. So, the cement content of this cement mixing pile is obtained relatively.

4 Conclusions

When the cement content is in the range of 0 to 100%, the curve of cement content and EDTA consumption is similar to hyperbolic distribution. A linear relationship between the reciprocal of the cement content and the reciprocal of the EDTA consumption is obtained. The cement content of the cement mixing pile is quickly determined by this linear relationship. This method is simple and easy to calculate, with a wide applicability.

References

1. Fu, H.C.: Using EDTA titration method to determine the cubic ratio of limes in ash foundation. Constr. Q. **29**(1), 25–27 (2011)
2. JGJ79-2002: Technical Code for Building Foundation Treatment. Chinese Building Industry Press, Beijing (2002)
3. Antiohos, S.K., Tsimas, S.: A novel way to upgrade the coarse part of a high calcium fly ash for reuse into cement systems. Waste Manag. **27**, 675–683 (2007)

Author Index

A

Abiodun, Abiola Ayopo, 773
Ahmed, Balqees A., 546
Airoldi, Sergio, 469, 473, 659
Aissaoui, Soufyane, 663
Aksoy, Yeliz Yükselen, 720
Ali, Haifa A., 546
Alpaydın, Şükran Gizem, 778
Al-Sammarraie, D., 730
Andò, Edward, 252
Apostolaki, Stefania, 619
Arroyo, Hiram, 123
Arroyo, Marcos, 274, 390
Assadollahi, Hossein, 668
Augarde, Charles, 858
Avinash, G. S. S., 826
Azzam, Rafig, 672

B

Bai, Bing, 291
Bao, Xiaohua, 478
Bauer, Erich, 191
Bersan, Silvia, 677
Bian, Hanliang, 685
Bilotta, Emilio, 324
Boháč, Jan, 578
Bossi, Giulia, 677
Bouaicha, Allaoua, 402
Bouazza, Abdelmalek, 822
Bretschnaider, Alberto, 473
Bretschneider, Alberto, 659
Bui Van, Duc, 782
Byrne, Byron, 725

C

Cabrera, Miguel Angel, 591
Cai, Guanghua, 787
Cai, Guojun, 685, 689, 699
Cai, Yuebo, 866
Cao, Wei, 3
Ceccato, Francesca, 296
Chang, Wen-Jong, 681
Che, N., 238
Che, Yuxuan, 128
Chen, Chen, 542
Chen, Cheng, 98
Chen, Chuan, 307
Chen, Jingcheng, 483
Chen, Jinjian, 336
Chen, Jyh-Fang, 681
Chen, Liang, 551
Chen, Ping, 13
Chen, Ren-Peng, 624
Chen, Shengshui, 534
Chen, Xing-wei, 136
Chen, You-liang, 423
Chen, Zuyu, 300
Cheng, Hongyang, 132
Cheng, Zhanlin, 653
Chibuzor Onyelowe, Kennedy, 782
Choo, Jinhyun, 312
Choosrithong, Kamchai, 316
Chou, Shih-Hsun, 681
Chow, Jun Kang, 233
Chu, Ya, 685
Ciantia, Matteo, 274
Çokça, E., 838

Cola, Simonetta, 677
Cui, Deshan, 229
Cui, Hongzhi, 478

D
Dafalias, Y. F., 225
Dai, B., 555
Dai, Guo-liang, 463
Dalvi, R. S., 805
Damiano, E., 648
Dang, Liet Chi, 792
Daniluk, Hubert, 89
Darban, R., 648
De Polo, Fabio, 677
Deng, Gang, 128
Derbin, Yury, 320
Develioglu, İnci, 488
Diano, Valentina, 597
Dinh, Anh Quan, 668
Dołżyk-Szypcio, Katarzyna, 140
Dong, Jungui, 492
Dong, Mei, 747
Dong, Tong, 8
Dong, Yijia, 415
Doose, Ralf, 615
Du, Guangyin, 787
Du, Xiuli, 52
Du, Yu, 699
Duan, Wei, 689

E
Elshafie, Mohammed Z. E. B., 694
Eric Sze, H. Y., 103

F
Fabozzi, Stefania, 324
Fabris, Carla, 328
Falsafizadeh, S. R., 496
Fan, HengHui, 797
Fan, Tingting, 501
Fang, Huanglei, 644
Feng, Feng-Cui, 355
Feng, Tugen, 506
Feng, Yongzhen, 510
Fern, E. James, 79
Fioravante, Vincenzo, 469, 473, 659
Fitzgerald, Breiffni, 340
Fleureau, Jean-Marie, 148
Fu, Hongyuan, 483
Fu, Longlong, 144

Fu, Peng, 332
Fu, Zhongzhi, 534

G
Galliková, Zuzana, 514
Gang, Li, 39
Gao, Qian-Feng, 148
Gao, Yan, 233
Gao, Yufeng, 74
Gates, Will P., 822
Geng, Xueyu, 628
Gens, Antonio, 274
Geutebrück, Ernst, 716
Girardi, Veronica, 296
Giretti, Daniela, 469, 473, 659
Gong, H., 238
Gong, Wei-ming, 463
Gong, Xiaonan, 747
Goodarzi, M., 730
Gu, Xiaoqiang, 153, 203
GuhaRay, Anasua, 801, 826
Guo, Aiguo, 94
Guo, Anbang, 519
Guo, Jun, 751
Guo, Peijun, 144
Guo, Xiaogang, 13
Guo, Xiaoxia, 65, 611
Guo, Zhen, 602
Gupta, Shashank, 801

H
Hai, Lu, 365
Haider, Abbas, 410
Han, Zhong, 116
Hattab, Mahdia, 148
He, Chuncan, 809
He, Jie, 174, 178
He, Jizhong, 128
He, S. S., 739
He, Xuzhen, 17, 157, 419
Herle, Ivo, 252, 607
Hicher, Pierre-Yves, 21, 439
Hirapure, Anurag S., 805
Ho, Yen-Te, 694
Hoang, Dinh Phuc, 782
Hong, Juntian, 161
Hong, Y., 524
Hosseininia, Ehsan Seyedi, 394
Hou, Chao, 743
Hou, W. S., 560

Hou, Yongmao, 336
Hu, Anfeng, 332
Hu, HaiJun, 797
Hu, Hui, 672
Hu, Minyun, 529
Hu, Qian-Qian, 165
Hu, Shengkun, 644
Hu, Xinli, 809
Hu, Yayuan, 31
Hua, Likun, 8
Huang, An-Bin, 681, 694
Huang, Biao, 336
Huang, Haiying, 199
Huang, Jianjun, 814
Huang, Jizhi, 26
Huang, Kai, 169
Hughes, Paul, 858
Hung, Wen-Yi, 694

I
Ighil Ameu, Lamine, 148
Igoe, David, 340
Irvine, K. N., 360

J
Jerman, Jan, 344
Ji, Enyue, 534, 870
Jia, Qi, 538
Jiang, Hong-Tao, 759
Jiang, Jianhong, 35
Jiang, Mingjing, 174, 178, 211, 238
Jiang, Xiang, 247
Jin, Yin-Fu, 112
Jin, Zhiyang, 478
Jin, Zhuang, 350
Jin, Zihao, 542
Jing, Xue-Ying, 186
Jostad, Hans Petter, 84
Jrad, Mohamad, 148

K
Kaliakin, Victor N., 35
Kallioglou, Polyxeni, 619
Kamel, Taous, 402
Kar, Arkamitra, 801, 826
Karkush, Mahdi O., 546
Khabbaz, Hadi, 792
Kiełczewski, Tomasz, 89
Kim, Jong Ryeol, 390, 419
Kim, Jong, 17, 157, 847
Kindler, Arne, 712
Klapperich, Herbert, 672
Komaravolu, V. P., 801
Kong, Gangqiang, 551

Kong, Liang, 8
Kong, Lingwei, 98
Kong, Yuxia, 818
Kotronis, Panagiotis, 350
Koulaouzidou, Kyriaki, 619
Kreiter, S., 730
Ku, Taeseo, 847
Kuang, Lianfei, 182, 767
Kuang, Lian-Fei, 39

L
Laouafa, Farid, 439
Lashkari, Ali, 61, 496
Laue, Jan, 538
Lav, M. A., 838
Lee, Sanghyun, 312
Li, Chao, 644
Li, Hui, 551
Li, Jian, 365
Li, Kuikui, 699
Li, Mingguang, 336
Li, Peng, 410
Li, Q., 555
Li, S. D., 560
Li, Shuai, 186
Li, Ting, 39
Li, Weichao, 586
Li, Wei-Hua, 355
Li, Weiyi, 203
Li, Xia, 233, 256, 265
Li, Yangyang, 360
Li, Yucheng, 569
Li, Zhaofeng, 233
Li, Zheng, 370
Liang, Hengchang, 767
Liang, Jingyu, 52
Liang, Weijian, 449
Liang, Wencheng, 699
Liao, Hongjian J., 582
Lim, Aswin, 373
Lin, Jia, 191
Lin, Jindi, 743
Ling, Hoe I., 35
Liu, Bangan, 703, 707, 743
Liu, Fang, 128
Liu, Feng, 381
Liu, Hanlong, 247
Liu, J. F., 739
Liu, Jian-kun, 377
Liu, Jun, 178, 444
Liu, Junzhe, 542
Liu, Li, 377
Liu, Lin, 108
Liu, Lingxiao, 510, 519, 636

Liu, Sihong, 207, 574
Liu, Si-hong, 43
Liu, Songyu, 685, 689, 699, 751, 787
Liu, Su-Ping, 759
Liu, Xin, 564
Liu, Yan, 365
Liu, Yang, 195
Liu, Yiming, 98
Long, Xing, 751
Loukidis, Dimitrios, 386
Lu, Dechun, 52
Lü, Xilin, 435
Lu, Yuke, 529
Luding, Stefan, 132
Luo, Shengmin, 569
Luo, Ting, 108
Lv, Hai-bo, 492

M

Ma, Chao, 52
Ma, Yanxia, 510, 519, 636
Ma, Yifei, 199
Magnanimo, Vanessa, 132
Mahawish, Aamir, 822
Mamirov, Maxat, 390
Mao, Hangyu, 574
Marshall, Alec M., 320
Martinez, Alejandro, 69
Mašín, David, 344
Mazhar, S., 826
Miao, Linchang, 644, 834
Minardo, A., 648
Ming, Haiyan, 478
Mohyla, Tomáš, 578
Monforte, Lluís, 390
Moon, Sung-Woo, 847
Mörz, T., 730
Mu, Qingyi Y., 582

N

Na, SeonHong, 57
Nalbantoglu, Zalihe, 773
Nasrollahi, Seyed Mehdi, 394
Ng, C. W. W., 582
Ngo, Anh Quan, 852
Nguyen Thi, Nu, 782
Nguyen, Huu Nam, 852
Nguyen, Huy Vuong, 852
Nguyen, Quoc Dung, 852
Nycz, Karolina, 712

O

O'Kelly, Brendan C., 586
Olivares, L., 648
Önal, Okan, 720
Onyelowe, Kennedy Chibuzor, 782
Ören, Ali Hakan, 720
Ou, Chang-Yu, 373
Özden, Gürkan, 720

P

Pachnicz, Michał, 216
Pan, Jiajun, 653
Pasuto, Alessandro, 677
Pei, H., 739
Peng, Chong, 398
Perrot, Arnaud, 307
Picarelli, L., 648
Pinzon, Gustavo, 591
Porcino, Daniela Dominica, 597
Prendergast, Luke J., 340
Priegnitz, Mike, 716
Pulat, Hasan Firat, 488
Pulko, Boštjan, 328

Q

Qi, Jilin, 94
Qian, Jiangu, 203
Qian, Qiuying, 506
Qian, Wenxun, 866
Qiao, Maowei, 743
Qin, Jie-Qiong, 755
Quan Tran, Van, 852

R

Račanský, Václav, 328
Rafa, Sid Ali, 402
Rahardjo, Harianto, 360
Rajczakowska, Magdalena, 216
Rangeard, Damien, 307
Rao, Dengyu, 291
Reiffsteck, Philippe, 663
Reinisch, Johannes, 734
Ren, W. G., 739
Rizvi, Zarghaam Haider, 405
Rojas, Eduardo, 123
Rouaz, Idriss, 402
Royston, Ronan, 725
Rui, Shengjie, 602

S

Sadler, Andrew, 858
Salimi, Mohammadjavad, 61
Sarkar, Saptarshi, 340
Sattari, Amir Shorian, 405
Saylan, Ali Alper, 720
Schenato, Luca, 677
Schweiger, Helmut F., 316, 328

Schwiteilo, Erik, 607
Shan, Yao, 144
Shao, Longtan, 65, 611
Sharma, L. K., 668
Sheil, Brian, 725
Shen, Chaomin, 207
Shen, Chao-min, 43
Shen, Kanmin, 602
Shen, Zhifu, 211
Shi, Bin, 759
Shougrakpam, Sangeeta, 830
Shuku, Takayuki, 132
Simonini, Paolo, 296, 677
Sirupa, R., 826
Šmilauer, Václav, 252
Sobótka, Maciej, 216
Song, Erxiang, 161, 410
Song, Yun-qi, 640
Stähler, F. T., 730
Stanski, T., 730
Stutz, Hans Henning, 69, 615
Su, Chao, 415
Sun, Songtai, 842
Sun, WaiChing, 57
Sun, Xiaohao, 834
Sun, Yi, 43
Sun, Yifei, 74
Sun, Zhiliang, 98
Süt-Ünver, İ., 838
Świtała, Barbara M., 79
Szypcio, Zenon, 220

T
Tamiolakis, Georgios-Pantelis, 386
Tan, Dao-Yuan, 755
Tan, Huiming, 814
Tang, Lin, 842
Tang, Xiaowu, 842
Tempone, Pamela, 132
Teng, Yining, 84
Tharaud, Blandine, 668
Theocharis, A. I., 225
Thoeni, Klaus, 132
Thurner, Robert, 734
Tomasello, Giuseppe, 597
Tong, Liyuan, 751
Trivedi, Ashutosh, 830
Tschuchnigg, Franz, 734
Tymiński, Wojciech, 89
Tyri, Danai, 619

V
Vairaktaris, E., 225
Van Dang, Phi, 782
Viggiani, Gioacchino, 252
Vinoth, Ganapathiraman, 847
Vu, Ba Thao, 852

W
Walker, James, 320
Wan, Liang-long, 116
Wanatowski, Dariusz, 320
Wang, Baotian, 870
Wang, Binglong, 763
Wang, C. S., 739
Wang, Chengcheng, 834
Wang, Dong, 458
Wang, Dongfang, 862
Wang, H. N., 238
Wang, Han-Lin, 624
Wang, Jianhua, 336
Wang, Jian-hua, 640
Wang, Jianye, 858
Wang, Kejia, 506
Wang, L. Z., 524
Wang, Lei, 247
Wang, Lizhong, 602
Wang, Pengyue, 365
Wang, R., 427
Wang, Rui, 3, 48, 165
Wang, Shengnian, 211
Wang, Shun, 229, 419
Wang, Su-ran, 423
Wang, W. D., 555
Wang, Xiaoxiao, 195
Wang, Xie-qun, 116
Wang, Xing, 8
Wang, Yu-Hsing, 233
Wang, Zhihua, 211
Wang, Zhijie, 242
Wei, Guang-Qing, 759
Wei, Hua, 866
Wei, Liming, 103
Wei, Xiao, 103
Wei, Yongbin, 703, 707, 743
Wiebicke, Max, 252
Wride, Natalie M., 628
Wu, Bangbiao, 632
Wu, Hongyu, 747
Wu, Kuen-Wei, 694
Wu, Nan, 291

Wu, Shuchong, 529
Wu, Shunchuan, 195
Wu, Wei, 13, 17, 94, 157, 191, 229, 398, 419, 782
Wu, Wenju, 636
Wu, Yongkang, 569, 862
Wu, Ze-Xiang, 350
Wuttke, Frank, 405, 615

X
Xia, Kaiwen, 381, 632
Xiang, Wei, 229
Xiao, Bin, 529
Xiao, Qiong, 256
Xiao, Sha, 431
Xiao, Yang, 247
Xie, Kanghe, 332
Xiong, Congcong, 763
Xu, Fei, 866
Xu, Guofang, 94, 98
Xu, Guoyuan, 26, 492
Xu, Li-yang, 261
Xu, Ming, 161
Xu, Ri-qing, 261
Xu, Tao, 291
Xu, Xiangchun, 751
Xu, Z. H., 555
Xu, Zhao, 355
Xue, Dawei, 435
Xue, Long, 48

Y
Yan, Xiao-ran, 261
Yan, YingJia, 797
Yang, B., 524
Yang, Dunshun, 265
Yang, Gang, 551
Yang, Guangqing, 242
Yang, Jie, 439
Yang, Jun, 103, 564
Yang, Junyan, 492
Yang, Longcai, 763
Yang, Qi, 542
Yang, Qing, 551
Yang, S., 427
Yang, Shuocheng, 153
Yang, XiuJuan, 797
Yao, Xiaoliang, 94
Yao, Yang-ping, 108
Ye, Guan-lin, 640
Yin, Jian-Hua, 755
Yin, Zhen-Yu, 112, 350, 439
You, Quan, 644

Yu, Fan, 94
Yu, Hai-Sui, 265, 283, 398
Yu, Haitao, 324
Yu, Yuzhen, 569, 862
Yu, Zhenhuan, 743
Yuan, Jun, 689
Yuan, Quan, 233
Yuan, Yong, 324
Yue, Zhong-qi Quentin, 136
Yue, Zhongqi, 431
Yukselen-Aksoy, Yeliz, 778

Z
Zadjaoui, Abdeldjalil, 663
Zeni, L., 648
Zhan, Liangtong, 13
Zhang, Canhong, 870
Zhang, Cheng-Cheng, 759
Zhang, Dichuan, 17, 157, 419
Zhang, Fuguang, 269
Zhang, Fuhai, 506
Zhang, Guoping, 569, 862
Zhang, Han, 809
Zhang, Hong-ru, 377
Zhang, Jian-Min, 3, 48, 165
Zhang, Jun-feng, 116
Zhang, Kaiyu, 381
Zhang, Lei, 648
Zhang, Lu, 797
Zhang, Ningning, 274
Zhang, Ruixuan, 842
Zhang, Shuo, 640
Zhang, Wuyu, 510, 519, 636
Zhang, Xilei, 763
Zhang, Ye, 763
Zhang, Youhu, 84
Zhang, Yuqin, 444
Zhang, Zhichao, 247
Zhao, Boya, 65, 611
Zhao, Guangsi, 767
Zhao, Hongyi, 453
Zhao, Jidong, 279, 449
Zhao, Wei, 743, 763
Zhao, Xiaodong, 767
Zhao, Xue-liang, 463
Zheng, Dong-Sheng, 453
Zheng, Jingbin, 458
Zheng, Wenbo, 809
Zheng, Yingren, 8
Zhou, C., 582
Zhou, Guoqing, 182, 767
Zhou, H. W., 427, 739
Zhou, Mi, 506
Zhou, Peijiao, 529

Author Index

Zhou, Shunhua, 144
Zhou, Wan-Huan, 186
Zhou, Wenjie, 602
Zhou, Yuefeng, 653
Zhou, Z. M., 560
Zhu, Fan, 279
Zhu, Hehua, 852
Zhu, Hua-Xiang, 186
Zhu, Jianfeng, 453
Zhu, Jungao, 534
Zhu, Liuwen, 699

Zhu, Qi-Yin, 39
Zhu, Wen-bo, 463
Zhu, Wenkai, 174
Zhu, Yanhua, 365
Zhu, Zhuo-Hui, 755
Zhuang, Pei-Zhi, 283
Zhuo, Z., 427
Ziegler, Martin, 242
Zou, Wei-lie, 116
Zou, Yazhou, 182
Zuo, Yongzhen, 653

Printed by Printforce, the Netherlands